CW00471684

Advances in Rotating Electric Machines

Advances in Rotating Electric Machines

Volume 1

Editor

Sérgio Cruz

MDPI • Basel • Beijing • Wuhan • Barcelona • Belgrade • Manchester • Tokyo • Cluj • Tianjin

Editor
Sérgio Cruz
University of Coimbra
Portugal

Editorial Office
MDPI
St. Alban-Anlage 66
4052 Basel, Switzerland

This is a reprint of articles from the Special Issue published online in the open access journal *Energies* (ISSN 1996-1073) (available at: https://www.mdpi.com/journal/energies/special_issues/Rotating_Electric_Machines).

For citation purposes, cite each article independently as indicated on the article page online and as indicated below:

LastName, A.A.; LastName, B.B.; LastName, C.C. Article Title. *Journal Name* **Year**, *Article Number*, Page Range.

Volume 1
ISBN 978-3-03936-780-1 (Hbk)
ISBN 978-3-03936-781-8 (PDF)

Volume 1-2
ISBN 978-3-03936-842-6 (Hbk)
ISBN 978-3-03936-843-3 (PDF)

Contents

About the Editor

Sérgio Cruz (Ph.D.) received his E.E. diploma and M.Sc. and Dr.Eng. degrees in electrical engineering from the University of Coimbra, Coimbra, Portugal, in 1994, 1999, and 2004, respectively. He is currently an Associate Professor and the Director of the Electric Machines Laboratory at the Department of Electrical and Computer Engineering, University of Coimbra. His teaching and research interests include power transformers, rotating electric machines, electric drives, and power electronic converters, with special emphasis on fault diagnosis, fault tolerance, and digital control.

Article

Finite Control Set Model Predictive Control of Six-Phase Asymmetrical Machines—An Overview

Pedro Gonçalves, Sérgio Cruz * and André Mendes

Department of Electrical and Computer Engineering, University of Coimbra, and Instituto de Telecomunicações, Pólo 2-Pinhal de Marrocos, P-3030-290 Coimbra, Portugal; pgoncalves@ieee.org (P.G.); amsmendes@ieee.org (A.M.)

* Correspondence: smacruz@ieee.org; Tel.: +351-239-796-272

Received: 29 October 2019; Accepted: 5 December 2019; Published: 10 December 2019

Abstract: Recently, the control of multiphase electric drives has been a hot research topic due to the advantages of multiphase machines, namely the reduced phase ratings, improved fault tolerance and lesser torque harmonics. Finite control set model predictive control (FCS-MPC) is one of the most promising high performance control strategies due to its good dynamic behaviour and flexibility in the definition of control objectives. Although several FCS-MPC strategies have already been proposed for multiphase drives, a comparative study that assembles all these strategies in a single reference is still missing. Hence, this paper aims to provide an overview and a critical comparison of all available FCS-MPC techniques for electric drives based on six-phase machines, focusing mainly on predictive current control (PCC) and predictive torque control (PTC) strategies. The performance of an asymmetrical six-phase permanent magnet synchronous machine is compared side-by-side for a total of thirteen PCC and five PTC strategies, with the aid of simulation and experimental results. Finally, in order to determine the best and the worst performing control strategies, each strategy is evaluated according to distinct features, such as ease of implementation, minimization of current harmonics, tuning requirements, computational burden, among others.

Keywords: multiphase electric drives; multiphase machines; six-phase machines; finite control set model predictive control; predictive current control; predictive torque control

---

1. Introduction

Rotating electrical machines with a number of phases higher than three ($n > 3$) are commonly referred to in the literature as multiphase machines [1]. Multiphase machines were first used in high power generation units during the 1920s due to the current limit of circuit breakers at that time and due to the size of the reactors needed to limit currents in the event of faults [2]. In the 1960s, it was demonstrated that an increase in the number of phases of electrical machines fed by voltage source inverters (VSIs) leads to an increase of the order of torque pulsations ($h = 2n$) and to a reduction of their magnitude [3]. Additionally, the increase in the number of phases also leads to a lower current or voltage per phase, decreasing the requirements of the power semiconductors ratings [4]. An electric drive with improved reliability based on a multiphase machine was first studied in the 1980s [5], where each phase was connected to an independent power converter and in the event of a fault in one or more phases, the drive could remain in operation with a reduced power rating. Since the decoupled control of flux and torque in multiphase machines only requires the regulation of two independent current components, regardless of the number of phases of the machine [6], multiphase machines provide additional degrees of freedom that can be used for several purposes without affecting the production of flux and torque [7]. The first works taking advantage of the additional degrees of freedom of multiphase machines were published in the 1990s, where the injection of current harmonics was used to enhance the torque developed by machines with concentrated windings [8,9]. Multimotor drives proposed in the 2000s is another application that takes advantage of the additional degrees of freedom, where a single n-phase VSI is able to drive independently up to $(n-1)/2$ machines if n is odd or up to $(n-2)/2$ machines if n is even, either connected in series or in parallel [10,11]. More recently, the additional degrees of freedom of multiphase machines are being used to provide: balancing of the dc-link capacitors of series-connected VSIs on the machine side [12]; unequal power sharing [13,14]; full-load test methods [15,16]; integrated battery charging for electric vehicles [17–19]; dynamic braking for non-regenerative electric drives [20,21]; and diagnosis of open-phase faults [22,23]. In addition to the reduced current or voltage ratings per phase, lower torque harmonics, improved fault-tolerant capabilities and additional degrees of freedom, multiphase machines also offer other advantages over their three-phase counterparts, namely: improved winding factors, reduced harmonic content in the magnetomotive force (MMF), lower rotor losses and lesser harmonics in the dc-link current [1,24–26]. Nowadays, electric drives based on multiphase machines are employed in a wide range of areas, such as aircraft [27,28], electric or hybrid vehicles [29], locomotive traction [30], high-speed elevators [31], ship propulsion [32], spacecraft [33] and wind energy applications [34–36].

In multiphase electric drives, n-phase machines are typically supplied by a n-phase VSI, whose power semiconductors are commanded by a control strategy in order to achieve variable speed operation [26]. Several control strategies have been reported in the literature over the years for multiphase electric drives, such as scalar or constant V/f control, field oriented control (FOC) and direct torque control (DTC) [25,26]. Scalar control regulates the speed of the machine by imposing a constant ratio between the amplitude and the frequency of the stator voltage [3]. Since the constant V/f control cannot control directly the currents, an unbalance in the machine can lead to the appearance of x-y current components with considerable magnitude [37]. Moreover, the reference voltages generated by scalar control are translated into command signals for the power semiconductors of the VSI using pulse width modulation (PWM) or space vector PWM (SV-PWM) techniques [7]. However, similarly to standard three-phase electric drives, scalar control cannot provide accurate control of the rotor speed of multiphase machines and leads to a poor dynamic performance [25]. On the other hand, both FOC and DTC schemes provide a decoupled control of the flux and torque, improving the control of the machine [38]. Typically, in FOC schemes the flux and torque of the machine are adjusted by regulating two independent current components with proportional-integral (PI) controllers, regardless of the number of phases of the machine, and the VSI control signals are synthesized with PWM or SV-PWM techniques [39,40]. In the case of DTC schemes, the flux and torque are controlled directly with hysteresis controllers and the control actuation is usually selected using a switching table [41,42].

In the last decade, finite control set model predictive control (FCS-MPC), along with control strategies such as FOC and DTC, has been proposed for the control of high-performance electric drives [43–45]. The main advantages of FCS-MPC over the classical control strategies are the improved dynamic performance, flexibility in the definition of control objectives and easy inclusion of constraints [46]. Since SV-PWM techniques can be hard to implement in multiphase drives [7], particularly for machines with a high number of phases or when multilevel converters are employed, FCS-MPC is also an attractive solution for multiphase drives since it does not require the use of a modulator [46]. FCS-MPC strategies use a discrete version of the system model to predict the future behavior of the controlled variables, considering a finite set of possible actuations of the power converters [47]. Typically, FCS-MPC strategies can be based on the application of a single switching state during a sampling period, referred to as optical switching vector MPC (OSV-MPC), or as an alternative, consider the application of an optimal switching sequence, known as optimal switching sequence (OSS-MPC) [48]. The control objectives of the FCS-MPC strategies are expressed in the form of a cost function, which evaluates the error between the controlled variables and their reference values. Hence, the optimal actuation is obtained by selecting, among the considered finite set of control actuations the one that leads to the minimum value of the cost function [46–48].

The FCS-MPC strategies available in the literature for multiphase drives are commonly classified according to their control objectives, such as predictive current control (PCC), predictive torque control (PTC) or predictive speed control (PSC) [49]. In the case of PCC schemes, the stator currents are the controlled variables, while the flux and torque are usually selected as the controlled variables in PTC strategies [50]. The PSC scheme eliminates the external PI speed loop present in PCC and PTC strategies although it requires the tuning of several weighting factors and depends on the mechanical parameters of the drive to estimate the load torque and predict the rotor speed [51]. Due to these limitations of PSC schemes, applications of both PCC and PTC strategies for multiphase drives are more popular among the research community and can be found in multiple publications [49,50,52–55].

Although several works have reported implementations of FCS-MPC strategies for electric drives based on multiphase machines in recent years, very few works attempted to review and compare these control strategies [50]. The publications [43,50,56,57] provide an overview of FCS-MPC strategies applied to five and six-phase machines, which are the simplest and the most addressed configurations in the literature [58]. However, these publications do not provide simulation or experimental results and lack a critical comparison among the considered FCS-MPC control strategies. On the other hand, a comparison between several FCS-MPC strategies applied to a six-phase PMSM drive was presented in Reference [49], although only simulation results were provided and the latest contributions in this field are missing.

This paper assembles in a single reference all published FCS-MPC strategies for electric drives based on six-phase machines. It presents a critical comparative study between the different FCS-MPC strategies, highlighting their advantages and drawbacks, being supported by a theoretical framework and by both simulation and experimental results obtained with a six-phase PMSM drive. Additionally, the paper includes a section providing an overview of the different topologies of multiphase electric drives and a section detailing the modeling of six-phase machine drives.

The paper is structured as follows—Section 2 provides an overview of the existing multiphase electric drives, Section 3 discusses the modeling of six-phase drives and Section 4 presents the theory behind the FCS-MPC strategies for the six-phase drives published so far. Moreover, Section 5 presents the simulation results of the reviewed FCS-MPC strategies, while Section 6 presents the experimental results for the same control strategies. Finally, Section 7 contains the main conclusions of this work.

2. Multiphase Electric Drives

Since n-phase machines are supplied by n-phase power converters in multiphase electric drives, the variable n is not restricted by the number of phases of the electric grid and can be selected according to the application [26]. Hence, this section provides an overview of the different types of n-phase machines and n-phase power converters reported in the literature. Since this paper provides an overview of the FCS-MPC strategies for six-phase drives ($n = 6$) in particular, this topology is analyzed in more detail in Section 2.3.

2.1. Types of Multiphase Electric Machines

The main difference between multiphase and standard three-phase machines is the configuration of the stator windings [58]. In multiphase machines, the stator windings are designed to have n phases and can be of distributed or concentrated type, depending on the number of stator slots per pole per phase [25]. Regarding the machine type in multiphase drives, the majority of the literature published in recent years has been focused on induction machines (IMs) and permanent magnet synchronous machines (PMSMs) [50,59]. In comparison to IMs, PMSMs provide a higher efficiency and power density, higher power factor and enhanced fault tolerance against open-phase faults [58,60].

Multiphase machines are typically classified according to the spatial displacement between phases and are denominated as symmetrical or asymmetrical machines [58]. In the symmetrical configuration, the stator windings of a n-phase machine are designed to have a displacement of $360/n$ electrical degrees between consecutive phases [61]. However, if n is an odd and non-prime number or if n

is an even number, that is, $n = \{6, 9, 12, 15, ...\}$, the stator windings of a n-phase machine can also be designed to have an asymmetrical configuration. In the asymmetrical configuration, the stator windings are associated in k sets of windings, each one with a phases ($n = a \cdot k$) spaced by $360/a$ electrical degrees and the k sets of windings are displaced by $\alpha = 180/n$ electrical degrees between them [62]. In the symmetrical configuration, the n-phases are usually wye-connected with a single neutral point, whereas in the asymmetrical configuration the a phases within each set of windings are wye-connected with k neutral points, which can be left isolated or connected to each other [59].

The asymmetrical configuration with k isolated neutral points is widely adopted since it restricts the circulation of zero-sequence currents (ZSC) [63] and provides isolation among the k sets of windings [26], although the number of independent currents is reduced to from $n-1$ to $n-k$ in comparison with the single neutral point case [25]. Since the k sets of windings are isolated from each other, the use of coupling inductors is not necessary as in the case of high-power three-phase machines, where several power converters are associated in parallel to achieve high power ratings [60]. In the case of a fault in either the converter or the machine, a simple fault-tolerant control strategy can be adopted for machines with $n = a \cdot k$ phases and k isolated neutrals by simply deactivating the affected set of windings, while the drive is maintained in operation with a reduction in the power rating of $1/k \times 100\%$ [26]. On the other hand, the asymmetrical configuration with a single neutral configuration has additional $k-1$ degrees of freedom, resulting in improved performance under fault-tolerant operation [62,64,65].

The majority of multiphase machines with an asymmetrical configuration are designed to have multiple sets of three-phase windings ($a = 3$) in order to maintain the compatibility with standard three-phase power converters [59,66]. Examples of application of these multiphase machines are the six-phase ($k = 2$) and twelve-phase generators ($k = 4$) used in wind energy applications [67], and the nine-phase machine used in high-speed elevators [31]. Although less common, the sets of windings can be arranged with a number of phases different from three, such as the fifteen-phase machine with three sets of five-phase windings ($a = 5$) reported in Reference [58] for a ship propulsion system.

2.2. Types of Power Converters

The n-phase power converters employed in multiphase drives can be of two types: n-phase VSIs or n-phase current source inverters (CSIs). Nowadays, n-phase VSIs are usually adopted in multiphase drives [58], while CSIs were used in some of the earlier multiphase drives [68,69]. Typically, two-level voltage source inverters (2L-VSIs) are used to drive multiphase machines in industrial applications, although multilevel topologies, such as the neutral-point-clamped (3L-NPC) converter can also be employed [7,70]. Other topologies of multiphase VSIs referred in the literature are the n-phase matrix converter and the n-phase H-bridge converter [26,66].

Multiphase power converters can be associated to supply the machine from one side or from both sides, being usually denominated as single or double-sided supply [7]. The single-sided configuration is the one typically employed in multiphase drives, since the stator windings in multiphase phase machines are commonly wye-connected into one or multiple stars [26]. In the double-sided configuration, the stator windings of the machine are supplied from both sides in an open winding configuration, increasing the number of levels of the phase voltage supplied by the VSIs and improving the fault-tolerant performance of the system [7,58]. However, the double-sided supply configuration requires twice the number of VSIs used in the single-sided configuration, increasing the complexity of the electric drive.

Typically, n-phase machines with multiple sets of a-phase windings ($n = a \cdot k$) are supplied from k VSIs, each one with a phases [59]. These k VSIs can be arranged into three configurations regarding the dc-link side: (i) a single dc-link; (ii) k isolated dc-links; (iii) k series-connected dc-links [58]. In healthy operation, both the single dc-link and the k isolated dc-links provide similar performance, the only difference is in fault-tolerant operation where the single dc-link provides better capabilities [71–73]. In spite of elevating the total dc-link voltage, the use of k series-connected dc-links requires a control strategy to guarantee the balance of the voltage of the dc-link capacitors and performs worse than the other configurations in fault-tolerant operation [12].

2.3. Particular Case: Six-Phase Drives

Among multiphase machines with multiple sets of three-phase windings, the asymmetrical six-phase machine with two isolated neutrals (2N) is the simplest and the most studied configuration [1,25,26,50,58,59,74]. In the literature, six-phase machines are also referred to as dual three-phase, dual-stator, double-star, quasi six-phase or split-phase machines [25,38]. The diagram of a typical six-phase drive is presented in Figure 1, where a six-phase machine (either an IM or a PMSM) with an asymmetrical winding configuration is supplied by two 2L-VSIs connected to a single dc-link.

Regarding the configuration of the stator windings of six-phase machines, the asymmetrical configuration is the most reported in the literature, where the two sets of three-phase windings are displaced by thirty electrical degrees ($\alpha = 30°$), as shown in Figure 1b [58]. The asymmetrical configuration provides a reduction of the MMF harmonic content and eliminates the torque harmonics of order $h = 6 \cdot m$, with m being an odd number [37,75,76]. Other values for the displacement between the two sets of windings, such as $\alpha = 0°$ and $\alpha = 60°$(symmetrical configuration), do not provide a reduction of the harmonic content of the MMF and torque [25,76,77].

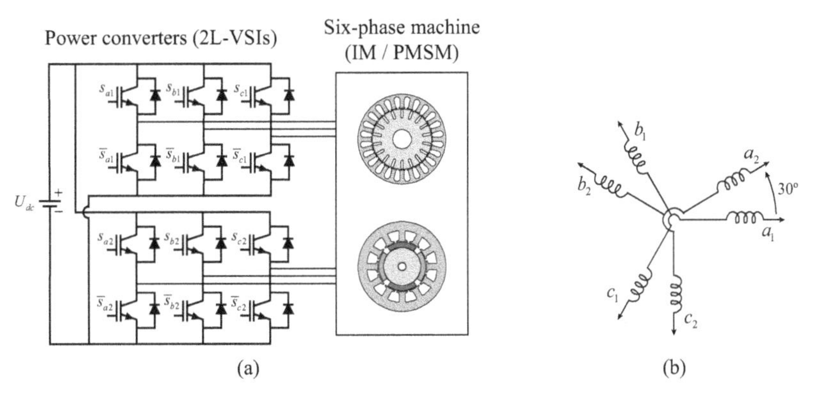

Figure 1. Six-phase asymmetrical drive: (**a**) power circuit; (**b**) winding arrangement of the six-phase asymmetrical machine with a 2N configuration.

Since in six-phase machines the two sets of windings are typically wye-connected, three neutral configurations are possible: (i) 2N; (ii) single isolated neutral (1N); and (iii) single neutral connected to the middle point of the dc-link bus or to an extra leg of the VSI, being termed as single non-isolated neutral (1NIN) in this paper [62]. The 2N configuration is often used since it provides a better dc-link voltage usage and avoids the circulation of ZSCs, leading to a better performance in steady-state operation, with lesser current harmonics, in comparison to the 1N and 1NIN configurations [58]. On the other hand, the 1N and 1NIN configurations proved to be advantageous in fault-tolerant operation [62,64,65,78], and in the enhancement of torque for the case of the 1NIN configuration [79,80].

3. System Model

This chapter presents the mathematical model of electric drives based on six-phase machines (both IMs and PMSMs) fed by two 2L-VSIs, which is required for the implementation of FCS-MPC strategies.

3.1. Introduction

In the literature, two distinct transformations are reported to model six-phase machines: the double *d-q* transformation and the variable space decomposition (VSD) transformation [38,50,81].

The double d-q transformation consists in the application of the Park transformation to both sets of windings [75,82] and was widely used in the first FOC and DTC strategies proposed for six-phase machines [25,38]. The VSD transformation is widely used nowadays not only in FCS-MPC but also in FOC and DTC strategies [50,58] since it is able to separate the current, flux linkage and voltage components responsible for the electromechanical energy conversion, mapped into the α-β subspace, from the remaining components, mapped into the x-y subspace, which can be used as additional degrees of freedom [7]. Moreover, the VSD transformation eliminates the coupling terms between the different subspaces in the model of six-phase machines, which are present when the double d-q transformation is used instead [81]. Additionally, the VSD transformation maps the current, flux and voltage harmonics of order $h = 12m \pm 1$ with $m = 1, 2, ...$ into the α-β subspace, while the harmonics of order $h = 6m \pm 1$ with $m = 1, 3, ...,$ are mapped into the x-y subspace [83].

3.2. Two-Level Voltage Source Inverters

Considering a six-phase machine with a 2N configuration, the phase voltages depend on the switching state vector $\mathbf{s}$ of the 2L-VSIs defined in (1) and are calculated with (2):

$$\mathbf{s} = \begin{bmatrix} s_{a1} & s_{b1} & s_{c1} & s_{a2} & s_{b2} & s_{c2} \end{bmatrix}^{\mathrm{T}}, \tag{1}$$

$$\begin{bmatrix} u_{a1s} \\ u_{b1s} \\ u_{c1s} \\ u_{a2s} \\ u_{b2s} \\ u_{c2s} \end{bmatrix} = \frac{U_{dc}}{3} \begin{bmatrix} 2 & -1 & -1 & 0 & 0 & 0 \\ -1 & 2 & -1 & 0 & 0 & 0 \\ -1 & -1 & 2 & 0 & 0 & 0 \\ 0 & 0 & 0 & 2 & -1 & -1 \\ 0 & 0 & 0 & -1 & 2 & -1 \\ 0 & 0 & 0 & -1 & -1 & 2 \end{bmatrix} \cdot \mathbf{s}, \tag{2}$$

where $s_u = \{0, 1\}$ is the switching state of phase u, with $u \in \{a_1, b_1, c_1, a_2, b_2, c_2\}$. If $s_u = 1$, the top insulated gate bipolar transistor (IGBT) of phase u is ON and the bottom IGBT is OFF, while the opposite is true when $s_u = 0$. By applying the VSD transformation [83] defined in (3) to the phase voltages given by (2), the stator voltage components of the six-phase machine in the α-β, x-y and z_1-z_2 subspaces are calculated by (4):

$$\mathbf{T}_{vsd} = \frac{1}{3} \begin{bmatrix} 1 & \cos\left(\frac{2\pi}{3}\right) & \cos\left(\frac{4\pi}{3}\right) & \cos\left(\frac{\pi}{6}\right) & \cos\left(\frac{5\pi}{6}\right) & \cos\left(\frac{9\pi}{6}\right) \\ 0 & \sin\left(\frac{2\pi}{3}\right) & \sin\left(\frac{4\pi}{3}\right) & \sin\left(\frac{\pi}{6}\right) & \sin\left(\frac{5\pi}{6}\right) & \sin\left(\frac{9\pi}{6}\right) \\ 1 & \cos\left(\frac{4\pi}{3}\right) & \cos\left(\frac{2\pi}{3}\right) & -\cos\left(\frac{\pi}{6}\right) & -\cos\left(\frac{5\pi}{6}\right) & -\cos\left(\frac{9\pi}{6}\right) \\ 0 & \sin\left(\frac{4\pi}{3}\right) & \sin\left(\frac{2\pi}{3}\right) & \sin\left(\frac{\pi}{6}\right) & \sin\left(\frac{5\pi}{6}\right) & \sin\left(\frac{9\pi}{6}\right) \\ 1 & 1 & 1 & 0 & 0 & 0 \\ 0 & 0 & 0 & 1 & 1 & 1 \end{bmatrix}, \tag{3}$$

$$\begin{bmatrix} u_{\alpha s} & u_{\beta s} & u_{xs} & u_{ys} & u_{z1s} & u_{z2s} \end{bmatrix}^{\mathrm{T}} = \mathbf{T}_{vsd} \cdot \begin{bmatrix} u_{a1s} & u_{b1s} & u_{c1s} & u_{a2s} & u_{b2s} & u_{c2s} \end{bmatrix}^{\mathrm{T}}. \tag{4}$$

The sixty-four different possibilities for the switching state vector $\mathbf{s}$ result in forty-nine distinct voltage vectors mapped into the α-β and x-y subspaces simultaneously, as shown in Figure 2. The projections of the voltage vectors in z_1-z_2 are not considered in the model of six-phase machines with a 2N configuration since ZSCs cannot circulate [84]. The index of the voltage vectors represented in Figure 2 is obtained by the conversion of the binary number of vector $\mathbf{s}$ into a decimal number.

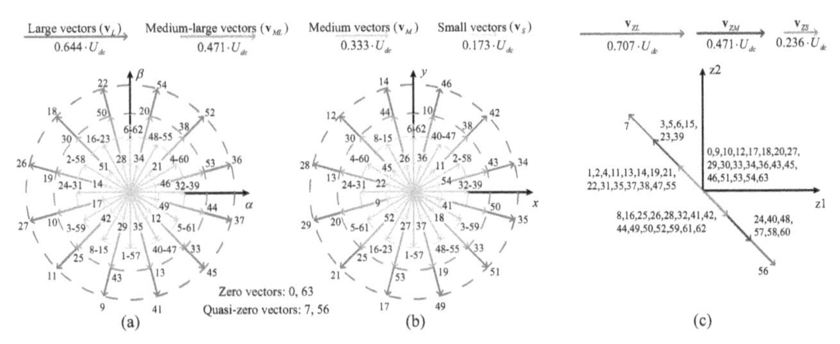

Figure 2. Voltage vectors of the six-phase machine in a stationary reference frame, mapped into the: (**a**) α-β; (**b**) x-y; (**c**) z1-z2 subspaces.

3.3. Six-Phase Induction Machine

In order to obtain the dynamic model of a six-phase IM in an arbitrary reference frame rotating at an angular speed ω_a, the following rotation matrix is used along with the VSD transformation:

$$\mathbf{R}(\theta_a) = \begin{bmatrix} \mathbf{T}_r(\theta_a) & \mathbf{0}_2 & \mathbf{0}_2 \\ \mathbf{0}_2 & \left(\mathbf{T}_r(\theta_a)\right)^{-1} & \mathbf{0}_2 \\ \mathbf{0}_2 & \mathbf{0}_2 & \mathbf{I}_2 \end{bmatrix}, \quad \mathbf{T}_r(\theta_a) = \begin{bmatrix} \cos(\theta_a) & \sin(\theta_a) \\ -\sin(\theta_a) & \cos(\theta_a) \end{bmatrix}, \tag{5}$$

where θ_a is the electrical angle of the arbitrary reference frame, with the d-axis aligned with the airgap, stator or rotor flux [38]. Matrix $R(\theta_a)$ rotates the α-β components in the counterclockwise direction in order to obtain the d-q components, while the x-y components are rotated in the clockwise direction in order to obtain the x'-y' components. This direction of rotation of the x-y components is adopted in recent works since it makes easier to control the unbalance of the machine [12,20,84,85].

Considering sinusoidally distributed windings, negligible saturation and symmetry between the different phases, the voltage equations of a six-phase IM in an arbitrary reference frame (rotating at an angular speed ω_a), obtained by the application of the VSD transformation along with (5) are given by [58,86]:

$$\begin{bmatrix} u_{ds} \\ u_{qs} \\ u_{xs'} \\ u_{ys'} \\ u_{z1s} \\ u_{z2s} \end{bmatrix} = R_s \begin{bmatrix} i_{ds} \\ i_{qs} \\ i_{xs'} \\ i_{ys'} \\ i_{z1s} \\ i_{z2s} \end{bmatrix} + \frac{\mathrm{d}}{\mathrm{dt}} \begin{bmatrix} \psi_{ds} \\ \psi_{qs} \\ \psi_{xs'} \\ \psi_{ys'} \\ \psi_{z1s} \\ \psi_{z2s} \end{bmatrix} + \omega_a \begin{bmatrix} 0 & -1 & 0 & 0 & 0 & 0 \\ 1 & 0 & 0 & 0 & 0 & 0 \\ 0 & 0 & 0 & 1 & 0 & 0 \\ 0 & 0 & -1 & 0 & 0 & 0 \\ 0 & 0 & 0 & 0 & 0 & 0 \\ 0 & 0 & 0 & 0 & 0 & 0 \end{bmatrix} \begin{bmatrix} \psi_{ds} \\ \psi_{qs} \\ \psi_{xs'} \\ \psi_{ys'} \\ \psi_{z1s} \\ \psi_{z2s} \end{bmatrix}, \tag{6}$$

$$\begin{bmatrix} u_{dr} \\ u_{qr} \end{bmatrix} = R_r \begin{bmatrix} i_{dr} \\ i_{qr} \end{bmatrix} + \frac{\mathrm{d}}{\mathrm{dt}} \begin{bmatrix} \psi_{dr} \\ \psi_{qr} \end{bmatrix} + (\omega_a - \omega_r) \begin{bmatrix} 0 & -1 \\ 1 & 0 \end{bmatrix} \begin{bmatrix} \psi_{dr} \\ \psi_{qr} \end{bmatrix}, \tag{7}$$

where $\{R_s, R_r\}$ are the equivalent resistance of the stator and rotor windings, respectively, ω_r is the rotor electric angular speed, the symbols u, i, ψ represent the voltage, current and flux linkage and indexes s and r stand for stator and rotor variables. The voltage equations of the six-phase IM can be written in the stationary or stator reference frame for $\omega_a = 0$, in the rotor reference frame for $\omega_a = \omega_r$, or in the synchronous reference frame for $\omega_a = \omega_s$, with ω_s being the synchronous angular speed.

Taking into account the effect of the mutual leakage inductance, which is non-negligible in six-phase machines with short-pitched windings [87], the relation between the flux linkage and the current components mapped into the d-q and x'-y' subspaces is given by [25,38]:

$$\begin{bmatrix} \psi_{ds} \\ \psi_{qs} \\ \psi_{xs'} \\ \psi_{ys'} \\ \psi_{z1s} \\ \psi_{z2s} \end{bmatrix} = \begin{bmatrix} L_s & 0 & 0 & 0 & 0 & 0 \\ 0 & L_s & 0 & 0 & 0 & 0 \\ 0 & 0 & L_{ls} & 0 & 0 & 0 \\ 0 & 0 & 0 & L_{ls} & 0 & 0 \\ 0 & 0 & 0 & 0 & L_0 & 0 \\ 0 & 0 & 0 & 0 & 0 & L_0 \end{bmatrix} \begin{bmatrix} i_{ds} \\ i_{qs} \\ i_{xs'} \\ i_{ys'} \\ i_{z1s} \\ i_{z2s} \end{bmatrix} + \begin{bmatrix} M_m & 0 \\ 0 & M_m \\ 0 & 0 \\ 0 & 0 \\ 0 & 0 \\ 0 & 0 \end{bmatrix} \begin{bmatrix} i_{dr} \\ i_{qr} \end{bmatrix}, \tag{8}$$

$$\begin{bmatrix} \psi_{dr} \\ \psi_{qr} \end{bmatrix} = \begin{bmatrix} M_m & 0 & 0 & 0 & 0 & 0 \\ 0 & M_m & 0 & 0 & 0 & 0 \end{bmatrix} \begin{bmatrix} i_{ds} \\ i_{qs} \\ i_{xs'} \\ i_{ys'} \\ i_{z1s} \\ i_{z2s} \end{bmatrix} + \begin{bmatrix} L_r & 0 \\ 0 & L_r \end{bmatrix} \begin{bmatrix} i_{dr} \\ i_{qr} \end{bmatrix}. \tag{9}$$

The inductance parameters in (8) and (9) are given by [88]:

$$\begin{cases} L_s = L_{ls} + 2L_{lm} + 3L_m \\ L_0 = L_{ls} + L_{lm} \\ L_r = L_{lr} + 3L_m \\ M_m = 3L_m \end{cases}, \tag{10}$$

where L_{ls} is the self leakage inductance of stator, L_{lm} is the mutual leakage inductance of the stator, L_m is the magnetizing inductance and L_{lr} is the self leakage inductance of the rotor. The torque developed by the six-phase IM is computed with [25]:

$$t_e = 3p\left(\psi_{ds} \cdot i_{qs} - \psi_{qs} \cdot i_{ds}\right) = -3p\left(\psi_{dr} \cdot i_{qr} - \psi_{qr} \cdot i_{dr}\right) = 3pM_m\left(i_{dr} \cdot i_{qs} - i_{qr} \cdot i_{ds}\right). \tag{11}$$

From (8) and (11), it becomes clear that the flux linkage components mapped into the x'-y' subspace do not contribute to the production of torque and only contribute to the stator leakage flux [87]. Since the equivalent impedance of the machine in the x'-y' subspace is very low, as it only depends on R_s and L_{ls}, it might lead to the circulation of large currents in this subspace, which contributes to the increase of the stator copper losses [89].

3.4. Six-Phase Permanent Magnet Synchronous Machine

Assuming sinusoidally distributed windings, negligible saturation and symmetry between the different phases, the dynamic model of a six-phase PMSM in the rotor reference frame (rotating at ω_r) obtained with the VSD transformation and rotation matrix (5) is defined by [58,76]:

$$\begin{bmatrix} u_{ds} \\ u_{qs} \\ u_{xs'} \\ u_{ys'} \\ u_{z1s} \\ u_{z2s} \end{bmatrix} = R_s \begin{bmatrix} i_{ds} \\ i_{qs} \\ i_{xs'} \\ i_{ys'} \\ i_{z1s} \\ i_{z2s} \end{bmatrix} + \frac{\mathrm{d}}{\mathrm{dt}} \begin{bmatrix} \psi_{ds} \\ \psi_{qs} \\ \psi_{xs'} \\ \psi_{ys'} \\ \psi_{z1s} \\ \psi_{z2s} \end{bmatrix} + \omega_r \begin{bmatrix} 0 & -1 & 0 & 0 & 0 & 0 \\ 1 & 0 & 0 & 0 & 0 & 0 \\ 0 & 0 & 0 & 1 & 0 & 0 \\ 0 & 0 & -1 & 0 & 0 & 0 \\ 0 & 0 & 0 & 0 & 0 & 0 \\ 0 & 0 & 0 & 0 & 0 & 0 \end{bmatrix} \begin{bmatrix} \psi_{ds} \\ \psi_{qs} \\ \psi_{xs'} \\ \psi_{ys'} \\ \psi_{z1s} \\ \psi_{z2s} \end{bmatrix}, \tag{12}$$

$$\begin{bmatrix} \psi_{ds} \\ \psi_{qs} \\ \psi_{xs'} \\ \psi_{ys'} \\ \psi_{z1s} \\ \psi_{z2s} \end{bmatrix} = \begin{bmatrix} L_d & 0 & 0 & 0 & 0 & 0 \\ 0 & L_q & 0 & 0 & 0 & 0 \\ 0 & 0 & L_x & 0 & 0 & 0 \\ 0 & 0 & 0 & L_y & 0 & 0 \\ 0 & 0 & 0 & 0 & L_{01} & 0 \\ 0 & 0 & 0 & 0 & 0 & L_{02} \end{bmatrix} \begin{bmatrix} i_{ds} \\ i_{qs} \\ i_{xs'} \\ i_{ys'} \\ i_{z1s} \\ i_{z2s} \end{bmatrix} + \begin{bmatrix} \psi_{ds,\mathrm{PM}} \\ \psi_{qs,\mathrm{PM}} \\ \psi_{xs',\mathrm{PM}} \\ \psi_{ys',\mathrm{PM}} \\ \psi_{z1s,\mathrm{PM}} \\ \psi_{z2s,\mathrm{PM}} \end{bmatrix}, \quad (13)$$

where $\{L_d, L_q, L_x, L_y, L_{01}, L_{02}\}$ are the equivalent inductances of the d, q, x', y', z1 and z2 axis, respectively and $\psi_{vs,\mathrm{PM}}$ is the v-component of the stator flux linkage due to permanent magnets (PMs), with $v \in \{d, q, x', y', z1, z2\}$. Considering only the fundamental component of the stator flux linkage due to the PMs, the flux components in the d-q, x'-y' and z1-z2 subspaces are given by:

$$\begin{bmatrix} \psi_{ds,\mathrm{PM}} & \psi_{qs,\mathrm{PM}} & \psi_{xs',\mathrm{PM}} & \psi_{ys',\mathrm{PM}} & \psi_{z1s,\mathrm{PM}} & \psi_{z2s,\mathrm{PM}} \end{bmatrix}^{\mathrm{T}} = \begin{bmatrix} \psi_{s,\mathrm{PM1}} & 0 & 0 & 0 & 0 & 0 \end{bmatrix}^{\mathrm{T}}, \quad (14)$$

where ψ_{PM1} is the peak value of the fundamental component of the stator flux linkage due to the PMs. The torque of a six-phase PMSM is calculated with (15) [81]:

$$t_e = 3p \left[\psi_{s,\mathrm{PM1}} i_{qs} + \left(L_d - L_q\right) i_{ds} i_{qs}\right]. \quad (15)$$

Considering a six-phase PMSM with surface-mounted PMs (SPMSM), the inductance parameters are given by:

$$\begin{cases} L_d = L_q = L_{dq} = L_{ls} + 2L_{lm} + 3L_m \\ L_x = L_y = L_{xy} = L_{ls} \\ L_{01} = L_{02} = L_0 = L_{ls} + L_{lm} \end{cases}, \quad (16)$$

and the torque expression is reduced to [76]:

$$t_e = 3p \left(\psi_{s,\mathrm{PM1}} i_{qs}\right). \quad (17)$$

Equations (12)–(17) show that only the d-q current components contribute to the production of torque in six-phase PMSMs with distributed windings and in the case of SPMSMs the torque depends only on the q-axis current component. On the other hand, the x'-y' current components are only limited by a small equivalent impedance, which can lead to the appearance of large x'-y' currents in six-phase PMSMs fed by VSIs [58].

4. Finite Control Set Model Predictive Control

Model predictive control (MPC) uses a model of the system to predict the future values of the output variables and selects a control actuation by minimizing a cost function, which defines the control objectives [47]. In the last decade, the increase in the computational power of real-time control platforms has made possible the application of MPC strategies to electric drives [46]. In the literature, MPC strategies are usually divided into two categories: CCS-MPC (continuous control set model predictive control) and FCS-MPC [43,44,57]. The FCS-MPC is usually preferred in the control of electric drives due to the easy inclusion of constraints and non-linearities in the cost function [48]. Due to the flexibility of FCS-MPC strategies, different control objectives can be set in the cost function, such as the reference tracking of current, torque, flux or speed. PCC and PTC are the most reported FCS-MPC variants for six-phase machine drives [50,58]. Although less common PSC was also proposed in Reference [51] to eliminate the speed PI controller present in PCC and PTC, although it requires the tuning of several weighting factors and depends on the mechanical parameters of the drive to estimate the load torque and predict the rotor speed. Hence, this paper is focused only on PCC and PTC variants.

4.1. Standard and Restrained Search Predictive Current Control

The standard predictive current control (S-PCC) strategy for electric drives based on six-phase IMs was introduced in References [90,91]. In order to predict the values of the stator currents for instant $k+h$, where h is the prediction horizon, the model of the six-phase IM (6)–(9) is discretized with the forward Euler method:

$$\begin{bmatrix} i_{ds}^{k+h} \\ i_{qs}^{k+h} \\ i_{xs'}^{k+h} \\ i_{ys'}^{k+h} \end{bmatrix} = \begin{bmatrix} 1-\frac{R_s T_s}{\sigma L_s} & \left(\frac{\omega_r}{\sigma}+\omega_k\right) T_s & 0 & 0 \\ -\left(\frac{\omega_r}{\sigma}+\omega_k\right) T_s & 1-\frac{R_s T_s}{\sigma L_s} & 0 & 0 \\ 0 & 0 & 1-\frac{R_s T_s}{L_{ls}} & -\omega_a T_s \\ 0 & 0 & \omega_a T_s & 1-\frac{R_s T_s}{L_{ls}} \end{bmatrix} \begin{bmatrix} i_{ds}^{k+h-1} \\ i_{qs}^{k+h-1} \\ i_{xs'}^{k+h-1} \\ i_{ys'}^{k+h-1} \end{bmatrix} + \begin{bmatrix} \frac{R_r M_m T_s}{\sigma L_r L_s} & \omega_r \frac{M_m T_s}{\sigma L_s} \\ -\omega_r \frac{M_m T_s}{\sigma L_s} & \frac{R_r M_m T_s}{\sigma L_r L_s} \\ 0 & 0 \\ 0 & 0 \end{bmatrix} \begin{bmatrix} i_{dr}^{k+h-1} \\ i_{qr}^{k+h-1} \end{bmatrix} + \begin{bmatrix} \frac{T_s}{\sigma L_s} & 0 & 0 & 0 \\ 0 & \frac{T_s}{\sigma L_s} & 0 & 0 \\ 0 & 0 & \frac{T_s}{L_{ls}} & 0 \\ 0 & 0 & 0 & \frac{T_s}{L_{ls}} \end{bmatrix} \begin{bmatrix} u_{ds}^{k+h-1} \\ u_{qs}^{k+h-1} \\ u_{xs'}^{k+h-1} \\ u_{ys'}^{k+h-1} \end{bmatrix}, \quad (18)$$

where $\sigma = 1 - M_m^2/(L_r L_s)$ and $\omega_k = \omega_a - \omega_r$. Since the rotor currents cannot be measured, they must be estimated either using an observer, such as the Luenberger observer or a Kalman filter [92,93] or using the past values of the measured variables [91,94]. In order to compensate the delay in the actuation, a prediction horizon of two samples ahead ($h = 2$) is usually selected in FCS-MPC strategies. Hence, the stator current components are predicted for instant $k+2$ using (18) (with $h = 2$), which depend on the rotor current components at instant $k+1$, given by:

$$\begin{bmatrix} i_{dr}^{k+1} \\ i_{qr}^{k+1} \end{bmatrix} = \begin{bmatrix} 1-\frac{R_r T_s}{\sigma L_r} & \left(\omega_a-\frac{\omega_r}{\sigma}\right) T_s \\ -\left(\omega_a-\frac{\omega_r}{\sigma}\right) T_s & 1-\frac{R_r T_s}{\sigma L_r} \end{bmatrix} \begin{bmatrix} i_{dr}^{k} \\ i_{qr}^{k} \end{bmatrix} + \begin{bmatrix} \frac{M_m R_s T_s}{\sigma L_r L_s} & -\frac{\omega_r M_m T_s}{\sigma L_r} \\ \frac{\omega_r M_m T_s}{\sigma L_r} & \frac{M_m R_s T_s}{\sigma L_r L_s} \end{bmatrix} \begin{bmatrix} i_{ds}^{k} \\ i_{qs}^{k} \end{bmatrix} \begin{bmatrix} -\frac{M_m T_s}{\sigma L_r} & 0 \\ 0 & -\frac{M_m T_s}{\sigma L_r} \end{bmatrix} \begin{bmatrix} u_{ds}^{k} \\ u_{qs}^{k} \end{bmatrix}. \quad (19)$$

Alternatively, if a six-phase PMSM is used instead, the predictions of the stator current for instant $k+h$ are computed with:

$$\begin{bmatrix} i_{ds}^{k+h} \\ i_{qs}^{k+h} \\ i_{xs'}^{k+h} \\ i_{ys'}^{k+h} \end{bmatrix} = \begin{bmatrix} 1-\frac{R_s T_s}{L_d} & \frac{\omega_r L_q T_s}{L_d} & 0 & 0 \\ -\frac{\omega_r L_d T_s}{L_q} & 1-\frac{R_s T_s}{L_q} & 0 & 0 \\ 0 & 0 & 1-\frac{R_s T_s}{L_x} & -\frac{\omega_r L_y T_s}{L_x} \\ 0 & 0 & \frac{\omega_r L_x T_s}{L_y} & 1-\frac{R_s T_s}{L_y} \end{bmatrix} \begin{bmatrix} i_{ds}^{k+h-1} \\ i_{qs}^{k+h-1} \\ i_{xs'}^{k+h-1} \\ i_{ys'}^{k+h-1} \end{bmatrix} + \begin{bmatrix} \frac{T_s}{L_d} & 0 & 0 & 0 \\ 0 & \frac{T_s}{L_q} & 0 & 0 \\ 0 & 0 & \frac{T_s}{L_x} & 0 \\ 0 & 0 & 0 & \frac{T_s}{L_y} \end{bmatrix} \begin{bmatrix} u_{ds}^{k+h-1} \\ u_{qs}^{k+h-1} \\ u_{xs'}^{k+h-1} \\ u_{ys'}^{k+h-1} \end{bmatrix} - \begin{bmatrix} 0 \\ \frac{\omega_r T_s}{L_q} \psi_{s,PM1} \\ 0 \\ 0 \end{bmatrix}, \quad (20)$$

The cost function of the S-PCC strategy is evaluated for forty-nine different voltage vectors (Figure 2) and is given by:

$$g_C = \left(i_{ds}^* - i_{ds}^{k+2}\right)^2 + \left(i_{qs}^* - i_{qs}^{k+2}\right)^2 + \lambda_{xy}\left[\left(i_{xs'}^* - i_{xs'}^{k+2}\right)^2 + \left(i_{ys'}^* - i_{ys'}^{k+2}\right)^2\right], \quad (21)$$

where λ_{xy} is the weighting factor that adjusts the relative importance of the reference tracking of the x'-y' current components over the d-q current components. The value of i_{ds}^* is regulated to impose rated flux in IMs, while in the case of SPMSMs, the value of i_{ds}^* is set to zero since the rated flux is produced by the PMs of the rotor [58]. For the operation above rated speed, the value of i_{ds}^* should be reduced in both cases in order to limit the level of the back-electromotive force (EMF), which increases proportionally with the rotor speed [78]. The value of i_{qs}^* can be set directly to regulate the torque of the machine or by a PI controller to regulate the speed of the machine. The voltage vector that minimizes the cost function (21) is selected for application during the next sampling period. Besides the last term of (21), which serves as a constraint to minimize the x-y current components, an additional constraint could be used to reduce the switching frequency, although it would require the tuning of a second weighting factor, which increases the complexity of the strategy. Although some PCC strategies consider the use of magnitude errors in the cost function, as in Reference [91], squared errors provide better reference tracking when the cost function has multiple terms, as stated in Reference [95].

In the case of IMs, the PCC strategies available in the literature use the model of the system in the stationary reference frame ($\omega_a = 0$) [90–92,94], while the PCC strategies for PMSMs use the model of the system in the synchronous reference frame ($\omega_a = \omega_r$), with the d-axis aligned with the flux due to the PMs [49,51]. It is important to note that although the control can be performed in both reference frames, using the synchronous reference frame can simplify the model of the system and avoid the extrapolation of current references to instant $k+2$ used in (21), since both i_{ds} and i_{qs} are constant quantities during steady-state conditions in this frame [67]. On the other hand, the use of the stationary reference frame reduces the number of rotational transformations required in PCC strategies, decreasing their computational burden.

The general diagram of the S-PCC strategy for a six-phase IM drive, considering a stationary reference frame, is given in Figure 3, where the inverse of $\mathbf{T}_r$ defined in (5) is used to obtain the current references in the stationary reference frame [90,91]. Vectors $\mathbf{i}_s^{\alpha\beta}$ and $\mathbf{u}_s^{\alpha\beta}$ are the stator current and voltage vectors in the stationary reference frame, respectively and are defined as:

$$\mathbf{i}_s^{\alpha\beta} = \begin{bmatrix} i_{\alpha s} & i_{\beta s} & i_{xs} & i_{ys} & i_{z1s} & i_{z2s} \end{bmatrix}^{\mathrm{T}}, \qquad \mathbf{u}_s^{\alpha\beta} = \begin{bmatrix} u_{\alpha s} & u_{\beta s} & u_{xs} & u_{ys} & u_{z1s} & u_{z2s} \end{bmatrix}^{\mathrm{T}}. \tag{22}$$

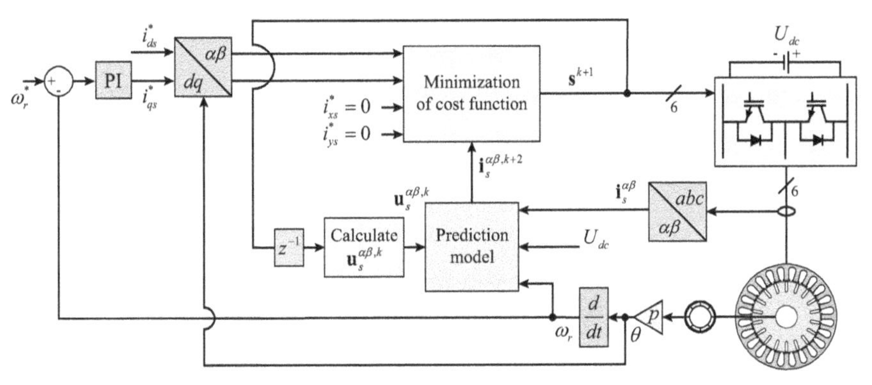

Figure 3. General diagram of the S-PCC strategy for six-phase IM drives.

The general diagram of the S-PCC strategy for six-phase PMSM drives, considering the model of the drive in the synchronous reference frame ($\omega_a = \omega_r$) is presented in Figure 4 [49]. Vectors $\mathbf{i}_s$ and $\mathbf{u}_s$ are the stator current and voltage vectors in the synchronous reference frame rotating at $\omega_a = \omega_r$, respectively and are defined as:

$$\mathbf{i}_s = \begin{bmatrix} i_{ds} & i_{qs} & i_{xs'} & i_{ys'} & i_{z1s} & i_{z2s} \end{bmatrix}^{\mathrm{T}}, \qquad \mathbf{u}_s = \begin{bmatrix} u_{ds} & u_{qs} & u_{xs'} & u_{ys'} & u_{z1s} & u_{z2s} \end{bmatrix}^{\mathrm{T}}. \tag{23}$$

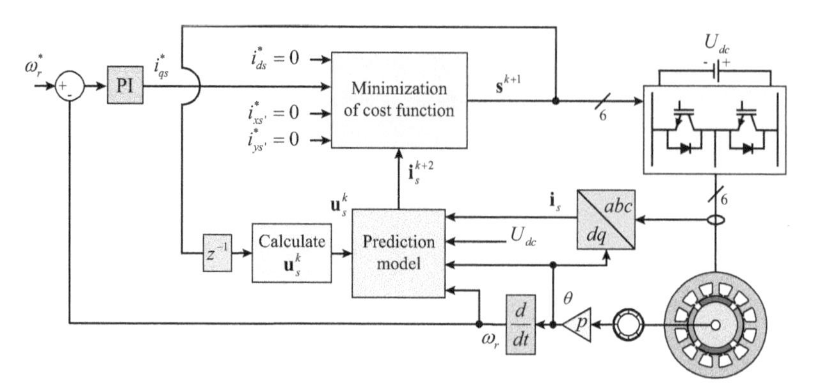

Figure 4. General diagram of the S-PCC strategy for six-phase PMSM drives.

The computational burden of the S-PCC strategy can be high for application in real-time platforms, especially if a multilevel VSI is employed [96]. The evaluation of only twelve large vectors in the S-PCC strategy was considered in References [90,91], however it led to the increase of the x'-y' current harmonics. As an alternative, a restrained search predictive current control (RS-PCC) strategy was proposed in Reference [97], which reduces the number of candidate switching states at each sampling instant. The RS-PCC algorithm imposes the following constraints: (i) the candidate voltage vectors can only generate commutations in two or less VSI legs; and (ii) A VSI leg cannot commute in two consecutive sampling periods. Thus, the RS-PCC algorithm reduces the number of voltage vectors for which the cost function (21) is evaluated from sixty-four to eleven voltage vectors if only two legs were commuted in the previous sampling period or to sixteen voltage vectors if only one leg was commuted [97].

4.2. Predictive Current Control Based on Pulse Width Modulation Schemes

A one-step modulation predictive current control (OSM-PCC) scheme was proposed in Reference [98], which optimizes the length of the voltage vectors in order to improve the performance of six-phase IMs at low speeds. This strategy considers only the twelve large voltage vectors in the α-β subspace and optimizes the length of the optimal vector by minimizing the function:

$$g_{cf}\left(\mathbf{v}_a, d_a\right) = \left(i_{ds}^* - d_a \cdot i_{ds,a}^{k+2} - (1-d_a) \cdot i_{ds,0}^{k+2}\right)^2 + \left(i_{qs}^* - d_a \cdot i_{qs,a}^{k+2} - (1-d_a) \cdot i_{qs,0}^{k+2}\right)^2 \quad (24)$$

where d_a is the duty cycle of the optimal voltage vector with $d_a \in [0,1]$, $\left\{i_{ds,0}^{k+2}, i_{qs,0}^{k+2}\right\}$ are the predicted d-q current components for instant $k+2$ due to the application of a zero vector during T_s and $\left\{i_{ds,a}^{k+2}, i_{qs,a}^{k+2}\right\}$ are the predicted d-q current components for instant $k+2$ due to the application of the optimal vector $\mathbf{v}_a$ during T_s. The minimization of (24) is performed by solving:

$$\frac{\partial g_{cf}\left(\mathbf{v}_a, d_a\right)}{\partial d_a} = 0, \quad (25)$$

which yields:

$$d_a = \frac{\left(i_{ds}^* - i_{ds,0}^{k+2}\right)\left(i_{ds}^{k+2} - i_{ds,0}^{k+2}\right) + \left(i_{qs}^* - i_{qs,0}^{k+2}\right)\left(i_{qs}^{k+2} - i_{qs,0}^{k+2}\right)}{\left(i_{ds}^{k+2} - i_{ds,0}^{k+2}\right)^2 + \left(i_{qs}^{k+2} - i_{qs,0}^{k+2}\right)^2}. \quad (26)$$

In order to minimize the x'-y' current harmonics and to provide a fixed switching frequency, a PWM-PCC strategy was proposed in Reference [99]. This strategy considers only thirteen voltage vectors (twelve large vectors in the α-β subspace and one zero vector) and since the PWM modulator is able to generate the optimal voltage vector with zero x-y components, the cost function is reduced to:

$$g_{cf} = \left(i_{ds}^{*} - i_{ds}^{k+2}\right)^2 + \left(i_{qs}^{*} - i_{qs}^{k+2}\right)^2. \tag{27}$$

However, it is important to mention that in order to generate zero x-y voltage components over a sampling period in the PWM-PCC strategy, the PWM modulator reduces the amplitude of the voltage vectors in the α-β subspace from $0.644 \cdot U_{dc}$ to $0.5 \cdot U_{dc}$, which is the limit of the linear modulation region [99]. An enhanced PWM-PCC (EPWM-PCC) was proposed in Reference [100], where the main difference in relation to Reference [99] is that the optimal voltage vector is firstly optimized in amplitude with (26), before being synthesized by the PWM modulator with zero voltage x-y components. Moreover, an extended range PWM-PCC (ERPWM-PCC) strategy that combines the EPWM-PCC approach for operation in the linear modulation region and the OSM-PCC method for operation in the overmodulation region was proposed in Reference [101] in order to improve the dc-link usage and the transient performance of six-phase machines.

The modulated PCC (M-PCC) proposed for six-phase IM drives in References [102–105] integrates a modulation technique in the control algorithm to reduce the x'-y' current components. This strategy considers that the α-β subspace is divided into forty-eight different sectors, which are defined by adjacent voltage vectors with the same amplitude. In order to calculate the duty cycles of the voltage vectors within each sector, the M-PCC strategy considers that the duty cycles of the zero and active vectors $\{d_z, d_i, d_j\}$ are inversely proportional to the value of the cost function (21) for the respective voltage vector, yielding [105]:

$$d_i = \frac{g_c(\mathbf{v}_z)\, g_c(\mathbf{v}_j)}{g_c(\mathbf{v}_z)\, g_c(\mathbf{v}_i) + g_c(\mathbf{v}_i)\, g_c(\mathbf{v}_j) + g_c(\mathbf{v}_z)\, g_c(\mathbf{v}_j)}, \tag{28}$$

$$d_j = \frac{g_c(\mathbf{v}_z)\, g_c(\mathbf{v}_i)}{g_c(\mathbf{v}_z)\, g_c(\mathbf{v}_i) + g_c(\mathbf{v}_i)\, g_c(\mathbf{v}_j) + g_c(\mathbf{v}_z)\, g_c(\mathbf{v}_j)}, \tag{29}$$

$$d_z = 1 - (d_i + d_j), \tag{30}$$

where $\{g_c(\mathbf{v}_z), g_c(\mathbf{v}_i), g_c(\mathbf{v}_j)\}$ are the values of the cost function (21) due to a zero voltage vector $\mathbf{v}_z$ and due to the active voltage vectors $\mathbf{v}_i$ and $\mathbf{v}_j$, respectively. The duty cycle d_z is equally divided among the two zero vectors $\mathbf{v}_0$ and $\mathbf{v}_{63}$, in order to achieve a fixed switching frequency. Finally, the M-PCC strategy determines the optimal sector by evaluating the cost function:

$$g_{cm} = g_c(\mathbf{v}_i)\, d_i + g_c(\mathbf{v}_j)\, d_j, \tag{31}$$

4.3. Predictive Current Control Based on Virtual Vectors

An innovative PCC strategy based on virtual vectors (VV-PCC) was proposed in References [106] to mitigate the current harmonics mapped into the x'-y' subspace of six-phase IM drives. The theory behind virtual vectors was initially introduced for the direct torque control (DTC) of five-phase [107,108] and six-phase [109] machines and consists in the creation of a new set of voltage vectors, denominated virtual vectors or synthetic vectors in the literature [110,111], with zero x-y voltage components. The twelve virtual vectors $\{\mathbf{v}_{v1}, \ldots, \mathbf{v}_{v12}\}$ with an amplitude of $0.598 \cdot U_{dc}$ shown in Figure 5 are created by the combination of one large and one medium-large vectors with the same phase in the α-β subspace (Figure 2), during a sampling period with the following duty cycles:

$$d_L = \sqrt{3} - 1 \approx 0.732, \qquad d_{MLI} = 1 - d_L \approx 0.268, \tag{32}$$

where d_L and d_{MLI} are the duty cycles of the large and medium-large vectors, respectively. Additionally, the zero virtual vector $\mathbf{v}_{v0}$ is obtained by the application of two zero vectors, $\mathbf{v}_0$ and

$\mathbf{v}_{63}$, with equal duty cycles. The VV-PCC strategy evaluates (27) for thirteen virtual vectors $\{\mathbf{v}_{v0}, ..., \mathbf{v}_{v12}\}$, and selects the virtual vector that minimizes the cost function for application during the next sampling period. The virtual vectors are synthesized with switching patterns centered to the middle of the sampling period as in References [110,111], in order to ease the implementation in digital controllers.

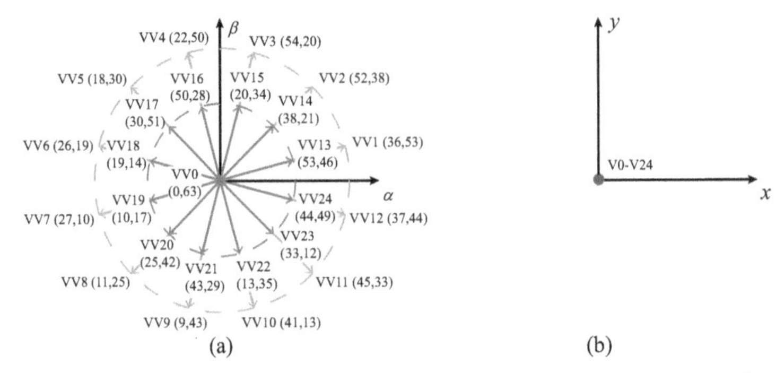

Figure 5. Virtual vectors in the stationary reference frame mapped into the: (**a**) α-β subspace; (**b**) x-y subspace.

To improve the performance of a six-phase PMSM drive at low speeds, a VV-PCC strategy based on an extended set of twenty-five virtual vectors (EVV-PCC) was proposed in Reference [112]. The extra twelve virtual vectors $\{\mathbf{v}_{v13}, ..., \mathbf{v}_{v24}\}$ shown in Figure 5 have an amplitude of $0.345 \cdot U_{dc}$ and are created by the combination of one medium-large and one small vector, with the same phase in the α-β subspace (Figure 2), during a sampling period with duty cycles given by:

$$d_{MLII} = \frac{\sqrt{3}}{3} \approx 0.577, \qquad d_S = 1 - d_{MLII} \approx 0.423, \tag{33}$$

where d_{MLII} and d_S are the duty cycles of the medium-large and small vectors, respectively. To maintain a reduced computational burden, the authors of Reference [112] use a deadbeat approach with the aim to reduce the number of candidate virtual vectors from twenty-four to only two.

A PCC strategy based on the optimal amplitude of virtual vectors (OAVV-PCC) was introduced in Reference [113] with the aim to reduce current/torque ripples at low speeds. The OAVV-PCC strategy computes (27) for twelve virtual vectors $\{\mathbf{v}_{v1}, ..., \mathbf{v}_{v12}\}$, selects the vector $\mathbf{v}_{va}$ that provides the minimum value for the cost function and minimizes (34) in order to obtain the duty cycle d_a of vector $\mathbf{v}_{va}$ with (35):

$$g_{cf}(\mathbf{v}_{va}, d_a) = \left(i_{ds}^* - d_a \cdot i_{ds,a}^{k+2} - (1 - d_a) \cdot i_{ds,0}^{k+2}\right)^2 + \left(i_{qs}^* - d_a \cdot i_{qs,a}^{k+2} - (1 - d_a) \cdot i_{qs,0}^{k+2}\right)^2, \tag{34}$$

$$d_a = \frac{\left(i_{ds}^* - i_{ds,0}^{k+2}\right)\left(i_{ds,a}^{k+2} - i_{ds,0}^{k+2}\right) + \left(i_{qs}^* - i_{qs,0}^{k+2}\right)\left(i_{qs,a}^{k+2} - i_{qs,0}^{k+2}\right)}{\left(i_{ds,a}^{k+2} - i_{ds,0}^{k+2}\right)^2 + \left(i_{qs,a}^{k+2} - i_{qs,0}^{k+2}\right)^2}, \tag{35}$$

where d_a is bounded to the interval $[0, 1]$ and $\left\{i_{ds,a}^{k+2}, i_{qs,a}^{k+2}\right\}$ are the predicted d-q current components for instant $k + 2$ considering the application of $\mathbf{v}_{va}$ during T_s. The OAVV-PCC strategy uses a centered switching pattern to apply $\mathbf{v}_{va}$ during $d_a \cdot T_s$ and $\mathbf{v}_{v0}$ during $(1 - d_a) \cdot T_s$. Since vector $\mathbf{v}_{v0}$ is obtained by the application of two zero vectors $\mathbf{v}_0$ and $\mathbf{v}_{63}$ with equal application times $((1 - d_a) \cdot T_s/2)$, a fixed switching frequency is obtained [113].

In order to improve the reference tracking of the d-q current components of six-phase machines, a VV-PCC strategy based on the application of two virtual vectors over a sampling period (VV2-PCC) is suggested in Reference [114]. This strategy evaluates the cost function (27) for vectors $\{\mathbf{v}_{v0}, \ldots, \mathbf{v}_{v12}\}$ and selects the two adjacent active virtual vectors or one active and one zero virtual vector $\{\mathbf{v}_{vi}, \mathbf{v}_{vj}\}$ that lead to the smallest values in the cost function. The optimal values for the duty cycles of vectors $\{\mathbf{v}_{vi}, \mathbf{v}_{vj}\}$ are obtained by minimizing:

$$g_{cf}\left(\mathbf{v}_{vi}, d_i, \mathbf{v}_{vj}, d_j\right) = \left(i_{ds}^{*} - d_i \cdot i_{ds,i}^{k+2} - d_j \cdot i_{ds,j}^{k+2}\right)^2 + \left(i_{qs}^{*} - d_i \cdot i_{qs,i}^{k+2} - d_j \cdot i_{qs,j}^{k+2}\right)^2, \quad (36)$$

where $\{d_i, d_j\}$ are both limited to the interval $[0, 1]$ and subjected to $d_i + d_j = 1$. The authors of Reference [114] evaluate (36) for a range of values of d_i from 0.5 to 1 with steps of 0.05 and with $d_j = 1 - d_i$, although an approach similar to that in References [113,115] can also be used to compute the optimal values for $\{d_i, d_j\}$. Finally, the VV2-PCC strategy has three switching possibilities: (i) application of only one active virtual vector (similarly to VV-PCC); (ii) application of a zero and an active virtual vector (similarly to OAVV-PCC); (iii) application of two active virtual vectors.

A PCC strategy based on virtual vectors with optimal amplitude and phase (OAPVV-PCC) that combines two active and one zero virtual vector during a sampling period is proposed in Reference [116]. This strategy applies an equivalent virtual vector optimized in both amplitude and phase to the machine, thus improving the reference tracking of the *d*-*q* current components in comparison to other PCC strategies based on virtual vectors. After selecting the two active virtual vectors $\{\mathbf{v}_{vi}, \mathbf{v}_{vj}\}$ from $\{\mathbf{v}_{v1}, \ldots, \mathbf{v}_{v12}\}$ that provide minimum values for (27), the OAPVV-PCC strategy optimizes first the phase and then the amplitude of the equivalent virtual vector to be applied. Considering that the equivalent virtual vector $\mathbf{v}_{vn}$ is defined as:

$$\mathbf{v}_{vn} = \mathbf{v}_{vi} \cdot d_i + \mathbf{v}_{vj} \cdot d_j, \quad (37)$$

with the duty cycles $\{d_i, d_j\}$ being subjected to the constraint $d_i + d_j = 1$, the minimization of (36) yields:

$$d_i = \frac{\left(i_{ds}^{*} - i_{ds,j}^{k+2}\right)\left(i_{ds,i}^{k+2} - i_{ds,j}^{k+2}\right) + \left(i_{qs}^{*} - i_{qs,j}^{k+2}\right)\left(i_{qs,i}^{k+2} - i_{qs,j}^{k+2}\right)}{\left(i_{ds,i}^{k+2} - i_{ds,j}^{k+2}\right)^2 + \left(i_{qs,i}^{k+2} - i_{qs,j}^{k+2}\right)^2}, \qquad d_j = 1 - d_i, \quad (38)$$

where $\{d_i, d_j\}$ are both limited to the interval $[0, 1]$. The amplitude of $\mathbf{v}_{vn}$ is then optimized by minimizing:

$$g_{cf}\left(\mathbf{v}_{vn}, d_n\right) = \left(i_{ds}^{*} - d_n \cdot i_{ds,n}^{k+2} - (1 - d_n) \cdot i_{ds,0}^{k+2}\right)^2 + \left(i_{qs}^{*} - d_n \cdot i_{qs,n}^{k+2} - (1 - d_n) \cdot i_{qs,0}^{k+2}\right)^2, \quad (39)$$

which gives the duty cycle d_n:

$$d_n = \frac{\left(i_{ds}^{*} - i_{ds,0}^{k+2}\right)\left(i_{ds,n}^{k+2} - i_{ds,0}^{k+2}\right) + \left(i_{qs}^{*} - i_{qs,0}^{k+2}\right)\left(i_{qs,n}^{k+2} - i_{qs,0}^{k+2}\right)}{\left(i_{ds,n}^{k+2} - i_{ds,0}^{k+2}\right)^2 + \left(i_{qs,n}^{k+2} - i_{qs,0}^{k+2}\right)^2}, \quad (40)$$

where $\left\{i_{ds,n}^{k+2}, i_{qs,n}^{k+2}\right\}$ are the predicted *d*-*q* current components for instant $k + 2$ due to the application of $\mathbf{v}_{vn}$ during T_s and $d_n \in [0, 1]$. Finally, the equivalent virtual vector with both optimal amplitude and phase is defined as:

$$\mathbf{v}_{vn}' = \mathbf{v}_{vi} \cdot d_i' + \mathbf{v}_{vj} \cdot d_j', \quad (41)$$

where the duty cycles $\left\{d_i', d_j'\right\}$ are given by:

$$d_i' = d_i \cdot d_n, \qquad d_j' = d_j \cdot d_n, \quad (42)$$

with $0 < d_i' + d_j' < 1$. The virtual vectors $\{\mathbf{v}_{vi}, \mathbf{v}_{vj}, \mathbf{v}_{v0}\}$ are synthesized during the next sampling period with the duty cycles $\{d_i', d_i', d_0\}$, where $d_0 = 1 - d_i' - d_j'$, using a centered switching pattern as in Reference [116], leading to a fixed switching frequency.

4.4. Bi-Subspace Predictive Current Control Based on Virtual Vectors

Although virtual vectors impose zero x-y voltage components over a sampling period, x'-y' currents with considerable magnitude may continue to circulate in the stator windings due to machine asymmetries, deadtime effects in the power switches of the VSIs or, in the case of PMSMs, the back-EMF harmonics due to the non-sinusoidal shape of PMs [117–119]. Since the elimination of these current harmonics requires the application of non-zero x-y voltages, the concept of dual virtual vectors was introduced in Reference [120]. In opposition to the standard virtual vectors, the dual virtual vectors only contain x-y voltage components, hence the control of the x'-y' currents can be performed without disturbing the reference tracking of the d-q current components, which regulate the flux and torque of the machine. The dual virtual vectors are created by the combination of a large and a medium-large vector with the same phase in the x-y subspace and the duty cycles given by (32), resulting in twelve dual virtual vectors with an amplitude of $0.598 \cdot U_{dc}$ in the x-y subspace (stationary reference frame), as shown in Figure 6.

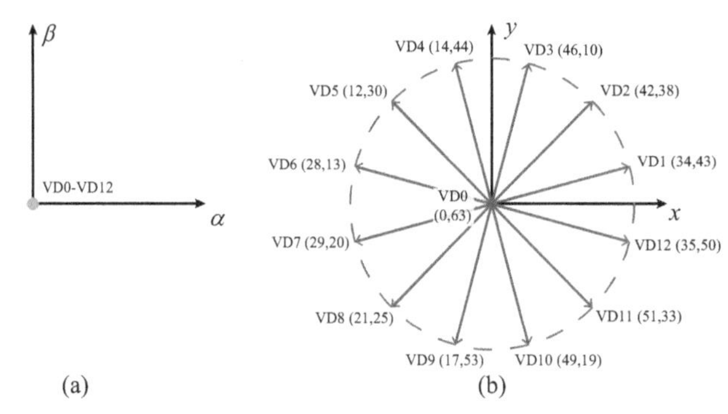

Figure 6. Dual virtual vectors in the stationary reference frame mapped into the: (**a**) α-β subspace; (**b**) x-y subspace.

The bi-subspace PCC strategy based on virtual vectors (BSVV-PCC) presented in Reference [120] aims to provide an accurate current control in both d-q and x'-y' subspaces. This strategy uses two FCS-MPC stages, where one regulates the d-q current components and the other regulates the x'-y' current components. The regulation of the d-q current components is performed as in the OAVV-PCC strategy, where the virtual vector $\mathbf{v}_{va}$ that minimizes (27) is optimized in amplitude by computing d_a with (35). Regarding the regulation of the x'-y' current components, the following cost function is evaluated for the twelve dual virtual vectors $\{\mathbf{v}_{dv1}, \ldots, \mathbf{v}_{dv12}\}$:

$$g_{cs} = \left(i_{xs'}^{*} - i_{xs'}^{k+2}\right)^2 + \left(i_{ys'}^{*} - i_{ys'}^{k+2}\right)^2, \quad (43)$$

where the values of $\left\{i_{xs'}^{*}, i_{ys'}^{*}\right\}$ are set to zero in order to minimize the x'-y' current components. Then, the duty cycle d_b of the optimal dual virtual vector $\mathbf{v}_{dvb}$ is obtained by minimizing:

$$g_{cs}\left(\mathbf{v}_{dvb}, d_b\right) = \left(i_{xs}^{*} - d_b \cdot i_{xs',b}^{k+2} - (1 - d_b) \cdot i_{xs',0}^{k+2}\right)^2 + \left(i_{ys}^{*} - d_b \cdot i_{ys',b}^{k+2} - (1 - d_b) \cdot i_{ys',0}^{k+2}\right)^2, \quad (44)$$

which results in:

$$d_b = \frac{\left(i^*_{xs'} - i^{k+2}_{xs',0}\right)\left(i^{k+2}_{xs',b} - i^{k+2}_{xs',0}\right) + \left(i^*_{ys'} - i^{k+2}_{ys',0}\right)\left(i^{k+2}_{ys',b} - i^{k+2}_{ys',0}\right)}{\left(i^{k+2}_{xs',b} - i^{k+2}_{xs',0}\right)^2 + \left(i^{k+2}_{ys',b} - i^{k+2}_{ys',0}\right)^2}, \quad (45)$$

where d_b belongs to the interval $[0,1]$ and $\left\{i^{k+2}_{xs',b}, i^{k+2}_{ys',b}\right\}$ are the predicted d-q current components for instant $k+2$ considering the application of $\mathbf{v}_{dvb}$ during T_s. Due to the voltage limitation of 2L-VSIs, the BSVV-PCC strategy imposes the following constraint to d_b:

$$\begin{cases} d'_b = 0, & d_b < 0 \\ d'_b = d_b, & 0 \le d_b \le 1 - d_a \\ d'_b = 1 - d_a, & d_b > 1 - d_a \end{cases}. \quad (46)$$

Finally, the vectors $\{\mathbf{v}_{va}, \mathbf{v}_{dvb}, \mathbf{v}_{v0}\}$ with duty cycles $\{d_a, d_b, d_0\}$ are applied to the machine in the next sampling period using centered switching patterns as described in Reference [120], thus leading to a fixed switching frequency.

4.5. Standard Predictive Torque Control

The standard predictive torque control (S-PTC) for six-phase IMs used in electric vehicles was presented in Reference [121]. Since in PTC schemes for IM drives the stator flux and torque are controlled directly in the stationary reference frame ($\omega_a = 0$), the stator current and rotor flux components are commonly selected as state variables [52,121]. Hence, from (6)–(9) and using the forward Euler discretization method, the following expressions are obtained:

$$\begin{bmatrix} i^{k+h}_{ds} \\ i^{k+h}_{qs} \\ i^{k+h}_{xs'} \\ i^{k+h}_{ys'} \end{bmatrix} = \begin{bmatrix} 1 - \frac{T_s}{\sigma\tau_s} - \frac{(1-\sigma)T_s}{\sigma\tau_r} & \omega_a T_s & 0 & 0 \\ -\omega_a T_s & 1 - \frac{T_s}{\sigma\tau_s} - \frac{(1-\sigma)T_s}{\sigma\tau_r} & 0 & 0 \\ 0 & 0 & 1 - \frac{R_s T_s}{L_{ls}} & -\omega_a T_s \\ 0 & 0 & \omega_a T_s & 1 - \frac{R_s T_s}{L_{ls}} \end{bmatrix} \begin{bmatrix} i^{k+h-1}_{ds} \\ i^{k+h-1}_{qs} \\ i^{k+h-1}_{xs'} \\ i^{k+h-1}_{ys'} \end{bmatrix} + \begin{bmatrix} \frac{R_r M_m T_s}{\sigma L_s L_r^2} & \frac{\omega_r M_m T_s}{\sigma L_s L_r} \\ -\frac{\omega_r M_m T_s}{\sigma L_s L_r} & \frac{R_r M_m T_s}{\sigma L_s L_r^2} \\ 0 & 0 \\ 0 & 0 \end{bmatrix} \begin{bmatrix} \psi^{k+h-1}_{dr} \\ \psi^{k+h-1}_{qr} \end{bmatrix} + \begin{bmatrix} \frac{T_s}{\sigma L_s} & 0 & 0 & 0 \\ 0 & \frac{T_s}{\sigma L_s} & 0 & 0 \\ 0 & 0 & \frac{T_s}{L_{ls}} & 0 \\ 0 & 0 & 0 & \frac{T_s}{L_{ls}} \end{bmatrix} \begin{bmatrix} u^{k+h-1}_{ds} \\ u^{k+h-1}_{qs} \\ u^{k+h-1}_{xs'} \\ u^{k+h-1}_{ys'} \end{bmatrix}, \quad (47)$$

$$\begin{bmatrix} \psi^{k+h}_{dr} \\ \psi^{k+h}_{qr} \end{bmatrix} = \begin{bmatrix} 1 - \frac{T_s}{\tau_r} & \omega_k T_s \\ -\omega_k T_s & 1 - \frac{T_s}{\tau_r} \end{bmatrix} \begin{bmatrix} \psi^{k+h-1}_{dr} \\ \psi^{k+h-1}_{qr} \end{bmatrix} + \begin{bmatrix} \frac{M_m T_s}{\tau_r} & 0 \\ 0 & \frac{M_m T_s}{\tau_r} \end{bmatrix} \begin{bmatrix} i^{k+h-1}_{ds} \\ i^{k+h-1}_{qs} \end{bmatrix}, \quad (48)$$

where $\tau_s = L_s/R_s$ and $\tau_r = L_r/R_r$. The stator flux components are obtained from the stator current and rotor flux components with:

$$\begin{bmatrix} \psi^{k+h}_{ds} \\ \psi^{k+h}_{qs} \\ \psi^{k+h}_{xs'} \\ \psi^{k+h}_{ys'} \end{bmatrix} = \begin{bmatrix} \sigma L_s & 0 & 0 & 0 \\ 0 & \sigma L_s & 0 & 0 \\ 0 & 0 & L_{ls} & 0 \\ 0 & 0 & 0 & L_{ls} \end{bmatrix} \begin{bmatrix} i^{k+h}_{ds} \\ i^{k+h}_{qs} \\ i^{k+h}_{xs'} \\ i^{k+h}_{ys'} \end{bmatrix} + \begin{bmatrix} \frac{M_m}{L_r} & 0 \\ 0 & \frac{M_m}{L_r} \\ 0 & 0 \\ 0 & 0 \end{bmatrix} \begin{bmatrix} \psi^{k+h}_{dr} \\ \psi^{k+h}_{qr} \end{bmatrix}, \quad (49)$$

In order to select the optimal voltage vector, the S-PTC scheme evaluates the following cost function for forty-nine distinct voltage vectors:

$$g_t = \left(\frac{t^*_e - t^{k+2}_e}{t_n}\right)^2 + \left(\frac{\psi^*_s - \psi^{k+2}_s}{\psi_{sn}}\right)^2 + C_{dq} + C_{xy}, \quad (50)$$

where ψ_{sn} is the rated stator flux, t_n is the rated torque and t_e^{k+2} is calculated by (11) using the predictions of the current and stator flux components for instant $k+2$. The term ψ_s^{k+2} is defined as:

$$\psi_s^{k+2} = \sqrt{\left(\psi_{ds}^{k+2}\right)^2 + \left(\psi_{qs}^{k+2}\right)^2}. \tag{51}$$

The terms C_{dq} and C_{xy} in (50) are overcurrent constraints that penalize currents above a certain magnitude in both d-q and x'-y' subspaces:

$$\begin{cases} C_{dq} = 0, & i_{s,dq}^{k+2} \leq i_{s,dq}^{max} \\ C_{dq} = 10^5, & i_{s,dq}^{k+2} > i_{s,dq}^{max} \end{cases}, \quad \begin{cases} C_{xy} = 0, & i_{s,xy}^{k+2} \leq i_{s,xy}^{max} \\ C_{xy} = 10^5, & i_{s,xy}^{k+2} > i_{s,xy}^{max} \end{cases}, \tag{52}$$

where $\left\{i_{s,dq}^{max}, i_{s,xy}^{max}\right\}$ are the maximum values for the current amplitude in both d-q and x'-y' subspaces and $\left\{i_{s,dq}^{k+2}, i_{s,xy}^{k+2}\right\}$ are defined as:

$$i_{s,dq}^{k+2} = \sqrt{\left(i_{ds}^{k+2}\right)^2 + \left(i_{qs}^{k+2}\right)^2}, \quad i_{s,xy}^{k+2} = \sqrt{\left(i_{xs'}^{k+2}\right)^2 + \left(i_{ys'}^{k+2}\right)^2}. \tag{53}$$

Finally, the voltage vector that minimizes (50) is applied to the six-phase IM during the next sampling period. As an example, the general diagram of the S-PTC strategy for six-phase IM drives, considering the model of the drive in the stationary reference frame ($\omega_a = 0$) is shown in Figure 7 [121].

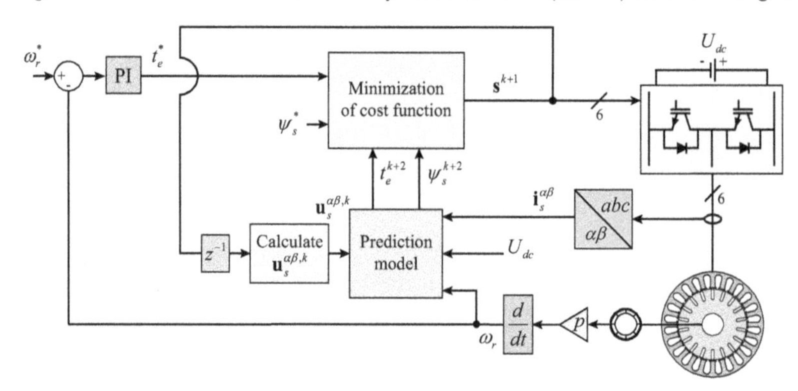

Figure 7. General diagram of the S-PTC strategy for six-phase IM drives.

An S-PTC strategy for six-phase PMSMs is presented in Reference [122], where the stator currents are predicted with (20) considering the synchronous reference frame ($\omega_a = \omega_r$) with the d-axis aligned with flux due to the PMs. The stator flux d-q and x'-y' components are calculated by:

$$\begin{bmatrix} \psi_{ds}^{k+h} \\ \psi_{qs}^{k+h} \\ \psi_{xs'}^{k+h} \\ \psi_{ys'}^{k+h} \end{bmatrix} = \begin{bmatrix} L_d & 0 & 0 & 0 \\ 0 & L_q & 0 & 0 \\ 0 & 0 & L_x & 0 \\ 0 & 0 & 0 & L_y \end{bmatrix} \begin{bmatrix} i_{ds}^{k+h} \\ i_{qs}^{k+h} \\ i_{xs'}^{k+h} \\ i_{ys'}^{k+h} \end{bmatrix} + \begin{bmatrix} \psi_{s,\mathrm{PM1}} \\ 0 \\ 0 \\ 0 \end{bmatrix}. \tag{54}$$

The S-PTC strategy in Reference [122] uses a pre-selection process to reduce the number of candidate voltage vectors from forty-nine to only three, based on the angle of the stator flux in the α-β and x-y subspaces (stationary reference frame) and on the signal of the torque error. The following cost function is evaluated for the three candidate voltage vectors:

$$g_{tf} = \left(t_e^* - t_e^{k+2}\right)^2 + \lambda_\psi \left(\psi_s^* - \psi_s^{k+2}\right)^2, \tag{55}$$

where λ_ψ is a weighting factor. The voltage vector that minimizes (55) is applied to the six-phase PMSM during the next sampling period. The general diagram of the S-PTC strategy for six-phase PMSM drives in the synchronous reference frame ($\omega_a = \omega_r$) is shown in Figure 8 [122].

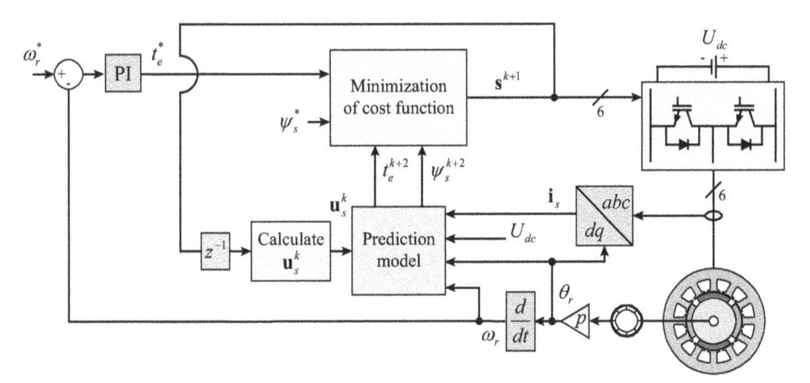

Figure 8. General diagram of the S-PTC strategy for six-phase PMSM drives.

4.6. Predictive Torque Control Based on the Duty Cycle Optimization of Voltage Vectors

An approach similar to the S-PTC, baptized as high robustness PTC (HR-PTC), which considers a discrete duty cycle optimization, is proposed in Reference [123]. Instead of considering the application of the optimal voltage vector $\mathbf{v}_a$ during the entire sampling period, this strategy finds the optimal duty cycle d_a of vector $\mathbf{v}_a$, with $d_a \in \{0, 0.2, 0.4, 0.6, 0.8, 1\}$, by minimizing:

$$g_{tf} = \left(t_e^* - d_a \cdot t_{e,a}^{k+2} - (1-d_a) \cdot t_{e,0}^{k+2}\right)^2 + \lambda_\psi \left(\psi_s^* - d_a \cdot \psi_{s,a}^{k+2} - (1-d_a) \cdot \psi_{s,0}^{k+2}\right)^2 \tag{56}$$

where $\left\{t_{e,a}^{k+2}, \psi_{s,a}^{k+2}\right\}$ are the predicted torque and stator flux for instant $k+2$ considering the application of vector $\mathbf{v}_a$ during T_s and $\left\{t_{e,a}^{k+2}, \psi_{s,a}^{k+2}\right\}$ are the predicted torque and stator flux for instant $k+2$ due to the application of a zero vector during T_s. The vector $\mathbf{v}_a$ is selected for application in the next sampling period during $d_a \cdot T_s$, where the value of d_a is selected from the minimization of (56).

A reduced cost function PTC (RCF-PTC) strategy was presented in Reference [124], where a deadbeat approach is used to determine the sector of the optimal voltage vector, reducing the number of candidates from forty-nine to only three. This strategy uses a model of the six-phase PMSM with stator current and stator flux components as state variables, hence the stator currents are predicted for instant $k+h$ with (20) and the d-q components of the stator flux are predicted for instant $k+h$ with:

$$\begin{bmatrix} \psi_{ds}^{k+h} \\ \psi_{qs}^{k+h} \end{bmatrix} = \begin{bmatrix} 1 & \omega_r T_s \\ -\omega_r T_s & 1 \end{bmatrix} \begin{bmatrix} \psi_{ds}^{k+h-1} \\ \psi_{qs}^{k+h-1} \end{bmatrix} + \begin{bmatrix} -R_s & 0 \\ 0 & -R_s \end{bmatrix} \begin{bmatrix} i_{ds}^{k+h-1} \\ i_{qs}^{k+h-1} \end{bmatrix} + \begin{bmatrix} 1 & 0 \\ 0 & 1 \end{bmatrix} \begin{bmatrix} u_{ds}^{k+h-1} \\ u_{qs}^{k+h-1} \end{bmatrix}. \tag{57}$$

In the RCF-PTC, the voltage components in the d-q subspace are obtained using a deadbeat approach, that is, considering $t_e^{k+2} = t_e^*$, and their angle in the stationary reference frame (α-β subspace) is used to select three candidate voltage vectors, one small, one medium-large and one large with the same phase. The amplitude of these three vectors is optimized by computing their duty cycle with:

$$d_a = \frac{\sqrt{\left(u_{\alpha s}^*\right)^2 + \left(u_{\beta s}^*\right)^2} \cdot \sqrt{\left(u_{\alpha s}^{k+1}\right)^2 + \left(u_{\beta s}^{k+1}\right)^2} \cdot \cos(\theta_v)}{\left(u_{\alpha s}^{k+1}\right)^2 + \left(u_{\beta s}^{k+1}\right)^2}, \tag{58}$$

where $\left\{u_{\alpha s}^{*}, u_{\beta s}^{*}\right\}$ are the α-β components of the stator voltage reference computed by the RCF-PTC strategy, $\left\{u_{\alpha s}^{k+1}, u_{\beta s}^{k+1}\right\}$ are the α-β components of the three candidate vectors and θ_v is the angle between the reference and candidate voltage vectors in the stationary reference frame (α-β subspace). Then, the voltage vector, among the three candidates, that provides minimal x'-y' current components is selected for application during the next sampling period. Thus, the cost function of the RCF-PTC is defined as [124]:

$$g_{fcs} = \left(\sqrt{i_{xs'}^{k+2} + i_{ys'}^{k+2}}\right)^2. \tag{59}$$

4.7. Predictive Torque Control Based on Virtual Vectors

In order to eliminate the stator flux weighting factor, a flux constrained PTC (FC-PTC) that calculates the stator flux references from the reference torque is presented in Reference [125]. Moreover, the FC-PTC strategy considers the use of virtual vectors $\{\mathbf{v}_{V0}, ..., \mathbf{v}_{V24}\}$ (Figure 5), hence the cost function for this strategy is defined as:

$$g_f = \left(\psi_{ds}^{*} - \psi_{ds}^{k+2}\right)^2 + \left(\psi_{qs}^{*} - \psi_{qs}^{k+2}\right)^2, \tag{60}$$

where the reference values of the d-q components of the stator flux $\left\{\psi_{ds}^{*}, \psi_{qs}^{*}\right\}$ are calculated with (61) considering $i_{ds}^{*} = 0$, which corresponds to maximum torque per ampere (MTPA) conditions in SPMSMs:

$$\begin{cases} \psi_{ds}^{*} = \psi_{s,\mathrm{PM1}} \\ \psi_{qs}^{*} = \dfrac{L_q t_e^{*}}{3p\psi_{s,\mathrm{PM1}}} \end{cases}. \tag{61}$$

As the computational burden of FC-PTC can be considerable for implementation in digital controllers, the authors of Reference [125] have used a look-up table in order to reduce the number of candidate virtual vectors from twenty-four to only six.

A multi-vector PTC (MV-PTC) scheme was proposed in Reference [126] with the aim to improve the steady-state operation of a six-phase PMSM drive. This strategy considers only twelve active virtual vectors $\{\mathbf{v}_{V1}, ..., \mathbf{v}_{V12}\}$ from Figure 5 and optimizes the amplitude of each one using:

$$T_a = \frac{t_e^{*} - t_e^{k+1} - \Delta t_{e,0} \cdot T_s}{\Delta t_{e,a} - \Delta t_{e,0}}, \tag{62}$$

where $\{\Delta t_{e,0}, \Delta t_{e,a}\}$ are the torque deviation due to the application of a zero and an active virtual vector, respectively and are defined as [126]:

$$\Delta t_{e,0} = t_{e,0}^{k+2} - t_e^{k+1}, \qquad \Delta t_{e,a} = t_{e,a}^{k+2} - t_e^{k+1}. \tag{63}$$

The MV-PTC strategy evaluates (55) for twelve virtual vectors with optimized amplitude and applies the optimal virtual vector in the next sampling period, combined with a zero virtual vector, leading to a fixed switching frequency.

5. Simulation Results

In order to assess and compare the performance of the different FCS-MPC strategies described in the previous section, several simulations results obtained with a six-phase PMSM drive are presented in this section. The 2L-VSIs were modelled in Matlab/Simulink using the ideal IGBT model from the Simscape Power Systems library and the six-phase PMSM was modeled using (6)–(11) with the parameters given in Table 1, where $\{P_s, U_s, I_s, n_n, t_n, \psi_{sn}\}$ are the rated values of the power, voltage, current, speed, torque and stator flux of the machine designed in Reference [76]. Since both the non-linearities of the power converters and the back-EMF harmonics contribute to the appearance of considerable x'-y' currents, the simulation model considers a deadtime of t_d = 2.2 μs in the power switches of the 2L-VSIs and also accounts for the 5th and 7th harmonics of the no-load flux linkage due to the PMs, whose amplitudes $\{\psi_{s,PM5}, \psi_{s,PM7}\}$ and phases $\{\phi_5, \phi_7\}$ are provided in Table 1.

Table 1. Parameters of the six-phase drive.

Parameter	Value	Parameter	Value	Parameter	Value	Parameter	Value
P_s (kW)	4	ψ_{sn} (mWb)	1013.8	$\psi_{s,PM1}$ (mWb)	980.4	U_{dc} (V)	650
U_s (V)	340	p	2	$\psi_{s,PM5}$ (mWb)	2.4	t_d (μs)	2.2
I_s (A)	3.4	R_s (Ω)	1.5	$\psi_{s,PM7}$ (mWb)	1.6	T_s (μs)	30, 40, 60, 100, 200
n_n (rpm)	1500	L_{dq} (mH)	53.8	ϕ_5 (deg)	1.3	λ_i	0.025
t_n (N.m)	28.4	L_{xy} (mH)	2.1	ϕ_7 (deg)	−12.7	λ_f	1000

To measure the performance of the six-phase PMSM drive under the considered FCS-MPC strategies, the following performance indicators are defined to quantify the reference tracking error of the current and stator flux components:

$$E_{i,v} = \frac{\frac{1}{N}\sum_{n=1}^{N} |i_{vs}^*(n) - i_{vs}(n)|}{\sqrt{2} \times I_s} \times 100\%, \tag{64}$$

$$E_{\psi,v} = \frac{\frac{1}{N}\sum_{n=1}^{N} |\psi_{vs}^*(n) - \psi_{vs}(n)|}{\psi_{sn}} \times 100\%, \tag{65}$$

where $v \in \{d, q, x', y'\}$ and N is the number of samples corresponding to a time window of 1 s. Moreover, the current harmonic distortion considering up to the fiftieth current harmonic is computed with:

$$\text{THD}_i = \frac{1}{6} \sum_{x=a1,\dots,c2} \frac{\sqrt{i_{xs,2}^2 + \dots + i_{xs,50}^2}}{i_{xs,1}} \times 100\%, \tag{66}$$

where $i_{xs,h}$ is the h-order harmonic of the x-phase current. In order to account for all harmonic content of currents, the total waveform distortion of current is defined as:

$$\text{TWD}_i = \frac{1}{6} \sum_{x=a1,\dots,c2} \frac{\sqrt{I_{xs}^2 - i_{xs,1}^2}}{i_{xs,1}} \times 100\%, \tag{67}$$

where I_{xs} is the rms value of the current in phase x. The total waveform ripple of torque is calculated with:

$$\text{TWR}_t = \frac{\sqrt{T_e^2 - \bar{t}_e^2}}{|\bar{t}_e|} \times 100\%, \tag{68}$$

where T_e is the torque rms value and $\bar{t}_e$ is the mean value of torque.

To compare the PCC strategies considered in Section 4, the six-phase drive is simulated in Matlab/Simulink environment for operation at a constant speed of 750 rpm and rated load condition (motoring mode), which is obtained by setting $i_{qs} = 4.8$ A. Different values of T_s were considered in the PCC strategies in order to obtain a mean switching frequency of around 5 kHz. A speed of 750 rpm was selected to show the difference in the performance of the strategies capable of applying multiple voltage vectors or multiple virtual vectors during a sampling period from the remaining, which provide a much better performance at speed levels below the rated value. The simulation results obtained for the steady-state operation of the six-phase PMSM drive under the considered PCC strategies are presented in Figures 9–11, while the respective performance indicators are summarized in Table 2.

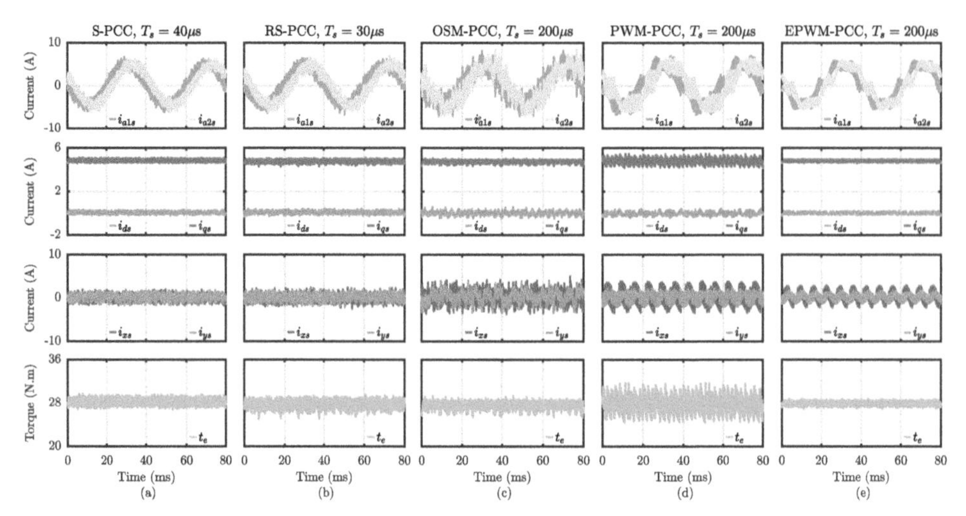

Figure 9. Simulation results for the PMSM drive operating at 750 rpm and rated load (motoring mode) for: (**a**) S-PCC; (**b**) RS-PCC; (**c**) OSM-PCC; (**d**) PWM-PCC and (**e**) EPWM-PCC.

The simulation results for the S-PCC strategy are presented in Figure 9a. Since the S-PCC only applies one out of sixty-four voltage vectors per sampling period and each voltage vector contains both α-β and x-y components, this strategy cannot completely suppress the x'-y' currents. Moreover, the value of λ_i could be increased to further minimize the x'-y' currents but this would degrade the reference tracking of the d-q currents, which regulate the flux and torque of the machine. The higher value obtained for the TWD$_i$ in comparison to the THD$_i$ in the case of the S-PCC (TWD$_i$ = 18.41% and THD$_i$ = 4.23%) shows that the observed distortion in the currents is mainly of high frequency and is mostly mapped into the x'-y' subspace. The RS-PCC strategy provides a reduced mean switching frequency in comparison to the S-PCC by limiting the number of candidate voltage vectors, giving a slightly deteriorated performance even with a smaller value of T_s. On the other hand, the OSM-PCC strategy optimizes the length of the applied voltage vector by combining it with two zero vectors ($\mathbf{v}_0$ and $\mathbf{v}_{63}$) over a sampling period, resulting in a fixed switching frequency of $\bar{f}_{sw} = 1/T_s$. Hence, the value of T_s is increased to 200 µs to obtain a fixed value of $\bar{f}_{sw} = 5.0$ kHz, which worsens the performance of the system in comparison to the S-PCC, as shown in Figure 9c but greatly reduces the computational requirements of digital control platforms for the execution of this control strategy. The use of a PWM technique in the PWM-PCC strategy avoids the injection of x-y voltage components and guarantees a fixed switching frequency, as in the case of the OSM-PCC strategy. Since no x-y voltage components are applied to the machine, the x'-y' currents components cannot be regulated, that is, they are left in open-loop. This leads to the appearance of low-frequency current harmonics in the x'-y' subspace, as shown in Figure 9d, caused by the deadtime effect of the power switches and by the back-EMF harmonics. The EPWM-PCC optimizes the amplitude of the applied voltage vector in the α-β subspace, while guaranteeing the application of zero x-y voltage components over a sampling period. Hence, the EPWM-PCC strategy improves the reference tracking of the d-q currents and reduces the value of TWR$_t$ in comparison to the S-PCC, OSM-PCC and PWM-PCC strategies. However, as in the case of PWM-PCC, the EPWM-PCC strategy is not able to regulate the x'-y' currents, giving a high value for the THD$_i$.

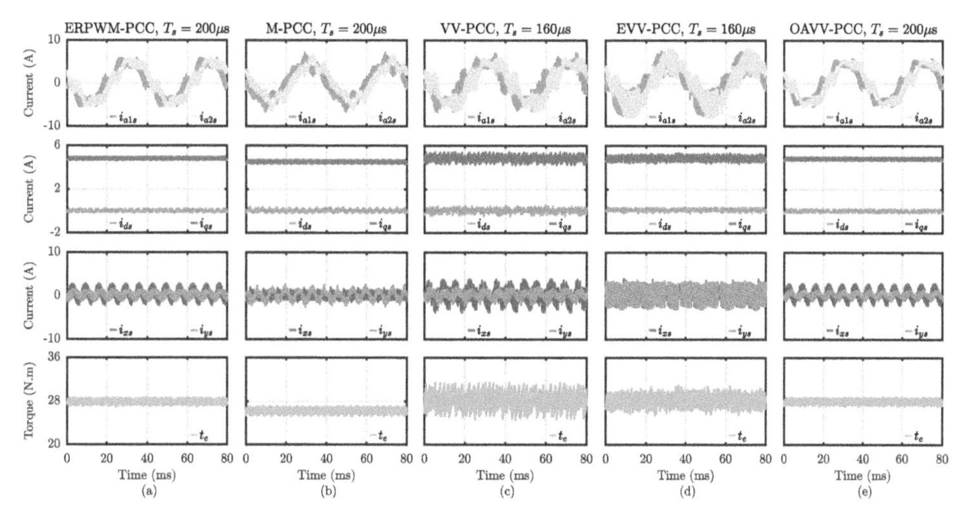

Figure 10. Simulation results for the PMSM drive operating at 750 rpm and rated load (motoring mode) for: (**a**) ERPWM-PCC; (**b**) M-PCC; (**c**) VV-PCC; (**d**) EVV-PCC and (**e**) OAVV-PCC.

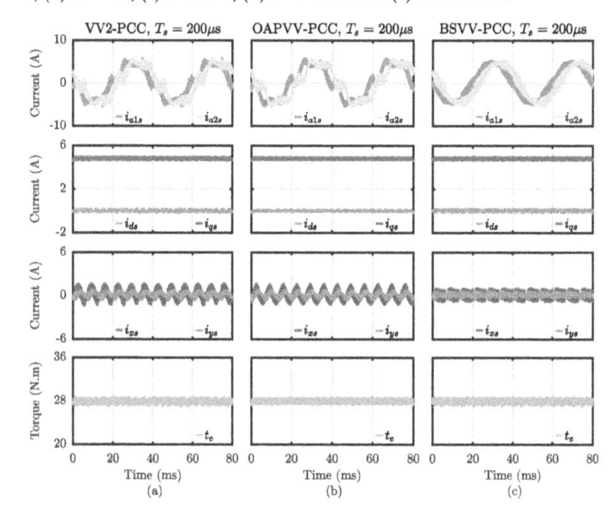

Figure 11. Simulation results for the PMSM drive operating at 750 rpm and rated load (motoring mode) for: (**a**) VV2-PCC; (**b**) OAPVV-PCC and (**c**) BSVV-PCC.

The simulation results for the drive operating with the ERPWM-PCC strategy are presented in Figure 10a. The performance of the drive in steady-state conditions for the considered point of operation is very similar to the one obtained with the EPWM-PCC strategy. However, the EPWM-PCC strategy can only apply a voltage vector with an amplitude of up to $0.5 \cdot U_{dc}$, which corresponds to the limit of the linear region of the PWM technique used. On the other hand, the ERPWM-PCC strategy is able to operate in both the linear and in the overmodulation regions (for an amplitude of the voltage vectors between $0.5 \cdot U_{dc}$ and $0.644 \cdot U_{dc}$), which not only improves the dc-link voltage usage but also the transient performance of the drive. The simulation results for a step in i_{qs}^* from 2.4 A to 4.8 A at $t = 10$ ms are shown in Figure 12 and validate the superior performance of the ERPWM-PCC strategy over the EPWM-PCC, obtaining a reduction of the rise time from 4 ms to 1.3 ms. However, when operating in the overmodulation region, the ERPWM-PCC strategy cannot guarantee the injection of zero x-y voltage components, as in the case of the operation in the linear region of modulation.

The simulation results for the steady-state operation of the drive under the M-PCC strategy are shown in Figure 10b. Differently from the PWM-PCC, EPWM-PCC and ERPWM-PCC strategies, the M-PCC strategy combines two active vectors and two zeros ($\mathbf{v}_0$ and $\mathbf{v}_{63}$) over a sampling period, which provides a fixed switching frequency but does not guarantee the application of zero x-y voltage components over a sampling period. The cost function of the M-PCC strategy evaluates the current errors in both subspaces and uses a weighting factor (λ_i) to determine the relative importance between the tracking of reference currents in both subspaces. Even when $\lambda_i = 0.025$ is selected, the current errors in the x-y subspace disturb the reference tracking of the d-q current components, as demonstrated by the increase in the values of $E_{i,d}$ and $E_{i,q}$ (Table 2), and a steady-state error is perceptible in both the q-axis current and torque, as shown in Figure 10b. An even smaller value for λ_i could be selected to reduce the steady-state errors in i_{qs} and in t_e but the amplitude of x-y current components would also increase. The VV-PCC strategy uses twelve active and one zero virtual vectors instead of standard fourty-nine voltage vectors to apply zero x-y voltage components to the machine. The results obtained for the VV-PCC strategy are presented in Figure 10c and are very similar to the ones obtained with the PWM-PCC strategy, however the virtual vectors have an amplitude of $0.598 \cdot U_{dc}$, which improves the dc-link voltage usage and the transient performance of the drive. The EVV-PCC strategy provides a decrease in the d-q currents errors and in the torque ripple in comparison to the VV-PCC strategy, as seen in Figure 10d, due to the addition of twelve small active virtual vectors with an amplitude of $0.345 \cdot U_{dc}$ to the control algorithm. The simulation results for the OAVV-PCC strategy are presented in Figure 10e and show a significant improvement in terms of torque ripple and d-q current errors during steady-state operation over the VV-PCC and EVV-PCC strategies. Since in the OAVV-PCC technique the selected virtual vector is combined with a zero virtual vector over a sampling period, the operation of the drive at low speeds is highly improved while maintaining a fixed switching frequency.

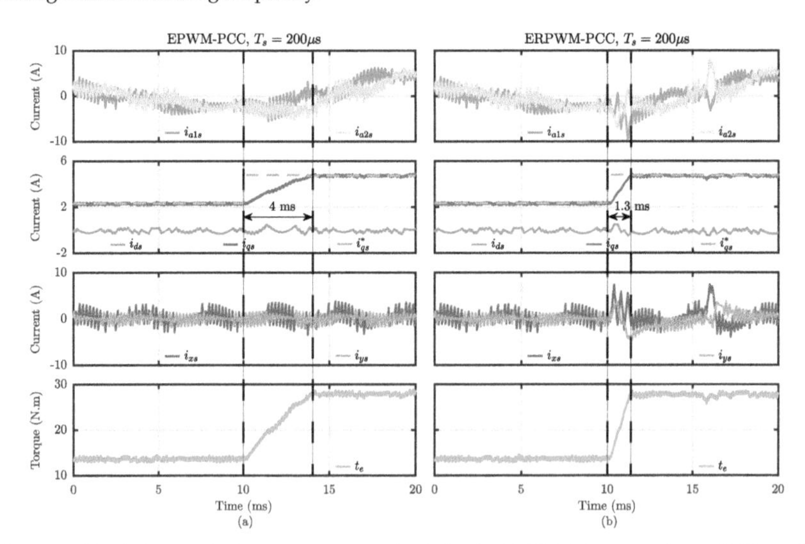

Figure 12. Simulation results for a step response in i^*_{qs} for the PMSM drive operating at 1300 rpm for: (**a**) EPWM-PCC and (**b**) ERPWM-PCC.

The simulation results for the drive operating under the VV2-PCC strategy are shown in Figure 11a. The obtained results show a performance similar to the one obtained with the OAVV-PCC strategy, however the torque ripple is slightly increased from to 1.18% to 1.33%. Although the VV2-PCC is able to apply one virtual vector and one zero virtual vector or two virtual vectors over a sampling period and in theory should provide lower current errors and lower torque ripple than the OAVV-PCC, this is not verified since the VV2-PCC strategy is only able to apply a finite set of values for the duty cycles of the two vectors, as discussed in Section 4.3. The simulation results for the drive operating under the OAPVV-PCC strategy are presented in Figure 11b and demonstrate a very good performance under steady-state operation in terms of tracking of the reference d-q current components and torque ripple. Since this strategy combines two active and one zero virtual vectors during a sampling period, the resultant voltage vector provides very low d-q current errors and the lowest value of TWR_t among the compared PCC strategies. As the PCC strategies based on PWM techniques, such as the EPWM-PCC and ERPWM-PCC, and the strategies based on virtual vectors, such as OAVV-PCC, VV2-PCC and OAPVV-PCC, do not apply x-y voltage components, those techniques cannot compensate the low frequency x'-y' current harmonics generated by the deadtime effects of the power switches and by the back-EMF harmonics. The simulation results for the BSVV-PCC strategy are presented in Figure 11c and show a significant reduction in the amplitude of the x'-y' current components. The BSVV-PCC strategy not only provides low current errors in both subspaces and low torque ripple but also provides the lowest values for the THD_i and TWD_i, among the compared control strategies.

Table 2. Performance indicators for the drive operating at 750 rpm and rated load (motoring mode) for the different PCC strategies.

Strategy	$E_{i,d}$ (%)	$E_{i,q}$ (%)	$E_{i,x}$ (%)	$E_{i,y}$ (%)	THD_i (%)	TWD_i (%)	TWR_t (%)	$\bar{f}_{sw}$ (kHz)
S-PCC	1.54	1.40	10.53	10.70	4.23	18.41	1.69	4.06
RS-PCC	1.93	2.77	11.16	10.22	4.71	19.64	2.53	3.32
OSM-PCC	3.50	2.89	25.58	22.91	27.47	43.60	2.22	5.00
PWM-PCC	2.95	4.30	18.35	10.89	18.64	27.42	5.15	5.00
EPWM-PCC	1.63	1.64	16.20	8.75	18.83	21.63	1.17	5.00
ERPWM-PCC	1.65	1.64	16.21	8.77	18.68	21.48	1.16	5.00
M-PCC	2.05	7.63	6.03	18.34	22.05	24.71	1.32	5.00
VV-PCC	3.00	3.88	18.77	9.87	20.65	26.38	4.74	5.27
EVV-PCC	1.72	2.98	14.97	15.99	14.29	29.32	3.57	5.00
OAVV-PCC	1.63	1.63	16.25	7.20	18.90	20.71	1.18	5.00
VV2-PCC	1.60	1.65	16.21	7.21	18.71	20.46	1.25	5.00
OAPVV-PCC	1.22	1.46	18.76	8.55	22.40	23.93	1.10	5.00
BSVV-PCC	1.34	1.55	5.47	2.71	3.66	9.37	1.18	5.00

The simulation results for the operation of the six-phase drive under PTC strategies are presented in Figure 13, while the corresponding performance indicators are given in Table 3. In comparison to the S-PCC, the S-PTC strategy provides lower torque ripple although with a higher current harmonic distortion, as seen in Figure 13a. The HR-PTC strategy is similar to the S-PTC but provides an optimization in amplitude of the selected voltage vector, by combining it with two zero vectors ($\mathbf{v}_0$ and $\mathbf{v}_{63}$). Since, each voltage vector contains both α-β and x-y current components, a large value for T_s leads to the appearance of large currents in the x-y subspace, thus a T_s = 60 µs was chosen. From Figure 13b, the HR-PTC strategy provides higher current distortion and higher torque ripple than the S-PTC strategy, even with a higher mean switching frequency ($f_{sw} = 13.26$ kHz). The RFC-PTC strategy, whose results are presented in Figure 13c, provides a lower torque ripple in comparison to the S-PTC and HR-PTC strategies, however it gives a higher value for the THD_i and leads to a high mean switching frequency ($f_{sw} = 10.0$ kHz). The simulation results for the FC-PTC strategy are presented in Figure 13d and show a reduction in the x'-y' flux errors due to the use of virtual vectors, even with a higher sampling period (T_s = 160 µs) in comparison to previous PTC strategies. The MV-PTC strategy improves the steady-state operation of the drive by optimizing the amplitude of the selected virtual vector, giving reduced flux errors and a low torque ripple for a fixed switching frequency of 5 kHz.

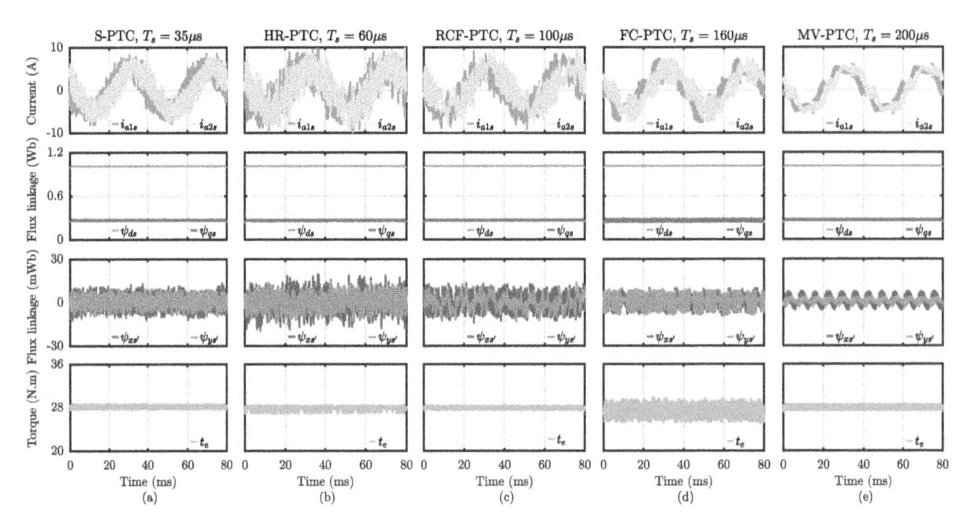

Figure 13. Simulation results for the PMSM drive operating at 750 rpm and rated load (motoring mode) for: (**a**) S-PTC; (**b**) HR-PTC; (**c**) RCF-PTC; (**d**) FC-PTC and (**e**) MV-PTC.

Table 3. Performance indicators for the drive operating at 750 rpm and rated load (motoring mode) for the different PTC strategies.

Strategy	$E_{\psi,d}$ (%)	$E_{\psi,q}$ (%)	$E_{\psi,x}$ (%)	$E_{\psi,y}$ (%)	**THD**$_i$ **(%)**	**THD**$_i$ **(%)**	**TWR**$_t$ **(%)**	$\bar{f}_{sw}$ **(kHz)**
S-PTC	0.44	0.24	0.39	0.27	12.05	41.74	0.69	5.87
HR-PTC	0.42	0.43	0.48	0.35	16.59	55.04	1.06	13.26
RFC-PTC	0.23	0.40	0.50	0.36	26.63	53.56	0.59	10.00
FC-PTC	0.50	1.05	0.25	0.16	16.57	28.69	3.42	5.11
MV-PTC	0.47	0.35	0.22	0.07	18.58	20.47	1.06	5.00

6. Experimental Results

6.1. Experimental Setup

The experimental results presented in this section were obtained with a six-phase PMSM drive, with the same parameters as the ones given in Table 1 in Section 5. The 4 kW six-phase asymmetrical PMSM is supplied by two 2L-VSIs by Semikron (SKiiP 132 GD 120), which are fed by a dc-bus with a voltage level of 650 V. The speed of the six-phase PMSM is regulated by a mechanically coupled 7.5 kW IM fed by a commercial variable frequency converter. The rotor position of the PMSM is measured with an incremental encoder with 2048 ppr. The PCC and PTC strategies are implemented in a digital control platform dS1103 by dSPACE and a cRIO-9066 by National Instruments is used to generate the switching patterns needed by the control strategies that: (i) optimize the amplitude of voltage vectors; (ii) require PWM techniques or (iii) consider the use of virtual vectors. In those control strategies, at the end of each sampling period, the dS1103 platform writes the six leg duty cycles in a digital port, which is read by the FPGA of the cRIO-9066. At the beginning of the next sampling period, the cRIO-9066 generates the switching signals for the 2L-VSIs with a symmetry to the middle of the sampling period. In order to maintain the processes of both platforms synchronized, an interrupt signal is generated at the beginning of each control cycle in the FPGA of the cRIO-9066, which determines the beginning of a new control cycle in the dS1103 platform. The experimental setup is shown in Figure 14.

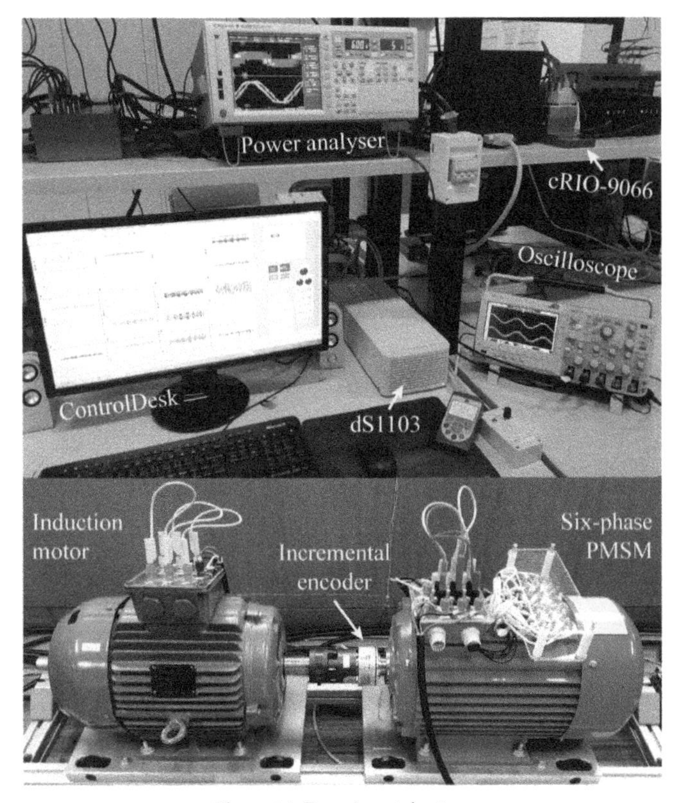

Figure 14. Experimental setup.

6.2. Obtained Results

The experimental results for the steady-state operation of the six-phase PMSM drive under the tested PCC strategies are shown in Figures 15–17, while the respective performance indicators are listed in Table 4, where $\bar{t}_{exe}$ is the mean execution time for each strategy. It is important to note that the execution times of the strategies that require the generation of custom switching patterns already contain the time required for the communication between the dS1103 and the cRIO-9066 platforms, which is around 15 µs. From Figure 15, it is shown that the RS-PCC strategy provides a worse performance than the S-PCC strategy in terms of current and torque ripples. However, the RS-PCC requires a lower execution time than the S-PCC strategy, which could be useful in digital control platforms with limited resources. As in the simulation results, both S-PCC and RS-PCC strategies give much higher values for the TWD_i, 27.87% and 39.30%, over the THD_i, 3.22% and 2.87%, meaning that the majority of the ripple observed in the phase currents is of high-frequency. The OSM-PCC strategy also gives a worse performance over previous strategies, increasing the TWD_i to 76.37%, but imposing a fixed switching frequency to the power switches of the inverters, which could ease the process of designing output filters for the six-phase machine. The use of a PWM technique in the PWM-PCC strategy leads to a reduction of the current ripple, mainly in the x'-y' currents, due to the imposition of mean zero x-y voltage components over a sampling period. However, since the PWM technique generates a fixed switching frequency of $\bar{f}_{sw} = 1/T_s$, the sampling period in the PWM-PCC strategy was set to T_s = 200 µs, which increases the d-q current errors and the torque ripple in comparison to previous strategies. The EPWM-PCC strategy provides a significant reduction in the torque ripple, that is, TWR_t decreased from 11.04% to 1.35%, due to the optimization in amplitude of the voltage vectors used by this strategy. Nonetheless, both PWM-PCC and EPWM-PCC strategies do not apply any x-y voltage components to the machine, meaning that the low order current harmonics mapped into the x'-y' subspace cannot be compensated.

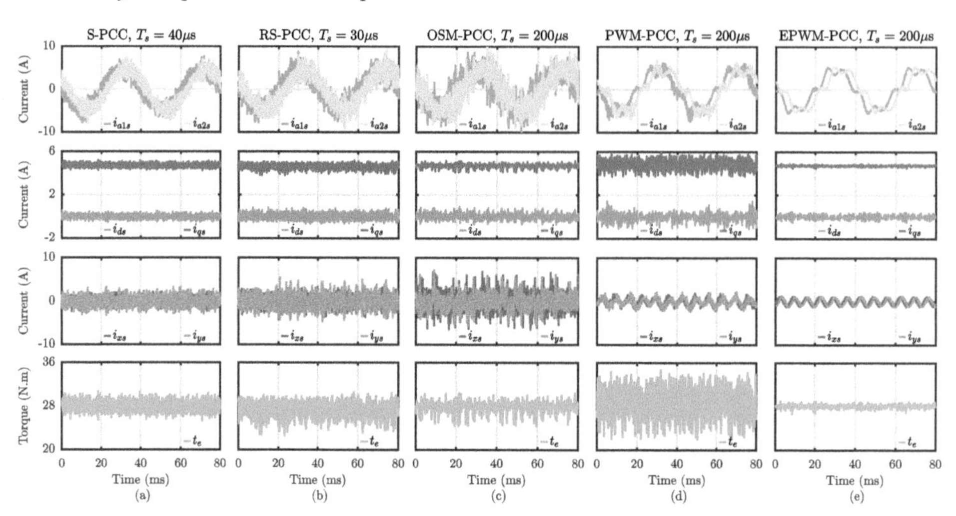

Figure 15. Experimental results for the PMSM drive operating at 750 rpm and rated load (motoring mode) for: (**a**) S-PCC; (**b**) RS-PCC; (**c**) OSM-PCC; (**d**) PWM-PCC and (**e**) EPWM-PCC.

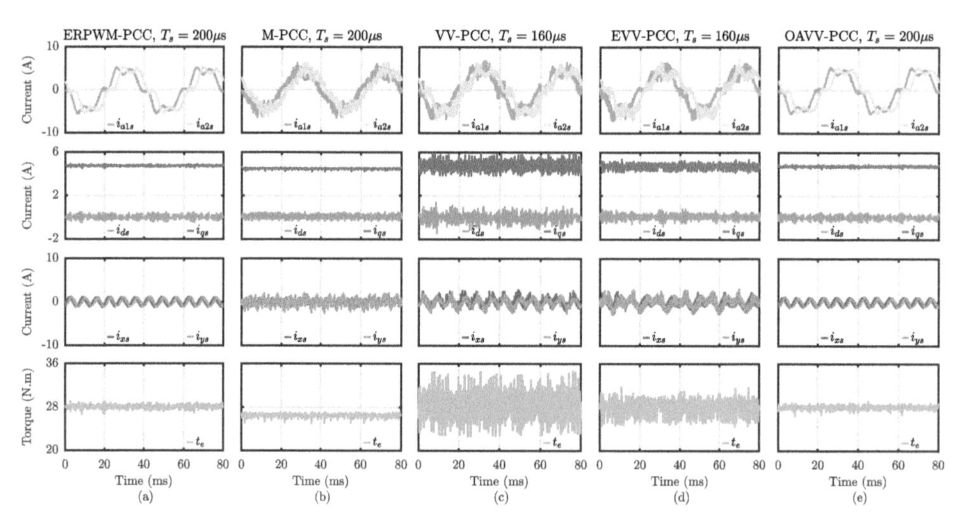

Figure 16. Experimental results for the PMSM drive operating at 750 rpm and rated load (motoring mode) for: (**a**) ERPWM-PCC; (**b**) M-PCC; (**c**) VV-PCC; (**d**) EVV-PCC and (**e**) OAVV-PCC.

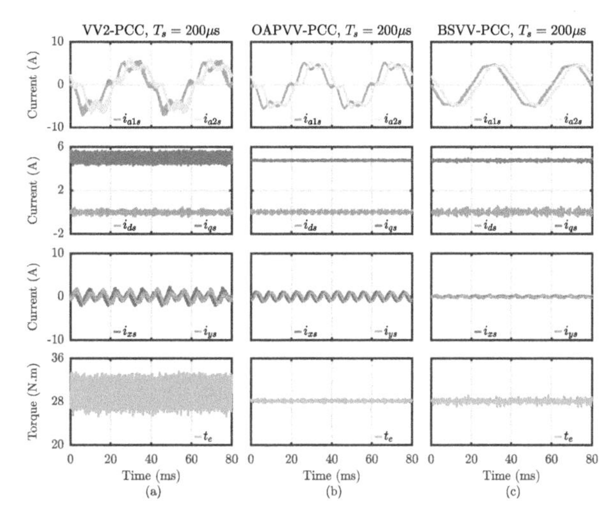

Figure 17. Experimental results for the PMSM drive operating at 750 rpm and rated load (motoring mode) for: (**a**) VV2-PCC; (**b**) OAPVV-PCC and (**c**) BSVV-PCC.

The experimental results for the operation of the six-phase drive under the strategies ERPWM-PCC, M-PCC, VV-PCC, EVV-PCC and OAVV-PCC are shown in Figure 16. In steady-state operation, the ERPWM-PCC strategy gives an equal performance to EPWM-PCC but since it is able to operate outside the linear modulation region (i.e., it can apply voltage vectors with a full amplitude of $0.644 \cdot U_{dc}$), the operation limits of the drive are increased and the performance of the drive during transients is enhanced. The M-PCC strategy integrates a different modulation strategy that combines two adjacent voltage vectors in the α-β subspace and two zero vectors ($\mathbf{v}_0$ and $\mathbf{v}_{63}$) in order to obtain a fixed switching frequency as with PWM-based PCC strategies. Although the M-PCC reduces the amplitude of the x'-y' current components, since the cost function of this strategy considers the reference current tracking errors in both subspaces, an optimal tracking of the current references in both subspaces is not possible and a steady-state error is observed in the q-axis current and in torque. In terms of computational requirements, the M-PCC strategy has a mean execution time of 78.0 µs, being the control strategy with higher computational requirements among the compared PCC techniques. The VV-PCC strategy considers the use of twelve large active and one zero virtual vectors, which avoids the application of x-y voltage components to the machine and presents a similar performance to the PWM-PCC strategy, although it provides higher dc-link voltage usage and leads to a mean switching frequency of 4.88 kHz. The EVV-PCC strategy manages to reduce the d-q current errors and torque ripple in comparison to the VV-PCC strategy, due to the inclusion of small virtual vectors. Although the EVV-PCC strategy is able to apply one out of twenty-five virtual vectors, this strategy uses a deadbeat approach to reduce the number of candidates to only two, thus providing a small execution time ($\bar{t}_{exe}$ = 28.28 µs). Since in the OAVV-PCC strategy the virtual vectors are optimized in amplitude, the d-q current errors and torque ripple are significantly reduced in comparison to VV-PCC and EVV-PCC strategies.

Table 4. Performance indicators for the drive operating at 750 rpm and rated load (motoring mode) for the PCC strategies.

Strategy	$E_{i,d}$ (%)	$E_{i,q}$ (%)	$E_{i,x}$ (%)	$E_{i,y}$ (%)	THD$_i$ (%)	TWD$_i$ (%)	TWR$_i$ (%)	$\bar{f}_{sw}$ (kHz)	$\bar{t}_{exe}$ (μs)
S-PCC	2.94	2.56	13.50	17.38	3.22	27.87	3.10	3.41	38.89
RS-PCC	4.22	4.61	18.83	22.64	2.87	39.30	4.75	3.77	25.25
OSM-PCC	5.96	3.35	50.37	36.17	23.77	76.37	4.51	5.00	35.22
PWM-PCC	7.52	9.29	15.34	15.00	22.76	29.37	11.04	5.00	29.99
EPWM-PCC	3.41	1.08	14.60	13.68	22.59	23.05	1.35	5.00	33.67
ERPWM-PCC	3.41	1.08	14.60	13.68	22.59	23.05	1.35	5.00	33.67
M-PCC	4.16	5.99	6.62	18.10	15.92	24.44	1.27	5.00	78.00
VV-PCC	7.14	8.47	18.01	15.39	24.09	31.87	10.00	4.88	29.54
EVV-PCC	4.71	4.41	15.28	20.62	25.35	32.04	5.41	4.85	28.28
OAVV-PCC	3.32	1.05	14.94	13.67	22.80	23.20	1.31	5.00	37.46
VV2-PCC	3.30	6.46	16.31	15.30	23.96	25.92	8.47	4.64	33.60
OAPVV-PCC	2.11	0.54	14.86	13.45	22.52	22.60	0.66	5.00	49.23
BSVV-PCC	3.28	1.03	2.16	3.40	4.61	6.38	1.30	5.00	35.45

The experimental results for the operation of the six-phase drive under strategies VV2-PCC, OAPVV-PCC and BSVV-PCC are shown in Figure 17. The VV2-PCC strategy combines two virtual vectors over a sampling period and offers a performance slightly worse than the one obtained with OAVV-PCC. The ripple in the d-q current components is due to a finite set of values that can be selected for the duty cycles of the two virtual vectors, as detailed in Section 4.3. The OAPVV-PCC strategy provides the lowest d-q current errors and torque ripple among the different PCC strategies. However, as in the case of PWM and virtual vector based PCC strategies, the low order harmonics in the x'-y' current components cannot be suppressed. On the other hand, the BSVV-PCC strategy is able to control both the d-q and x'-y' current components and provides the lowest x'-y' current errors and the lowest current harmonic distortion (TWD$_i$ = 6.38%) among all tested PCC strategies.

The experimental results for the operation of the six-phase drive under strategies S-PTC, HR-PTC, RFC-PTC, FC-PTC and MV-PTC are shown in Figure 18, while the corresponding performance indicators are given in Table 5. The obtained results show that although the S-PTC strategy provides a higher current harmonic distortion over the S-PCC strategy, 67.17% versus 27.87%, it gives a smaller torque ripple, 1.71% versus 3.10%. The HR-PTC strategy optimizes the amplitude of the optimal voltage vector, which would improve the steady-state performance of the drive. However, since it also increases the number of commutations of the power switches, the sampling period was set to T_s = 60 μs. Even with a higher switching frequency of 12.83 kHz, the HR-PTC gives the worst results in terms of current waveform distortion (TWD$_i$ = 89.36%) among the compared PTC strategies. The RFC-PTC strategy leads to a lower torque ripple in comparison to the previous PTC strategies, however it produces a high current distortion (TWD$_i$ = 77.02%), even with a high switching frequency of 10 kHz. On the other hand, the FC-PTC strategy reduces the ripple of the phase currents due to the use of virtual vectors. Moreover, the MV-PTC strategy optimizes the amplitude of the selected virtual vector, giving the lowest torque ripple (TWR$_t$ = 0.46%) for the compared PTC strategies, while maintaining a fixed switching frequency of 5 kHz.

Table 5. Performance indicators for the drive operating at 750 rpm and rated load (motoring mode) for all the PTC strategies.

Strategy	$E_{\psi,d}$ (%)	$E_{\psi,q}$ (%)	$E_{\psi,x}$ (%)	$E_{\psi,y}$ (%)	THD$_i$ (%)	TWD$_i$ (%)	TWR$_t$ (%)	$\bar{f}_{sw}$ (kHz)	$\bar{t}_{exe}$ (μs)
S-PTC	0.85	0.40	0.52	0.47	13.66	67.17	1.71	5.03	18.51
HR-PTC	0.97	0.66	0.72	0.59	17.00	89.36	2.87	12.83	43.35
RFC-PTC	0.52	0.23	0.64	0.51	23.00	77.02	1.02	10.00	34.34
FC-PTC	1.03	1.63	0.20	0.24	25.85	31.42	7.14	4.97	29.90
MV-PTC	0.77	0.10	0.19	0.18	22.96	23.20	0.46	5.00	34.01

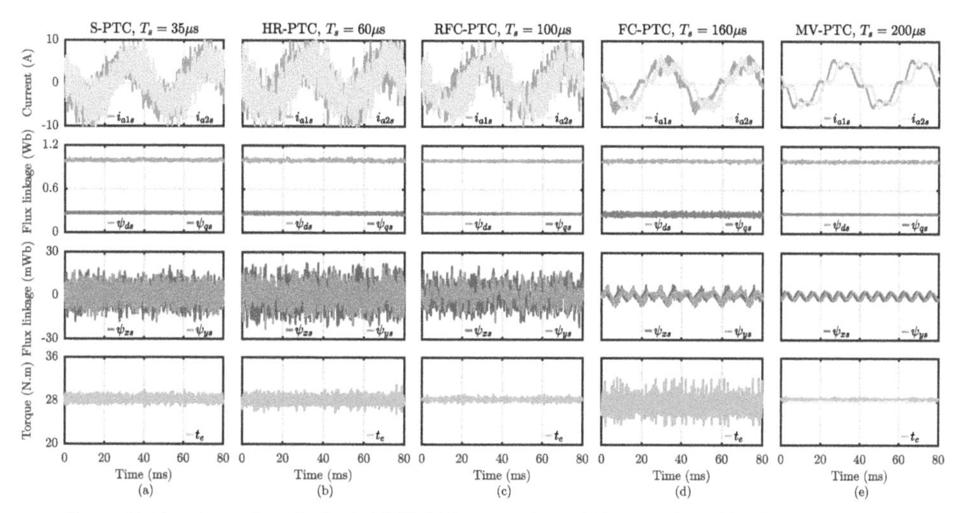

Figure 18. Experimental results for the PMSM drive operating at 750 rpm and rated load (motoring mode) for: (**a**) S-PTC; (**b**) HR-PTC; (**c**) RCF-PTC; (**d**) FC-PTC and (**e**) MV-PTC.

6.3. Comparison of Tested PCC and PCC Strategies

In order to summarize the merits and demerits of all tested PCC and PTC strategies applied to an electric drive based on a six-phase SPMSM, a comparison between these control strategies is given in Table 6. The following statements are defined to evaluate each control strategy:

- S1: The concept of the control strategy is simple and of easy implementation.
- S2: The control strategy produces a fixed switching frequency.
- S3: The high-frequency content of the phase currents (mapped into the x'-y' subspace) is minimized due to the use of a PWM technique or virtual vectors.
- S4: The low-frequency order harmonics of the phase currents (mapped into the x'-y' subspace) due to deadtime effects and back-EMF harmonics are suppressed by the control strategy.
- S5: No weighting factors need to be tuned.
- S6: The computational burden of the control strategy is low.
- S7: The control strategy gives a good performance at low speeds.
- S8: The control strategy provides full dc-bus usage, that is, it is able to apply a voltage vector with an amplitude up to $0.644 \cdot U_{dc}$ in the α-β subspace.
- S9: A separate and fast digital control platform (e.g. an FPGA) is not required to generate switching patterns for the power switches of the 2L-VSIs.

Each statement listed above is classified in Table 6 with a '+', when it is verified for the control strategy and with a '−' when the statement is not verified. Additionally, symbol '0' is employed in the case when the statement is not completely verified. For instance, control strategy M-PCC is not able to eliminate the high-frequency ripple of phase currents as other strategies based on PWM techniques or virtual vectors but still provides less current waveform distortion than S-PCC, RS-PCC, OSM-PCC, S-PTC, HR-PTC and RFC-PTC strategies. Moreover, in the case of statement S8, the control strategies based on virtual vectors are classified with a symbol '0' since the maximum amplitude of virtual vectors in the α-β subspace is $0.598 \cdot U_{dc}$, which is 7.14% smaller than the length of the large voltage vectors (Figure 2).

According to Table 6, the BSVV-PCC strategy is the best among the tested PCC and PTC strategies, since it verifies seven out of the nine statements given above. This strategy is simple and intuitive, gives a fixed switching frequency, minimizes both high-frequency and low-frequency harmonics of the x'-y' currents, does not require the tuning of weighting factors and provides good performance at low speeds. Moreover, the ERPWM-PCC and OAVV-PCC strategies are classified in second place fulfilling six out of nine statements, since they are not capable of suppressing the low-frequency harmonics of x'-y' currents and the ERPWM-PCC strategy requires a x-y weighting factor for the operation in the overmodulation region. Additionally, the MV-PTC strategy is the best strategy among the PTC strategies, verifying five out of nine statements. This strategy loses to BSVV-PCC in the complexity of the algorithm and in the inability to eliminate low order x'-y' current harmonics. On the other hand, the control strategies that provided the worst performance were the HR-PTC and RFC-PTC strategies, which did not comply with five out of nine statements given above. When testing these two strategies, lower values for the sampling time were used, giving high values for the mean switching frequency, in order to avoid excessive high-frequency ripple in the phase currents of the machine.

Table 6. Comparison between the PCC and PTC strategies applied to an electric drive based on a six-phase SPMSM.

Strategy	S1	S2	S3	S4	S5	S6	S7	S8	S9
S-PCC	+	−	−	0	−	+	−	+	+
RS-PCC	+	−	−	0	−	+	−	+	+
OSM-PCC	+	+	−	−	−	+	0	+	−
PWM-PCC	+	+	+	−	+	+	−	−	−
EPWM-PCC	+	+	+	−	+	+	+	−	−
ERPWM-PCC	+	+	+	−	0	+	+	+	−
M-PCC	−	+	0	0	−	−	+	+	−
VV-PCC	+	−	+	−	+	+	−	0	−
EVV-PCC	+	−	+	−	+	+	0	0	−
OAVV-PCC	+	+	+	−	+	+	+	0	−
VV2-PCC	0	−	+	−	+	0	+	0	−
OAPVV-PCC	0	+	+	−	+	0	+	0	−
BSVV-PCC	+	+	+	+	+	+	+	0	−
S-PTC	+	−	−	0	−	+	−	+	+
HR-PTC	+	−	−	−	−	0	0	+	−
RFC-PTC	−	−	−	−	+	+	0	+	−
FC-PTC	−	−	+	−	+	+	0	0	−
MV-PTC	−	+	+	−	+	+	+	0	−

Each statement is classified with: '+' (verified); '0' (not completely verified); and '−' (not verified).

6.4. Parameter Sensitivity

Since FCS-MPC strategies use a machine model to predict the future behavior of the controlled variables, the accuracy of these predictions depends on the equivalent parameters of the machine [127]. As these parameters can vary for different operating conditions, any parameter mismatch will cause an error in the predictions of the FCS-MPC algorithm and will lead to a deteriorated performance of the drive [128,129]. For the case of six-phase IM drives, only Reference [130] has studied the parameter sensitivity of the S-PCC strategy, while the parameter sensitivity of FCS-MPC strategies for six-phase PMSM drives remains uncovered. Hence, the parameter sensitivity of the control strategies BSVV-PCC, OAVV-PCC and MV-PTC to variations of ±30% in the values of parameters R_s and L_{dq} is analyzed in this section. Additionally, since the BSVV-PCC is able to control the x'-y' current components, the analysis of this strategy to variations in the parameter L_{xy} is also considered.

The parameter sensitivity of the OAVV-PCC strategy is tested experimentally for variations of ±30% in the parameters R_s and L_{dq}, for the operation of the six-phase PMSM drive at 750 rpm and rated load (motoring mode) and the obtained performance indicators are given in Table 7. The OAVV-PCC provides a slightly worse performance for errors of ±30% in the value of R_s, since the current errors, torque ripple and current harmonic distortion are marginally increased. In the case of an error of −30% in L_{dq}, the OAVV-PCC strategy still provides an acceptable performance, although in the case of a +30% error both the d-q current errors and torque ripple are heavily increased.

The performance indicators obtained for the MV-PTC strategy considering variations of ±30% in the parameters R_s and L_{dq} are given in Table 8. As in the previous case, only an error of +30% in the parameter L_{dq} provides a considerable degradation of the drive performance in terms of d-q current errors and torque ripple. Moreover, when an error of -30% is considered in L_{dq}, the d-axis flux error increases substantially in comparison to the normal case, while the indicators for the remaining cases of Table 8 only change marginally.

Table 9 contains the performance indicators for the BSVV-PCC strategy considering variations of $\pm30\%$ in the parameters R_s, L_{dq} and L_{xy}. Similarly to what was observed with the previous strategies, a mismatch in the value of R_s has a small impact on the performance of the six-phase drive. On the other hand, an error of $+30\%$ in L_{dq} negatively influences the performance of the drive, as shown by the high values of the d-q current errors and TWR_t. The errors in L_{xy} slightly increase the x-y current errors and consequently give higher values for the current harmonic distortion in comparison to the normal case. It is important to note that errors in L_{xy} do not affect significantly the value of TWR_t, since the x-y current components do not contribute to the production of torque.

Table 7. Parameter sensitivity of the OAVV-PCC strategy.

Strategy	$E_{i,d}$ (%)	$E_{i,q}$ (%)	$E_{i,x}$ (%)	$E_{i,y}$ (%)	THD_i (%)	TWD_i (%)	TWR_t (%)
Normal parameters	3.32	1.05	14.94	13.67	22.80	23.20	1.31
$0.7 \cdot R_s$	3.47	1.25	15.12	14.17	23.50	23.96	1.36
$1.3 \cdot R_s$	3.50	1.07	15.14	14.30	23.38	23.73	1.35
$0.7 \cdot L_{dq}$	4.13	0.96	14.99	14.05	23.21	23.28	1.11
$1.3 \cdot L_{dq}$	13.90	5.91	15.12	15.57	22.56	31.50	7.22

Table 8. Parameter sensitivity of the MV-PTC strategy.

Strategy	$E_{\psi,d}$ (%)	$E_{\psi,q}$ (%)	$E_{\psi,x}$ (%)	$E_{\psi,y}$ (%)	THD_i (%)	TWD_i (%)	TWR_t (%)
Normal parameters	0.77	0.10	0.19	0.18	22.96	23.20	0.46
$0.7 \cdot R_s$	3.57	0.12	0.19	0.17	21.85	22.10	0.31
$1.3 \cdot R_s$	0.86	0.13	0.20	0.18	23.24	23.52	0.47
$0.7 \cdot L_{dq}$	3.57	0.12	0.19	0.17	21.85	22.10	0.31
$1.3 \cdot L_{dq}$	2.38	0.70	0.19	0.19	23.09	25.04	3.24

Table 9. Parameter sensitivity of the BSVV-PCC strategy.

Strategy	$E_{i,d}$ (%)	$E_{i,q}$ (%)	$E_{i,x}$ (%)	$E_{i,y}$ (%)	THD_i (%)	TWD_i (%)	TWR_t (%)
Normal parameters	3.28	1.03	2.16	3.40	4.61	6.38	1.30
$0.7 \cdot R_s$	3.42	1.17	2.40	3.38	4.68	6.39	1.35
$1.3 \cdot R_s$	3.44	1.08	2.42	3.47	4.75	6.66	1.36
$0.7 \cdot L_{dq}$	3.55	0.89	2.22	3.41	4.66	5.89	1.09
$1.3 \cdot L_{dq}$	5.09	1.95	2.83	3.62	4.51	9.15	2.43
$0.7 \cdot L_{xy}$	3.62	1.15	2.62	3.96	5.41	6.61	1.41
$1.3 \cdot L_{xy}$	3.59	1.10	2.46	3.16	4.22	6.77	1.40

7. Conclusions

This paper has presented a critical comparative study of the FCS-MPC strategies available in the literature for electric drives based on six-phase asymmetrical machines, including a comprehensive overview of the theoretical background of these FCS-MPC strategies. It also assembles in a single reference the mathematical models of the six-phase drive topology, based on either IMs or PMSMs.

A total of thirteen PCC and five PTC strategies applied to a six-phase PMSM drive were compared side-by-side, with the aid of simulation and experimental results and their merits and shortcomings were discussed. In general, the PCC strategies favor a reduced harmonic content in the phase currents of the machine, while the PTC strategies produce a smaller torque ripple. The control strategies based on virtual vectors provide less high-frequency harmonic content in the x'-y' currents. Additionally, the control strategies based on virtual vectors optimized in amplitude or optimized in both amplitude and phase provide the lowest current or flux errors in the d-q subspace and improve the performance of the drive at low speeds. The low-frequency components of the x'-y' currents, due to deadtime effects in the power switches and due to the back-EMF harmonics in the case of six-phase PMSMs, were only suppressed by the BSVV-PCC strategy.

In the authors' opinion, this paper is useful to introduce FCS-MPC to control engineers or researchers working in the area of control of multiphase electric drives. Additionally, the paper can also help those who are already engaged in this field to select the best FCS-MPC strategy for their application, considering the merits and shortcomings of each strategy.

Author Contributions: Conceptualization, S.C. and P.G.; methodology, P.G. and S.C.; software, P.G.; validation, P.G.; formal analysis, P.G.; investigation, P.G.; resources, S.C. and A.M.; data curation, P.G.; writing—original draft preparation, P.G. and S.C.; writing—review and editing, P.G., S.C., and A.M.; visualization, S.C. and A.M.; supervision, S.C. and A.M.; project administration, P.G. and S.C.; funding acquisition, S.C. and A.M.

Funding: This work is funded by FCT/MEC through national funds and when applicable co-funded by FEDER—PT2020 partnership agreement under the project UID/EEA/50008/2019. P.G. also acknowledges the financial support of the Portuguese Foundation for Science and Technology (FCT) under the scholarship SFRH/BD/129286/2017.

Conflicts of Interest: The authors declare no conflict of interest. The funders had no role in the design of the study; in the collection, analyses, or interpretation of data; in the writing of the manuscript, or in the decision to publish the results.

Abbreviations

1N	single isolated neutral
1NIN	single non-isolated neutral
2L-VSI	two-level voltage source inverter
2N	two isolated neutrals
3L-NPC	three-level neutral-point-clamped
BSVV-PCC	bi-subspace predictive current control based on virtual vectors
CSI	current source inverter
CCS-MPC	continuous control set model predictive control
DTC	direct torque control
EMF	electromotive force
EPWM-PCC	enhanced pulse width modulation predictive current control
ERPWM-PCC	extended range pulse width modulation predictive current control
EVV-PCC	predictive current control based on a extended set of virtual vectors
FC-PTC	flux constrained predictive torque control
FCS-MPC	finite control set model predictive control
FOC	field oriented control
FPGA	field-programmable gate array
HR-PTC	high robustness predictive torque control
IGBT	insulated gate bipolar transistor
IM	induction machine
M-PCC	modulated predictive current control
MMF	magnetomotive force
MPC	model predictive control
MTPA	maximum torque per ampere
MV-PTC	multi-vector predictive torque control
OAVV-PCC	predictive current control based on virtual vectors with optimal amplitude
OAPVV-PCC	predictive current control based on virtual vectors with optimal amplitude and phase
OSM-PCC	one step modulation model predictive control
OSS-MPC	optimal switching sequence model predictive control
OSV-MPC	optimal switching vector model predictive control
PCC	predictive current control
PI	proportional-integral
PM	permanent magnet
PMSM	permanent magnet synchronous machine
PSC	predictive speed control
PTC	predictive torque control
PWM	pulse width modulation
PWM-PCC	pulse width modulation predictive current control
RCF-PTC	reduced cost function predictive torque control
RS-PCC	restrained search predictive current control

S-PCC	standard predictive current control
S-PTC	standard predictive torque control
SPMSM	permanent magnet synchronous machine with surface-mounted permanent magnets
SV-PWM	space vector pulse width modulation
THD	total harmonic distortion
TWD	total waveform distortion
TWR	total waveform ripple
VV-PCC	predictive current control based on virtual vectors
VV2-PCC	predictive current control based on the application of two virtual vectors
VSD	variable space decomposition
VSI	voltage source inverter
ZSC	zero-sequence current

List of Symbols

General

C_{dq}, C_{xy}	overcurrent hard constraints for the *d-q* and *x-y* subspaces
d_i	duty cycles of vectors $\mathbf{v}_i$, $\mathbf{v}_{vi}$ or $\mathbf{v}_{dvi}$
$E_{i,d}$, $E_{i,q}$, $E_{i,x}$, $E_{i,y}$,	current error of d, q, x and y-axis components
$E_{\psi,d}$, $E_{\psi,q}$, $E_{\psi,x}$, $E_{\psi,y}$,	flux linkage error of d, q, x and y-axis components
$\bar{f}_{sw}$	mean switching frequency
g_c, g_{cf}, g_f	cost functions for PTC strategies
g_{tf}, g_{fcs}, g_{cm}, g_{cs}	cost functions for PCC strategies
i_{dr}, i_{qr}	*d-q* rotor current components
i_{ds}, i_{qs}, $i_{xs'}$, $i_{ys'}$, i_{z1s}, i_{z2s}	*d-q*, x'-y' and z1-z2 stator current components
$i_{ds,0}$, $i_{qs,0}$, $i_{xs',0}$, $i_{ys',0}$	*d-q* and x'-y' stator current components due to a zero vector
$i_{ds,i}$, $i_{qs,i}$, $i_{xs',i}$, $i_{ys',i}$	*d-q* and x'-y' stator current components due to vector $\mathbf{v}_i$, $\mathbf{v}_{vi}$ or $\mathbf{v}_{dvi}$
$\mathbf{i}_s$	stator current vector in a synchronous reference frame
$\mathbf{i}_s^{\alpha\beta}$	stator current vector in a stationary reference frame
$i_{s,dq}$, $i_{s,xy}$	current amplitude in the *d-q* and *x-y* subspaces
L_d, L_q, L_x, L_y, L_{z1}, L_{z2}	d, q, x, y, z1 and z2-axis inductances
L_{dq}, L_{xy}, L_0	*d-q*, *x-y* and z1-z2 subspace inductances
L_{lm}	stator mutual leakage inductance
L_{ls}, L_{lr}	stator and rotor self leakage inductances
L_s, L_r, L_m	stator, rotor and magnetizing inductances
p	number of pole-pairs
$\mathbf{s}$	switching state vector
s_{a1s}, s_{b1s}, s_{c1s}, s_{a2s}, s_{b2s}, s_{c2s}	phase switching states
t_e	electromagnetic torque
$t_{e,0}$	electromagnetic torque due to a vector $\mathbf{v}_{v0}$
$t_{e,i}$	electromagnetic torque due to a vector $\mathbf{v}_{vi}$
$\bar{t}_{exe}$	mean execution time
T_s	sampling period
THD_i, TWD_i	current total harmonic distortion and total waveform distortion
TWR_r	torque waveform ripple
$\mathbf{u}_s$	stator voltage vector in a synchronous reference frame
$\mathbf{u}_s^{\alpha\beta}$	stator voltage vector in a stationary reference frame
u_{a1s}, u_{b1s}, u_{c1s}, u_{a2s}, u_{b2s}, u_{c2s}	phase stator voltages
u_{dr}, u_{qr}	*d-q* rotor voltage components
u_{ds}, u_{qs}, $u_{xs'}$, $u_{ys'}$, u_{z1s}, u_{z2s}	*d-q*, x'-y' and z1-z2 stator voltage components
$u_{\alpha s}$, $u_{\beta s}$, u_{xs}, u_{ys}	α-β, *x-y* stator voltage components
$\mathbf{v}_i$	voltage vector with index i
$\mathbf{v}_{dvi}$	dual virtual vector with index i
$\mathbf{v}_{vi}$	virtual vector with index i
θ	rotor electrical position
λ_i, λ_f	weighting factors of current and flux
ψ_{dr}, ψ_{qr}	*d-q* rotor flux linkage components
ψ_{ds}, ψ_{qs}, $\psi_{xs'}$, $\psi_{ys'}$, ψ_{z1s}, ψ_{z2s}	*d-q*, x'-y' and z1-z2 stator flux linkage components
$\psi_{ds,PM}$, $\psi_{qs,PM}$, $\psi_{xs',PM}$, $\psi_{ys',PM}$, $\psi_{z1s,PM}$, $\psi_{z2s,PM}$	*d-q*, x'-y' and z1-z2 stator flux linkage components due to the PMs

ψ_s	flux linkage amplitude in the *d*-*q* subspace
$\psi_{s,0}$	flux linkage amplitude in the *d*-*q* subspace due to a zero vector
$\psi_{s,i}$	flux linkage amplitude in the *d*-*q* subspace due to a vector $\mathbf{v}_{vi}$
$\psi_{s,\mathrm{PMi}}$	*i*-order harmonic component of the flux linkage due to the PMs
ω_a	electrical angular speed of an arbitrary reference frame
ω_s, ω_r	stator and rotor electrical angular speeds
$\mathbf{R}$	rotation matrix
$\mathbf{T}_{vsd}$	VSD transformation
U_{dc}	dc-link voltage
Subscripts	
$d,q,x',y',z1,z2$	d, q, x, y, z1 and z2-axis quantities
s, r	stator and rotor quantities
n	rated value
α,β,x,y	d, q, x, y, z1 and z2-axis quantities
Superscripts	
$*$	reference value
$k+2$	predicted quantity for instant $k+2$
$k+h$	predicted quantity for instant $k+h$

References

1. Singh, G. Multi-phase induction machine drive research—A survey. *Elect. Power Syst. Res.* **2002**, *61*, 139–147. [CrossRef]
2. Alger, P.L.; Freiburghouse, E.; Chase, D. Double windings for turbine alternators. *Trans. Am. Inst. Electr. Eng.* **1930**, *49*, 226–244. [CrossRef]
3. Ward, E.; Harer, H. Preliminary investigation of an invertor-fed 5-phase induction motor. *Proc. Inst. Electr. Eng.* **1969**, *116*, 980–984. [CrossRef]
4. Schiferl, R.; Ong, C. Six phase synchronous machine with AC and DC stator connections, Part I: Equivalent circuit representation and steady-state analysis. *IEEE Trans. Power App. Syst.* **1983**, *102*, 2685–2693. [CrossRef]
5. Jahns, T.M. Improved reliability in solid-state AC drives by means of multiple independent phase drive units. *IEEE Trans. Ind. Appl.* **1980**, *16*, 321–331. [CrossRef]
6. Parsa, L. On advantages of multi-phase machines. In Proceedings of the 31st Annual Conference of IEEE Industrial Electronics Society (IECON), Raleigh, NC, USA, 6–10 November 2005; pp. 1–6.
7. Levi, E. Advances in converter control and innovative exploitation of additional degrees of freedom for multiphase machines. *IEEE Trans. Ind. Electron.* **2016**, *63*, 433–448. [CrossRef]
8. Toliyat, H.A.; Lipo, T.A.; White, J.C. Analysis of a concentrated winding induction machine for adjustable speed drive applications. I. Motor analysis. *IEEE Trans. Energy Convers.* **1991**, *6*, 679–683. [CrossRef]
9. Toliyat, H.A.; Lipo, T.A.; White, J.C. Analysis of a concentrated winding induction machine for adjustable speed drive applications. II. Motor design and performance. *IEEE Trans. Energy Convers.* **1991**, *6*, 684–692. [CrossRef]
10. Gataric, S. A polyphase Cartesian vector approach to control of polyphase AC machines. In Proceedings of the IEEE Industry Application Society (IAS) Annual Meeting, Rome, Italy, 8–12 October 2000; Volume 3, pp. 1648–1654.
11. Jones, M.; Vukosavic, S.N.; Levi, E. Parallel-connected multiphase multidrive systems with single inverter supply. *IEEE Trans. Ind. Electron.* **2009**, *56*, 2047–2057. [CrossRef]
12. Che, H.S.; Levi, E.; Jones, M.; Duran, M.J.; Hew, W.P.; Rahim, N.A. Operation of a six-phase induction machine using series-connected machine-side converters. *IEEE Trans. Ind. Electron.* **2014**, *61*, 164–176. [CrossRef]
13. Zoric, I.; Jones, M.; Levi, E. Arbitrary power sharing among three-phase winding sets of multiphase machines. *IEEE Trans. Ind. Electron.* **2018**, *65*, 1128–1139. [CrossRef]
14. Subotic, I.; Dordevic, O.; Gomm, B.; Levi, E. Active and Reactive Power Sharing Between Three-Phase Winding Sets of a Multiphase Induction Machine. *IEEE Trans. Energy Convers.* **2019**, *34*, 1401–1410. [CrossRef]
15. Zabaleta, M.; Levi, E.; Jones, M. A novel synthetic loading method for multiple three-phase winding electric machines. *IEEE Trans. Energy Convers.* **2019**, *34*, 70–78. [CrossRef]
16. Abduallah, A.A.; Dordevic, O.; Jones, M.; Levi, E. Regenerative test for multiple three-phase machines with even number of neutral points. *IEEE Trans. Ind. Electron.* **2020**, *67*, 1684–1694. [CrossRef]

17. Subotic, I.; Bodo, N.; Levi, E.; Jones, M.; Levi, V. Isolated chargers for EVs incorporating six-phase machines. *IEEE Trans. Ind. Electron.* **2016**, *63*, 653–664. [CrossRef]
18. Subotic, I.; Bodo, N.; Levi, E.; Dumnic, B.; Milicevic, D.; Katic, V. Overview of fast on-board integrated battery chargers for electric vehicles based on multiphase machines and power electronics. *IET Elect. Power Appl.* **2016**, *10*, 217–229. [CrossRef]
19. Subotic, I.; Bodo, N.; Levi, E. Integration of six-phase EV drivetrains into battery charging process with direct grid connection. *IEEE Trans. Energy Convers.* **2017**, *32*, 1012–1022. [CrossRef]
20. Duran, M.J.; Gonzalez-Prieto, I.; Barrero, F.; Levi, E.; Zarri, L.; Mengoni, M. A simple braking method for six-phase induction motor drives with unidirectional power flow in the base-speed region. *IEEE Trans. Ind. Electron.* **2017**, *64*, 6032–6041. [CrossRef]
21. Gonzalez-Prieto, I.; Duran, M.J.; Barrero, F.J. Fault-tolerant control of six-phase induction motor drives with variable current injection. *IEEE Trans. Power Electron.* **2017**, *32*, 7894–7903. [CrossRef]
22. Duran, M.J.; Gonzalez-Prieto, I.; Rios-Garcia, N.; Barrero, F. A simple, fast, and robust open-phase fault detection technique for six-phase induction motor drives. *IEEE Trans. Power Electron.* **2018**, *33*, 547–557. [CrossRef]
23. Gonzalez-Prieto, I.; Duran, M.J.; Rios-Garcia, N.; Barrero, F.; Martin, C. Open-switch fault detection in five-phase induction motor drives using model predictive control. *IEEE Trans. Ind. Electron.* **2018**, *65*, 3045–3055. [CrossRef]
24. Williamson, S.; Smith, S. Pulsating torque and losses in multiphase induction machines. *IEEE Trans. Ind. Appl.* **2003**, *39*, 986–993. [CrossRef]
25. Levi, E.; Bojoi, R.; Profumo, F.; Toliyat, H.; Williamson, S. Multiphase induction motor drives—A technology status review. *IET Elect. Power Appl.* **2007**, *1*, 489–516. [CrossRef]
26. Levi, E. Multiphase electric machines for variable-speed applications. *IEEE Trans. Ind. Electron.* **2008**, *55*, 1893–1909. [CrossRef]
27. Cao, W.; Mecrow, B.C.; Atkinson, G.J.; Bennett, J.W.; Atkinson, D.J. Overview of electric motor technologies used for more electric aircraft (MEA). *IEEE Trans. Ind. Electron.* **2012**, *59*, 3523–3531.
28. Bojoi, R.; Cavagnino, A.; Tenconi, A.; Vaschetto, S. Control of shaft-line-embedded multiphase starter/generator for aero-engine. *IEEE Trans. Ind. Electron.* **2016**, *63*, 641–652. [CrossRef]
29. Bojoi, R.; Rubino, S.; Tenconi, A.; Vaschetto, S. Multiphase electrical machines and drives: A viable solution for energy generation and transportation electrification. In Proceedings of the IEEE International Conference and Exposition on Electrical and Power Engineering (EPE), Iasi, Romania, 20–22 October 2016; pp. 632–639.
30. Bojoi, R.; Cavagnino, A.; Cossale, M.; Tenconi, A. Multiphase starter generator for a 48-V mini-hybrid powertrain: Design and testing. *IEEE Trans. Ind. Appl.* **2016**, *52*, 1750–1758.
31. Jung, E.; Yoo, H.; Sul, S.K.; Choi, H.S.; Choi, Y.Y. A nine-phase permanent-magnet motor drive system for an ultrahigh-speed elevator. *IEEE Trans. Ind. Appl.* **2012**, *48*, 987–995. [CrossRef]
32. Liu, Z.; Wu, J.; Hao, L. Coordinated and fault-tolerant control of tandem 15-phase induction motors in ship propulsion system. *IET Electr. Power Appl.* **2017**, *12*, 91–97. [CrossRef]
33. Moraes, T.D.S.; Nguyen, N.K.; Semail, E.; Meinguet, F.; Guerin, M. Dual-multiphase motor drives for fault-tolerant applications: Power electronic structures and control strategies. *IEEE Trans. Power Electron.* **2017**, *33*, 572–580. [CrossRef]
34. Zhu, Z.; Hu, J. Electrical machines and power-electronic systems for high-power wind energy generation applications: Part II–power electronics and control systems. *COMPEL Int. J. Comput. Math. Elect. Electron. Eng.* **2012**, *32*, 34–71. [CrossRef]
35. Yaramasu, V.; Wu, B.; Sen, P.C.; Kouro, S.; Narimani, M. High-power wind energy conversion systems: State-of-the-art and emerging technologies. *Proc. IEEE* **2015**, *103*, 740–788. [CrossRef]
36. Prieto-Araujo, E.; Junyent-Ferré, A.; Lavernia-Ferrer, D.; Gomis-Bellmunt, O. Decentralized control of a nine-phase permanent magnet generator for offshore wind turbines. *IEEE Trans. Energy Convers.* **2015**, *30*, 1103–1112. [CrossRef]
37. Abbas, M.A.; Christen, R.; Jahns, T.M. Six-phase voltage source inverter driven induction motor. *IEEE Trans. Ind. Appl.* **1984**, *IA-20*, 1251–1259. [CrossRef]
38. Bojoi, R.; Farina, F.; Profumo, F.; Tenconi, A. Dual-three phase induction machine drives control—A survey. *IEEJ Trans. Ind. Appl.* **2006**, *126*, 420–429. [CrossRef]

39. Bojoi, R.; Lazzari, M.; Profumo, F.; Tenconi, A. Digital field-oriented control for dual three-phase induction motor drives. *IEEE Trans. Ind. Appl.* **2003**, *39*, 752–760. [CrossRef]
40. Singh, G.K.; Nam, K.; Lim, S. A simple indirect field-oriented control scheme for multiphase induction machine. *IEEE Trans. Ind. Electron.* **2005**, *52*, 1177–1184. [CrossRef]
41. Bojoi, R.; Farina, F.; Griva, G.; Profumo, F.; Tenconi, A. Direct torque control for dual three-phase induction motor drives. *IEEE Trans. Ind. Appl.* **2005**, *41*, 1627–1636. [CrossRef]
42. Hatua, K.; Ranganathan, V. Direct torque control schemes for split-phase induction machine. *IEEE Trans. Ind. Appl.* **2005**, *41*, 1243–1254. [CrossRef]
43. Wang, F.; Mei, X.; Rodriguez, J.; Kennel, R. Model predictive control for electrical drive systems-an overview. *CES Trans. Elect. Mach. Syst.* **2017**, *1*, 219–230.
44. Liu, C.; Luo, Y. Overview of advanced control strategies for electric machines. *Chin. J. Elect. Eng.* **2017**, *3*, 53–61.
45. Wang, F.; Zhang, Z.; Mei, X.; Rodríguez, J.; Kennel, R. Advanced control strategies of induction machine: Field oriented control, direct torque control and model predictive control. *Energies* **2018**, *11*, 120. [CrossRef]
46. Kouro, S.; Perez, M.A.; Rodriguez, J.; Llor, A.M.; Young, H.A. Model predictive control: MPC's role in the evolution of power electronics. *IEEE Ind. Electron. Mag.* **2015**, *9*, 8–21. [CrossRef]
47. Bordons, C.; Montero, C. Basic principles of MPC for power converters: Bridging the gap between theory and practice. *IEEE Ind. Electron. Mag.* **2015**, *9*, 31–43. [CrossRef]
48. Vazquez, S.; Rodriguez, J.; Rivera, M.; Franquelo, L.G.; Norambuena, M. Model predictive control for power converters and drives: Advances and trends. *IEEE Trans. Ind. Electron.* **2017**, *64*, 935–947. [CrossRef]
49. Gonçalves, P.F.; Cruz, S.M.; Mendes, A.M. Comparison of Model Predictive Control Strategies for Six-Phase Permanent Magnet Synchronous Machines. In Proceedings of the 44th Annual Conference of the IEEE Industrial Electronics Society (IECON), Washington, DC, USA, 21–23 October 2018; pp. 5801–5806.
50. Barrero, F.; Duran, M.J. Recent advances in the design, modeling, and control of multiphase machines—Part I. *IEEE Trans. Ind. Electron.* **2016**, *63*, 449–458. [CrossRef]
51. Ye, D.; Li, J.; Qu, R.; Lu, H.; Lu, Y. Finite set model predictive mtpa control with vsd method for asymmetric six-phase pmsm. In Proceedings of the IEEE International Electric Machines and Drives Conference (IEMDC), Miami, FL, USA, 21–24 May 2017; pp. 1–7.
52. Riveros, J.A.; Barrero, F.; Levi, E.; Durán, M.J.; Toral, S.; Jones, M. Variable-speed five-phase induction motor drive based on predictive torque control. *IEEE Trans. Ind. Electron.* **2013**, *60*, 2957–2968. [CrossRef]
53. Lim, C.S.; Levi, E.; Jones, M.; Rahim, N.A.; Hew, W.P. FCS-MPC-based current control of a five-phase induction motor and its comparison with PI-PWM control. *IEEE Trans. Ind. Electron.* **2014**, *61*, 149–163. [CrossRef]
54. Wu, X.; Song, W.; Xue, C. Low-Complexity Model Predictive Torque Control Method Without Weighting Factor for Five-Phase PMSM Based on Hysteresis Comparators. *Trans. Emerg. Sel. Top. Power Electron.* **2018**, *6*, 1650–1661. [CrossRef]
55. Prieto, I.G.; Zoric, I.; Duran, M.J.; Levi, E. Constrained Model Predictive Control in Nine-phase Induction Motor Drives. *IEEE Trans. Energy Convers.* **2019**, *34*, 1881–1889. [CrossRef]
56. Liu, Z.; Li, Y.; Zheng, Z. A review of drive techniques for multiphase machines. *CES Trans. Elect. Mach. Syst.* **2018**, *2*, 243–251. [CrossRef]
57. Tenconi, A.; Rubino, S.; Bojoi, R. Model Predictive Control for Multiphase Motor Drives—A Technology Status Review. In Proceedings of the IEEE International Power Electronic Conference (IPEC), Niigata, Japan, 20–24 May 2018; pp. 732–739.
58. Duran, M.J.; Levi, E.; Barrero, F. Multiphase electric drives: Introduction. In *Wiley Encyclopedia of Electrical and Electronics Engineering*; Webster, J., Ed.; John Wiley and Sons, Inc.: Hoboken, NJ, USA, 2017; pp. 1–26.
59. Duran, M.J.; Barrero, F. Recent advances in the design, modeling, and control of multiphase machines—Part II. *IEEE Trans. Ind. Electron.* **2016**, *63*, 459–468. [CrossRef]
60. Yaramasu, V.; Dekka, A.; Durán, M.J.; Kouro, S.; Wu, B. PMSG-based wind energy conversion systems: Survey on power converters and controls. *IET Elect. Power Appl.* **2017**, *11*, 956–968. [CrossRef]
61. Yepes, A.G.; Malvar, J.; Vidal, A.; López, O.; Doval-Gandoy, J. Current harmonics compensation based on multiresonant control in synchronous frames for symmetrical n-phase machines. *IEEE Trans. Ind. Electron.* **2015**, *62*, 2708–2720. [CrossRef]

62. Yepes, A.G.; Doval-Gandoy, J.; Baneira, F.; Perez-Estevez, D.; Lopez, O. Current harmonic compensation for n-phase machines with asymmetrical winding arrangement and different neutral configurations. *IEEE Trans. Ind. Appl.* **2017**, *53*, 5426–5439. [CrossRef]
63. Betin, F.; Capolino, G.A.; Casadei, D.; Kawkabani, B.; Bojoi, R.I.; Harnefors, L.; Levi, E.; Parsa, L.; Fahimi, B. Trends in electrical machines control: Samples for classical, sensorless, and fault-tolerant techniques. *IEEE Ind. Electron. Mag.* **2014**, *8*, 43–55. [CrossRef]
64. Che, H.S.; Duran, M.J.; Levi, E.; Jones, M.; Hew, W.P.; Rahim, N.A. Postfault operation of an asymmetrical six-phase induction machine with single and two isolated neutral points. *IEEE Trans. Power Electron.* **2014**, *29*, 5406–5416. [CrossRef]
65. Munim, W.N.W.A.; Duran, M.J.; Che, H.S.; Bermúdez, M.; González-Prieto, I.; Rahim, N.A. A unified analysis of the fault tolerance capability in six-phase induction motor drives. *IEEE Trans. Power Electron.* **2017**, *32*, 7824–7836. [CrossRef]
66. Reusser, C. Power Converter Topologies for Multiphase Drive Applications. In *Electric Power Conversion*; Gaiceanu, M., Ed.; IntechOpen: London, UK, 2018; pp. 1–22.
67. Yaramasu, V.; Wu, B. *Model Predictive Control of Wind Energy Conversion Systems*; IEEE: Hoboken, NJ, USA, 2017.
68. Andersen, E. 6-phase induction motors for current source inverter drives. In Proceedings of the 16th Annual Meeting IEEE Industry Applications Society, Philadelphia, PA, USA, 5–9 October 1981; pp. 607–618.
69. Gopakumar, K.; Sathiakumar, S.; Biswas, S.; Vithayathil, J. Modified current source inverter fed induction motor drive with reduced torque pulsations. *IEE Proc. B* **1984**, *131*, 159–164. [CrossRef]
70. Levi, E.; Bodo, N.; Dordevic, O.; Jones, M. Recent advances in power electronic converter control for multiphase drive systems. In Proceedings of the IEEE Workshop on Electrical Machines Design, Control and Diagnosis (WEMDCD), Paris, France, 11–12 March 2013; pp. 158–167.
71. Duran, M.J.; Prieto, I.G.; Bermudez, M.; Barrero, F.; Guzman, H.; Arahal, M.R. Optimal fault-tolerant control of six-phase induction motor drives with parallel converters. *IEEE Trans. Ind. Electron.* **2016**, *63*, 629–640. [CrossRef]
72. Gonzalez-Prieto, I.; Duran, M.J.; Che, H.; Levi, E.; Bermúdez, M.; Barrero, F. Fault-tolerant operation of six-phase energy conversion systems with parallel machine-side converters. *IEEE Trans. Power Electron.* **2016**, *31*, 3068–3079. [CrossRef]
73. Gonzalez-Prieto, I.; Duran, M.J.; Barrero, F.; Bermudez, M.; Guzman, H. Impact of postfault flux adaptation on six-phase induction motor drives with parallel converters. *IEEE Trans. Power Electron.* **2017**, *32*, 515–528. [CrossRef]
74. Fuchs, E.; Rosenberg, L. Analysis of an alternator with two displaced stator windings. *IEEE Trans. Power App. Syst.* **1974**, *93*, 1776–1786. [CrossRef]
75. Nelson, R.; Krause, P. Induction machine analysis for arbitrary displacement between multiple winding sets. *IEEE Trans. Power App. Syst.* **1974**, *PAS-93*, 841–848. [CrossRef]
76. Gonçalves, P.F.; Cruz, S.M.; Mendes, A.M. Design of a six-phase asymmetrical permanent magnet synchronous generator for wind energy applications. *J. Eng.* **2019**, *2019*, 4532–4536. [CrossRef]
77. Eldeeb, H.M.; Abdel-Khalik, A.S.; Kullick, J.; Hackl, C.M. Pre and Post-fault Current Control of Dual Three-Phase Reluctance Synchronous Drives. *IEEE Trans. on Ind. Electron.* **2019**. [CrossRef]
78. Eldeeb, H.M.; Abdel-Khalik, A.S.; Hackl, C.M. Post-Fault Full Torque-Speed Exploitation of Dual Three-Phase IPMSM Drives. *IEEE Trans. Ind. Electron.* **2019**, *66*, 6746–6756. [CrossRef]
79. Hu, Y.; Zhu, Z.; Odavic, M. Torque capability enhancement of dual three-phase PMSM drive with fifth and seventh current harmonics injection. *IEEE Trans. Ind. Appl.* **2017**, *53*, 4526–4535. [CrossRef]
80. Wang, K. Effects of harmonics into magnet shape and current of dual three-phase permanent magnet machine on output torque capability. *IEEE Trans. Ind. Electron.* **2018**, *65*, 8758–8767. [CrossRef]
81. Hu, Y.; Zhu, Z.; Odavic, M. Comparison of two-individual current control and vector space decomposition control for dual three-phase PMSM. *IEEE Trans. Ind. Appl.* **2017**, *53*, 4483–4492. [CrossRef]
82. Lipo, T. A d-q model for six phase induction machines. *Proc. Int. Conf. Elect. Mach.* **1980**, *2*, 860–867.
83. Zhao, Y.; Lipo, T.A. Space vector PWM control of dual three-phase induction machine using vector space decomposition. *IEEE Trans. Ind. Appl.* **1995**, *31*, 1100–1109. [CrossRef]
84. Eldeeb, H.M.; Abdel-Khalik, A.S.; Hackl, C.M. Dynamic modeling of dual three-phase IPMSM drives with different neutral configurations. *IEEE Trans. Ind. Electron.* **2019**, *66*, 141–151. [CrossRef]

85. Duran, M.J.; González-Prieto, I.; González-Prieto, A.; Barrero, F. Multiphase energy conversion systems connected to microgrids with unequal power-sharing capability. *IEEE Trans. Energy Convers.* **2017**, *32*, 1386–1395. [CrossRef]
86. Che, H.S.; Levi, E.; Jones, M.; Hew, W.P.; Rahim, N.A. Current control methods for an asymmetrical six-phase induction motor drive. *IEEE Trans. Power Electron.* **2014**, *29*, 407–417. [CrossRef]
87. Hadiouche, D.; Razik, H.; Rezzoug, A. On the modeling and design of dual-stator windings to minimize circulating harmonic currents for VSI fed AC machines. *IEEE Trans. Ind. Appl.* **2004**, *40*, 506–515. [CrossRef]
88. Che, H.S.; Abdel-Khalik, A.S.; Dordevic, O.; Levi, E. Parameter estimation of asymmetrical six-phase induction machines using modified standard tests. *IEEE Trans. Ind. Electron.* **2017**, *64*, 6075–6085. [CrossRef]
89. Tessarolo, A.; Bassi, C. Stator harmonic currents in VSI-fed synchronous motors with multiple three-phase armature windings. *IEEE Trans. Energy Convers.* **2010**, *25*, 974–982. [CrossRef]
90. Arahal, M.; Barrero, F.; Toral, S.; Duran, M.; Gregor, R. Multi-phase current control using finite-state model-predictive control. *Control Eng. Pract.* **2009**, *17*, 579–587. [CrossRef]
91. Barrero, F.; Arahal, M.R.; Gregor, R.; Toral, S.; Durán, M.J. A proof of concept study of predictive current control for VSI-driven asymmetrical dual three-phase AC machines. *IEEE Trans. Ind. Electron.* **2009**, *56*, 1937–1954. [CrossRef]
92. Recalde, R.I.G. The asymmetrical dual three-phase induction machine and the mbpc in the speed control. In *Induction Motors: Modelling and Control*; Araujo, R., Ed.; IntechOpen: London, UK, 2012; pp. 385–400.
93. Rodas, J.; Gregor, R.; Takase, Y.; Moreira, H.; Riveray, M. A comparative study of reduced order estimators applied to the speed control of six-phase generator for a WT applications. In Proceedings of the 39th Annual Conference of the IEEE Industrial Electronics Society, Vienna, Austria, 10–13 November 2013; pp. 5124–5129.
94. Dasika, J.D.; Qin, J.; Saeedifard, M.; Pekarek, S.D. Predictive current control of a six-phase asymmetrical drive system based on parallel-connected back-to-back converters. In Proceedings of the IEEE Energy Conversion Congress and Exposition (ECCE), Raleigh, NC, USA, 15–20 September 2012; pp. 137–141.
95. Rodriguez, J.; Cortes, P. *Predictive Control of Power Converters and Electrical Drives*; John Wiley & Sons: Hoboken, NJ, USA, 2017.
96. Durán, M.J.; Barrero, F.; Prieto, J.; Toral, S. Predictive current control of dual three-phase drives using restrained search techniques and multi level voltage source inverters. In Proceedings of the IEEE International Symposium on Industrial Electronics (ISIE), Bari, Italy, 4–7 July 2010; pp. 3171–3176.
97. Duran, M.J.; Prieto, J.; Barrero, F.; Toral, S. Predictive current control of dual three-phase drives using restrained search techniques. *IEEE Trans. Ind. Electron.* **2011**, *58*, 3253–3263. [CrossRef]
98. Barrero, F.; Arahal, M.R.; Gregor, R.; Toral, S.; Durán, M.J. One-step modulation predictive current control method for the asymmetrical dual three-phase induction machine. *IEEE Trans. Ind. Electron.* **2009**, *56*, 1974–1983. [CrossRef]
99. Gregor, R.; Barrero, F.; Toral, S.; Duran, M.; Arahal, M.; Prieto, J.; Mora, J. Predictive-space vector PWM current control method for asymmetrical dual three-phase induction motor drives. *IET Elect. Power Appl.* **2010**, *4*, 26–34. [CrossRef]
100. Barrero, F.; Prieto, J.; Levi, E.; Gregor, R.; Toral, S.; Durán, M.J.; Jones, M. An enhanced predictive current control method for asymmetrical six-phase motor drives. *IEEE Trans. Ind. Electron.* **2011**, *58*, 3242–3252. [CrossRef]
101. Prieto, J.; Barrero, F.; Lim, C.S.; Levi, E. Predictive current control with modulation in asymmetrical six-phase motor drives. In Proceedings of the International Power Electronics and Motion Control Conference (EPE/PEMC), Novi Sad, Serbia, 4–6 September 2012; pp. 1–8.
102. Ayala, M.; Gonzalez, O.; Rodas, J.; Gregor, R.; Rivera, M. Predictive control at fixed switching frequency for a dual three-phase induction machine with Kalman filter-based rotor estimator. In Proceedings of the IEEE International Conference on Automatica (ICA-ACCA), Curico, Chile, 19–21 October 2016; pp. 1–6.
103. Ayala, M.; Rodas, J.; Gregor, R.; Doval-Gandoy, J.; Gonzalez, O.; Saad, M.; Rivera, M. Comparative study of predictive control strategies at fixed switching frequency for an asymmetrical six-phase induction motor drive. In Proceedings of the IEEE International Electrical Machines and Drives Conference (IEMDC), Miami, FL, USA, 21–24 May 2017.

104. Gonzalez, O.; Ayala, M.; Rodas, J.; Gregor, R.; Rivas, G.; Doval-Gandoy, J. Variable-speed control of a six-phase induction machine using predictive-fixed switching frequency current control techniques. In Proceedings of the IEEE International Symposium on Power Electron. for Distributed Generation Syst. (PEDG), Charlotte, NC, USA, 25–28 June 2018; pp. 1–6.
105. Gonzalez, O.; Ayala, M.; Doval-Gandoy, J.; Rodas, J.; Gregor, R.; Rivera, M. Predictive-Fixed Switching Current Control Strategy Applied to Six-Phase Induction Machine. *Energies* **2019**, *12*, 2294. [CrossRef]
106. Gonzalez-Prieto, I.; Duran, M.J.; Aciego, J.J.; Martin, C.; Barrero, F. Model predictive control of six-phase induction motor drives using virtual voltage vectors. *IEEE Trans. Ind. Electron.* **2018**, *65*, 27–37. [CrossRef]
107. Gao, L.; Fletcher, J.E.; Zheng, L. Low-speed control improvements for a two-level five-phase inverter-fed induction machine using classic direct torque control. *IEEE Trans. Ind. Electron.* **2011**, *58*, 2744–2754. [CrossRef]
108. Zheng, L.; Fletcher, J.E.; Williams, B.W.; He, X. A novel direct torque control scheme for a sensorless five-phase induction motor drive. *IEEE Trans. Ind. Electron.* **2011**, *58*, 503–513. [CrossRef]
109. Hoang, K.D.; Ren, Y.; Zhu, Z.Q.; Foster, M. Modified switching-table strategy for reduction of current harmonics in direct torque controlled dual-three-phase permanent magnet synchronous machine drives. *IET Elect. Power Appl.* **2015**, *9*, 10–19. [CrossRef]
110. Ren, Y.; Zhu, Z.Q. Reduction of both harmonic current and torque ripple for dual three-phase permanent-magnet synchronous machine using modified switching-table-based direct torque control. *IEEE Trans. Ind. Electron.* **2015**, *62*, 6671–6683. [CrossRef]
111. Pandit, J.K.; Aware, M.V.; Nemade, R.V.; Levi, E. Direct torque control scheme for a six-phase induction motor with reduced torque ripple. *IEEE Trans. Power Electron.* **2017**, *32*, 7118–7129. [CrossRef]
112. Luo, Y.; Liu, C. Elimination of harmonic currents using a reference voltage vector based-model predictive control for a six-phase PMSM motor. *IEEE Trans. Power Electron.* **2019**, *34*, 6960–6972. [CrossRef]
113. Gonçalves, P.; Cruz, S.; Mendes, A. Predictive current control based on variable amplitude virtual vectors for six-phase permanent magnet synchronous machines. In Proceedings of the 20th International Conference on Industrial Technology (ICIT), Melbourne, Australia, 13–15 February 2019; pp. 310–316.
114. Aciego, J.J.; Prieto, I.G.; Duran, M.J. Model predictive control of six-phase induction motor drives using two virtual voltage vectors. *Trans. Emerg. Sel. Top. Power Electron.* **2019**, *7*, 321–330. [CrossRef]
115. Gonçalves, P.; Cruz, S.; Mendes, A. Fixed and Variable Amplitude Virtual Vectors for Model Predictive Control of Six-Phase PMSMs with Single Neutral Configuration. In Proceedings of the 20th International Conference on Industrial Technology (ICIT), Melbourne, Australia, 13–15 February 2019; pp. 310–316.
116. Gonçalves, P.; Cruz, S.; Mendes, A. Predictive Current Control of Six-Phase Permanent Magnet Synchronous Machines Based on Virtual Vectors with Optimal Amplitude and Phase. In Proceedings of the 2nd International Conference on Smart Energy Systems and Technologies (SEST), Porto, Portugal, 9–11 September 2019; pp. 1–6.
117. Hu, Y.; Zhu, Z.Q.; Liu, K. Current control for dual three-phase permanent magnet synchronous motors accounting for current unbalance and harmonics. *IEEE Trans. Emerg. Sel. Top. Power Electron.* **2014**, *2*, 272–284.
118. Karttunen, J.; Kallio, S.; Peltoniemi, P.; Silventoinen, P. Current harmonic compensation in dual three-phase PMSMs using a disturbance observer. *IEEE Trans. Ind. Electron.* **2016**, *63*, 583–594. [CrossRef]
119. Karttunen, J.; Kallio, S.; Honkanen, J.; Peltoniemi, P.; Silventoinen, P. Partial current harmonic compensation in dual three-phase PMSMs considering the limited available voltage. *IEEE Trans. Ind. Electron.* **2017**, *64*, 1038–1048. [CrossRef]
120. Gonçalves, P.F.; Cruz, S.M.; Mendes, A.M. Bi-subspace predictive current control of six-phase PMSM drives based on virtual vectors with optimal amplitude. *IET Elect. Power Appl.* **2019**, *13*, 1672–1683. [CrossRef]
121. Prieto, J.; Riveros, J.A.; Bogado, B.; Barrero, F.; Toral, S.; Cortés, P. Electric propulsion technology based in predictive direct torque control and asymmetrical dual three-phase drives. In Proceedings of the 13th International Conference on Intelligent Transportation Systems (ITSC), Madeira Island, Portugal, 19–22 September 2010; pp. 397–402.
122. Luo, Y.; Liu, C. A simplified model predictive control for a dual three-phase PMSM with reduced harmonic currents. *IEEE Trans. Ind. Electron.* **2018**, *65*, 9079–9089. [CrossRef]
123. Luo, Y.; Liu, C. Model Predictive Control for a Six-Phase PMSM with High Robustness Against Weighting Factor Variation. *IEEE Trans Ind. Appl.* **2019**, *55*, 2781–2791. [CrossRef]

124. Luo, Y.; Liu, C. Model predictive control for a six-phase PMSM motor with a reduced-dimension cost function. *IEEE Trans. Ind. Electron.* **2020**, *67*, 969–979. [CrossRef]
125. Luo, Y.; Liu, C. A flux constrained predictive control for a six-phase PMSM motor with lower complexity. *IEEE Trans. Ind. Electron.* **2019**, *66*, 5081–5093. [CrossRef]
126. Luo, Y.; Liu, C. Multi-Vectors Based Model Predictive Torque Control for a Six-Phase PMSM Motor with Fixed Switching Frequency. *IEEE Trans. Energy Convers.* **2019**, *34*, 1369–1379. [CrossRef]
127. Martín, C.; Bermúdez, M.; Barrero, F.; Arahal, M.R.; Kestelyn, X.; Durán, M.J. Sensitivity of predictive controllers to parameter variation in five-phase induction motor drives. *Control Eng. Pract.* **2017**, *68*, 23–31. [CrossRef]
128. Siami, M.; Khaburi, D.A.; Rodriguez, J. Torque ripple reduction of predictive torque control for PMSM drives with parameter mismatch. *IEEE Trans. Ind. Electron.* **2017**, *32*, 7160–7168. [CrossRef]
129. Abdelrahem, M.; Hackl, C.M.; Kennel, R.; Rodriguez, J. Efficient Direct-Model Predictive Control with Discrete-Time Integral Action for PMSGs. *IEEE Trans. Energy Convers.* **2019**, *34*, 1063–1072. [CrossRef]
130. Bogado, B.; Barrero, F.; Arahal, M.; Toral, S.; Levi, E. Sensitivity to electrical parameter variations of predictive current control in multiphase drives. In Proceedings of the 39th Annual Conference of the IEEE Industrial Electronics Society (IECON), Vienna, Austria, 10–13 November 2013; pp. 5215–5220.

Article

Direct Torque Control of PMSM with Modified Finite Set Model Predictive Control

GuangQing Bao [1,2,3], WuGang Qi [1,2,3,*] and Ting He [4]

[1] College of Electrical Engineering and Information Engineering, Lanzhou University of Technology, Lanzhou 730050, China; gqbao@lut.cn
[2] Key Laboratory of Gansu Advanced Control for Industrial Processes, Lanzhou University of Technology, Lanzhou 730050, China
[3] National Demonstration Center for Experimental Electrical and Control Engineering Education, Lanzhou University of Technology, Lanzhou 730050, China
[4] Gansu Natural Energy Research Institute, Lanzhou 730046, China; 13679261064@163.com
* Correspondence: qwg_0218@163.com

Received: 7 November 2019; Accepted: 24 December 2019; Published: 3 January 2020

Abstract: A direct torque control (DTC) with a modified finite set model predictive strategy is proposed in this paper. The eight voltage space vectors of two-level inverters are taken as the finite control set and applied to the model predictive direct torque control of a permanent magnet synchronous motor (PMSM). The duty cycle of each voltage vector in the finite set can be estimated by a cost function, which is designed based on factors including the torque error, maximum torque per ampere (MTPA), and stator current constraints. Lyapunov control theory is introduced in the determination of the weight coefficients of the cost function to guarantee stability, and thus the optimal voltage vector reference value of the inverter is obtained. Compared with the conventional finite control set model predictive control (FCS-MPC) method, the torque ripple is reduced and the robustness of the system is clearly improved. Finally, the simulation and experimental results verify the effectiveness of the proposed control scheme.

Keywords: direct torque control; finite control set mode predictive control; duty cycle; maximum torque per ampere; permanent magnet synchronous motor

---

1. Introduction

Permanent magnet synchronous motor (PMSM) direct torque control (DTC) has been widely used in industry due to its simple control structure, fast dynamic response, and high efficiency [1]. However, the traditional DTC control results in large torque ripple due to the insufficient switching frequency of the inverters and two nonlinear hysteresis comparators [2,3].

To solve the problem of torque ripple, many scholars have put forward many improved methods. An improved method is calculating the effective voltage vector action time in real time to ensure the minimum torque ripple by the current torque error [4–6]. This method reduces the torque ripple to a certain extent, but the calculation process is complicated. At present, voltage space vector modulation (SVM) is also introduced into DTC [7,8]. This method can effectively reduce the torque ripple, but it cannot eliminate the steady-state error of the torque. At the same time, the calculation process is highly dependent on the parameters of the motor and has poor robustness.

As a real-time online optimization control method, model predictive control (MPC) has received extensive attention in the field of electric drives and power converters due to its high dynamics and resistance to parameter disturbance [9,10]. The method of finite control set MPC (FCS-MPC) directly utilizes the discreteness of inverter output voltage and the finiteness of switching state; at the same time, it does not need modulation and has a small amount of computation, which makes it a hot research topic in the field of power electronics [11,12]. In [13], on the premise that the flux linkage and torque of the motor are limited within the target range, a longer prediction step is selected to optimize the switching frequency of the inverter. A strategy of separate control of transient and steady state are adopted to reduce the switching frequency of inverter [14]. Mayne [15] proposed an FCS-MPC algorithm with model error compensation which enhances the robustness of the system. In [16], FCS-MPC method is applied in a grid-connected converter, and the stability of the control system is proved by Lyapunov stability theory. In addition, the virtual voltage vector is also integrated into FCS-MPC architecture, and the optimal voltage vector is preselected by using the results of Lyapunov stability theory. This method reduces the number of possible voltage vectors from 38 to 10, thus reducing the computational burden of the controller.

The cost function of FCS-MPC is the key to realizing the optimal selection of the voltage vector, which can be established according to the control targets, the system model based predictive variables, and the reference variables [17]. Depending on the control situation, it is also necessary to include system constraints. The cost function can contain multiple control objectives, variables, and constraints. Correct design of reasonable weight coefficients is very important for selecting the voltage vectors and governing the switching sequence of the inverter based on predictive current control of the PMSM with the FCS model [18,19].

The main contribution of this paper is to combine Lyapunov function theory with FCS-MPC to reduce torque ripple. Among them, the operating conditions of MTPA and torque tracking are guaranteed by using cost function. The Lyapunov function theory is introduced in the calculation of the finite set voltage vector duty cycle to obtain the optimal voltage vector. The simulation and experimental results show that the method achieves fast torque response and torque ripple minimization.

2. Discrete Mathematical Model of PMSM and Drive

The main circuit of the electric drive system under consideration is illustrated in Figure 1, which also corresponds to the experimental system layout. The PMSM has the characteristics of being multivariable and nonlinear and has strong coupling [20]. In order to simplify the mathematical model of PMSM, the following assumptions are made: (1) Y-shaped connection of stator windings, symmetrical distribution of three-phase windings, and space difference of each winding is 120°. (2) Eddy current loss, hysteresis loss, and the change of motor parameters are neglected. (3) It is assumed that permanent magnets on the rotor generate a main magnetic field in the stator–rotor airgap (the magnetic field is distributed sinusoidally along the circumference of the airgap) 22. So the continuous time model for PMSM in the d–q coordinate system can be described as:

$$L_d \frac{di_d}{dt} = (u_d - Ri_d + \omega L_q i_q), \quad (1)$$

$$L_q \frac{di_q}{dt} = (u_q - Ri_q - \omega L_d i_d - \omega \psi_f), \quad (2)$$

$$T_e = 1.5p(\psi_f + (L_d - L_q) i_d) i_q, \quad (3)$$

where u_d, u_q, i_d and i_q are the stator voltages and currents, R is the stator winding resistance, L_d and L_q are the d- and q-axes inductances, ψ_f is the flux linkage of the rotor, p is the number of pairs of poles, ω is the mechanical angular speed, and T_e is the electromagnetic torque. i_q is proportional to the electromagnetic torque and i_d is proportional to the reactive power. A predictive current control scheme is thus formed in which the reference current is generated by an external speed control loop.

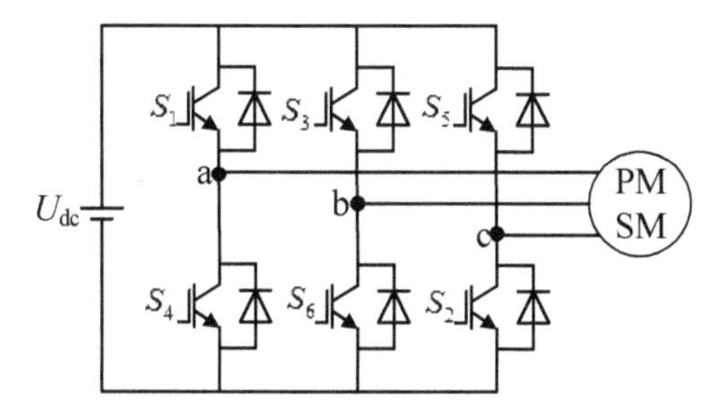

Figure 1. The main circuit of the permanent magnet synchronous motor (PMSM) drive.

Discretize the mathematical model of PMSM to obtain the necessary conditions for model prediction. Since the sampling time T_s is sufficiently small, the Euler approximation is used for the stator current derivative di/dt of the sampling time T_s.

$$\frac{di}{dt} = \frac{i(k+1) - i(k)}{T_s}, \tag{4}$$

where T_s is the sampling period. Then, the discrete current model of the PMSM is obtained:

$$L_d i_d^{k+1} = (L_d - T_s R) i_d^k + T_s \omega L_q i_q^k + T_s u_d^k, \tag{5}$$

$$L_q i_q^{k+1} = (L_q - T_s R) i_q^k - T_s \omega L_d i_d^k - T_s \omega \psi_f + T_s u_q^k, \tag{6}$$

$$T_e^{k+1} = 1.5 p i_q^{k+1} [\psi_f + (L_d - L_q) i_d^{k+1}]. \tag{7}$$

The commutation process of two-level voltage source inverters is accomplished by use of DC bus. The switching state can be represented by the switching signals S_a and S_b and switching on and off on different bridge arms. The upper and lower switches of each bridge arm of the inverter cannot be turned on at the same time to prevent short circuit S_c, as shown in Table 1:

Table 1. Inverter switching states.

	S_a	S_b	S_c
0	S_1 on S_4 off	S_3 on S_6 off	S_2 on S_5 off
1	S_1 off S_4 on	S_3 off S_6 on	S_2 off S_5 on

From the above table, the vector form function of the switching state of the three-phase bridge arm (S) is as follows:

$$S = \frac{2}{3}(S_a + aS_b + a^2 S_c), \tag{8}$$

where $a = e^{j2\pi/3}$. The relationship between output voltage vector with the switch state can be defined as:

$$U_{out} = SU_{dc} = \frac{2}{3} U_{dc}(S_a + aS_b + a^2 S_c), \tag{9}$$

where U_{dc} is the DC source voltage and U_{out} is the output voltage of the inverter.

Considering the different switching states of the inverters, eight different voltage vectors are obtained, as shown in Table 2. It is noted that $V0 = V7$, resulting in a finite set of seven different voltage vectors in the plane.

Table 2. Switching state and voltage vector.

S_a	S_b	S_c	Voltage Vector U_{out}
0	0	0	$V0 = 0$
0	0	1	$V1 = \frac{2}{3}U_{dc}e^{j\frac{4\pi}{3}}$
0	1	0	$V2 = \frac{2}{3}U_{dc}e^{j\frac{2\pi}{3}}$
0	1	1	$V3 = \frac{2}{3}U_{dc}e^{j\pi}$
1	0	0	$V4 = \frac{2}{3}U_{dc}$
1	0	1	$V5 = \frac{2}{3}U_{dc}e^{j\frac{5\pi}{3}}$
1	1	0	$V6 = \frac{2}{3}U_{dc}e^{j\frac{\pi}{3}}$
1	1	1	$V7 = 0$

3. Predictive Control Based on Duty Cycle

Within one sampling interval, the cost function values that correspond to the seven voltage vectors are calculated by the FCS. The switching state that minimizes the cost function is chosen as the switching state of the inverter. The process of model prediction optimization is shown in Figure 2, where x, x_p, and x^* are the torque of the real response, the predicted value, and the reference value, respectively. If the corresponding $x(k)$ is the best value at time k, the cost function value $x(k + 1)$ corresponding to the seven voltage vectors is calculated at time $(k + 1)$, and the seven calculated cost functions values are compared with reference values, where the closest is to be the optimal solution at the moment $(k + 1)$. V_4 is the control signal of the voltage vector corresponding to the optimal value of the time $(k + 1)$. In the same way, V_3 should be selected as the optimal control signal at time $(k + 2)$ [12].

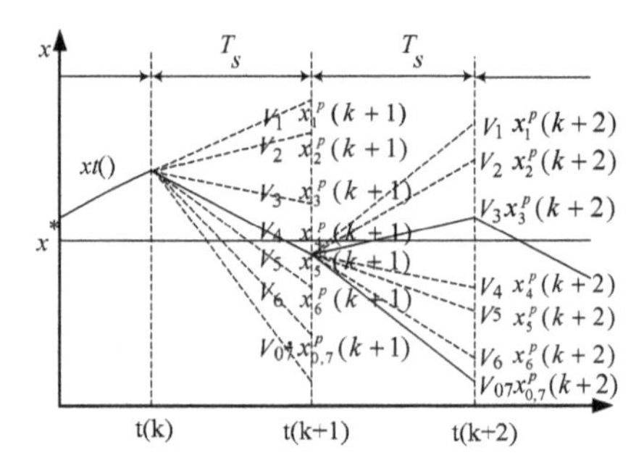

Figure 2. Optimization process diagram of model predictive control (MPC).

3.1. Design of Cost Function

The maximum torque/current control (MTPA) consumes the minimum stator current at the same electromagnetic torque output, which is usually used in interior PMSM systems. The space vector analysis of PMSM in the d–q coordinate system is shown in Figure 3. Assuming that the stator current vector i_s leads q axis β, the current component in the d–q coordinate system is as follows:

$$\begin{cases} i_d = -|i_s| \sin\beta \\ i_q = |i_s| \cos\beta \\ i_s = \sqrt{i_d^2 + i_q^2} \end{cases}, \tag{10}$$

$$T_e = 1.5p[\psi_f i_s \cos\beta - (Ld - Lq) i_s^2 \cos\beta \sin\beta]. \tag{11}$$

If the amplitude of the stator current of the PMSM is a constant value, it can be concluded that the electromagnetic torque of the motor is only related to β. In order to find the maximum value of the torque, that is, to satisfy the MTPA condition, and calculate Equation (11) with β as a variable, the following relationship is obtained:

$$\frac{dT_e}{d\beta} = -\psi_f is \sin\beta - (L_d - L_q) i_s [\cos\beta^2 - \sin\beta^2] = 0. \tag{12}$$

Because i_s and β satisfy the Pythagorean theorem, the relationship of i_d and i_q under the MTPA condition can thus be derived as follows:

$$i_d + \frac{L_d - L_q}{\psi_f}(i_d^2 - i_q^2) = 0. \tag{13}$$

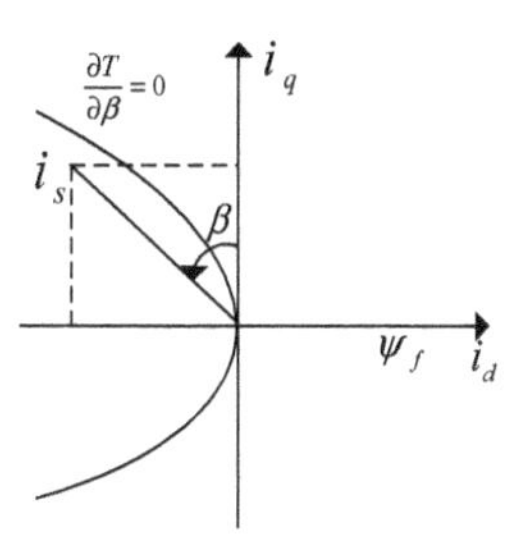

Figure 3. Space vector analysis of the PMSM.

The most important part of MPC is the cost function design. The cost function is expressed in orthogonal coordinates and considering the torque tracking, MTPA condition, and current constraint:

$$J(k) = K_T J_T(k) + K_M J_M(k) + K_C(J_{C1}(k) + J_{C2}(k)), \tag{14}$$

where K_T, K_M, and K_C are the weighting factors of the cost function and are positive real numbers. The main aim of the MPC is to minimize the torque error. Thus, the cost for the error $J_T(k)$ is:

$$J_T(k) = \left(T_e^* - T_e^{k+1}\right)^2. \tag{15}$$

Since high currents lead to large losses, $J_M(k)$ is to have a low absolute current:

$$J_M(k) = \left(i_d^k + \frac{L_d - L_q}{\psi_F}\left(\left(i_d^k\right)^2 - \left(i_q^k\right)^2\right)\right)^2. \tag{16}$$

Tertiary control goals can be added in order to ensure the operational safety of the system. $J_{C1}(k)$ denotes the current limit, and the current amplitude of the system must be smaller than the maximum allowable value $I_{\max}$.

$$J_{C1}(k) = \begin{cases} 0 & \left(I_{\max} \geq \sqrt{\left(i_d^k\right)^2 + \left(i_q^k\right)^2}\right) \\ \left(I_{\max} - \sqrt{\left(i_d^k\right)^2 + \left(i_q^k\right)^2}\right)^2 & \left(I_{\max} < \sqrt{\left(i_d^k\right)^2 + \left(i_q^k\right)^2}\right) \end{cases} \tag{17}$$

To make the point of the cost function converge to the parabola on the left side of the i_q axis, as shown in Figure 3. Adding current constraints $J_{C2}(k)$ to the cost function:

$$J_{C2}(k) = \begin{cases} 0, \left(i_d^k \leq 0\right) \\ \left(i_d^k\right)^2, \left(i_d^k > 0\right) \end{cases}. \tag{18}$$

To ensure current constraints and avoid control variables converging to the wrong MTPA curve, $K_C \gg K_T, K_M$ should be satisfied. For the choice of K_T and K_M, $K_T + K_M = 1$ can be made. A larger value of K_T leads to faster convergence of the torque, while a larger value of K_M indicates that the MTPA state converges quickly. $K_T > K_M$ can be selected [21] to achieve a fast torque response.

3.2. Duty Cycle Calculation

The duty cycle of each voltage vector in a finite set is calculated by Lyapunov function to eliminate torque and current ripple. Without considering the current limitation, that is, ignoring $J_{C1}(k)$ and $J_{C2}(k)$, the Lyapunov function is expressed as follows:

$$V(\mathrm{k}+1) = k_T J_T(k+1) + k_M J_M(k+1). \tag{19}$$

In order to optimize the operation of the PMSM, the condition $V(k+1) = 0$ must be fulfilled to ensure $T_e^{k+1} = T_e^*$ and its current operating point is on the MTPA curve. In addition, the condition $dV(k+1)/dt = 0$ indicates that the PMSM is operating at the optimum state.

Lemma 1. *Defining the function of the current:*

$$f(i_d, i_q) = i_d + (L_d - L_q)(i_d^2 - {}_q^2)/\psi_f \tag{20}$$

If the torque of the PMSM is kept at a nonzero constant value, the function $f(i_d, i_q)$ is a strict monotonic function along the constant torque curve. The correlative poof is introduced in the Appendix A.

With the help of the function $f(i_d, i_q)$, the stator current trajectory is shown in Figure 4. Point A is the initial point; the trajectory first reaches the point B along the constant torque curve and then reaches the point C along the MTPA curve with the condition $V(k) = 0$. When the current trajectory is along $\overrightarrow{AB}$ to $\overrightarrow{BC}$, the Lyapunov function is strictly decreasing.

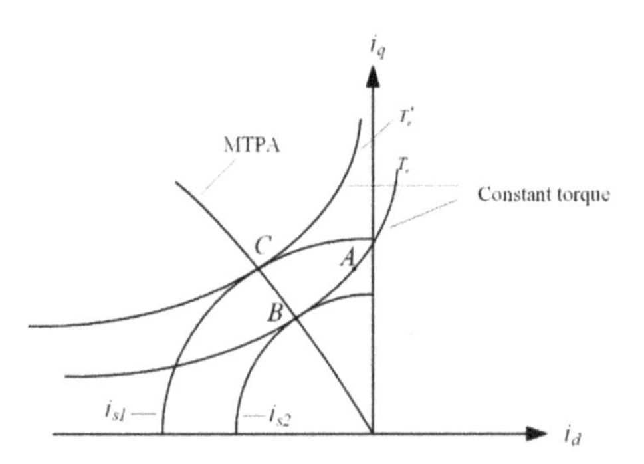

Figure 4. Stator current vector trajectory.

Lemma 2. *Under the initial condition of the PMSM, if the back EMF of the PMSM is within the voltage limit range of the inverter, there exists a feasible voltage vector u_{dq}^* that fulfills:*

$$dV(k+1)/dt \leq 0. \tag{21}$$

Lemma 3. *For a given current dynamics:*

$$\frac{di_{dq}^{k+1}}{dt} = Ai_{dq}^{k+1} + Bu_{dq}^{k+1} + E, \tag{22}$$

where $A = \begin{bmatrix} -\frac{R}{L_d} & \frac{\omega L_q}{L_d} \\ -\frac{\omega L_d}{L_q} & -\frac{R}{L_q} \end{bmatrix}$, $B = \begin{bmatrix} \frac{1}{L_d} & 0 \\ 0 & \frac{1}{L_q} \end{bmatrix}$, $E = \begin{bmatrix} 0 \\ -\frac{-\omega\psi_f}{L_q} \end{bmatrix}$.

If the back EMF of the PMSM is within the voltage limit range of the inverter, then at least one voltage vector is satisfied:

$$\frac{dV(k+1)}{dt} = \frac{\partial V(k+1)}{\partial i_{dq}^{k+1}}(Ai_{dq}^{k+1} + Bu_{dq}^{k+1} + E) \leq 0. \tag{23}$$

The proofs of Lemma 2 and Lemma 3 are introduced in the Appendix A.

In order to minimize the ripple of the PMSM current and torque, the Lyapunov function is expected to keep $V = 0$ and $dV(k+1)/dt = 0$ in a steady state. The duty cycle of each voltage vector satisfying the desired requirement is calculated as follows:

$$T^{j}_{duty,k+1} = \begin{cases} T\sigma\,, & \frac{dV(K+1)}{dt} > 0 \\ 0\,, & \frac{dV(K+1)}{dt} = 0 \\ \frac{-V(K+1)}{dV(k+1)/dt}\,, & \frac{dV(K+1)}{dt} < 0 \end{cases}, \tag{24}$$

Here, $T_\sigma \ll T_s$ is a time constant. The voltage vector with $dV(k+1)/dt > 0$ has to be taken into account in the FCS-MPC, and then the current reaches the limit, T_σ is introduced, and the current constraint plays a major role in the cost function. When there is disturbance in the PMSM system, such as machine parameter drift and voltage error in the inverter, we also need to select a smaller T_σ to compromise the disturbance at this point. When the PMSM model and the prediction are accurate, $T_\sigma = 0$ is the ideal value for the ripple reduction. All voltage vectors $u^{j}_{dq,k+1}$ with $dV(k+1)/dt < 0$ are candidates to minimize the cost function.

3.3. Finite Control Set Model Prediction

Finally, the optimum output voltage which can be realized by the SVPWM is obtained by Equation (23). Only when voltage $u^{j}_{dq,k+1}$ acts on the PMSM will it affect the torque, that is, to ensure the implementation of MTPA and $dV(k+1)/dt = 0$. It is noteworthy that setting the optimal duty cycle T_{duty}^{k+1} equal to T_s in the transient state of the proposed control strategy can minimize the cost function. Therefore, under the same sampling frequency, the torque response of the new strategy is as fast as that of the traditional FCS-MPC. In steady state, $dV(k+1)/dt = 0$ can be maintained. Therefore, current ripple and torque ripple can be minimized. Figure 5 is a block diagram of the predictive direct torque control algorithm of the FCS-MPC proposed in this paper.

$$\begin{bmatrix} u_d^{k+1} \\ u_q^{k+1} \end{bmatrix} = (1 - T_{duty}^{k+1}) \begin{vmatrix} Ri_d^{k+1} - \omega L_q i_q^{k+1} \\ Ri_q^{k+1} - \omega L_d i_d^{k+1} + \omega\psi_F \end{vmatrix} + T_{duty}^{k+1} \begin{vmatrix} u_d^{j,k+1} \\ u_q^{j,k+1} \end{vmatrix} \tag{25}$$

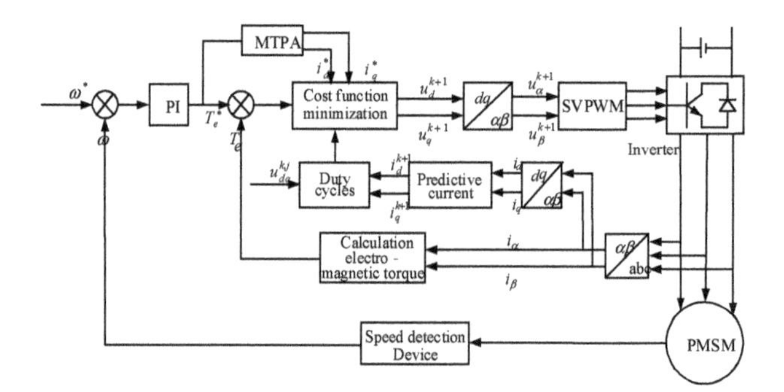

Figure 5. Block diagram of the proposed MPC.

4. Simulation Results

In order to investigate the effectiveness and feasibility of the proposed method, the FCS-MPC of the surface-mounted PMSM based on MATLAB/Simulink software was built. Then, it was compared with that of the classical FCS-MPC method. The parameter settings are shown in Table 3.

Table 3. Surface-mounted permanent magnet synchronous motor (SPMSM) parameters for simulation.

Variable	Parameter	Value
P_N	Rated power	6.5 kW
n_N	Rated speed	1500 rpm
U_N	Rated voltage	306 V
I_N	Rated current	12.3 A
R	Stator resistance	1.01 Ω
L	L_d, Lq	15 mH
p	Number of pairs of poles	4
J	Rotary inertia	0.01535 kg·m^2
ψ_f	Flux linkage of the rotor	0.175 Wb

4.1. Steady-State Operation

Figure 6 shows the simulation results of traditional FCS-MPC and improved FCS-MPC under steady-state conditions. The reference value of motor speed is set to 500 rpm, and the load torque is 20 N·m. As can be seen from the figure, compared with conventional FCS-MPC, the torque ripple of the improved method is significantly reduced. In Figure 6, the average values of the torque ripples of proposed (0.9 N·m) are smaller compared to those of conventional FCS-MPC (3 N·m) under the same conditions. It is noted that, in this paper, torque ripples are calculated by the following equation [22]:

$$T_{rip} = \sqrt{\frac{1}{N}\sum_{i=1}^{N}(T_e - T_{ave})^2}, \tag{26}$$

where N is the number of samples and T_{ave} is the average value of the torque.

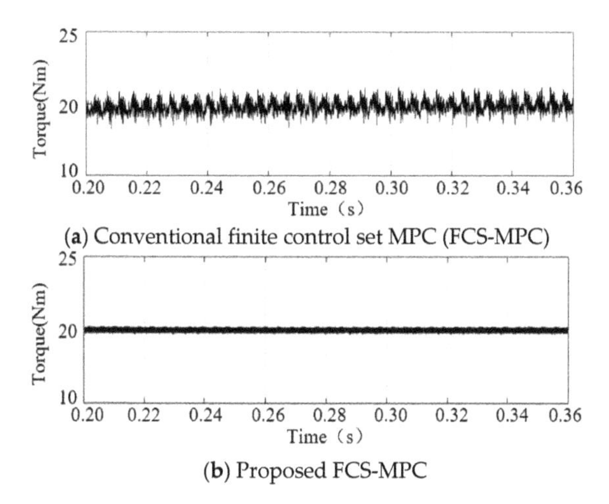

(**a**) Conventional finite control set MPC (FCS-MPC)

(**b**) Proposed FCS-MPC

Figure 6. Torque steady-state response simulation.

4.2. *Dynamic Response*

In order to compare the torque dynamic, a magnified view of the torque response is shown in Figure 7. From Figure 7, we can see that the torque suddenly increases from 10 to 20 N·m in only 0.003 s, and in the process of torque mutation, the conventional FCS-MPC method has obvious overshoot. Compared with the conventional FCS-MPC, the improved method has faster dynamic response because the motor adopts optimized duty cycle modulation in the whole control cycle.

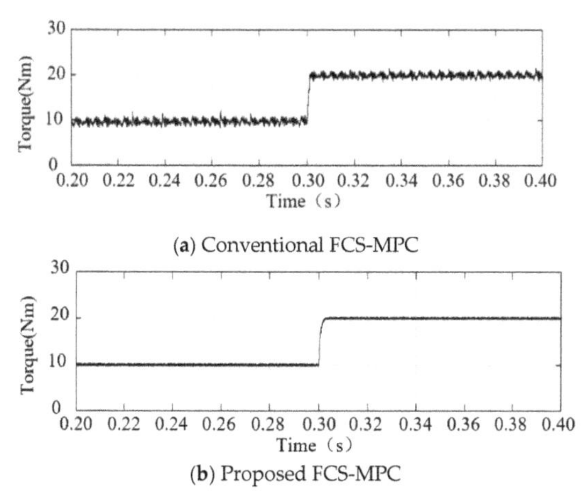

(**a**) Conventional FCS-MPC

(**b**) Proposed FCS-MPC

Figure 7. Torque dynamic simulation at t = 0.3 s.

4.3. *Motor Parameter Robustness*

As we all know, a temperature rise of the motor will cause the stator internal resistance to increase. The influence of internal resistance on the torque is discussed in this paper. The torque waveforms of the two control methods in the condition of rated stator resistance (the blue solid line) and 1.5-times the rated value (the red dashed line) are given in Figure 8a,b, respectively. It is noticeable that slight torque ripples occur under the conventional FCS-MPC method when the resistance increases, while the proposed FCS-MPC control has better robustness in which the torque is insensitive to the resistance change. However, when entering a steady state, the two control systems are not affected by internal resistance changes.

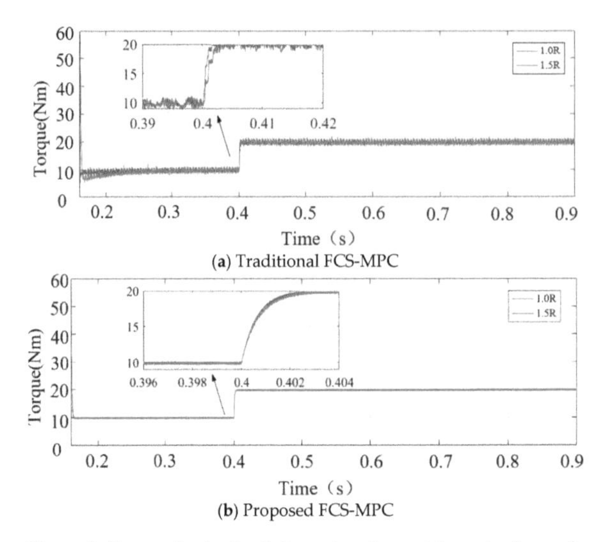

Figure 8. Torque ripple simulation when the resistance is changed.

Almost all permanent magnet devices, including permanent magnet motors, will exhibit the demagnetization phenomenon in varying degrees due to factors such as the working environment and usage time. The effect of changes in the motor flux on torque is also discussed. Figure 9a,b shows the torque waveforms of the two control methods when the rotor flux is 0.9-times the rated value, equal to the rated value, and 1.1-times the rated value, to simulate permanent magnet motors that have been used for a long time, used normally, and newly manufactured. As shown in Figure 9, the blue point line is 0.9-times the rated flux, the red solid line is the rated flux, and the green dotted line is 1.5-times the rated flux. Evidently, the conventional FCS-MPC for the change of flux is the change of the maximum torque. However, the improved predictive control is insensitive to the flux change.

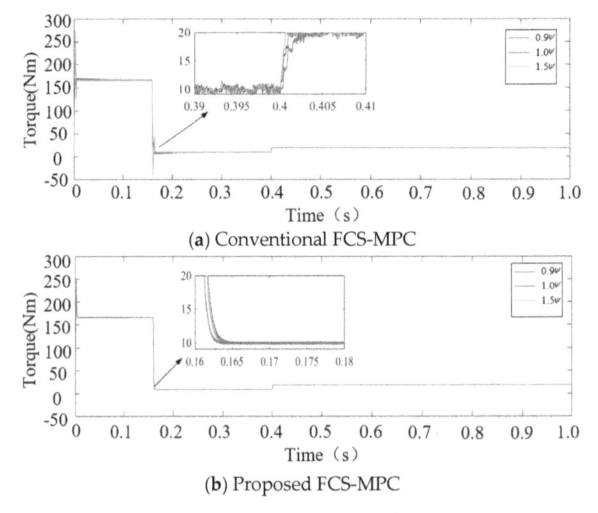

Figure 9. Torque ripple simulation when the flux is changed.

For a mechanical load, the oscillation of torque is very bad for the shaft and load of the motor. Therefore, it can be seen from the simulation waveforms of the above two sections that the robustness of predictive control to parameter changes is better than that of conventional FCS-MPC. In theory, the switch meter of direct torque control is fixed in advance and cannot be adjusted according to the running state of the motor. Therefore, the robustness of direct predictive control to parameter changes is better than that of direct torque control, which is also consistent with the previous prediction results.

5. Experimental Test

The experiment aimed to verify the torque performance of this method with surface-mounted permanent magnet synchronous motor (SPMSM) and interior permanent magnet synchronous motor (IPMSM). The experimental platform as shown in Figure 10 was built by using dSPACE1102. The load of the PMSM is provided by the load machine.

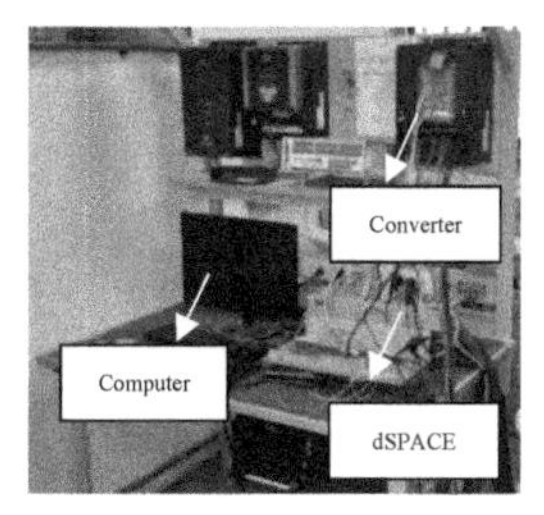

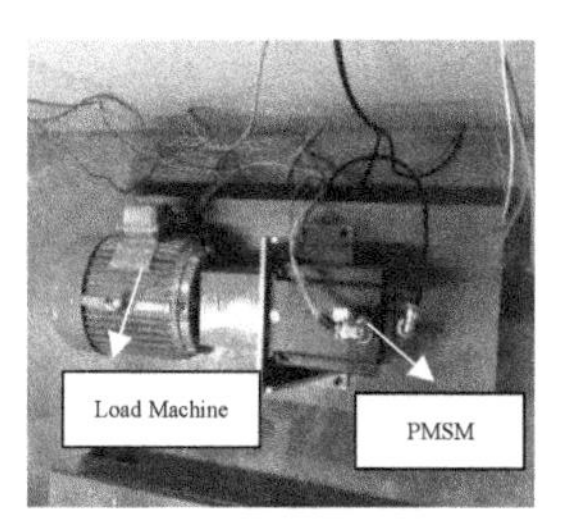

Figure 10. The controlled 6.5 kW PMSM drive test rig.

5.1. Experiment on SPMSM

The parameters of the SPMSM are shown in Table 3. The steady-state performance test results of conventional FCS-MPC and proposed FCS-MPC are shown in Figure 11. In the experiment, the reference load torque was 20 N·m, and the results show that the proposed FCS-MPC significantly reduced the torque ripple under the same average switching frequency, while the conventional FCS-MPC had a lot of torque ripples (6 N·m) compared to proposed FCS-MPC (2.8 N·m). This is consistent with the simulation results.

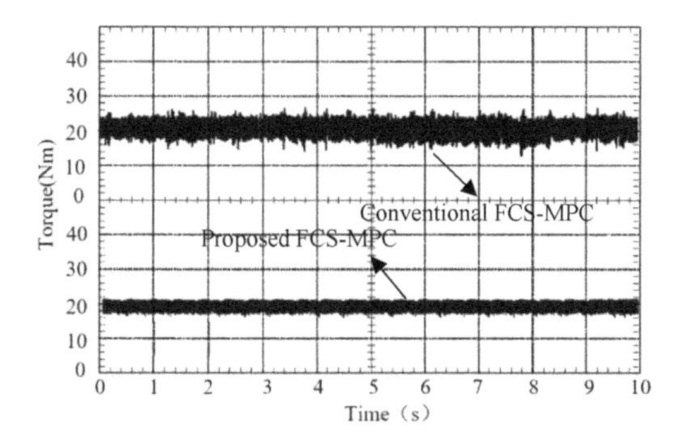

Figure 11. Torque steady-state responses (SPMSM).

In order to test the dynamic performance of the proposed FCS-MPC method, a load torque of 25 N·m was suddenly added when the steady speed of the motor was 500 rpm. Figure 12 shows that the response times of proposed FCS-MPC and conventional FCS-MPC are approximately the same; however, the proposed FCS-MPC strategy torque ripple is smaller than that of the conventional FCS-MPC. Meanwhile, the proposed FCS-MPC method enables the motor to quickly recover to the reference speed value compared to conventional FCS-MPC.

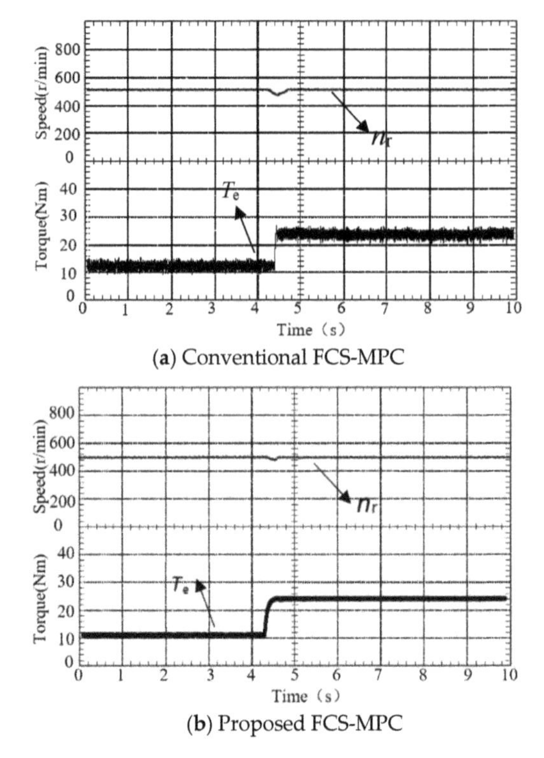

Figure 12. Speed dynamic and torque dynamic responses (SPMSM).

5.2. Experiment on IPMSM

Some parameters of the experiment motor are shown in Table 4. The following experiments were carried out in three aspects: torque steady-state responses, speed dynamics, and torque dynamics responses.

Table 4. Interior permanent magnet synchronous motor (IPMSM) parameters for simulation.

Variable	Parameter	Value
P_N	Rated power	5.2 kW
n_N	Rated speed	1500 rpm
U_N	Rated voltage	380 V
R	Stator resistance	0.39 Ω
L	Ld	0.88 mH
L	Lq	1.62 mH
p	Number of pairs of poles	4
ψ_f	Flux linkage of the rotor	0.163 Wb

Figure 13 shows electromagnetic torque waveforms of two methods, with the load torque reference value set to 30 N·m. As can be seen from Figure 13, the method of proposed FCS-MPC can significantly reduce torque ripple by comparing torque waveforms; the average torque ripple of proposed method is only 3 N·m while that of conventional FCS-MPC is 7 N·m.

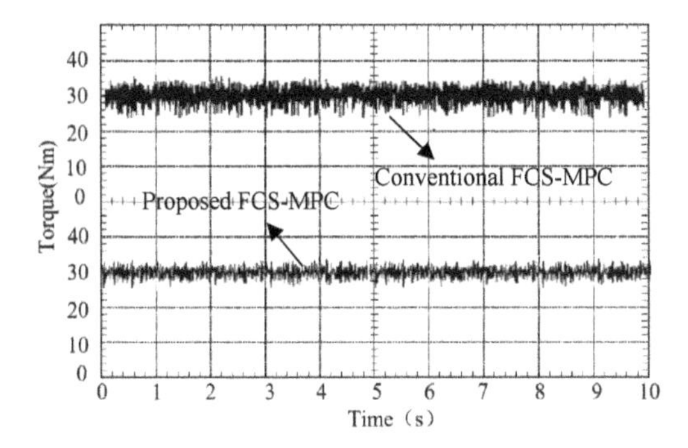

Figure 13. Torque steady-state responses (IPMSM).

Figure 14 shows the waveforms of speed and electromagnetic torque of two methods when the starting moment of the motor and step change of load torque with load torque increasing from 10 to 20 N·m. It can be seen from Figure 14 that at the moment of starting the motor, the speed and electromagnetic torque of the conventional and proposed FCS-MPC methods increase sharply, reaching the given reference value quickly; however, compared with the conventional FCS-MPC, the strategy of proposed FCS-MPC has smaller torque ripple. When the load torque changes abruptly, the proposed FCS-MPC can also track the change of torque quickly, and the corresponding speed is faster than conventional FCS-MPC.

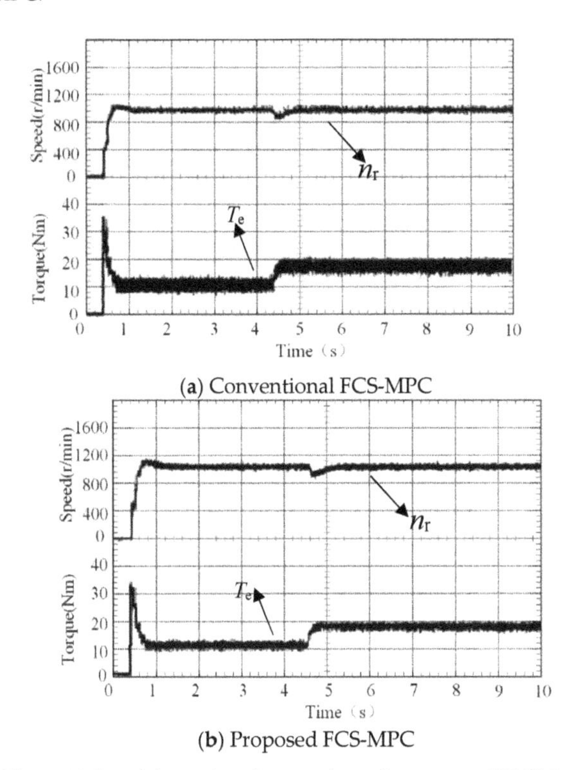

Figure 14. Speed dynamic and torque dynamic responses (IPMSM).

Through the analysis of the above experiments, compared with the conventional FCS-MPC control system, the proposed FCS-MPC in this paper effectively reduces the torque ripple and improves the following performance of the motor torque.

6. Summary

In this paper, a new scheme of direct torque control for PMSM based on a finite control set (FCS) model is proposed. The eight voltage vectors of the two-level converter are utilized as an FCS for the torque prediction of the PMSM. The cost function is used to estimate the duty cycle of each voltage vector. Thus, the optimal voltage vector can be obtained from eight voltage vectors and their duty cycles. Compared with the classical FCS-MPC method, the proposed method has smaller torque ripple and excellent dynamic performance.

Author Contributions: Conceptualization and methodology, G.B. and W.Q.; software, T.H.; validation, G.B. and W.Q.; writing—original draft preparation, T.H.; writing—review and editing, G.B. and W.Q.; funding acquisition, G.B. All authors have read and agreed to the published version of the manuscript.

Funding: This research was funded by supported by State Key Laboratory of Large Electric Drive System and Equipment Technology (SKLLDJJ032016018) and National Natural Science Foundation of China (51967012) and Scientific Research and Innovation Team Project of Gansu Education Department (2018C-09).

Conflicts of Interest: The authors declare no conflict of interest.

Appendix A

Proof of Lemma 1. Finding partial derivatives of $f(i_d, i_q)$ and $T_e = 1.5pi_q[\psi_f + (L_d - L_q)i_d]$ based on current vectors i_d and i_q, respectively:

$$\frac{\partial f(i_d, i_q)}{\partial i_{dq}} = [1 + \frac{2(L_d - L_q)}{\psi_f} i_d, -\frac{2(L_d - L_q)}{\psi_f} i_q] \tag{A1}$$

$$\frac{\partial T_e}{\partial i_{dq}} = [1.5p(L_d - L_q)i_q, 1.5p(\psi_f + (L_d - L_q)i_q)] \tag{A2}$$

Considering the torque T_e as constant, the trajectory of the operating point moves along the direction of the α which fulfills $(T_e/i_{dq}) \cdot \alpha = 0$. The vector α can be described as:

$$\alpha = \varepsilon[-1.5p(\psi_f + (L_d - L_q)i_q), 1.5p(L_d - L_q)i_q)]^T, \tag{A3}$$

where ε is a positive real number.

$$\frac{\partial f(i_d, i_q)}{\partial i_{dq}} \cdot \alpha = 1.5p\varepsilon\left[\psi_F + 3(L_d - L_q)i_d\right] - \frac{3p\varepsilon(L_d - L_q)^2}{\psi_f}(i_d^2 + i_q^2) < 0. \tag{A4}$$

Therefore, the function $f(i_d, i_q)$ is a strict monotonic function along the constant torque curve. □

Proof of Lemma 2. The Lyapunov function is derived as:

$$\frac{dV(k+1)}{dt} = \frac{\partial V(k+1)}{\partial i_{dq}^{k+1}} \cdot \frac{di_{dq}^{k+1}}{dt}. \tag{A5}$$

Therefore, the trajectory of the current makes the value of Lyapunov function decrease and there exists a Δi_{dq}^{k+1} which fulfills:

$$\frac{\partial V(k+1)}{di_{dq}^{k+1}} \Delta i_{dq}^{k+1} \leq 0. \tag{A6}$$

Therefore, there exists a current derivative:

$$\frac{di_{dq}^{k+1}}{dt} = \mu \cdot \Delta i_{dq}^{k+1}, \tag{A7}$$

so that:

$$\frac{dV(k+1)}{dt} = \frac{\partial V(k+1)}{\partial i_{dq}^{k+1}} \cdot \mu \cdot \Delta i_{dq}^{k+1} \leq 0. \tag{A8}$$

Here, μ can be a very small positive constant. □

Proof of Lemma 3. Any reference vector $u_{dq}^{*}(x)$ within the region of feasibility [having a magnitude of less than $(2/3)U_{dc}$] is contained within one of the six nonzero switching sectors of width $(\pi/3)$ with vertices (v_d^0, v_q^0), (v_d^i, v_q^i), and (v_d^j, v_q^j), where $i, j \in \{1, \ldots, 6\}$ are the nonzero switching states to the left and right of the reference vector and (v_d^0, v_q^0) is one of the two zero vectors. Containment within a switching sector ensures the existence of coefficients γ and η satisfying $\gamma, \eta \geq 0$ and $\gamma + \eta \leq 1$ such that the reference vector is expressible as a convex combination of the realizable inputs, given by:

$$u_{dq}^{*}(x) = \gamma v_{dq}^{i} + \eta v_{dq}^{j} + (1 - \gamma - \eta) v_{dq}^{0}. \tag{A9}$$

Plugging Equation (A10) into Equation (21) and noting that the system is control affine, we see that [23]:

$$\begin{aligned}\frac{dV(k+1)}{dt} &= \frac{\partial V(k+1)}{\partial i_{dq}^{k+1}}(Ai_{dq}^{k+1} + Bu_{dq}^{*} + E) = \frac{\partial V(k+1)}{\partial i_{dq}^{k+1}}(Ai_{dq}^{k+1} + B(\gamma v_{dq}^{i} + \eta v_{dq}^{j} + (1-\gamma-\eta)v_{dq}^{0}) + E) \\ &= \gamma \frac{\partial V(k+1)}{\partial i_{dq}^{k+1}}(Ai_{dq}^{k+1} + Bv_{dq}^{i}) + \eta \frac{\partial V(k+1)}{\partial i_{dq}^{k+1}}(Ai_{dq}^{k+1} + Bv_{dq}^{j}) \\ &+ (1-\gamma-\eta)\frac{\partial V(k+1)}{\partial i_{dq}^{k+1}}(Ai_{dq}^{k+1} + Bv_{dq}^{0}) + \frac{\partial V(k+1)}{\partial i_{dq}^{k+1}}E\end{aligned} \tag{A10}$$

Because γ, η, and $(1 - \gamma - \eta)$ are all nonnegative, the following inequalities hold [12]:

$$\begin{cases} \frac{\partial V(k+1)}{\partial i_{dq}^{k+1}}(Ai_{dq}^{k+1} + Bv_{dq}^{i}) \leq 0 \\ \frac{\partial V(k+1)}{\partial i_{dq}^{k+1}}(Ai_{dq}^{k+1} + Bv_{dq}^{j}) \leq 0 \\ \frac{\partial V(k+1)}{\partial i_{dq}^{k+1}}(Ai_{dq}^{k+1} + Bv_{dq}^{0}) \leq 0 \end{cases}, \tag{A11}$$

which completes the proof in Lemma 3. The theorem also guarantees the stability of the proposed control scheme. □

References

1. Yang, J.F.; Hu, Y.W. Optimal direct torque control of permanent magnet synchronous motor. *Proc. Chin. Soc. Electr. Eng.* **2011**, *31*, 109–115.
2. Depenbrock, M. Direct self-control (DSC) of inverter fed induction machine. *IEEE Trans. Power Electron.* **1988**, *3*, 420–429. [CrossRef]
3. Uddin, M.N.; Zou, H.; Azevedo, F. Online loss-minimization-based adaptive flux observer for direct torque and flux control of PMSM drive. *IEEE Trans. Ind. Appl.* **2016**, *52*, 425–431. [CrossRef]
4. Swierczynski, D.; Kazmierkowski, M.P. Direct torque control of permanent magnet synchronous motor (PMSM) using space vector modulation (DTC-SVM)-simulation and experimental results. In Proceedings of the Conference of the IEEE 2002 28th Annual Conference of the Industrial Electronics Society, Sevilla, Spain, 5–8 November 2002.

5. Tang, L.; Zhong, L.; Rahman, M.; Hu, Y. A novel direct torque controlled interior permanent magnet synchronous machine drive with low ripple in flux and torque and fixed switching frequency. *IEEE Trans. Power Electron.* **2004**, *19*, 346–354. [CrossRef]
6. Cho, Y.; Lee, K.B.; Song, J.H.; Lee, Y. Torque-ripple minimization and fast dynamic scheme for torque predictive control of permanent magnet synchronous motors. *IEEE Trans. Power Electron.* **2015**, *30*, 2182–2190. [CrossRef]
7. Vafaie, M.H.; Dehkordi, B.M.; Moallem, P.; Kiyoumarsi, A. A new predictive direct torque control method for improving both steady-state and transient-state operations of the PMSM. *IEEE Trans. Power Electron.* **2016**, *31*, 3738–3753. [CrossRef]
8. Zhang, Y.; Zhu, J. Direct torque control of permanent magnet synchronous motor with reduced torque ripple and commutation frequency. *IEEE Trans. Power Electron.* **2011**, *26*, 235–248. [CrossRef]
9. Rodriguez, J.; Kazmierkowski, M.P.; Espinoza, J.R.; Zanchetta, P.; Abu-Rub, H.; Young, H.A.; Rojas, C.A. State of the art of finite control set model predictive control in power electronics. *IEEE Trans. Ind. Inform.* **2012**, *9*, 1003–1016. [CrossRef]
10. Rodriguez, J.; Pontt, J.; Silva, C.; Salgado, M.; Rees, S.; Ammann, U.; Lezana, P.; Huerta, R.; Cortes, P. Predictive control of three-phase inverter. *Electron. Lett.* **2004**, *40*, 561–562. [CrossRef]
11. Preindl, M.; Bolognani, S. Model predictive direct torque control with finite control set for PMSM drive systems, part 1: Maximum torque per ampere operation. *IEEE Trans. Ind. Inform.* **2013**, *9*, 1912–1921. [CrossRef]
12. Liu, Q.; Hameyer, K. A finite control set model predictive direct torque control for the PMSM with MTPA operation and torque ripple minimization. In Proceedings of the 2015 IEEE International Electric Machines & Drives Conference (IEMDC), Coeur d'Alene, ID, USA, 10–13 May 2015; pp. 804–810.
13. Geyer, T.; Papafotiou, G.; Morari, M. Model predictive direct torque control; part I: Concept, algorithm, and analysis. *IEEE Trans. Ind. Electron.* **2009**, *56*, 1894–1905. [CrossRef]
14. Preindl, M.; Schaltz, E.; Thogersen, P. Switching frequency reduction using model predictive direct current control for high-power voltage source inverters. *IEEE Trans. Ind. Electron.* **2010**, *58*, 2826–2835. [CrossRef]
15. Mayne, D.Q.; Rawlings, J.B.; Rao, C.V.; Scokaert, P.O.M. Constrained model predictive control: Stability and optimality. *Automatica* **2000**, *36*, 789–814. [CrossRef]
16. Alam, K.S.; Akter, M.P.; Xiao, D.; Zhang, D.; Rahman, M.F. Asymptotically Stable Predictive Control of a Grid-Connected Converter Based on Discrete Space Vector Modulation. *IEEE Trans. Ind. Inform.* **2018**, *15*, 2775–2785. [CrossRef]
17. Aguilera, R.P.; Quevedo, D.E. On stability and performance of finite control set MPC for power converters. In Proceedings of the Workshop Predictive Control Electrical Drives Power Electronics, Munich, Germany, 14–15 October 2011; pp. 55–62.
18. Lee, J.S.; Lorenz, R.D.; Valenzuela, M.A. Time-optimal and loss minimizing deadbeat-direct torque and flux control for interior permanent magnet synchronous machines. *IEEE Trans. Ind. Appl.* **2014**, *50*, 1880–1890. [CrossRef]
19. Xu, W.; Lorenz, R.D. Dynamic loss minimization using improved deadbeat-direct torque and flux control for interior permanent-magnet synchronous machines. *IEEE Trans. Ind. Appl.* **2014**, *50*, 1053–1065. [CrossRef]
20. Chen, Y.; Sun, D.; Lin, B.; Ching, T.W.; Li, W. Dead-beat direct torque and flux control based on sliding-mode stator flux observer for PMSM in electric vehicles. In Proceedings of the 41st Annual Conference of the IEEE Industrial Electronics Society, Yokohama, Japan, 9–12 November 2015; pp. 2270–2275.
21. Barcaro, M.; Fornasiero, E.; Bianchi, N.; Bolognani, S. Design procedure of IPM motor drive for railway traction. In Proceedings of the IEEE International Electric Machines & Drives Conference (IEMDC), Niagara Falls, ON, Canada, 15–18 May 2011; pp. 983–988.
22. Bianchi, N.; Bolognani, S.; Ruzojcic, B. Design of a 1000 HP permanent magnet synchronous motor for ship propulsion. In Proceedings of the 13th European Conference on Power Electronics and Applications (EPE'09), Barcelona, Spain, 8–10 September 2009; pp. 1–8.
23. Prior, G.; Krstic, M. Quantized-input control Lyapunov approach for permanent magnet synchronous motor drives. *IEEE Trans. Control Syst. Technol.* **2013**, *21*, 1784–1794. [CrossRef]

Article

New Modulation Technique to Mitigate Common Mode Voltage Effects in Star-Connected Five-Phase AC Drives

Markel Fernandez [1,*], Andres Sierra-Gonzalez [2], Endika Robles [1], Iñigo Kortabarria [1], Edorta Ibarra [1] and Jose Luis Martin [1]

[1] Derpartment of Electronic Technology, University of the Basque Country (UPV/EHU), Plaza Ingeniero Torres Quevedo 1, 48013 Bilbao, Spain; endika.robles@ehu.eus (E.R.); inigo.kortabarria@ehu.eus (I.K.); edorta.ibarra@ehu.eus (E.I.); joseluis.martin@ehu.eus (J.L.M.)

[2] Tecnalia Research and Innovation, C. Mikeletegi 7, 20009 Donostia, Spain; andres.sierra@tecnalia.com

* Correspondence: markel.fernandez@ehu.eus

Received: 29 November 2019; Accepted: 29 January 2020; Published: 31 January 2020

Abstract: Star-connected multiphase AC drives are being considered for electromovility applications such as electromechanical actuators (EMA), where high power density and fault tolerance is demanded. As for three-phase systems, common-mode voltage (CMV) is an issue for multiphase drives. CMV leads to shaft voltages between rotor and stator windings, generating bearing currents which accelerate bearing degradation and produce high electromagnetic interferences (EMI). CMV effects can be mitigated by using appropriate modulation techniques. Thus, this work proposes a new Hybrid PWM algorithm that effectively reduces CMV in five-phase AC electric drives, improving their reliability. All the mathematical background required to understand the proposal, i.e., vector transformations, vector sequences and calculation of analytical expressions for duty cycle determination are detailed. Additionally, practical details that simplify the implementation of the proposal in an FPGA are also included. This technique, HAZSL5M5-PWM, extends the linear range of the AZSL5M5-PWM modulation, providing a full linear range. Simulation results obtained in an accurate multiphase EMA model are provided, showing the validity of the proposed modulation approach.

Keywords: multiphase electric drives; CMV; modulation techniques; PWM

1. Introduction

AC electric drives are used in a wide variety of industrial applications such as in compressors [1], in electric vehicle propulsion systems [2,3] and in more electric aircraft (MEA) [4], among others. Although three-phase systems dominate the AC drive market, multiphase solutions are gaining popularity [5–7]. Multiphase systems are preferable for applications where high fault tolerance is required [8], such as for MEA applications, where electromechanical actuators (EMA) for control surfaces, fuel pumps, landing gears, environmental control systems and starter-generators need to be operated [9–11]. Apart from their intrinsic fault tolerance, other benefits of multiphase drives include a reduced current per phase (reducing copper losses and increasing efficiency) [4], noise and electromagnetic interference (EMI) minimization [12,13], higher power density and lower torque ripple [14], making them attractive for transport electrification. Among the multiphase topologies available in the scientific literature, star-connected five-phase technologies (Figure 1) can be highlighted, as they provide a good trade-off between system complexity and fault tolerance [15,16]. Specifically, multiphase permanent magnet synchronous machines (PMSM) are being considered for aircrafts due to their superior power density [17,18].

In general, AC electric drives can experience issues due to the common-mode voltage (CMV) [19] and common mode currents (CMC) [20]. CMV variations are generated by the commutation of the power converter devices, producing EMI [21] and bearing currents that can compromise the integrity of the electric machine [22]. Such voltage variations create new capacitive paths through the motor bearings, leading to premature aging. Capacitive currents, electrostatic discharge machine (EDM) currents, circulating currents and rotor-to-ground currents can flow through the bearings [23,24] (Figure 2), and their harmful effects depend on the type of bearing, size of the machine and how the machine is used.

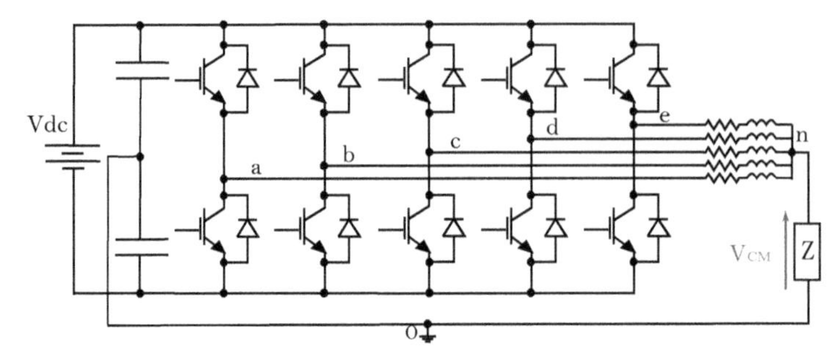

Figure 1. CMV in a five-phase power system.

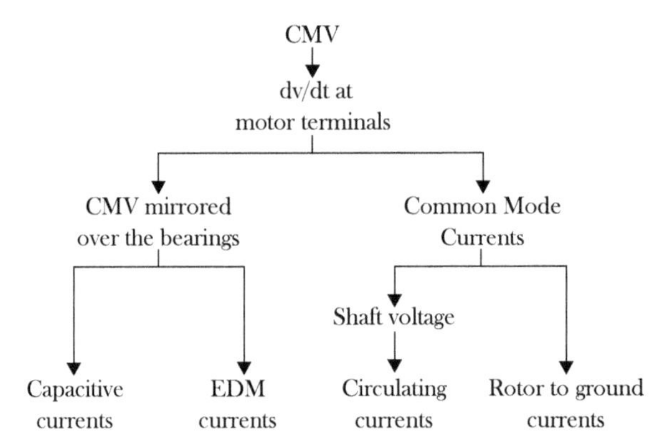

Figure 2. Cause-and-effect chain of common mode voltage (CMV) (adapted from [25]).

Operating at higher switching frequencies can entail more severe CMV related issues (additional EMI generation and larger number of *dv/dt*) [26]. As considerable efforts are being carried out to widespread the usage of wide bandgap (WBG) devices in AC drives, much higher switching frequencies are expected in the future, making the investigation on CMV mitigation a popular topic [13,19,20]. Thus, a wide variety of solutions have been proposed in the literature. Such solutions can be classified as passive or active. Passive solutions are those which mitigate or eliminate the harmful effects generated by CMV, while active solutions are intended to reduce or totally avoid CMV generation. Among passive solutions, Faraday shielding [27,28], ceramic and hybrid bearings [22,29], shielded cables [23,30] and shaft grounding rings [27,31] can be highlighted. On the other hand, modulation techniques and new inverter topologies such as multilevel inverters [32], single-phase transformerless inverters [33,34] and three-phase inverters [35] among others [36] are the most common active solutions. Among all these solutions, modulation algorithms can be considered for CMV reduction in star-connected five-phase AC drives due to their ease of implementation, low cost, and because no additional hardware is needed.

In [37], the authors initially proposed a CMV reduction modulation technique for five-phase inverters, named AZSL5M5-PWM. However, the proposal has been only considered for passive loads (star-connected RL loads) and solely validated in open-loop. From the obtained results, it has been concluded that the linear range of the original AZSL5M5-PWM is limited, which can prevent the utilization of this technique in electric drives where operation close to the base speed (without entering in field weakening region) is desirable, as is the case in most EMA systems. Thus, a hybrid AZSL5M5-PWM technique (HAZSL5M5-PWM) that provides the same linear range as conventional space vector PWM (SV-PWM) is proposed in this work, and its performance is evaluated in an EMA system.

This manuscript is organized as follows. First of all, conventional SV-PWM for star-connected five-phase power systems is presented, where the harmonic projection of the stator voltages into their corresponding orthogonal subspaces by means of Clarke transformation is mathematically justified. After that and considering the third harmonic elimination constraint, it is shown how CMV variations are generated in the multiphase drive. Secondly, the most relevant reduced common-mode voltage PWM (RCMV-PWM) modulation techniques are briefly described, focusing on their limitations. After that, the proposed Hybrid AZSL5M5-PWM modulation technique is presented providing the required tools for duty cycle calculation, and validated by means of simulation. The target of the proposed modulation technique is to effectively reduce CMV in star connected multiphase systems, while the hybridization is performed to cover the whole operation range of the drive. Open-loop and detailed five-phase EMA simulations are conducted to perform the validation, where not only CMV reduction is verified, but other figures such as total harmonic distortion (THD) and efficiency are evaluated in order to demonstrate that the achieved CMV reduction does not significantly penalize other relevant drive figures.

2. Influence of the SV-PWM Technique in the CMV of a Star-Connected Five-Phase AC Drive

SV-PWM is one of the most used modulation techniques in three-phase and multiphase power systems thanks to its easy digital implementation and optimum DC bus voltage utilization. As a star-connected five-phase system has four degrees of freedom, stator voltages and currents can be represented into two separated two-dimensional planes, α-β and x-y, and one homopolar component by means of the following amplitude invariant Clarke transformation [38]:

$$\begin{bmatrix} v_\alpha \\ v_\beta \\ v_x \\ v_y \\ v_0 \end{bmatrix} = \frac{2}{5} \begin{bmatrix} 1 & cos(2\pi/5) & cos(4\pi/5) & cos(6\pi/5) & cos(8\pi/5) \\ 0 & sin(2\pi/5) & sin(4\pi/5) & sin(6\pi/5) & sin(8\pi/5) \\ 1 & cos(4\pi/5) & cos(8\pi/5) & cos(12\pi/5) & cos(16\pi/5) \\ 0 & sin(4\pi/5) & sin(8\pi/5) & sin(12\pi/5) & sin(16\pi/5) \\ \frac{1}{2} & \frac{1}{2} & \frac{1}{2} & \frac{1}{2} & \frac{1}{2} \end{bmatrix} \begin{bmatrix} v_a \\ v_b \\ v_c \\ v_d \\ v_e \end{bmatrix}. \quad (1)$$

The Clarke transformation allows us to decouple the 5-dimensional voltage vector in the *abcde* reference frame into three orthogonal subspaces (α-β, x-y and 0). For a surface-mounted permanent magnet synchronous machine (SM-PMSM), this decoupling is done through the diagonalization of the inductance matrix **L** (2).

$$\mathbf{L} = \begin{bmatrix} L_{11} & L_{12} & L_{13} & L_{14} & L_{15} \\ L_{21} & L_{22} & L_{23} & L_{24} & L_{25} \\ L_{31} & L_{32} & L_{33} & L_{34} & L_{35} \\ L_{41} & L_{42} & L_{43} & L_{44} & L_{45} \\ L_{51} & L_{52} & L_{53} & L_{54} & L_{55} \end{bmatrix}. \quad (2)$$

For SM-PMSMs, the elements of **L** can be considered invariant with respect to the rotor angular position, as the surface placed magnets have a permeability near that the one of the air. Therefore, a SM-PMSM behaves like a non-salient pole synchronous machine [39]. As the windings in each phase are manufactured identically, the mutual inductances between any pair of phases separated with the same electrical angle are equal, i.e., $L_{12} = L_{15} = L_{21} = L_{23} = L_{51} = L_{jk}$ (if $|j-k| = 1$) or $L_{13} = L_{31} = L_{25} = L_{52} = L_{jk}$ (if $|j-k| = 2$). Similarly, all the self-inductances are equal ($L_{11} = L_{22} = \cdots = L_{55}$). This type of matrix is known as a circulant matrix, and it has some special properties [40]. For example, it guarantees that **L** is orthogonally diagonalizable by a transformation represented by a 5×5 real matrix [41,42].

The circulant matrices are diagonalized by the Fourier Matrix [40,43]. Therefore, the Clarke transformation decomposes the 5-dimensional vectors according to their harmonic components. In the α-β sub-space, the $h = 5(l-1) \pm 1$ harmonic components are projected while, in the x-y sub-space, the $h = 5(l-1) \pm 3$ ones are projected, being $l \in \{1,3,5,...\}$ [41,44]. The harmonic components of order $h = 5l$ are projected into the zero-sequence or homopolar sub-space. In Table 1, the odd harmonics associated with each sub-space according to the Clarke transformation of (1) are presented for a 5-phase machine.

Table 1. Sub-space harmonics mapping for a five-phase machine.

Sub-Space	Harmonics
$\alpha - \beta$	$h = 1, 9, 11, 19...$
$x - y$	$h = 3, 7, 13, 17...$
zero-sequence	$h = 0, 5, 15, 25...$

The number of possible switching states or space vectors is 2^5, where 30 are active vectors and two are zero vectors (Figure 3). Active vectors can be classified depending on their magnitude as:

- Large vectors, where $|V_l| = 4/5V_{DC}\, cos(\pi/5)$, which correspond to the outer decagon of Figure 3.
- Medium vectors, where $|V_m| = 2/5V_{DC}$, which correspond to the middle decagon of Figure 3.
- Small vectors, where $|V_s| = 4/5V_{DC}\, cos(2\pi/5)$, which correspond to the inner decagon of Figure 3.

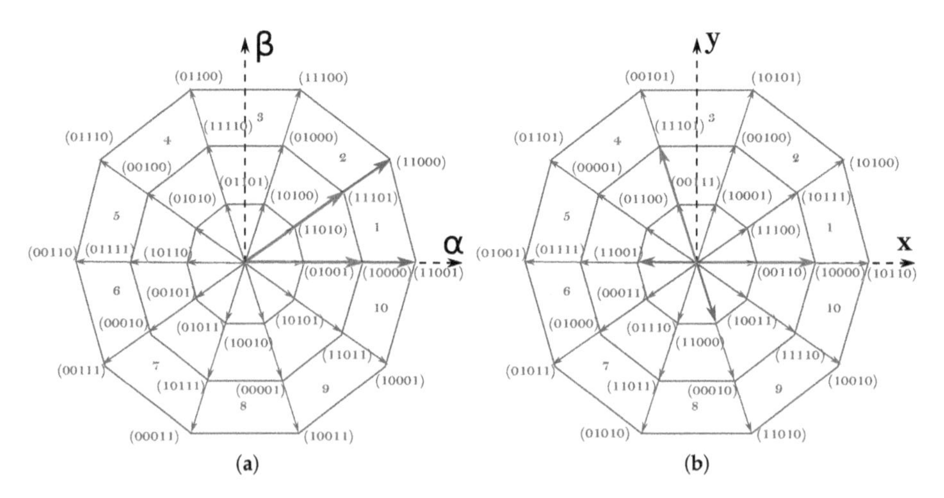

Figure 3. Five-phase SV-PWM: α-β and x-y planes with their corresponding space vectors and switching states (a '1' in a switching state represents that the top switch of a given phase is activated, while a '0' represents that its complementary switch is activated). **(a)** α-β vector plane. **(b)** x-y vector plane.

For an n-phase system, $n-1$ active vectors must be applied at each commutation period in order to achieve a sinusoidal output [45]. Thus, four active vectors must be used in a star-connected five-phase system. Although various possible active vector combinations are possible to produce a given output voltage vector, the most common alternative consists on using two large and two medium adjacent vectors. As an illustrative example, Figure 3 shows the vectors used to synthesize a given reference voltage vector located in the first sector of the α-β plane, being the following the application sequence that minimizes switching losses: 00000, 10000, 11000, 11001, 11101, 11111, 11101, 11001, 11000, 10000, 00000. If the third harmonic component needs to be eliminated, the application-time ratio between medium and large vectors must satisfy (3), as with this ratio the sum of the applied vectors in the x-y plane is zero (Figure 3) [46].

$$\frac{t_{large}}{t_{medium}} = 1.618. \tag{3}$$

As a result, the maximum achievable output voltage following this modulation approach is:

$$V_{o_{max}} = \frac{4}{5}\cos\left(\frac{\pi}{5}\right)\cos\left(\frac{\pi}{10}\right) = 0.6155 V_{DC}, \tag{4}$$

and the CMV generated in the five-phase system is:

$$V_{CM}(t) = \frac{1}{5}\left[V_{a0}(t) + V_{b0}(t) + V_{c0}(t) + V_{d0}(t) + V_{e0}(t)\right]. \tag{5}$$

When using SV-PWM and applying the vector sequence that corresponds to the first sector of the α-β plane, the CMV waveform of Figure 4 is obtained. From (5), it can be deduced that all large and short vectors generate CMV levels of $\pm 0.3V_{DC}$, while CMV levels are of $\pm 0.1V_{DC}$ for medium vectors and of $\pm 0.5V_{DC}$ for null vectors. When evaluating the impact of the CMV, the difference between the maximum and minimum CMV levels (Δ_{CMV}, Figure 4) must be considered, and the number of CMV variations for each commutation period (N_{CMV}) must also be taken into account.

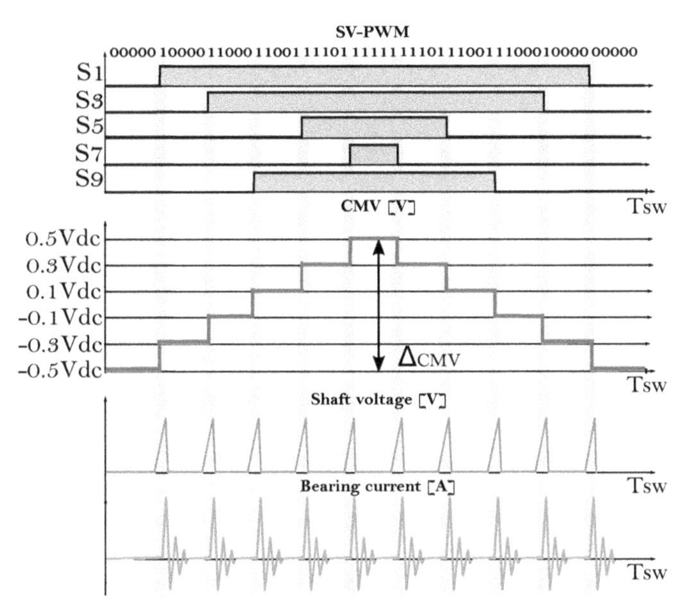

Figure 4. CMV waveform of SV-PWM technique (adapted from [23]).

3. RCMV-PWM Techniques

As zero vectors are responsible for generating the maximum CMV levels (Figure 4), most of the RCMV-PWM techniques avoid the application of these vectors to reduce Δ_{CMV} and N_{CMV}. In [38], an extension of the three-phase active zero state PWM (AZS-PWM) [47] modulation technique to the five-phase scenario is proposed. This technique replaces zero vectors by applying two active vectors with the opposite phase at the same time. In this work, this technique will be named AZSL2M2-PWM as, apart from the active vectors that substitute zero vectors, two large (L2) and two medium (M2) vectors are used at each modulation period. This technique shows a good harmonic performance and DC bus utilization, being its linear range $0 \leq m \leq 1$. However, Δ_{CMV} is not greatly reduced, as only $\pm 0.5V_{DC}$ CMV levels are avoided (Table 2). Similarly, a modulation algorithm that employs four large active vectors in conjunction with two active vectors with opposite phases (AZSL4-PWM) is proposed in [38]. This technique has the same linear range as SV-PWM and AZSL2M2-PWM, and considerably reduces Δ_{CMV}, as only applies large vectors. Nonetheless, N_{CMV} remains as for SV-PWM (Table 2).

M5-PWM [48] and L5-PWM [49] techniques completely eliminate Δ_{CMV} and N_{CMV} by only using odd or even medium (M5-PWM), or odd or even large (L5-PWM) active vectors. However, this is achieved at the cost of introducing additional power losses, significantly reducing the linear range up to 0.5257 for M5-PWM (Table 2), and generating high harmonic distortion for L5-PWM, making them inappropriate for many industrial applications. Authors in [48,49] also propose variants that use ten medium (M10-PWM) or ten large (L10-PWM) vectors. These techniques enhance the linear range by increasing the available vectors, but do not reduce Δ_{CMV} and N_{CMV} as much as with M5-PWM and L5-PWM (Table 2).

Table 2. Summary of the most relevant features of SV-PWM and RCMV-PWM techniques.

Modulation	Δ_{CMV} [V]	N_{CMV}	Δ_{CMV} Reduction [%]	N_{CMV} Reduction [%]	v_{CM} Waveform	Linear Range
SV-PWM	V_{DC}	10	-	-	$0.5V_{DC}$, $0.3V_{DC}$, $0.1V_{DC}$, $-0.1V_{DC}$, $-0.3V_{DC}$, $-0.5V_{DC}$	$0 \leq m \leq 1$
AZSL2M2-PWM	$0.6V_{DC}$	6	−40%	−40%	$0.3V_{DC}$, $0.1V_{DC}$, $-0.1V_{DC}$, $-0.3V_{DC}$	$0 \leq m \leq 1$
AZSL4-PWM	$0.2V_{DC}$	10	−80%	0%	$0.1V_{DC}$, $-0.1V_{DC}$	$0 \leq m \leq 1$
M5-PWM	0	0	−100%	−100%	$-0.3V_{DC}$	$0 \leq m \leq 0.5257$
M10-PWM	$0.6V_{DC}$	2	−40%	−80%	- 0 3VD, G 0 3VD	$0 \leq m \leq 0.618$
L5-PWM	0	0	−100%	−100%	$0.1V_{DC}$	$0 \leq m \leq 0.8507$
L10-PWM	$0.2V_{DC}$	6	−80%	−40%	$0.1V_{DC}$, $-0.1V_{DC}$	$0 \leq m \leq 1$
AZSL5M5-PWM	$0.4V_{DC}$	2	−60%	−80%	$0.3V_{DC}$, 10.- V_{DC}	$0 \leq m \leq 0.8507$

Among the reviewed techniques, AZSL2M2-PWM and L10-PWM best suit for industrial applications, as they keep the linear range with a reasonable THD while effectively reducing Δ_{CMV}. However, N_{CMV} reduction by means of such modulation algorithms is limited. Thus, a new RCMV-PWM technique that further reduces N_{CMV} while keeping an extended linear range is proposed in the following section.

4. Proposed RCMV-PWM Technique

4.1. Active Zero State L5M5 PWM Technique (AZSL5M5-PWM)

The main part of the proposed RCMV-PWM technique, named active zero state L5M5 PWM (AZSL5M5-PWM), is based on the AZSL2M2-PWM technique [38]. However, and unlike AZSL2M2-PWM, the proposed scheme only uses odd or even vectors to further reduce the CMV voltage variations. For example, if odd vectors are only considered, this leads to the sector distribution of Figure 5a. Thus, five medium vectors and five large vectors are exclusively used to synthesize the reference voltage (Figure 5), and CMV varies between $-0.3V_{DC}$ and $0.1V_{DC}$ (if only even vectors are used, CMV varies between $-0.1V_{DC}$ and $0.3V_{DC}$). Consequently, $\Delta_{CMV} = 0.4V_{DC}$ and $N_{CMV} = 2$ (Table 2).

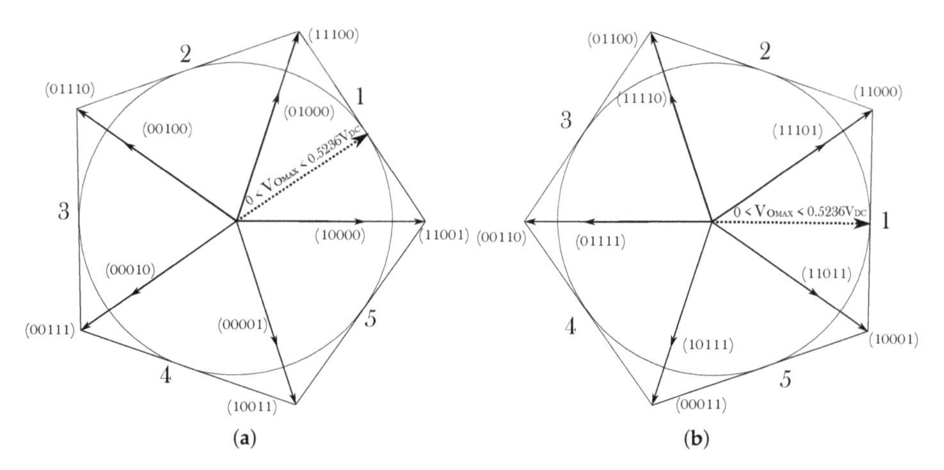

Figure 5. Sector distribution of AZSL5M5-PWM modulation scheme in the α-β plane. (**a**) AZSL5M5-PWM implemented with odd vectors. (**b**) AZSL5M5-PWM implemented with even vectors.

For the sake of simplicity, all the following analyses are conducted considering the AZSL5M5-PWM variant which exclusively uses odd vectors. The procedure is analogous for even vectors.

The application time of each vector can be easily calculated by solving the following system:

$$\begin{bmatrix} V_\alpha^* \\ V_\beta^* \\ V_x^* \\ V_y^* \end{bmatrix} = \begin{bmatrix} V_{m_{r\alpha}} & V_{l_{l\alpha}} & V_{l_{r\alpha}} & V_{m_{l\alpha}} \\ V_{m_{r\beta}} & V_{l_{l\beta}} & V_{l_{r\beta}} & V_{m_{l\beta}} \\ V_{m_{rx}} & V_{l_{lx}} & V_{l_{rx}} & V_{m_{lx}} \\ V_{m_{ry}} & V_{l_{ly}} & V_{l_{ry}} & V_{m_{ly}} \end{bmatrix} \begin{bmatrix} \delta_1 \\ \delta_2 \\ \delta_3 \\ \delta_4 \end{bmatrix}, \tag{6}$$

where V_α^*, V_β^*, V_x^* and V_y^* are the reference voltage projections in the α-β and x-y planes, δ_1, δ_2, δ_3 and δ_4 are the duty cycles for each vector, and the 4×4 matrix is composed of the magnitudes of the vectors to be applied in each sector, where V_{m_r} refers to the modulus of the medium vector on the right side of the reference vector (V_{ref}) (Figure 6, ③), V_{m_l} refers to the modulus of the medium vector on the left (Figure 6, ④), and V_{l_l} and V_{l_r} refer to the modulus of the large vectors on both left and right sides, respectively (Figure 6, ② and ①). Consequently, a 4×4 matrix should be defined for each sector. In general, V_x^* and V_y^* are set to zero in order to cancel the voltage third harmonic. From (6), it is possible to explicitly determine the values of the duty cycles δ_1, δ_2, δ_3 and δ_4 with respect to the reference voltages V_α^* and V_β^*:

$$\begin{aligned}
\delta_1 &= \alpha_1 V_\alpha^*/V_{DC} + \beta_1 V_\beta^*/V_{DC} = -a_1 sin\left[\frac{2s\pi}{5}\right] V_\alpha^*/V_{DC} + a_1 cos\left[\frac{2s\pi}{5}\right] V_\beta^*/V_{DC}, \\
\delta_2 &= \alpha_2 V_\alpha^*/V_{DC} + \beta_2 V_\beta^*/V_{DC} = -a_2 sin\left[\frac{2(s-1)\pi}{5}\right] V_\alpha^*/V_{DC} + a_2 cos\left[\frac{2(s-1)\pi}{5}\right] V_\beta^*/V_{DC}, \\
\delta_3 &= \alpha_3 V_\alpha^*/V_{DC} + \beta_3 V_\beta^*/V_{DC} = a_2 sin\left[\frac{2s\pi}{5}\right] V_\alpha^*/V_{DC} - a_2 cos\left[\frac{2s\pi}{5}\right] V_\beta^*/V_{DC}, \\
\delta_4 &= \alpha_4 V_\alpha^*/V_{DC} + \beta_4 V_\beta^*/V_{DC} = a_1 sin\left[\frac{2(s-1)\pi}{5}\right] V_\alpha^*/V_{DC} - a_1 cos\left[\frac{2(s-1)\pi}{5}\right] V_\beta^*/V_{DC},
\end{aligned} \tag{7}$$

where being $s = \{1,2...5\}$ the corresponding sector in the $\alpha\beta$ plane, and being $a_1 = (-5+\sqrt{5})/\sqrt{2(5+\sqrt{5})}$ and $a_2 = \sqrt{(10/(5+\sqrt{5})}$.

Regarding the practical implementation of the proposed technique, the 2×4 matrix $\mathbf{M}_s$ can be defined as in (8), where the elements of such matrix can be precalculated for each sector and stored into look-up tables (LUT). In this way, the computational burden and implementation complexity of the algorithm are greatly reduced. Table 3 summarizes the values of $\mathbf{M}_s$ for each sector $s = \{1,2...5\}$.

$$\mathbf{M_s} = \begin{bmatrix} \alpha_1 & \beta_1 \\ \alpha_2 & \beta_2 \\ \alpha_3 & \beta_3 \\ \alpha_4 & \beta_4 \end{bmatrix}. \tag{8}$$

Table 3. Values of $\mathbf{M}_s$ depending on the $\alpha\beta$ plane sector.

Sector 1 (M_1)	Sector 2 (M_2)	Sector 3 (M_3)	Sector 4 (M_4)	Sector 5 (M_5)
$\begin{bmatrix} 0.691 & -0.224 \\ 0 & 1.176 \\ 1.118 & -0.363 \\ 0 & 0.726 \end{bmatrix}$	$\begin{bmatrix} 0.427 & 0.588 \\ -1.118 & 0.363 \\ 0.691 & 0.951 \\ -0.691 & 0.224 \end{bmatrix}$	$\begin{bmatrix} -0.427 & 0.588 \\ -0.691 & -0.951 \\ -0.691 & 0.951 \\ -0.427 & -0.588 \end{bmatrix}$	$\begin{bmatrix} -0.691 & -0.224 \\ 0.691 & -0.951 \\ -1.118 & -0.363 \\ 0.427 & -0.588 \end{bmatrix}$	$\begin{bmatrix} 0 & -0.726 \\ 1.118 & 0.363 \\ 0 & -1.176 \\ 0.691 & 0.224 \end{bmatrix}$

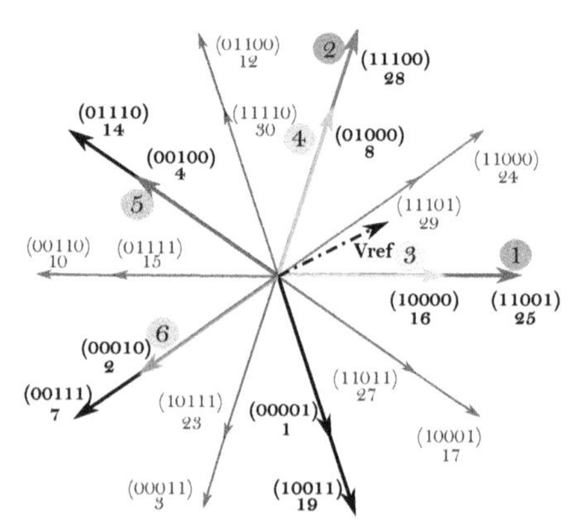

Figure 6. AZSL5M5-PWM modulation vector sequence (①–⑥) for sector 1 when only odd vectors are applied.

The main difference of the AZSL5M5-PWM technique over the AZSL2M2-PWM one is that, there are no strictly phase-opposite vector pairs, as only odd/even vectors can be used (Figure 6). To solve this problem, three active vectors (two medium and one large) are used to replace a zero vector.

First, the large vector on the right side of the sector (Figure 6, ①) is applied during $t_0/3$, and each medium vector (Figure 6, ⑤ and ⑥) is applied during $t_0/3$.

Since the applied vector sequence has a great impact on the CMV and on the switching losses, a sequence with minimum commutations has been chosen for each sector. For instance, when the reference voltage vector lays in sector 1, the next vector sequence is applied: 11001, 11100, 10000, 01000, 00100, 00010, 01000, 10000, 11100 and 11001 (Figure 6). Odd vector variant AZSL5M5-PWM vector sequences depending on the reference voltage sector are given in Table 4. It is important to note that, for AZSL5M5-PWM, more than one commutation is produced at each vector change.

Table 4. AZSL5M5-PWM vector sequences (odd vectors).

AZSL5M5-PWM	Vector Sequence
Sector 1	11001 11100 10000 01000 00100 00010 01000 10000 11100 11001
Sector 2	11100 01110 01000 00100 00010 00001 00100 01000 01110 11100
Sector 3	01110 00111 00100 00010 00001 10000 00010 00100 00111 01110
Sector 4	00111 10011 00010 00001 10000 01000 00001 00010 10011 00111
Sector 5	10011 11001 00001 10000 01000 00100 10000 00001 11001 10011

As the α-β plane is divided into five sectors instead of ten (Figure 5), the linear range of AZSL5M5-PWM is slightly reduced, being the maximum achievable output voltage:

$$V_{O_{MAX}} = \frac{4}{5}\cos\left(\frac{\pi}{5}\right)\cos\left(\frac{\pi}{5}\right) = 0.5236V_{DC}. \tag{9}$$

For applications where achieving full linear range is mandatory, an hybrid modulation that extends AZSL5M5-PWM's linear range is proposed in the following.

4.2. Hybridization of the Proposed Modulation Algorithm

Three operation areas have been differentiated in Figure 7 to carry out the hybrid modulation algorithm and extend the linear range of AZSL5M5-PWM. The AZSL5M5-PWM hybrid variant (HASZL5M5-PWM) that uses odd vectors (white area) has been chosen as the main modulation scheme. When V_{ref} steps over the boundaries of the white pentagon, two things may occur. On the one hand, V_{ref} might remain within the shadowed boundaries. In such case, AZSL5M5-PWM with even vectors would be applied. On the other hand, if V_{ref} is out of the limits of AZSL5M5-PWM with even or odd vectors, SV-PWM can be used to fulfill the remaining area, marked with lines. This modification extends the linear range of AZSL5M5-PWM up to 26.8%. Further variants with greater CMV reduction could also be considered if a full range RCMV-PWM technique, such as AZSL2M2-PWM, is used instead of SV-PWM.

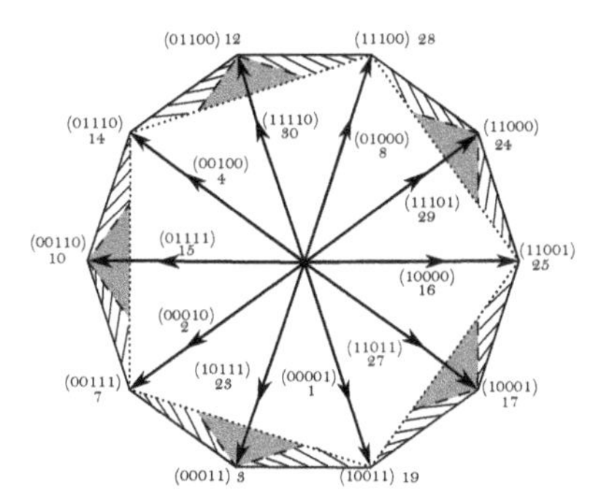

Figure 7. HAZSL5M5-PWM. White: AZSL5M5-PWM with odd vectors; shadowed: AZSL5M5-PWM with even vectors; lines: SV-PWM.

5. Simulation Results

In order to validate the proposal, two simulation platforms have been implemented in Matlab/Simulink. On the one hand, an open-loop model has been created to evaluate the HAZSL5M5-PWM technique and compare its performance with other existing ones regardless of the influence of a control algorithm. On the other, a detailed five-phase EMA model has been implemented to evaluate the proposal in the context of a variable speed AC drive. Without losing generality and in order not to significantly increase the computational burden of the model, ideal switch models that do not consider switching transients nor dead-time effects have been adopted in both simulation platforms. The obtained results and their discussion are provided in the following.

5.1. Open-Loop Model Simulation Results

Figure 8 shows the open-loop model block diagram. SimPowerSystem blocks have been used to model the power elements. The battery has been modeled as an ideal DC voltage source. The power-converter block includes a two-level five-phase voltage source inverter, where each switching device includes a detailed loss and thermal model, allowing an accurate estimation of inverter losses. In this work, a loss model of the International Rectifier AUIRGPS4067D1 IGBT has been implemented for each switch, whose main parameters are detailed in Table 5. The loss and thermal model follows the same approach as the one presented by the authors in [50]. The analytical approach used in this work to estimate the instantaneous conduction and switching losses is commonly used by the scientific community [51] and by the industry [52]. On the other hand, the adopted 1D thermal modeling approach has been verified by the authors in [53], where it has been compared to 3D finite element method (FEM) simulation, obtaining almost the same results. Finally, a passive star-connected five-phase RL load has been included. The most significant parameters of the open-loop model are collected in Table 6.

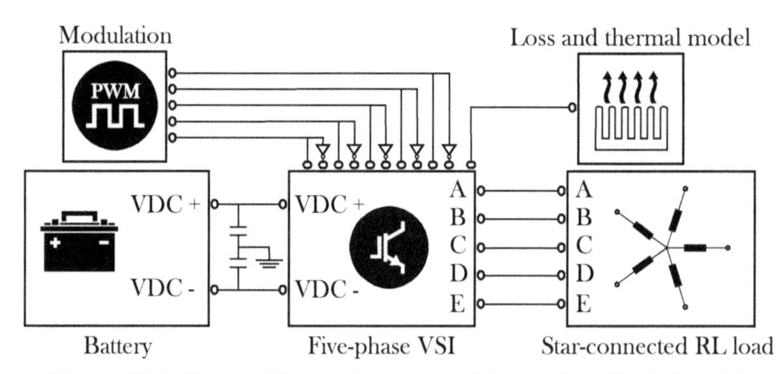

Figure 8. Block diagram of the constituting parts of the open-loop Simulink model.

Table 5. Most relevant parameters of the simulated International Rectifier AUIRGPS4067D1 IGBT.

Parameter	Value	Unit
Nominal current per switch	120	A
Maximum blocking voltage	600	V
Typical IGBT collector-emitter voltage	1.7	V
Typical diode forward voltage	1.7	V
Typical IGBT turn-on switching loss	8.2	mJ
Typical IGBT turn-off switching loss	2.9	mJ
Typical diode reverse recovery	2.4	mJ
IGBT thermal resistance	0.2	°C/W
Diode thermal resistance	0.25	°C/W
Allowable junction temperature	−55 to 175	°C

Table 6. The most significant parameters of the open-loop simulation platform.

Variable	Symbol	Value	Unit
Load resistance	R_{Load}	0.001	Ω
Load inductance	L_{Load}	1	mH
Battery voltage	V_{DC}	320	V
Modulator frequency	f_{mod}	50	Hz
Switching frequency	f_{sw}	10000	Hz

Figure 9 shows the THD and the efficiencies obtained for the proposed algorithm and for other techniques for all the linear range. As it was expected, RCMV-PWM modulations show greater harmonic content when compared to SV-PWM due to the use of phase-opposite vectors. However, for high modulation index values, all the studied modulations produce a similar THD. On the other hand, while AZSL2M2-PWM and SV-PWM have similar efficiencies, the HAZSL5M5-PWM has a slightly lower efficiency, which increases for low modulation indexes.

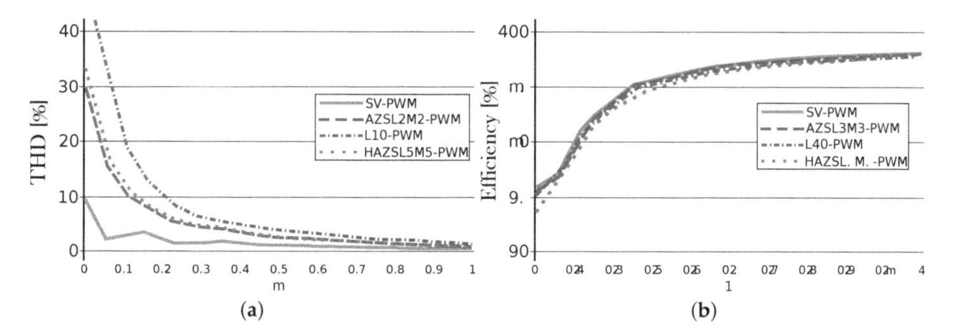

Figure 9. Total harmonic distortion (THD) and efficiency of studied modulation techniques for static operation points. (**a**) THD. (**b**) Efficiency.

Power losses can be seen in more detail in Figure 10. As mentioned before, HAZSL5M5-PWM requires more commutations at each vector change which entails an increase of switching losses (Figure 10a). However, conduction losses are almost equal in all modulations (Figure 10b). On the other hand, Figure 10c shows the load power as a function of the modulation index.

Regarding CMV mitigation, the proposed HAZSL5M5-PWM technique reduces Δ_{CMV} and N_{CMV} by a 60% and 80%, respectively, when $m \leq 0.8507$. These percentages are reduced while m gets close to 1. The worst case scenario, when modulation index is 1, AZSL5M5-PWM is active 29.78% of the simulated time while SV-PWM is active the 70.22% of the simulated time. In such a case, the Δ_{CMV} is reduced by 17.86% and N_{CMV} is reduced by 23.82%. In addition, when applying the operation condition equivalent to maximum torque ($T_{em_{max}}$ = 26 Nm) and maximum speed (ω_{max} = 105 rpm) that allows this particular application (modulation index = 0.96), AZSL5M5-PWM is active 49.8% of the simulated time while SV-PWM is active the 50.2% of the simulated time, reducing the Δ_{CMV} by 29.88% and N_{CMV} by 39.84%. So, even when the most torque and speed values are considered, HAZSL5M5-PWM reduces the N_{CMV} as much as AZSL2M2-PWM and L10-PWM techniques.

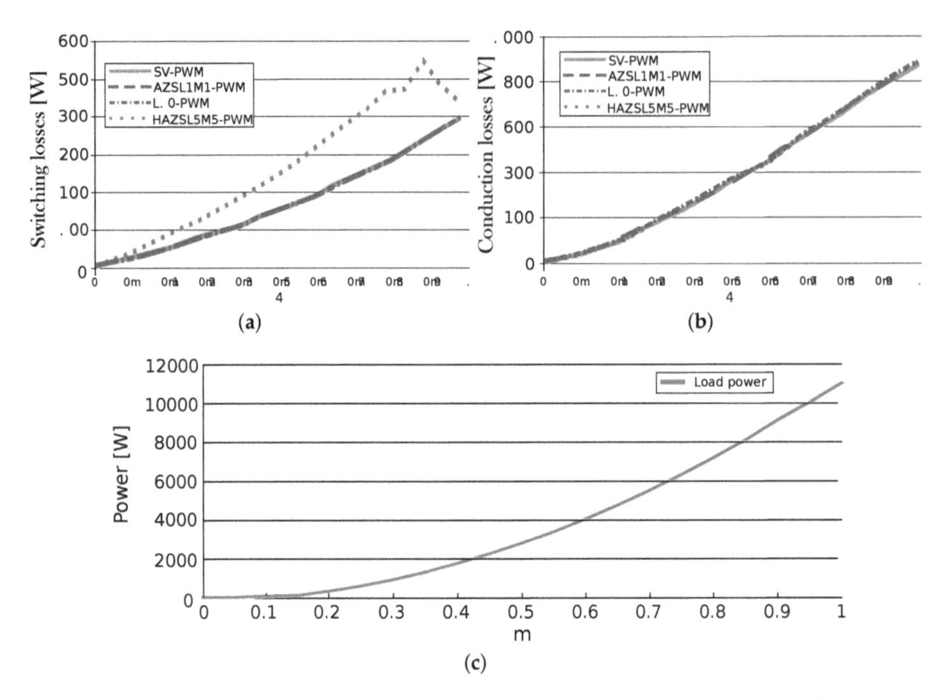

Figure 10. Distribution of power losses for the of studied modulation techniques for static operation points. (**a**) Switching losses. (**b**) Conduction losses. (**c**) Load power.

5.2. EMA Model Simulation Results

Figure 11 shows the block diagram of the implemented EMA model where, as in the open-loop model, the same power loss model based on the International Rectifier AUIRGPS4067D1 IGBT has been implemented. The EMA incorporates a star-connected five-phase PMSM, which third harmonic back-EMF component is negligible. Table 7 shows the main parameters of the simulated EMA.

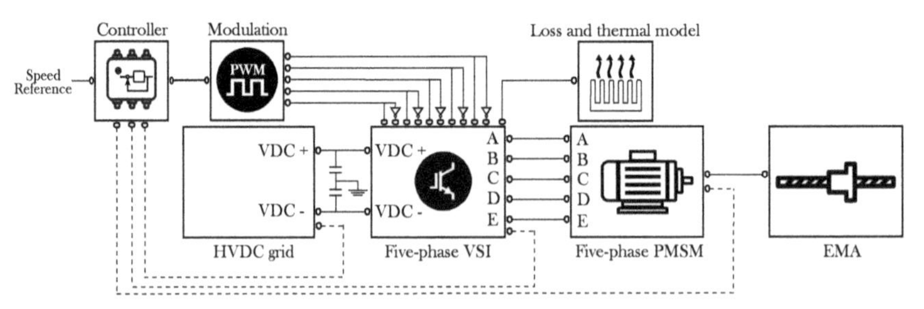

Figure 11. General block diagram of the electromechanical actuator (EMA) simulation platform.

The five-phase PMSM stator voltages are given by:

$$\mathbf{V} = \mathbf{RI} + \mathbf{L}\frac{d\mathbf{I}}{dt} + \frac{d\mathbf{\Psi}_{PM}}{dt}, \tag{10}$$

where **V** and **I** are five-dimensional vectors whose element (v_j and i_j, $j \in [a, b, ..., e]$) are the per-phase voltages and currents, respectively. **R** is a 5×5 diagonal matrix, where each diagonal element represents the phase resistance. **L** is the 5×5 stator inductance matrix, where each element L_{ij} ($i, j \in [a, b, ..., e]$) represent the self- ($i = j$) and mutual-inductances ($i \neq j$) between phases i

and j. Being the value of the mutual-inductances very low and magnetic saturation phenomena negligible, mutual-inductances have been considered zero and self-inductances have been considered constant in the implemented electric machine model. A perfectly balanced stator has been considered. These assumptions have been done without losing generality for the evaluation of the proposed modulation algorithm as, if any non-ideality is present in the electric drive, torque and speed loops are responsible of their compensation, while the proposed algorithm synthesizes the commanded reference voltages and minimizes CMV. The term $\mathbf{\Psi}_{PM}$ is the five-dimensional flux linkage vector ($\mathbf{\Psi}_{PM} = [\Psi_{PMa}, \Psi_{PMb}, \ldots, \Psi_{PMe}]^T$) produced due to the permanent magnets.

Table 7. Most significant parameters of the simulated EMA.

Parameter	Symbol	Value	Unit
Rated power	P_{nom}	1.51	kW
Rated torque	T_{nom}	12.1	Nm
Rated speed	ω_{nom}	1200	RPM
Pole-pair number	N_p	9	–
Stator resistance	R_s	1.5	Ω
Stator self-inductance	L_s	9.6	mH
PM flux linkage	Ψ_{PM}	0.13	Wb
HVDC grid voltage	V_{DC}	270	V
Switching frequency	f_{sw}	10000	Hz

The torque produced by the motor is given by:

$$T_{em} = \mathbf{I}^T \frac{d\mathbf{\Psi}_{PM}}{d\theta_m}, \tag{11}$$

where θ_m is the angular mechanical rotor position. The dynamics of the rotational movement are given by:

$$T_{em} - T_l = J\frac{d\omega_m}{dt} + B\omega_m, \tag{12}$$

where T_l is the load torque produced by the EMA, J is the total inertia moment of the rotating masses, including EMA and motor, ω_m is the rotational speed of the rotor and B is the viscous friction coefficient.

Figure 12 shows the detailed diagram of the controller. The controller consists of two control loops. The outer one regulates the rotational speed of the motor. This loop has a proportional-integral (PI) controller tuned in z. For this application, the damping factor has been set to $\xi = 0.707$, while the settling-time has been set to $T_s = 50$ ms. The inner loop tracks the current references through a vector controller [54]. Again, $\xi = 0.707$ for the current regulator, while $T_s = 5$ ms. In this particular case, only two PI controllers are required to control the first harmonic components (i_{d1}, i_{q1}), as there is no third harmonic back-EMF component and the proposed PWM technique intrinsically regulates to zero the third harmonic voltages (V_x^* and V_y^* are imposed to be zero). It must be taken into account that, for this particular control approach, a conventional microcontroller sine-triangle PWM peripheral cannot be used due to the modulation algorithm computational requirements. Thus, an FPGA should be incorporated to implement the modulation algorithm, while implementing the speed and current loops in a fixed-point or floating point DSP.

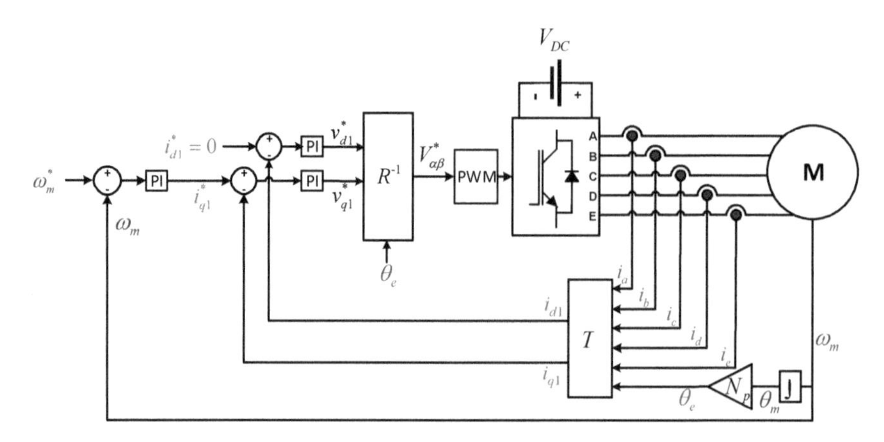

Figure 12. Block diagram of the EMA speed and torque controller.

Therefore, the following *abcde* to *d*1-*q*1 transformation (*T*) is used in the controller:

$$T = \frac{2}{5}\begin{bmatrix} cos(\theta_e) & cos(\theta_e - 2\pi/5) & cos(\theta_e - 4\pi/5) & cos(\theta_e - 6\pi/5) & cos(\theta_e - 8\pi/5) \\ -sin(\theta_e) & -sin(\theta_e - 2\pi/5) & -sin(\theta_e - 4\pi/5) & -sin(\theta_e - 6\pi/5) & -sin(\theta_e - 8\pi/5) \end{bmatrix}, \tag{13}$$

this being the matrix product of the transformation in (1) with the following rotational matrix:

$$R = \begin{bmatrix} cos(\theta_e) & sin(\theta_e) & 0 & 0 & 0 \\ -sin(\theta_e) & cos(\theta_e) & 0 & 0 & 0 \end{bmatrix}, \tag{14}$$

where θ_e is the electrical rotor position of the motor, being $\theta_e = N_p\theta_m$.

Once the inner loop PI controllers provide the voltage references (v^*_{d1}, v^*_{q1}), such references are transformed into the $\alpha\beta$ frame by applying the R^{-1} (pseudo-inverse of *R*) matrix and fed to the PWM block. Matrix R^{-1} is the classical counter-clockwise rotation transformation [42]:

$$R^{-1} = \begin{bmatrix} cos(\theta_e) & -sin(\theta_e) \\ sin(\theta_e) & cos(\theta_e) \end{bmatrix}, \tag{15}$$

Several simulations have been performed for various torque and speed conditions that cover the whole operation range of the EMA in order to evaluate the figures of the HAZSL5M5-PWM algorithm compared to other techniques.

Figures 13a–c show the efficiency results and the distribution between conduction and switching losses. Similar results as in open-loop simulations have been obtained for the EMA platform. As expected, switching losses increase when applying HAZSL5M5-PWM. However, such losses are not linear since hybrid AZSL5M5-PWM also includes SV-PWM algorithm and, when high modulation indexes are required, SV-PWM and HAZSL5M5-PWM techniques operate together reducing commutation losses. In terms of overall system efficiency, it is only reduced for about 1% when compared HAZSL5M5-PWM to SV-PWM. However, Δ_{CMV} and N_{CMV} are significantly reduced thanks to the proposed technique. In addition, in this particular application, AZSL5M5-PWM is operating all the time in all the simulated operation points except the one described in the previous section (T_{em} = 26 Nm and ω = 105 rpm). Consequently, the benefits of the AZSL5M5-PWM are fully exploited in the vast majority of the operation points of this application.

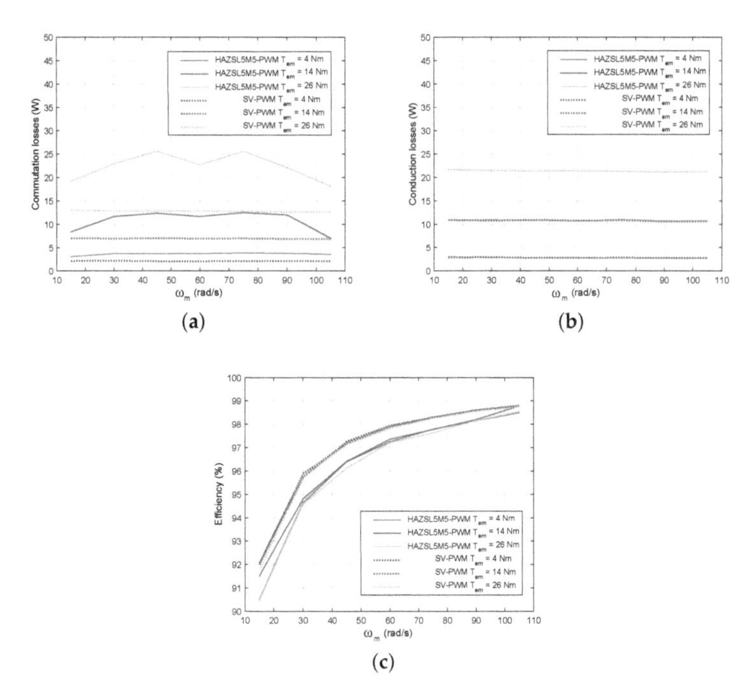

Figure 13. Power losses and efficiency of the proposed HAZSL5M5-PWM algorithm compared to other modulation techniques. (**a**) EMA switching losses for the studied modulation techniques. (**b**) EMA conduction losses for the studied modulation techniques. (**c**) EMA system efficiency with for the studied modulation techniques.

6. Conclusions

This paper introduces the CMV issue in multiphase electric drives and, more precisely, in EMA drives for MEA applications. In this context, the HAZSL5M5-PWM modulation technique is proposed. The basis of this technique merges the usage of phase-opposite vectors and the use of only odd or even active vectors to further reduce the CMV. The THD and efficiency characteristics of the proposed RCMV-PWM algorithm are evaluated and compared with SV-PWM and other RCMV-PWM techniques in an open-loop and in an EMA drive Simulink models. It is shown that the proposed algorithm achieves better THD performance when high modulation indexes are required. In addition, this technique reduces CMV variations (N_{CMV}) up to 80% when compared to SV-PWM, which directly implies a reduction in the leakage currents that affect the bearings. Therefore, the EMA reliability is improved. In exchange, with the proposed modulation the system efficiency slightly decreases. Thus, the investigation of other hybridization alternatives for the HAZSL5M5-PWM modulation technique that also consider THD and efficiency can be considered for future research.

Author Contributions: M.F. concept for the article, writing, proposed modulation technique development, open-loop and closed-loop models simulations and analysis. A.S-G. EMA model development and support on closed-loop model simulations. E.R. IGBT and thermal model development, support on open-loop simulations and review. I.K. support on modulation technique development, conceptual support, review and supervision. E.I. simulation platform development, review and supervision. J.L.M. review. All authors have read and agreed to the published version of the manuscript.

Funding: This work has been supported in part by the Government of the Basque Country within the fund for research groups of the Basque University system IT978-16 and in part by the Government of the Basque Country within the research program ELKARTEK as the project ENSOL (KK-2018/00040).

Conflicts of Interest: The authors declare no conflict of itnerest.

Abbreviations

The following abbreviations are used in this manuscript:

AC	Alternating current
AZS-PWM	Active Zero-State Pulse Width Modulation
AZSL2M2-PWM	Active Zero-State Two Large Two Medium
AZSL4-PWM	Active Zero-State Four Large
AZSL5M5-PWM	Active Zero-State Five Large Five Medium
CMC	Common mode current
CMV	Common mode voltage
DC	Direct current
EDM	Electric discharge machining
EMA	Electromechanical actuator
EMI	Electromagnetic interferences
FEM	Finite element method
GaN	Gallium nitride
HVDC	High-voltage direct current
IGBT	Insulated gate bipolar transistor
L5-PWM	Five Large Pulse Width Modulation
L10-PWM	Ten Large Pulse Width Modulation
M5-PWM	Five Medium Pulse Width Modulation
M10-PWM	Ten Medium Pulse Width Modulation
MEA	More Electric Aircrafts
PI	Proportional Integral
PMSM	Permanent Magnet Synchronous Machine
PWM	Pulse Width Modulation
RCMV-PWM	Reduced Common Mode Voltage Pulse Width Modulation
SiC	Silicon Carbide
SV-PWM	Space Vector Pulse Width Modulation
THD	Total Harmonic Distortion
VSI	Voltage Source Inverter
WBG	Wide Bandgap

References

1. Uzhegov, N.; Smirnov, A.; Park, C.H.; Ahn, J.H.; Heikkinen, J.; Pyrhönen, J. Design Aspects of High-Speed Electrical Machines With Active Magnetic Bearings for Compressor Applications. *IEEE Trans. Ind. Electron.* **2017**, *64*, 8427–8436. [CrossRef]
2. Riba, J.R.; López-Torres, C.; Romeral, L.; Garcia, A. Rare-earth-free propulsion motors for electric vehicles: A technology review. *Renew. Sustain. Energy Rev.* **2016**, *57*, 367–379. [CrossRef]
3. Kumar, M.S.; Revankar, S.T. Development scheme and key technology of an electric vehicle: An overview. *Renew. Sustain. Energy Rev.* **2017**, *70*, 1266–1285. [CrossRef]
4. Riveros, J.A.; Barrero, F.; Levi, E.; Durán, M.J.; Toral, S.; Jones, M. Variable-Speed Five-Phase Induction Motor Drive Based on Predictive Torque Control. *IEEE Trans. Ind. Electron.* **2013**, *60*, 2957–2968. [CrossRef]

5. Negahdari, A.; Yepes, A.G.; Doval-Gandoy, J.; Toliyat, H.A. Efficiency Enhancement of Multiphase Electric Drives at Light-Load Operation Considering Both Converter and Stator Copper Losses. *IEEE Trans. Power Electron.* **2019**, *34*, 1518–1525. [CrossRef]
6. Liu, Z.; Li, Y.; Zheng, Z. A review of drive techniques for multiphase machines. *CES Trans. Electr. Mach. Syst.* **2018**, *2*, 243–251. [CrossRef]
7. Diana, M.; Ruffo, R.; Guglielmi, P. PWM Carrier Displacement in Multi-N-Phase Drives: An Additional Degree of Freedom to Reduce the DC-Link Stress. *Energies* **2018**, *11*, 443. [CrossRef]
8. Zheng, P.; Sui, Y.; Zhao, J.; Tong, C.; Lipo, T.A.; Wang, A. Investigation of a Novel Five-Phase Modular Permanent-Magnet In-Wheel Motor. *IEEE Trans. Magn.* **2011**, *47*, 4084–4087.10.1109/TMAG.2011.2150207. [CrossRef]
9. Cao, W.; Mecrow, B.C.; Atkinson, G.J.; Bennett, J.W.; Atkinson, D.J. Overview of Electric Motor Technologies Used for More Electric Aircraft (MEA). *IEEE Trans. Ind. Electron.* **2012**, *59*, 3523–3531. [CrossRef]
10. Bozhko, S.; Hill, C.I.; Yang, T. More-Electric Aircraft: Systems and Modeling. In *Wiley Encyclopedia of Electrical and Electronics Engineering*; American Cancer Society: New York, NY, USA, 2018; pp. 1–31. [CrossRef]
11. Wheeler, P.W.; Clare, J.C.; Trentin, A.; Bozhko, S. An overview of the more electrical aircraft. *J. Aerosp. Eng.* **2012**, *227*, 578–585. [CrossRef]
12. Deng, Q.; Wang, Z.; Chen, C.; Czarkowski, D.; Kazimierczuk, M.K.; Zhou, H.; Hu, W. Modeling and Control of Inductive Power Transfer System Supplied by Multiphase Phase-Controlled Inverter. *IEEE Trans. Power Electron.* **2019**, *34*, 9303–9315. [CrossRef]
13. Belkhode, S.; Jain, S. Optimized Switching PWM Technique With Common-Mode Current Minimization for Five-Phase Open-End Winding Induction Motor Drives. *IEEE Trans. Power Electron.* **2019**, *34*, 8971–8980. [CrossRef]
14. Liu, Z.; Zheng, Z.; Xu, L.; Wang, K.; Li, Y. Current Balance Control for Symmetrical Multiphase Inverters. *IEEE Trans. Power Electron.* **2016**, *31*, 4005–4012. [CrossRef]
15. Prieto, B. Design and Analysis of Fractional-Slot Concentrated-Winding Multiphase Fault-Tolerant Permanent Magnet Synchronous Machines. Ph.D. Thesis, Tecnum Universidad de Navarra, Donostia, Spain, 2015.
16. Iqbal, A.; Rahman, K.; Abdallah, A.A.; Moin, K.A.S.; Abdellah, K. Current Control of a Five-phase Voltage Source Inverter. In Proceedings of the International Conference on Power Electronics and their Applications (ICPEA), Elazig, Turkey, 27 September 2013.
17. Bennett, J.W.; Atkinson, G.J.; Mecrow, B.C.; Atkinson, D.J. Fault-Tolerant Design Considerations and Control Strategies for Aerospace Drives. *IEEE Trans. Ind. Electron.* **2012**, *59*, 2049–2058. [CrossRef]
18. Bojoi, R.; Cavagnino, A.; Tenconi, A.; Tessarolo, A.; Vaschetto, S. Multiphase electrical machines and drives in the transportation electrification. In Proceedings of the IEEE 1st International Forum on Research and Technologies for Society and Industry Leveraging a better tomorrow (RTSI), Torino, Italy, 18 September 2015. [CrossRef]
19. Takahashi, S.; Ogasawara, S.; Takemoto, M.; Orikawa, K.; Tamate, M. Common-Mode Voltage Attenuation of an Active Common-Mode Filter in a Motor Drive System Fed by a PWM Inverter. *IEEE Trans. Ind. Appl.* **2019**, *55*, 2721–2730. [CrossRef]
20. Karampuri, R.; Jain, S.; Somasekhar, V.T. Common-Mode Current Elimination PWM Strategy Along with Current Ripple Reduction for Open-Winding Five-Phase Induction Motor Drive. *IEEE Trans. Power Electron.* **2019**, *34*, 6659–6668. [CrossRef]
21. Espina, J.; Balcells, J.; Arias, A.; Ortega, C. Common mode EMI model for a direct matrix converter. *IEEE Trans. Ind. Electron.* **2011**, *58*, 5049–5056. [CrossRef]
22. Muetze, A.; Binder, A. Don't lose your bearings. *IEEE Ind. Appl. Mag.* **2006**, *12*, 22–31. [CrossRef]
23. Hadden, T.; Jiang, J.W.; Bilgin, B.; Yang, Y.; Sathyan, A.; Dadkhah, H.; Emadi, A. A Review of Shaft Voltages and Bearing Currents in EV and HEV Motors. In Proceedings of the Industrial Electronics Society (IECON), Florence, Italy, 24 October 2016; pp. 1578–1583.
24. Mütze, A. Thousands of hits: On inverter-induced bearing currents, related work, and the literature. *Elektrotechnik Inf.* **2011**, *128*, 382–388. [CrossRef]
25. Muetze, A. On a New Type of Inverter-Induced Bearing Current in Large Drives With One Journal Bearing. *IEEE Trans. Ind. Appl.* **2010**, *46*, 240–248. [CrossRef]

26. Morya, K.; Gardner, M.C.; Anvari, B.; Liu, L.; Yepes, A.G.; Doval-Gandoy, J.; Toliyat, H.A. Wide Bandgap Devices in AC Electric Drives: Opportunities and Challenges. *IEEE Trans. Transp. Electrif.* **2019**, *5*, 3–20. [CrossRef]
27. Oh, W.; Willwerth, A. Shaft Grounding—A Solution to Motor Bearing Currents. *Am. Soc. Heating Refrig. Air Cond. Eng. Trans.* **2008**, *114*, 246–251.
28. Muetze, A.; Binder, A. Calculation of influence of insulated bearings and insulated inner bearing seats on circulating bearing currents in machines of inverter-based drive systems. *IEEE Trans. Ind. Appl.* **2006**, *42*, 965–972. [CrossRef]
29. Muetze, A. Bearing Currents in Inverter-Fed AC-Motors. Ph.D. Thesis, Der Technischen Universitaet Darmstadt, Darmstadt, Germany, 2004.
30. Schiferl, R.F.; Melfi, M.J. Bearing current remediation options. *IEEE Ind. Appl. Mag.* **2004**, *10*, 40–50. [CrossRef]
31. Muetze, A.; Oh, W. Application of Static Charge Dissipation to Mitigate Electric Discharge Bearing Currents. *IEEE Trans. Ind. Appl.* **2008**, *44*, 135–143. [CrossRef]
32. Nguyen, N.; Nguyen, T.; Lee, H. A reduced siwtching loss PWM strategy to eliminate common-mode voltage in multilevel inverters. *IEEE Trans. Power Electron.* **2015**, *30*, 5425–5438. [CrossRef]
33. Syed, A.; Kalyani, S. Evaluation of single phase transformerless photovoltaic inverters. *Electr. Electron. Eng. Int. J.* **2015**, *4*, 25–39. [CrossRef]
34. Freddy, T.K.S.; Rahim, N.A.; Hew, W.P.; Che, H.S. Comparison and Analysis of Single-Phase Transformerless Grid-Connected PV Inverters. *IEEE Trans. Power Electron.* **2014**, *29*, 5358–5369. [CrossRef]
35. Freddy, T.K.S.; Rahim, N.A.; Hew, W.P.; Che, H.S. Modulation Techniques to Reduce Leakage Current in Three-Phase Transformerless H7 Photovoltaic Inverter. *IEEE Trans. Ind. Electron.* **2015**, *62*, 322–331. [CrossRef]
36. Kouro, S.; Leon, J.I.; Vinnikov, D.; Franquelo, L.G. Grid-Connected Photovoltaic Systems: An Overview of Recent Research and Emerging PV Converter Technology. *IEEE Ind. Electron. Mag.* **2015**, *9*, 47–61. [CrossRef]
37. Fernandez, M.; Robles, E.; Kortabarria, I.; Andreu, J.; Ibarra, E. Novel modulation techniques to reduce the common mode voltage in multiphase inverters. In Proceedings of the IEEE Industrial Electronics Society Conference (IECON), Lisabon, Portugal, 14 October 2019; pp. 1898–1903.
38. Durán, M.J.; Prieto, J.; Barrero, F.; Riveros, J.A.; Guzman, H. Space-Vector PWM with Reduced Common-Mode Voltage for Five-Phase Induction Motor Drives. *IEEE Trans. Ind. Electron.* **2013**, *60*, 4159–4168. [CrossRef]
39. Rahman, A. *Power Electronics and Motor Drives*; CRC Press: Boca Raton, FL, USA, 2016; pp. 5–10.
40. Gray, R. Toeplitz and Circulant Matrices: A Review. *Found. Trends Commun. Inf. Theory* **2006**, *2*, 155–239. [CrossRef]
41. Semail, E; Bouscayrol, A.; Hautier, P.J. Vectorial formalism for analysis and design of polyphase synchronous machines. *Eur. Phys. J. Appl. Phys.* **2003**, *22*, 207–220. [CrossRef]
42. Tang, K.T. *Mathematical Methods for Engineers and Scientists 1*; Springer: Berlin, Germany, 2006.
43. Davis, P. *Circulant Matrices*; John Wiley & Sons Inc.: Hoboken, NJ, USA, 1979.
44. Zhar, H.; Gong, J.; Semail, E.; Scuiller, F. Comparison of Optimized Control Strategies of a High-speed Traction Machine with Five Phases and Bi-Harmonic Electromotive Force. *Energies* **2016**, *9*, 952. [CrossRef]
45. Kelly, W.; Strangas, E.G.; Miller, J.M. Multiphase Space Vector Pulse Width Modulation. *IEEE Power Eng. Rev.* **2002**, *22*, 53–53. [CrossRef]
46. Iqbal, A.; Levi, E. Space vector modulation schemes for a five-phase voltage source inverter. In Proceedings of the European Conference on Power Electronics and Applications (ECPEA), Barcelona, Spain, 10 September 2005; pp. 1–12. [CrossRef]
47. Yen-Shin, L.; Po-Sheng, C.; Hsiang-Kuo, L.; Chou, J. Optimal common-mode voltage reduction PWM technique for inverter control with consideration of the dead-time effects-part II: Applications to IM drives with diode front end. *IEEE Trans. Ind. Appl.* **2004**, *40*, 1613–1620. [CrossRef]
48. Iqbal Alammari, R.; Mosa, M.; Abu-Rub, H. Finite set model predictive current control with reduced and constant common mode voltage for a five-phase voltage source inverter. In Proceedings of the International Symposium on Industrial Electronics (ISIE), Istambul, Turkey, 4 June 2014; pp. 479–484. [CrossRef]

49. Munim, N.A.; Ismail, M.F.; Abidin, A.F.; Haris, H.M. Multi-phase inverter Space Vector Modulation. In Proceedings of the International Power Engineering and Optimization Conference (PEOCO), Langkawi, Malaysia, 22 July 2013; pp. 149–154. [CrossRef]
50. Robles, E.; Fernandez, M.; Ibarra, E.; Andreu, J.; Kortabarria, I. Mitigation of Common Mode Voltage Issues in Electric Vehicle Drive Systems by Means of an Alternative AC-Decoupling Power Converter Topology. *Energies* **2019**, *12*, 3349. [CrossRef]
51. Sadigh, A.K.; Dargahi, V.; Corzine, K. Analytical determination of conduction power loss and investigation of switching power loss for modified flying capacitor multicell converters. *IET Power Electron.* **2016**, *9*, 175–187. [CrossRef]
52. Wintrich, A.; Nicolai, U.; Tursky, W.; Reimann, T. *Application Manual Power Semiconductors*; Semikron: Nürnberg, Germany, 2017.
53. Matallana, A.; Robles, E.; Ibarra, E.; Andreu, J.; Delmonte, N.; Cova, P. A methodology to determine reliability issues in automotive SiC power modules combining 1D and 3D thermal simulations under driving cycle profiles. *Microelectron. Reliab.* **2019**, *102*, 1–9. [CrossRef]
54. Parsa, L.; Toliyat, H. Five-Phase Permanent-Magnet Motor Drives. *IEEE Trans. Ind. Appl.* **2005**, *41*, 30–37. [CrossRef]

Article

A Novel Vector Control Strategy for a Six-Phase Induction Motor with Low Torque Ripples and Harmonic Currents

Hamidreza Heidari [1,*], Anton Rassõlkin [1], Toomas Vaimann [1], Ants Kallaste [1], Asghar Taheri [2], Mohammad Hosein Holakooie [2] and Anouar Belahcen [1,3]

1 Department of Electrical Power Engineering and Mechatronics, Tallinn University of Technology, 19086 Tallinn, Estonia; anton.rassolkin@taltech.ee (A.R.); toomas.vaimann@taltech.ee (T.V.); ants.kallaste@taltech.ee (A.K.); anouar.belahcen@aalto.fi (A.B.)
2 Department of Electrical Engineering, University of Zanjan, Zanjan 45371-38791, Iran; taheri@znu.ac.ir (A.T.); hosein.holakooie@znu.ac.ir (M.H.H.)
3 Department of Electrical Engineering, Aalto University, 11000 Aalto, Finland
* Correspondence: haheid@taltech.ee; Tel.:+372-5613-9797

Received: 25 January 2019; Accepted: 19 March 2019; Published: 21 March 2019

Abstract: In this paper, a new vector control strategy is proposed to reduce torque ripples and harmonic currents represented in switching table-based direct torque control (ST-DTC) of a six-phase induction motor (6PIM). For this purpose, a new set of inputs is provided for the switching table (ST). These inputs are based on the decoupled current components in the synchronous reference frame. Indeed, using both field-oriented control (FOC) and direct torque control (DTC) concepts, precise inputs are applied to the ST in order to achieve better steady-state torque response. By applying the duty cycle control strategy, the loss subspace components are eliminated through a suitable selection of virtual voltage vectors. Each virtual voltage vector is based on a combination of a large and a medium vector to make the average volt-seconds in loss subspace near to zero. Therefore, the proposed strategy not only notably reduces the torque ripples, but also suppresses the low frequency current harmonics, simultaneously. Simulation and experimental results clarify the high performance of the proposed scheme.

Keywords: direct torque control; duty cycle control; harmonic currents; six-phase induction motor; torque ripple

1. Introduction

With the emergence of power electronic devices and adjustable-speed drives, multiphase machines have attracted wide attention for special applications in the naval, automotive, and aerospace industries [1,2]. The key features of these machines are high reliability, fault tolerant operation, low rate of power switches, low torque pulsation and low dc-link voltage [3–7]. Among multiphase motors, multiple three-phase winding motors have received more interest due to their advantage of compatibility with conventional three-phase technology. Considering this benefit, the asymmetrical 6PIM, which is composed of two sets of three-phase windings spatially shifted by 30 electrical degrees, seems desirable for many applications.

Direct Torque Control (DTC) is a simple and powerful scheme for variable speed 6PIM drives that provides high-performance torque and stator flux control [8]. However, it suffers from some serious drawbacks, including high torque ripples and low-frequency current harmonics, which can strikingly degrade the performance of the drive system [9,10]. A great deal of effort has been invested in alleviating DTC high torque ripples, which mostly includes the modification of the hysteresis controller [11,12], amending the switching table (ST) [13,14], or replacing hysteresis controllers with other control strategies to provide Pulse Width Modulation (PWM)-based DTC [15]. A global minimum torque ripple using modified switching pattern has been proposed in [16]. In [17], the torque ripples have been reduced by applying active and zero voltage vectors in each sampling period using a predictive DTC. With regard to a large number of voltage vectors in six-phase voltage source inverter (VSI), elimination of loss subspace components seems possible through the vector space decomposition (VSD) model. Aiming to reduce harmonic currents, elimination of $z_1 - z_2$ (loss subspace) components using the duty cycle concept in DTC is proposed in [9,12] for a five-phase induction motor, and a six-phase permanent magnet motor, respectively. Moreover, harmonic currents have been reduced by adding an inductance filter to the 6PIM drive system [18]. On the other hand, structure reconfiguration of 6PIM to minimize both harmonic currents and torque ripples has been done in [4]. On the other hand, field-oriented control (FOC) can be easily applied to many types of electrical machines [19].

The main focus of this paper is on the parallel torque ripple and harmonic current reduction in vector control of a 6PIM. To achieve these goals, a new vector control scheme is proposed, using a combination of DTC FOC concepts to moderate steady-state torque ripples. To reduce harmonic currents, twelve virtual voltage vectors are introduced by combination of large and single medium voltage vectors (e.g., 48 and 57 in Figure 1a, respectively). The duration of each voltage vector is determined such that the average volt-seconds in the $z_1 - z_2$ subspace becomes near to zero. Consequently, low-frequency current harmonics experience a considerable reduction, current will be much smoother, and the efficiency of the drive system will be increased.

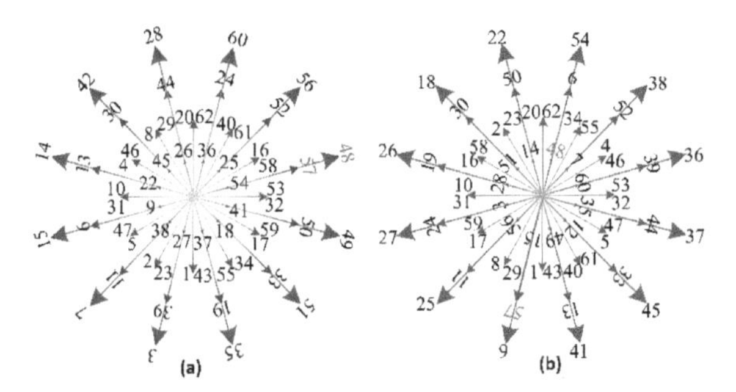

Figure 1. 6PIM model in (**a**) $\alpha - \beta$, (**b**) $z_1 - z_2$ bspaces.

The rest of this paper is organized as follows: in Section 2, 6PIM modeling is presented. The conventional DTC and its drawbacks such as torque ripples and harmonic currents are discussed in Section 3. The proposed method to reduce torque ripples and harmonic currents is presented in Section 4.

2. 6PIM Modelling

There are two common methods for modelling of 6PIMs: double $d - q$ [20] and VSD [21]. According to the VSD approach, the machine's parameters are mapped into active and loss subspaces, which makes the control strategy more convenient and efficient. Moreover, the VSD method can easily be extended to other types of motors. In this study, the 6PIM modelling is based on the VSD strategy. According to this modelling technique, a 6PIM with near-sinusoidal distributed windings is modelled in three orthogonal subspaces, which are commonly named as the $\alpha - \beta$, $z_1 - z_2$, and $o_1 - o_2$ subspaces. The produced voltage space vectors by switching states of a two-level six-phase voltage source inverter in the $\alpha - \beta$, $z_1 - z_2$ subspaces are shown in Figure 1. Among these subspaces, only the $\alpha - \beta$ components share useful electromechanical energy conversion, while $z_1 - z_2$ and $o_1 - o_2$ components do not generate any electromechanical energy in the air-gap and just produce losses. The fundamental components and also harmonics of the order $12n \pm 1 (n = 1, 2, 3 \ldots)$ are mapped into the $\alpha - \beta$ subspace. The losses of 6PIM are mapped into the $z_1 - z_2$ and $o_1 - o_2$ subspaces which include harmonics by the order of $6n \pm 1(n = 1,3,5\ldots)$ and $3n(n = 1,2,3\ldots)$, respectively. By the assumption that the stator mutual leakage inductance is ignored, the components of $z_1 - z_2$ and $o_1 - o_2$ subspaces have the same form [21]. Since the active and loss components are investigated separately, it is clear that the control of 6PIM will be more efficient by using VSD. Additionally, isolation of the neutral points of two three-phase windings, makes the $o_1 - o_2$ subspace losses become zero [21].

2.1. 6PIM Model in $\alpha - \beta$ Subspace

As already mentioned, only the $\alpha - \beta$ subspace components contribute to the electrical energy conversion. Using the VSD strategy, the normal six-dimensional electrical components of the 6PIM are mapped into the $\alpha - \beta$, $z_1 - z_2$, and $o_1 - o_2$ subspaces by an appropriate matrix named T_6, which is presented in Appendix A. The $\alpha - \beta$ voltage equations in the stationary reference frame are as follows:

$$\begin{cases} \overline{V_s} = R_S\overline{I_s} + \rho\overline{\Psi_s} \\ 0 = R_r\overline{I_r} + \rho\overline{\Psi_r} - j\omega_r\overline{\Psi_r} \end{cases} \tag{1}$$

where, $\overline{V}_s = v_{s\alpha} + jv_{s\beta}$, $\overline{I}_s = i_{s\alpha} + ji_{s\beta}$, $\overline{I}_r = i_{r\alpha} + ji_{r\beta}$, $\overline{\Psi}_s = \psi_{s\alpha} + j\psi_{s\beta}$, $\overline{\Psi}_r = \psi_{r\alpha} + j\psi_{r\beta}$, R_S is the stator resistance, R_r is the rotor resistance, ω_r is the angular speed, and ρ is the derivative operator. The stator flux linkage ($\overline{\Psi}_s$) and rotor flux linkage ($\overline{\Psi}_r$) can be expressed as:

$$\begin{bmatrix} \overline{\Psi}_s \\ \overline{\Psi}_r \end{bmatrix} = \begin{bmatrix} L_s & M \\ M & L_r \end{bmatrix} \begin{bmatrix} \overline{I}_s \\ \overline{I}_r \end{bmatrix} \tag{2}$$

where, L_s, L_r, and M are the stator, rotor and magnetizing inductances, respectively.

2.2. 6PIM Model in $z_1 - z_2$ Subspace

The 6PIM model in the $z_1 - z_2$ subspace behaves as a passive resistor–inductor (R–L) circuit as:

$$\begin{bmatrix} V_{sz_1} \\ V_{sz_2} \end{bmatrix} = \begin{bmatrix} R_s + \rho L_{ls} & 0 \\ 0 & R_s + \rho L_{ls} \end{bmatrix} \begin{bmatrix} I_{sz_1} \\ I_{sz_2} \end{bmatrix} \tag{3}$$

where, L_{ls} is the stator leakage inductance. In this paper, 6PIM is applied with two isolated neutral points, with which this structure prevents the zero sequence currents. Hence, the $o_1 - o_2$ components can be neglected.

3. Conventional DTC of 6PIM

In a six-phase voltage source inverter (VSI), there are $2^6 = 64$ switching states. Each state produces a voltage space vector (defined as V_k) in the $\alpha - \beta$ or $z_1 - z_2$ subspaces, shown in Figure 1. As can be seen, there are 12 large (e.g., 48), 12 single medium (e.g., 57), 24 double medium (e.g., 53), 12 small (e.g., 54), and 4 null voltage vectors. The block diagram of conventional DTC is shown in Figure 2.

The stator flux in this approach is obtained as:

$$\overline{\Psi}_s = \int (\overline{V}_s - R_s\overline{I}_s)dt \tag{4}$$

The electromagnetic torque can be calculated using the stator flux and current as:

$$T_e = 1.5\ P(\overline{\Psi}_s \cdot \overline{I}_s^*) \tag{5}$$

where, P is the number of pole pairs. The reference values of the stator flux and electromagnetic torque are compared with the estimated ones, and the errors are applied to the hysteresis controller.

The outputs of the hysteresis regulators denote the signs of torque and flux change. In order to minimize the errors, the optimum vector is selected through ST, which is tabulated in Table 1.

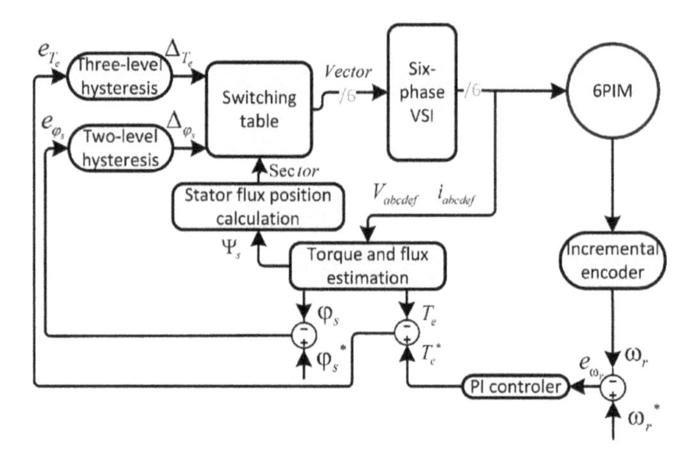

Figure 2. Conventional DTC block diagram.

In this table, k is the number of the sector. V_k is the applied voltage vector to the inverter, which is defined as binary numbers in switching states of VSI as in Table 2.

Table 1. ST of conventional DTC.

Ψ_s/sector k	Δ_{T_e}=1	Δ_{T_e}=0	Δ_{T_e}=−1
$\Delta_{\phi_s} = 1$	V_{k+1}	V_0	V_{k+10}
$\Delta_{\phi_s} = -1$	V_{k+4}	V_0	V_{k+7}

Table 2. Selected vectors in ST of conventional DTC.

V_0	V_1	V_2	V_3	V_4	V_5	V_6	V_7	V_8	V_9	V_{10}	V_{11}	V_{12}
0, 21, 42, 63	48	56	60	28	12	14	15	7	3	35	57	49

In the conventional DTC, only the large voltage vectors in the $\alpha - \beta$ subspace are applied to the 6PIM to maximize the utilization of the dc-link. From Equation (4), it can be seen that the stator flux variations and the applied voltage vectors have the same direction. Hence, the changes in the stator flux depend on the applied voltage vectors. Compared to the stator time constant, the rotor time constant is very large. Therefore, the rotor flux linkage changes are negligible and it can be assumed constant during short transients [22]. By the application of the active voltage vectors, stator flux linkage vector will be moved away from rotor flux linkage vector and the angle between them will be greater. This leads to changes in torque according to Equation (5).

4. Harmonic Currents Reduction by Duty Cycle Control Strategy

From Figure 1, it can be seen that each voltage vector in the $\alpha - \beta$ subspace has a corresponding vector in the $z_1 - z_2$ subspace with different position and magnitude. It is recommended to make the average volt-second outcome in the $z_1 - z_2$ subspace near to zero. Accordingly, two voltage vectors have been applied to the inverter in each sampling period. Active voltage vectors should be in a same direction (in order to have high effect on torque) and their correspondents in $z_1 - z_2$ subspace should be in an opposite direction (in order to have less losses). Therefore, the selected vectors will produce high outcome in $\alpha - \beta$ subspace and low outcome in the $z_1 - z_2$ subspace.

The applied voltage vectors in the $\alpha-\beta$ subspace are expressed as:

$$\begin{cases} V_{M_{\alpha-\beta}} = \frac{\sqrt{2}}{3}V_{dc} \\ V_{L_{\alpha-\beta}} = \frac{\sqrt{6}+\sqrt{2}}{6}V_{dc} \end{cases} \tag{6}$$

where, $V_{M_{\alpha-\beta}}$ and $V_{L_{\alpha-\beta}}$ are single-medium and large voltage vectors in the $\alpha-\beta$ subspace. A suitable duty ratio is calculated as:

$$\begin{cases} \left|V_{M_{\alpha-\beta}}T_{M_{\alpha-\beta}}\right| = \left|V_{L_{\alpha-\beta}}T_{L_{\alpha-\beta}}\right| \\ T_{M\alpha-\beta} + T_{L_{\alpha-\beta}} = T_s \end{cases} \Rightarrow \begin{cases} T_{L_{\alpha-\beta}} = 0.73\ T_s \\ T_{M_{\alpha-\beta}} = 0.27\ T_s \end{cases} \tag{7}$$

where, $T_{L_{\alpha-\beta}}$, and $T_{M_{\alpha-\beta}}$ are the duration of the large and single-medium voltage vectors application in the $\alpha-\beta$ subspace, and T_s is the sampling period. In this way, the losses in the z_1-z_2 subspace are reduced strikingly, while the reduction in the electromagnetic components is subtle. For instance, vectors number 48 and 57 have the same direction in the $\alpha-\beta$ subspace and the opposite direction in the z_1-z_2 subspace. These voltage vectors are applied to the motor as illustrated in Figure 3, where g_n is the number of legs in six-phase VSI, and K_v is the duty ratio defined as:

$$K_v = 1 - \frac{\left|V_{M_{\alpha-\beta}}\right|}{\left|V_{L_{\alpha-\beta}}\right|} = 0.27 \tag{8}$$

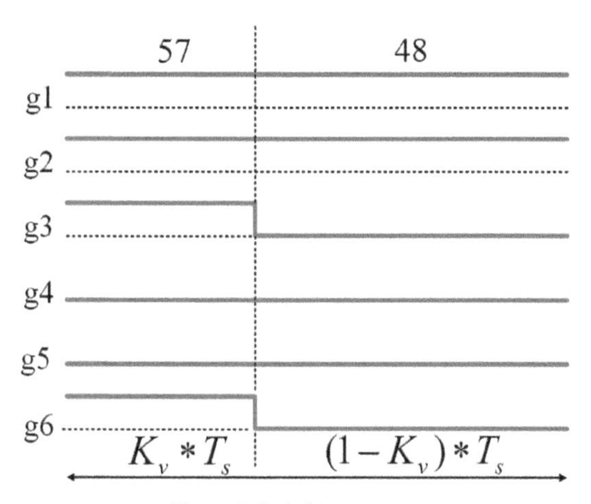

Figure 3. Switching pattern.

If $g_n = 1$ the upper switch is on and the lower switch is off. On the contrary, when $g_n = 0$, the lower switch becomes on and the upper switch turns off. The compound of these vectors in each sampling period is named virtual vector, shown in Figure 4.

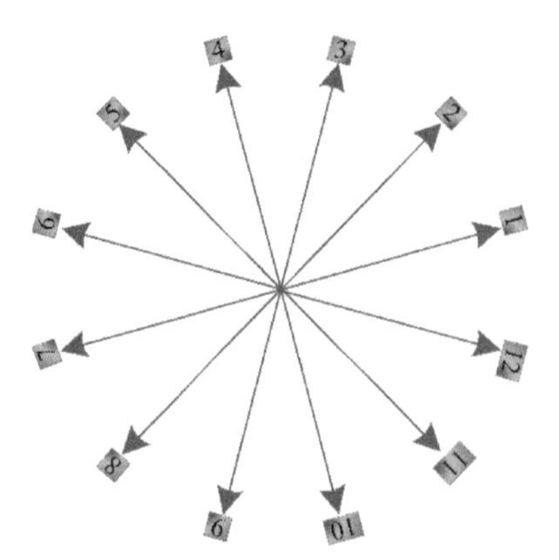

Figure 4. Virtual vectors in α-β subspace.

Figure 3 shows that in two legs (among the six legs) of the inverter, the switches' status has been changed. Therefore, by this method has more switching frequency against DTC. This increase in switching frequency is less than twice. The vectors in the $\alpha - \beta$ subspace are replaced by virtual vectors in ST shown in Table 2.

5. Proposed Control Algorithm for the 6PIM Drive

Using FOC framework [23], 6PIM's mathematical equations are transformed to the synchronous reference frame $(d - q)$, which creates possibility of decoupled control of the torque and flux as a permanent-magnet separated-excitation *dc* motor. In this reference frame, the stator flux vector is located on *d*-axis which is shown in Figure 5.

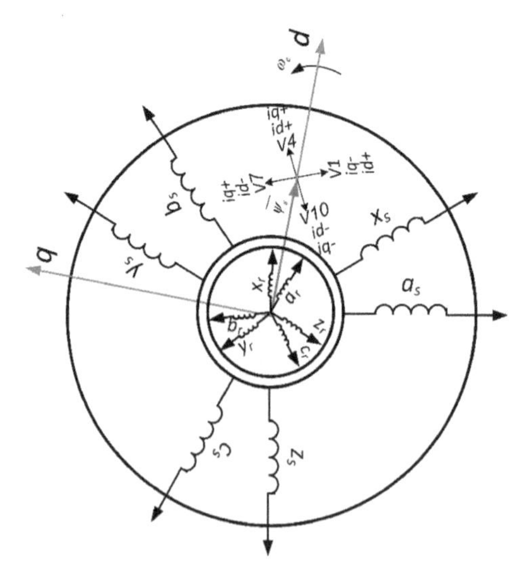

Figure 5. 6PIM structure and stator flux vector in 3rd sector and applied inverter voltage vectors.

Orthogonal currents of the 6PIM are mapped into the synchronous reference frame using Park transformation based on the flux vector position, which is achieved by field orientation process. For the

6PIM control, i_q is a torque -and i_d is a flux- producing components. Hence, the equations of the torque and flux are related to stator currents in the $d-q$ frame as follows:

$$\begin{cases} T_e \propto i_q \\ \lambda_s \propto i_d \end{cases} \tag{9}$$

In order to decrease the torque ripples in the 6PIM, a new approach is employed by modifying the ST's inputs. Since the inputs of the ST in the classical ST-DTC are the errors between command and actual values of the electromagnetic torque and the stator flux, it seems effective to use the errors between the set and actual values of the i_q, i_d, instead. In order to redesign the ST-DTC method to use these inputs, the inputs of ST in the conventional DTC are used as inputs for PI regulators. The outputs of PI regulators are the command values of the currents in the $d-q$ axis. Replacing Δ_{T_e}, Δ_{ϕ_s} by Δi_q, Δi_d in ST-DTC, respectively, the proposed method provides better inputs to the same ST presented in Table 1 which leads to a better performance in 6PIM. Table 3 shows that through defining virtual vectors, the ST is the same with conventional DTC with different inputs.

Table 3. The switching table of the proposed scheme.

Ψ_s/Sector k	Δ_{i_q}=1	Δ_{i_q}=0	Δ_{i_q}=−1
$\Delta_{i_d}=1$	V_{k+1}	V_0	V_{k+10}
$\Delta_{i_d}=-1$	V_{k+4}	V_0	V_{k+7}

Δ_{i_q} and Δ_{i_d} imply that changing the signs of the i_q and i_d is required. If i_q needs to be increased, then $\Delta_{i_q}=1$. If there is no i_q requirement, then $\Delta_{i_q}=0$. Also, $\Delta_{i_q}=-1$ denotes the decrease of i_q. All the states are defined as the same for the notation of Δ_{i_d}. These digital output signals of the hysteresis controllers are described as:

$$\begin{gathered} \Delta_{i_q}=1 \quad if \quad i_q \leq i_q{}^{\star}-|hysteresis\ band| \\ \Delta_{i_q}=0 \quad if \quad i_q=i_q{}^{\star} \\ \Delta_{i_q}=-1 \quad if \quad i_q \geq i_q{}^{\star}+|hysteresis\ band| \end{gathered} \tag{10}$$

Similarly, for the changes required for the d-axis of the stator current, Δi_d is described as:

$$\begin{gathered} \Delta_{i_d}=1 \quad if \quad i_d \leq i_d{}^{\star}-|hysteresis\ band| \\ \Delta_{i_d}=0 \quad if \quad i_d \geq i_d{}^{\star}+|hysteresis\ band| \end{gathered} \tag{11}$$

The block diagram of the proposed control strategy is shown in Figure 6. To concurrently achieve low Total Harmonics Distortion)THD(of the motor currents and low torque ripples, the proposed vector control scheme is synthesized with the duty cycle control strategy. The ST applies two voltage vectors in each sampling period in order to eliminate z_1-z_2 subspace components. In comparison with the conventional DTC, the switching frequency of the proposed scheme is increased (less than twice according to Figure 3.) because two voltage vectors are applied in each sampling period. In contrary, both harmonic currents and torque ripples are reduced. Moreover, the proposed scheme has fast dynamic response, similar to the conventional DTC, and does not need any PWM modulator that creates complexity and time delay.

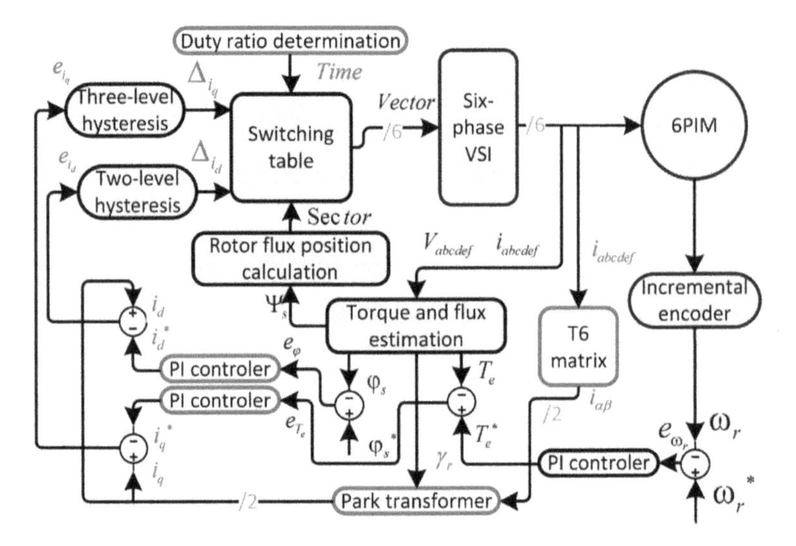

Figure 6. Block diagram of the proposed control scheme for 6PIM.

6. Simulation Results

The proposed and duty cycle control methods are simulated in MATLAB/Simulink. The sampling period of both methods are set to 100 μs. All the parameters are assumed to be constant, although each of them can be changed under the thermal effect, which is not within the scope of this essay. The simulations are carried out based on real specifications for 6PIM. The 6PIM parameters are specified in Table 4.

Table 4. 6PIM parameters.

Parameter, Unit	Value	Parameter, Unit	Value
Rated power, W	700	L_m, mH	588
Rated voltage, V	200	L_s, mH	603.3
Rated current, A	2	L_r, mH	604.4
Rated speed, rpm	1400	R_s, Ω	15.0
Frequency, Hz	50	R_r, Ω	7.91

The simulation results for the duty cycle and the proposed DTC strategies under load change from 0 to about 3.5 Nm at $t = 0.5$ s, speed command of 100 rad/s, and flux command of 0.5 Wb are shown in Figures 7 and 8, a speed torque and stator flux reference signals are shown with red dashed lines.

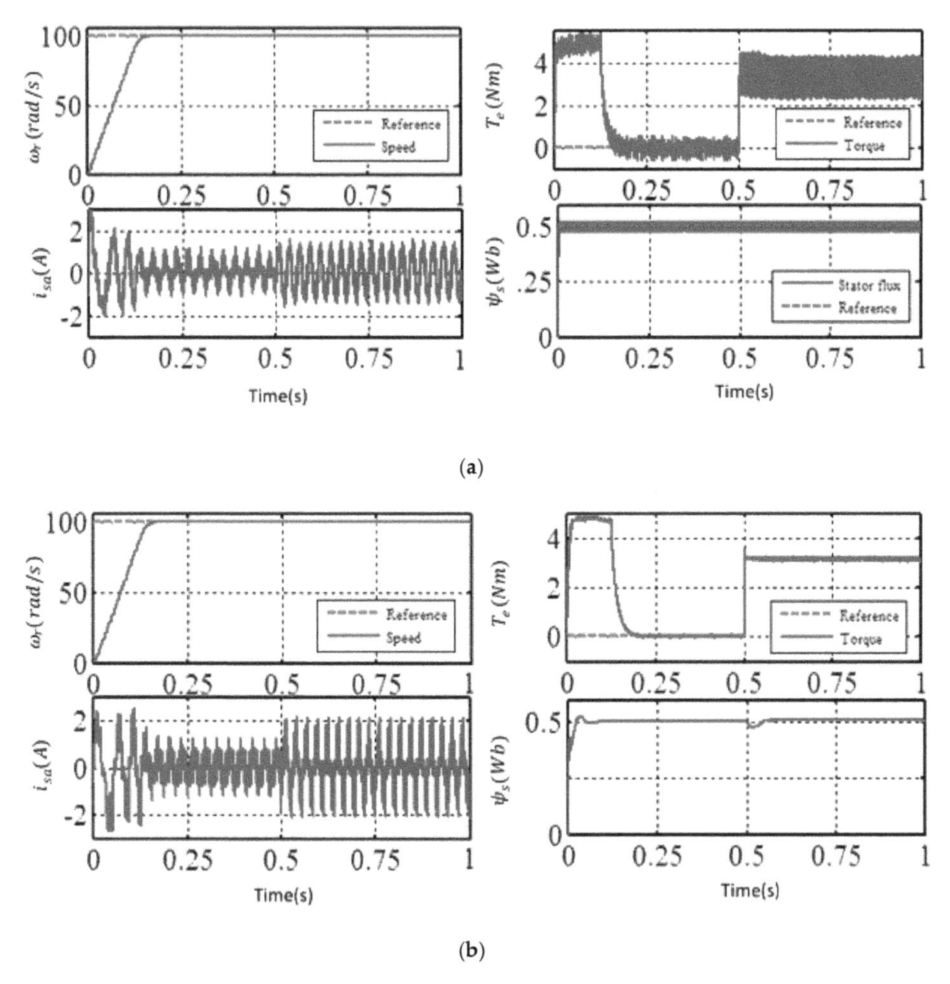

Figure 7. Load change condition of the 6PIM controlled by the duty cycle control strategy (**a**) and proposed method (**b**).

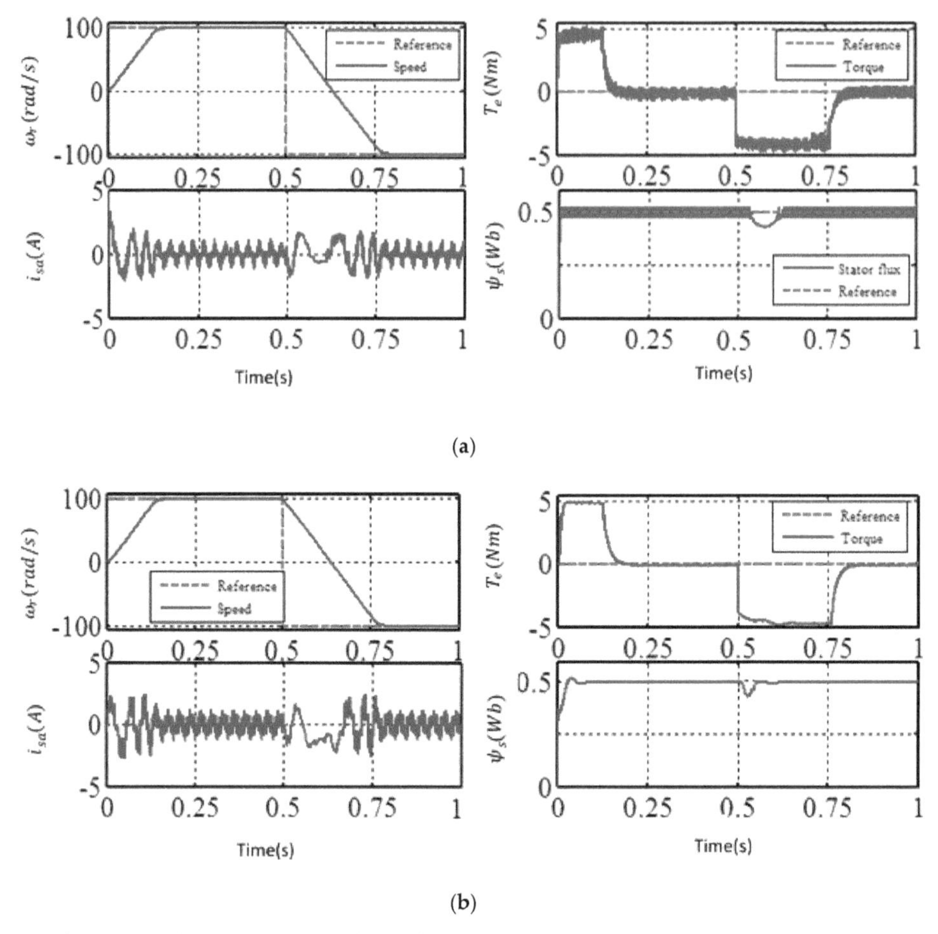

Figure 8. Speed direction change condition of the 6PIM controlled by the duty cycle control strategy (**a**) and proposed method (**b**).

As it can be seen from the both simulations (Figures 7 and 8), the torque ripples and stator flux fluctuations of proposed method is lower in compared to duty cycle control strategy. However, due to additional PI controller is used for stator flux, Equation (4), running time is higher.

7. Experimental Setup

In addition to the simulations, the performance of the proposed method is validated experimentally. Figure 9 shows the experimental setup, which contains the 6PIM and its coupled load, the main processor, two three-phase VSIs, current and voltage transducers, shaft encoder, and single-phase bridge rectifier.

The applied processor used in the driver is an eZDSP F28335 based on the floating point TMS320F28335 chip. The motor speed is measured by an Autonics incremental shaft encoder (Autonics, Busan, South Korea) mechanically coupled to the 6PIM with resolution of 2500 P/R. LEM LTS6np current transducers are implemented to measure all the phases' currents in order to be used in the estimation and the control processes. The DC-link voltage is also measured using $LV\ 25 - p$ voltage transducer. A DC generator is applied as load machine and a PCI-1716 data acquisition card (DAQ, Advantech, Milpitas, CA, USA) as an A/D converter. A 700-W, 24-stator slots three-phase squirrel-cage induction motor, which has been rewound to construct a 4-pole asymmetrical 6PIM is also tested for the proposed method performance. The MATLAB/Embedded Coder is used to generate usable code for the code composer studio development environment. The digital motor control and *IQmath* libraries along with *IQ17* data type are employed. The sampling period is set to $T_s = 100$ µs with a dead-time of 2 µs.

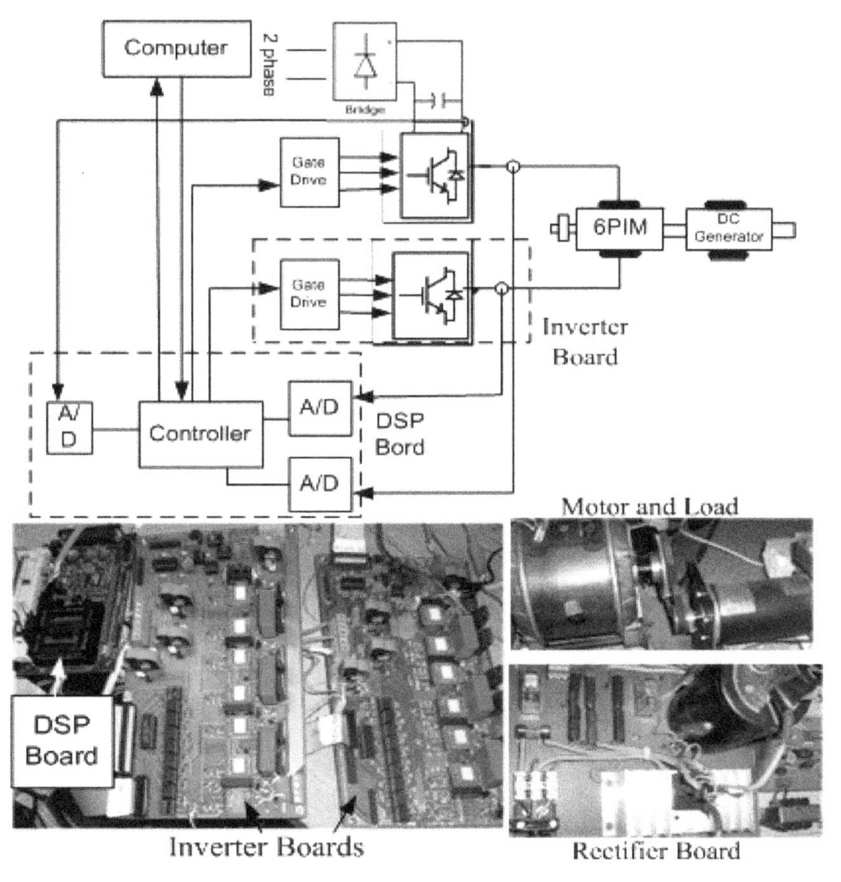

Figure 9. Experimental setup.

To demonstrate the torque ripples reduction precisely, the torque figures are shown within a short time frame in Figure 10.

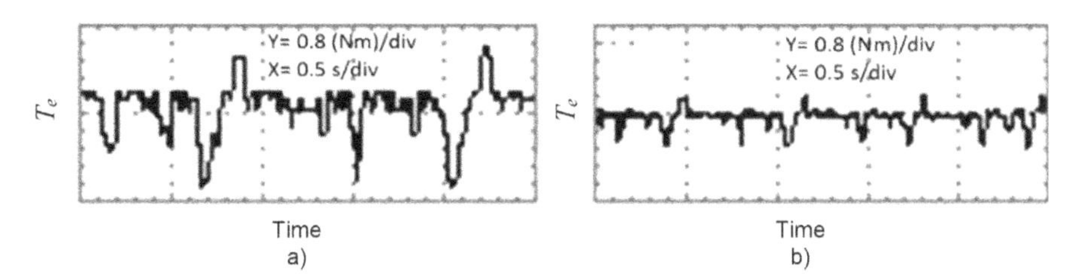

Figure 10. The torque response of the 6PIM in steady state with 4 Nm load, derived by (**a**) duty cycle control strategy; (**b**) proposed method.

The experimental results of the duty cycle control strategy and the proposed method under the load changing from 0 to about 3.5 N/m are shown in Figure 11. In this test, the speed and flux commands are 100 rad/s and 0.5 Wb, respectively. The provided tests illustrate the alleviation of torque ripples in the proposed method compared with the duty cycle control strategy.

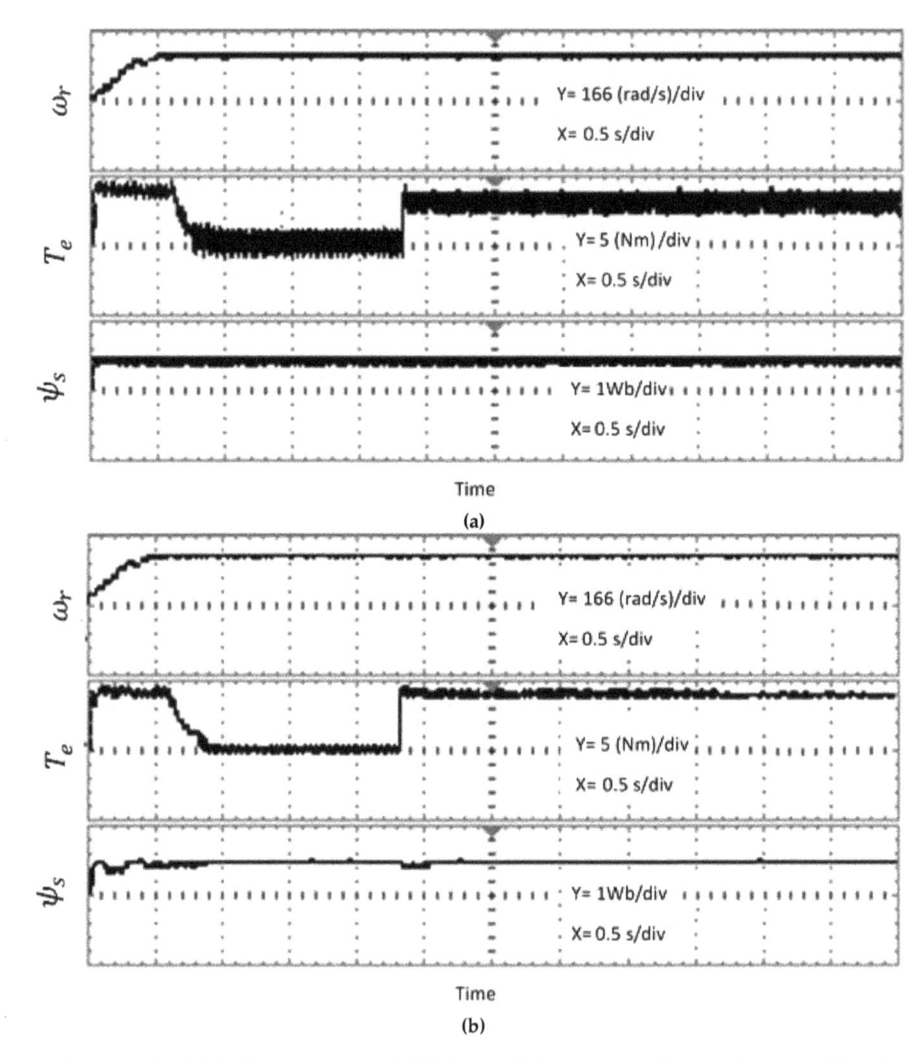

Figure 11. Load injection experiment of 6PIM, controlled by duty cycle control strategy (**a**) and proposed method (**b**).

In Table 5, the torque ripples in the no load condition are investigated to show the differences between the proposed method and the conventional DTC and the duty cycle control-based DTC strategies. It is clear that the torque ripples for the 6PIM is effectively decreased for the proposed method in comparison with other two methods. Furthermore, as it is seen from current THD in Figure 12, the low order harmonics of the stator currents, especially the fifth and seventh harmonics, are considerably reduced for the proposed control method.

Table 5. Torque ripples in three different conditions of 6PIM driving by three different methods.

Method	No-Load	Rated Load	50% Load
Conventional DTC	%45	%49	%47
Duty cycle control	%41	%37	%39
Proposed method	%9	%12	%10

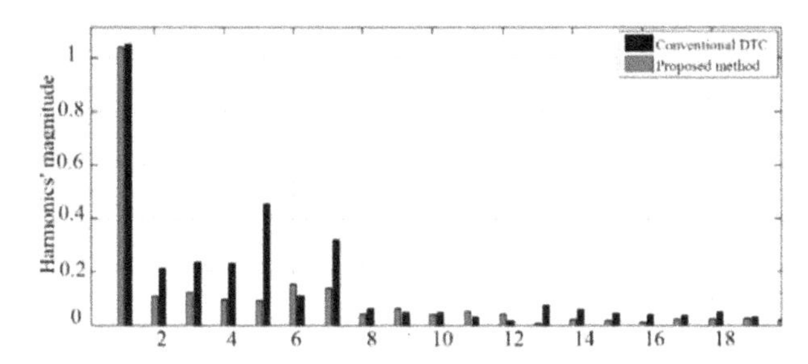

Figure 12. Current THD of the conventional DTC and proposed method.

8. Conclusions

In this paper the performance of the 6PIM was improved by a new vector control strategy. Using a new set of inputs for the ST in DTC method and applying the duty cycle control strategy leads to decrease in both torque ripples and harmonic currents. From a complexity viewpoint, the proposed technique falls between the DTC and FOC. This method is more simple compared with FOC strategy due to the absence of PWM algorithm, and has a fast dynamic similar to DTC strategy. The main limitation of the proposed technique is variable switching frequency compared with FOC. The effectiveness of the proposed control strategy was confirmed using both simulation and experimental tests.

Author Contributions: Methodology, validation and formal analysis, H.H.; writing—original draft H.H., A.T., M.H.H.; writing—review & editing, A.R., T.V. and A.K.; supervision, A.B.

Funding: This work was supported by the Estonian Research Council under Grants PUT1260.

Conflicts of Interest: The authors declare no conflict of interest.

Appendix A

$$T = \frac{1}{\sqrt{3}} \begin{bmatrix} 1 & \cos(4\gamma) & \cos(8\gamma) & \cos(\gamma) & \cos(5\gamma) & \cos(9\gamma) \\ 1 & \sin(4\gamma) & \sin(8\gamma) & \sin(\gamma) & \sin(5\gamma) & \sin(9\gamma) \\ 1 & \cos(8\gamma) & \cos(4\gamma) & \cos(5\gamma) & \cos(\gamma) & \cos(9\gamma) \\ 1 & \sin(8\gamma) & \sin(4\gamma) & \sin(5\gamma) & \sin(\gamma) & \sin(9\gamma) \\ 1 & 1 & 1 & 0 & 0 & 0 \\ 0 & 0 & 0 & 1 & 1 & 1 \end{bmatrix}$$

where, $\gamma = \alpha = \frac{\pi}{6}$

References

1. Levi, E. Multiphase electric machines for variable-speed applications. *IEEE Trans. Ind. Electron.* **2008**, *55*, 1893–1909. [CrossRef]
2. Holakooie, M.H.; Ojaghi, M.; Taheri, A. Direct Torque Control of Six-Phase Induction Motor with a Novel MRAS-Based Stator Resistance Estimator. *IEEE Trans. Ind. Electron.* **2018**, *65*, 7685–7696. [CrossRef]

3. Che, H.S.; Duran, M.J.; Levi, E.; Jones, M.; Hew, W.P.; Rahim, N.A. Postfault operation of an asymmetrical six-phase induction machine with single and two isolated neutral points. *IEEE Trans. Power Electron.* **2014**, *29*, 5406–5416. [CrossRef]
4. Jahns, T.M. Improved reliability in solid-state drives for large asynchronous ac machines by means of multiple independent phase-drive units. *IEEE Trans. Ind. Appl.* **1980**, *IA-16*, 321–331. [CrossRef]
5. Fnaiech, M.A.; Betin, F.; Capolino, G.A.; Fnaiech, F. Fuzzy logic and sliding-mode controls applied to six-phase induction machine with open phases. *IEEE Trans. Ind. Electron.* **2010**, *57*, 354–364. [CrossRef]
6. Parsa, L.; Toliyat, H.A.; Goodarzi, A. Five-phase interior permanent-magnet motors with low torque pulsation. *IEEE Trans. Ind. Appl.* **2007**, *43*, 40–46. [CrossRef]
7. Marouani, K.; Baghli, L.; Hadiouche, D.; Kheloui, A.; Rezzoug, A. A new PWM strategy based on a 24-sector vector space decomposition for a six-phase VSI-Fed dual stator induction motor. *IEEE Trans. Ind. Electron.* **2008**, *55*, 1910–1920. [CrossRef]
8. Pandit, J.K.; Aware, M.V.; Nemade, R.V.; Levi, E. Direct Torque Control Scheme for a Six-Phase Induction Motor with Reduced Torque Ripple. *IEEE Trans. Power Electron.* **2017**, *32*, 7118–7129. [CrossRef]
9. Zheng, L.; Fletcher, J.E.; Williams, B.W.; He, X. A novel direct torque control scheme for a sensorless five-phase induction motor drive. *IEEE Trans. Ind. Electron.* **2011**, *58*, 503–513. [CrossRef]
10. Lokriti, A.; Salhi, I.; Doubabi, S. IM Direct Torque Control with no flux distortion and no static torque error. *ISA Trans.* **2015**, *59*, 256–267. [CrossRef]
11. Ambrozic, A.; Buja, G.S.; Menis, R. Band-Constrained Technique for Direct Torque Control of Induction Motor. *IEEE Trans. Ind. Electron.* **2004**, *51*, 776–784. [CrossRef]
12. Ren, Y.; Zhu, Z.Q. Enhancement of steady-state performance in direct-torque-controlled dual three-phase permanent-magnet synchronous machine drives with modified switching table. *IEEE Trans. Ind. Electron.* **2015**, *62*, 3338–3350. [CrossRef]
13. Taheri, A.; Rahmati, A.; Kaboli, S. Comparison of efficiency for different switching tables in six-phase induction motor DTC drive. *J. Power Electron.* **2012**, *12*, 128–135. [CrossRef]
14. Singh, B.; Jain, S.; Dwivedi, S. Torque ripple reduction technique with improved flux response for a direct torque control induction motor drive. *IET Power Electron.* **2013**, *6*, 326–342. [CrossRef]
15. Lai, Y.S.; Chen, J.H. A new approach to direct torque control of induction motor drives for constant inverter switching frequency and torque ripple reduction. *IEEE Trans. Energy Convers.* **2001**, *16*, 220–227.
16. Shyu, K.K.; Lin, J.K.; Pham, V.T.; Yang, M.J.; Wang, T.W. Global minimum torque ripple design for direct torque control of induction motor drives. *IEEE Trans. Ind. Electron.* **2010**, *57*, 3148–3156. [CrossRef]
17. Abad, G.; Rodriguez, M.A.; Poza, J. Two-Level VSC Based Predictive Direct Torque Control of the Doubly Fed Induction Machine with Reduced Torque and Flux Ripples at Low Constant Switching Frequency. *IEEE Trans. Power Electron.* **2008**, *23*, 1050–1061. [CrossRef]
18. Wang, T.; Fang, F.; Wu, X.; Jiang, X. Novel filter for stator harmonic currents reduction in six-step converter fed multiphase induction motor drives. *IEEE Trans. Power Electron.* **2013**, *28*, 498–506. [CrossRef]
19. Autsou, S.; Saroka, V.; Karpovich, D.; Rassolkin, A.; Gevorkov, L.; Vaimann, T.; Kallaste, A.; Belahcen, A. Comparative study of field-oriented control model in application for induction and synchronous reluctance motors for life-cycle analysis. In Proceedings of the 2018 25th International Workshop on Electric Drives: Optimization in Control of Electric Drives (IWED), Moscow, Russia, 31 January–2 February 2018.
20. Nelson, R.H.; Krause, P.C. Induction machine analysis for arbitrary displacement between multiple winding sets. *IEEE Trans. Power Appar. Syst.* **1974**, *PAS-93*, 841–848. [CrossRef]
21. Zhao, Y.; Lipo, T.A. Space Vector PWM Control of Dual Three-phase Induction Machine Using Vector Space Decomposition. *IEEE Trans. Ind. Appl.* **1995**, *31*, 1100–1109. [CrossRef]
22. Binder, A.; Muetze, A. Scaling effects of inverter-induced bearing currents in AC machines. *IEEE Trans. Ind. Appl.* **2008**, *44*, 769–776. [CrossRef]
23. Holakooie, M.H.; Taheri, A.; Sharifian, M.B.B. MRAS based speed estimator for sensorless vector control of a linear induction motor with improved adaptation mechanisms. *J. Power Electron.* **2015**, *15*, 1274–1285. [CrossRef]

Article

Comparative Analysis of High Frequency Signal Injection Based Torque Estimation Methods for SPMSM, IPMSM and SynRM

Maria Martinez *, David Reigosa, Daniel Fernandez and Fernando Briz

Electrical, Electronic, Computers and Systems Engineering, University of Oviedo, 33204 Oviedo, Spain; diazdavid@uniovi.es (D.R.); fernandezalodaniel@uniovi.es (D.F.); fbriz@uniovi.es (F.B.)

* Correspondence: martinezgmaria@uniovi.es; Tel.: +34-985-103000 (ext. 6436)

Received: 31 December 2019; Accepted: 20 January 2020; Published: 28 January 2020

Abstract: Torque estimation in permanent magnet synchronous machines and synchronous reluctance machines is required in many applications. Torque produced by a permanent magnet synchronous machine depends on the permanent magnets' flux and *dq*-axes inductances, whereas torque in synchronous reluctance machines depends on the *dq*-axes inductances. Consequently, precise knowledge of these parameters is required for proper torque estimation. The use of high frequency signal both for permanent magnets' flux and *dq*-axes inductances estimation has been recently shown to be a viable option. This paper reviews the physical principles, implementation and performance of high-frequency signal injection based torque estimation for permanent magnet synchronous machines and synchronous reluctance machines.

Keywords: torque estimation; online parameters estimation; permanent magnet synchronous machines; synchronous reluctance machines; high frequency signal injection

---

1. Introduction

The use of Permanent Magnet Synchronous Machines (PMSMs) has substantially increased in industrial applications, electric traction, renewables energies, etc. due to their superior performance compared with other types of electrical machines, such as Induction Machines (IM), in terms of power density, torque density, dynamic response, wide speed range, simplicity of the control and efficiency. However, the high and often unpredictable price of rare-earth materials and the risk of demagnetisation due to excessive operating currents/temperatures are a concern for this type of machines. Synchronous Reluctance Machines (SynRM) have gained popularity over the last years as a viable alternative thanks to their lower cost and high tolerance to overcurrents and/or overheating and overall increased robustness, resulting from the absence of magnets in the rotor [1–7].

Torque production capability of PMSMs depends on the magnetisation state of the permanent magnets (PMs) and the saliency ratio, i.e., the difference between *d*- and *q*-axes inductances. Magnet strength and inductances can change during normal operation of the machine due to changes in the fundamental current and/or PM temperature [8–10]. An increase of the PM temperature reduces the PM remanent flux (i.e., magnetisation state), and consequently the machine torque for a given stator current. In addition, PM remanent flux variation changes the *d*-axis saturation level (*d*-axis assumed to be aligned with PM flux), and therefore the *d*-axis inductance [11]. Also injection of fundamental current will change the saturation level, resulting therefore into inductance variations [11,12]. Torque in SynRMs is function of *d* and *q*-axes inductances exclusively, which can vary significantly with the current level due to saturation of the core material.

Precise knowledge of the torque produced by the machine is required in many applications. Torque measurement is problematic. Torque transducers based on strain gauges are likely the preferred option [13–17]. However, this type of sensor can introduce resonances into the system, are highly sensitive to electromagnetic interference, requires precise mounting and calibration to ensure accuracy and their cost could account for a significant portion of the drive cost [18]. Less popular options are torque measurement systems based on torsional displacement [19] or magnetoelastic effect [20]. Torque measurement systems based on torsional displacement are less sensitive to electromagnetic noise but use optical probes, which are expensive and require accurate calibration [19]. Torque measurement based on magneto-elastic sensors is simpler and uses non-contact technology without the need of calibration, but it requires special shaft materials and can incur in torque measurement errors due to shaft thermal expansion. Regardless of the method being used, precise torque measurement is expensive, and requires extra elements (sensor, cables, connectors, ...), what can reduce the reliability of the drive. These concerns have boosted the interest in torque estimation methods.

Torque estimation methods can be roughly separated into (a) methods based on the torque equation [21,22] and (b) indirect estimation methods [23–31]. The first type can use the General Torque Equation (GTE) assuming constant motor parameters [21], Flux estimation with Compensation Scheme (FCS) where dq-flux linkage is estimated by the measurement of stator voltage, currents and rotor position [21] or look-up-tables, which are used to adjust the machine parameters according to machine operating conditions [22]. A large number on indirect torque estimation methods have been proposed, which go from a simple power balance with known electric power and rotor speed [23], observer based methods [24–29] (e.g., sliding mode observers [24], model reference adaptive systems [25,26], model reference observers, reduced order observers [27], recursive least square parameters estimation [28] or affine projection algorithms parameters estimation [29]), methods requiring additional sensors, e.g., giant magnetoresistance effect (GMR) based methods [30], or neural networks based methods [31]. All these methods [21–31] require previous knowledge of certain machine parameters (e.g., resistances or inductances) and/or its operating condition (e.g., temperature).

Injection of a high-frequency (HF) signal has been recently proposed for on-line estimation of the machine parameters used by the torque equation. Consequently, they belong to category (a) discussed above [12,32,33]. The HF signal is injected via inverter on top of the fundamental voltage responsible of torque production, meaning that it does not interfere with the normal operation of the machine. Furthermore, no additional sensors or cabling are needed. The HF signal can be either a current or a voltage signal and have different shapes: pulsating, square-wave, sinusoidal, etc. Although the physical principles are the same in all the cases, this choice can result in significant differences in the implementation. Injection of a voltage signal is easier in principle as the inverter is a voltage source. On the contrary, current injection will require the use of current controllers, but will be shown to improve the reliability of the method. Note that instead of periodic HF signal, PWM pulses could also be used, either transient currents or neutral voltage following each inverter pulse being measured and processed in this case. However, its implementation implies changes in the hardware (e.g., additional sensors, access to the neutral of the machine, etc.), increasing the implementation cost and difficulty.

This paper reviews the use of HF signal injection based parameter estimation methods aimed to improve the accuracy of torque estimation for Interior PMSMs (IPMSMs), Surface PMSMs (SPMSMs) and SynRMs. Pros and cons of each method will be discussed in terms of implementation requirements, accuracy and suitability depending on the machine design. A key aspect for the proposed method will be modelling the relationship between the incremental inductances obtained from the injected HF signal and the apparent inductance used in conventional torque equation. The paper is organised as follows. The fundamental model of the synchronous machine (SM) is presented in Section 2, while the HF model is presented in Section 3. Parameter identification and torque estimation using different form of HF signal injections is discussed in Section 4. Finally, experimental results are provided in Section 5.

2. Fundamental Model of a Synchronous Motor

The fundamental model of a synchronous machine in a reference frame synchronous with the rotor is given by (1) [8], where R_d, R_q, L_d and L_q are the d and q-axes resistances and inductances, respectively; ω_r is the mechanical rotational speed; λ_{pm} is PM flux linkage; and p is the differential operator. In the nomenclature used in this paper, superscripts "s" and "r" indicate the stationary and rotor synchronous reference frames, respectively; subscripts "s" and "r" indicate stator and rotor variables, respectively; and subscripts "dq" and "HF" indicate fundamental and HF components respectively. Finally, "*" is used to indicate commanded values, and "^" indicates estimated values.

$$\begin{bmatrix} v_{sd}^r \\ v_{sq}^r \end{bmatrix} = \begin{bmatrix} R_d + pL_d & -\omega_r L_q \\ \omega_r L_d & R_q + pL_q \end{bmatrix} \begin{bmatrix} i_{sd}^r \\ i_{sq}^r \end{bmatrix} + \begin{bmatrix} 0 \\ \lambda_{pm}\omega_r \end{bmatrix} \tag{1}$$

The output torque in the steady state can be expressed by (2), where P stands for the number of poles [19]. The first term on the left-hand side of (2), T_{syn} is the electromagnetic/synchronous torque due to the PM flux linking magnet and stator coils, whereas the second term on the right-hand side of (2), T_{rel}, is the reluctance torque due to saliency of the machine, i.e., the difference between d- and q-axis inductances.

$$T_{out} = \frac{3}{2}\frac{P}{2}\left[\lambda_{pm} i_{sq}^r + \left(L_d - L_q\right) i_{sd}^r i_{sq}^r\right] = T_{syn} + T_{rel} \tag{2}$$

It is observed from (2) that estimation of the synchronous torque T_{syn} requires knowledge of λ_{pm}, whereas estimation of the reluctance torque T_{rel} requires knowledge of the differential inductance $(L_d - L_q)$.

As already mentioned, this paper will address torque estimation for three machine designs: IPMSM, SPMSM and SynRM. Their schematic cross sections and main characteristics being shown in Figure 1 and Table 1, respectively. From the combined analysis of (2) and Table 1, the mechanisms for torque production of each machine design becomes evident.

Torque produced by IPMSMs and SPMSMs is determined both by the magnetic torque due to the PMs (synchronous torque) and the reluctance torque due to different L_d and L_q inductances, i.e., saliency ratio. Note that for the case of SPMSM, the saliency ratio is often negligible. Torque produced by SynRM will be determined only by the saliency ratio. Proper torque estimation will, therefore, require precise estimation of λ_{pm} and/or L_d and L_q.

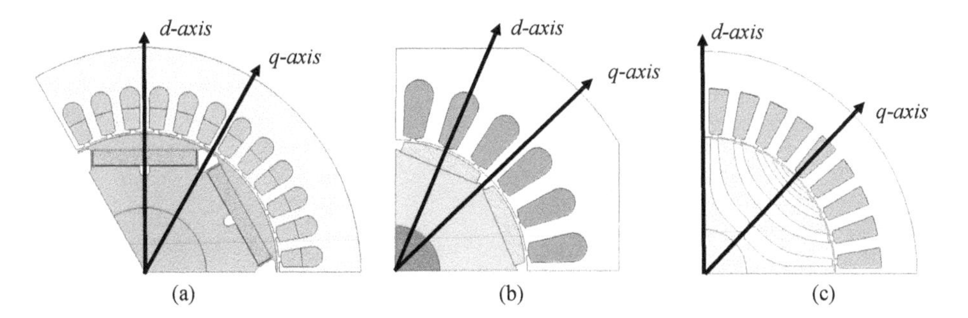

Figure 1. Partial cross-sectional view of the motors being analysed: (**a**) IPMSM, (**b**) SPMSM and (**c**) SynRM.

Table 1. Comparison of motors parameters.

	IPMSM	SPMSM	SynRM
Magnetic flux linkage	$\lambda_{pm} > 0$	$\lambda_{pm} > 0$	$\lambda_{pm} = 0$
d and q inductances	$L_d < L_q$	$L_d \approx L_q$	$L_d >> L_q$

Parameters Variation with Operating Conditions of the Machine

Torque equation is function of inductances and magnet strength, which can change with the operating point of the machine. q- and d-axes inductances can be modelled as (3) and (4).

$$L_{q(I_{sd},I_{sq})} = L_{q0}\left(1 + \alpha_{I_{sd}} I^r_{sd} + \alpha_{I_{sq}} I^r_{sq}\right) \tag{3}$$

$$L_{d(I_{sd},I_{sq},T_r)} = L_{d0}\left(1 + \alpha_{I_{sd}} I^r_{sd} + \alpha_{I_{sq}} I^r_{sq} + \alpha_{T_r}\left(T_r - T_{r0}\right)\right) \tag{4}$$

From Equation (3), q-axis inductance is seen to be function of the fundamental component of the d-axis current due to cross-coupling, and of the q-axes current due to saturation [11]. Similarly, d-axis inductance (4) is function of the fundamental d- and q-axes currents, but also of the PMs remanent flux, which varies with temperature T_r [8,11,12]. Variation of d-axis inductance with PM flux λ_{pm} has been shown to be almost linear, and can therefore be modelled as (5) [12], where λ_{pm0} and L_{d0} are the base value of PM flux and d-axis inductance, i.e., at room temperature (T_{r0}) and when there is no fundamental current, and $L_{d(I_{sd},I_{sq},T_r)}$ is the d-axis inductance for a given magnet temperature T_r and current I_{sdq}.

$$\lambda_{pm} = \lambda_{pm0}\frac{L_{d0}}{L_{d(I_{sd},I_{sq},T_r)}} \tag{5}$$

Substituting (3)–(5) into (2), the general torque equation can be written as (6), which evidences the dependence of torque with the operating condition of the machine, and highlights the incorrectness of assuming constant inductances. Parameters estimation using HF signal injection is discussed following.

$$T_{out} = \frac{3}{2}\frac{P}{2}\left[\left(\lambda_{pm0}\frac{L_{d0}}{L_{d(I_{sd},I_{sq},T_r)}}\right) i^r_{sq} + \left(L_{d(I_{sd},I_{sq},T_r)} - L_{q(I_{sd},I_{sq})}\right) i^r_{sd} i^r_{sq}\right] \tag{6}$$

3. HF Model of the Synchronous Machine

The model representing the behaviour of a synchronous machine when the stator is fed at a frequency sufficiently higher than the rotor frequency (7) can be deduced from (1) by neglecting magnet flux, as it does not contain any HF component [12].

$$\begin{bmatrix} v^r_{sdHF} \\ v^r_{sqHF} \end{bmatrix} = \begin{bmatrix} R_{dHF} + pL_{dHF} & -\omega_r L_{qHF} \\ \omega_r L_{dHF} & R_{qHF} + pL_{qHF} \end{bmatrix} \begin{bmatrix} i^r_{sdHF} \\ i^r_{sqHF} \end{bmatrix} \tag{7}$$

As the HF signal being applied is a voltage, Equation (7) can be solved for the high frequency current as

$$\begin{bmatrix} i^r_{sdHF} \\ i^r_{sqHF} \end{bmatrix} = \frac{1}{(R_{dHF}+pL_{dHF})(R_{qHF}+pL_{qHF})+\omega_r^2 L_{dHF}L_{qHF}} \begin{bmatrix} R_{qHF} + pL_{qHF} & \omega_r L_{qHF} \\ -\omega_r L_{dHF} & R_{dHF} + pL_{dHF} \end{bmatrix} \begin{bmatrix} v^r_{sdHF} \\ v^r_{sqHF} \end{bmatrix} \tag{8}$$

Torque in Equation (6) is a function of absolute (apparent) inductances. However, the inductances of the HF model (7) and (8) are incremental inductances [34], i.e., inductances estimated by means of HF signal injection will be incremental. It is needed therefore the establish the relationship between absolute and incremental inductances.

The inductance of a stator winding is defined as the relationship between flux linkage λ divided by the stator current I producing that flux [35], where N is the number of turns of the stator winding.

$$L = \frac{\lambda}{I} = \frac{N\phi}{I} \tag{9}$$

Flux ϕ in the core which can be defined as (10), R being the reluctance of the magnetic circuit.

$$\phi = \frac{NI}{R} \tag{10}$$

Combining (9) and (10), the inductance can be written as (11), where μ is the permeability of the material ($\mu = \mu_0\mu_r$); A is the mean cross-sectional area of the magnetic circuit and l is the mean length of the magnetic circuit.

$$L = \frac{N^2}{R} = \frac{N^2\mu A}{l} \tag{11}$$

It is seen from (11) that the inductance is proportional to the permeability [35].

$$L(H) = k\mu(H) \tag{12}$$

Static and dynamic permeabilities can be used to analyse the B-H curve of ferromagnetic materials (see Figure 2). The static permeability is defined as the slope of a straight line from the origin to the actual operating point A, i.e., the ratio of flux density (B) vs. field intensity (H) at every operating point (13). The static permeability gives the absolute inductance of the machine.

$$\mu_s = \frac{1}{\mu_0}\frac{B}{H} \tag{13}$$

On the other hand, the slope (AB) of the B–H curve (14) is denoted as relative, differential or dynamic permeability.

$$\mu_d = \frac{1}{\mu_0}\frac{\partial B}{\partial H} \tag{14}$$

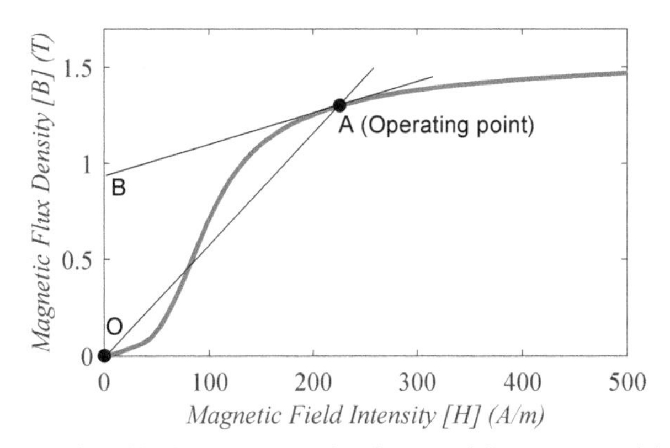

Figure 2. BH curve for NSSMC 50H470 material, and static and dynamic permeability definition.

As an example, Figure 3 shows the BH curve and corresponding static and dynamic permeabilities μ_s and μ_d for the core material NSSMC 50H470. It is observed that for very low excitation levels static and dynamic permeabilities have a ratio close to one, whereas for high current levels the ratio is almost constant (~15 in the figure, note the logarithmic scale). It was deduced from (12) that absolute and incremental inductances will mirror the behaviour of static and dynamic permeabilities, i.e., the relationship between absolute and incremental inductances can be modeled as (15), with k_μ being defined as (16).

$$L_{dq(I_{sd},I_{sq},T_r)} = k_\mu L_{dqHF\left(I_{sd},I_{sq},T_r\right)} \tag{15}$$

$$k_\mu = \frac{\mu_s}{\mu_d} \tag{16}$$

It is apparent from Figure 4 that at high excitation levels, the relationship between both inductances is almost constant. However, to model k_μ in the whole operating range, some type of polynomial function would be needed.

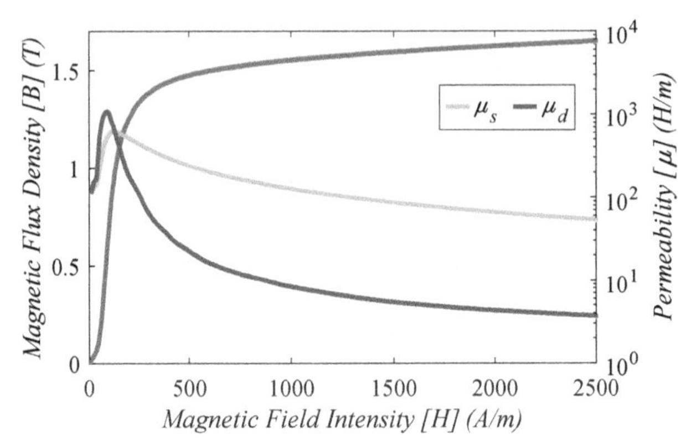

Figure 3. BH curve of ferromagnetic material and corresponding static and dynamic permeabilities (see Figure 2).

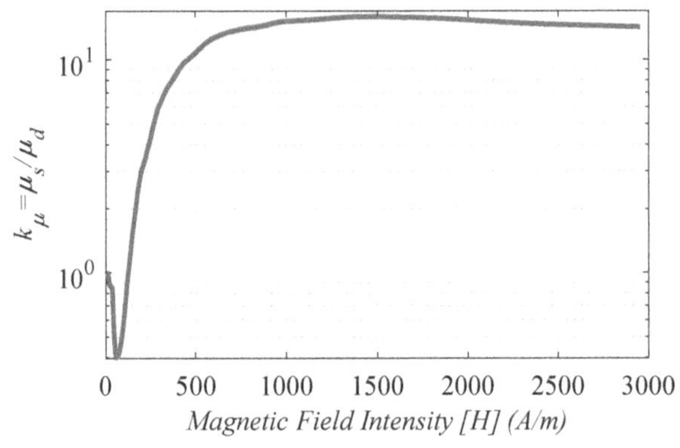

Figure 4. Ratio between static and dynamic permeabilities of the NSSMC 50H470 core material in Figure 3.

λ_{pm} variation shown in (5) can be therefore rewritten as (17), the output torque can be finally estimated combining (15) and (17) as (18).

$$\lambda_{pm} = \lambda_{pm0} \frac{L_{dHF_0}}{L_{dHF\left(I_{sd},I_{sq},T_r\right)}} \tag{17}$$

$$\hat{T}_{out} = \frac{3}{2}\frac{P}{2}\left[\left(\lambda_{pm0}\frac{L_{dHF_0}}{L_{dHF(I_{sd},I_{sq},T_r)}}\right)i^r_{sq} + \left(k_\mu L_{dHF(I_{sd},I_{sq},T_r)} - k_\mu L_{qHF(I_{sd},I_{sq})}\right)i^r_{sd}i^r_{sq}\right] \quad (18)$$

4. HF Inductances Estimation

Different forms of HF excitation can be used to estimate the HF inductances. Generally speaking, voltage injection results in easier implementations as the inverter is a voltage source, but will be sensitive to speed, also being affected by the non-purely inductive machine behaviour. Those drawbacks can be overcome by injecting a HF current, at the price of an increase in the complexity of implementation. Rotating Voltage Injection, Pulsating Voltage Injection and Pulsating dq-axes Current Injection are discussed in the next subsections, followed by a comparative analysis. It has to be noted that other types of periodic high frequency signal injection (e.g., square wave) could be used. These options have not been discussed in this paper due to room constrains.

4.1. Pulsating Voltage Injection

HF inductances L_{dHF} and L_{qHF} can be estimated by injecting a voltage HF signal in a predefined direction referred to the d-axes. This form of excitation is commonly referred as pulsating voltage.

The HF model in the synchronous reference frame (7) can be written as function of the mean ΣL (20) and differential ΔL (21) inductances (19). Note that the resistive component has been neglected, as at high frequency it is significantly smaller than the inductive terms and the p operator in (7) has been replaced by $j\omega_{HF}$. Furthermore, rotor speed dependent terms in (7) have been neglected assumed that the HF voltage signal has a frequency much higher than the fundamental rotating frequency (i.e., $\omega_{HF} >> \omega_r$).

$$\begin{bmatrix} v^r_{sdHF} \\ v^r_{sqHF} \end{bmatrix} = \begin{bmatrix} j\omega_{HF}(\Sigma L - \Delta L) & 0 \\ 0 & j\omega_{HF}(\Sigma L + \Delta L) \end{bmatrix} \begin{bmatrix} i^r_{sdHF} \\ i^r_{sqHF} \end{bmatrix} \quad (19)$$

$$\Sigma L = \frac{L_{dHF} + L_{qHF}}{2} \quad (20)$$

$$\Delta L = \frac{L_{dHF} - L_{qHF}}{2} \quad (21)$$

The currents induced in the stator terminals expressed in the stator reference frame can be derived from the inverse matrix as (22)

$$\begin{bmatrix} i^s_{sdHF} \\ i^s_{sqHF} \end{bmatrix} = \frac{1}{j\omega_{HF}\left(\Sigma L^2 - \Delta L^2\right)} \begin{bmatrix} \Sigma L + \Delta L\cos(2\theta_r) & \Delta L\sin(2\theta_r) \\ \Delta L\sin(2\theta_r) & \Sigma L - \Delta L\cos(2\theta_r) \end{bmatrix} \begin{bmatrix} v^s_{sdHF} \\ v^s_{sqHF} \end{bmatrix} \quad (22)$$

If a pulsating HF voltage is injected in the stator terminals of the machine (23), the resulting HF current can be derived substituting in (23) into (22) as (26),

$$\begin{bmatrix} v^{s*}_{sdHF} \\ v^{s*}_{sqHF} \end{bmatrix} = V^*_{HF}\sin(\theta_{HF}) \begin{bmatrix} \cos(\hat{\theta}_r) \\ \sin(\hat{\theta}_r) \end{bmatrix} \quad (23)$$

$$\theta_{HF} = \omega_{HF}t \quad (24)$$

$$\hat{\theta}_r = \theta_r + \varphi \quad (25)$$

$$\begin{bmatrix} i^s_{sdHF} \\ i^s_{sqHF} \end{bmatrix} = \frac{V^*_{HF}\sin(\theta_{HF})}{j\omega_{HF}\left(\Sigma L^2 - \Delta L^2\right)} \begin{bmatrix} \Sigma L\cos(\hat{\theta}_r) + \Delta L\cos(2\theta_r - \hat{\theta}_r) \\ \Sigma L\sin(\hat{\theta}_r) + \Delta L\sin(2\theta_r - \hat{\theta}_r) \end{bmatrix} \quad (26)$$

where V^*_{HF} is the magnitude of the injected HF signal, θ_{HF} is the phase of the HF signal (24), ω_{HF} is the frequency of the HF signal, $\hat{\theta}_r$ is the injection angle of the pulsating HF voltage (25) and φ is an arbitrary angle; e.g., if $\varphi = 0$, the pulsating HF voltage will be injected in the d-axis of the machine, whereas if $\varphi = \pi/2$, it will be injected in the q-axis.

By synchronization with the injection reference frame "$\hat{\theta}_r$", the stator currents (26) are transformed into (27)

$$\begin{bmatrix} i_{sdHF}^{\hat{\theta}_r} \\ i_{sqHF}^{\hat{\theta}_r} \end{bmatrix} = \frac{V_{HF}^* \sin(\theta_{HF})}{j\omega_{HF}\left(\Sigma L^2 - \Delta L^2\right)} \begin{bmatrix} \Sigma L + \Delta L \cos(2(\theta_r - \hat{\theta}_r)) \\ \Delta L \sin(2(\theta_r - \hat{\theta}_r)) \end{bmatrix} \quad (27)$$

Finally, if the HF is injected between d and q-axes, i.e., $\hat{\theta}_r = \theta_r + \pi/4$, (27) can be simplified into (28) [32]. From (28) the HF inductances (29) and (30) are readily obtained.

$$\begin{bmatrix} i_{sdHF}^{\hat{\theta}_r} \\ i_{sqHF}^{\hat{\theta}_r} \end{bmatrix} = \frac{V_{HF}^* \sin(\theta_{HF})}{j\omega_{HF}\left(\Sigma L^2 - \Delta L^2\right)} \begin{bmatrix} \Sigma L + \Delta L \cos(-\pi/2) \\ \Delta L \sin(-\pi/2) \end{bmatrix} = \frac{V_{HF}^* \sin(\theta_{HF})}{j\omega_{HF}\left(\Sigma L^2 - \Delta L^2\right)} \begin{bmatrix} \Sigma L \\ -\Delta L \end{bmatrix} \quad (28)$$

$$L_{dHF} = \frac{V_{HF}^* \sin(\theta_{HF})}{j\omega_{HF}\left(i_{sdHF}^{\hat{\theta}_r} - i_{sqHF}^{\hat{\theta}_r}\right)} \quad (29)$$

$$L_{qHF} = \frac{V_{HF}^* \sin(\theta_{HF})}{j\omega_{HF}\left(i_{sdHF}^{\hat{\theta}_r} + i_{sqHF}^{\hat{\theta}_r}\right)} \quad (30)$$

Figure 5 shows the inverter control block diagram and the signal processing needed for torque estimation when pulsating HF voltage injection at an arbitrary angle of injection, $\hat{\theta}_r$, is used for dq-axes HF inductance estimation. The HF voltage v_{sdqHF1}^{r*}, is injected in open-loop. A High-Pass Filter (HPF) is needed to isolate the HF components of the overall stator currents. Inputs to the signal processing block are the commanded HF voltage v_{sdqHF1}^{r*}, the induced HF currents i_{sdqHF1}^{r} and the fundamental commanded current I_{sdq}^{r*}. A Band Pass Filter (BPF) is used to isolate the positive sequence component of the HF induced current. The d and q-axis HF inductances are estimated using (29) and (30), the PM flux is estimated using (17) and the output torque, $\hat{T_{out}}$, is finally estimated using (18). As the estimated dq-axes inductances already reflect the effects of temperature on PM magnetisation state as well as the effects of fundamental current, knowledge of PM temperature is not needed.

Appealing properties of HF pulsating voltage injection are its simplicity and the fact that a single frequency allows the estimation of d- and q-axes HF inductances.

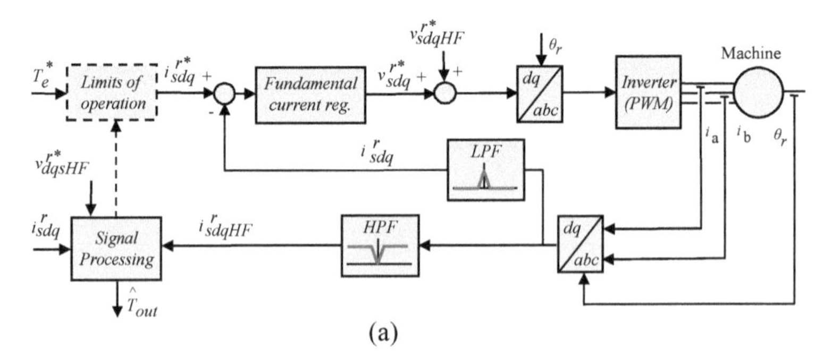

Figure 5. *Cont.*

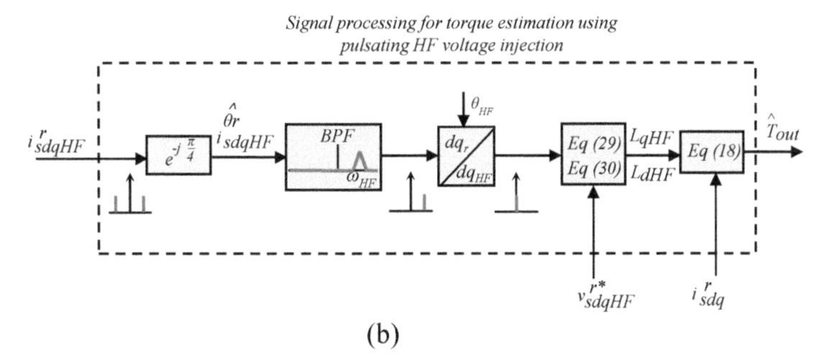

Figure 5. Implementation using pulsating HF voltage injection at an arbitrary angle of injection, $\hat{\theta}_r$. (**a**) Power converter control. (**b**) Signal processing for torque estimation.

4.2. Rotating Voltage Injection

While the assumption of pure inductive behaviour at high frequencies can be acceptable for most PMSMs designs, it can be arguable for SynRMs. In this case, the use of a HF pulsating voltage at 45° could incur in large errors in the estimated inductances, and consequently in the estimated torque. The use of a rotating HF voltage can be advantageous in this case.

When a rotating HF voltage (31) is injected in a synchronous machine, the HF currents induced in stator windings can be obtained by inserting (31) into (8).

$$v^{r*}_{sdqHF} = \begin{bmatrix} \bar{V}^{r*}_{sdHF} \\ \bar{V}^{r*}_{sqHF} \end{bmatrix} = \begin{bmatrix} V^*_{HF} \cos(\omega_{HF} t) \\ V^*_{HF} \sin(\omega_{HF} t) \end{bmatrix} = V^*_{HF} e^{j(\omega_{HF} t)} = v^{r*}_{sdqHFpc} \tag{31}$$

By solving (8), the stator HF currents in the rotor synchronous reference frame are obtained as

$$i^r_{sdHF} = \frac{R_{qHF} + j\omega_{HF} L_{qHF}}{(R_{dHF} + j\omega_{HF} L_{dHF})(R_{qHF} + j\omega_{HF} L_{qHF}) + \omega_r^2 L_{dHF} L_{qHF}} \left[v^r_{sdHF} + \frac{\omega_r L_{qHF} v^r_{sqHF}}{R_{qHF} + j\omega_{HF} L_{qHF}} \right] \tag{32}$$

$$i^r_{sqHF} = \frac{R_{dHF} + j\omega_{HF} L_{dHF}}{(R_{dHF} + j\omega_{HF} L_{dHF})(R_{qHF} + j\omega_{HF} L_{qHF}) + \omega_r^2 L_{dHF} L_{qHF}} \left[v^r_{sqHF} - \frac{\omega_r L_{dHF} v^r_{sdHF}}{R_{dHF} + j\omega_{HF} L_{dHF}} \right] \tag{33}$$

It can be seen from (32) and (33) that obtaining L_{dHF} and L_{qHF} is not straightforward due to cross-coupling between dq-axes. However, if the frequency of the injected HF signal is sufficiently higher than the rotor frequency, the rotor speed dependent terms can be safely neglected, the HF currents induced in the stator simplifying to (34) and (35). An orientative value for this assumption can be $\omega_{HF} > \omega_r + 2 \cdot \pi \cdot 500$ rad/s [12].

$$i^r_{sdHF} = \frac{v^r_{sdHF}}{(R_{dHF} + j\omega_{HF} L_{dHF})} = \frac{V^*_{HF} \cos(\omega_{HF} t)}{(R_{dHF} + j\omega_{HF} L_{dHF})} \tag{34}$$

$$i^r_{sqHF} = \frac{v^r_{sqHF}}{(R_{qHF} + j\omega_{HF} L_{qHF})} = \frac{V^*_{HF} \sin(\omega_{HF} t)}{(R_{dHF} + j\omega_{HF} L_{dHF})} \tag{35}$$

Estimation of L_{dHF} and L_{qHF} can be obtained from the imaginary part of the dq-axes impedance (36) and (37).

$$Z_{dHF} = R_{dHF} + j\omega_{HF} L_{dHF} = \frac{V^*_{HF} \cos(\omega_{HF} t)}{i^r_{sdHF}} \tag{36}$$

$$Z_{qHF} = R_{qHF} + j\omega_{HF} L_{qHF} = \frac{V_{HF}^{*} \sin(\omega_{HF} t)}{i_{sqHF}^{r}} \tag{37}$$

Figure 6 shows the inverter control and signal processing for torque estimation using rotating HF voltage injection. As for the implementation shown in Figure 5, the commanded HF voltage, v_{sdqHF}^{r*} is injected in open-loop, with no additional controllers required. A High-Pass Filter (HPF) is also needed to isolate the HF components of the overall stator currents. Inputs of the signal processing block are the commanded HF voltage v_{sdqHF}^{r*}, the induced HF currents i_{sdqHF}^{r}, and the fundamental commanded current I_{sdq}^{r*}. The d and q-axis HF inductances are estimated using (36) and (37), the PM flux is estimated using (17) and the output torque, $\hat{T_{out}}$, is finally estimated using (18).

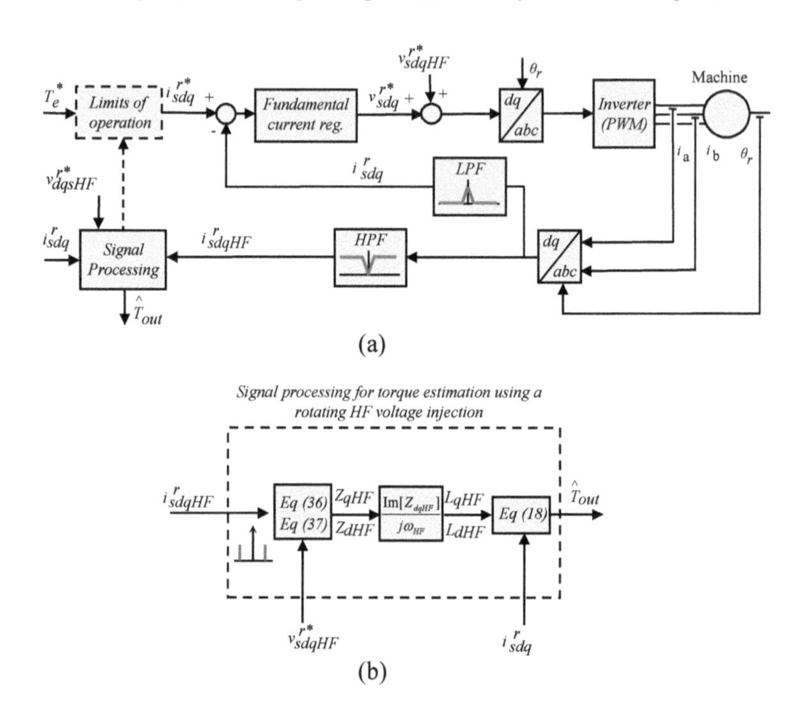

Figure 6. Implementation using rotating HF voltage injection. (**a**) Power converter control. (**b**) Signal processing for torque estimation.

The frequency of the HF signal should be sufficiently higher than the rotor frequency to safely neglect the rotor speed dependent (i.e., cross-coupling) terms. This can be problematic in the case of IPMSM designed to operate at very high-speeds. Furthermore, as the HF voltage is injected in a reference frame synchronous with the rotor, the effective frequency of the HF frequency signal will be $\omega_r + \omega_{HF}$. It must be guaranteed that it is smaller than half of the switching frequency (Nyquist frequency).

4.3. Pulsating Dq-Axes Current Injection

Injection of a pulsating HF current has been proposed as a viable mean to avoid issues due to the assumption of a purely inductive behaviour of the windings, as well as the problems with high speed machines discussed in the preceding subsections. HF inductances can be estimated by injecting two pulsating HF currents at different frequencies into the d- and q-axis. HF current controllers are needed for this purpose [12]. The process to estimate L_{dHF} is described following, and identical procedure is used to obtain L_{qHF}.

If a HF current is injected into the d-axis of the machine and the q-axis current is force to be equal to zero (38), the HF voltages commanded by the HF current controller will be of the form shown in (39). A fictitious HF voltage vector can be defined consisting only of the d-axis component $v^{r'}_{sdqHF1}$ (40). Both HF currents and voltages in (38) and (40) can be separated into positive sequence ($i^{r*}_{sdqHF1pc}$ and $v^{r'}_{sdqHF1pc}$) and negative sequence ($i^{r*}_{sdqHF1nc}$ and $v^{r'}_{sdqHF1nc}$) components, (41) and (42), of magnitude equal to half of the original signal. The d-axis HF impedance (43) can be obtained from the positive or negative sequence indistinctly. The d-axis HF inductance is finally obtained as the imaginary part of (43) and (44).

$$i^{r*}_{sdqHF1} = \begin{bmatrix} \bar{I}^{r*}_{sdHF1} \\ \bar{I}^{r*}_{sqHF1} \end{bmatrix} = \begin{bmatrix} I^{*}_{HF} \cos(\omega_{dHF} t) \\ 0 \end{bmatrix} \tag{38}$$

$$v^{r*}_{sdqHF1} = \begin{bmatrix} \bar{V}^{r*}_{sdHF1} \\ \bar{V}^{r*}_{sqHF1} \end{bmatrix} = \begin{bmatrix} (R_{dHF} + j\omega_{dHF} L_{dHF})\, \bar{I}^{r}_{sdHF1} \\ \omega_r L_{dHF} \bar{I}^{r}_{sdHF1} \end{bmatrix} \tag{39}$$

$$v^{r'}_{sdqHF1} = \begin{bmatrix} \bar{V}^{r*}_{sdHF1} \\ 0 \end{bmatrix} = \begin{bmatrix} (R_{dHF} + j\omega_{dHF} L_{dHF})\, \bar{I}^{r}_{sdHF1} \\ 0 \end{bmatrix} = \begin{bmatrix} \bar{V}^{r'}_{sdqHF1} \cos(\omega_{dHF} t + \varphi_{Z_d}) \\ 0 \end{bmatrix} \tag{40}$$

$$i^{r*}_{sdqHF1} = \frac{I^{*}_{HF}}{2} e^{j\omega_{dHF} t} + \frac{I^{*}_{HF}}{2} e^{-j\omega_{dHF} t} = i^{r*}_{sdqHF1pc} + i^{r*}_{sdqHF1nc} \tag{41}$$

$$v^{r'}_{sdqHF1} = \frac{\left|v^{r'}_{sdqHF1}\right|}{2} e^{j(\omega_{dHF} t - \varphi_{Zd})} + \frac{\left|v^{r'}_{sdqHF1}\right|}{2} e^{j(-\omega_{dHF} t + \varphi_{Zd})} = v^{r'}_{sdqHF1pc} + v^{r'}_{sdqHF1nc} \tag{42}$$

$$Z_{dHF} = R_{dHF} + j\omega_{dHF} L_{dHF} = \frac{v^{r'}_{sdqHF1pc}}{i^{r*}_{sdqHF1pc}} = \frac{v^{r'}_{sdqHF1nc}}{i^{r*}_{sdqHF1nc}} \tag{43}$$

$$L_{dHF} = \frac{Im\left[Z_{dHF}\right]}{j\omega_{dHF}} \tag{44}$$

Figure 7 shows the inverter control block diagram and the signal processing needed for torque estimation when pulsating dq-axes HF current signals are used for dq-axes HF inductance estimation. Two HF resonant current controllers (45) are used to inject the HF currents, where K_p is the proportional gain, ω_{HF} is the resonant frequency, and c is the zero position of the controller.

$$PR(s) = K_p \frac{(s+c)^2 + \omega_{HF}^2}{s^2 + \omega_{HF}^2} \tag{45}$$

Inputs to the signal processing block are the commanded high frequency resonant currents i^{r*}_{sdqHF1} and i^{r*}_{sdqHF2}, the output voltage of the resonant controllers $v^{r'}_{sdqHF1}$ and $v^{r'}_{sdqHF2}$ and the fundamental commanded current I^{r*}_{sdq}. Two band stop filters, $BSF1$ and $BSF2$, are used to remove the negative sequence components of the HF currents and voltages. The d and q-axis HF impedances are estimated using (43), the d and q-axis HF inductances are estimated using (44), the PM flux is estimated using (17) and, the output torque, $\hat{T}_{out}$, is finally estimated using (18).

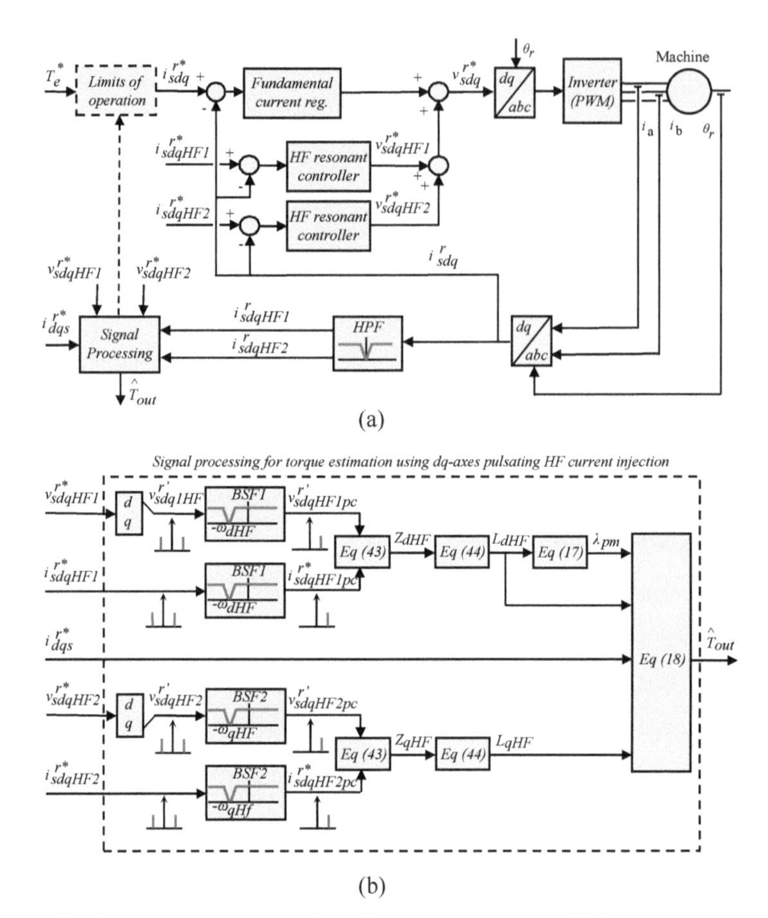

Figure 7. HF inductance estimation using HF pulsating current injection. (**a**) Power converter control. (**b**) Signal processing for torque estimation.

Note that in this case a pure inductive behaviour is not assumed, cross-coupling terms do not affect to the estimations either. However, the fact that there are two HF signals might have a larger impact on machine performance (noise, vibration, ...) compared to methods using a single HF signal, the computational burden also being larger. Finally, Table 2 summarises the main characteristics of the methods analysed in this section.

Table 2. Comparison between HF Signal injection methods for SM parameter estimation.

	Pulsating dq-axes HF Current Injection	Pulsating HF Voltage Injection	Rotating HF Voltage Injection
Pure inductive HF model can be assumed	✓	✗	✓
Sensitive machine speed	✓	✗	✗
Need of additional controllers	✗	✓	✓
Number of injected signals	2	1	1
Computational burden	✗	✓	✓

5. Experimental Results

The proposed methods have been tested in the machines designs shown in Figure 1; the corresponding parameters can be found in Table 3. A schematic representation of the test bench as well as pictures of the different elements are shown in Figure 8. The inverter feeding the machines under test (Inverter 2 in Figure 8a) uses 1200 V, 100 A IGBT power modules, with a switching frequency of 10kHz. The load machine is a 40kW axial PMSM machines (EMRAX 228 [36]) driven by a BAMOCAR-PG-D3 power converter [37] (Inverter 1 in Figure 8a). Torque is measured using a T5 Interface Torque transducer [13], with 12-Bit resolution, 10 kHz, ±100 Nm, 0.2% combined error. Currents in the machines being tested are measured using standard 1% Hall-effect based-current sensors and 12-bit analog-to-digital converters [38]. For the sake of completeness, a different type of HF excitation has been used with each machine design.

Table 3. Test machines parameters.

	P_{rated} [kW]	I_{rated}[A]	L_{d0} [mH]	L_{q0} [mH]	λ_{pm0}[Wb]	ω_{rated} [rpm]	Poles
IPMSM	4	14	10.5	23	0.64	1000	6
SPMSM	5.3	15	5.54	6.81	0.59	1000	8
SynRM	1.5	3.9	410	100	-	1500	4

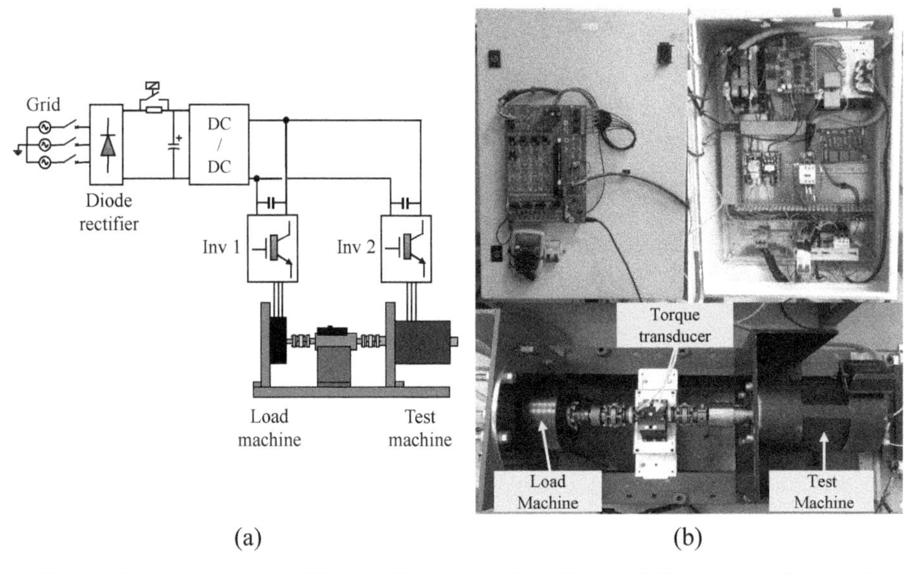

Figure 8. Experimental set-up: (**a**) Schematic representation. (**b**) Top left: Control board (Inverter 2). Top right: Power converter (Inverter 2). Bottom: Picture of the test bench.

5.1. IPMSM Torque Estimation

Two pulsating HF currents have been used in this case, of magnitude 0.05 pu and frequency of $\omega_{dHF} = 2 \cdot \pi \cdot 500$ rad/s and $\omega_{qHF} = 2 \cdot \pi \cdot 1000$ rad/s, respectively (38). The *HPF* needed to isolate the HF current components (see Figure 7) has a bandwidth of 5 Hz. Band stop filters *BSF*1 and *BSF*2 used to remove the negative sequence components of the HF currents and voltages have a bandwidth of $2 \cdot \pi \cdot 10$ rad/s.

Due to rotor magnets, the machine will work at high saturation levels even for no-load conditions. The coefficient linking the HF estimated inductances (incremental inductances) and the absolute inductances is assumed to be constant, $k_\mu = 14$ (see Figure 4).

Figure 9a shows the estimated torque from (2) ($\hat{T}_{out_{conv}}$) assuming constant parameters; the estimated torque when the machine parameters are estimated from the injected HF current (18) ($\hat{T}_{out_{HF}}$); and the measured torque using the torque transducer shown in Figure 8 (T_{out}), when the magnitude of the fundamental current I^r_{sdq} changes from 0p.u. to 1p.u. following a Maximum Torque Per Ampere (MTPA) trajectory. Figure 9b shows the estimation error using both general torque equation and the proposed method. It can be observed that torque estimation error is reduced when the machine parameters are estimated using HF signal injection; the improvement being more relevant at higher current levels. This is an expected result since the dq-axes inductances values will differ more to their base values as the saturation level increases.

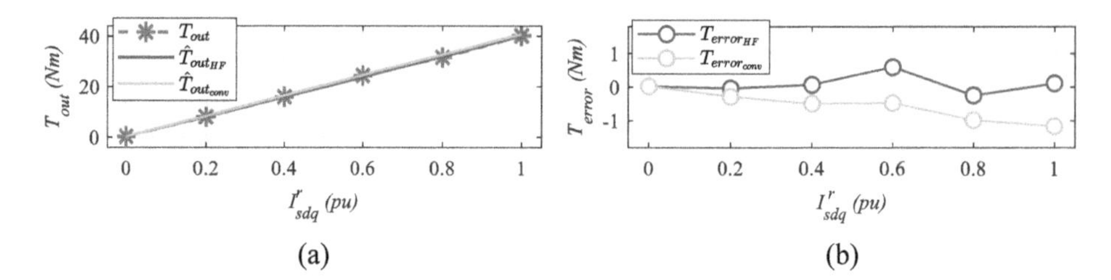

Figure 9. Experimental results: IPMSM & pulsating current injection: (**a**) Estimated and measured torque. (**b**) Estimated torque error. $0 < I^r_{sdq} < 1$ p.u., $\omega_r = 50Hz$, $I_{HF} = 0.05$ p.u., $\omega_{dHF} = 2 \cdot \pi \cdot 500$ rad/s , $\omega_{qHF} = 2 \cdot \pi \cdot 1000$ rad/s.

5.2. SPMSM Torque Estimation

HF inductances have been estimated in this case using pulsating voltage injection at 45° as described in Section 4.1. A HF voltage of 10 V and 250 Hz has been used (23). A band pass filter of 100 Hz was used to isolate the HF currents. Similar for the case of the IPMSM, the machine will work at high saturation levels even at no-load conditions due to the magnets. Therefore, also, in this case the coefficient linking the estimated HF inductances and the absolute inductances has been considered constant $k_\mu = 14$.

Similarly to Figure 9, Figure 10a shows the estimated torque assuming constant parameters and adapting machine parameters using HF voltage injection. The fundamental current,I^r_{sdq} was varied from 0 p.u. to 1 p.u. following a MTPA trajectory. Figure 10b shows the estimation error for both methods. As for the IPMSM case, torque estimation error is also reduced when the machine parameters are estimated using HF signal injection, the improvement being more noticeable than for the IPMSMs.

Finally, Figure 11 shows the actual and estimated torque when there is a step-like change in the q-axis current command from 0 to 1 p.u.. It can be observed from the error shown in Figure 11b that the torque estimator responds in the range of ms.

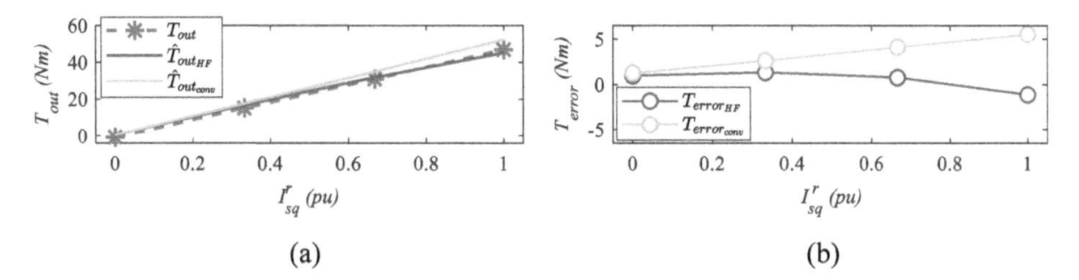

Figure 10. Experimental results: SPMSM & pulsating voltage injection: (**a**) Estimated and measured torque. (**b**) Estimated torque error. $0 < I^r_{sdq} < 1$ p.u., $\omega_r = 16$ Hz, $V_{HF} = 10$ V, $\omega_{HF} = 2 \cdot \pi \cdot 250$ rad/s.

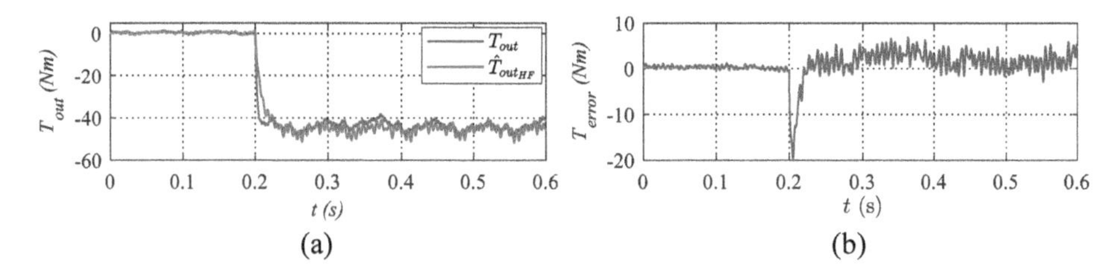

Figure 11. Experimental results: SPMSM & pulsating voltage injection: (**a**) Estimated and measured torque. (**b**) Estimated torque error. Transient response to a step-like change in q-axis current command from 0 to 1 p.u., $\omega_r = 16$ Hz, $V_{HF} = 10$ V, $\omega_{HF} = 2 \cdot \pi \cdot 250$ rad/s.

5.3. SynRM Torque Estimation

Torque estimation for the SynRM has been performed using rotating HF voltage injection. A HF voltage of 40 V and 500 Hz has been used (31). Figure 12 shows experimental results when the magnitude of the fundamental current, I^r_{sq}, changes from 0p.u. to 1p.u. following a MTPA trajectory. In a first approach, the machine has been considered to be working at low-middle saturation levels. Therefore, the coefficient linking the HF estimated inductances and the absolute inductances k_μ, has been considered to be constant and equal to 1 (i.e., the incremental inductance has been assumed to be approximately equal to the absolute inductance, see Figure 3). However, from the torque estimation error shown in Figure 12b, it is deduced that once the machine begins to saturate this approach is not longer valid. This suggests that the relation between the HF estimated inductances and the absolute inductances has to be adjusted using at least a second order polynomial. This is a subject of ongoing research.

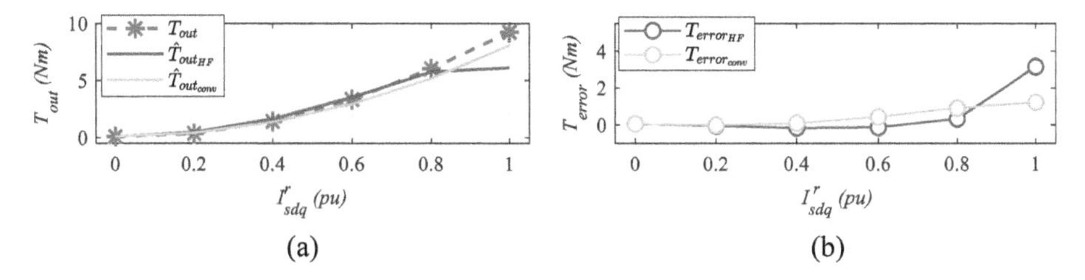

Figure 12. Experimental results: SynRM & rotating voltage injection: (**a**) Estimated and measured torque. (**b**) Estimated torque error. $0 < I^r_{sdq} < 1$ p.u., $\omega_r = 16$ Hz, $V_{HF} = 50$ V, $\omega_{HF} = 2 \cdot \pi \cdot 500$ rad/s.

6. Conclusions

Parameter estimation using HF signal injection aimed to improve the accuracy of torque estimation methods has been addressed in this paper, with the aim of making the estimation robust against variations in the operating conditions of the machine. This implies a reformulation of the torque equation, which will be function of the HF (incremental) inductances instead of the absolute inductances. Accurate modelling of the relationship between the incremental and absolute inductances will be therefore of paramount importance. Additionally, estimation of the PM flux is based on the linear relation with the d-axis inductance.

Three different types of HF signal injection have been considered: Pulsating Voltage Injection, Rotating Voltage Injection and Pulsating dq-axes Current Injection. In all the cases, the signal is superposed on top of the fundamental excitation applied by the inverter, not interfering therefore with the normal operation of the drive. It is concluded that pulsating current injection is advantageous as it is insensitive to the resistive components of the HF model and to cross-coupling effect. In change, its implementation is slightly more difficult due to the need of HF current controllers. In all the cases, no modification of the hardware is required.

Experimental verification using IPMSM, a SPMSM and a SynRM have been presented, which confirm the viability of the proposed methods.

Author Contributions: All the authors contributed in formulating the problem and designing the research proposal. M.M. wrote the paper, collected the experimental data and analysed the data. D.R. suggested the research topic, guided M.M. to complete the research and helped to analyse the data. D.F. developed the set up for the experimental tests. F.B. helped to write and finish the paper. All authors discussed the results and contributed to revised the final manuscript. All authors have read and agreed to the published version of the manuscript.

Funding: This work was funded in part by the Research, Technological Development and Innovation Programs of the Spanish Ministry Economy and Competitiveness, under grant MINECO-17-ENE2016-80047-R, by the Government of Asturias under project IDI/2018/000188 and FEDER funds and by the University of Oviedo under grant PAPI 2018-PF-12.

Conflicts of Interest: The authors declare no conflicts of interest. The funders had no role in the design of the study; in the collection, analyses, or interpretation of data; in the writing of the manuscript; or in the decision to publish the results.

Abbreviations

The following abbreviations are used in this manuscript:

HF	High Frequency
IPMSMs	Interior Permanent Magnet Synchronous Machines
MTPA	Maximum Torque Per Ampere
SMs	Synchronous Machines
SPMSMs	Surface Permanent Magnet Synchronous Machines
SynRMs	Synchronous Reluctance Machines

References

1. Matsuo, T.; Lipo, T.A. Field oriented control of synchronous reluctance machine. In Proceedings of the IEEE Power Electronics Specialist Conference—PESC '93, Seattle, WA, USA, 20–24 June 1993; pp. 425–431. [CrossRef]
2. Maroufian, S.S.; Pillay, P. Torque Characterization of a Synchronous Reluctance Machine Using an Analytical Model. *IEEE Trans. Transp. Electrif.* **2018**, *4*, 506–516. [CrossRef]
3. Shi, R.; Toliyat, H.A.; El-Antably, A. A DSP-based direct torque control of five-phase synchronous reluctance motor drive. In Proceedings of the APEC 2001, Sixteenth Annual IEEE Applied Power Electronics Conference and Exposition (Cat. No.01CH37181), Anaheim, CA, USA, 4–8 March 2001; Volume 2, pp. 1077–1082, doi:10.1109/APEC.2001.912500. [CrossRef]

4. Shinke, A.; Hasegawa, M.; Matsui, K. Torque estimation for synchronous reluctance motors using robust flux observer to magnetic saturation. In Proceedings of the 2009 IEEE International Symposium on Industrial Electronics, Seoul, Korea, 5–8 July 2009; pp. 1569–1574. [CrossRef]
5. Yousefi-Talouki, A.; Pescetto, P.; Pellegrino, G.; Boldea, I. Combined Active Flux and High-Frequency Injection Methods for Sensorless Direct-Flux Vector Control of Synchronous Reluctance Machines. *IEEE Trans. Power Electron.* **2018**, *33*, 2447–2457. [CrossRef]
6. Senjyu, T.; Kinjo, K.; Urasaki, N.; Uezato, K. High efficiency control of synchronous reluctance motors using extended Kalman filter. *IEEE Trans. Ind. Electron.* **2003**, *50*, 726–732. [CrossRef]
7. Kim, S.; Im, J.; Go, S.C.; Bae, J.; Kim, W.; Kim, K.; Kim, C.; Lee, J. Robust Torque Control of DC Link Voltage Fluctuation for SynRM Considering Inductances With Magnetic Saturation. *IEEE Trans. Magn.* **2010**, *46*, 2005–2008. [CrossRef]
8. Gieras, J.F. *Permanent Magnet Motor Technology: Design and Applications*, 3rd ed.; CRC Press: Boca Raton, FL, USA, 2009.
9. Limsuwan, N.; Kato, T.; Akatsu, K.; Lorenz, R.D. Design and Evaluation of a Variable-Flux Flux-Intensifying Interior Permanent-Magnet Machine. *IEEE Trans. Ind. Appl.* **2014**, *50*, 1015–1024. [CrossRef]
10. Jung, H.; Park, D.; Kim, H.; Sul, S.; Berry, D.J. Non-Invasive Magnet Temperature Estimation of IPMSM Based on High-Frequency Inductance With a Pulsating High-Frequency Voltage Signal Injection. *IEEE Trans. Ind. Appl.* **2019**, *55*, 3076–3086. [CrossRef]
11. Reigosa, D.; Fernández, D.; Martínez, M.; Guerrero, J.M.; Diez, A.B.; Briz, F. Magnet Temperature Estimation in Permanent Magnet Synchronous Machines Using the High Frequency Inductance. *IEEE Trans. Ind. Appl.* **2019**, *55*, 2750–2757. [CrossRef]
12. Martinez, M.; Reigosa, D.; Fernandez, D.; Guerrero, J.M.; Briz, F. Enhancement of Permanent Magnet Synchronous Machines Torque Estimation Using Pulsating High Frequency Current Injection. *IEEE Trans. Ind. Appl.* **2019**, 1–9. [CrossRef]
13. Interface. Torque Sensors: T5. 2019. Available online: https://interfaceforce.co.uk/ (accessed on 1 September 2019).
14. HBM. HBM Torque Sensors. Available online: https://www.hbm.com/en/0264/torque-transducers-torque-sensorstorque-meters/ (accessed on 1 September 2019).
15. TE Connectivity. Measure Reaction and Rotating Torque. Available online: https://www.te.com/global-en/products/sensors/torque-sensors.html (accessed on 16 October 2019).
16. Futek. Futek Torque Sensors. Available online: https://www.futek.com/store/Torque%20Sensors (accessed on 16 October 2019).
17. Magtrol Motors Testing and Sensors. Magtrol Torque Transducers. Available online: https://www.magtrol.com/product-category/torque-transducers/ (accessed on 16 October 2019).
18. Heins, G.; Thiele, M.; Brown, T. Accurate Torque Ripple Measurement for PMSM. *IEEE Trans. Instrum. Meas.* **2011**, *60*, 3868–3874. [CrossRef]
19. Sue, P.; Wilson, D.; Farr, L.; Kretschmar, A. High precision torque measurement on a rotating load coupling for power generation operations. In Proceedings of the 2012 IEEE International Instrumentation and Measurement Technology Conference Proceedings, Graz, Austria, 13–16 May 2012; pp. 518–523. [CrossRef]
20. Zakrzewski, J. Combined magnetoelastic transducer for torque and force measurement. *IEEE Trans. Instrum. Meas.* **1997**, *46*, 807–810. [CrossRef]
21. Yeo, K.C.; Heins, G.; Boer, F.D. Comparison of torque estimators for PMSM. In Proceedings of the 2008 Australasian Universities Power Engineering Conference, Sydney, NSW, Australia, 14–17 December 2008; pp. 1–6.
22. Cheng, B.; Tesch, T.R. Torque Feedforward Control Technique for Permanent-Magnet Synchronous Motors. *IEEE Trans. Ind. Electron.* **2010**, *57*, 969–974. [CrossRef]
23. Jukic, F.; Sumina, D.; Erceg, I. Comparison of torque estimation methods for interior permanent magnet wind power generator. In Proceedings of the 2017 19th International Conference on Electrical Drives and Power Electronics (EDPE), Dubrovnik, Croatia, 4–6 October 2017; pp. 291–296. [CrossRef]

24. Xu, J.X.; Panda, S.K.; Pan, Y.J.; Lee, T.H.; Lam, B.H. Improved PMSM pulsating torque minimization with iterative learning and sliding mode observer. In Proceedings of the 2000 26th Annual Conference of the IEEE Industrial Electronics Society, IECON 2000, 2000 IEEE International Conference on Industrial Electronics, Control and Instrumentation, 21st Century Technologies, Nagoya, Japan, 22–28 October 2000; Volume 3, pp. 1931–1936. [CrossRef]
25. Lam, B.H.; Panda, S.K.; Xu, J.X. Torque ripple minimization in PM synchronous motors an iterative learning control approach. In Proceedings of the IEEE 1999 International Conference on Power Electronics and Drive Systems, PEDS'99 (Cat. No.99TH8475), Hong Kong, China, 27–29 July 1999; Volume 1, pp. 144–149. [CrossRef]
26. Chung, S.K.; Kim, H.S.; Kim, C.G.; Youn, M.J. A new instantaneous torque control of PM synchronous motor for high-performance direct-drive applications. *IEEE Trans. Power Electron.* **1998**, *13*, 388–400. [CrossRef]
27. Dong, X.; Tianmiao, W.; Hongxing, W. Comparison between model reference observer and reduced order observer of PMSM torque. In Proceedings of the 2011 6th IEEE Conference on Industrial Electronics and Applications, Beijing, China, 21–23 June 2011; pp. 663–667. [CrossRef]
28. Liu, Q.; Hameyer, K. High-Performance Adaptive Torque Control for an IPMSM With Real-Time MTPA Operation. *IEEE Trans. Energy Convers.* **2017**, *32*, 571–581. [CrossRef]
29. Mohamed, Y.A.I.; Lee, T.K. Adaptive self-tuning MTPA vector controller for IPMSM drive system. *IEEE Trans. Energy Convers.* **2006**, *21*, 636–644. [CrossRef]
30. Traoré, W.F.; McCann, R. Torque measurements in synchronous generators using giant magnetoresistive sensor arrays via the Maxwell stress tensor. In Proceedings of the 2013 IEEE Power Energy Society General Meeting, Vancouver, BC, Canada, 21–25 July 2013; pp. 1–5. [CrossRef]
31. Lin, Z.; Reay, D.S.; Williams, B.W.; He, X. Online Modeling for Switched Reluctance Motors Using B-Spline Neural Networks. *IEEE Trans. Ind. Electron.* **2007**, *54*, 3317–3322. [CrossRef]
32. Reigosa, D.; Kang, Y.G.; Martínez, M.; Fernández, D.; Guerrero, J.M.; Briz, F. SPMSMs Sensorless Torque Estimation Using High Frequency Signal Injection. In Proceedings of the 2019 IEEE Energy Conversion Congress and Exposition (ECCE), Baltimore, MD, USA, 29 September–3 October 2019; pp. 2388–2393. [CrossRef]
33. Martinez, M.; Laborda, D.F.; Reigosa, D.; Fernández, D.; Guerrero, J.M.; Briz, F. SynRM Sensorless Torque Estimation Using High Frequency Signal Injection. In Proceedings of the 2019 IEEE 10th International Symposium on Sensorless Control for Electrical Drives (SLED), Turin, Italy, 9–10 September 2019; pp. 1–5. [CrossRef]
34. Xu, W.; Lorenz, R.D. High-Frequency Injection-Based Stator Flux Linkage and Torque Estimation for DB-DTFC Implementation on IPMSMs Considering Cross-Saturation Effects. *IEEE Trans. Ind. Appl.* **2014**, *50*, 3805–3815. [CrossRef]
35. Hanselman, D.C. *Brushless Permanent-Magnet Motor Design*; McGraw-Hill Inc.: New York, NY, USA, 1994.
36. Emrax e-Motors. Standard Motors: Emrax 228. 2019. Available online: https://emrax.com/products/emrax-228 (accessed on 16 October 2019).
37. UniTek. Bamocar-D3. 2019. Available online: https://www.unitek-industrie-elektronik.de/bamocar-d3 (accessed on 16 October 2019).
38. Texas Instrument. Microcontrollers(MCU): TMS320F28335. 2019. Available online: http://www.ti.com/product/TMS320F28335 (accessed on 16 October 2019).

Article

Sensorless Control for IPMSM Based on Adaptive Super-Twisting Sliding-Mode Observer and Improved Phase-Locked Loop

Shuo Chen [1], Xiao Zhang [1,*], Xiang Wu [1], Guojun Tan [1] and Xianchao Chen [2]

[1] School of Electrical and Power Engineering, China University of Mining and Technology, Xuzhou 221116, China; ts17130047a3@cumt.edu.cn (S.C.); zb13060003@cumt.edu.cn (X.W.); gjtan@cumt.edu.cn (G.T.)

[2] Xuzhou Yirui Construction Machinery Co. Ltd., Xuzhou 221000, China; xinchao623@163.com

* Correspondence: zhangxiao@cumt.edu.cn; Tel.: +86-137-0521-7567

Received: 6 March 2019; Accepted: 26 March 2019; Published: 29 March 2019

Abstract: In traditional sensorless control of the interior permanent magnet synchronous motors (IPMSMs) for medium and high speed domains, a control strategy based on a sliding-mode observer (SMO) and phase-locked loop (PLL) is widely applied. A new strategy for IPMSM sensorless control based on an adaptive super-twisting sliding-mode observer and improved phase-locked loop is proposed in this paper. A super-twisting sliding-mode observer (STO) can eliminate the chattering problem without low-pass filters (LPFs), which is an effective method to obtain the estimated back electromotive forces (EMFs). However, the constant sliding-mode gains in STO may cause instability in the high speed domain and chattering in the low speed domain. The speed-related adaptive gains are proposed to achieve the accurate estimation of the observer in wide speed range and the corresponding stability is proved. When the speed of IPMSM is reversed, the traditional PLL will lose its accuracy, resulting in a position estimation error of 180°. The improved PLL based on a simple strategy for signal reconstruction of back EMF is proposed to ensure that the motor can realize the direction switching of speed stably. The proposed strategy is verified by experimental testing with a 60-kW IPMSM sensorless drive.

Keywords: interior permanent magnet synchronous motor (IPMSM); sensorless control; adaptive algorithm; super-twisting sliding mode observer (STO); phase-locked loop (PLL)

1. Introduction

Recently, interior permanent magnet synchronous motors (IPMSMs) have been extensively utilized in the fields of electromechanical drives, electric vehicles, and numerical control servo systems due to their robustness, high efficiency, high power density, and compactness [1–4]. The usage of position sensors decreases the reliability and increases the cost and volume of IPMSM drives. In order to overcome these shortcomings caused by the use of mechanical position sensors, sensorless control technology has become one of the important research directions in related fields [5,6]. Generally, sensorless control strategies can be divided into two categories. The first one is called signal injection methods [7–9]. This method is based on the salient pole effect of the motor, which is mainly used in zero and low speed domains. The second one is called back EMF based methods [10–19], which utilizes the estimated back EMF signals to obtain the position information of the motor. Because the magnitude of back EMF is in proportion to the speed of the motor, the performance of back EMF based methods at ultra-low and zero speed is extremely poor [11]. Hence, back EMF based methods and signal injection methods are usually combined to achieve sensorless control for a whole speed range [12–14]. Back EMF based methods primarily includes the model adaptive method (MRAS) [16], the Kalman filtering method (EKF) [17], and the sliding mode observer (SMO) [2,18,19], etc. Compared with MRAS and EKF, SMO has simpler structure and stronger robustness. Hence, SMO is extensively applied in sensorless control strategy [19].

The signum function used in traditional SMO can introduce high frequency harmonics into the estimated signals, which eventually lead to the inevitable chattering phenomenon. Therefore, low-pass filters (LPFs) are commonly utilized to smooth the estimated signals. However, the LPFs in turn bring the disadvantages of phase delay of estimated signals. In [20], signum function is utilized to reduce the SMO chattering phenomenon caused by sigmoid function. In [21], an adaptive filter is proposed to reduce the negative effects of LPFs. However, these methods cannot completely avoid phase delay caused by LPFs. In [22,23], the super-twisting algorithm is proposed to eliminate the chattering phenomenon caused by signum function. The super-twisting sliding mode observer (STO) can effectively eliminate the sliding-mode chattering phenomenon without compromising robustness and avoid the use of LPFs. In [24], the stability of STO is further analyzed by using the Lyapunov function and the corresponding stability conditions are given. In [25], the sensorless control strategy based on STO and resistance identification is proposed for SPMSM. Resistance identification enhances the robustness of the super-twisting sliding mode observer. Although STO performs well in reducing chattering, there is still a problem to be solved. When the constant sliding-mode gains are adopted in this method, the sliding-mode gains should be big enough to meet the stability condition in the wide speed range. But the large sliding-mode gains will lead to a large chattering phenomenon, especially in a low speed domain [19].

Traditionally, the position information is obtained by the estimated back electromotive forces through arc-tangent method directly. However, the arc-tangent function makes position information susceptible to harmonics and noises. In order to improve estimation performance, the quadrature phase-locked loop algorithm is proposed in [6], which is called the traditional PLL in this paper. High-order harmonics can be filtered out due to the special structure of PLL. When the speed of IPMSM is reversed, the traditional PLL will lose its accuracy, resulting in a position estimation error of 180°. The reason for such drawback is that the sign of the back EMFs has an effect on the sign of the equivalent position error [26,27]. To solve the aforementioned problem, Refs. [26,27] proposed a kind of PLL, which constructs the equation of the equivalent position error based on tangent function. Such a scheme may overcome the problem, but it brings complexity to the algorithm and it is vulnerable to harmonics and noises due to the introduction of a tangent function.

In this paper, a new strategy based on adaptive super-twisting sliding mode observer and improved PLL for IPMSM sensorless control is proposed to overcome aforementioned limitations. Super-twisting sliding-mode observer is utilized to obtain the estimated back electromotive forces. Moreover, speed-related adaptive gains are proposed to achieve accurate estimation in a wide speed domain so that they widen the speed range of the super-twisting sliding-mode observer. On the basis of existing stability conditions in [24], the stability of the proposed adaptive STO is proved in this paper. To improve the shortcomings of the above-mentioned two kinds of PLL, a simple strategy for signal reconstruction of back EMF is proposed. Based on this strategy, the improved PLL can overcome the limitation of speed reversal existing in traditional PLL without the introduction of tangent function. Besides, the improved PLL has simple structure, great steady performance, and transient response. Finally, the proposed strategy based on adaptive STO and improved PLL is verified by experimental testing with a 60-kW IPMSM sensorless drive.

2. Adaptive Super-Twisting Sliding-Mode Observer

For the sake of convenience, magnetic saturation is neglected and it is assumed that the flux linkage distribution is perfectly sinusoidal. The model of IPMSM is shown in Figure 1. The ABC, $\alpha\beta$ and dq frames represent the natural, the stationary, and the rotating reference frames, respectively.

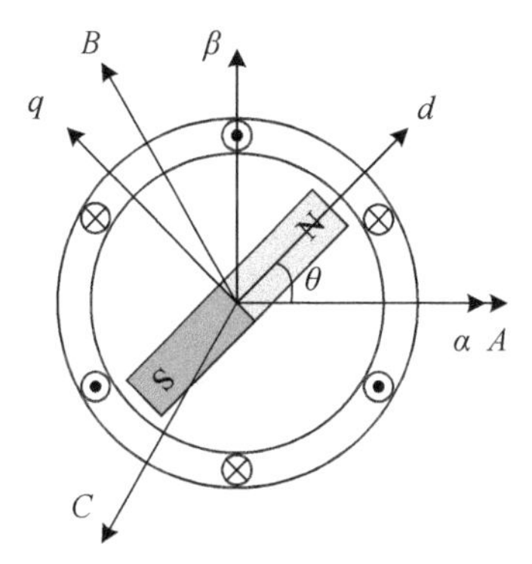

Figure 1. The model of interior permanent magnet synchronous motors (IPMSM).

The mathematic model of IPMSM in $\alpha\beta$ stationary reference frame is expressed as

$$u_\alpha = Ri_\alpha + L_d\frac{di_\alpha}{dt} + \omega_e(L_d - L_q)i_\beta + e_\alpha \tag{1}$$

$$u_\beta = Ri_\beta + L_q\frac{di_\beta}{dt} - \omega_e(L_d - L_q)i_\alpha + e_\beta \tag{2}$$

where u_α, u_β are stator voltages; i_α, i_β are stator currents; R is stator resistance; ω_e is electrical rotor speed; ψ_f is PM flux linkage; and L_d,L_q are stator inductances. e_α and e_β are the $\alpha\beta$-axis back EMFs of IPMSM, satisfying $e_\alpha = -Esin\theta$ and $e_\beta = Ecos\theta$. θ is the rotor position and E is the amplitude of back EMF [28], satisfying

$$E = (L_d - L_q)\left(\omega_e i_d - \frac{di_q}{dt}\right) + \omega_e\psi_f \tag{3}$$

2.1. Super-Twisting Algorithm

A. Levant proposed the super-twisting algorithm to eliminate the chatter caused by the signum function in [23,29]. The fundamental form of this algorithm is written as follows:

$$\frac{d\hat{x}_1}{dt} = -k_1|\hat{x}_1 - x_1|\text{sign}(\hat{x}_1 - x_1) + \hat{x}_2 + \rho_1 \tag{4}$$

$$\frac{d\hat{x}_2}{dt} = -k_2\text{sign}(\hat{x}_1 - x_1) + \rho_2 \tag{5}$$

where x_i, $\hat{x}_i$, k_i, sign(), and ρ_i are state variables, estimation of state variables, sliding-mode gains, signum function, and perturbation terms, respectively. The corresponding conditions of the stability of the super-twisting algorithm have been educed in [24]. If ρ_1 and ρ_2 in Equations (6) and (7) satisfy the following conditions:

$$\rho_1 \le \delta_1|x_1|^{\frac{1}{2}},\ \rho_2 = 0 \tag{6}$$

where δ_1 is a positive constant and the sliding-mode gains k_1 and k_2 meet the condition:

$$k_1 > 2\delta_1,\ k_2 > k_1\frac{5\delta_1 k_1 + 4\delta_1^2}{2(k_1 - 2\delta_1)} \tag{7}$$

the stability of the system can be guaranteed.

2.2. Super-Twisting Sliding Mode Observer for IPMSM Sensorless Control

To estimate the back EMFs conveniently, the mathematic mode of IPMSM shown in Equations (3) and (4) is organized into the current model:

$$\frac{di_\alpha}{dt} = -\frac{R}{L_d}i_\alpha - \omega_e\frac{L_d - L_q}{L_d}i_\beta + \frac{u_\alpha}{L_d} - \frac{e_\alpha}{L_d} \tag{8}$$

$$\frac{di_\beta}{dt} = -\frac{R}{L_d}i_\beta + \omega_e\frac{L_d - L_q}{L_d}i_\alpha + \frac{u_\beta}{L_d} - \frac{e_\beta}{L_d} \tag{9}$$

The estimated currents are taken as state variables in Equations (4) and (5), then the STO for IPMSM sensorless control be represented as

$$\frac{d\hat{i}_\alpha}{dt} = -\frac{R}{L_d}\hat{i}_\alpha - \hat{\omega}_e\frac{L_d - L_q}{L_d}\hat{i}_\beta + \frac{u_\alpha}{L_d} - \frac{k_1}{L_d}|\bar{i}_\alpha|^{\frac{1}{2}}\text{sign}(\bar{i}_\alpha) - \frac{1}{L_d}\int k_2\text{sign}(\bar{i}_\alpha)dt \tag{10}$$

$$\frac{d\hat{i}_\beta}{dt} = -\frac{R}{L_d}\hat{i}_\beta + \hat{\omega}_e\frac{L_d - L_q}{L_d}\hat{i}_\alpha + \frac{u_\beta}{L_d} - \frac{k_1}{L_d}|\bar{i}_\beta|^{\frac{1}{2}}\text{sign}(\bar{i}_\beta) - \frac{1}{L_d}\int k_2\text{sign}(\bar{i}_\beta)dt \tag{11}$$

where $\bar{i}_\alpha = \hat{i}_\alpha - i_\alpha$, $\bar{i}_\beta = \hat{i}_\beta - i_\beta$ and ^ represents the estimated variable. It should be noticed that, differently from the STO for SPMSM sensorless control in [26], the perturbation term ρ_1 in Equation (4) for IPMSM sensorless control is replaced by $-\frac{R}{L_d}\hat{i}_\alpha - \hat{\omega}_e\frac{L_d-L_q}{L_d}\hat{i}_\beta + \frac{u_\alpha}{L_d}$ and $-\frac{R}{L_d}\hat{i}_\beta + \hat{\omega}_e\frac{L_d-L_q}{L_d}\hat{i}_\alpha + \frac{u_\beta}{L_d}$, respectively.

By substituting the perturbation terms into Equation (6) and taking estimated currents as state variables, Equation (6) can be reformulated as

$$-\frac{R}{L_d}\hat{i}_\alpha - \hat{\omega}_e\frac{L_d - L_q}{L_d}\hat{i}_\beta + \frac{u_\alpha}{L_d} \le \delta_1|\hat{i}_\alpha|^{\frac{1}{2}} \tag{12}$$

$$-\frac{R}{L_d}\hat{i}_\beta + \hat{\omega}_e\frac{L_d - L_q}{L_d}\hat{i}_\alpha + \frac{u_\beta}{L_d} \le \delta_1|\hat{i}_\beta|^{\frac{1}{2}} \tag{13}$$

If δ_1 is large enough, the stable conditions can be guaranteed easily. By subtracting Equations (8) and (9) from Equations (10) and (11) respectively, the state equations of the current estimation errors can be obtained:

$$\frac{d\bar{i}_\alpha}{dt} = -\frac{R}{L_d}\bar{i}_\alpha - \frac{L_d - L_q}{L_d}(\hat{\omega}_e\hat{i}_\beta - \omega_e i_\beta) - \frac{k_1}{L_d}|\bar{i}_\alpha|^{\frac{1}{2}}\text{sign}(\bar{i}_\alpha) - \frac{1}{L_d}\int k_2\text{sign}(\bar{i}_\alpha)dt + \frac{e_\alpha}{L_d} \tag{14}$$

$$\frac{d\bar{i}_\beta}{dt} = -\frac{R}{L_d}\bar{i}_\beta + \frac{L_d - L_q}{L_d}(\hat{\omega}_e\hat{i}_\alpha - \omega_e i_\alpha) - \frac{k_1}{L_d}|\bar{i}_\beta|^{\frac{1}{2}}\text{sign}(\bar{i}_\beta) - \frac{1}{L_d}\int k_2\text{sign}(\bar{i}_\beta)dt + \frac{e_\beta}{L_d} \tag{15}$$

when STO reaches the sliding surface, it is approximately considered that the estimated value is equal to the actual value ($\hat{\omega}_e \approx \omega_e$, $\hat{i}_\alpha \approx i_\alpha$ and $\hat{i}_\beta \approx i_\beta$). Then the equivalent control law of the back EMFs is expressed as

$$\hat{e}_\alpha = k_1|\bar{i}_\alpha|^{\frac{1}{2}}\text{sign}(\bar{i}_\alpha) + \int k_2\text{sign}(\bar{i}_\alpha)dt \tag{16}$$

$$\hat{e}_\beta = k_1|\bar{i}_\beta|^{\frac{1}{2}}\text{sign}(\bar{i}_\beta) + \int k_2\text{sign}(\bar{i}_\beta)dt \tag{17}$$

The linear term $k_1|\bar{i}_\alpha|^{\frac{1}{2}}\text{sign}(\bar{i}_\alpha)$ determines the convergence rate of the STO and the integral term $\int k_2\text{sign}(\bar{i}_\alpha)dt$ is related to the suppression of chattering phenomena. Hence, k_2 usually has a large value.

2.3. Adaptive Super-Twisting Sliding Mode Observer for IPMSM Sensorless Control

Although STO performs well in reducing chattering, there is still a problem to be solved. When the constant sliding-mode gains are adopted in this method, the sliding-mode gains should be large enough to meet the stable conditions when the IPMSM runs at high speed. However, due to the excessive sliding mode gains, the performance of the STO in the low speed domain will be seriously deteriorated [19]. In order to extract accurate rotor position in wide speed range, the STO for IPMSM with speed-related adaptive gains is proposed in this paper. The speed-related adaptive gains k_1 and k_2 are adopted as

$$k_1 = l_1\omega_e^*,\ k_2 = l_2\omega_e^{*2} \tag{18}$$

$$\omega_e^* = \begin{cases} \omega_{emin} & 0 \le \hat{\omega}_e < \omega_{emin} \\ \mathrm{LPF}(\hat{\omega}_e) & \omega_{emin} \le \hat{\omega}_e \le \omega_{emax} \\ \omega_{emax} & \hat{\omega}_e > \omega_{emax} \end{cases} \tag{19}$$

where l_1 and l_2 are adaptive coefficients, ω_{emax} is the maximum electrical rotor speed of motor, ω_{emin} is the minimum electrical rotor speed allowed by the STO for back EMFs observation. The first-order LPF in the STO is utilized to smooth the gain variations and improve the robustness of the observer in the transient process. Its cut-off frequency is determined according to ω_{emax} and switching frequency. The stability of adaptive STO is proved as follows:

In Equations (12) and (13), compared with $\frac{u_\alpha}{L_d}$ and $\frac{u_\beta}{L_d}$, $\frac{R}{L_d}\hat{i}_\alpha$, $\hat{\omega}_e\frac{L_d-L_q}{L_d}\hat{i}_\beta$, $\frac{R}{L_d}\hat{i}_\beta$ and $\hat{\omega}_e\frac{L_d-L_q}{L_d}\hat{i}_\alpha$ can be neglected. Then, the perturbation terms can be simplified as

$$\rho_1(i_\alpha) \approx \frac{u_\alpha}{L_d},\ \rho_1(i_\beta) \approx \frac{u_\beta}{L_d} \tag{20}$$

then, Equation (6) can be rewritten as

$$|\rho_1(i_\alpha)| \approx \left|\frac{u_\alpha}{L_d}\right| \approx \frac{\omega_e\psi_f}{L_d} \le \delta_1|\hat{i}_\alpha|^{\frac{1}{2}} \tag{21}$$

when STO reaches the sliding surface, $|\hat{i}_\alpha|^{\frac{1}{2}}$ is in a certain range and $\omega_e^* \approx \omega_e$. δ_1 is replaced by $\lambda\omega_e$ in Equation (21), Equation (21) can be rewritten as

$$|\rho_1(i_\alpha)| \approx \left|\frac{u_\alpha}{L_d}\right| \approx \frac{\omega_e\psi_f}{L_d} \le \lambda|\hat{i}_\alpha|^{\frac{1}{2}}\omega_e \tag{22}$$

This formula can be satisfied by choosing a large λ. Substituting $\delta_1 = \lambda\omega_e$, $k_1 = l_1\omega_e$ and $k_2 = l_2\omega_e^2$ into Equation (7), Equation (7) can be rewritten as

$$k_1 = l_1\omega_e > 2\delta_1 = 2\lambda\omega_e \tag{23}$$

$$k_2 = l_2{\omega_e}^2 > k_1\frac{5\delta_1 k_1 + 4\delta_1^2}{2(k_1 - 2\delta_1)} = l_1\frac{5\lambda l_1 + 4\lambda^2}{2l_1 - 4\lambda}{\omega_e}^2 \tag{24}$$

It is obvious that when the adaptive coefficients l_1 and l_2 satisfy the condition $l_1 > 2\lambda$ and $l_2 > l_1\frac{5\lambda l_1+4\lambda^2}{2l_1-4\lambda}$, the stability conditions of adaptive STO can be satisfied. The black diagram of adaptive STO for IPMSM sensorless control is shown in Figure 2.

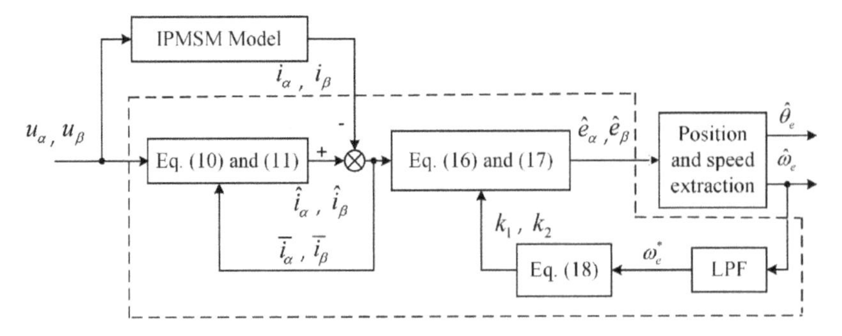

Figure 2. The black diagram of adaptive super-twisting sliding-mode observer (STO) for IPMSM sensorless control.

3. Acquisition of Position Information

Traditionally, the position information is obtained by the estimated back electromotive forces through arc-tangent method directly.

$$\hat{\theta}_e = -\arctan\cdot(\frac{\hat{e}_\alpha}{\hat{e}_\beta}) \tag{25}$$

The electrical rotor speed can be calculated by $\hat{\omega}_e = \frac{d\hat{\theta}_e}{dt}$. However, the estimated position and speed is susceptible to noise and harmonics because of the usage of arc-tangent method. Especially when $\hat{e}_\beta$ crosses zero, the obvious estimation errors may be produced. Ref. [6] proposed the quadrature phase-locked loop algorithm to mitigate the adverse effect. In this paper, this algorithm is called the traditional PLL.

3.1. Traditional PLL

The transfer function of the traditional PLL can be written as

$$G(s) = \frac{\hat{\theta}_e}{\theta_e} = \frac{EK_p s + EK_i}{s^2 + EK_p s + EK_i} \tag{26}$$

where K_p is the proportional gain, K_i is the integral gain. The structure of the traditional PLL is represented in Figure 3.

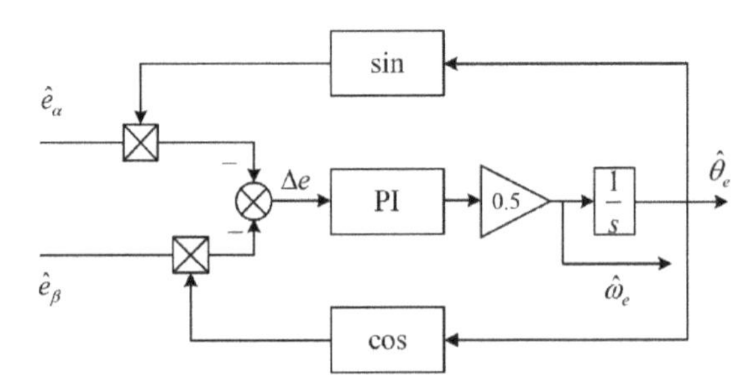

Figure 3. The structure of the traditional phase-locked loop (PLL).

The bode diagram of Equation (26) with different E is shown in Figure 4. As shown in Figure 4, E varies with the rotor speed, so the bandwidth of the PLL is influenced by the operating frequency of motor. This could make the design of system parameters more difficult and deteriorate the accuracy of the position estimation. Therefore, the normalization of the back EMFs is necessary.

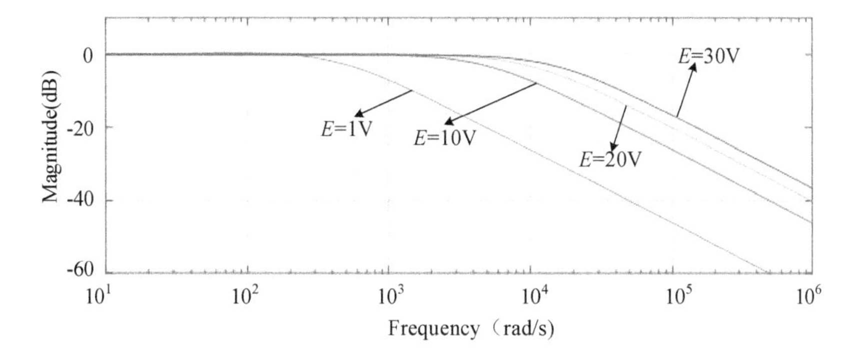

Figure 4. Bode diagram of the traditional PLL transfer function with different E.

By normalizing the estimated back EMF, the equivalent position error Δe can be written as

$$\begin{aligned}\Delta e &= \frac{1}{\sqrt{\hat{e}_\alpha^2+\hat{e}_\beta^2}}\left[-\hat{e}_\alpha \cos(\hat{\theta}_e) - \hat{e}_\beta \sin(\hat{\theta}_e)\right] \\ &= -\hat{e}_{\alpha n} \cos(\hat{\theta}_e) - \hat{e}_{\beta n} \sin(\hat{\theta}_e) \\ &= \sin(\theta_e)\cos(\hat{\theta}_e) - \sin(\hat{\theta}_e)\cos(\theta_e) \\ &= \sin(\theta_e - \hat{\theta}_e) \approx \theta_e - \hat{\theta}_e\end{aligned} \tag{27}$$

where $\hat{e}_{\alpha n}$ and $\hat{e}_{\beta n}$ are the normalized back EMFs, and the closed-loop transfer function of the traditional PLL with back EMF normalization can be obtained by

$$G(s) = \frac{\hat{\theta}_e}{\theta_e} = \frac{K_p s + K_i}{s^2 + K_p s + K_i} \tag{28}$$

The traditional PLL has the characteristics of LPF. High-order harmonics can be filtered out due to the special structure of phase-locked loop. However, when the speed of IPMSM is reversed, the traditional PLL will lose its accuracy, resulting in a position estimation error of 180°. When the parameters of PLL are set for one direction of rotation, the estimation of rotor position is correct for this direction only and an error of 180° will be produced in the other direction. Such a drawback makes the traditional PLL not suitable for applications where the motor needs to switch the direction of rotation. The theoretical analysis of the above problem is shown in Section 3.3.

3.2. Tangent-Based PLL

To solve the aforementioned problem, Refs. [26,27] proposed a kind of PLL scheme, which constructs the equivalent position error equation based on tangent function.

$$\begin{aligned}\Delta e &= \frac{\frac{\hat{e}_\alpha}{\hat{e}_\beta} - \frac{\sin(\frac{\hat{\theta}_e}{2})}{\cos(\frac{\hat{\theta}_e}{2})}}{1+\frac{\hat{e}_\alpha}{\hat{e}_\beta}\cdot\frac{\sin(\frac{\hat{\theta}_e}{2})}{\cos(\frac{\hat{\theta}_e}{2})}} = \frac{\tan(\theta_e)-\tan(\frac{\hat{\theta}_e}{2})}{1+\tan(\theta_e)\cdot\tan(\frac{\hat{\theta}_e}{2})} \\ &= \tan\left(\theta_e - \frac{\hat{\theta}_e}{2}\right)\end{aligned} \tag{29}$$

The structure of the tangent-based PLL is shown in Figure 5. When the system achieves the steady point, rotor position can be calculated as

$$\theta_e = \frac{\hat{\theta}_e}{2} \tag{30}$$

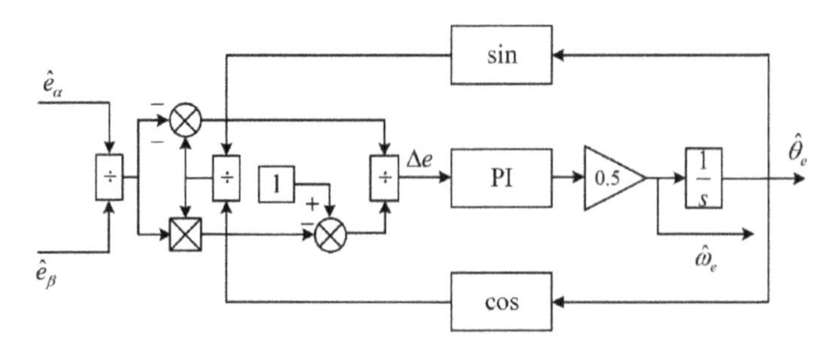

Figure 5. The structure of the tangent-based PLL.

This kind of PLL can solve the reversal problem. However, it increases the complexity of the algorithm. And it is vulnerable to harmonic and noise interference due to the introduction of tangent function. Especially, during $\hat{e}_\beta$ crosses zero and the rotor position crosses $\pm\frac{\pi}{2}$, the obvious estimation error may occur.

3.3. Improved PLL

The improved PLL is based on a simple EMF signals reconstruction strategy. The structure of the improved PLL is depicted in Figure 6 and the equation of the equivalent position error in the proposed scheme can be expressed as

$$\begin{aligned}\Delta e &= -\hat{e}_{\alpha n}\hat{e}_{\beta n}\cos(2\hat{\theta}_e) + \frac{(\hat{e}_{\alpha n}{}^2 - \hat{e}_{\beta n}{}^2)}{2}\sin(2\hat{\theta}_e) \\ &= \frac{1}{2}[\sin(2\theta_e)\cos(2\hat{\theta}_e) - \sin(2\hat{\theta}_e)\cos(2\theta_e)] \\ &= \frac{1}{2}\sin(2(\theta_e - \hat{\theta}_e))\end{aligned} \tag{31}$$

when the system reaches the stable point, Δe can be derived as

$$\begin{aligned}\Delta e = &\ \frac{1}{2}\sin(2(\theta_e - \hat{\theta}_e)) \\ &\approx \theta_e - \hat{\theta}_e\end{aligned} \tag{32}$$

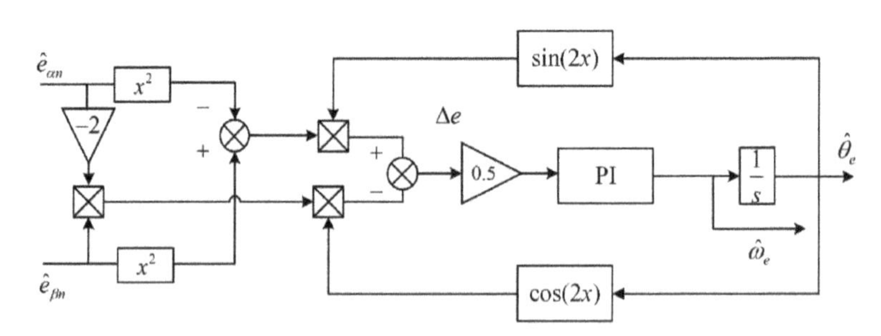

Figure 6. The structure of the improved PLL.

In the positive speed range of the motor,

$$\hat{e}_{\alpha n} = -\sin(\theta_e),\ \hat{e}_{\beta n} = \cos(\theta_e) \tag{33}$$

and the dynamic equations of the traditional PLL with back EMF normalization are represented as

$$\frac{de_\theta}{dt} = e_\omega \tag{34}$$

$$\frac{de_\omega}{dt} = -K_p \cos(e_\theta)e_\omega - K_I \sin(e_\theta) \tag{35}$$

where $e_\theta = \theta_e - \hat{\theta}_e$, $e_\omega = \omega_e - \hat{\omega}_e$. The phase trajectory of the traditional PLL for positive speed is shown in Figure 7a. As shown in Figure 7a, there are three equilibrium points in the system, which are (0,0), (π,0) and ($-\pi$,0). Among the three equilibrium points, only (0,0) is stable point. The others are saddle points. That means the trajectories in the phase trajectory of traditional PLL for positive speed will move to the origin. In other words, e_θ and e_ω can converge to (0,0) in limited time, which meets the requirements of estimation performance.

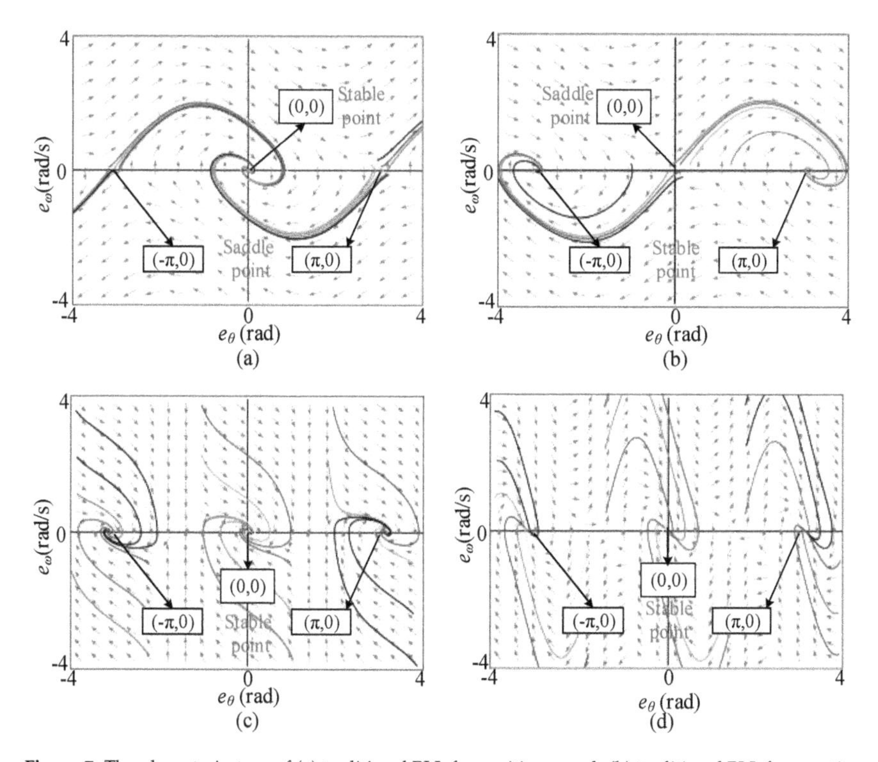

Figure 7. The phase trajectory of (**a**) traditional PLL for positive speed. (**b**) traditional PLL for negative speed. (**c**) tangent-based PLL for both positive and negative speed. (**d**) improved PLL for both positive and negative speed.

But when the direction of rotation is reversed, the symbols of the back EMF change and the same symbolic change can be detected on the equivalent position error signal Δe:

$$\hat{e}_{\alpha n} = \sin(\theta_e),\ \hat{e}_{\beta n} = -\cos(\theta_e) \tag{36}$$

$$\begin{aligned} \Delta e &= -\hat{e}_{\alpha n}\cos(\hat{\theta}_e) - \hat{e}_{\beta n}\sin(\hat{\theta}_e) \\ &= -\sin(\theta_e)\cos(\hat{\theta}_e) + \sin(\hat{\theta}_e)\cos(\theta_e) \\ &= \sin(\hat{\theta}_e - \theta_e) \approx -(\theta_e - \hat{\theta}_e) \end{aligned} \tag{37}$$

And the dynamic equations of the traditional PLL are rewritten as

$$\frac{de_\theta}{dt} = e_\omega \tag{38}$$

$$\frac{de_\omega}{dt} = K_p \cos(e_\theta)e_\omega + K_I \sin(e_\theta) \tag{39}$$

The phase trajectory of the traditional PLL for negative speed is given in Figure 7b. The system has the same three equilibrium points, which are (0,0), (π,0) and ($-\pi$,0). However, (0,0) changes into saddle point and ($\pm\pi$,0) become stable points. The trajectories in the nonlinear system depart from (0,0) to reach the stable points ($\pm\pi$,0) so that the system produce a position estimation error of 180°. Although this problem can be solved by resetting the gains of the PI controller, it is difficult to implement in real-time control system. Therefore, the traditional PLL cannot meet the requirements of applications where the motor needs to switch the direction of rotation.

The phase trajectory of the tangent-based PLL for both positive and negative speed is shown in Figure 7c. More details can be found in [26,27]. In this kind of PLL system, (0,0), (π,0), and ($-\pi$,0) are three stable points. By setting the proper parameters of PI regulator, e_θ and e_ω can converge to (0,0). That means the tangent-based PLL can solve the reversal problem. But due to the introduction of tangent function, it is vulnerable to harmonic and noise interference. Especially when $\hat{e}_\beta$ crosses zero and the position crosses $\pm\frac{\pi}{2}$, the obvious estimation errors will be produced. This algorithm is difficult to adopt in practice.

Compared with the traditional PLL and the tangent-based PLL, the improved PLL makes the speed reversal of motor not cause the symbolic change of the equivalent position error Δe by using a simple back EMF signals reconstruction strategy without tangent function. The dynamic equations are the same for both positive and negative speed and can be represented as

$$\frac{de_\theta}{dt} = e_\omega \tag{40}$$

$$\frac{de_\omega}{dt} = \frac{1}{2}\left[-K_p \cos(2e_\theta)2e_\omega - K_I \sin(2e_\theta)\right] \tag{41}$$

There are five equilibrium points in the system, which are (0,0), ($\pm\pi$,0) and ($\pm\frac{\pi}{2}$,0). In order to confirm the properties of equilibrium points in the system conveniently, the nonlinear equation of state is linearized. The Jacobian matrix $J(e_\theta, e_\omega)$ for (40) and (41) is represented as

$$J(e_\theta, e_\omega) = \begin{bmatrix} 0 & 1 \\ 2K_p \sin(2e_\theta)e_\omega - K_I \cos(2e_\theta) & -K_p \cos(2e_\theta) \end{bmatrix} \tag{42}$$

Substituting $(e_\theta, e_\omega) = (0,0)$ and $(e_\theta, e_\omega) = (\pm\pi, 0)$ into (42) respectively, the expression is the same at these points:

$$J(e_\theta, e_\omega)_{(e_\theta, e_\omega)=(0,0),(\pm\pi,0)} = \begin{bmatrix} 0 & 1 \\ -K_I & -K_p \end{bmatrix} \tag{43}$$

The eigenvalues of (43) can be expressed as

$$\lambda_1 = \frac{-K_p + \sqrt{K_p^2 - 4K_I}}{2}, \ \lambda_2 = \frac{-K_p - \sqrt{K_p^2 - 4K_I}}{2} \tag{44}$$

Because $K_p > 0$ and $K_I > 0$, λ_1 and λ_2 have negative real parts. That means (0,0) and ($\pm\pi$,0) are stable points.

Substituting $(e_\theta, e_\omega) = (\pm\frac{\pi}{2}, 0)$ into Equation (42) respectively, the expression is the same at these points:

$$J(e_\theta, e_\omega)_{(e_\theta, e_\omega)=(\pm\frac{\pi}{2},0)} = \begin{bmatrix} 0 & 1 \\ K_I & K_p \end{bmatrix} \tag{45}$$

The eigenvalues of Equation (45) can be expressed as

$$\lambda_1 = \frac{K_p + \sqrt{K_p^2 + 4K_I}}{2} > 0, \lambda_2 = \frac{K_p - \sqrt{K_p^2 + 4K_I}}{2} < 0 \tag{46}$$

Because $\lambda_1 > 0$ and $\lambda_2 < 0$, $(\pm\frac{\pi}{2},0)$ are saddle points in the system. In summary, among the five equilibrium points, (0,0) and $(\pm\pi,0)$ are stable points and $(\pm\frac{\pi}{2},0)$ are saddle points. The phase trajectory of the improved PLL for both positive and negative speed is shown in Figure 7d. Similar to the tangent-based PLL, each of these stable points is a focal point that the neighborhood phase trajectories will be attracted to. Moreover, because there is no introduction of the arctangent function, this method has better robustness than the tangent-based PLL. By selecting the appropriate gains of the PI regulator, e_θ and e_ω will converge to the origin. That means the motor can switch the speed direction steadily by adopting the proposed PLL.

4. Experimental Results

The control diagram of proposed sensorless control strategy for IPMSM based on adaptive STO and improved PLL is shown in Figure 8. The double closed-loop vector control is adopted. The details of the adaptive STO and the improved PLL are shown in Figures 2 and 6, respectively.

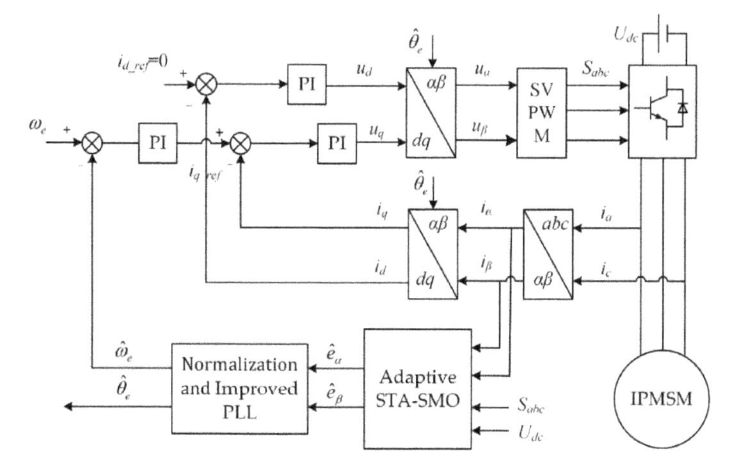

Figure 8. The control diagram of proposed sensorless control strategy for IPMSM.

An experimental prototype is shown in Figure 9 and the corresponding experimental platform was established as shown in Figure 10. The platform is mainly composed of two water-cooled IPMSMs, one rectifier, two inverters, and three controllers. The motor 1 is connected with inverter 1, and the proposed strategy is implemented by the controller 1. The motor 2 is a load motor which is controlled by the inverter 2, which is controlled by controller 2. Table 1 lists the parameters of the IPMSM. A 540 V dc-link voltage is obtained by the PWM rectifier for testing and verifying the performance of the proposed strategy. The rectifier is controlled by controller 3. In the experiment, TMS320F2812 DSP is adopted to carry out the new sensorless control strategy. All signals are converted by a digital-to-analog chip (TLV5610) and displayed on a digital oscilloscope. The traditional two-level inverter topology is adopted [30]. Switching frequency of the inverter and sampling frequency of the control system are set to 10 kHz. A rotary decoder (PGA411-Q1) is employed to obtain the actual position and speed of the motor, which are used for comparing and verifying the performance of the proposed strategy.

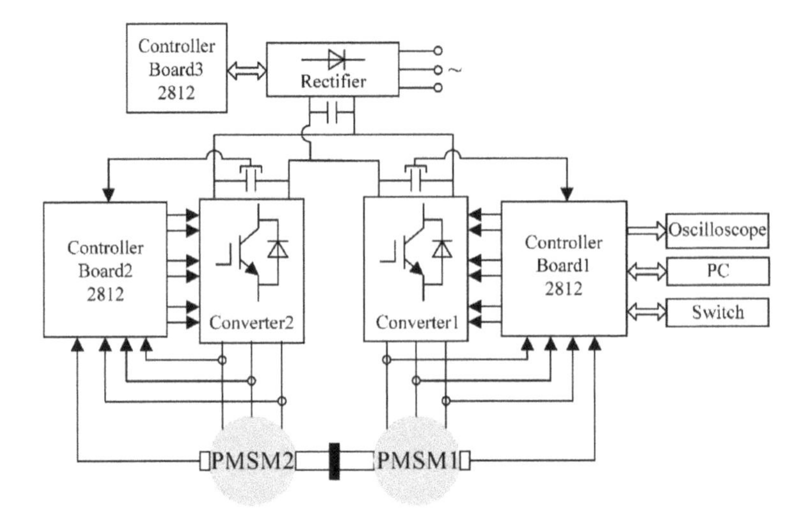

Figure 9. The experimental prototype.

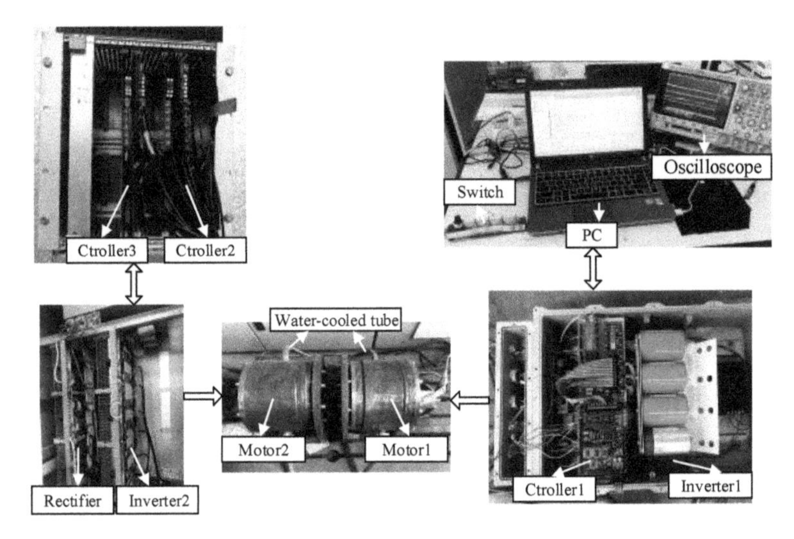

Figure 10. The experimental platform.

Table 1. Parameters of the IPMSM.

Parameter	Value
Flux linkage	0.225 Wb
d/q-axis inductor	0.95/2.05 mH
Resistance	0.1 Ω
Pole pairs	4
Rated power	60 kW
Rated speed	3000 rpm

4.1. Experimental Results of Adaptive Super-Twisting Sliding Mode Observer

The performances of the STO with constant sliding-mode gains in different speed ranges are presented in Figures 11 and 12. The parameters of the STO are $k_1 = 15$ and $k_2 = 60,000$ and the parameters of the PI regulator in the PLL are $K_p = 250$ and $K_i = 20,000$. Since the STO is based on the back electromotive forces model, the performance of STO is unreliable in ultra-low and zero speed domains. In this paper, IF control is adopted to ensure the start-up for IPMSM sensorless control. The threshold of speed that transiting from IF control to sensorless control is set to 300 r/min. The Figure 11 shows the performance of STO with no load from 0 to 1000 rpm.lo The IPMSM starts up in open-loop by using IF control at 1 s and switches to sensorless control at 2 s. Obviously, the estimation errors are large in the process of start-up and it takes about 1 s for the observer to get accurate rotor position information. When the IPMSM operates at 1000 r/min under sensorless control, the speed estimation error is within ±8 r/min and the position estimation error is between 1.08° and 7.2°. The estimated back EMFs have good sinusoidal properties. This means the STO with $k_1 = 15$ and $k_2 = 60,000$ can operate perfectly at 1000 r/min.

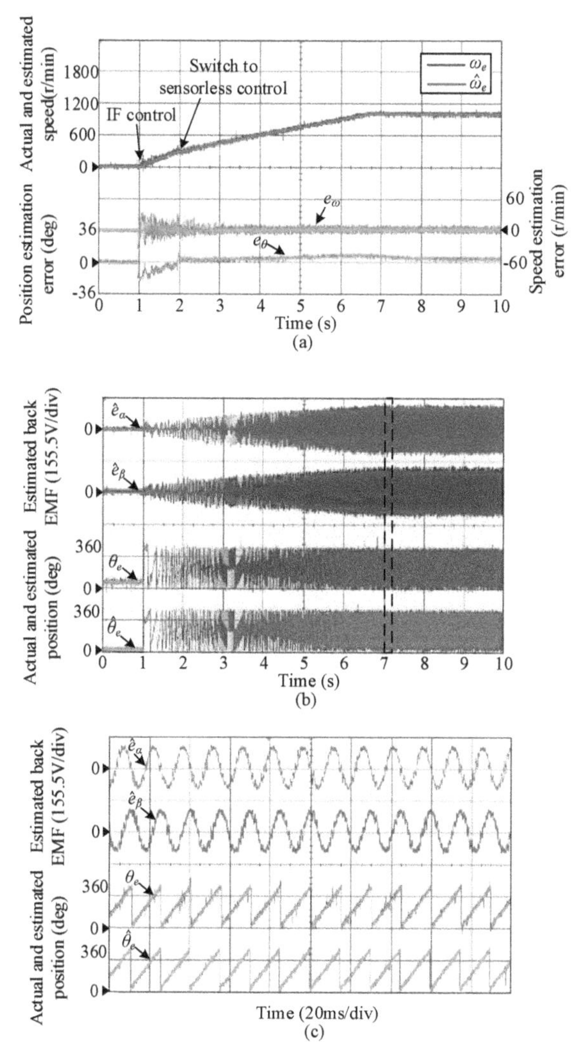

Figure 11. The performance of STO with no load from 0 to 1000 rpm. (**a**) Actual and estimated speed, speed estimation error, and position estimation error. (**b**) Estimated back electromotive forces (EMFs) and Actual and estimated position. The waveforms in (**b**) at 1000 r/min are zoomed in (**c**).

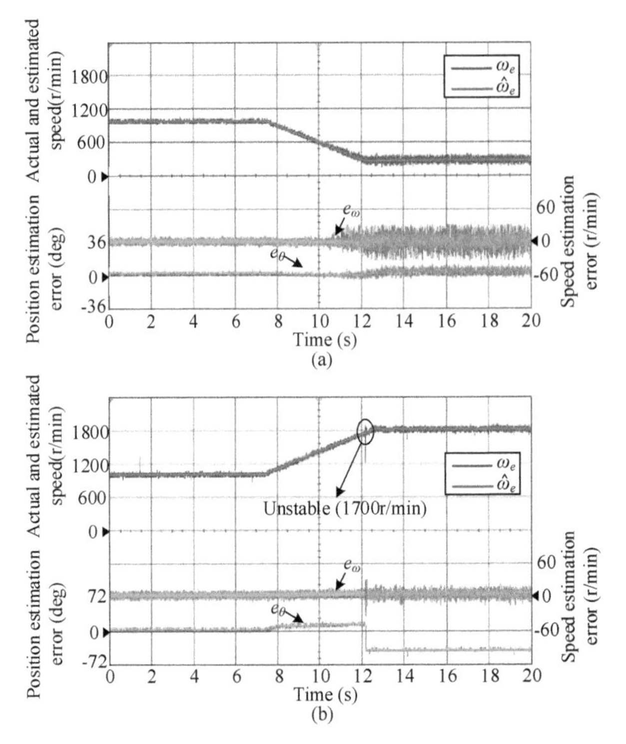

Figure 12. The performance of STO in wide speed range. (**a**) The performance of STO with no load from 1000 r/min to 300 r/min in closed-loop. (**b**) The performance of STO with no load from 1000 r/min to 1800 r/min in open-loop.

The performances of the STO with $k_1 = 15$ and $k_2 = 60,000$ from 1000 r/min to 300 r/min in closed-loop and from 1000 r/min to 1800 r/min in open-loop are shown in the Figure 12. In the process of motor speed decreasing from 1000 r/min to 300 r/min, the error of speed and position estimation increases significantly. That is because excessive sliding-mode gains lead to the large chattering of the estimated signals, resulting in severe chattering of the motor. It is dangerous to test the STO for the IPMSM in high speed range and closed-loop, so the speed is raised from 1000 r/min to 1800 r/min in open-loop. The corresponding performance is given in Figure 12b. The STO becomes unreliable at about 1700 r/min. At about 1700 r/min, the position estimation error jumps abruptly from 10.8° to −40° and the estimated speed has a large flutter. This means the IPMSM cannot operate at high speed over 1700 r/min in closed-loop. That is because the sliding-mode gains are too small to meet the stability conditions of STO. Experimental results presented in Figure 12 illustrate that the performance of STO in low and high speed range is limited by the constant sliding-mode gains and it is necessary to adopt speed-related adaptive sliding-mode gains.

The adaptive coefficients of the observer can be calculated by $l_1 = \frac{k_1}{\omega_e}$ and $l_2 = \frac{k_2}{\omega_e^2}$. The STO with $k_1 = 15$ and $k_2 = 60,000$ can operate perfectly at 1000 r/min ($\omega_e \approx$418.9 rad/s). So in this paper, $l_1 = \frac{15}{418.9} \approx 0.036$ and $l_2 = \frac{60,000}{418.9^2} \approx 0.342$. After applying the proposed adaptive STO, the IPMSM works well in wide speed range and closed-loop as shown in Figure 13. Throughout the operation, the speed estimation error is within ±10 r/min and the position estimation error is less than 10.8°. It is obvious that the position and speed estimation errors are significantly lower than the observer with constant sliding-mode gains, when the IPMSM runs in low and high speed range.

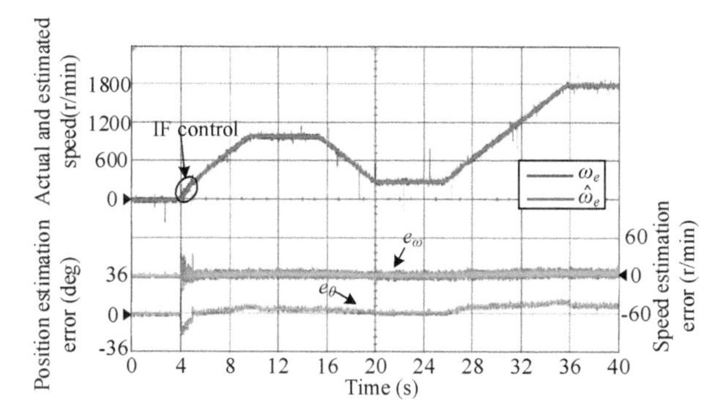

Figure 13. The performance of adaptive STO with no load in closed-loop under variable speed: raises from 0 r/min to 1000 r/min, drops to 300 r/min, and raises to 1800 r/min.

The dynamic performance of adaptive STO at 1800 r/min is shown in Figure 14. A 40 N·m load is enabled at 3 s and disabled at 6.2 s. The estimated speed can track the actual speed accurately and the estimated position error is less than 10.8° in the course of operation. The DC error of the position estimation increases by about 5° after loading and this is due to the mismatch of parameters caused by the increase of current after loading [12,31]. Hence, the performance of the adaptive STO could be verified.

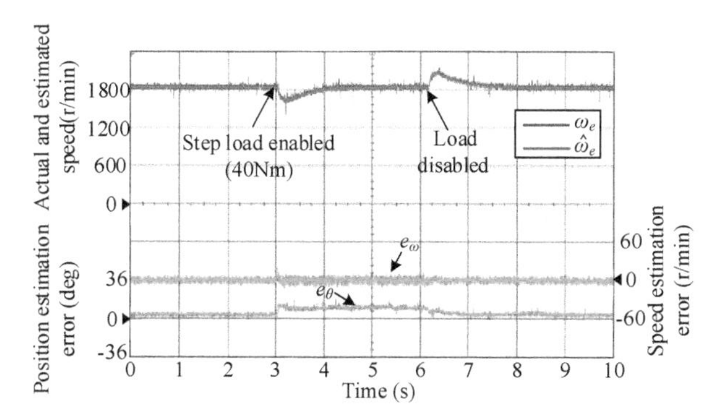

Figure 14. The dynamic performance of adaptive STO at 1800 r/min.

4.2. Experimental Results of the Proposed Improved PLL

The performances of traditional PLL, tangent-based PLL, and proposed improved PLL when the IPMSM turns from positive speed to reverse speed in open-loop are shown in Figure 15. For comparative purposes, three kinds of PLL operate under the same conditions: $K_p = 250$ and $K_i = 20,000$. The speed command is turned from 600 r/min to -600 r/min at 0.6 s.

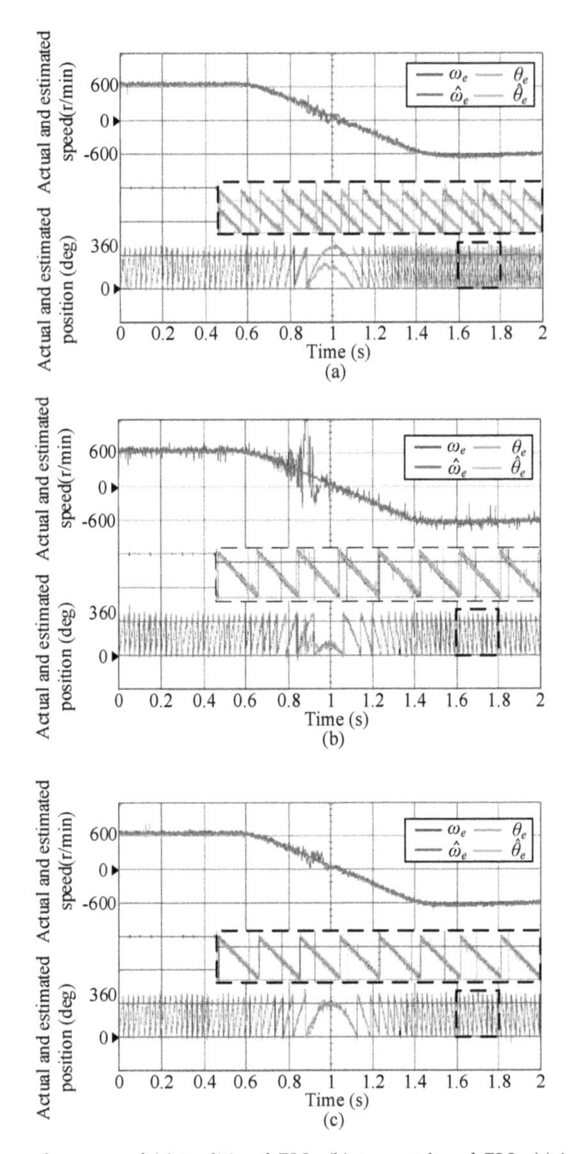

Figure 15. The performance of (**a**) traditional PLL, (**b**) tangent-based PLL, (**c**) improved PLL in open-loop from 600 r/min to −600 r/min.

As shown in Figure 15a, The estimated speed follows the actual speed accurately, when the rotation direction of the motor is positive. But when the speed of IPMSM is reversed, the conventional PLL loses its accuracy and produces a large position estimation error (180°). This prevents the motor from turning from positive speed to reverse speed in closed-loop. The performance of tangent-based PLL is shown in Figure 15b. Although tangent-based PLL can solve the speed reversal problem, the introduction of division and tangent functions increases the complexity of the algorithm and makes the tangent-based PLL vulnerable to harmonic and noise, especially when the back EMF crosses zero and the position crosses $\pm\frac{\pi}{2}$ where an obvious estimation error may occur. Excessive speed and position chattering shown in Figure 15b means the algorithm cannot be adopted in practice. The performances of the proposed improved PLL in open-loop and closed-loop are shown in Figures 15c and 16, respectively. It is clearly that the improved PLL has great performance when the IPMSM turns from positive speed to reverse speed. Thus, the effectiveness of the proposed improved PLL can be verified.

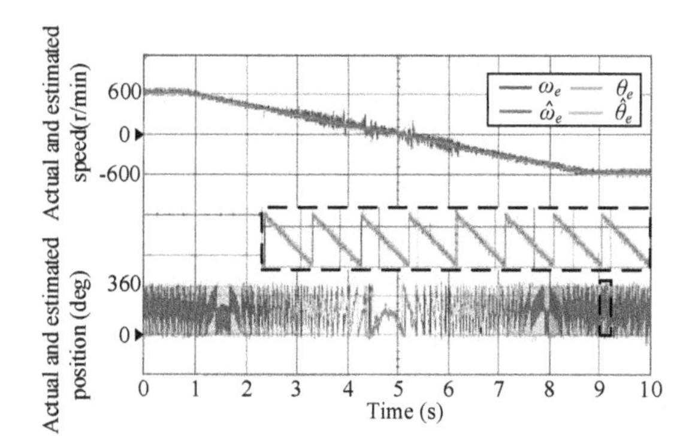

Figure 16. The performance of proposed improved PLL in closed-loop from 600 r/min to −600 r/min.

5. Conclusions

A new strategy for IPMSM sensorless control based on adaptive STO and improved PLL is proposed in this paper. STO is utilized to obtain the estimated back electromotive forces and the speed-related adaptive gains are proposed to achieve the accurate estimation of the observer in wide speed range. Moreover, the improved PLL based on a simple strategy for signal reconstruction of back EMF is proposed to overcome the limitation of speed reversal existing in traditional PLL without the introduction of tangent function. The experimental results show that the speed range of the super-twisting sliding-mode observer can be widened by adopting the proposed adaptive algorithm and the improved PLL has great performance so that IPMSM can realize the direction switching of speed stably.

Author Contributions: S.C. proposed the new sensorless control strategy. S.C. and X.W. performed the experiments and analyzed the data. S.C. wrote the first draft and X.W., X.Z., G.T. and X.C. guided and revised the manuscript.

Funding: The project is supported by the National Natural Science Foundational of China under Award U1610113.

Conflicts of Interest: The authors declare no conflict of interest.

References

1. Li, S.; Zhou, X. Sensorless Energy Conservation Control for Permanent Magnet Synchronous Motors Based on a Novel Hybrid Observer Applied in Coal Conveyer Systems. *Energies* **2018**, *11*, 2554. [CrossRef]
2. Wang, Y.; Wang, X.; Xie, W.; Dou, M. Full-Speed Range Encoderless Control for Salient-Pole PMSM with a Novel Full-Order SMO. *Energies* **2018**, *11*, 2423. [CrossRef]
3. Wang, G.; Ding, L.; Li, Z.; Lin, Z.; Xu, J.; Zhang, G.; Zhan, H.; Zhan, H.; Ni, R.; Xu, D. Enhanced Position Observer Using Second-Order Generalized Integrator for Sensorless Interior Permanent Magnet Synchronous Motor Drives. *IEEE Trans. Energy Convers.* **2014**, *29*, 486–495.
4. Wang, M.-S.; Tsai, T.-M. Sliding Mode and Neural Network Control of Sensorless PMSM Controlled System for Power Consumption and Performance Improvement. *Energies* **2017**, *10*, 1780. [CrossRef]
5. Joo, K.J.; Park, J.S.; Lee, J. Study on Reduced Cost of Non-Salient Machine System Using MTPA Angle Pre-Compensation Method Based on EEMF Sensorless Control. *Energies* **2018**, *11*, 1425. [CrossRef]
6. Wang, G.; Li, Z.; Zhang, G.; Yu, Y.; Xu, D. Quadrature PLL-Based High-Order Sliding-Mode Observer for IPMSM Sensorless Control with Online MTPA Control Strategy. *IEEE Trans. Energy Convers.* **2013**, *28*, 214–224. [CrossRef]

7. Tian, L.; Zhao, J.; Sun, J. Sensorless Control of Interior Permanent Magnet Synchronous Motor in Low-Speed Region Using Novel Adaptive Filter. *Energies* **2016**, *9*, 1084. [CrossRef]
8. Liu, J.; Zhu, Z. Novel Sensorless Control Strategy with Injection of High-Frequency Pulsating Carrier Signal into Stationary Reference Frame. *IEEE Trans. Ind. Appl.* **2014**, *50*, 2574–2583. [CrossRef]
9. Yoon, Y.; Sul, S.; Morimoto, S.; Ide, K. High-Bandwidth Sensorless Algorithm for AC Machines Based on Square-Wave-Type Voltage Injection. *IEEE Trans. Ind. Appl.* **2011**, *47*, 1361–1370. [CrossRef]
10. Jung, T.-U.; Jang, J.-H.; Park, C.-S. A Back-EMF Estimation Error Compensation Method for Accurate Rotor Position Estimation of Surface Mounted Permanent Magnet Synchronous Motors. *Energies* **2017**, *10*, 1160. [CrossRef]
11. Cho, Y. Improved Sensorless Control of Interior Permanent Magnet Sensorless Motors Using an Active Damping Control Strategy. *Energies* **2016**, *9*, 135. [CrossRef]
12. Tuovinen, T.; Hinkkanen, M. Adaptive Full-Order Observer with High-Frequency Signal Injection for Synchronous Reluctance Motor Drives. *IEEE J. Emerg. Sel. Top. Power Electron.* **2014**, *2*, 181–189. [CrossRef]
13. Yousefi-Talouki, A.; Pescetto, P.; Pellegrino, G.; Ion, B. Combined Active Flux and High Frequency Injection Methods for Sensorless Direct Flux Vector Control of Synchronous Reluctance Machines. *IEEE Trans. Power Electron.* **2018**, *33*, 2447–2457. [CrossRef]
14. Yousefi-Talouki, A.; Pescetto, P.; Pellegrino, G. Sensorless Direct Flux Vector Control of Synchronous Reluctance Motors Including Standstill, MTPA and Flux Weakening. *IEEE Trans. Ind. Appl.* **2017**, *53*, 3598–3608. [CrossRef]
15. Wang, G.; Yang, R.; Xu, D. DSP-Based Control of Sensorless IPMSM Drives for Wide-Speed-Range Operation. *IEEE Trans. Ind. Electron.* **2013**, *60*, 720–727. [CrossRef]
16. Zhao, Y.; Qiao, W.; Wu, L. Improved Rotor Position and Speed Estimators for Sensorless Control of Interior Permanent-Magnet Synchronous Machines. *IEEE J. Emerg. Sel. Top. Power Electron.* **2014**, *2*, 627–639. [CrossRef]
17. Park, J.B.; Wang, X. Sensorless Direct Torque Control of Surface-Mounted Permanent Magnet Synchronous Motors with Nonlinear Kalman Filtering. *Energies* **2018**, *11*, 969. [CrossRef]
18. Qiao, Z.; Shi, T.; Wang, Y.; Yan, Y.; Xia, C.; He, X. New Sliding-Mode Observer for Position Sensorless Control of Permanent-Magnet Synchronous Motor. *IEEE Trans. Ind. Electron.* **2013**, *60*, 710–719. [CrossRef]
19. Lin, S.; Zhang, W. An Adaptive Sliding-Mode Observer with a Tangent function-based PLL Structure for Position Sensorless PMSM Drives. *Int. J. Electr. Power Energy Syst.* **2017**, *88*, 63–74. [CrossRef]
20. Kim, H.; Son, J.; Lee, J. A High-Speed Sliding-Mode Observer for the Sensorless Speed Control of a PMSM. *IEEE Trans. Ind. Electron.* **2011**, *58*, 4069–4077.
21. Cascella, G.L.; Salvatore, N.; Salvatore, L. Adaptive Sliding-Mode Observer for Field Oriented Sensorless Control of SPMSM. In Proceedings of the 2003 IEEE International Symposium on Industrial Electronics (Cat. No. 03TH8692), Rio de Janeiro, Brazil, 9–11 June 2003; Volume 2, pp. 1137–1143.
22. Levant, A. Principles of 2-sliding Mode Design. *Automatica* **2007**, *43*, 576–586. [CrossRef]
23. Levant, A. Sliding Order and Sliding Accuracy in Sliding Mode Control. *Int. J. Control* **1993**, *58*, 1247–1263. [CrossRef]
24. Moreno, J.A.; Osorio, M. A Lyapunov Approach to Second-Order Sliding Mode Controllers and Observers. In Proceedings of the 47th IEEE Conference on Decision and Control, Cancun, Mexico, 9–11 December 2008; pp. 2856–2861.
25. Liang, D.; Li, J.; Qu, R. Sensorless Control of Permanent Magnet Synchronous Machine Based on Second-Order Sliding-Mode Observer with Online Resistance Estimation. *IEEE Trans. Ind. Appl.* **2017**, *53*, 3672–3682. [CrossRef]
26. Olivieri, C.; Tursini, M. A Novel PLL Scheme for a Sensorless PMSM Drive Overcoming Common Speed Reversal Problems. In Proceedings of the IEEE International Symposium on Power Electronics, Electrical Drives, Automation and Motion, Sorrento, Italy, 20–22 June 2012.
27. Olivieri, C.; Parasiliti, F.; Tursini, M. A Full-Sensorless Permanent Magnet Synchronous Motor Drive with an Enhanced Phase-Locked Loop Scheme. In Proceedings of the IEEE International Conference on Electrical Machines, Marseille, France, 2–5 September 2012; pp. 2202–2208.
28. Chen, Z.; Tomita, M.; Doki, S.; Okuma, S. An Extended Electromotive Force Model for Sensorless Control of Interior Permanent-Magnet Synchronous Motors. *IEEE Trans. Ind. Electron.* **2007**, *43*, 576–586.
29. Levant, A. Robust Exact Differentiation via Sliding Mode Technique. *Automatica* **1998**, *34*, 379–384. [CrossRef]

30. Wu, X.; Tan, G.; Ye, Z.; Liu, Y.; Xu, S. Optimized Common-Mode Voltage Reduction PWM for Three-Phase Voltage Source Inverters. *IEEE Trans. Power Electron.* **2016**, *31*, 2959–2969. [CrossRef]
31. Li, Y.; Zhu, Z.; Howe, D.; Bingham, C. Improved Rotor Position Estimation in Extended Back-EMF Based Sensorless PM Brushless AC Drives with Magnetic Saliency. In Proceedings of the IEEE International Electric Machines & Drives Conference, Antalya, Turkey, 3–5 May 2007; pp. 214–229.

Article

A Super-Twisting Sliding-Mode Stator Flux Observer for Sensorless Direct Torque and Flux Control of IPMSM

Junlei Chen, Shuo Chen, Xiang Wu, Guojun Tan * and Jianqi Hao

School of Electrical and Power Engineering, China University of Mining and Technology, Xuzhou 221116, China
* Correspondence: gjtan@cumt.edu.cn; Tel.: +86-138-0521-9335

Received: 10 June 2019; Accepted: 1 July 2019; Published: 3 July 2019

Abstract: The scheme based on direct torque and flux control (DTFC) as well as active flux is a good choice for the interior permanent magnet synchronous motor (IPMSM) sensorless control. The precision of the stator flux observation is essential for this scheme. However, the performance of traditional observers like pure integrator and the low-pass filter (LPF) is severely deteriorated by disturbances, especially dc offset. Recently, a sliding-mode stator flux observer (SMFO) was proposed to reduce the dc offset in the estimated stator flux. However, it cannot eliminate the dc offset totally and will cause the chattering problem. To solve these problems, a novel super-twisting sliding-mode stator flux observer (STSMFO) is proposed in this paper. Compared with SMFO, STSMFO can reduce the chattering and fully eliminate the dc offset without any amplitude and phase compensation. Then, the precision of the stator flux and rotor position can be greatly improved over a wide speed region. The detailed mathematical analysis has been given for comparing it with another three traditional observers. The numerical simulations and experimental testing with an IPMSM drive platform have been implemented to verify the capability of the proposed sensorless scheme.

Keywords: interior permanent magnet synchronous motor (IPMSM); active flux; sensorless control; stator flux observation; super-twisting sliding-mode stator flux observer (STSMFO)

---

1. Introduction

Interior permanent magnet synchronous motor (IPMSM) has been utilized in wide industrial fields because of its advantages like high torque density, fast response, and low torque ripple [1–3]. Direct torque control (DTC) is one of the most popular strategies for ac machines for its fast torque response and strong robustness [4,5]. However, the traditional DTC relies on the hysteresis comparators and the switching table, which will result in the unfixed switching frequency and the inaccurate compensation of the flux magnitude and the torque error. Furthermore, the ripple will be caused by the hysteresis comparators in the stator flux, and ripple can deteriorate the performance of the IPMSM drive [6]. To overcome this problem, the space vector modulation (SVM) is introduced into the DTC algorithm and the DTC-SVM is proposed [7]. The reference voltage is selected via three variables which are torque error, reference stator flux amplitude, and feedback flux vector, respectively. The introduction of the SVM can fix the switching frequency and reduce the ripple. However, the robustness is poor due to the single torque closed-loop structure of the DTC-SVM. In [8], the direct torque and flux control (DTFC) is proposed to enhance the robustness of the DTC-SVM. Compared with the DTC-SVM, one extra proportional-integral (PI) regulation is utilized to control flux and it can realize the flux to track the flux reference more precisely.

For the sake of realizing the high-performance control of IPMSM, the accurate position information is essential. However, position sensors not only decrease the reliability of the IPMSM system but also increase the cost. Therefore, research of the sensorless strategies for PMSMs have been paid much attention. In general, sensorless methods are divided into two categories which are based on signal injection and machine model according to speed region. At low speeds, the signal injection is generally adopted to realize rotor estimation [9–11]. However, it can cause the extra loss and torque ripple. With the speed increasing to the medium domain, model-based schemes are widely utilized to observe the position [12–16]. Due to its ease of implementation, plenty of research is focused on the designation of the back-emf observer like the extended Kalman filter [12] and the sliding-mode observer (SMO) [13]. However, the capability of the back-emf-based sensorless scheme is poor at low speeds because of the non-negligible noise. In [17], the active flux concept is proposed to transform salient-pole motors into virtual nonsalient-pole motors. The active flux is independent of speed and the direction aligns with rotor direction. Therefore, the active flux-based sensorless scheme can be readily implemented. Even at low speeds, the sensorless control can be carried out without signal injection [18]. Moreover, the stator flux can be observed by the DTFC and the corresponding robustness is enhanced due to the regulation of the DTFC. Therefore, the active flux-based sensorless scheme is suitable for the DTFC-based IPMSM drive system

Based on the IPMSM model, the stator flux is the integral of the back-emf and it is critical for flux observers to consider the influence of disturbances, especially dc offset [19]. When the pure integrator is adopted as a flux observer, the flux dc offset will increase with time, eventually resulting in the saturation of the observer [20]. In [21], a disturbance observer is designed to eliminate the dc component, but it is difficult to implement due to its complex structure. In [22], the dc component of the back-emf is filtered out by a low-pass filter (LPF). Actually, the LPF is a pure integrator cascaded with a high-pass filter. However, the amplitude attenuation and the phase delay are inevitable because of the introduction of the LPF. Therefore, the accurate compensation is essential for the sensorless control scheme. In [23], the SMFO is proposed to observe the stator flux without compensation, which is robust against disturbances and can reduce the dc offset through its compensation term. However, the dc disturbance rejection capability of the SMFO is limited and the chattering problem of the traditional SMFO is inevitable. It is worth mentioning that even a little dc offset can cause serious speed ripples. In [24], super-twisting algorithm (STA) is proposed to decrease the chattering of the SMO without reducing the robustness and it has been widely utilized in observers and controllers [25–28].

This paper proposed a novel STSMFO for the active flux-based IPMSM sensorless scheme. The stator flux observer based on STA is designed to overcome the disadvantages of the traditional observer, including the saturation effect, the position compensation, and flux dc offset. The corresponding Lyapunov stability of the STSMFO is proved based on references [29,30]. Then, the precision of the flux and position estimation can be greatly enhanced over a wide speed region. Moreover, the proposed sensorless method is better than most model-based sensorless strategies at low speeds due to the fact that it can be successfully utilized at low speeds without signal injection and compensation. Finally, the capability of the scheme based on the novel STSMFO is confirmed through numerical simulation and experimental testing with an IPMSM drive platform.

This paper is organized as follows: Section 2 gives the IPMSM model and the concept of DTFC. Section 3 introduces the active flux-based sensorless strategy and the mathematical analysis of three traditional observers. Section 4 introduces the proposed STSMFO and its analysis. The evaluation of the simulation and the experiment is given in Section 5. Section 6 draws the conclusions.

2. Direct Torque and Flux Control

Figure 1 shows the relationship between the stator flux-oriented coordinate (x-y), rotating reference coordinate (d-q), and stationary reference coordinate (α-β). The IPMSM stator flux and voltage equations in the d-q axis are presented as:

$$\begin{cases} u_d = R_s i_d + \frac{d\psi_d}{dt} - \omega_e \psi_q \\ u_q = R_s i_q + \frac{d\psi_q}{dt} + \omega_e \psi_d \end{cases} \tag{1}$$

$$\begin{cases} \psi_q = L_q i_q + \psi_f \\ \psi_q = L_q i_q \end{cases} \tag{2}$$

where, u_{dq}, i_{dq}, and ψ_{dq} are the stator voltages, currents, and flux linkages in the d-q axis, respectively. R_S is the resistance, L_{dq} are the stator inductances, ω_e is the rotor electrical angular velocity, ψ_f is the rotor flux linkage, δ is the load angle, θ_e is the rotor electrical position, and θ_s is the stator flux angle.

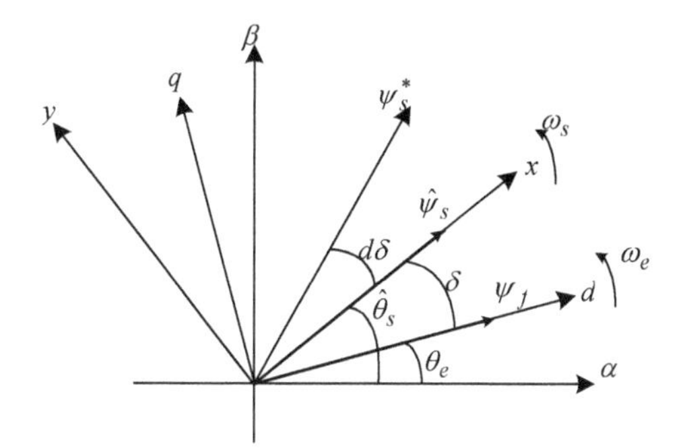

Figure 1. The relationship between various reference frames.

Transforming (1) to an x-y reference frame and combining the relationship between load angle and torque [31], the voltage equation is found as:

$$\begin{cases} u_x = R_s i_x + \frac{d|\psi_s|}{dt} \\ u_y = R_s i_y + \omega_s|\psi_s| = R_s i_y + \omega_e|\psi_s| + \frac{1}{K}\cdot\frac{dT_e}{dt} \end{cases} \tag{3}$$

where, $K = \frac{3p_n}{2L_dL_q}\left[\psi_f L_q cos\delta - |\psi_s|\left(L_q - L_d\right)cos2\delta\right]$, p_n is the number of pole pairs, ψ_s is the stator flux vector, and u_{xy} and i_{xy} are the stator voltages and currents in the x-y frame, respectively.

The IPMSM stator flux and torque equations in the α-β axis can be given as:

$$\begin{cases} \psi_\alpha = \int(u_\alpha - R_s i_\alpha)dt \\ \psi_\beta = \int\left(u_\beta - R_s i_\beta\right)dt \end{cases} \tag{4}$$

$$|\psi_s| = \sqrt{\psi_\alpha^2 + \psi_\beta^2} \tag{5}$$

$$T_e = \frac{3}{2}p_n\left(\psi_\alpha i_\beta - \psi_\beta i_\alpha\right) \tag{6}$$

where, $u_{\alpha\beta}$, $i_{\alpha\beta}$, and $\psi_{\alpha\beta}$ are the α-β axis stator voltages, currents, and flux linkages respectively, and T_e is torque.

3. Active Flux-Based Sensorless Control

To observe the rotor position of IPMSMs easily, the active flux is proposed to transform IPMSMs into virtual nonsalient-pole machines. The d-q axis active flux equation in reference [17] is given as:

$$\begin{cases} \psi_{dAF} = \psi_d - L_q i_d = \psi_f + \left(L_q - L_d\right)\cdot i_d \\ \psi_{qAF} = \psi_q - L_q i_q = 0 \end{cases} \tag{7}$$

where, ψ_{dqAF} are the active fluxes in the d-q axis. Equation (7) shows that the active flux aligns with the rotor direction, therefore active flux in the α-β frame results in:

$$\begin{cases} \psi_{\alpha AF} = \psi_\alpha - L_q i_\alpha = \psi_{dAF} cos\theta_e \\ \psi_{\beta AF} = \psi_\beta - L_q i_\beta = \psi_{dAF} sin\theta_e \end{cases} \tag{8}$$

where, $\psi_{\alpha\beta AF}$ are the active fluxes in the α-β axis. Equation (8) shows that the rotor position can be observed via the active flux vector. One way to extract it is a phase-locked loop. Moreover, it is worth mentioning that the active flux is obtained on the basis of stator flux. Therefore, the accuracy of the rotor position observation directly relies on the estimation accuracy of stator flux. In practice, a dc component in back-emf is inevitable and it is the main reason for the deteriorating performance of the flux observer. Considering the dc component, the stator flux equation is presented as:

$$\psi_s = \int (u_s - R_s i_s + A_0) dt = \int e_s dt \tag{9}$$

where u_s and i_s are the stator voltage and current, A_0 is the dc component, and e_s is the back-emf. For the convenience of analysis, e_s is expressed as:

$$e_s = A_0 + A_1 sin(\omega_1 t + \varphi_1) \tag{10}$$

where, $A_1 sin(\omega_1 t + \varphi_1)$ is the fundamental component. A_1, ω_1, and φ_1 are amplitude, angular frequency, and initial angel of the fundamental component, respectively. Taking the Laplace transformation of back-emf, it can be found as:

$$E_s(s) = \frac{A_0}{s} + A_1 \frac{s \cdot sin\varphi_1 + \omega_1 cos\varphi_1}{s^2 + \omega_1^2} \tag{11}$$

where, $E_s(s)$ is the Laplacian form of e_s and s is the Laplacian operator.

3.1. Pure Integrator-Based Observer

The Laplace transform of pure integrator is given as:

$$\psi_{s\_I}(s) = \frac{E_s(s)}{s} \tag{12}$$

where, $\psi_{s\_I}(s)$ is the Laplacian form of ψ_{s_I}, and ψ_{s_I} is the pure integrator-observed stator flux. Taking (11) into (12) then results in [19]:

$$\psi_{s\_I} = A_0 t + \frac{A_1 cos\varphi_1}{\omega_1} + \frac{A_1}{\omega_1} sin\left(\omega_1 t + \varphi_1 - \frac{\pi}{2}\right) \tag{13}$$

As seen, two disturbance terms exist in the estimated stator flux by pure integrator. One is a component increasing with time linearly, which is caused by a dc component in the back-emf and it can eventually result in the saturation of the integrator. Another one is a dc offset determined by the fundamental wave and it will change at various initial positions. These two terms can cause serious distortions of the estimated stator flux. Furthermore, the precise rotor position estimation cannot be achieved due to the distorted active flux.

3.2. Low-Pass Filter (LPF)-Based Observer

For the sake of filtering out the dc component in the back-emf, a high-pass filter is used to cascade with pure integrator. The Laplace transform of the observer is found as:

$$\psi_{s\_LPF}(s) = \frac{1}{s} \cdot \frac{s}{s + \omega_c} \cdot E_s(s) = \frac{E_s(s)}{s + \omega_c} \tag{14}$$

which shows that the observer is actually a LPF. In Equation (14), $\psi_{s\_LPF}(s)$ is the Laplace transform of ψ_{s_LPF}, ψ_{s_LPF} and ω_c are the estimated stator flux and cutoff frequency of LPF, respectively. ω_c is always selected much lower than ω_1 because it is, in effect, the cutoff frequency of the high-pass filter. Taking (11) into (14) then results in [19]:

$$\psi_{s\_LPF} = \frac{A_0}{\omega_c} - \frac{A_0}{\omega_c}e^{-\omega_c t} + \frac{A_1 cos(\varphi_1 + \theta_1)}{\sqrt{\omega_1^2 + \omega_c^2}}e^{-\omega_c t} + \frac{A_1}{\sqrt{\omega_1^2 + \omega_c^2}} sin\left(\omega_1 t + \varphi_1 - \frac{\pi}{2} + \theta_1\right) \quad (15)$$

where, $\theta_1 = tan^{-1}(\omega_c/\omega_1)$ is the phase delay caused by the LPF. Compared with Equation (13), the component causing integrator saturation is removed. The two main distorted terms decrease exponentially to zero with time. Although the dc offset still cannot be eliminated totally, it is already decreased greatly by LPF. However, the phase delay θ_1 and amplitude attenuation $A_1/\sqrt{\omega_1^2 + \omega_c^2}$ of the fundamental wave is inevitable because of the introduction of the LPF. Therefore, the estimated rotor position by LPF needs precise compensation to ensure good performance of the sensorless control.

3.3. First-Order Sliding-Mode Stator Flux Observer (SMFO)

Obviously, no matter whether a pure integrator-based observer or LPF-based observer is used, it cannot achieve high-performance sensorless control. Consequently, the SMFO is proposed to observe the stator flux. Figure 2 shows the diagram of the SMFO.

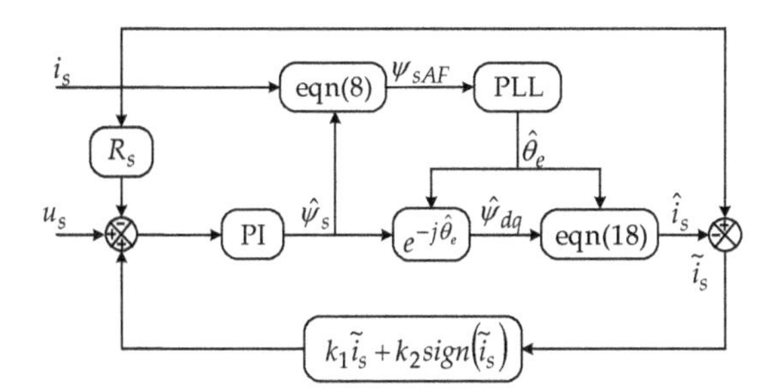

Figure 2. The diagram of the first-order sliding-mode stator flux observer (SMFO).

The SMFO, considering both the current and the voltage model of IPMSM, is designed. Moreover, the former is essentially utilized to compensate the latter. The stator flux and current are defined as the state variable and the output, respectively. On the basis of the IPMSM model, equations of the SMFO result in:

$$\frac{d\psi_s}{dt} = u_s - R_s i_s + k_1\tilde{i}_s + k_2 sign(\tilde{i}_s) \quad (16)$$

$$\begin{bmatrix} \hat{\psi}_d \\ \hat{\psi}_q \end{bmatrix} = \begin{bmatrix} cos\hat{\theta}_e & sin\hat{\theta}_e \\ -sin\hat{\theta}_e & cos\hat{\theta}_e \end{bmatrix} \begin{bmatrix} \hat{\psi}_\alpha \\ \hat{\psi}_\beta \end{bmatrix} \quad (17)$$

$$\begin{bmatrix} \hat{i}_\alpha \\ \hat{i}_\beta \end{bmatrix} = \begin{bmatrix} cos\hat{\theta}_e & -sin\hat{\theta}_e \\ sin\hat{\theta}_e & cos\hat{\theta}_e \end{bmatrix} \begin{bmatrix} \frac{\hat{\psi}_d - \psi_f}{L_d} \\ \frac{\hat{\psi}_q}{L_q} \end{bmatrix} \quad (18)$$

where, $\tilde{i}_s = i_s - \hat{i}_s$ is the stator current estimation error, "^" is the estimated value, and sign () is the sign function. k_1 and k_2 are linear and nonlinear gains of the SMFO which affect the dynamic performance and robustness, respectively. The Lyapunov stability has been proved in reference [8]. Considering the dc offset, (16) can be found as:

$$\frac{d\psi_s}{dt} = u_s - R_s i_s + A_0 + k_1\tilde{i}_s + k_2 sign(\tilde{i}_s) \quad (19)$$

Actually, the SMFO is used to observe the dc offset according to the current estimation error. Further, the estimated dc offset is fed back to the back-emf to realize the accurate estimation of the stator flux. However, the SMFO cannot remove flux dc offset totally. In the steady state, $A_0 + k_1\widetilde{i_s} + k_2 sign(\widetilde{i_s}) = 0$ and the effect of the dc component on the flux estimation is eliminated. However, in the process of reaching a steady state, the estimated flux dc offset is already accumulated. The estimated flux dc offset can be given as:

$$\psi_{s0} = \int_0^{t_1} (A_0 + k_1\widetilde{i_s} + k_2 sign(\widetilde{i_s}))dt \tag{20}$$

where, ψ_{s0} is the flux dc offset, and t_1 is the time to steady state. In general, ψ_{s0} can be omitted because A_0 and t_1 are small, but if new dc disturbances interfere with the system, it will cause the accumulation of flux estimation dc offset and eventually have a non-negligible impact on the stator flux observation.

4. Super-Twisting Sliding-Mode Stator Flux Observer

The estimation precision of the stator flux dramatically affects the capability of the DTFC-based active flux sensorless scheme. The first-order SMFO can observe the stator flux without phase shift and amplitude attenuation. However, the conventional first-order SMFO is still sensitive to dc disturbances and suffers from the chattering problem. To overcome the limitations of the first-order SMFO, the well-known STA was utilized to design an observer for estimating stator flux in this paper. With the STA, the chattering can be reduced and the dc offset can be eliminated effectively.

4.1. Super-Twisting Algorithm

To improve the chattering problem of the traditional SMO, the super-twisting algorithm was proposed in reference [24]. The equation of STA considering perturbation terms can be presented as:

$$\begin{cases} \frac{dx_1}{dt} = -k_1|\widetilde{x_1}|^{0.5} sign(\widetilde{x_1}) + x_2 + \rho_1(x_1, t) \\ \frac{dx_2}{dt} = -k_2 sign(\widetilde{x_1}) + \rho_2(x_2, t) \end{cases} \tag{21}$$

where, x_i, $\widetilde{x_i}(\widetilde{x_i} = \hat{x}_i - x_i)$, k_i and ρ_i are state variables, state variables estimation error, gains of the STA, and perturbation terms, respectively. The Lyapunov stability has been proved in references [29,30]. According to [29,30], the observer is stable when sliding-mode gains and the perturbation terms satisfy (22):

$$\begin{cases} k_1 > 2\delta_1, k_2 > k_1 \frac{5\delta_1 k_1 + 4\delta_1^2}{2(k_1 - 2\delta_1)} \\ |\rho_1| \leq \delta_1 |x_1|^{0.5}, \rho_2 = 0 \end{cases} \tag{22}$$

where, δ_1 is a positive constant.

4.2. Super-Twisting Sliding-Mode Stator Flux Observer

To solve problems of chattering and the dc offset in the traditional SMFO, a stator flux observer is proposed based on STA. Substituting $x_1 = \hat{\psi}_s$ into (21), then Equation (21) can be rewritten as:

$$\frac{d\hat{\psi}_s}{dt} = -k_1|\widetilde{\psi_s}|^{0.5} sign(\widetilde{\psi_s}) - k_2 \int sign(\widetilde{\psi_s})dt + \rho_1(\hat{\psi}_s, t) \tag{23}$$

where, $\rho_1(\hat{\psi}_s, t)$ is:

$$\rho_1(\hat{\psi}_s, t) = u_s - R_s i_s + A_0 \tag{24}$$

Taking (24) into (22) results in:

$$|u_s - R_s i_s + A_0| \leq \delta_1 |\psi_s|^{0.5} \tag{25}$$

which shows that the system can converge in finite time to sliding surface for a large enough δ_1. By subtracting (23) from (4), the flux error equation is found as:

$$\frac{d\widetilde{\psi}_s}{dt} = k_1|\widetilde{\psi}_s|^{0.5} sign(\widetilde{\psi}_s) + k_2 \int sign(\widetilde{\psi}_s)dt - A_0 \tag{26}$$

At the starting point, the stator flux estimation value is different from the actual value due to the existence of dc offset A_0. Then, the sliding-mode terms $k_1|\widetilde{\psi}_s|^{0.5} sign(\widetilde{\psi}_s) + k_2 \int sign(\widetilde{\psi}_s)dt$ are used to estimate the dc component. Furthermore, the estimated dc component is fed back to compensate the back-emf. At a steady state, the flux error is on the sliding surface. Taking $\widetilde{\psi}_s = 0$ into (26) results in:

$$A_0 = k_1|\widetilde{\psi}_s|^{0.5} sign(\widetilde{\psi}_s) + k_2 \int sign(\widetilde{\psi}_s)dt \tag{27}$$

which shows that the effect of the dc component has been removed. In addition, in the process of reaching the steady state, the accumulated flux estimation dc offset can be given as:

$$\psi_{s0\_STA} = \int \left(A_0 - k_1|\widetilde{\psi}_s|^{0.5} sign(\widetilde{\psi}_s) - k_2 \int sign(\widetilde{\psi}_s)dt \right) dt \tag{28}$$

As we all know, the integral term $k_2 \int sign(\widetilde{\psi}_s)dt$ will remain until the ψ_{s0_STA} is zero. Therefore, compared with the SMFO, the chattering caused by the sign function can be reduced greatly and the estimated flux dc offset can be eliminated completely due to the integral term and a large gain k_2 of the STSMFO. In addition, it is worth mentioning that the stator flux amplitude reference value and estimated flux angle are utilized to replace the actual stator flux vector. The diagram of the STSMFO is shown in Figure 3.

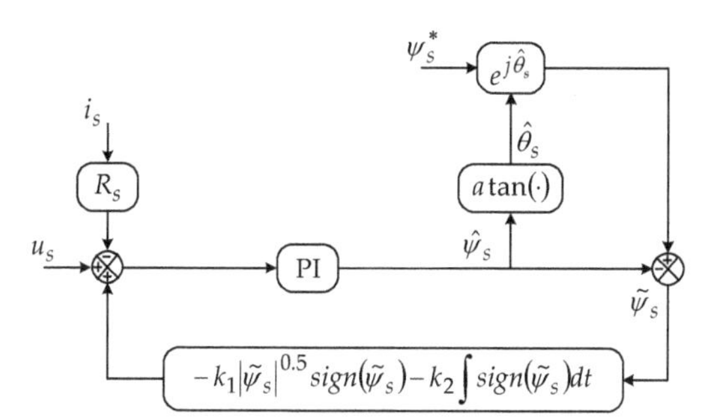

Figure 3. The diagram of the super-twisting sliding-mode stator flux observer (STSMFO).

In conclusion, two disturbance terms by pure integrator can cause serious distortions of the estimated stator flux. The LPF can eliminate the dc offset at the cost of amplitude attenuation and phase shift. The SMFO can reduce the dc offset greatly without any influence on the fundamental wave. However, the SMFO will accumulate the small disturbance and cause chattering. Finally, the proposed STSMFO can remove the dc offset completely and smooth the estimated stator flux.

5. Evaluation via Simulation and Experiment

The IPMSM control scheme is given in Figure 4. Both the simulation and the experiment based on Figure 4 were carried out to verify the capability of the STSMFO. The parameters of the IPMSM are listed in Table 1.

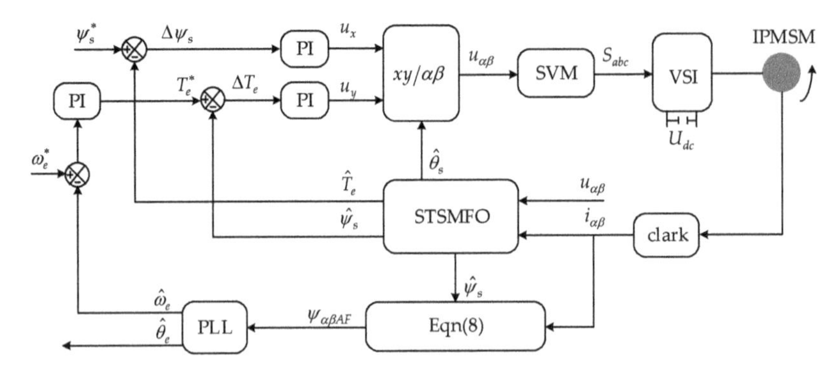

Figure 4. The proposed interior permanent magnet synchronous motor (IPMSM) sensorless scheme.

Table 1. IPMSM Parameters.

Parameter	Value
rated speed	3000 rpm
rated power	60 kW
rated voltage	380 V
rated current	130 A
rotational inertia	0.2 kg·m^2
p_n	4
R_s	0.1 Ω
L_d/L_q	0.95/2.05 mH
ψ_f	0.225 Wb

5.1. Simulation Results

MATLAB/Simulink software was utilized to numerically demonstrate the capability of the STSMFO. The simulation results of α-axis estimated stator fluxes at 300 rpm and 1500 rpm are given in Figure 5.

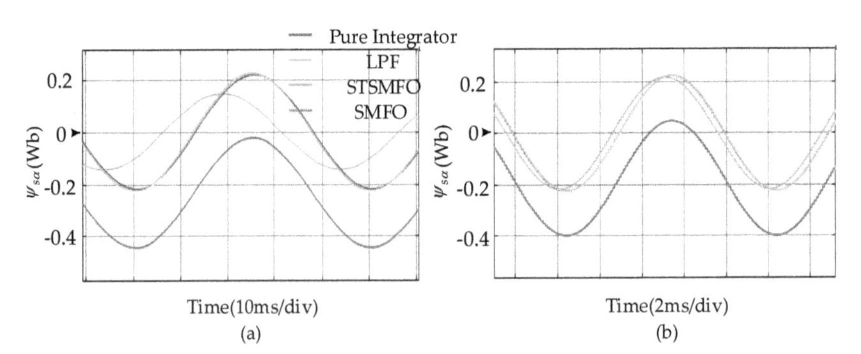

Figure 5. Simulation results of α-axis estimated stator fluxes: (**a**) The stator fluxes at 300 rpm, (**b**) The stator fluxes at 1500 rpm.

The pure integrator estimated stator flux deviates from the ideal value due to the dc offset. Figure 5a,b shows that the LPF will delay the phase and attenuate the amplitude of flux estimation. Moreover, as previously analyzed, the problems have been improved with the increase of the frequency of the fundamental wave. The simulation result of the rotor position observation at 300 rpm is shown in Figure 6 which shows that the pure integrator cannot estimate the rotor position correctly. In addition, the estimation error of the STSMFO is 0.01 rad which is lower than 0.06 rad of the SMFO and much lower than 0.85 rad of the LPF.

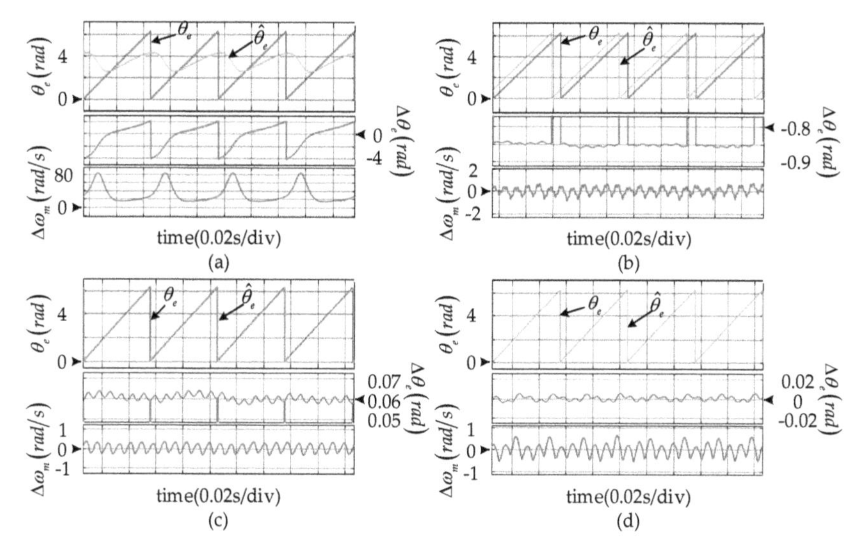

Figure 6. The simulation result of the position observation at 300 rpm: (**a**) Integrator, (**b**) low-pass filter (LPF), (**c**) The SMFO, (**d**) The STSMFO.

Although both the SMFO and the STSMFO can precisely estimate the stator flux, the SMFO cannot totally eliminate the dc offset. To prove it further, Figure 7 plots the simulation results of the flux estimation under the 9 V dc offset (added to the $u_{s\alpha}$). From the figure, it is clear that the SMFO cannot eliminate the dc offset, while the STSMFO removes the dc offset completely due to the integral term of the STA. Furthermore, the flux dc offset ψ_{s0} is inverse in proportion to the SMFO gains. However, there is a tradeoff between the ψ_{s0} and the SMFO gains because the larger SMFO gains will deteriorate the chattering problem.

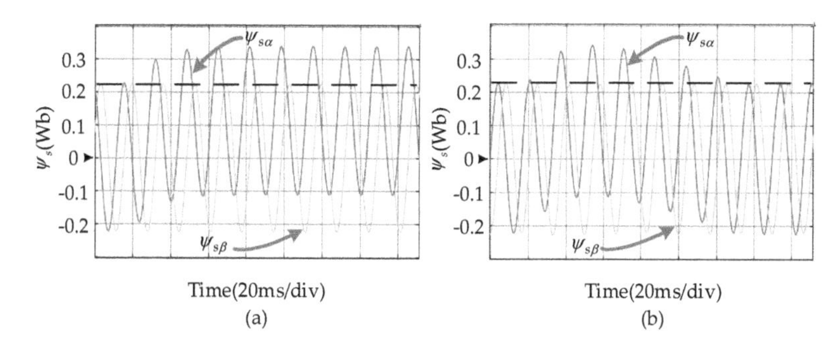

Figure 7. Simulation results of estimated stator fluxes under the 9 V dc offset: (**a**) The SMFO, (**b**) The STSMFO.

5.2. Experimental Results

Experiments on a 2-level IPMSM drive system [32] were subsequently carried out to further verify the effectiveness of the proposed observer. The composition of the IPMSM test platform is given in Figure 8, where a DSP TMS320F2812 was chosen as the MCU and output signals of a D/A chip (TLV5610) were displayed by an oscilloscope to monitor system variables in real-time. The three-phase

PWM inverter [33] is composed of IGBTs (FF650R17IE4) and a 540V dc-link voltage was obtained by the PWM rectifier. The sampling and switching periods were set to 100 μs. The sensors of current (CHB-500SG) and voltage (LV25-P) were utilized to measure ab-phase currents and dc-link voltage. The actual rotor position was detected by a rotary decoder (PGA411-Q1) for comparison. The main parameters of the IPMSM were the same as those utilized in the simulation. Moreover, the parameters of the PI regulator in phase locked loop (PLL) were $K_p = 15$ and $K_i = 250$. In addition, the parameters of the STSMFO were $K_1 = 2.5$ and $K_2 = 5000$.

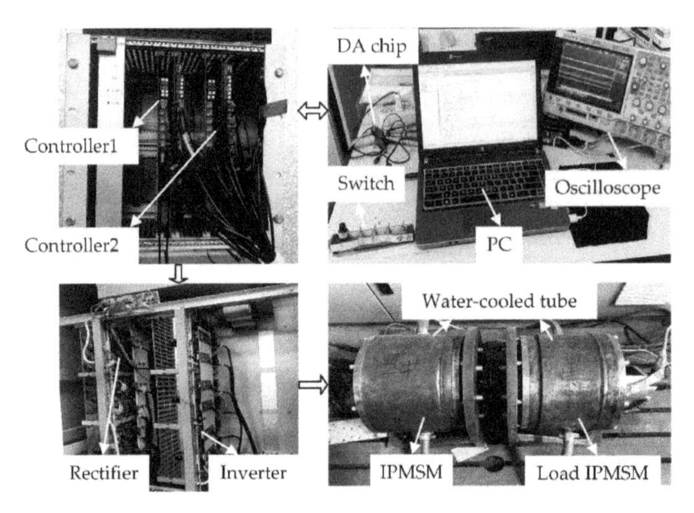

Figure 8. The IPMSM experimental platform.

5.2.1. Steady-State Performance

Figures 9 and 10 show the estimated stator flux trajectories and rotor positions at 300 rpm, respectively. The result in Figure 9a confirms that the pure integrator will introduce a monotonously increasing dc offset in the flux estimation with respect to time. It can really deteriorate the performance of the observer, resulting in the pure integrator not being able to estimate the correct position at all. The error of estimation is approximately equal to 300 rpm. The dc offset can be removed by the LPF, but the amplitude is attenuated from the flux amplitude reference which is 0.225 Wb to 0.17 Wb. Moreover, the LPF also generates a phase delay and causes a position error of about 30°. Although both the SMFO and the STSMFO can estimate the stator flux accurately, the SMFO cannot remove the dc offset totally and results in a larger position error than the STSMFO.

Besides, Figures 11 and 12 plot the estimated stator flux trajectories and rotor positions at 1200 rpm. As same as the situation at 300 rpm, the pure integrator cannot observe the flux correctly and the flux locus still deviates from the ideal locus. As described in Equation (15), the problems of amplitude attenuate, and phase delay caused by the LPF are improved with the increasing speed. Compared with the performance at 300 rpm, the amplitude of the estimated flux increases from 0.175 Wb to 0.205 Wb and the position error decreases to 10°. However, the LPF still cannot meet the requirements of a high-performance sensorless control. The performance of the SMFO and the STSMFO are similar to the situation at 300 rpm. They can estimate the stator flux accurately. Moreover, the position error caused by the STSMFO is still a bit lower than the SMFO.

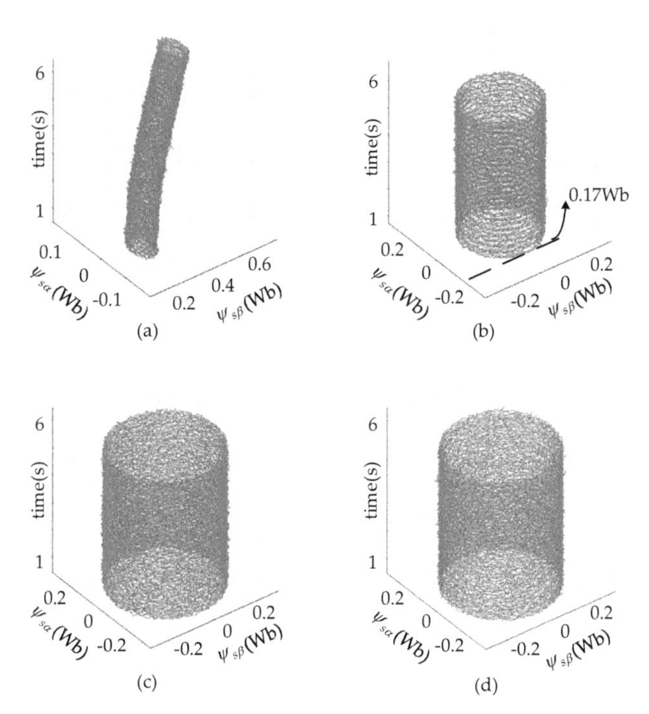

Figure 9. The estimated stator flux loci at 300 rpm: (**a**) Integrator, (**b**) LPF, (**c**) The SMFO, (**d**) The STSMFO.

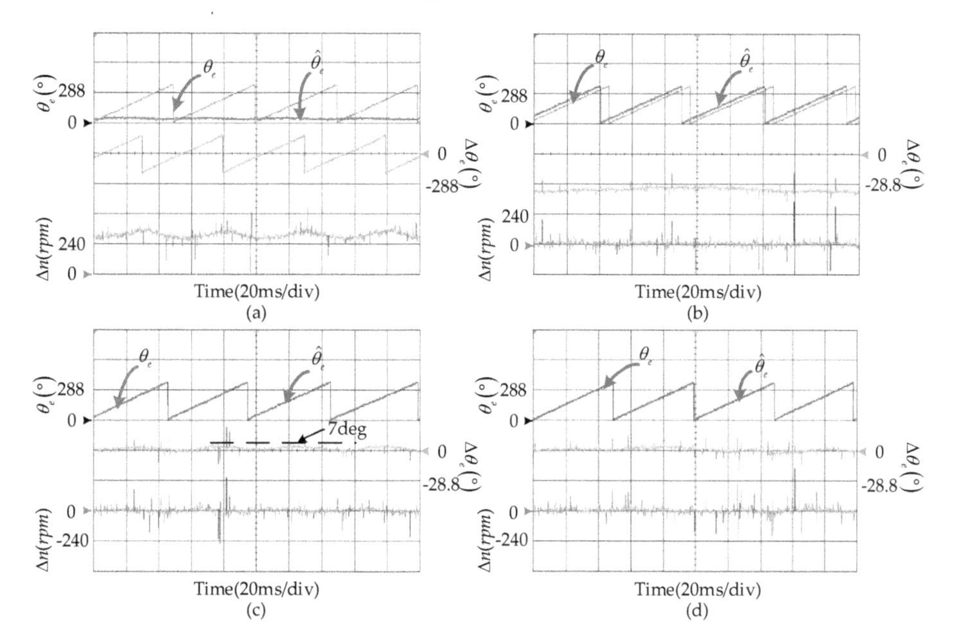

Figure 10. The performance of the position observation at 300 rpm: (**a**) Integrator, (**b**) LPF, (**c**) The SMFO, (**d**) The STSMFO.

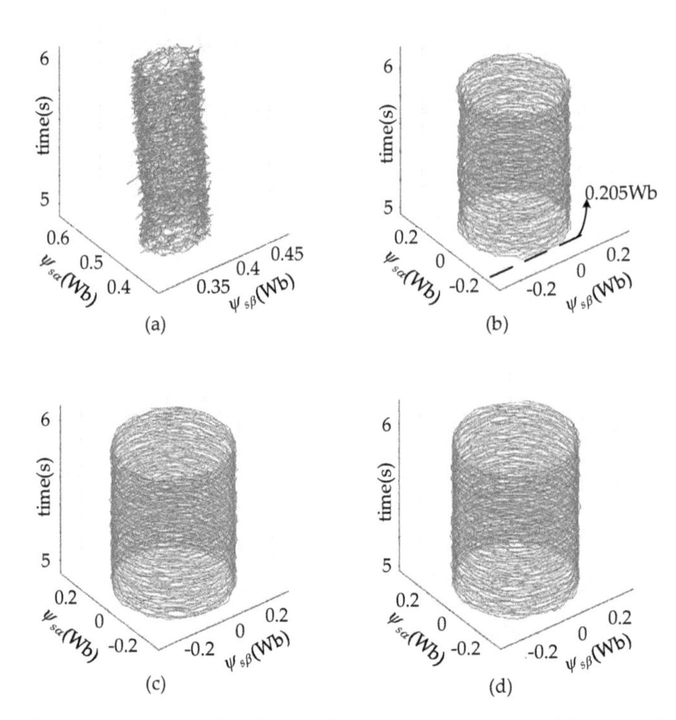

Figure 11. The estimated stator flux loci at 1200 rpm: (**a**) Integrator, (**b**) LPF, (**c**) The SMFO, (**d**) The STSMFO.

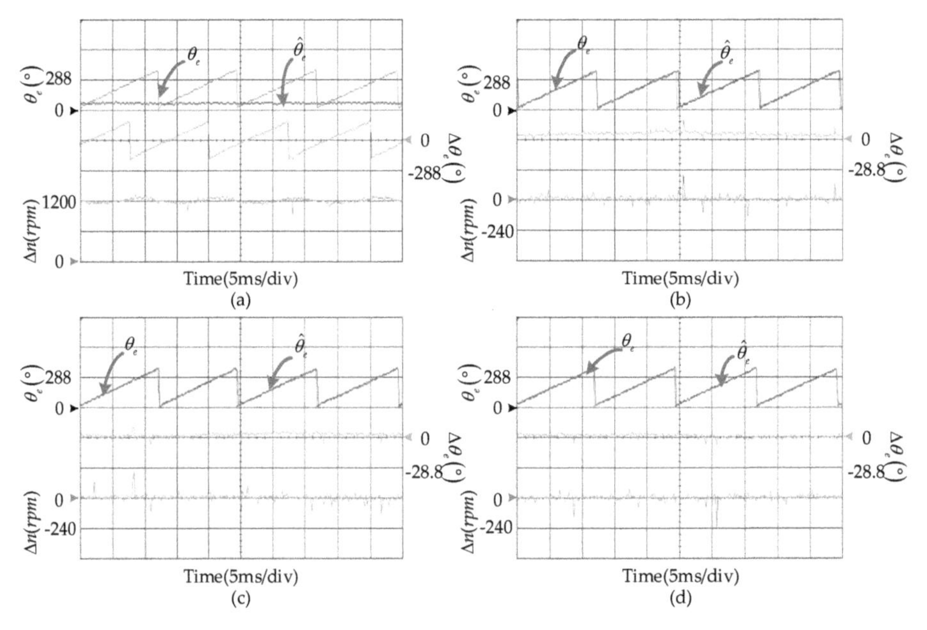

Figure 12. The performance of the position observation at 1200 rpm: (**a**) Integrator, (**b**) LPF, (**c**) The SMFO, (**d**) The STSMFO.

The experimental results of sensorless control which is based on the STSMFO at the 150 rpm are presented in Figure 13. The stator flux and its amplitude are given in Figure 13a. Figure 13b plots the actual and estimated position, estimated speed, and position error. As can be seen in the figure, the stator flux can be observed accurately and the chattering is small. Moreover, the figure shows a good position tracking ability of the STSMFO with a position error being controlled within 7°. Therefore, the STSMFO can be carried out at a low speed region.

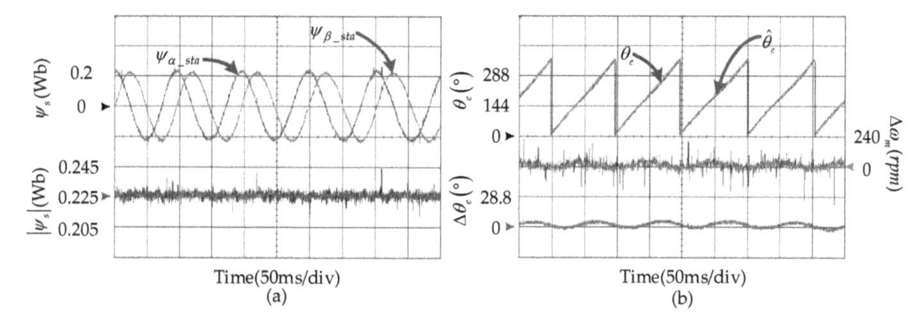

Figure 13. The sensorless drive by the STSMFO at 150 rpm: (**a**) The stator flux estimation, (**b**) The rotor position estimation.

5.2.2. DC Disturbance Rejection

To evaluate the dc disturbance rejection capability of the STSMFO, experiments with added dc disturbance signals are also implemented and the results are illustrated in Figures 14 and 15. The 20 V dc offset is added to the $u_{s\alpha}$ and the estimated fluxes by two observers are presented in the figure. As can be seen, there is a 0.125 Wb flux dc offset on the estimated stator flux by the SMFO since it cannot totally eliminate the dc disturbance. The flux observed by the STSMFO is also subject to dc offset, but it can recover to normal value within two sample times. The corresponding three-dimensional trajectories in Figure 14 illustrate this more intuitively.

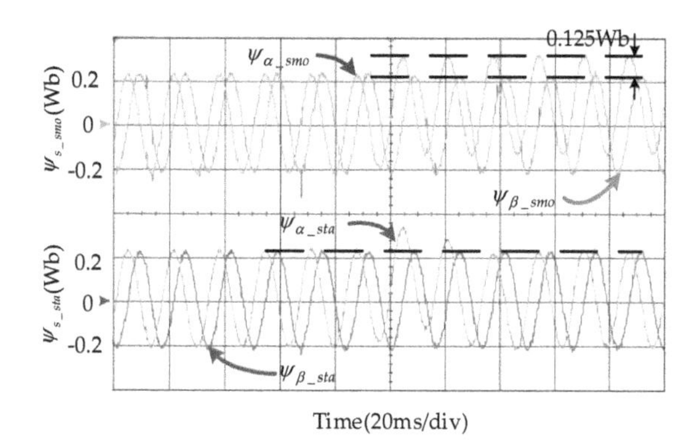

Figure 14. The experimental results of estimated stator fluxes under the 20 V dc offset.

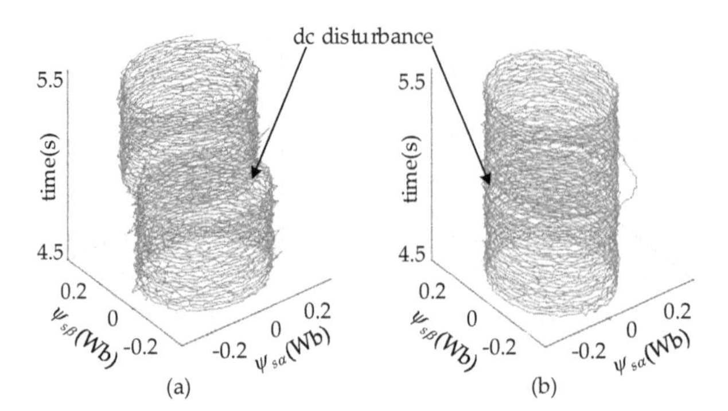

Figure 15. Three-dimensional stator flux trajectories under dc disturbance: (**a**) The SMFO, (**b**) The STSMFO.

The sensorless drive results under the 20V dc offset are illustrated in Figure 16. The picture plots the position error, the stator flux, and its amplitude. Compared with the open loop estimation, the estimated flux dc offset of the closed-loop estimation is smaller due to the flux regulation of the DTFC. However, a small flux dc offset can still cause large chattering in the speed and it can deteriorate the performance of the sensorless control. As for the STSMFO, the flux dc offset can be eliminated within one sample time and the chattering problem of the estimated rotor position can also be solved accordingly. As can be seen, the STSMFO has stronger disturbance rejection performance than the SMFO.

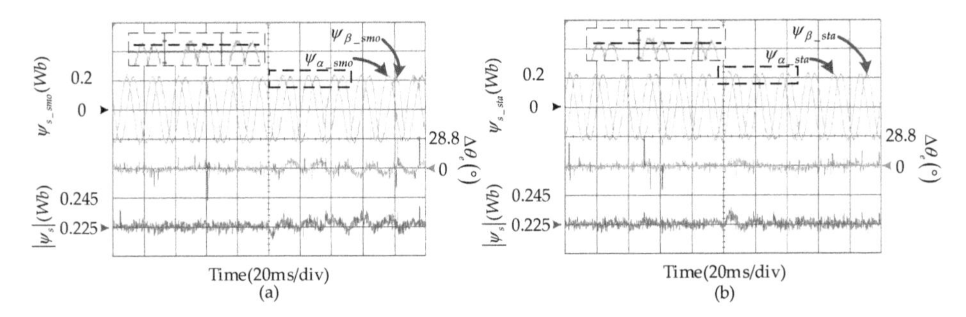

Figure 16. The sensorless drive results under the 20 V dc offset. (**a**) The SMFO, (**b**) The STSMFO.

5.2.3. Dynamic Capability

To further verify the performance of the STSMFO, the result of speed variation between 300 rpm and 1,500 rpm is given in Figure 17, which shows that the observer can accurately estimate the speed during the variable speed and control position error within 8°. Figure 18 plots the experimental result of the proposed sensorless method at 900 rpm under a 30 N·m load disturbance. The estimated speed, estimated position error, estimated speed error, and load torque are all shown in the figure. The results indicate a good speed tracking ability of the STSMFO under a load step, with a rotor position error being about 14.4°.

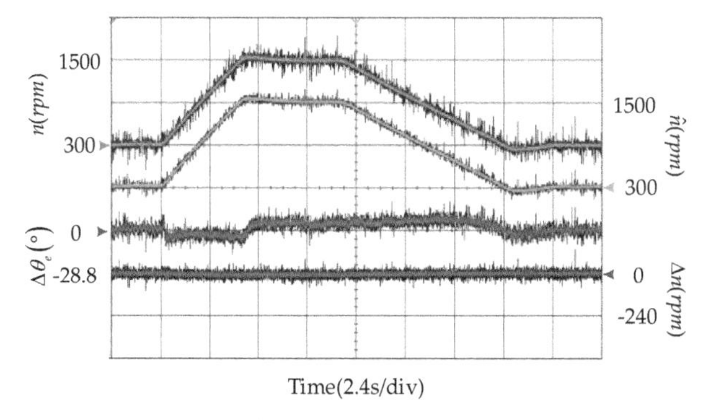

Figure 17. The STSMFO-based sensorless drive during variable speed operation.

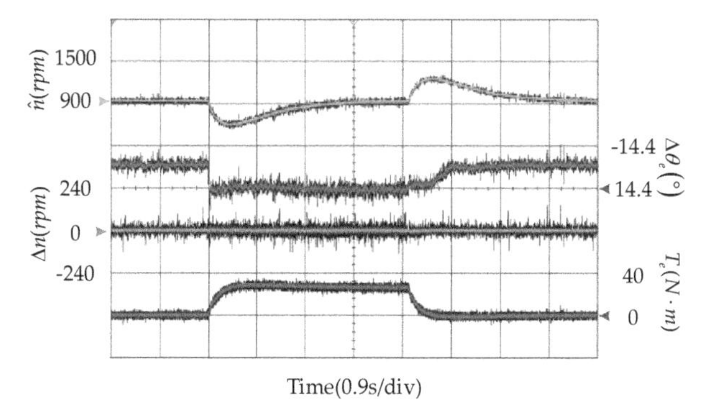

Figure 18. The STSMFO-based sensorless drive with 30 N·m load step at 900 rpm.

To validate the capability of the STSMFO at low speeds, the experimental result of sensorless drive during speed variation form 60 rpm to 300 rpm is presented in Figure 19. The actual and estimated speed, speed, and position estimation error are presented in the figure. It shows that the estimated rotor position error chatters a little around 14° at 60 rpm, but the speed estimation maintains a relatively high tracking accuracy over the whole speed domain, thus ensuring a stable motor operation.

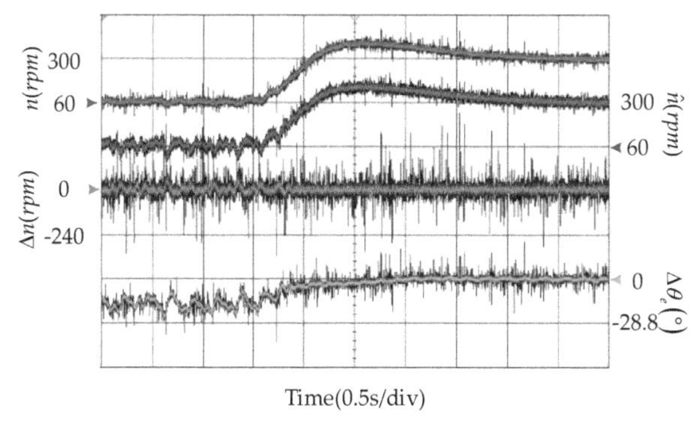

Figure 19. The STASFO-based sensorless drive results at a low-speed region.

6. Conclusions

This paper proposed a novel STSMFO for the active flux-based sensorless scheme. To demonstrate its effectiveness, a detailed mathematical analysis was given for comparing it with traditional observers, i.e., the pure integrator, the LPF, and the SMFO. It shows that the STSMFO can easily solve the problems caused by dc offset, unknown integral initial value, and saturation. Moreover, the robustness of STSMFO is much better than SMFO. Then, the precision of the stator flux estimation can be enhanced greatly without any amplitude and phase compensation. Furthermore, the rotor position can be observed accurately via the active flux over a wide speed domain. The numerical simulations and experimental tests have confirmed the capability of the steady state, the dynamic response, and the disturbance rejection.

Author Contributions: J.C. proposed the new scheme. J.C., S.C., and X.W. performed the simulations, experiments, and data analysis. J.C. wrote the first draft and S.C., X.W., G.T., and J.H. guided and revised the manuscript.

Funding: The project is supported by the National Natural Science Foundational of China (Grant number: U1610113) and National Key R&D Program of China (Grant number: 2016YFC0600804).

Conflicts of Interest: The authors declare no conflict of interest.

References

1. Li, J.; Huang, X.; Niu, F.; You, C.; Wu, L.; Fang, Y. Prediction Error Analysis of Finite-Control-Set Model Predictive Current Control for IPMSMs. *Energies* **2018**, *11*, 2051. [CrossRef]
2. Lara, J.; Xu, J.; Chandra, A. Effects of Rotor Position Error in the Performance of Field Oriented Controlled PMSM Drives for Electric Vehicle Traction Applications. *IEEE Trans. Ind. Electron.* **2016**, *63*, 4738–4751. [CrossRef]
3. Rahman, M.M.; Uddin, M.N. Third Harmonic Injection Based Nonlinear Control of IPMSM Drive for Wide Speed Range Operation. *IEEE Trans. Ind. Appl.* **2014**, *55*, 3174–3184. [CrossRef]
4. Takahashi, I.; Noguchi, T. A New Quick-Response and High-Efficiency Control Strategy of an Induction Motor. *IEEE Trans. Ind. Appl.* **1986**, *22*, 820–827. [CrossRef]
5. Ouanjli, N.E.; Derouich, A.; Ghzizal, A.E.; Motahhir, S.; Chebabhi, A.; Mourabit, Y.E.; Taoussi, M. Modern improvement techniques of direct torque control for induction motor drives—A review. *Prot. Control Mod. Power Syst.* **2019**, *4*, 136–147. [CrossRef]
6. Choi, Y.; Choi, H.H.; Jung, J. Feedback Linearization Direct Torque Control with Reduced Torque and Flux Ripples for IPMSM Drives. *IEEE Trans. Power Electron.* **2016**, *31*, 3728–3737. [CrossRef]
7. Tang, L.; Zhong, L.; Rahman, M.F.; Hu, Y. A Novel Direct Torque Controlled Interior Permanent Magnet Synchronous Machine Drive with Low Ripple in Flux and Torque and Fixed Switching Frequency. *IEEE Trans. Power Electron.* **2004**, *19*, 346–354. [CrossRef]
8. Foo, G.; Rahman, M.F. Sensorless Direct Torque and Flux-Controlled IPM Synchronous Motor Drive at Very Low Speed Without Signal Injection. *IEEE Trans. Ind. Electron.* **2010**, *57*, 395–403. [CrossRef]
9. Chen, J.; Yang, S.; Tu, J. Comparative Evaluation of a Permanent Magnet Machine Saliency-Based Drive with Sine-Wave and Square-Wave Voltage Injection. *Energies* **2018**, *11*, 2189. [CrossRef]
10. Yousefi-Talouki, A.; Pescetto, P.; Pellegrino, G.; Boldea, I. Combined Active Flux and High Frequency Injection Methods for Sensorless Direct Flux Vector Control of Synchronous Reluctance Machines. *IEEE Trans. Power Electron.* **2018**, *33*, 2447–2457. [CrossRef]
11. Yoon, Y.; Sul, S.; Morimoto, S.; Ide, K. High-Bandwidth Sensorless Algorithm for AC Machines Based on Square-Wave-Type Voltage Injection. *IEEE Trans. Ind. Appl.* **2011**, *47*, 1361–1370. [CrossRef]
12. Shi, Y. Online Identification of Permanent Magnet Flux Based on Extended Kalman Filter for IPMSM Drive with Position Sensorless Control. *IEEE Trans. Ind. Electron.* **2012**, *59*, 4169–4178. [CrossRef]
13. Chen, S.; Zhang, X.; Wu, X.; Tan, G.; Chen, X. Sensorless Control for IPMSM Based on Adaptive Super-Twisting Sliding-Mode Observer and Improved Phase-Locked Loop. *Energies* **2019**, *12*, 1225. [CrossRef]
14. Vieira, R.P.; Gastaldini, C.C.; Azzolin, R.Z.; Gründling, H.A. Sensorless Sliding-Mode Rotor Speed Observer of Induction Machines Based on Magnetizing Current Estimation. *IEEE Trans. Ind. Electron.* **2014**, *61*, 4573–4582. [CrossRef]

15. Fan, Y.; Zhang, L.; Cheng, M.; Chau, K.T. Sensorless SVPWM-FADTC of a New Flux-Modulated Permanent-Magnet Wheel Motor Based on a Wide-Speed Sliding Mode Observer. *IEEE Trans. Ind. Electron.* **2015**, *62*, 3143–3151. [CrossRef]
16. Gadoue, S.M.; Giaouris, D.; Finch, J.W. MRAS Sensorless Vector Control of an Induction Motor Using New Sliding-Mode and Fuzzy-Logic Adaptation Mechanisms. *IEEE Trans. Energy Convers.* **2010**, *25*, 394–402. [CrossRef]
17. Boldea, I.; Paicu, M.C.; Andreescu, G. Active Flux Concept for Motion-Sensorless Unified AC Drives. *IEEE Trans. Power Electron.* **2008**, *23*, 2612–2618. [CrossRef]
18. Boldea, I.; Paicu, M.C.; Andreescu, G.; Blaabjerg, F. "Active Flux" DTFC-SVM Sensorless Control of IPMSM. *IEEE Trans. Energy Convers.* **2009**, *24*, 314–322. [CrossRef]
19. Zhao, R.; Xin, Z.; Loh, P.C.; Blaabjerg, F. A Novel Flux Estimator Based on Multiple Second-Order Generalized Integrators and Frequency-Locked Loop for Induction Motor Drives. *IEEE Trans. Power Electron.* **2017**, *32*, 6286–6296. [CrossRef]
20. Xu, W.; Jiang, Y.; Mu, C.; Blaabjerg, F. Improved Nonlinear Flux Observer-Based Second-Order SOIFO for PMSM Sensorless Control. *IEEE Trans. Power Electron.* **2019**, *34*, 565–579. [CrossRef]
21. Choi, J.; Nam, K.; Bobtsov, A.A.; Pyrkin, A.; Ortega, R. Robust Adaptive Sensorless Control for Permanent-Magnet Synchronous Motors. *IEEE Trans. Power Electron.* **2017**, *32*, 3989–3997. [CrossRef]
22. Lin, T.C.; Zhu, Z.Q.; Liu, J.M. Improved Rotor Position Estimation in Sensorless-Controlled Permanent-Magnet Synchronous Machines Having Asymmetric-EMF with Harmonic Compensation. *IEEE Trans. Ind. Electron.* **2015**, *62*, 6131–6139. [CrossRef]
23. Roberto, M.; Edmundo, B.; Marco, A.A.; Hernández, C. Sensorless Predictive DTC of a Surface-Mounted Permanent-Magnet Synchronous Machine Based on Its Magnetic Anisotrogy. *IEEE Trans. Ind. Electron.* **2013**, *60*, 3016–3024.
24. Levant, A. Sliding order and sliding accuracy in sliding mode control. *Int. J. Control* **1993**, *58*, 1247–1263. [CrossRef]
25. Liang, D.; Li, J.; Qu, R. Sensorless Control of Permanent Magnet Synchronous Machine Based on Second-Order Sliding-Mode Observer with Online Resistance Estimation. *IEEE Trans. Ind. Appl.* **2017**, *53*, 3672–3682. [CrossRef]
26. Liang, D.; Li, J.; Qu, R.; Kong, W. Adaptive Second-Order Sliding-Mode Observer for PMSM Sensorless Control Considering VSI Nonlinearity. *IEEE Trans. Power Electron.* **2018**, *33*, 8994–9004. [CrossRef]
27. Li, Z.; Zhou, S.; Xiao, Y.; Wang, L. Sensorless Vector Control of Permanent Magnet Synchronous Linear Motor Based on Self-adaptive Super-twisting Sliding Mode Controller. *IEEE Access* **2019**, *7*, 44998–45011. [CrossRef]
28. Holakooie, M.H.; Ojaghi, M.; Taheri, A. Modified DTC of a Six-Phase Induction Motor with a Second-Order Sliding-Mode MRAS-Based Speed Estimator. *IEEE Trans. Power Electron.* **2019**, *34*, 600–611. [CrossRef]
29. Moreno, J.A.; Osorio, M. A Lyapunov approach to second-order sliding mode controlers and observer. In Proceedings of the 47th IEEE Conference on Decision and Control, Cancun, Mexico, 9–11 December 2008; pp. 2856–2861.
30. Moreno, J.A.; Osorio, M. Strict Lyapunov Functions for the Super-Twisting Algorithm. *IEEE Trans. Autom. Control* **2012**, *57*, 1035–1040. [CrossRef]
31. Zhong, L.; Rahman, M.F.; Hu, W.; Lim, K.W. Analysis of Direct Torque Control in Permanent Magnet Synchronous Motor Drives. *IEEE Trans. Power Electron.* **1997**, *12*, 528–536. [CrossRef]
32. Wu, X.; Tan, G.; Ye, Z.; Liu, Y.; Xu, S. Optimized Common-Mode Voltage Reduction PWM for Three-Phase Voltage Source Inverters. *IEEE Trans. Power Electron.* **2016**, *31*, 2959–2969. [CrossRef]
33. Xia, Y.; Gou, B.; Xu, Y. Anew ensemble-based classifier for IGBT open-circuit fault diagnosis in three-phase PWM converter. *Prot. Control Mod. Power Syst.* **2018**, *3*, 364–372. [CrossRef]

Article

An Improved Torque Control Strategy of PMSM Drive Considering On-Line MTPA Operation

Zhanqing Zhou [1], Xin Gu [2], Zhiqiang Wang [1], Guozheng Zhang [2,*] and Qiang Geng [2,*]

[1] School of Artificial Intelligence, Tianjin Polytechnic University, Tianjin 300387, China
[2] School of Electrical Engineering and Automation, Tianjin Polytechnic University, Tianjin 300387, China
* Correspondence: zhanggz@tju.edu.cn (G.Z.); gengqiang@tju.edu.cn (Q.G.); Tel.: +86-1382-133-1262 (G.Z.); +86-1382-020-8856 (Q.G.)

Received: 12 July 2019; Accepted: 30 July 2019; Published: 31 July 2019

Abstract: An improved direct torque control with space-vector modulation (DTC-SVM) scheme is presented in this paper. In the conventional DTC-SVM scheme, torque control performance is affected by the load conditions, due to the inappropriate linearization of the relationship between the flux angle and electromagnetic torque. Different from the conventional method, a torque controller with load angle estimation (TC-LAE) is proposed and the change rate of torque is regulated according to the variation of the load conditions, which could ensure the rapidity and consistency of torque performance at different load conditions. Meanwhile, an online permanent magnet synchronous motor and maximum torque per ampere (PMSM-MTPA) operation strategy based on the fitting solving method is proposed instead of the traditional two-dimensional look-up table, and the reference value of flux amplitude is calculated online to meet the MTPA requirement with the proposed method. The improved strategy is applied on a 6 kW PMSM, and the simulation and experimental results verified the effectiveness and the feasibility of the proposed strategy.

Keywords: direct torque control (DTC); permanent magnet synchronous motor (PMSM); maximum torque per ampere (MTPA) operation; DTC with space-vector modulation (DTC-SVM)

---

1. Introduction

A lot of work has been done to improve dynamic torque performance and to optimize the output efficiency of the torque of permanent magnet synchronous motors (PMSMs) in recent years [1–8]. Additionally, various optimal torque control strategies have been proposed, such as direct torque control (DTC) [4,5], predictive torque control [6,7] and nonlinear control strategies [8], etc. The DTC strategy combined with space vector pulse width modulation (PWM) [9], which used continuous rotated voltage vector to regulate the flux of the motor. The torque control performance was improved compared with conventional DTC [10–13].

The DTC-SVM scheme usually consists of two parts [14]: one is the selection of the flux reference based on the two-dimensional look-up table offline. In this part, the reference value of the current is achieved using the torque–current table based on the maximum torque per ampere (MTPA) criterion, and then, the reference value of flux is calculated according to the relationship between flux and current of the motor. Hence, the MTPA operation of the DTC-SVM scheme could be achieved [15]. The other is the calculation of the reference flux angle based on the proportional integral (PI) controller. The relationship between the electromagnetic torque and the flux angle is linearized approximately, which resulted in excellent torque performance which could be maintained at different load conditions.

The research of DTC-SVM scheme always focuses on two aspects: one is to reduce the impact on the MTPA operation brought about by the change of parameters of the motor. The other is to improve the control/precision of the torque against variations in the load conditions.

The validity of the data in the torque–current two-dimensional table depends on the accuracy of the motor's parameters in the selection mechanism of the reference flux amplitude. Slow variation of the parameters is inevitable because of copper loss and magnetic saturation. Hence, the operation's conditions might deviate from the MTPA [16]. Therefore, the online MTPA control is an ideal solution to these problems. At present, the online solution of MTPA could be classified into the direct solving method [17] and engineering optimization method [18,19]. The MTPA criterion is a fourth-order equation about the stator current. The direct solving method is to solve this fourth-order equation online using the Ferrari method. Then, the reference value of the current could be obtained. The engineering optimization method is to change the fourth-order equation to an online optimization problem. The direct solving relationship between the torque and MTPA criterion is established. The voltage limitations of the inverter, the extreme current of the motor and the operational conditions are used as the boundary criterions. The stator current, which meets the MTPA criterion, is the optimization object. Then, the solving of the current reference value is realized. After that, the reference value of the stator flux amplitude for the DTC-SVM scheme could be obtained with this reference value of the current [20]. It is worth illustrating that the impacts of parameter variation on the online MTPA operation can be eliminated using certain parameter identification or self-adaptive methods.

The linearization of the relationship between the flux angle and electromagnetic torque is used as the control object for the torque control of the conventional DTC-SVM scheme, thus, the torque loop can be regarded as a second-order system with a PI controller. However, the dampening of this torque loop will be affected by the load conditions, which could lead to different torque adjustable performances of the conventional DTC-SVM scheme. There are three kinds of control strategies to improve the performance of torque control caused by inappropriate linearization:

1. Variable parameter PI control. A third-order characteristic curve of the relationship between the parameters of the PI controller and the load torque is established by the interpolation fitting method. The parameters of the PI controller will be adjusted according to the curve mentioned before to eliminate the impact of load torque on the performance of PI control [21].
2. Nonlinear control. Back-stepping control [22], variable structure control [23], sliding mode control [24,25] and other nonlinear controllers are utilized to realize torque control. The nonlinear controller has the advantages of a rapid dynamic response and good adaptability against external disturbances and nonlinearity of the parameters. The dynamic performance of the motor is rapid and consistent under different load conditions.
3. Deadbeat torque control [26–28] and predictive torque control [29–31]. The deviation of the torque is used as the input, and the required stator voltage for torque control can be obtained by the predictive mechanism. In these kinds of strategies, the stator voltage is adjusted online to eliminate the impact of load on the performance of torque by the predictive/deadbeat controller according to the mathematical model and load condition of the motor.

To improve the performance of the conventional DTC-SVM, a novel online MTPA method based on Lagrange interpolation and an improved torque controller with load angle estimation (TC-LAE) are proposed in this paper. Different from the existing MTPA scheme, the proposed MTPA scheme takes the stator flux linkage as a variable instead of the stator current. Furthermore, the direct selection of the reference flux amplitude satisfied with the MTPA criterion could be realized on-line by Lagrange interpolation. Besides, a P-type torque controller with load angle estimation is adopted instead of the inappropriate linearization PI controller, so that the parameters of torque controller could be adjusted online according to the actual load angle to improve the control performance of torque under different load conditions.

2. Examination of Conventional DTC-SVM Scheme

The structure diagram of the conventional DTC-SVM scheme is shown in Figure 1.

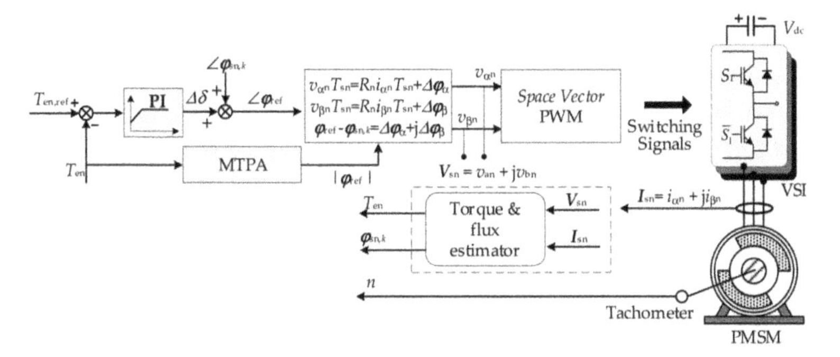

Figure 1. The conventional direct torque control with space-vector modulation (DTC-SVM) scheme.

2.1. Selection of Reference Flux Amplitude Based on the MTPA Criterion

When operating below the rated speed of a PMSM, the MTPA operation should be satisfied for high-efficiency operation, which means the relationship between the electromagnetic torque and the stator current might meet the requirements of the following equations [15].

$$\begin{cases} i_{\mathrm{dn}} = \frac{1}{2(L_{\mathrm{qn}}-L_{\mathrm{dn}})} - \sqrt{\frac{1}{4(L_{\mathrm{qn}}-L_{\mathrm{dn}})^2} + i_{\mathrm{qn}}^2} \\ T_{\mathrm{en}} = i_{\mathrm{qn}} + (L_{\mathrm{dn}} - L_{\mathrm{qn}})i_{\mathrm{dn}}i_{\mathrm{qn}} \end{cases} \quad (1)$$

During mathematical derivation of this paper, the per-unit value is employed for generality. The base value is selected as follows: $i_{\mathrm{b}} = i_{\mathrm{N}}$; $L_{\mathrm{b}} = \varphi_{\mathrm{b}}/i_{\mathrm{b}}$; $\varphi_{\mathrm{b}} = \varphi_{\mathrm{r}}$; $T_{\mathrm{eb}} = 1.5p\varphi_{\mathrm{b}}i_{\mathrm{b}}$; $T_{\mathrm{sb}} = 1/\omega_{\mathrm{N}}$.

The reference value of i_{dn} and i_{qn} that meet the requirement of the MTPA criterion are obtained using look-up table, and the equation of the flux linkage after normalization can be obtained by:

$$\begin{cases} \varphi_{\mathrm{dn}} = L_{\mathrm{dn}}i_{\mathrm{dn}} + \varphi_{\mathrm{rn}} = \varphi'_{\mathrm{dn}} + 1 \\ \varphi_{\mathrm{qn}} = L_{\mathrm{qn}}i_{\mathrm{qn}} \end{cases} \quad (2)$$

where, $\varphi'_{\mathrm{dn}} = L_{\mathrm{dn}}i_{\mathrm{dn}}$. Furthermore, the reference value of stator flux amplitude is:

$$|\varphi_{\mathrm{sn}}| = \sqrt{(L_{\mathrm{dn}}i_{\mathrm{dn}} + 1)^2 + (L_{\mathrm{qn}}i_{\mathrm{qn}})^2} \quad (3)$$

It can seen from Equations (1)–(3) that the MTPA operation of the PMSM depends on the accuracy of the motor's parameters. Hence, the online MTPA operation method could be used to reduce the impact of the parameters on the operation of the MTPA. The essence is changing the offline look-up table which meets the requirement of Equation (1) to solve the fourth-order equation online, which takes i_{d} as the independent variable and T_{en} as the parameter. Finally, the reference amplitude could be obtained using the square root operation, as shown in Equation (3).

2.2. Torque Control Based on the PI Controller

The electromagnetic torque can be written as the expression of the flux amplitude and load angle, which is [14]:

$$T_{\mathrm{en}} = \frac{|\varphi_{\mathrm{sn}}|}{L_{\mathrm{qn}}}\left[\rho \sin\delta - \frac{1}{2}|\varphi_{\mathrm{sn}}|(\rho - 1)\sin 2\delta\right] \quad (4)$$

where, $\rho = L_{\mathrm{qn}}/L_{\mathrm{dn}}$, $\rho > 1$. δ represents the load angle, which is the angle between the stator voltage vector and the flux vector. The incremental quantity of the load angle $\Delta\delta$ is related to the phase angle of the motor's stator flux vector.

$$\angle\varphi_{\mathrm{ref}} = \angle\varphi_{\mathrm{s},k} + \Delta\delta \quad (5)$$

The approximate linearization processing of Equation (4) is taken when the load angle is δ_0

$$T_{\text{en}} = k_{\text{T}}(\delta - \delta_0) + T_0 \tag{6}$$

where,

$$k_{\text{T}} = \frac{|\varphi_{\text{sn}}|}{L_{\text{qn}}}\left[\rho \cos\delta_k - |\varphi_{\text{sn}}|(\rho - 1)\cos 2\delta_k\right] \tag{7}$$

$$T_0 = T_{\text{en}}\big|_{\delta_k = \delta_0} \tag{8}$$

The PI controller is used to realize torque control for the conventional DTC-SVM scheme, according to Equation (6). The structure block diagram of the torque loop is shown in Figure 2. The close-loop transfer function of the torque control link could be obtained from this figure, and the damping ζ and natural characteristic frequency ω_{n} can be derived as:

$$\zeta = \frac{K_{\text{p}}}{2}\sqrt{\frac{k_{\text{T}}}{K_{\text{i}}T_{\text{s}}}}, \quad \omega_{\text{n}} = \sqrt{\frac{K_{\text{i}}k_{\text{T}}}{T_{\text{s}}}} \tag{9}$$

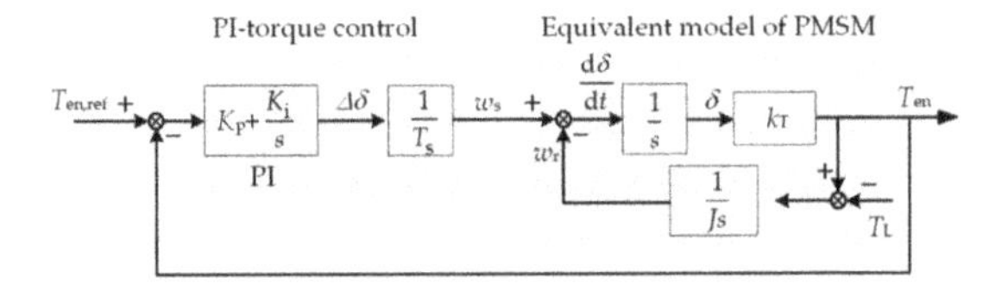

Figure 2. The control block of the torque loop.

In the block diagram, as shown in Figure 2, the tuning processing of the parameters of the PI controller is as follows: firstly, the expectation value of ζ is 0.707 in engineering practice. Secondly, considering the regulation performance and the disturbance immunity of the control system, T_{i} can usually be chosen as 15 T_{s} to 25 T_{s} in the digital control system, where $T_{\text{i}} = K_{\text{p}}/K_{\text{i}}$ [32]. At last, the change rate of torque k_{T} corresponding certain load conditions is selected to calculate the parameters of the PI controller generally.

However, k_{T} varies for different parameters, stator flux amplitudes and load angles of the motor, and because of the nonlinear characteristic of Equation (7), it is hard to select a particular k_{T} to tune the PI parameters. For example, the change curve of k_{T} is calculated according to (7), where the stator flux amplitude changes from 0.6 to 1.0 (per-unit value) and the load angle changes from 0.0 to 1.5 (per-unit value), as shown in Figure 3.

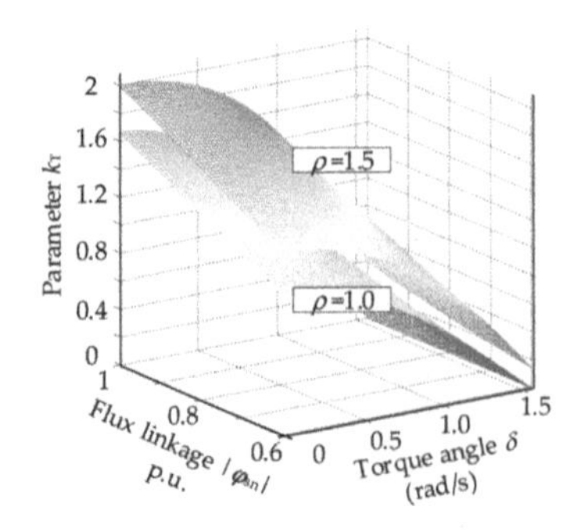

Figure 3. The relationship of the torque change rate k_{T} with the flux amplitude and load angle.

As shown in Figure 3, k_T is mainly affected by the load angle of the motor. The heavier the load is, the smaller the value of k_T is, inversely. So if k_T under the rated load condition is used to calculate the PI parameters, the k_T would be increased under light load conditions. Thus, it can be known from equation (9), ζ will be larger than 0.707 for light load operations, so the adjustment time of the transiente torque will be longer. Conversely, if k_T under the no-load condition is used, k_T will be decreased when operating under rated load conditions. Then, ζ is smaller than 0.707, which is possible to cause an oscillation process of the torque regulation.

3. An Improved DTC-SVM Scheme

The block diagram of the proposed improved DTC-SVM scheme is presented in Figure 4. For the reference calculation of the flux amplitude, a novel MTPA criterion expressed by the stator flux linkage is constructed, and the reference value of flux amplitude can be obtained with Lagrange interpolation online directly. For the reference calculation of flux phase angle, this section puts forward a novel P-type torque controller with load angle estimation (TC-LTE), which could regulate the flux phase angle as load angle variation. Moreover, by adding the relevant correction, compensation and limitation blocks for the incremental quantity of the load angle $\Delta\delta$, the impact of the voltage limitation circle, rotation of the permanent magnet and load angle stability on the torque control performance could be depressed.

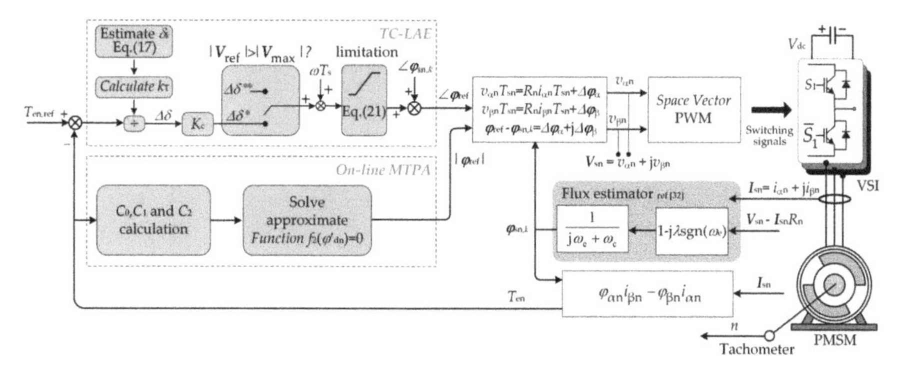

Figure 4. The schematic diagram of the proposed DTC-SVM scheme.

3.1. Novel Online MPTA Scheme

By substituting (2) into (1), the following equation can be derived:

$$a\varphi'^4_{\mathrm{dn}} + b\varphi'^3_{\mathrm{dn}} + c\varphi'^2_{\mathrm{dn}} + d\varphi'_{\mathrm{dn}} + eT^2_{\mathrm{en}}L^2_{\mathrm{qn}} = 0 \tag{10}$$

where, $a = (1-\rho)^3\rho^2$; $b = 3\rho^2(1-\rho)^2$; $c = 3\rho^2(1-\rho)$; $d = \rho^2$; $e = -(1-\rho)$.

Equation (10) is a novel MTPA criterion expressed by flux linkage; it can determine the reference flux amplitude that satisfied MTPA criterion under different load conditions directly.

In order to solve (10), Lagrange interpolation is adopted to fit the left polynomial of (10). Then, the feasible solution of (10) is equivalent to the zero point of the fitting polynomial. At last, the reference value of the flux amplitude can be determined with simple calculation. Assuming:

$$f_1(\varphi'_{\mathrm{dn}}) = a\varphi'^4_{\mathrm{dn}} + b\varphi'^3_{\mathrm{dn}} + c\varphi'^2_{\mathrm{dn}} + d\varphi'_{\mathrm{dn}} + eT^2_{\mathrm{en}}L^2_{\mathrm{qn}} \tag{11}$$

According to the theory of Lagrange interpolation [33], there are two necessary steps to fit the above polynomial. The first step is to confirm the solution region of (10). The second step is to select the samples and calculate the remainder of the interpolation.

Calculating the derivative of φ'_{dn} in (11), and making this derivation equal to zero, that is:

$$4(1-\rho)^3\varphi'^3_{dn} + 9(1-\rho)^2\varphi'^2_{dn} + 6(1-\rho)\varphi'_{dn} + 1 = 0 \tag{12}$$

The above equation has a single real root $\frac{1}{4(\rho-1)}$ and a double real root $\frac{1}{\rho-1}$. Taking the second order derivative at these two points, we can get $\left.\frac{d^2 f_1}{d\varphi'^2_{dn}}\right|_{\varphi'_{dn}=\frac{1}{4(\rho-1)}} < 0$ and $\left.\frac{d^2 f_1}{d\varphi'^2_{dn}}\right|_{\varphi'_{dn}=\frac{1}{(\rho-1)}} < 0$. Obviously, $\left[\frac{1}{4(\rho-1)}, f_1\left(\frac{1}{4(\rho-1)}\right)\right]$ and $\left[\frac{1}{\rho-1}, f_1\left(\frac{1}{\rho-1}\right)\right]$ are the maximum points of f_1. Besides, considering the intercept of the $\varphi'_{dn} - f_1$ plot, $f_1(0) = eT_{en}{}^2L_{qn}{}^2 > 0$, it can be concluded that (10) has one positive solution and one negative solution. Furthermore, as shown in (1), $i_{dn} < 0$. Hence, the negative solution could be the unique feasible solution of (10). The schematic diagram of the interpolation trajectory and its remainder are drawn in Figure 5.

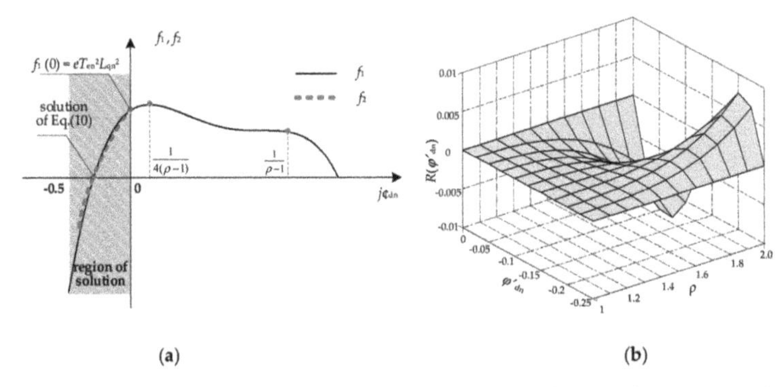

Figure 5. The schematic diagram of the interpolation trajectory and remainder. (**a**) Interpolation; (**b**) remainder of interpolation.

To avoid the irreversible demagnetization of the permanent magnet, always keep $\varphi'_{dn} > -0.5$ [34]. Therefore, the solution region of (10) can be determined as (−0.5, 0).

In general, the value of φ'_{dn} is always small under the MTPA operation, then the interplotion samples could be selected in [−0.25, 0]. This section adopted Lagrange parabolic interpolation to fit the curve of f_1, and the interplotion samples are chosen as (0, C_0), (−0.15, C_1) and (−0.25, C_2), where $C_0 = f_1(0) = eT^2_{en}L^2_{qn}$, $C_1 = f_1(\varphi'_{dn})\big|_{\varphi'_{dn}=-0.15}$ and $C_2 = f_1(\varphi'_{dn})\big|_{\varphi'_{dn}=-0.25}$, respectively. Based on the interpolation formula [33], The final interpolation polynomial can be expressed as follows:

$$f_2(\varphi'_{dn}) = (26.67C_0 - 66.67C_1 + 26.67C_2)\varphi'^2_{dn} + (10.67C_0 - 16.67C_1 + 4C_2)\varphi'_{dn} + C_0 \tag{13}$$

and the interpolation remainder for (13) is:

$$R_n(\varphi'_{dn}) = \frac{\varphi'_{dn}f_1'''(\varphi'_{dn})}{6}(\varphi'_{dn} + 0.15)(\varphi'_{dn} + 0.25) \tag{14}$$

For $\varphi'_{dn} \in [-0.25, 0]$ and $\rho \in [1.0, 2.0]$, the numerical analysis results of $R_n(\varphi'_{dn})$ is shown in Figure 5b. It can be seen from this figure that $-0.01 < R_n(\varphi'_{dn}) < 0.01$, thus the solution of $f_2(\varphi'_{dn}) = 0$ could be regarded as the solution of (10), approximately. This means that we can determine the reference flux amplitude with the solution of $f_2(\varphi'_{dn}) = 0$ online, and the complex process for solving (10) directly can be avoided. Particularly, the following extra conditions must be satisfied during the determination of the reference flux amplitude.

1. To ensure stable operation of the PMSM, the reference value of the flux amplitude [20],

$$\left|\varphi_{\text{sn,ref}}\right| < \frac{\rho}{\rho - 1} \tag{15}$$

2. If the PMSM works under no-load conditions, the electromagnetic torque and stator current are almost zero. According to (3),

$$\left|\varphi_{\text{sn,ref}}\right| = 1 \tag{16}$$

3.2. Torque Controller with Load Angle Estimation (TC-LAE)

The control block of the proposed TC-LAE is shown Figure 4. The estimation equation of the load angle can be expressed with stator current and flux, which is:

$$\delta_k = \arctan\left[\frac{1.5pL_{\text{qn}}T_{\text{en}}}{\left|\varphi_{\text{sn},k}\right|^2 - L_{\text{qn}}(\varphi_\alpha i_\alpha + \varphi_\beta i_\beta)}\right] \tag{17}$$

by substituting the estimated load angle into (7), we can predict the value of k_{T} in real-time. Furthermore, the incremental quantity of the load angle at the next control instant, which is denoted as $\Delta\delta$, can be obtained by taking the difference operation on both sides of (4), that is:

$$\Delta\delta = \frac{1}{k_{\text{T}}}\Delta T_{\text{en}} \tag{18}$$

In addition, a P-type controller is employed in the TC-LAE for $\Delta\delta$ trimming, to depress the impact of several disturbance factors, such as sampling error and parameter mismatches, on torque performance. The control parameter of this P-type controller is K_{c}, and $K_{\text{c}} > 0$.

The conventional DTC-SVM takes a constant k_{T} for the parameter tuning of the PI controller. Differening from the conventional method, the proposed TC-LAE adjusts $\Delta\delta$ with the appropriate k_{T}, which is calculated based on the actual load angle. With the aid of TC-LAE, the improved DTC-SVM can achieve a fast and consistent torque response under different load conditions.

3.3. Correction, Compensation and Limitations of $\Delta\delta$

3.3.1. Correction

In DTC-SVM strategies, during large torque demands, the torque controller will give an output that demands the selection of voltage vectors to increase the torque. However, once the reference voltage vector tip point lies outside the hexagon, the space-vector PWM yields a negative time length, resulting in an inevitable volt-seconds error [35,36]. A voltage vector on the hexagon boundary (the modified reference voltage vector) must be selected and at least one back step has to be taken to recalculate the vector time lengths that generate the modified reference voltage vector. Shown in Figure 6, the two popular modified reference vector choices are the minimum magnitude error PWM (MMEPWM) method (also called the one-step-optimal method), and the minimum phase error PWM (MPEPWM) method. However, the MMEPWM and MPEPWM could not ensure the stable output of the load angle and flux amplitude simultaneously at the transient instant [37]. Hence, a special voltage vector correction algorithm is proposed in this section, which is shown in Figure 6.

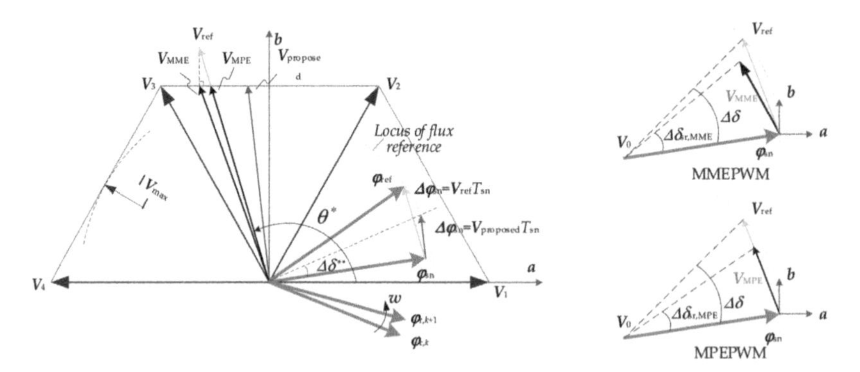

Figure 6. The correction algorithm of the incremental value of the load angle.

Assuming the reference torque is increased at the instant kT_s, and $\varphi_{sn,k}$ donotes the current stator flux vector; φ_{ref} denotes the reference stator flux vector; V_{ref} denotes the reference vetor of the stator voltage obtained by (17), which hopes to make the stator flux vector ($\varphi_{sn,k+1}$) equal to φ_{ref} at the instant $(k+1)T_s$. However, it can be seen from Figure 6 that the actual voltage vector is V_{act} due to the existance of the voltage limitation circle. Consequently, $\varphi_{sn,k+1}$ could not follow φ_{ref} under the effects of V_{act}, it will lead to an ampltide error for flux control. To make sure $|\varphi_{sn,k+1}| = |\varphi_{ref}|$, and fullly ulitize the voltage capablity of the VSI at the same time, we should revise $\Delta\delta$ when the amplitude of V_{ref} is beyond the voltage limitation circle. On the basis of the vector raltionship in Figure 6, and with the help of cosine theorem, the revised $\Delta\delta^{**}$ can be derived as:

$$\Delta\delta^{**} = \arccos\left(\frac{|V_{ref}T_{sn}|^2 - |\varphi_{ref}|^2 - |\varphi_{sn,k}|^2}{-2|\varphi_{ref}||\varphi_{sn,k}|}\right) \tag{19}$$

where, T_{sn} denotes the per-unit value of the control period, and the base value of time is selected as $1/\omega_N$, ω_N is the rated electrical angular frequency of the PMSM; V_{ref} denotes the voltage vector with the $\Delta\delta$ revising algorithm, its amplitude equals the maximum value of the output voltage of the VSI.

3.3.2. Compensation

The rotor permanent magnet of the PMSM keeps rotating during normal operation. Assuming the rotor rotates counterclockwise, and ω denotes the rotor electrical angular frequency, it can be seen from Figure 6 that the value of $\Delta\delta$ obtained by controller is ωT_s less than the actual required value because of the rotation of the permanent magnet. This angular deviation will result in offsets for torque control during high speed operations, hence it is necessary to compensate the angular deviation, that is:

$$\Delta\delta^* = \Delta\delta + \omega T_s \tag{20}$$

3.3.3. Limitation

Load angle stability must be ensured when the PMSM is operating under heavy load conditions, hence, the limitation block should be utilized for $\Delta\delta$ adjustment. Taking the derivative of δ in (7), and making this derivation equal to zero, the maximum load angle δ_m can be obtained:

$$\delta_m = \arccos\left[\frac{\rho - \sqrt{\rho^2 + 8|\varphi_{sn}|^2(\rho-1)^2}}{4|\varphi_{sn}|(\rho-1)}\right] \tag{21}$$

Therefore, the maximum allowable vaule of $\Delta\delta$ (denoted as $\Delta\delta_{\mathrm{m}}$) can be obtained:

$$\Delta\delta_{\mathrm{m}} = \delta_{\mathrm{m}} - \delta_k \tag{22}$$

4. Simulation Results

In order to study the control performance of the proposed TC-LAE, numerical simulations have been carried out using Matlab/Simulink. The parameters of the control system are presented in Table 1. It should be illustrated that the speed loop consisting of a PI controller is added out of the torque loop. The maximum value of the output torque, which is restricted by the PI controller of the speed loop, is $1.2T_{\mathrm{N}}$, which is 230 Nm.

Table 1. Parameters of the control system.

Parameter	Symbol	Value
Number of pole-pairs	p	8
Permanent magnet flux of PMSM	ϕ_{ρ}	0.9031 Wb
Stator resistance of PMSM	R_{s}	0.76 Ω
d-axis inductance of PMSM	L_{d}	23.05 mH
q-axis inductance of PMSM	L_{q}	24.26 mH
Rated speed of PMSM	n_{N}	300 r/min
Rated torque of PMSM	T_{N}	192 Nm
Rated voltage of PMSM	v_{N}	380 V
Rated current of PMSM	I_{N}	11.8 A
DC voltage of VSI	V_{dc}	560 V
Switching frequency of VSI	f_{sw}	5 kHz
Sampling period of controller	T_{s}	200 μs
Proportional coefficient of speed PI controller	K_{pw}	100
Integral coefficient of speed PI controller	K_{iw}	50 T_{s}

4.1. The Correction and Compensation for $\Delta\delta$

The simulation waveforms with/without the correction algorithm for $\Delta\delta$ are shown in Figure 7. The motor operates at 100 r/min with no load. When t = 1.5 s, the reference value of speed n_{ref} is set to 200 r/min, the speed PI controller reaches the positive limitation. Without the correction algorithm, $\Delta\delta$ increased rapidly, resulting from the sudden increase of T_{ref}. The amplitude of the stator voltage vector is increased correspondingly according to the analysis in Section 3.3. As can be seen from Figure 7, taking the voltage amplitude limitation of the SVM into consideration, when the reference voltage (v_{α}, v_{β})is beyond the range of the voltage limitation circle, the amplitude of the stator flux linkage $|\varphi_{\mathrm{s}}|$ slides for a short time and a dynamic deviation will appear between the reference value and the actual value of the stator flux linkage. So, the stator current will increase rapidly($i_{\max}$ = 23 A). With the correction algorithm, the tracking ability of the flux control is improved, and the dynamic current is reduced effectively when the reference torque is changed suddenly($i_{\mathrm{a,max}}$ = 21 A).

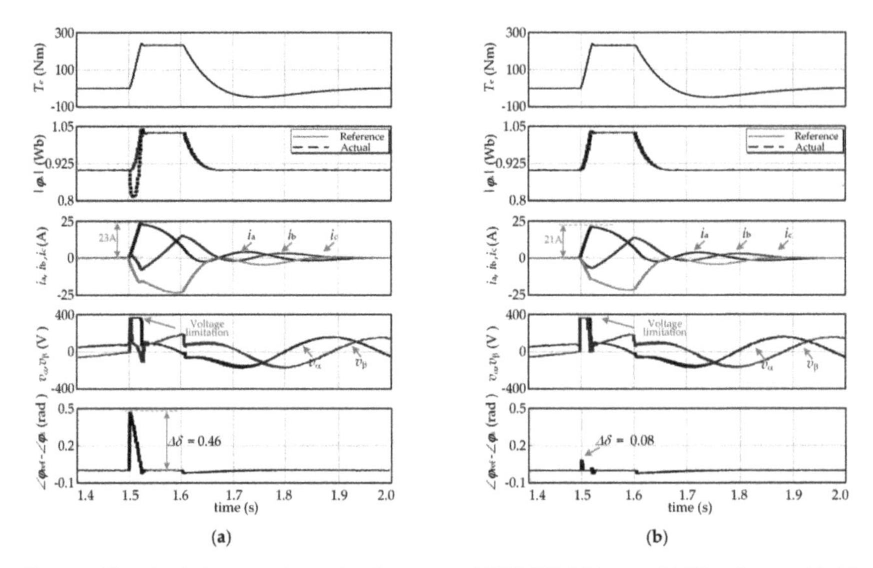

Figure 7. The simulation waveforms for the proposed DTC-SVM Scheme. (**a**) Waveforms with $\Delta\delta$ correction; (**b**) waveforms with $\Delta\delta$ correction.

4.2. Selection of the Torque Controller Parameter, K_c

The impact of the selection of K_c on the static and dynamic performance is analyzed in this section. The standard deviation is used to evaluate the extent of the torque ripple in the anaylsis process. That is:

$$\sigma_T = \sqrt{\frac{1}{n-1}\sum_{i=1}^{n}\left(T_e(i) - \overline{T_e}\right)^2} \tag{23}$$

where, $\overline{T_e} = \frac{1}{n}\sum_{i=1}^{n} T_e(i)$, n is the number of samples and $n = 1000$.

In the simulation, the motor is operated stably at 100 r/min with 50 Nm. When $t = 0.1$ s, n_{ref} is set to 200 r/min. When K_c is between 0.2 and 2.5, the variation rules of the static torque ripple σ_T and the regulating time of electromagnetic torque t_d are shown in Figure 8:

1. When the value of K_c is less than 2.0, the torque ripple of the motor is small. When the value of K_c is greater than 2.0, the static oscillation of the electromagnetic torque appears.
2. When the value of K_c is between 0.5 and 2.0, the dynamic performance of the electromagnetic torque is basically consistent. When the value of K_c is less than 0.5, the regulating time of the torque becomes longer. So, the dynamic performance of the system is degraded.

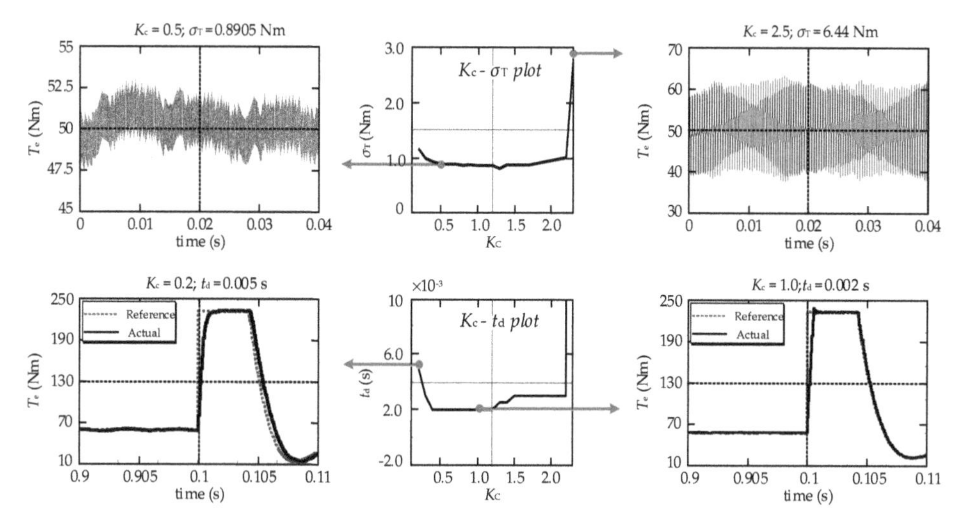

Figure 8. Steady and transient torque performance of the proposed DTC-SVM scheme under different values of K_c.

4.3. Dynamic Characteristic of the Motor Control System

The dynamic simulation waveforms of the improved DTC-SVM scheme and the conventional DTC-SVM scheme are shown in Figure 9. In the simulation, for the conventional DTC-SVM scheme, the parameters of the speed PI controller are consistent with Table 1, the torque PI controller are $K_p = 6.25 \times 10^{-4}$ and $T_i = 20\ T_s$. The motor is operated at a steady-state of 100 r/min with 50 Nm. When $t = 0.1$ s, n_{ref} is set to −100 r/min and the motor is rotating in reverse. When $t = 0.2$ s, n_{ref} is set to 100 r/min again and the motor rotates normally. When $t = 0.3$ s, the load is suddenly increased to 150 Nm.

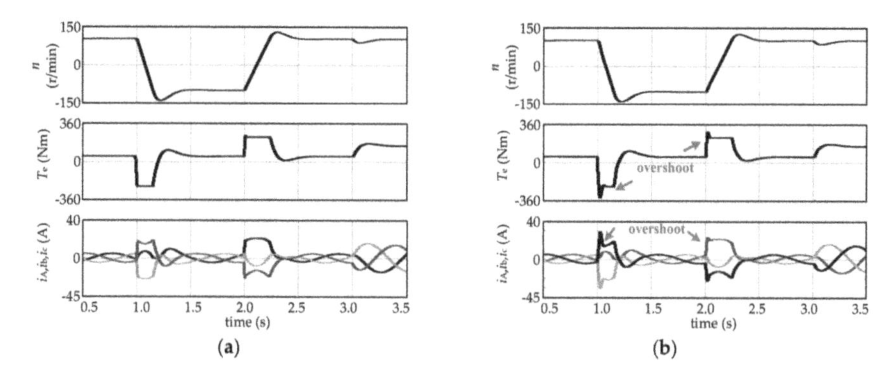

Figure 9. Speed dynamic simulation waveforms. (**a**) The improved DTC-SVM scheme; (**b**) the conventional DTC-SVM scheme.

As can be seen from Figure 9, the improved DTC-SVM scheme inherits the advantages of the conventional scheme, which has a rapid torque response and excellent stator currents. Meanwhile, because the torque loop of the conventional DTC-SVM is a second-order system, overshoot during torque regulation will inevitably occur. However, the improved DTC-SVM regulates the electromagnetic torque using k_T varied with the current load condition, so overshoots of the electromagnetic torque has been restrained to some extent.

5. Experimental Results

To verify the feasibility and effectiveness of the improved DTC-SVM scheme, experiments have been carried out on a 6 kW PMSM. The parameters of the experimental setup is consistant with the simulation, which is shown in Table 1. In the experimental setup, which is shown in Figure 10, a TMS320F28335 digital signal processor (DSP) is employed for the control strategy; the stator currents are measured by a LA-50P Hall sensor produced by LEM® (Geneva, Switzerland), and the DC-side voltage is measured by the VSM025A Hall sensor, and the sampling tasks of the stator currents and DC-side voltage are accomplished by the DSP; the angular velocity is obtained from the incremental mode optical shaft angle encoder; the electromagnetic torque is estimated with the mathematical model of the PMSM. Besides, the sampling and control period of the DSP is 200 μs.

Figure 10. Photograph of the experimental setup.

5.1. MTPA Operation

In the simulations and experiments, the motor operated at 100 r/min. At first, the motor is operated at 20 Nm, and the load is added at 20 Nm per time, until the load reaches 200 Nm. The actual electromagnetic torque, amplitude of stator flux linkage, average value of d-/q-axis currents and RMS value of the phase current under each load condition are measured, and the experimental data are plotted in Figure 11.

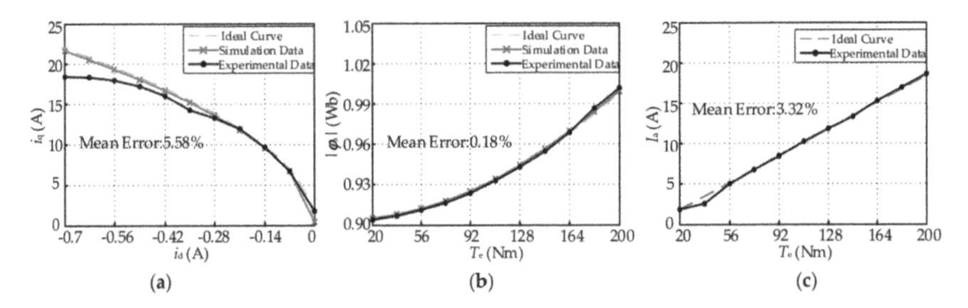

Figure 11. The performance of proposed on-line maximum torque per ampere (MTPA) method. (**a**) $i_d - i_q$ plot; (**b**) $T_e - |_s|$ plot; (**c**) $T_e - I_A$ plot.

It can be seen from Figure 11 that the MTPA trajectory obtained by simulation and experiments with the proposed online MTPA method almost coincide with the theoretical MTPA trajectory based on Equation (1).

5.2. Torque Control Performance

Figures 12 and 13 give the experimental waveforms with the conventional DTC-SVM scheme using two different parameters of torque PI controllers, which are calculated with k_T under no-load and rated load conditions, respectively. Figure 14 gives the experimental waveforms with the improved DTC-SVM scheme. In the experiments, firstly, the motor is operated at 100 r/min with no load and

50 Nm load, respetively. Next, the reference speed is increased to 200 r/min. Besides, to assure the incremental quantities of electromagnetic torque are consistent for different load conditions, the limitation of the speed PI controller is set to 100 Nm when the motor is operating under the no load condition, and 150 Nm for the 50 Nm load, correspondingly.

In Figure 12, the parameters of the torque controller of the conventional DTC-SVM scheme are calculated by the constant k_T corresponding to the no load condition. Then, it can be obtained that $K_p = 6.25 \times 10^{-4}$ and $T_i = 20\ T_s$. It can be seen from Figure 12 that the motor has favorable torque performance when operating under the no load condition. However, the dampening of the torque loop will be decreased on account of the decreased k_T when the motor operates with a 50 Nm load, and this will cause na oscillation process during the transiente torque.

In Figure 13, the parameters of the torque controller of the conventional DTC-SVM scheme are calculated by the constant k_T corresponding to the rated load condition. Then, it can be obtained $K_p = 2.0 \times 10^{-4}$ and $T_i = 20\ T_s$. It can be seen from Figure 13 that the torque response is favorable when the motor is operated with a 50 Nm load. However, the damping of the torque loop will be increased because of the increment of k_T when the motor is operating under the no load condition, resulting in a longer regualtion time of the eletromagnetic torque than the no load condition.

It can be seen from Figure 14 that the motor has a rapid and consistent torque response when operated under different load conditions for the proposed DTC-SVM scheme.

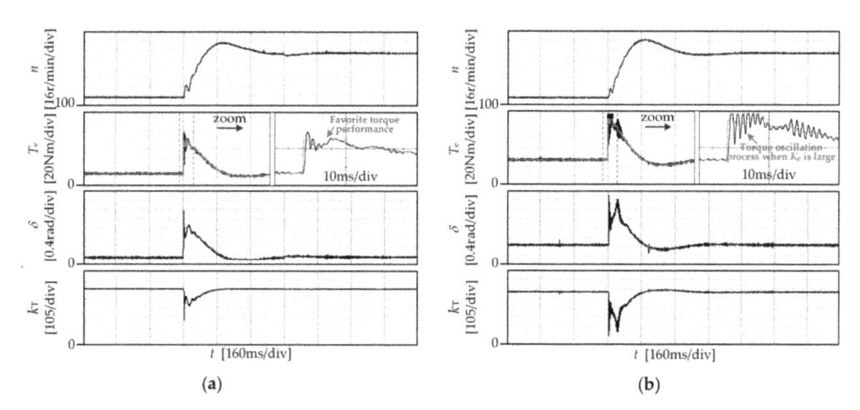

Figure 12. Experimental waveforms of the conventional DTC-SVM scheme during transient operation at $K_p = 6.25 \times 10^{-4}$; $T_i = 20\ T_s$. (**a**) No load; (**b**) 50 Nm load.

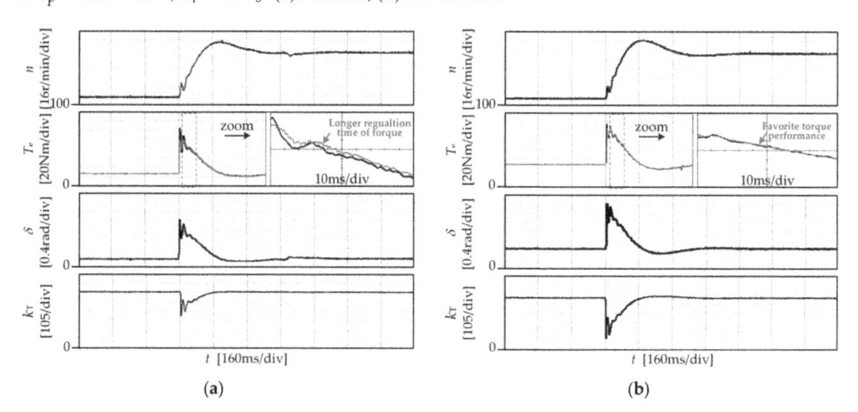

Figure 13. Experimental waveforms of the conventional DTC-SVM scheme during transient operation at $K_p = 2.0 \times 10^{-4}$; $T_i = 20\ T_s$. (**a**) No load; (**b**) 50 Nm load.

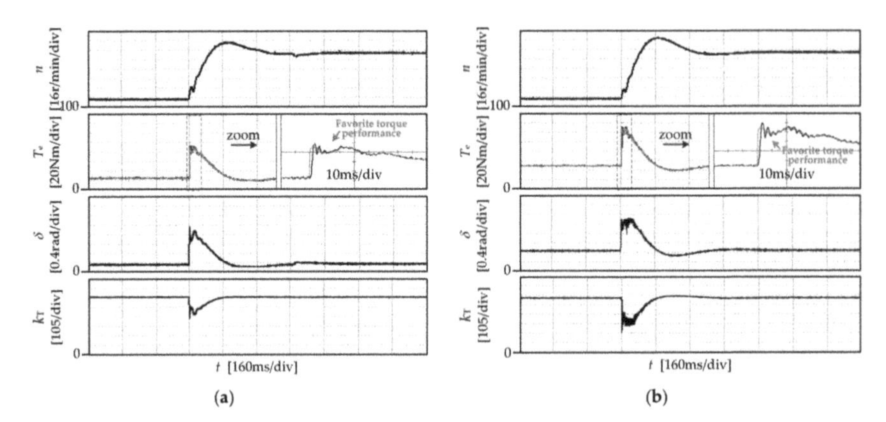

Figure 14. Experimental waveforms of the improved and conventional DTC-SVM scheme during transient operation. (**a**) No load; (**b**) 50 Nm load.

5.3. Characteristics of the Torque Controller

Figure 15 gives experimental waveforms with/without the compensation algorithm mentioned in Section 3.3. In the experiment, the motor is operated at 200 r/min with no load steadily. As shown in Figure 15, with the compensation algorithm, the electromagnetic torque could track its reference value without static error. While without the compensation algorithm, there are deviations between the electromagnetic torque and reference torque.

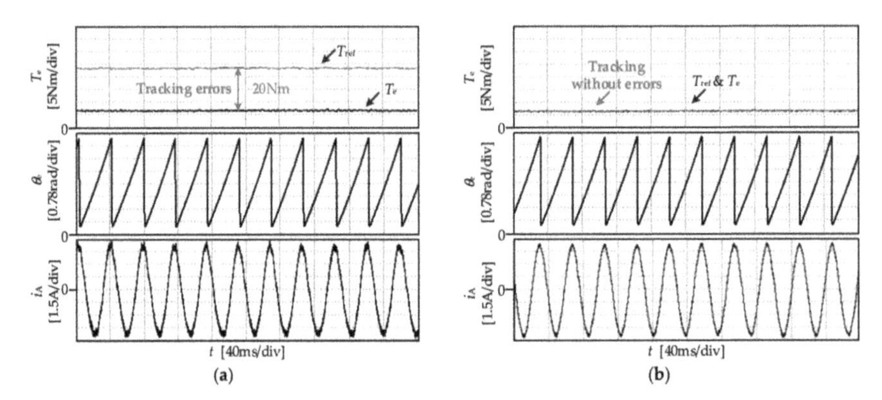

Figure 15. Experimental waveforms with/without compensation for the proposed DTC-SVM. (**a**) Without compensation; (**b**) with compensation.

Figure 16 gives the experimental waveforms of torque response for the value of K_c equal to 0.2, 1.0 and 2.0, respectively. In the experiments, firstly, the motor is operated at 100 r/min with no load. Next, the load is increased to 50 Nm. As can be seen from Figure 16, with the increase of K_c, the ripple amplitude of the static torque is also increased, but the motor has a faster torque dynamic response; with the decrease of K_c, although the ripple amplitude of the static torque is decreased, the dynamic performance of the motor deteriorated. Obviously, the above experimental results are consistent with the simulation.

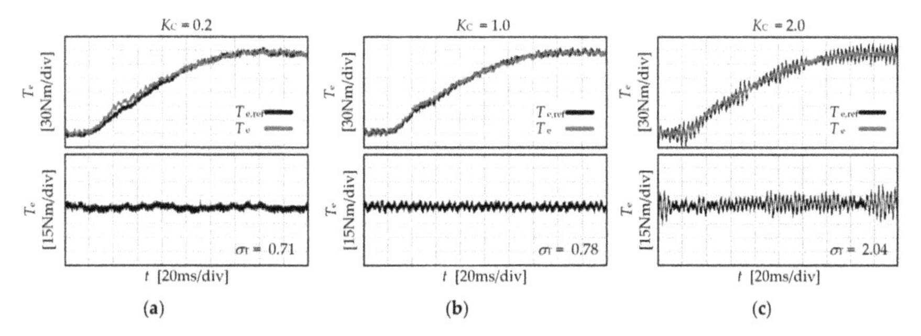

Figure 16. Transient and steady torque performance of the proposed DTC-SVM scheme with different values of K_c. (**a**) $K_c = 0.2$; (**b**) $K_c = 1.0$; (**c**) $K_c = 2.0$.

6. Conclusions

The constant torque change rate is used to regulate the electromagnetic torque in the conventional DTC-SVM, and the variation of the torque change rate has not been taken into consideration. Therefore, this kind of control mode will lead to a phenomenon where dampening of the torque control changes with the variation of the output torques. So, the dynamic performances of the electromagnetic torque are different under different output torque conditions. In order to solve this problem, this paper puts forward an improved DTC-SVM scheme. Compared with the conventional scheme, the proposed scheme adopts a torque controller with torque angle estimation (TC-LAE). With this torque controller, the torque change rate is adjusted in real-time according to the variation of the output electromagnetic torques. The dynamic performance of the torque control is improved. Meanwhile, for the determination of the reference flux amplitude, the MTPA criterion expressed by the flux linkage is established and the Lagrange interpolation fitting method is used to realize the on-line MTPA operation of the PMSM. With the proposed online MTPA method, we can determine the reference flux amplitude directly that could ensure the PMSM-MTPA operation, instead of utilizing the traditional two-dimensional look-up table. The simulation and experimental results of the improved and conventional scheme were researched using a 6 kW PMSM. The conclusions are as follows:

1. The improved scheme inherits the advantages of the conventional DTC-SVM scheme. The torque control ability under different load conditions has been improved; the impact of load variation on torque performance has been eliminated.
2. The reference value of the stator voltage given by the torque controller maybe exceed the range of the voltage limitation circle when the electromagnetic torque is changed suddenly. Then, the tracking deviation of the flux occurs and causes an impulse of the current. With the corresponding correction algorithm, the impact of the voltage limitation circle on the performance of torque control is eliminated.
3. To eliminate the deviation of torque control resulting from the rotation of the permanent magnet; a compensation term was added to the reference value of the flux phase angle in the improved scheme, and a no-error control of the electromagnetic torque was realized.
4. The selection of the torque control parameter K_c will impact the torque control performance in the improved scheme. The selection of K_c is analyzed by the simulation and experiment. The analysis results show that the torque ripple becomes greater when the K_c is large; the dynamic performance of the torque is degraded when the K_c is small.

Author Contributions: Conceptualization, Z.Z. and Z.W.; methodology, Z.Z.; software, Z.Z.; validation, X.G. and G.Z.; formal analysis, Z.Z. and Q.G.; writing—original draft preparation, Z.Z.; writing—review and editing, Z.Z. and G.Z., and; funding acquisition, Z.Z.

Funding: This research was funded by "The Science & Technology Development Fund of Tianjin Education Commission for Higher Education, grant number 2018KJ207".

Conflicts of Interest: The authors declare no conflict of interest.

Nomenclature

$v, i, \varphi_s, \varphi_r$	Stator voltage, current and flux, permanent magnet flux
$\boldsymbol{\varphi}_s, \boldsymbol{V}_s, \boldsymbol{I}_s$	Stator flux, voltage and current vector
ω_m, ω	Mechanical and electrical angular velocity
T_e, δ, ρ	Electromagnetic torque, load angle and salient rate of the motor
p, L, R	Number of pole pairs, stator inductance, stator resistance
k_T, K_c	Change rate of torque, control parameter of load angle controller
K_p, K_i, T_s	Proportionality coefficient, integral coefficient and control period
ABC, αβ, dq	Three-phase, two-phase stationary frame and rotating reference frame
$\lvert\cdot\rvert, \angle, \Delta$	Amplitude and phase angle of vector, Incremental quantity of variables
b, n, N (subscript)	Base value, per-unit value and rated value
k, (subscript)	Variables at the instant kT_s
ref, 0 (subscript)	Reference value and balance point value

References

1. Wang, D.; Yuan, T.; Wang, X.; Wang, X.; Ni, Y. Performance improvement of servo control system driven by dovel PMSM-DTC based on fixed sector division criterion. *Energies* **2019**, *12*, 2154. [CrossRef]
2. Wang, F.; Zhang, Z.; Mei, X.; Rodríguez, J.; Kennel, R. Advanced control strategies of induction machine: Field oriented control direct torque control and model predictive control. *Energies* **2018**, *11*, 120. [CrossRef]
3. Pavel, K.; Jiri, L. Induction motor drive direct torque control and predictive torque control comparison based on switching pattern analysis. *Energies* **2018**, *11*, 1793.
4. Wang, D.; Yuan, T.; Wang, X.; Wang, X.; Li, W. A Composite vectors modulation strategy for PMSM DTC systems. *Energies* **2018**, *11*, 2729. [CrossRef]
5. Milutin, P.; Nebojša, M.; Vojkan, K.; Bojan, B. An improved scheme for voltage sag override in direct torque controlled induction motor drives. *Energies* **2017**, *10*, 663. [CrossRef]
6. Li, H.; Lin, J.; Lu, Z. Three vectors model predictive torque control without weighting factor based on electromagnetic torque feedback compensation. *Energies* **2019**, *12*, 1393. [CrossRef]
7. Xu, Y.; Shi, T.; Yan, Y.; Gu, X. Dual-vector predictive torque control of permanent magnet synchronous motors based on a candidate vector table. *Energies* **2019**, *12*, 163. [CrossRef]
8. Song, Q.; Li, Y.; Jia, C. A novel direct torque control method based on asymmetric boundary layer sliding mode control for PMSM. *Energies* **2018**, *11*, 657. [CrossRef]
9. Hoang, K.D.; Ren, Y.; Zhu, Z.Q.; Foster, M. Modified switching-table strategy for reduction of current harmonics in direct torque controlled dual-three phase permanent magnet synchronous machine drives. *IET Electr. Power Appl.* **2015**, *9*, 10–19. [CrossRef]
10. Xia, C.; Zhao, J.; Yan, Y.; Shi, T. A novel direct torque control of matrix converter-fed PMSM drives using duty cycle control for torque ripple reduction. *IEEE Trans. Ind. Electron.* **2014**, *61*, 2700–2713. [CrossRef]
11. Yan, Y.; Zhao, J.; Xia, C.; Shi, T. Direct torque control of matrix converter-fed permanent magnet synchronous motor drives based on master and slave vectors. *IEEE Trans. Power Electron.* **2015**, *8*, 288–296. [CrossRef]
12. Zhu, Z.Q.; Ren, Y.; Liu, J. Improved torque regulator to reduce steady-state error of torque response for direct torque control of permanent magnet synchronous machine drives. *IET Trans. Electr. Power Appl.* **2014**, *8*, 108–116. [CrossRef]
13. Buja, G.S.; Kazmierkowski, M.P. Direct torque control of PWM inverter-fed AC motors-a survey. *IEEE Trans. Ind. Electron.* **2004**, *51*, 744–757. [CrossRef]
14. Tang, L.; Zhong, L.; Rahman, M.F.; Hu, Y. Novel direct torque controlled interior permanent magnet synchronous machine drive with low ripple in flux and torque and fixed switching frequency. *IEEE Trans. Power Electron.* **2004**, *19*, 346–354. [CrossRef]
15. Inoue, Y.; Morimoto, S.; Sanada, M. Control method suitable for direct-torque-control-based motor drive system satisfying voltage and current limitations. *IEEE Trans. Ind. Appl.* **2012**, *48*, 970–976. [CrossRef]
16. Krishnan, B. *Permanent Magnet Synchronous and Brushless DC Motor Drives*; CRC Press: Boca Raton, FL, USA, 2009; Chapter 7.

17. Jung, S.Y.; Hong, J.; Nam, K. Current minimizing torque control of the IPMSM using Ferrari's method. *IEEE Trans. Power Electron.* **2014**, *28*, 5603–5617. [CrossRef]
18. Preindl, M.; Bolognani, S. Optimal state reference computation with constrained MTPA criterion for PM motor drives. *IEEE Trans. Power Electron.* **2015**, *30*, 4524–4535. [CrossRef]
19. Lemmens, J.; Vanassche, P.; Driesen, J. PMSM drive current and voltage limiting as a constraint optimal control problem. *IEEE J. Emerg. Sel. Top. Power Electron.* **2015**, *3*, 326–338. [CrossRef]
20. Rahman, M.F.; Zhang, L.; Lim, K.W. A direct torque-controlled interior permanent magnet synchronous motor drive incorporating field weakening. *IEEE Trans. Ind. Appl.* **1998**, *34*, 1246–1253. [CrossRef]
21. Inoue, Y.; Morimoto, S.; Sanada, M. Examination and linearization of torque control system for direct torque controlled IPMSM. *IEEE Trans. Ind. Appl.* **2010**, *46*, 159–166. [CrossRef]
22. Foo, G.; Rahman, M.F. Direct torque and flux control of an IPM synchronous motor drive using a backstepping approach. *IET Trans. Electr. Power Appl.* **2009**, *3*, 413–421. [CrossRef]
23. Chen, S.Z.; Cheung, N.C.; Wong, K.C.; Wu, J. Integral variable structure direct torque control of doubly fed induction generator. *IET Trans. Electr. Power Appl.* **2011**, *5*, 18–25. [CrossRef]
24. Liu, Y. Suppressing stick-slip oscillation in underactuated multibody dril-strings with parametric uncertainties using sliding-mode control. *IET Trans. Control Theory Appl.* **2015**, *9*, 91–102. [CrossRef]
25. Lascu, C.; Trzynadlowski, A.M. Combining the principles of sliding mode, direct torque control, and space-vector modulation in a high-performance sensorless AC drive. *IEEE Trans. Ind. Appl.* **2004**, *40*, 170–177. [CrossRef]
26. Xu, W.; Lorenz, R.D. Dynamic loss minimization using improved deadbeat-direct torque and flux control for interior permanent magnet synchronous machines. *IEEE Trans. Ind. Appl.* **2014**, *50*, 1053–1065. [CrossRef]
27. Zhu, H.; Xiao, X.; Li, Y. Torque ripple reduction of the torque predictive control scheme for permanent-magnet synchronous motors. *IEEE Trans. Ind. Electron.* **2012**, *59*, 871–877. [CrossRef]
28. Song, W.; Ma, J.; Zhou, L.; Feng, X. Deadbeat predictive power control of single phase three level neutral-point-clamped converters using space-vector modulation for electric railway traction. *IEEE Trans. Power Electron.* **2015**, *31*, 721–732. [CrossRef]
29. Cho, Y.; Lee, K.; Song, J.; Lee, Y. Torque-ripple minimization and fast dynamic scheme for torque predictive control of permanent-magnet synchronous motors. *IEEE Trans. Power Electron.* **2015**, *30*, 2182–2190. [CrossRef]
30. Lee, J.S.; Choi, C.H.; Seok, J.K.; Lorenz, R.D. Deadbeat direct torque and flux control of interior permanent magnet machines with discrete time stator current and stator flux linkage observer. *IEEE Trans. Ind. Appl.* **2011**, *47*, 1749–1758. [CrossRef]
31. Preindl, M.; Bolognani, S. Model predictive direct torque control with finite control set for PMSM drive systems, part 1: Maximum torque per ampere operation. *IEEE Trans. Ind. Inform.* **2013**, *9*, 1912–1921. [CrossRef]
32. Doncker, R.; Pulle, D.; Veltman, A. *Advanced Electrical Drives Analysis, Modeling, Control*; Springer: New York, NY, USA, 2010; Chapter 4.
33. Burden, R.; Faires, J. *Numerical Analysis*; Cengage Learning: Boston, MA, USA, 2010; Chapter 3.
34. Tang, R. *Modern Permanent Magnet Machines: Theory and Design*; China Machine Press: Beijing, China, 2000; Chapters 3–4.
35. Hinkkanen, M.; Luomi, J. Modified integrator for voltage model flux estimation of induction motors. *IEEE Trans. Ind. Electron.* **2003**, *50*, 818–820. [CrossRef]
36. Luukko, J.; Pyrhonen, O.; Niemela, M.; Pyrhonen, J. Limitation of the load angle in a direct-torque-controlled synchronous machine drive. *IEEE Trans. Ind. Electron.* **2004**, *51*, 793–798. [CrossRef]
37. Jidin, A.; Idris, N.; Yatim, A.; Sutikno, T.; Elbuluk, M. An optimized switching strategy for quick dynamic torque control in DTC-hysteresis-based induction machines. *IEEE Trans. Ind. Electron.* **2011**, *58*, 3391–3400. [CrossRef]

Article

An Improved UDE-Based Flux-Weakening Control Strategy for IPMSM

Xin Gu [1], Tao Li [1], Xinmin Li [1], Guozheng Zhang [1] and Zhiqiang Wang [2,*]

[1] School of Electrical Engineering and Automation, Tiangong University, Tianjin 300387, China; guxin@tjpu.edu.cn (X.G.); litaoauto@foxmail.com (T.L.); lixinmin@tju.edu.cn (X.L.); zhanggz@tju.edu.cn (G.Z.)

[2] School of Artificial Intelligence, Tiangong University, Tianjin 300387, China

* Correspondence: wangzhq@tju.edu.cn; Tel.: +86-1392-098-9991

Received: 5 September 2019; Accepted: 25 October 2019; Published: 25 October 2019

Abstract: Interior permanent magnet synchronous motors (IPMSMs) are usually used in electric vehicle drives and in other applications. In order to enlarge the speed range of IPMSMs, the flux-weakening control method is adopted. The traditional flux-weakening control strategy degrades the control performance because of parameter mismatches caused by variation of motor parameters. An improved uncertainty and disturbance estimator (UDE)-based flux-weakening control strategy is proposed for IPMSM drives in this paper. The parameter tuning method in the UDE-based control is improved. In addition, a flux-weakening adjusting factor is put forward to reduce the torque fluctuation when the operation point switches between the constant torque region and the flux-weakening region. This factor can be adjusted online by a lookup table. Finally, the validity of proposed method is verified by the simulation and experimental results. The results show that the proposed control strategy can effectively enhance the robustness of the system in the flux-weakening region, and make the system switch more smoothly between the constant torque region and the flux-weakening region.

Keywords: IPMSM; uncertainty and disturbance estimator; flux-weakening control

---

1. Introduction

Interior permanent magnet synchronous motors (IPMSMs) have been widely used in the electric vehicle (EV) drives due to their simple structure, wide speed range, high power, and torque density. IPMSMs often run above the rated speed in some applications, such as EVs, which is the maximum speed that the motor can obtain in the constant torque region. The flux-weakening control is usually adopted in an IPMSM system for acquiring higher speed and meeting the application demand. A negative d-axis current is injected to the stator windings in the flux-weakening control strategy. The air gap magnetic field of an IPMSM is reduced under the direct axis armature reaction caused by negative d-axis current. Then, the motor speed increases as the magnetic field reduction. The copper and iron losses of IPMSMs increase with an increase in speed, which results in the temperature rising. The parameters of IPMSMs, such as stator resistance and inductance, vary nonlinearity with the temperature rising. The control performance of the IPMSM system is degraded under uncertain parameter variation. Meanwhile, the torque fluctuation caused by switching between the constant torque region and the flux-weakening region bring adverse effects to the smoothness of IPMSMs.

An improved feedforward control strategy is proposed to reduce the impacts from parameter variation caused by flux-weakening control in [1–4]. The reference values of d-axis and q-axis currents are given by a lookup table. The parameter identification is adopted in flux-weakening control to enhance the systemic robustness in [5,6]. The parameters of the motor can be identified online using this method. The influence of magnetic saturation and stator resistance are considered in [7], and an optimal control strategy is proposed. However, the flux observation is limited by the rotor position in this strategy. A linearized and constrained model predictive control is put forward to the flux-weakening control in [8]. However, this method involves a large amount of calculation and is extremely sensitive to parameter variation. In order to suppress the influence of parameter variation on the system, a voltage feedback control strategy is applied in the flux-weakening region. A flux-weakening current output by a proportional integral (PI) controller is added to the d-axis current reference. The input of the PI controller is the difference between the output voltage amplitude and the maximum available voltage amplitude of inverter. This method is simple, independent of motor parameters, and has good robustness, but the dynamic performance needs to be improved. The conventional voltage feedback method is improved in [9]. The difference between DC link voltage and output voltage of current controller is used to calculate the phase angle of reference current space vector by an adaptive algorithm. However, the global stability of the IPMSM system cannot be guaranteed. A single current regulator is proposed to improve the voltage utilization of the DC bus in [10]. This method eliminates the poor effect caused by the coupling of the d and q axis current, but the decrease of efficiency and stability is still not to be ignored. A line modulation-based flux-weakening control was proposed to maximize the DC bus voltage utilization in [11].

The difference between the output voltage amplitude and the maximum voltage amplitude of the inverter is regarded as the judgment of whether to enter the flux-weakening region to reduce the switching fluctuation caused by different control algorithms in constant torque region and flux-weakening region in [2,8,12,13]. A variable coefficient is used to adjust the stator flux linkage which contributes to generating the maximum and most suitable torque and to achieve smooth switching [3]. The d-axis current reference is modified by comparing the switching period and summation of active switching times for inverter pulse width modulation control in the flux-weakening region, thus extending the hexagon of the space vector modulation in [14]. The smooth switching between the constant torque region and the flux-weakening region is realized by using the amplitude of inverter output voltage and d-axis current as input of set-reset flip-flop in [15].

Recently, an uncertainty and disturbance estimator (UDE)-based control has attracted much attention and has been applied to unknown time delay systems, nonlinear systems, and power converters [16–22]. The algorithm assumes that an unknown continuous signal can be estimated by an appropriate filter and compensated to the control system. It is helpful for solving the problem that the control performance deteriorates due to the parameter variations in flux-weakening control and has strong robustness. The UDE-based method is applied to the inverter for improving the output voltage quality of the inverter and reducing the total harmonic distortion in [16,17], and this method improves the stability of the system in nonaffine and nonlinear systems [18]. The UDE-based control is applied to surface permanent magnet synchronous motor and proposes a simple parameter tuning algorithm in [19], but does not analyze the operating condition in flux-weakening regions. The influence of different filters on the system has been discussed in detail [20]. The asymptotic tracking and disturbance suppression are realized under different types of reference values.

In this paper, an improved UDE-based flux-weakening control strategy for an IPMSM system is proposed to solve the poor robustness caused by parameter variation in the flux-weakening region, and the torque fluctuation when the operation point switches between the constant torque region and flux-weakening region. The motor model and flux-weakening control strategy are presented in Section 2. The UDE-based control is introduced and the parameter tuning method is improved in Section 3. The flux-weakening adjusting factor and its lookup table method are proposed in Section 4. The analysis of parameter mismatches is carried out in Section 5. The experimental validation of proposed method is carried out in Section 6. Finally, Section 7 comprises the summary.

2. Mode and Flux-Weakening Control Strategy

The model of an IPMSM in the synchronous rotating coordinate system is expressed as

$$u_d = R_s i_d + L_d \frac{di_d}{dt} - \omega_e L_q i_q, \quad (1)$$

$$u_q = R_s i_q + L_q \frac{di_q}{dt} + \omega_e L_d i_d + \omega_e \psi_f, \quad (2)$$

$$T_e = \frac{3}{2} p \left[\psi_f i_q + \left(L_d - L_q \right) i_d i_q \right], \quad (3)$$

where R_s, L_d, and L_q are the stator resistance, d-axis, and q-axis inductance, respectively; ψ_f is permanent magnet flux linkage; p is pole pairs; i_d and i_q are d-axis and q-axis currents; ω_e is the electric angular of rotor; and T_e is the electromagnetic torque.

The maximum torque per ampere (MTPA) control strategy is often used in IPMSMs to achieve the optimal configuration of the given current when the IPMSM runs in the constant torque region. This strategy makes full use of the reluctance torque due to the difference between L_d and L_q. It can minimize the stator current at the same electromagnetic torque. In the MTPA, the d-axis current $i_d^*{}_{\cdot MTPA}$ is shown as

$$i^*_{d.MTPA} = \frac{\psi_f - \sqrt{4\left(L_q - L_d\right)^2 (i^*_q)^2 - \psi_f{}^2}}{2\left(L_q - L_d\right)}, \quad (4)$$

where i_q^* stands for the q-axis current reference which is obtained by the output of the speed controller.

A valid operating point of the IPMSM is limited by the voltage and current constraints. Considering the influence of the stator resistance, the voltage and current constraints are shown as follows:

$$(L_d i_d + \psi_f)^2 + (L_q i_q)^2 \le \left(\frac{U_{smax} - R_s I_{smax}}{\omega_e}\right)^2, \quad (5)$$

$$i_d^2 + i_q^2 \le I_{smax}^2, \quad (6)$$

where U_{smax} and I_{smax} represent the maximum voltage and maximum current, and they are limited by the inverter and motor.

The reference voltage output obtained by the current controller can reach or exceed the maximum value of the inverter with the increase of motor speed. Thus, the MTPA control strategy cannot further improve the speed of the IPMSM. The flux-weakening control is usually adopted to solve this problem. The negative d-axis current is further increased in flux-weakening region. According to the Equations (5) and (6), the d-axis current $i^*_{d.FW}$ in flux-weakening control is expressed as follows:

$$i^*_{d.FW} = \frac{-\psi_f + \sqrt{\left(\frac{U_{smax} - R_s I_{smax}}{\omega_e}\right)^2 - \left(L_q i^*_q\right)^2}}{L_d}. \quad (7)$$

The MTPA control strategy is adopted when the motor is running in the constant torque region, and the motor is switched to flux-weakening control when the motor speed is higher than the maximum speed of the constant torque region. The traditional control strategy based on PI controller is shown in Figure 1.

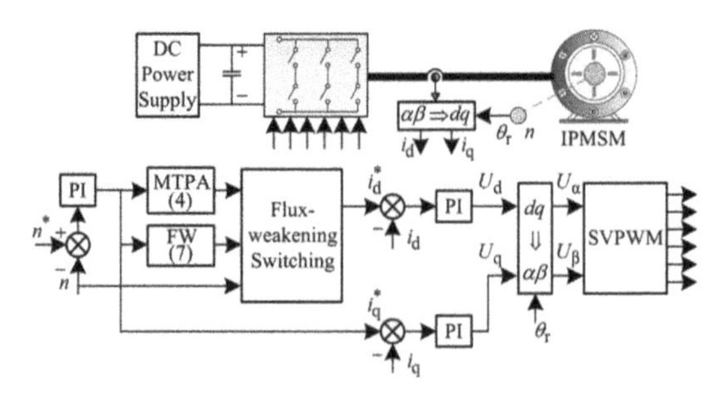

Figure 1. The traditional control strategy based on PI controller.

It can be seen from Figure 1 that the d-axis current is calculated by formulas. The parameters of PI controller cannot match well with the variable motor parameter caused by flux weakening. This leads to the performance decline of the IPMSM system. In addition, the fluctuation caused by the switching between the constant torque region and flux-weakening region at a fixed speed is necessary to be solved.

3. The Design of UDE-Based Control

The uncertainty and disturbance are regarded as unknown signal in a UDE-based control system. A stable reference model is used to satisfy the desired tracking performance of the closed-loop system. Uncertainties and disturbances are compensated to the controller by using an appropriate filter. Thus, the control law of UDE-based control is obtained.

The motor parameter variations and systemic random disturbances are regarded as an unknown signal in the IPMSM control system. The current-loop model of UDE-based control under the synchronous rotating coordinate system is obtained as

$$\frac{di_d}{dt} = -\frac{R_s}{L_d} i_d + \omega_e \frac{L_q}{L_d} i_q + \frac{1}{L_d} u_d + f_d + D_d, \tag{8}$$

$$\frac{di_q}{dt} = -\frac{R_s}{L_q} i_q - \omega_e \frac{L_d}{L_q} i_d + \frac{1}{L_q} u_q - \frac{1}{L_q} \omega_e \psi_f + f_q + D_q, \tag{9}$$

where D_d and D_q are the d-axis and q-axis systemic random disturbances; f_d and f_q are the d-axis and q-axis uncertainties caused by the variations of motor resistance, inductances, and flux linkage. The f_d and f_q are expressed as the average current ripple vector in sample period (0, T_s] and (T_s, $2T_s$] are defined as follows:

$$\begin{cases} f_d = -\frac{\Delta R_s}{L_d} i_d - \frac{\Delta L_d}{L_d} \frac{di_d}{dt} + \omega_e \frac{\Delta L_q}{L_d} i_q \\ f_q = -\frac{\Delta R_s}{L_q} i_q - \frac{\Delta L_q}{L_q} \frac{di_q}{dt} - \omega_e \frac{\Delta L_d}{L_q} i_d - \omega_e \frac{\Delta \psi_f}{L_q} \end{cases}. \tag{10}$$

Equations (8) and (9) can be rewritten in matrix from as

$$\dot{x}(t) = Ax(t) + Bu(t) + d_0(t) + f(t) + D(t), \tag{11}$$

where $x(t) = [i_d \; i_q]^T$, $u(t) = [u_d \; u_q]^T$, $f(t) = [f_d \; f_q]^T$, $D(t) = [D_d \; D_q]^T$,

$$d_0(t) = \begin{bmatrix} 0 \\ -\frac{\omega_e \psi_f}{L_q} \end{bmatrix}, A = \begin{bmatrix} -\frac{R_s}{L_d} & \omega_e \frac{L_q}{L_d} \\ -\omega_e \frac{L_d}{L_q} & -\frac{R_s}{L_q} \end{bmatrix}, B = \begin{bmatrix} \frac{1}{L_d} & 0 \\ 0 & \frac{1}{L_q} \end{bmatrix}.$$

3.1. The Reference Model

A stable reference model is used to achieve asymptotic tracking of the motor reference current in UDE-based control. The selection of reference model is related to control object. According to the structure of the IPMSM control system, the following linear reference model is adopted.

$$\dot{x}_{\rm m}(t) = A_{\rm m}x_{\rm m}(t) + B_{\rm m}c(t), \tag{12}$$

where $x_{\rm m} = [i_{\rm dm}\ i_{\rm qm}]^{\rm T}$ is the d-axis and q-axis current reference state vector, $c = [i_{\rm d}^{*}\ i_{\rm q}^{*}]^{\rm T}$ is the d-axis and q-axis current reference command vector.

$x_{\rm m}(t)$ and $c(t)$ are the second-order matrices selected in the reference model for achieving the desired tracking performance of the control system. Due to the difference between the d-axis and q-axis inductances in IPMSMs, the coefficient matrix of reference model is given as

$$A_{\rm m} = \begin{bmatrix} -\alpha & 0 \\ 0 & -\beta \end{bmatrix},\ B_{\rm m} = \begin{bmatrix} \alpha & 0 \\ 0 & \beta \end{bmatrix}, \tag{13}$$

where α and β are two independent positive real numbers.

Based on the above reference model, the currents can asymptotically track the reference state vector of the d-axis and q-axis current by controlling the input voltage $u(t)$ of the control system. The tracking error and dynamic error are expressed as

$$\mathrm{e}(t) = x_{\rm m}(t) - x(t), \tag{14}$$

$$\dot{\mathrm{e}}(t) = (A_{\rm m} + K)\mathrm{e}(t), \tag{15}$$

where K is an error feedback gain matrix of the reference model which is used to guarantee the system stability.

When K is a zero matrix, the current state vector can be expressed as follows by combining Equations (12), (14) and (15).

$$\dot{x}(t) = A_{\rm m}x(t) + B_{\rm m}c(t), \tag{16}$$

since matrix B and $B_{\rm m}$ are invertible as they are both diagonal matrices. By substituting (16) into (11), the input $u(t)$ can be expressed as

$$u(t) = B^{-1}[A_{\rm m}x(t) + B_{\rm m}c(t) - Ax(t) - d_0(t) - f(t) - D(t)]. \tag{17}$$

If (17) is used as the input of the control system, by substituting (17) into (15), the error dynamics of current is obtained:

$$\dot{\mathrm{e}}(t) = A_{\rm m}\mathrm{e}(t). \tag{18}$$

The eigenvalues $-\alpha$ and $-\beta$ of the matrix $A_{\rm m}$ are negative, so the dynamic tracking error of current is asymptotically stable according to Lyapunov stability theory. The error feedback gain matrix K should be selected as a zero matrix. From Equation (17), the control input $u(t)$ can maintain stability of the reference model tracking the current. However, $u(t)$ cannot be directly used in the control system because of the parameter uncertainties and systemic random disturbances.

3.2. The UDE-Based Control Law

In [22], the uncertainty and disturbance are estimated and compensated in the linear time invariant system by a filter with desired bandwidth, and this method improves the robustness of the system. The parameter uncertainties and systemic random disturbances of motor can be defined as

$$u_{\rm de}(t) = f(t) + D(t). \tag{19}$$

According to the current state Equation (11), Equation (19) is further written as

$$u_{\mathrm{de}}(t) = \dot{x}(t) - Ax(t) - Bu(t) - d_0(t). \tag{20}$$

From Equation (20), it can be seen that the motor parameter uncertainties and the systemic random disturbances can be represented by the known state variable $x(t)$ and the control variable $u(t)$. The filter g_f(t) is designed with unity gain and zero phase shifts for estimating the parameters uncertainties and the systematic random disturbances. The estimating value can be expressed as

$$\hat{u}_{\mathrm{de}}(t) = [f(t) + D(t)] * g_f(t) = \left[\dot{x}(t) - Ax(t) - Bu(t) - d_0(t)\right] * g_f(t), \tag{21}$$

where "*" is the convolution operator.

The design of the filter is mainly related to the suppression ability of the parameter uncertainties and systemic random disturbances. A detailed discussion is made for the step and sinusoidal disturbances in [20], and the corresponding filter is designed to achieve asymptotic tracking. Considering that the disturbance of IPMSM control system is mainly step signals, it is sufficient to choose a first-order low-pass filter to meet the actual working conditions. The frequency domain transfer function of filter is

$$G_f(s) = \frac{1}{Ts+1} = \frac{\gamma}{s+\gamma}, \tag{22}$$

where $\gamma = 1/T$ and γ is the bandwidth for the selected first-order low-pass filter.

Therefore, the frequency domain form of the parameter uncertainties and systematic random disturbances can be expressed as

$$U_{\mathrm{de}}(s) = \left[\dot{X}(s) - AX(s) - BU(s) - D_0(s)\right]G_f(s). \tag{23}$$

The UDE-based control law can be obtained by substituting (23) into the frequency domain form of (17).

$$U(s) = B^{-1}\left[\frac{1}{1-G_f(s)}(A_{\mathrm{m}}X_{\mathrm{m}}(s) + B_{\mathrm{m}}C(s)) - AX(s) - D_0(s) - \frac{sG_f(s)}{1-G_f(s)}X(s)\right] \tag{24}$$

By simplification, Equation (24) can be rewritten as follows:

$$U(s) = B^{-1}\left[A_{\mathrm{m}}X_{\mathrm{m}}(s) + B_{\mathrm{m}}C(s) - (AX(s) + D_0(s)) + K_{\mathrm{P}}E(s) + K_{\mathrm{I}}\frac{1}{s}E(s)\right], \tag{25}$$

where

$$K_{\mathrm{P}} = (\gamma I - A_{\mathrm{m}}) = \begin{bmatrix} \alpha+\gamma & 0 \\ 0 & \beta+\gamma \end{bmatrix}, K_{\mathrm{I}} = -\gamma A_{\mathrm{m}} = \begin{bmatrix} \alpha\times\gamma & 0 \\ 0 & \beta\times\gamma \end{bmatrix} = \begin{bmatrix} M_{\mathrm{d}} & 0 \\ 0 & M_{\mathrm{q}} \end{bmatrix}.$$

From the Equation (25), the traditional UDE-based control without parameter tuning is shown in Figure 2. This structure is mainly divided into three parts, including differential feedforward, PI regulator, and model inversion.

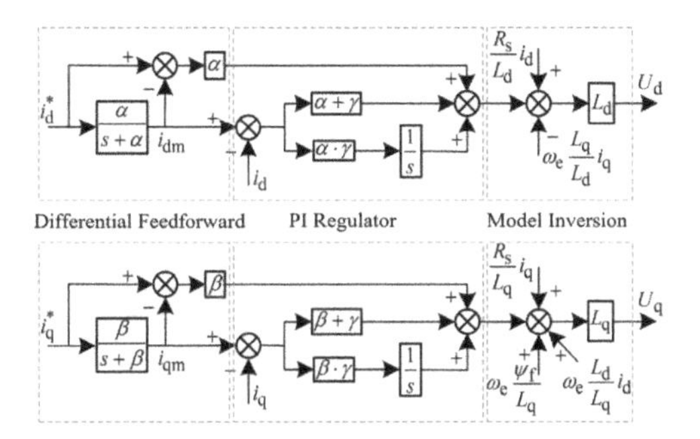

Figure 2. The traditional uncertainty and disturbance estimator (UDE)-based control without parameter tuning.

The eigenvalues $-\alpha$ and $-\beta$ of the matrix A_m are influenced by the system itself, so the adjustable range of A_m is small and γ plays an important role in asymptotic reference tracking. A simple dual-loop parameter tuning algorithm for PMSMs is adjusted by the current reference tracking performance in [19]. This method avoids the trial-and-error procedure, but is not suitable for IPMSMs.

3.3. Parameter Tuning Method

The rapidity of system response depends on K_P, while the following performance depends on K_I. The K_P and K_I can be adjusted by tuning α, β, and γ in order to achieve the asymptotic following of the current reference in UDE-based control. The tracking performance is not only related to known disturbances, but also influenced by unknown external disturbances. Therefore, the following performance can be further improved by tuning K_I online when K_P is determined. The K_I is represented by intrinsic components M_d and M_q and variable components N_d and N_q in this paper. The new integral gain $\widetilde{K_I}$ is shown in (26).

$$\widetilde{K_I} = \begin{bmatrix} M_d + N_d & 0 \\ 0 & M_q + N_q \end{bmatrix} \tag{26}$$

The intrinsic components M_d and M_q can be determined by previous analysis. The variable components N_d and N_q can be obtained by the out of proportional component whose inputs are the d-axis and q-axis current tracking errors. Thus, the online tuning of K_I can be achieved. The d-axis current is negative, and the q-axis current is positive in the control process. Therefore, the d-axis current tracking error is obtained by subtracting the given value from the feedback value. On the contrary, the q-axis current tracking error is obtained by subtracting the feedback value from the given value. This method can guarantee the positive output of proportional component during the bigger current errors. The coefficients K_{Id} and K_{Iq} are used to represent variation of N_d and N_q. Equation (26) is rewritten as follows:

$$\begin{cases} M_d + N_d = \alpha \times \gamma + K_{Id}\left(i_d - i_d^*\right) \\ M_q + N_q = \beta \times \gamma + K_{Iq}\left(i_q^* - i_q\right) \end{cases}. \tag{27}$$

Therefore, the structure of the improved UDE-based control with parameter tuning is shown in Figure 3.

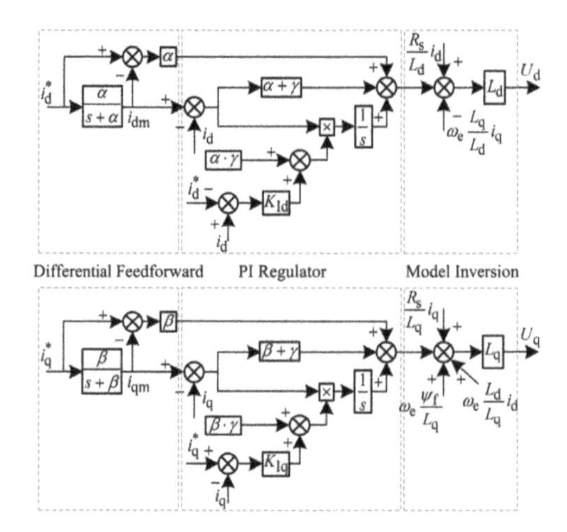

Figure 3. The improved UDE-based control with parameter tuning.

4. The Flux-Weakening Switching Control

4.1. Analysis of Switching Point

The two different d-axis current calculation methods need to be switched when the motor is switched between the constant torque region and the flux-weakening region. Considering the motor parameter variations, Equations (4) and (7) can be expressed as

$$i^*_{\mathrm{d.MTPA}} = \frac{\widetilde{\psi}_{\mathrm{f}} - \sqrt{4\left(\widetilde{L}_{\mathrm{q}} - \widetilde{L}_{\mathrm{d}}\right)^2 (i^*_{\mathrm{q}})^2 - \widetilde{\psi}_{\mathrm{f}}{}^2}}{2\left(\widetilde{L}_{\mathrm{q}} - \widetilde{L}_{\mathrm{d}}\right)}, \tag{28}$$

$$i^*_{\mathrm{d.FW}} = \frac{-\widetilde{\psi}_{\mathrm{f}} + \sqrt{\left(\frac{U_{\mathrm{smax}} - \widetilde{R}_{\mathrm{s}} I_{\mathrm{smax}}}{\omega_{\mathrm{e}}}\right)^2 - \left(\widetilde{L}_{\mathrm{q}} i^*_{\mathrm{q}}\right)^2}}{\widetilde{L}_{\mathrm{d}}}, \tag{29}$$

where "~" represents a variable scalar.

It can be seen from (28) and (29) that the d-axis current of the smooth switching point fluctuates due to the motor parameter variations. The currents at the selected switching points are usually smaller than that of the ideal switching points for ensuring the stability. The voltage and current constraints are represented in the plane of d-q axis current, as shown in Figure 4.

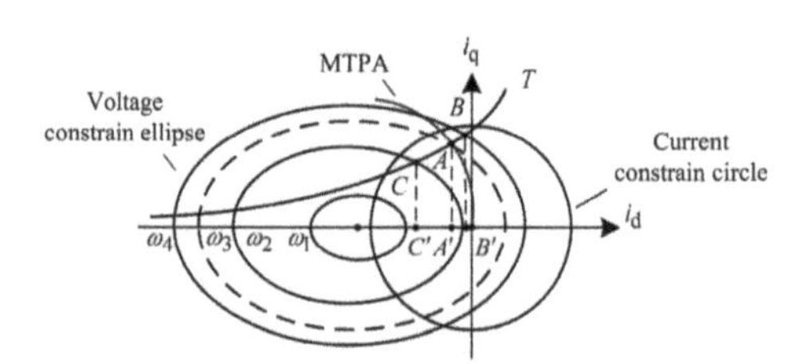

Figure 4. The voltage and current constrains.

The maximum speed of the MTPA control strategy can reach point A in the constant torque region, that is, the theoretical switching point, and the corresponding d-axis current is A'. The optimal switching point is the point which does not lead any fluctuation during the switching. The optimal switching point may deviate from A to B or C, which leads to corresponding current deviation from A' to B' or C' because of the motor parameter variations. Thus, the deviation of flux-weakening current may result in fluctuations during switching. The d-axis current can be adjusted with the size of voltage limit ellipse in a small range.

4.2. Flux-Weakening Adjusting Factor

From the above analysis, it is known that the d-axis current can be regulated by changing the size of voltage limit ellipse in a small range so that the flux-weakening current can reach the optimal switching point.

The d-axis current is calculated by the (7) in flux-weakening region. The voltage-drop on the stator resistance $R_s I_{smax}$ is very small, so the $R_s I_{smax}/\omega_e$ can be ignored with the increase of motor speed. A flux-weakening adjusting factor k is introduced before the speed in (7) and the d-axis current in flux-weakening region $\widetilde{i^*_{d.FW}}$ can be written as

$$\widetilde{i^*_{d.FW}} = \frac{-\psi_f + \sqrt{\left(\frac{U_{smax}}{k \cdot \omega_e}\right)^2 - \left(L_q i^*_q\right)^2}}{L_d}. \quad (30)$$

It can be seen from (30) that the size of voltage limit ellipse can be regulated by adjusting the factor k, so that the d-axis current reaches the optimal switching point.

4.3. Online Control Method

The error between the d-axis current reference under MTPA control and the d-axis current reference under flux-weakening control is defined as the reference error of d-axis current.

$$i^*_{d.error} = i^*_{d.MTPA} - i^*_{d.FW} \quad (31)$$

The d-axis current under different methods can be reflected by the reference error of d-axis current considering the motor parameter variations, and the load torque can be regarded as the function of q-axis current reference. The flux-weakening adjusting factor k can be determined by the reference error of d-axis current $i^*_{d.error}$ and the q-axis current reference i_q^*. Therefore, the $i^*_{d.error}$ and i_q^* can reflect the switching situation in real time.

Due to the transience of switching period, it is significant to obtain the factor k in a current control cycle. The lookup table method is effective for getting the k value as quickly as possible. Figure 5 shows a 3D table of the factor k varies with $i^*_{d.error}$ and i_q^*. The DC bus voltage range is from 270 to 320 V, with an interval of 5 V, and the given speed ramp range is from 0 to 1.1, with an interval of 0.1 in the table. The given speed ramp is defined as reached speed per unit time, where 1 represents the speed, that reaches 3000 r/min per second. The base value of $i^*_{d.error}$ and i_q^* is 62.5 A.

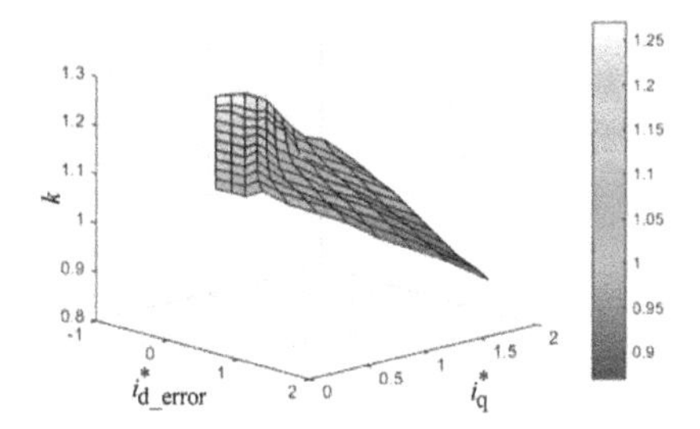

Figure 5. The 3D table of k varying with $i^*_{d.error}$ and i_q^*.

From the above analysis, the improved UDE-based control strategy for IPMSMs is shown in Figure 6.

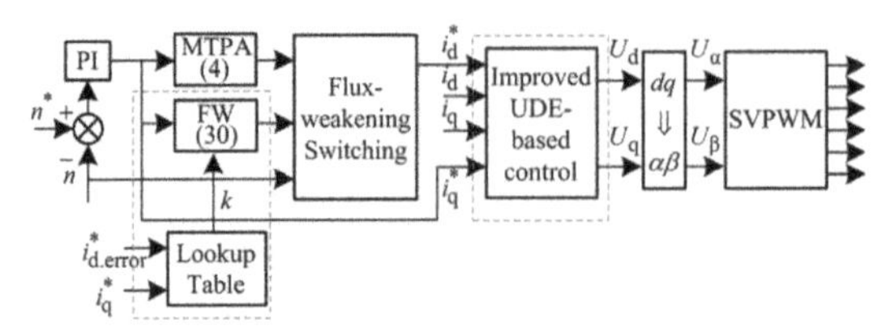

Figure 6. The improved UDE-based control strategy for interior permanent magnet synchronous motors (IPMSMs).

5. Analysis of Parameter Mismatches

The d-axis inductance, q-axis inductance, and flux linkage vary with the increase in speed, which results in parameter mismatches. The parameter mismatches between the controller and the motor reduce the systemic control performance. The mismatch of flux linkage is usually ignored in flux-weakening control. The robust analysis of parameter mismatches is implemented by combining with Bode plots of the transfer function in the traditional PI controller, the traditional UDE-based control without parameter tuning, and the improved UDE-based control with parameter tuning. The nominal parameters of an IPMSM in simulation are listed in Table 1, which are the same as for the experimental motor.

Table 1. Motor parameters.

Parameter	Value	Parameter	Value
Rated power	20 kW	Stator resistance	11.4 mΩ
Rated voltage	320 V	d-axis inductance	0.2 mH
Rated current	62.5 A	q-axis inductance	0.555 mH
Rated speed	3000 r/min	Flux linkage	0.07574 Wb
Rated torque	64 Nm	Pole pairs	4
Rated frequency	200 Hz		

The variation of d-axis and q-axis inductance with 20% nominal value is utilized to analyze the robustness of parameter mismatches in the flux-weakening region. The speed of the motor is 6000 r/min, the angular velocity is 2513 rad/s, the q-axis current is 84 A, and the d-axis current is obtained using Formula (7). Figure 7 are the Bode plots of parameter mismatches in the traditional control strategy-based PI controller, the traditional UDE-based control without parameter tuning, and the improved UDE-based control with parameter tuning. The parameters in controller are nominal L_d and L_q. The actual motor parameters are L_d' and L_q'.

It can be seen that the Bode plot of the traditional control strategy-based PI controller masks the Bode plot of the traditional UDE-based control without parameter tuning, which illustrates that the essence of traditional UDE-based control is same as the traditional control strategy. The improved UDE-based control optimizes the phase margin in the operation area compared with the traditional PI controller and traditional UDE-based control. The Bode plots of nominal parameters is shown by a black line in Figure 7, and the Bode plots considering the parameter mismatches are shown by green and red lines. The phase margin is reduced when the controller parameters are larger than the motor parameters, which is not conducive to the stability of the system. The phase margin is improved when the parameters of the controller are less than those of the motor, which is beneficial to the stability of the system.

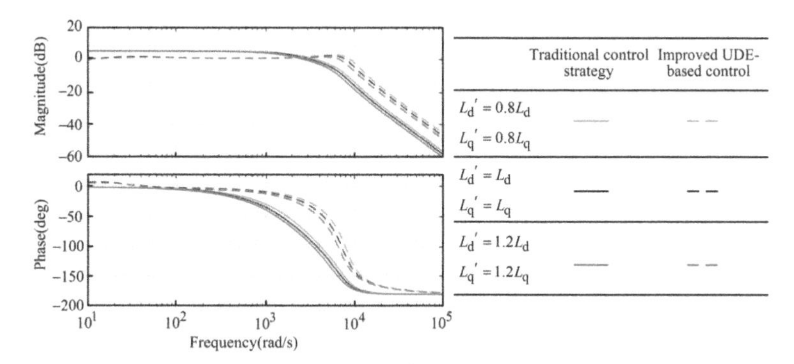

Figure 7. Bode plots of current loops.

6. Experimental Results and Analysis

Some experiments are carried out to verify the validity of the improved UDE-based flux-weakening control strategy for IPMSMs. The experimental system is shown in Figure 8. In the experimental system, a two-level voltage source inverter is utilized, and TMS320F28335DSP is used as the core controller. The speed and position signal are obtained by a photoelectric encoder. The control frequency is 10 kHz, which is the same as pulse width modulation (PWM) frequency. The nominal parameters of the test IPMSM in the experiments are listed in Table 1.

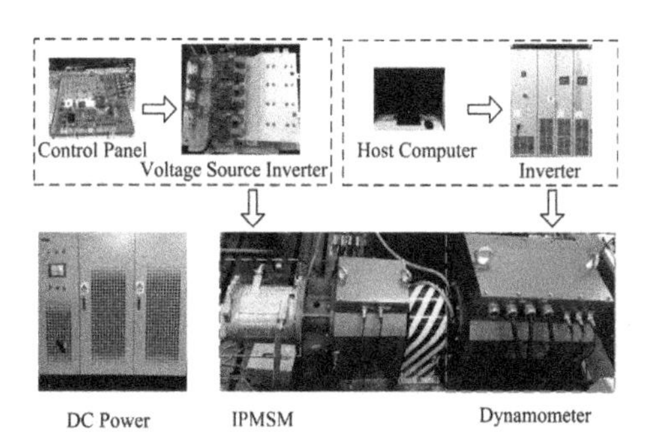

Figure 8. Experimental platform.

6.1. Tracking Performance in Constant Torque Region

The tracking performance experiments in constant torque region are carried out under the traditional PI control, the traditional UDE-based control without parameter tuning. The waveforms of d-axis current, q-axis current, speed, and torque under the three methods are respectively shown in Figure 9a–c. The given speed ramp is set to 1500 r/min per second, and the motor runs without loads. The parameters of reference model and filter are $\alpha = 5100$, $\beta = 8500$, and $\gamma = 0.00007$ in the traditional UDE-based control without parameter tuning and improved UDE-based control with parameter tuning. The coefficients of variable components in the improved UDE-based control are $K_{\text{Id}} = 0.0009$ and $K_{\text{Iq}} = 0.0004$. The variable components N_d, N_q are shown in Figure 10.

It can be seen from above results that the traditional PI controller and the improved UDE-based control with parameter tuning meet the tracking performance of d-axis and q-axis current in the constant torque region. However, the traditional UDE-based control without parameter tuning cannot follow the reference currents closely and eliminate steady errors. Meanwhile, the improved UDE-based control avoids turning the parameters α, β, and γ by adjusting K_{Id} and K_{Iq}. The complexity of parameter turning is effectively reduced.

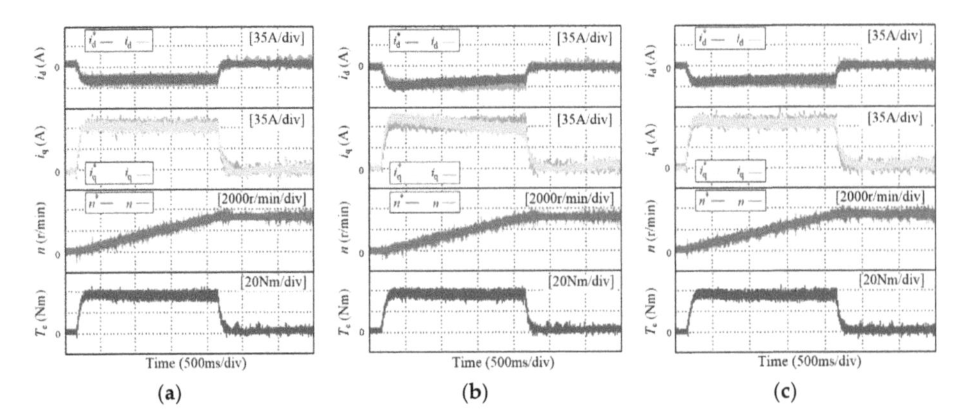

Figure 9. Tracking performance in the constant torque region: (**a**) The traditional control strategy-based PI controller; (**b**) The traditional UDE-based control without parameter tuning; (**c**) The improved UDE-based control with parameter tuning.

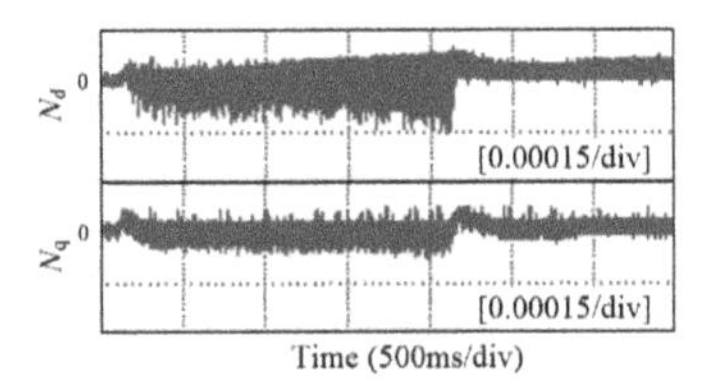

Figure 10. Experimental waveform of N_d, N_q in the improved UDE-based control strategy.

Thus, the traditional UDE-based control without parameter tuning is not suitable for the IPMSM drives. The traditional control strategy-based PI controller in Figure 1 and the improved UDE-based control strategy in Figure 6 are used to compare the tracking performance in the flux-weakening region of IPMSMs.

6.2. Tracking Performance in the Flux-Weakening Region

The tracking waveforms of the traditional control strategy-based PI controller and the improved UDE-based control strategy in the flux-weakening region are shown in Figure 11.

The given speed ramp is set to 600 r/min per second and the motor runs without loads. The speed of switching point is set at 4500 r/min which is the same as the simulation. The above K_{Id} and K_{Iq} are needed to adjust to suitable for the flux-weakening region. The adjusted K_{Id} and K_{Iq} are 0.0012 and 0.016. It indicates that increasing coefficients of variable components is helpful to improve the tracking performance.

It can be seen from Figure 11 that the two methods generate d-axis and q-axis current fluctuations in flux-weakening region, which leads to the increase in torque ripple. However, the d-axis current and torque fluctuation under the proposed method is obviously suppressed. The d-axis current fluctuation is reduced from 63 to 41 A, and the torque ripple is reduced from 30 to 19.75 Nm at 6000 r/min. The steady-state current fluctuation and torque ripple are lower after reaching 6000 r/min.

The standard deviation (SD) analysis of above experimental results is shown in Table 2. It can be seen that the standard deviation of steady-state d-axis current, q-axis current, speed, and torque are reduced by 24%, 41.24%, 0.785%, and 22.32% respectively. Thus, the proposed method reduces the current and torque ripples effectively in the flux-weakening region, which is consistent with the above theoretical analysis.

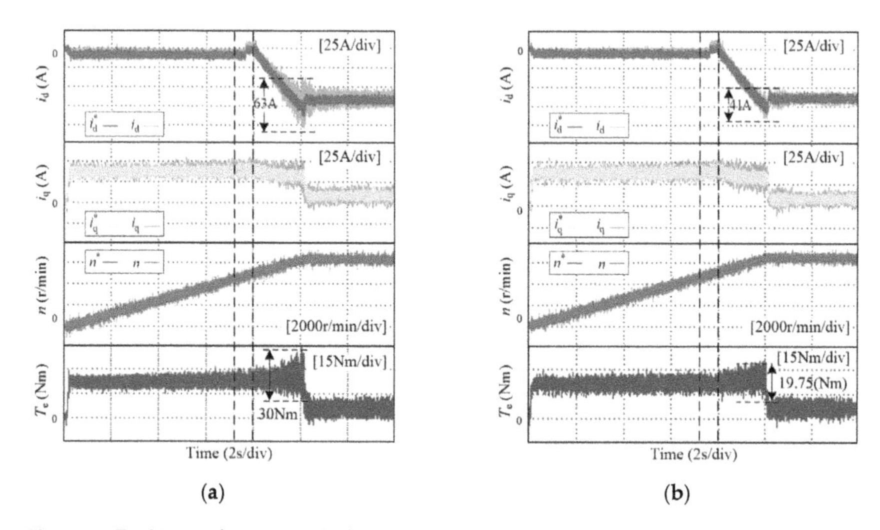

Figure 11. Tracking performance in the flux-weakening region: (**a**) The traditional control strategy-based PI controller; (**b**) The improved UDE-based control strategy.

Table 2. The comparison of SD without parameter mismatches.

Steady-State Value	Traditional Method	Proposed Method	Reduction Rate
d-axis current	4.3504	2.5564	41.24%
q-axis current	3.1577	2.3998	24.00%
Speed	83.56	82.904	0.79%
Torque	2.29716	1.7844	22.32%

Figure 12 shows the loading and unloading waveforms of d-axis current, q-axis current, speed, and torque in the flux-weakening region at 6000 r/min by using the above two methods. The motor runs in 6000 r/min and 8 Nm load conditions. The dynamometer suddenly loads 20 Nm to the motor. After a period of stable operation, the dynamometer suddenly reduces to 20 Nm load to reach stability.

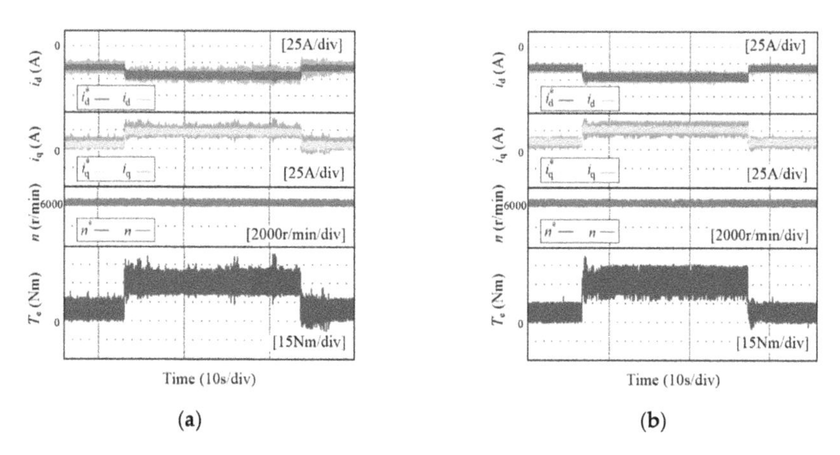

Figure 12. The loading and unloading waveforms in the flux-weakening region: (**a**) The traditional control strategy-based PI controller; (**b**) The improved UDE-based control strategy.

It can be seen from the Figure 12 that the improved UDE-based control strategy improves the tracking performance of current loop and reduces the current and torque fluctuations during sudden loading and unloading. In the meantime, the improved UDE-based control strategy has a certain ability to suppress the random interference in the system.

6.3. The Flux-Weakening Switching

It is evident that the switching fluctuation of d-axis current and torque is existent between the constant torque region and the flux-weakening region in Figure 11. The flux-weakening switching control-based lookup table method is adopted in the improved UDE-based control strategy for suppressing the switching fluctuations.

The waveforms of d-axis current, q-axis current, speed, and torque under different load torques are shown in Figure 13. The switching point is also set at 4500 r/min. The initial load torques in Figure 13a–c are 35, 26, and 21 Nm, respectively. The corresponding given speed ramps are set to 1500, 1000, and 750 r/min per second. The values of k are 0.85, 0.82, and 0.8, which are given by the 3D table shown in Figure 5.

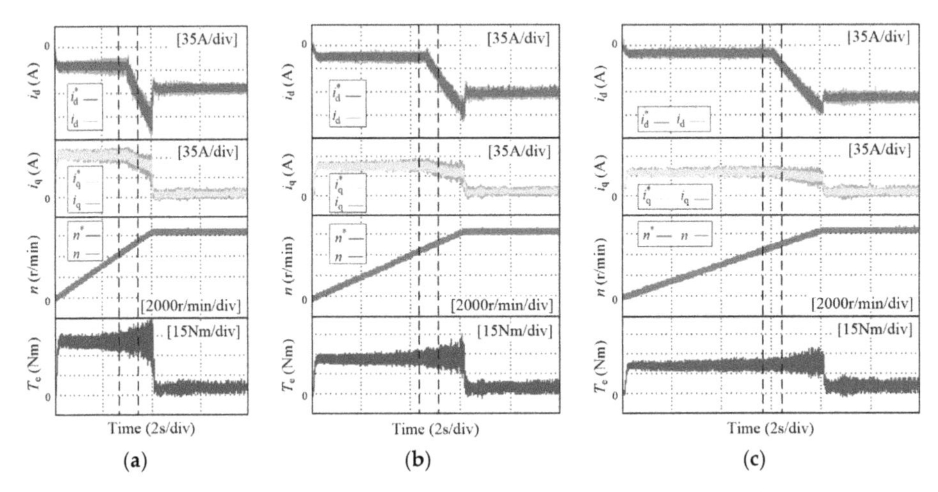

Figure 13. The flux-weakening switching waveforms under different load torque: (**a**) T_L = 35 Nm; (**b**) T_L = 26 Nm; (**c**) T_L = 21 Nm.

Compared with the d-axis current and torque in Figures 11 and 13, just as shown in the dotted line frame, the flux-weakening switching control-based lookup table method realizes smooth switching under different load torques. Since the essence of the proposed flux-weakening switching control based on lookup table is to control the reference error of d-axis current, the proposed switching control method thus makes the system switch more smoothly between the constant torque region and the flux-weakening region. The proposed switching method is simple compared with the traditional method. The lookup table method does not add additional computation to the control system. It is only necessary to establish a parameter table for the same type of motor, so as to avoid the complexity of establishing multiple parameter tables.

6.4. Analysis of Parameter Mismatches

Figures 11 and 13 show the experimental waveforms with nominal parameters of IPMSMs in the controller. It can be seen that the control performance has degraded when the motor runs in the flux-weakening region. According to the analysis of parameter mismatches, the variation of d-axis inductance and q-axis inductance with 20% nominal value in controller is utilized to verify the robustness of parameter mismatches in flux-weakening control. The waveforms with $L_{d0} = 1.2L_d$ and $L_{q0} = 1.2L_q$ are shown in Figure 14. It can be seen that the traditional control strategy based on PI controller further increase the current and torque ripples because of parameter mismatches in Figure 14a. The improved UDE-based control strategy still effectively depresses the ripples in steady and transient states in Figure 14b. When the speed reaches 6000 r/min, the steady-state torque ripple decreases from 42 to 25 Nm, and the maximum transient-state torque ripple is reduced from 50 to 40 Nm.

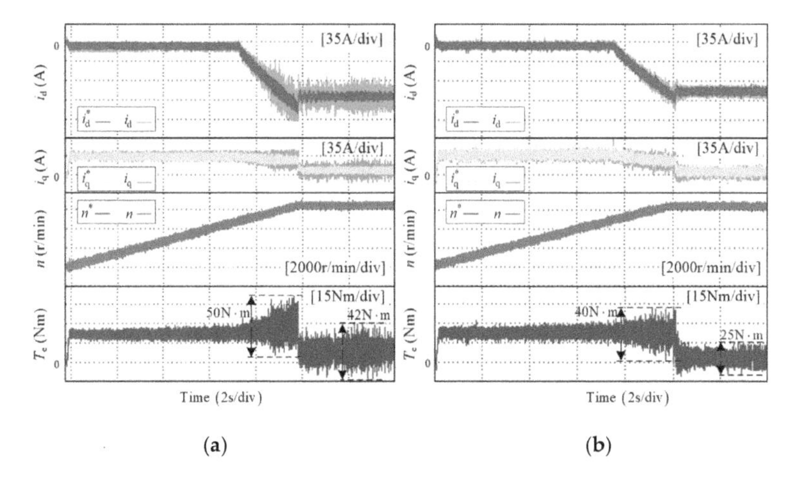

Figure 14. Experimental waveforms with $L_{d0} = 1.2L_d$, $L_{q0} = 1.2L_q$: (**a**) The traditional control strategy-based PI controller; (**b**) The improved UDE-based control strategy.

The SD analysis of above experimental results is listed in Table 3 considering the parameter mismatches.

Table 3. The comparison of SD with parameter mismatches.

Steady-State Value	Traditional Method	Proposed Method	Reduction Rate
d-axis current	12.59763	4.07657	67.64%
q-axis current	8.67091	3.59451	58.55%
Speed	76.6672	73.5504	4.07%
Torque	4.73176	2.77968	41.89%

It can be seen from Figures 11 and 14 that the fluctuations of d-axis current, q-axis current, and torque increase in both the traditional control strategy and proposed control strategy. This is the reason that the power of the motor reaches its limit in the flux-weakening region with an increase of rotational speed. The influences of dead time effect, high harmonics, and low carrier ratio on the system are serious. These factors play an important role in the fluctuation of motor in the flux-weakening region compared to the constant torque region. Therefore, the fluctuations of both traditional control strategy and proposed control strategy increase. However, the proposed control strategy reduces d-axis current, q-axis current, and torque fluctuations greatly in the case of parameter mismatches according to Table 3. Hence, the robustness of control system is improved effectively.

7. Conclusions

The two existing problems in the flux-weakening control for IPMSMs are discussed in this paper, including the decline of control performance with parameter mismatches and the switching fluctuation between the constant torque region and the flux-weakening region. An improved UDE-based flux-weakening control strategy for IPMSMs is proposed to solve them. It not only improves the tracking performance and reduces the torque ripple in flux-weakening region, but also realizes smooth switching between the constant torque region and the flux-weakening region. Compared to the computational burden, the traditional control strategy needs 23.13 μs and the proposed control strategy needs 30.99 μs when a control cycle contains 100 μs. It can be seen that there is not much difference between the execution time of the two control strategies and better control performance can be obtained. The proposed method also enhances the robustness of system effectively under parameter mismatches caused by the variation of parameters.

Author Contributions: Conceptualization, X.G., X.L. and Z.W.; methodology, X.G. and T.L.; software, X.L.; validation, X.L. and G.Z.; formal analysis, G.Z. and T.L.; writing—original draft preparation, T.L.; writing—review and editing, X.G. and Z.W., and; funding acquisition, X.G. and Z.W.

Funding: This research was funded by "The Natural Science Foundation of Tianjin, grant number 19JCYBJC21800 and 18JCQNJC74400" and "The National Natural Science Foundation of China, grant number 51977150".

Conflicts of Interest: The authors declare no conflict of interest.

References

1. Hoang, K.D.; Aorith, H.K.A. Online Control of IPMSM Drives for Traction Applications Considering Machine Parameter and Inverter Nonlinearities. *IEEE Trans. Transp. Electrif.* **2015**, *1*, 312–325. [CrossRef]
2. Nalepa, R.; Orlowska-Kowalska, T. Optimum Trajectory Control of the Current Vector of a Nonsalient-Pole PMSM in the Field-Weakening Region. *IEEE Trans. Ind. Electron.* **2012**, *59*, 2867–2876. [CrossRef]
3. Chen, Y.; Huang, X.; Wang, J.; Niu, F.; Zhang, J.; Fang, Y.; Wu, L. Improved Flux-Weakening Control of IPMSMs Based on Torque Feedforward Technique. *IEEE Trans. Power Electron.* **2018**, *33*, 10970–10978. [CrossRef]
4. Liu, H.; Zhu, Z.; Mohamed, E.; Fu, Y.; Qi, X. Flux-Weakening Control of Nonsalient Pole PMSM Having Large Winding Inductance, Accounting for Resistive Voltage Drop and Inverter Nonlinearities. *IEEE Trans. Power Electron.* **2012**, *27*, 942–952. [CrossRef]
5. Schoonhoven, G.; Uddin, M.N. MTPA- and FW-Based Robust Nonlinear Speed Control of IPMSM Drive Using Lyapunov Stability Criterion. *IEEE Trans. Ind. Appl.* **2016**, *52*, 4365–4374. [CrossRef]
6. Uddin, M.N.; Chy, M.M.I. Online Parameter-Estimation-Based Speed Control of PM AC Motor Drive in Flux-Weakening Region. *IEEE Trans. Ind. Appl.* **2008**, *44*, 1486–1494. [CrossRef]
7. Jo, C.; Seol, J.Y.; Ha, I.J. Flux-Weakening Control of IPM Motors with Significant Effect of Magnetic Saturation and Stator Resistance. *IEEE Trans. Ind. Electron.* **2008**, *55*, 1330–1340. [CrossRef]
8. Mynar, Z.; Vesely, L.; Vaclavek, P. PMSM Model Predictive Control with Field-Weakening Implementation. *IEEE Trans. Ind. Electron.* **2016**, *63*, 5156–5166. [CrossRef]
9. Bolognani, S.; Calligaro, S.; Petrella, R. Adaptive Flux-Weakening Controller for Interior Permanent Magnet Synchronous Motor Drives. *IEEE J. Emerg. Sel. Top. Power Electron.* **2014**, *2*, 236–248. [CrossRef]
10. Ekanayake, S.; Dutta, R.; Rahman, M.F.; Rathnayake, R.M.H.M. A modified single-current-regulator control scheme for deep flux- weakening operation of interior permanent magnet synchronous motors. In Proceedings of the IECON 2016-42nd Annual Conference of the IEEE Industrial Electronics Society Florence, Florence, Italy, 23–26 October 2016; pp. 2624–2629. [CrossRef]
11. Wang, W.; Zhang, J.; Cheng, M. Line-modulation-based flux-weakening control for permanent-magnet synchronous machines. *IET Power Electron.* **2018**, *11*, 930–936. [CrossRef]
12. Chaoui, H.; Khayamy, M.; Aljarboua, A.A. Adaptive Interval Type-2 Fuzzy Logic Control for PMSM Drives with a Modified Reference Frame. *IEEE Trans. Ind. Electron.* **2017**, *64*, 3786–3797. [CrossRef]
13. Dong, Z.; Yu, Y.; Li, W.; Wang, B.; Xu, D. Flux-Weakening Control for Induction Motor in Voltage Extension Region: Torque Analysis and Dynamic Performance Improvement. *IEEE Trans. Ind. Electron.* **2018**, *65*, 3740–3751. [CrossRef]
14. Lin, P.Y.; Lai, Y.S. Voltage Control Technique for the Extension of DC-Link Voltage Utilization of Finite-Speed SPMSM Drives. *IEEE Trans. Ind. Electron.* **2012**, *59*, 3392–3402. [CrossRef]
15. Stojan, D.; Drevensek, D.; Plantic, Ž.; Grcar, B.; Stumberger, G. Novel Field-Weakening Control Scheme for Permanent-Magnet Synchronous Machines Based on Voltage Angle Control. *IEEE Trans. Ind. Appl.* **2012**, *48*, 2390–2401. [CrossRef]

16. Gadelovits, S.Y.; Zhong, Q.C.; Kadirkamanathan, V.; Kuperman, A. UDE-Based Controller Equipped with a Multi-Band-Stop Filter to Improve the Voltage Quality of Inverters. *IEEE Trans. Ind. Electron.* **2017**, *64*, 7433–7443. [CrossRef]
17. Zhong, Q.C.; Wang, Y.; Ren, B. UDE-Based Robust Droop Control of Inverters in Parallel Operation. *IEEE Trans. Ind. Electron.* **2017**, *64*, 7552–7562. [CrossRef]
18. Ren, B.; Zhong, Q.C.; Chen, J. Robust Control for a Class of Nonaffine Nonlinear Systems Based on the Uncertainty and Disturbance Estimator. *IEEE Trans. Ind. Electron.* **2015**, *62*, 5881–5888. [CrossRef]
19. Ren, J.; Ye, Y.; Xu, G.; Zhao, Q.; Zhu, M. Uncertainty-and-Disturbance -Estimator-Based Current Control Scheme for PMSM Drives with a Simple Parameter Tuning Algorithm. *IEEE Trans. Power Electron.* **2017**, *32*, 5712–5722. [CrossRef]
20. Ren, B.; Zhong, Q.C.; Dai, J. Asymptotic Reference Tracking and Disturbance Rejection of UDE-Based Robust Control. *IEEE Trans. Ind. Electron.* **2017**, *64*, 3166–3176. [CrossRef]
21. Harnefors, L.; Nee, H.P. Model-based current control of AC machines using the internal model control method. *IEEE Trans. Ind. Appl.* **1998**, *34*, 133–141. [CrossRef]
22. Zhong, Q.C.; Rees, D. Control of uncertain LTI systems based on an uncertainty and disturbance estimator. *ASMEJ. Dyn. Syst. Meas. Control* **2004**, *126*, 905–910. [CrossRef]

Article

High Frequency Square-Wave Voltage Injection Scheme-Based Position Sensorless Control of IPMSM in the Low- and Zero- Speed Range

Shuang Wang, Jianfei Zhao * and Kang Yang

School of Mechatronic Engineering and Automation, Shanghai University, Baoshan District, Shanghai 200444, China; wang-shuang@shu.edu.cn (S.W.); yangk@shu.edu.cn (K.Y.)
* Correspondence: jfzhao@shu.edu.cn; Tel.: +86-021-6613-0935

Received: 5 November 2019; Accepted: 12 December 2019; Published: 14 December 2019

Abstract: In this paper, a new sensorless control scheme with the injection of a high-frequency square-wave voltage of an interior permanent-magnet synchronous motor (IPMSM) at low- and zero-speed operation is proposed. Conventional schemes may face the problems of obvious current sampling noise and slow identification in the process of magnetic polarity detection at zero speed operation, and the effects of inverter voltage error on the rotor position estimation accuracy at low speed operation. Based on the principle analysis of d-axis magnetic circuit characteristics, a method for determining the direction of magnetic polarity of d-axis two-opposite DC voltage offset by uninterruptible square-wave injection is proposed, which is fast in convergence rate of magnetic polarity detection and more distinct. In addition, the strategy injects a two-opposite high-frequency square-wave voltage vectors other than the one voltage vector into the estimated synchronous reference frame (SRF), which can reduce the effects of inverter voltage error on the rotor position estimation accuracy. With this approach, low-pass filter (LPF) and band-pass filter (BPF), which are used to obtain the fundamental current component and high-frequency current response with rotor position information respectively in the conventional sensorless control, are removed to simplify the signal process for estimating the rotor position and further improve control bandwidth. Finally, the experimental results on an IPMSM drive platform indicate that the rotor position with good steady state and dynamic performance can be obtained accurately at low-and zero-speed operation with the sensorless control strategy.

Keywords: sensorless control; high frequency square-wave voltage; interior permanent-magnet synchronous motor (IPMSM); magnetic polarity detection; rotor position estimation

1. Introduction

Permanent-magnet synchronous motors (PMSMs) have been widely applied in industrial fields [1–4] for the excellent features of high reliability, high efficiency, high torque density, good dynamic performance, etc. According to the structure of the permanent magnet of the rotor, the PMSM can be divided into interior permanent-magnet synchronous motor (IPMSM) and surface-mounted permanent-magnet synchronous motor (SPMSM). Compared with the SPMSM, the IPMSMs have attracted much attention in industrial fields recently due to their greatly improved overload capacity, power density and speed regulation range, which ascribes to the permanent magnet of the IPMSM is located inside the rotor.

To take advantage of these features, the information of rotor position acquired generally by the mechanical sensors, i.e., resolver or encoder, is necessarily required when the field-oriented control (FOC) scheme [5,6] is adopted. However, these mechanical sensors [7,8] mounted on the shaft of a PMSM bring several disadvantages such as extra cost, extra volume, low reliability, etc. In order to overcome these disadvantages, various kinds of position-sensorless control strategies that estimate the rotor position information without a mechanical sensor have been proposed in past decades [9–12]. Conventional sensorless control methods are mainly divided into two groups: (1) For medium- to high-speed operation, the schemes based on the estimation of back electromotive force (EMF), which contains the position information, (2) for low-speed operation, usually under 5% of the rated speed, the schemes based on the machine saliency injects an additional high-frequency (HF) signal to the PMSM which generates a response containing information of the rotor position.

This paper investigates the position-sensorless control strategy using the HF signal injection method in the low- and zero-speed operation, and according to the types of the injected HF signals, the signal can be mainly divided into rotating sinusoidal voltage injection (RSVI) [13], pulsating sinusoidal voltage injection (PSVI) [14], and pulsating square-wave voltage injection (PUVI) [15]. The first HF rotating sinusoidal voltage injection method was proposed by Lorenz R.D in the early time, and in this method, the balanced rotating voltage signals were injected into the stationary reference frame (SRF) and then the induced currents were extracted to obtain the rotor position information. However, this method injects an additional HF voltage into the SRF system, which will lead to the torque ripple caused by the fluctuation of q-axis current. Moreover, the saliency will be reduced under heavy load and the detection accuracy will be worse due to the magnetic saturation effect. For the improvement of the RSVI, pulsating sinusoidal voltage injection and pulsating square-wave voltage injection are proposed, which inject HF voltages in the estimated d-axis of the rotational reference frame (RRF) other than the SRF. To overcome these problems, a PSVI method [16] injects HF voltage into the estimated d-axis reference frame. Similar to RSVI method, the injection voltage frequency of PSVI method is usually about 1/10 of the carrier frequency. Therefore, this method should use low-pass filter (LPF) and band-pass filter (BPF) to separate the fundamental current component and HF currents, respectively, which decreases the control bandwidth and the dynamic performance. To improve the dynamic performance, reference [17] proposed the PUVI method, in which the HF square-wave injected into the estimated d-axis and increased the frequency of the injected signal as high as possible to the pulse width modulation (PWM) carrier frequency. Since the frequency of the injected HF square-wave voltage signal is much higher than the cut-off frequency of current loop, the LPFs in the current feedback loop can be omitted [18], which improves the response speed to some extent. Seung et al. [19] proposed a method that the frequency of the injected HF square-wave voltage signal is the same as the PWM switching frequency, in which two current samples were taken in a square-wave period and three arithmetic operations were performed to obtain the HF induced current signal. Besides, the dynamics of the position-sensorless control can be improved and the acoustic noise can be remarkably reduced. However, this method still needs to use a BPF for obtaining the feedback current signal. In [20], the fundamental current signal of the stator winding is obtained by arithmetic operation, without using a filter, but it is still necessary to sample the current twice in one square-wave period.

The conventional square-wave injection method has many advantages, such as low current noise, high bandwidth, and high steady state performance, etc. However, the voltage error caused by the nonlinearity of the inverter still needs to be taken seriously and solved. In addition, the magnetic polarity detection, which is one of the main problems to be improved, is essential for smooth startup and robust control of IPMSM. The conventional method [21] proposes to detect the initial position of the rotor by pulse voltage vector method, which injects a series of pulse voltage vectors with the same amplitude and different direction into the motor stator winding, and estimates the rotor pole position by comparing the magnitude of the response current based on the nonlinear saturation characteristics of the motor stator core. However, the current amplitude does not change much as the voltage vector approaches the rotor pole position. At the same time, the phase current measurement error and the inverter nonlinearity will affect the measurement accuracy [16].

In order to solve the problem of initial position detection in conventional sensorless control of IPMSM, a rotating voltage injection sensorless method of estimating the initial rotor position of a direct torque controlled IPMSM drive is proposed in [22], which injects a HF voltage to the windings and extracts the amplitude of the corresponding stator current components based on motor salient effect. However, this method still has some problems, such as obvious current sampling noise, influence by motor parameter variations, and special position interference. The improvement of the RSVI, PUVI, which has been proven to be the best injection type for the low- and zero-speed position sensorless control [23], can rarely be found for the initial rotor position detection in the sensorless control system. Xie et al. [24] proposed a PUVI method in which the two opposite voltage vectors were injected to reduce the effects of inverter voltage error on the position estimation accuracy. However, this method only studies the low-speed range, and does not conduct research and analysis on the zero-speed range, especially the lack of research on magnetic polarity detection, which is essential for smooth startup and robust control of IPMSM. In [25], the positive and negative test pulse voltages are injected into the estimated d-axis, and the positive direction of the magnetic polarity is judged by the time when the currents at different magnetic polarity are attenuated from the steady state value to 0. However, the position estimation is stopped in the process of polarity identification, which makes the structure of the code more complicated. This problem also exists in [26] and [27], and the interval of signal injection must be existed, resulting in long execution time and poor stability. Then, [28] uses the difference in current measured at zero time of each PWM period to determine the magnetic polarity. The advantage is that no additional injection voltage is required, and the convergence speed is fast. The disadvantage of this method is that the position estimation depends on the accuracy of the current. Once the current signal-to-noise ratio is too small, the rotor position cannot be accurately obtained. Based on the above references and the improvement points of this article, the comparison of Table 1 is obtained.

Table 1. List of the research contents covered by the articles.

Literature	Remove LPF	Remove BPF/HPF	Reduce Audible Noise	Nonlinearity of the Inverter	Magnetic Polarity Judgment
Ref. [4]	×	√	×	×	×
Ref. [8]	√	×	×	×	√
Ref. [10]	×	×	√	×	×
Ref. [11]	×	×	√	×	×
Ref. [12]	√	√	√	×	×
Ref. [13]	√	×	×	×	√
Ref. [15]	√	√	×	×	×
Ref. [16]	×	√	×	×	√
Ref. [18]	√	√	×	√	×
Ref. [23]	√	√	×	×	√
Ref. [24]	√	√	×	√	×
This paper	√	√	×	√	√

Therefore, in view of the existing sensorless estimation methods for rotor position, a HF square-wave voltage injection scheme-based rotor position estimation of IPMSM in low-speed range and a new method of magnetic polarity detection in zero-speed range are proposed. Based on the principle analysis of d-axis magnetic circuit characteristics, a method for determining the direction of magnetic polarity of d-axis two-opposite DC voltage offset by uninterruptible square-wave injection is proposed, which is simple and can quickly converge by comparing the absolute value of the peak-to-peak value in the d-axis high-frequency current response. Therefore, the method of magnetic polarity identification proposed in this paper is not only suitable for the case where the motor is stationary, but also suitable for the free running condition of the motor. At the same time, the strategy injects a two-opposite HF square-wave voltage vectors into the estimated SRF, which considers the effects of inverter voltage error on the rotor position estimation accuracy. With this approach, the LPF and the BPF are removed to simplify the signal process for estimating the rotor position and further improve control bandwidth.

This paper is organized as follows. First, the rotor position estimation strategy based on conventional HF square-wave voltage injection is analyzed in Section 2. In Section 3, the rotor position estimation strategy based on improved HF square-wave voltage injection and proposed magnetic polarity detection are investigated. Then, in Section 4, comprehensive simulation and experimental setup are introduced and experiments are provided to prove the effectiveness of the improved sensorless control strategy and proposed magnetic polarity detection method. Finally, Section 5 concludes this paper.

2. Analysis of Rotor Position Estimation Strategy Based on Conventional Square-Wave Voltage Injection

2.1. Mathematical Model of IPMSM

Assuming that the IPMSM operates in an unsaturated state with negligible hysteresis loss and eddy current loss, the mathematical model of the IPMSM in the α-β stationary reference frame is given as:

$$\begin{aligned}\begin{bmatrix} u_{\alpha s} \\ u_{\beta s} \end{bmatrix} = R_s \begin{bmatrix} i_{\alpha s} \\ i_{\beta s} \end{bmatrix} + \begin{bmatrix} L^r_{sum} + L^r_{dif}\cos 2\theta_r & L^r_{dif}\sin 2\theta_r \\ L^r_{dif}\sin 2\theta_r & L^r_{sum} - L^r_{dif}\cos\theta_r \end{bmatrix} \frac{d}{dt}\begin{bmatrix} i_{\alpha s} \\ i_{\beta s} \end{bmatrix} + \\ 2L^r_{dif}\omega_r \begin{bmatrix} -\sin 2\theta_r & \cos 2\theta_r \\ \cos 2\theta_r & \sin\theta_r \end{bmatrix}\begin{bmatrix} i_{\alpha s} \\ i_{\beta s} \end{bmatrix} + \omega_r\psi_f \begin{bmatrix} -\sin\theta_r \\ \cos\theta_r \end{bmatrix}\end{aligned} \tag{1}$$

where $u_{\alpha,\beta s}$, $i_{\alpha,\beta s}$ are the α- and β-axes stator voltage and current, respectively. R_s is the stator resistance; ω_r is the rotor speed; ψ_r is the linkage magnetic flux, and d/dt represents derivative operator. $L^r{}_{sum}$, $L^r{}_{dif}$ are average inductance and differential inductance, respectively, and are defined as $L^r{}_{sum} = (L^r_d + L^r_q)/2$ and $L^r{}_{dif} = (L^r_d - L^r_q)/2$, respectively, where $L^r{}_d$ and $L^r{}_q$ are d-and q-axis inductance, respectively. θ_r is the actual rotor position, and the physical model of actual and estimated rotor reference frames is shown in Figure 1.

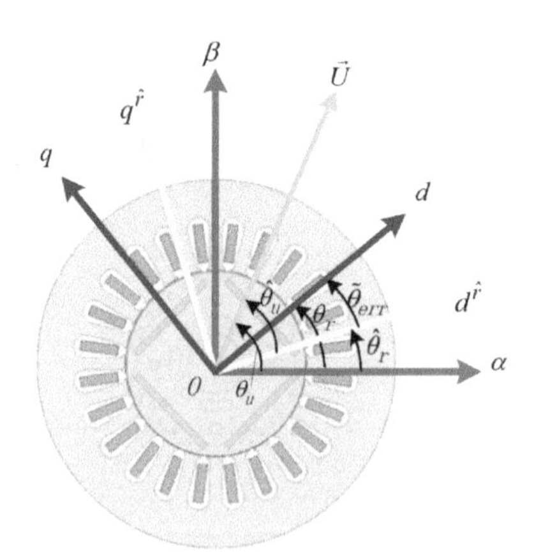

Figure 1. Physical model of the actual and estimated rotor reference frames.

According to Euler's formula, the voltage and current vector can be described as

$$\begin{cases} \vec{U}_{\alpha s} = U_{\alpha s} e^{j\theta_r} \\ \vec{U}_{\beta s} = U_{\beta s} j e^{j\theta_r} \\ \vec{I}_{\alpha s} = I_{\alpha s} e^{j\theta_r} \\ \vec{I}_{\beta s} = I_{\beta s} j e^{j\theta_r} \end{cases} \tag{2}$$

where $U_{\alpha s}$, $I_{\alpha s}$, $U_{\beta s}$, and $I_{\beta s}$ are the magnitudes of $\vec{U}_{\alpha s}$ and $\vec{I}_{\beta s}$ on the α- and β-axes, respectively; e is the natural constant; j is the imaginary unit.

Under the premise that IPMSM operates at zero- and low-speed range, the product term related to ω_r can be omitted because the speed is close to 0. Therefore, (1) can be simplified as

$$\begin{cases} \vec{U}_{\alpha s} = R_s \vec{I}_{\alpha s} + \left(L^r_{sum} + L^r_{dif} \cos 2\theta_r\right) \frac{d\vec{I}_{\alpha s}}{dt} - jL^r_{dif} \sin 2\theta_r \frac{d\vec{I}_{\beta s}}{dt} \\ \vec{U}_{\beta s} = R_s \vec{I}_{\beta s} + jL^r_{dif} \sin 2\theta_r \frac{d\vec{I}_{\alpha s}}{dt} + \left(L^r_{sum} - L^r_{dif} \cos 2\theta_r\right) \frac{d\vec{I}_{\beta s}}{dt} \end{cases} \tag{3}$$

By adding the two equations in (3), (3) is rewritten as

$$\begin{aligned} \vec{U}_{\alpha s} + \vec{U}_{\beta s} = {} & R_s\left(\vec{I}_{\alpha s} + \vec{I}_{\beta s}\right) + \frac{(L^r_d + L^r_q)}{2}\left(\frac{d\vec{I}_{\alpha s}}{dt} + \frac{d\vec{I}_{\beta s}}{dt}\right) \\ & + \frac{(L^r_d - L^r_q)}{2} \cos 2\theta_r \left(\frac{d\vec{I}_{\alpha s}}{dt} - \frac{d\vec{I}_{\beta s}}{dt}\right) + j\frac{(L^r_d - L^r_q)}{2} \sin 2\theta_r \left(\frac{d\vec{I}_{\alpha s}}{dt} - \frac{d\vec{I}_{\beta s}}{dt}\right) \end{aligned} \tag{4}$$

2.2. *Signal-Process Method in the Estimated Rotor Reference Frame*

Figure 2 shows the control system scheme for obtaining position information with the conventional square-wave injection method. By reasonably selecting the injection voltage, the voltage generated on the stator resistance in (4) can be neglected. After Euler transform, (4) is derived as

$$d\vec{I}_{\alpha\beta s} = \frac{L^r_d(\vec{U}_{\alpha\beta s} - \vec{U}^*_{\alpha\beta s} e^{j2\theta_r}) + L^r_q(\vec{U}_{\alpha\beta s} + \vec{U}^*_{\alpha\beta s} e^{j2\theta_r})}{2L^r_d L^r_q} dt \tag{5}$$

where $\vec{U}^*_{\alpha\beta s}$ is the conjugate vector of $\vec{U}_{\alpha\beta s}$. Since $d\vec{I}_{\alpha\beta s}/dt$ can be regarded as $\Delta\vec{I}_{\alpha\beta s}/\Delta t$ during one PWM switching period, where $\Delta\vec{I}_{\alpha\beta s}$ is the α- and β-axes current variation in the stationary reference frame. Therefore, the current variation can be expressed as

$$\Delta\vec{I}_{\alpha\beta s} = \left(\frac{L^r_d + L^r_q}{2L^r_d L^r_q} U_{\alpha\beta s} e^{j\theta_u} - \frac{L^r_d - L^r_q}{2L^r_d L^r_q} U_{\alpha\beta s} e^{j(2\theta_r - \theta_u)}\right)\Delta t \tag{6}$$

where θ_u is the angle of the voltage vector in the stationary reference frame. By converting the current variation in the α-β stationary reference frame to the estimated $\hat{d} - \hat{q}$ rotating reference frame, the current variation can be rewritten as

$$\Delta\vec{I}_{\hat{d}\hat{q}s} = \left(\frac{L^r_d + L^r_q}{2L^r_d L^r_q} - \frac{L^r_d - L^r_q}{2L^r_d L^r_q} e^{j2(\theta_r - \hat{\theta}_u - \hat{\theta}_r)}\right)\Delta t \cdot \vec{U}_{\hat{d}\hat{q}s} \tag{7}$$

where $\hat{\theta}_r$ is the estimated position angle. As shown in Figure 1, the relationship between different angles are given as

$$\begin{cases} \theta_u = \hat{\theta}_u + \hat{\theta}_r \\ \theta_r = \tilde{\theta}_{err} + \hat{\theta}_r \end{cases} \tag{8}$$

where $\widetilde{\theta}_{err}$ and $\hat{\theta}_u$ are the error angle of the actual rotating reference frame and the estimated rotating reference frame, the angle of the voltage vector in the estimated rotating reference frame, respectively. If $d^{\hat{r}}$-axis is selected as the injection axis, the form of square-wave voltage is shown as

$$U_{inj}(t) = \begin{cases} U_{inj}e^{j2\hat{\theta}_u}, 0 < t_m(T) < \frac{T}{2} \\ 0, \frac{T}{2} < t_m(T) < T \end{cases} \tag{9}$$

where T and $t_m(T)$ are the square-wave period, the remainder of t divided by T, respectively.

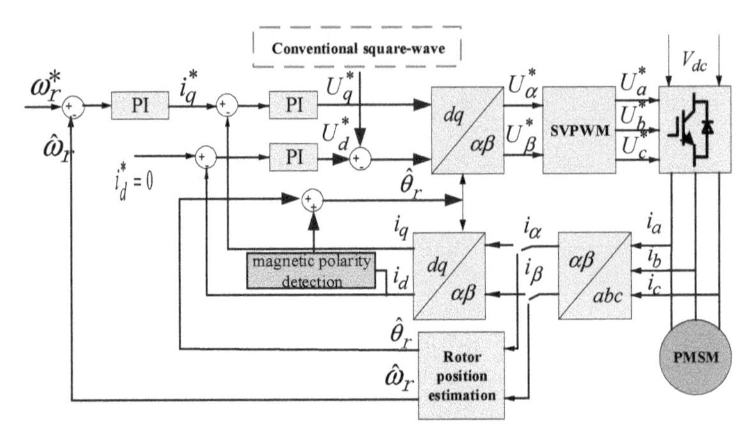

Figure 2. Control system scheme for obtaining position information with the conventional square-wave injection method.

The current variation in the estimated $d^{\hat{r}} - q^{\hat{r}}$ rotating reference frame after injecting the conventional square-wave voltage can be calculated as

$$\Delta\vec{I}_{d^{\hat{r}}q^{\hat{r}}s} = \frac{\Delta t(L_d^r + L_q^r)}{2L_d^r L_q^r}U_{inj} - \frac{\Delta t(L_d^r - L_q^r)}{2L_d^r L_q^r}U_{inj}\cos(2\widetilde{\theta}_{err}) - \frac{\Delta t(L_d^r - L_q^r)}{2L_d^r L_q^r}U_{inj}\sin(2\widetilde{\theta}_{err})j \tag{10}$$

It can be seen from (10) that the error angle can be directly extracted from the imaginary part of the current variation when the error angle is small, which is obtained as

$$\mathrm{Im}(\Delta\vec{I}_{d^{\hat{r}}q^{\hat{r}}s}) = \Delta\vec{I}_{q^{\hat{r}}s} = C \cdot \sin(2\widetilde{\theta}_{err}) \approx 2C \cdot \widetilde{\theta}_{err} \tag{11}$$

where $C = \frac{\Delta t(L_d^r - L_q^r)}{2L_d^r L_q^r}u_{inj}$, which is a constant. The estimated rotor position $\hat{\theta}_r$ is obtained by signal-processing in the sensorless control, as shown in Figure 3. The full name of PLL in Figure 3 is phase-locked loop.

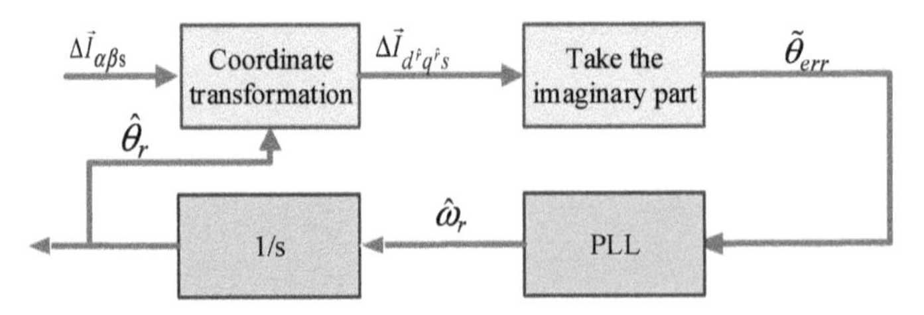

Figure 3. Signal processing diagram of the conventional square-wave voltage injection method for IPMSM.

2.3. Voltage Vector Injection Scheme

Since the motor speed is extremely low, i.e., less than 5% of rated speed and the motor state changes minimally during one PWM switching period, the performance of motor is basically the same when the PI controller integrates once every two PWM switching periods and integrates once for each PWM switching period. In the subsequent experiments, one switching frequency of 10 kHz was used, and the injection frequency was 1/2 of the switching frequency. Although the system of rotor position estimation can extract the position information of the rotor and obtain an estimated value, the estimated value may be consistent with the actual position, or may be offset by π rad, so that the positive direction of the actual d-axis cannot be determined. Therefore, the magnetic polarity must be judged before the motor runs at low speeds. The conventional method to solve the problem is usually to inject voltage pulses of equal width in both positive and negative directions into the d-axis in the estimated rotating reference frame. Considering that the duty period of the PWM calculated in the ARM controller will be updated in the next period, the conventional control sequence of injection voltage and sampling current under position-sensorless control at zero- and low-speed is shown in Figure 4.

As can be seen from Figure 4, current sampling was performed at the beginning of each current action period, and the current variation obtained by the difference between the measured currents in the two periods can be used for position estimation. The PI controller of the current loop only acts after the current sampling step in the FOC control period, so by separating the FOC control period and the voltage injection period, an additional low-pass filter for extracting the estimated rotor position can be omitted.

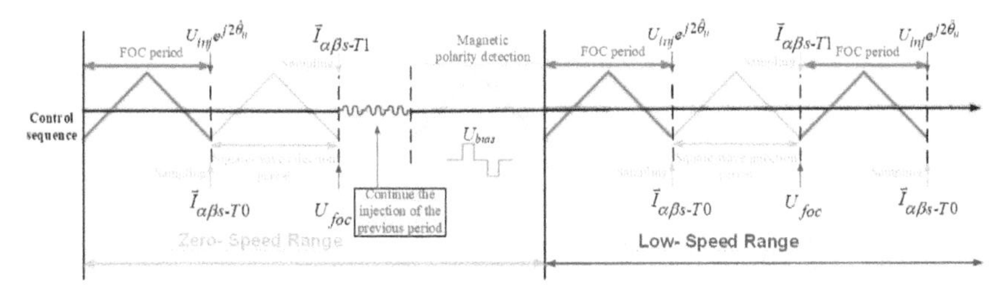

Figure 4. Conventional control sequence of injection voltage and sampling current at zero- and low-speed.

3. Analysis of Rotor Position Estimation Strategy Based on Improved Square-Wave Voltage Injection

3.1. Improved Signal-Process Method in the Estimated Rotor Reference Frame

At low speeds, various position-sensorless driving methods are generally affected by the nonlinear voltage error of the inverter. The nonlinear factors in the voltage-type inverter cause the HF response current to be distorted, which results in the rotor position estimation error and affects the stability of sensorless control. If the error is not compensated, the injected voltage vector cannot be injected into the target axis accurately, resulting in the observation position offset. In (6), $\left|\frac{L_d^r+L_q^r}{2L_d^rL_q^r}\right| = \left|-\frac{L_d^r-L_q^r}{2L_d^rL_q^r}\right|$, when square-wave voltage is injected to the estimated $d^{\hat{r}}$ -axis, the rotor position angle error signal is obtained from the term $\left[-\frac{\Delta t(L_d^r-L_q^r)}{2L_d^rL_q^r}u_{inj}\sin(2\widetilde{\theta}_{err})\right]$ in the imaginary part of $\Delta\vec{I}_{d^{\hat{r}}q^{\hat{r}}s}$ However, due to the nonlinearity of the voltage-source inverters, the square-wave voltage cannot be injected into the target position of $d^{\hat{r}}$ -axis accurately, so a small error will be enlarged from the term $\left[\frac{\Delta t(L_d^r-L_q^r)}{2L_d^rL_q^r}u_{inj}\right]$ in the real part of $\Delta\vec{I}_{d^{\hat{r}}q^{\hat{r}}s}$ Therefore, when extracting the rotor position angle error signal, the term $\left[\frac{\Delta t(L_d^r-L_q^r)}{2L_d^rL_q^r}u_{inj}\right]$ will not be completely eliminated, which will seriously affect the accuracy of conventional square-wave injection for estimating the rotor position angle. In the conventional method, the observation value of rotor position can be obtained by injecting only one voltage vector in the injection period, but the voltage error affected by the nonlinearity of the voltage-source inverters is not compensated. In addition, the conventional method first calculates the error and then compensates the error, which is not only complicated to operate, but also the digital control systems, e.g., DSP and dSPACE, have a delay of switching periods. Therefore, the voltage error calculated in each switching period will only be compensated in the next switching period, which leads to inaccurate voltage error compensation. Therefore, on the basis of the above injection method, it is important to study a simple and accurate inverter voltage error compensation method to improve the accuracy of rotor position estimation and the control performance of IPMSM.

In order to realize the compensation of the voltage error caused by the nonlinear factor of the inverter, e.g., dead-time of switches and turn-on and turn-off voltage drop of switches, another voltage vector with the same amplitude and opposite direction can be injected in the next switching period of injecting the positive voltage vector on the basis of separating the FOC control period and the square-wave injection period, at which time the PI controller of current loop acts and updates every three switching periods. Figure 5 shows the inverter single-phase bridge arm structure, and the improved physical model of actual and estimated rotor reference frames is shown in Figure 6.

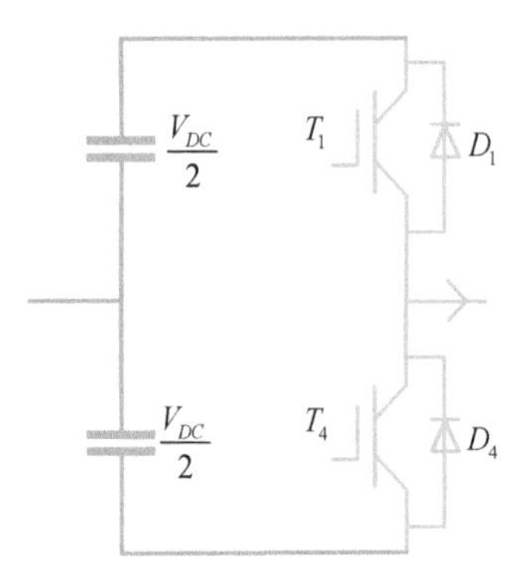

Figure 5. Inverter single-phase bridge arm structure.

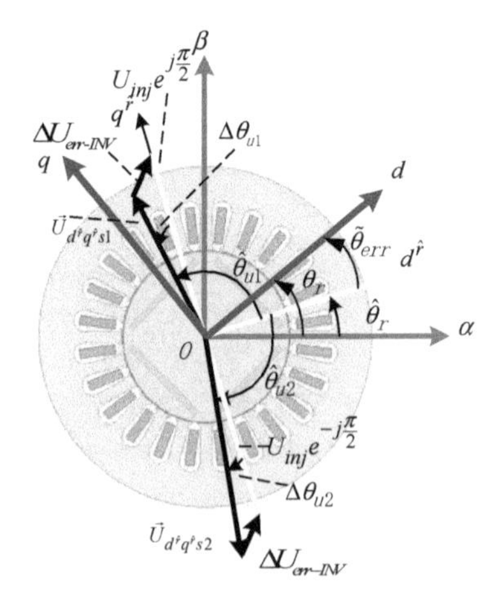

Figure 6. The improved physical model of actual and estimated rotor reference frames.

3.2. Improved Voltage Vector Injection Scheme

Similarly, the details of the ARM processor update in the next period after calculating the PWM duty period need to be considered. Figure 7 shows the improved control sequence of injection voltage and sampling current under position-sensorless control at zero- and low-speed. As shown in Figure 7, when the IPMSM operates at zero-speed, the initial position detection needs to be performed first, and then the magnetic polarity detection needs to be considered. When the IPMSM operates at low-speed, in the first PWM carrier period, the sensorless control system performs FOC control, without superimposing any high frequency vector. The current response generated by the FOC vector acts as the control current of three PWM periodic current loops. Then, before the beginning of the second PWM carrier period, a positive square-wave voltage is injected into the forward direction of $q^{\hat{r}}$, and then a negative square-wave voltage of the same magnitude is injected into the reverse direction of $q^{\hat{r}}$, before the beginning of the third PWM carrier period. Finally, the currents acquired by the second and third PWM periods are compared with the current acquired by the previous PWM period. By making a difference, two varying currents are obtained to calculate the rotor position angle.

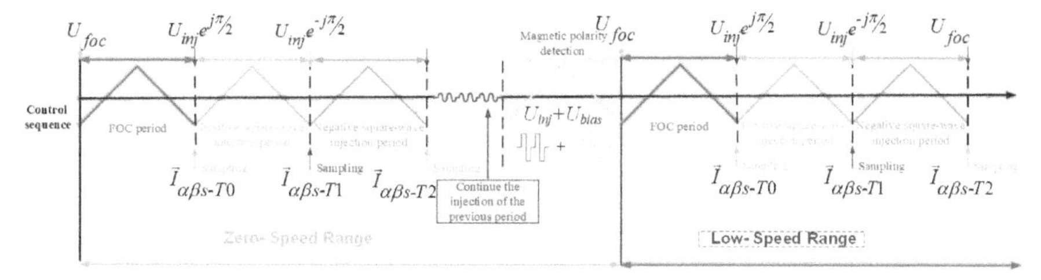

Figure 7. The improved control sequence of injection voltage and sampling current at zero- and low-speed.

Define the inverter voltage error as $\Delta U_{err-INV}$. Since the motor speed is very low and the switching period is very short, assuming that $\Delta U_{err-INV}$ does not change during the two switching periods, the voltage error caused by the nonlinearity of the inverter can be expressed as

$$\begin{cases} \vec{U}_{d^{\hat{r}}q^{\hat{r}}s1} = U_{inj}e^{j\hat{\theta}_{u1}} \\ \quad \approx U_{inj}e^{\ j\pi/2} - \Delta\vec{U}_{err-INV} \\ \vec{U}_{d^{\hat{r}}q^{\hat{r}}s2} = U_{inj}e^{j\hat{\theta}_{u2}} \\ \quad \approx U_{inj}e^{-j\pi/2} - \Delta\vec{U}_{err-INV} \end{cases} \tag{12}$$

By substituting (12) into (7), the simplified equation is rewritten as (13) and (14)

$$\begin{aligned} \Delta\vec{I}_{d^{\hat{r}}q^{\hat{r}}s1} &= \frac{\Delta t(L_d^r+L_q^r)}{2L_d^rL_q^r}(U_{inj}e^{\ j\pi/2} - \Delta\vec{U}_{err-INV}) - \frac{\Delta t(L_d^r-L_q^r)}{2L_d^rL_q^r}U_{inj}e^{j(2\tilde{\theta}_{err}-\hat{\theta}_{u1})} \\ &= \Delta t\left(\frac{L_d^r+L_q^r}{2L_d^rL_q^r}U_{inj}e^{\ j\pi/2} - \frac{L_d^r+L_q^r}{2L_d^rL_q^r}\Delta\vec{U}_{err-INV} - \frac{L_d^r-L_q^r}{2L_d^rL_q^r}U_{inj}e^{j(2\tilde{\theta}_{err}-\hat{\theta}_{u1})}\right) \end{aligned} \tag{13}$$

$$\begin{aligned} \Delta\vec{I}_{d^{\hat{r}}q^{\hat{r}}s2} &= \frac{\Delta t(L_d^r+L_q^r)}{2L_d^rL_q^r}(U_{inj}e^{\ -j\pi/2} - \Delta\vec{U}_{err-INV}) - \frac{\Delta t(L_d^r-L_q^r)}{2L_d^rL_q^r}U_{inj}e^{j(2\tilde{\theta}_{err}-\hat{\theta}_{u2})} \\ &= \Delta t\left(\frac{L_d^r+L_q^r}{2L_d^rL_q^r}U_{inj}e^{\ -j\pi/2} - \frac{L_d^r+L_q^r}{2L_d^rL_q^r}\Delta\vec{U}_{err-INV} - \frac{L_d^r-L_q^r}{2L_d^rL_q^r}U_{inj}e^{j(2\tilde{\theta}_{err}-\hat{\theta}_{u2})}\right) \end{aligned} \tag{14}$$

Thus, the rotor position estimation scheme can be realized by an improved physical model of actual and estimated rotor reference frames, as shown in Figure 6. In Figure 6, the relationship between angles can be described as

$$\begin{cases} \hat{\theta}_{u1} = \pi/2 + \Delta\theta_{u1} \\ \hat{\theta}_{u2} = -\pi/2 - \Delta\theta_{u2} \\ \Delta\theta_{u1} \approx \Delta\theta_{u2} \end{cases} \tag{15}$$

Since $\Delta\theta_{u1}$ is very close to 0, combining (13), (14), and (15), the obtained equation is calculated as (16).

$$\mathrm{Re}(\Delta\vec{I}_{d^{\rho}q^{\rho}s1} - \Delta\vec{I}_{d^{\rho}q^{\rho}s2}) = -\frac{\Delta t(L_d^r - L_q^r)}{L_d^r L_q^r} U_{inj} \cos\Delta\theta_{u1} \cdot \sin 2\widetilde{\theta}_{err} = 2C \cdot \sin 2\widetilde{\theta}_{err} \approx 4C \cdot \widetilde{\theta}_{err} \tag{16}$$

Similarly, if the injection voltage from the d^{ρ}-axis is selected, the estimated rotor position angle can be expressed as

$$\mathrm{Im}(\Delta\vec{I}_{d^{\rho}q^{\rho}s1} - \Delta\vec{I}_{d^{\rho}q^{\rho}s2}) = 2C \cdot \sin 2\widetilde{\theta}_{err} \approx 4C \cdot \widetilde{\theta}_{err} \tag{17}$$

It is observed that the (16) and (17) can finally estimate the rotor position angle after considering the voltage error caused by the nonlinearity of the inverter. Thus, the rotor position-sensorless estimation scheme can be realized by an improved square-wave injection method, as shown in Figure 8.

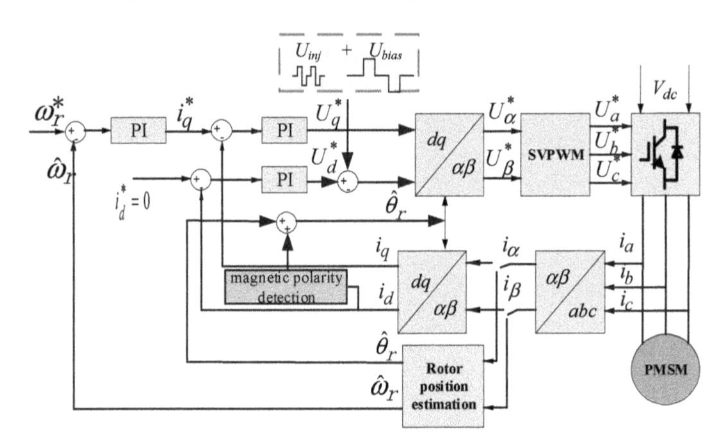

Figure 8. Control system scheme for obtaining position information with the improved square-wave injection method.

3.3. Determining the Direction of Magnetic Polarity

In the conventional method, after the square-wave voltage is injected, the magnetic polarity is determined by injecting two square-wave pulses of opposite directions and equal durations. However, in the process of software algorithm implementation, the switching task of the algorithm state machine is additionally increased, which makes the structure of the code more complicated, and the noise of the current sampling affects the accuracy of the judgment, which may lead to the magnetic polarity judgment error.

In this paper, a fast-initial position identification method is proposed. After the positive and negative square-wave injection, the given d-axis bias voltage U_{bias} is added and the direction of the bias voltage is changed on the basis of the uninterrupted square-wave injection. By comparing the absolute value of the peak-to-peak value of the d-axis HF current response, the magnetic polarity identification is completed, which is relatively simple.

At the same time, the bias current is not directly given in this paper, but the bias current is generated by a given bias voltage. Figure 9 shows the characteristics curve of the d-axis magnetic circuit and the high-frequency current response diagram. The incremental inductance at $X1$ and $X2$ can be defined as

$$\begin{cases} L_1 = \frac{\partial \psi}{\partial i}\Big|_{X_1} \\ L_2 = \frac{\partial \psi}{\partial i}\Big|_{X_2} \end{cases} \tag{18}$$

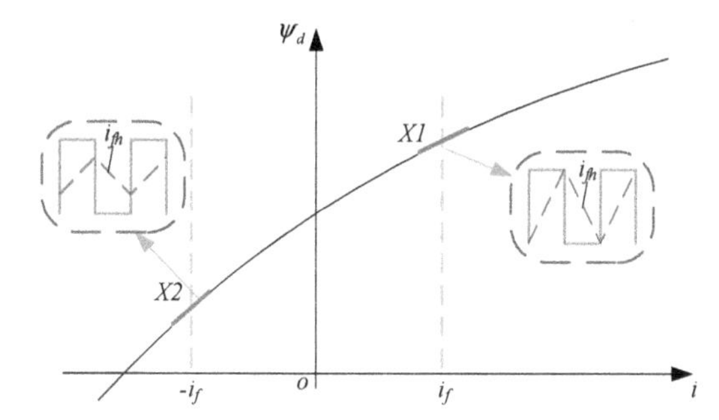

Figure 9. The characteristics curve of the d-axis magnetic circuit and the high-frequency current response diagram.

As shown in Figure 9, the incremental inductance $L2$ is larger than $L1$. Therefore, when the same flux is changed, the current at $X1$ changes greatly and the current at $X2$ changes little. Since the square-wave is not interrupted during the initial position identification process, the convergence speed of magnetic polarity identification is fast. In addition, it is necessary to set U_{bias} to 0 at an intermediate time period of a given bias voltage $\pm U_{bias}$, so that the fundamental current returns to the initial state. When the d-axis DC bias voltage is the same as the magnetic polarity of the rotor ($X1$), the stator flux saturation is increased, the incremental inductance is decreased, and the absolute value of the d-axis HF response current peak-to-peak value is increased. When the voltage is opposite to the magnetic polarity of the rotor ($X2$), the saturation of the stator flux is weakened, the incremental inductance is increased, and the absolute value of the peak-to-peak value of the d-axis HF response current is decreased.

Therefore, the magnetic polarity identification of the rotor can be realized by comparing the absolute values of the peak-to-peak value of the HF current response generated by the HF voltage under the given bias of the positive and negative d-axis voltage. If the peak-to-peak value of forward HF current is greater than peak-to-peak value of the reverse HF current, the estimated position angle direction is directed to the N pole; i.e., the estimated position angle is the actual position angle of the rotor. In addition, if the peak-to-peak value of forward HF current is less than the peak-to-peak value of reverse HF current, the estimated position angle direction is directed to the S pole, which means the rotor position angle needs to be compensated for π. Figure 10 shows the flow charts of two methods for magnetic polarity identification of the rotor.

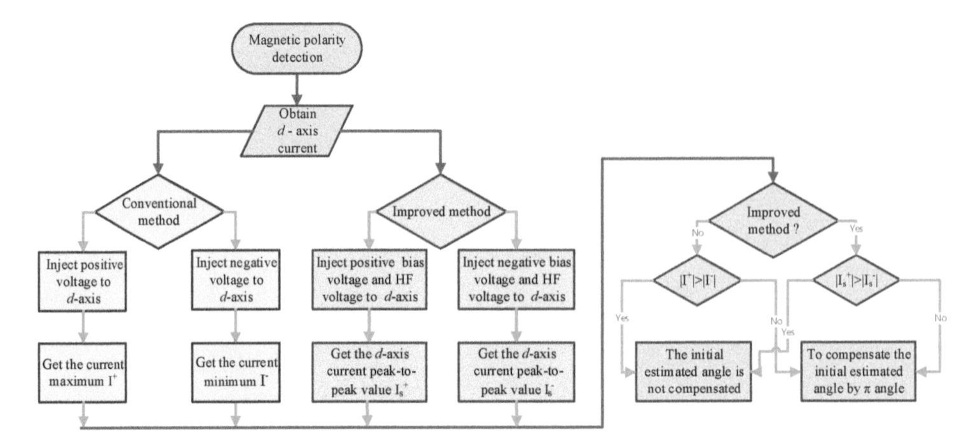

Figure 10. The flow charts of two methods for magnetic polarity identification of the rotor.

4. Simulation and Experimental Results

4.1. Simulationl Results

Because of the need to implement discrete simulation models in Matlab/Simulink, this paper uses the M function in Simulink and the data storage unit to save the discrete data for the next period of calculation. Before conducting the experiment, it is first determined by simulation that the square-wave injection is performed by injecting a HF voltage into the d-axis or a HF voltage into the q-axis. In the simulation, the bus voltage is 310 V, and the amplitude of the injected HF voltage vector is 70 V.

Figure 11 shows the current response when the q-axis is injected with positive and negative pulse voltages for position estimation. Figure 12 shows the current response when the d-axis is injected with positive and negative pulse voltages for position estimation. According to the torque equation of an IPMSM, the q-axis current has a predominant influence on the torque. As shown in Figure 11b, the HF current response with high amplitude is generated in the q-axis, which has a great influence on the torque ripple. However, as shown in Figure 12b, it can be seen that the amplitude of the HF current response in the q-axis is small, which is more conducive to the stability of the control system. Therefore, in subsequent experiments, the rotor position estimation is performed by injecting a HF voltage vector in the d-axis instead of the q-axis injection.

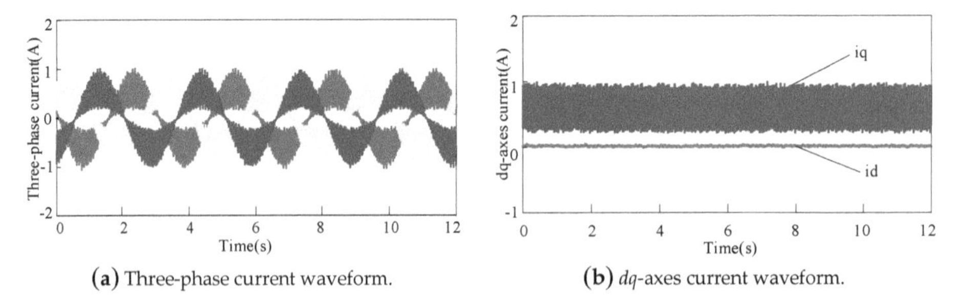

(**a**) Three-phase current waveform. (**b**) dq-axes current waveform.

Figure 11. Current response when the q-axis is injected with positive and negative pulse voltages for position estimation.

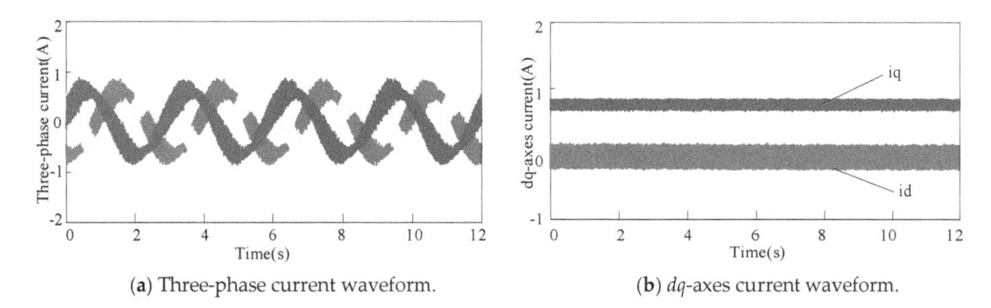

(a) Three-phase current waveform. (b) dq-axes current waveform.

Figure 12. Current response when the d-axis is injected with positive and negative pulse voltages for position estimation.

4.2. Experimental Platform

In order to verify the feasibility of this method in engineering and the accuracy of rotor position estimation, the proposed position-sensorless control scheme was verified on the platform with a 400W IPMSM, as shown in Figure 13. As shown in Figure 13, the experimental platform mainly includes two parts: the tested system and the load system.

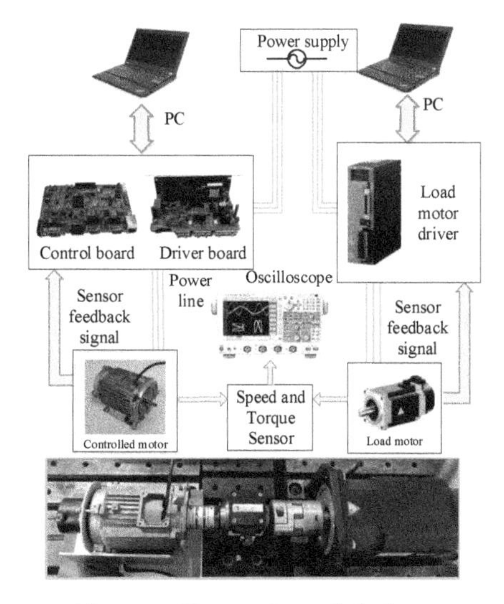

Figure 13. The experimental platform.

The tested system can be mainly divided into three groups: (1) the upper computer for status monitoring, (2) the built-in pump motor as the controlled object, and (3) the control and drive system with the Infineon XMC4500 chip as the core. The load system can be mainly divided into three groups: (1) the upper computer for command transmission and status monitoring, (2) the industrial servo drive, and (3) the high-performance servo motor as load. In addition, the controlled motor and the load motor are connected by speed and torque sensors. The IPMSM and system parameters are shown in Table 2.

Table 2. IPMSM and system parameters for experiment.

Parameter	Quantity	Unit
Pole pairs	2	poles
Resistance	1.6	[Ω]
d-axis inductance	15	[mH]
q-axis inductance	18.8	[mH]
Rated speed	3000	[rpm]
Rated power	400	[W]
Rated voltage	220	[V]
Rated current	2.28	[A]
Rated torque	1.27	[N·m]
PWM switching frequency	10	[kHz]
Injection voltage magnitude	70	[V]

4.3. Initial Rotor Position Estimation

In the experiment, the PWM switching frequency remains unchanged at 10 kHz. The rotor position estimated by the conventional HF square-wave voltage injection method and the improved HF square-wave voltage injection method are compared, and the obtained experimental waveforms are shown in Figures 14 and 15.

The experimental conditions are as follows: the conventional HF square-wave injection frequency is 10 kHz, and the rotor position is estimated every two PWM switching periods. The improved HF square-wave injection frequency is 10 kHz, and rotor position estimation is performed every three PWM switching periods. The initial position angle is artificially fixed at 30°, 60°, 120°, and 150° in advance by acquiring the angle of the encoder. Figure 16 shows the offset angles of the estimated and actual angles of the two methods. The initial position detection in Figure 14c,d and Figure 15c,d are both before the magnetic polarity judgment, and the magnetic polarity judgment are analyzed in the subsequent experiments.

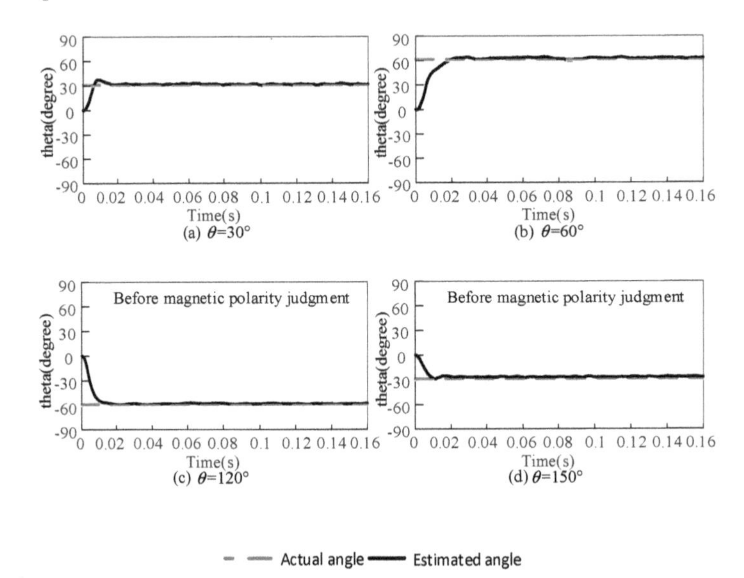

Figure 14. Initial position estimation response curve under conventional HF square-wave voltage injection.

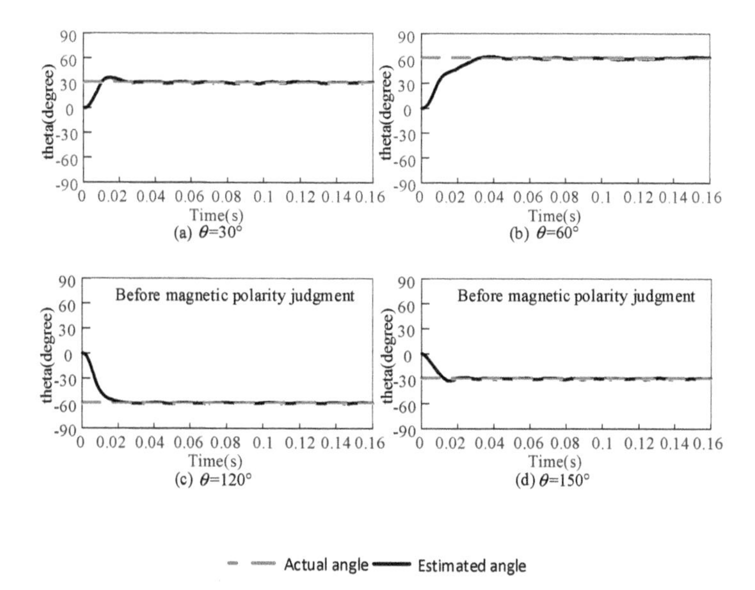

Figure 15. Initial position estimation response curve under improved HF square-wave voltage injection.

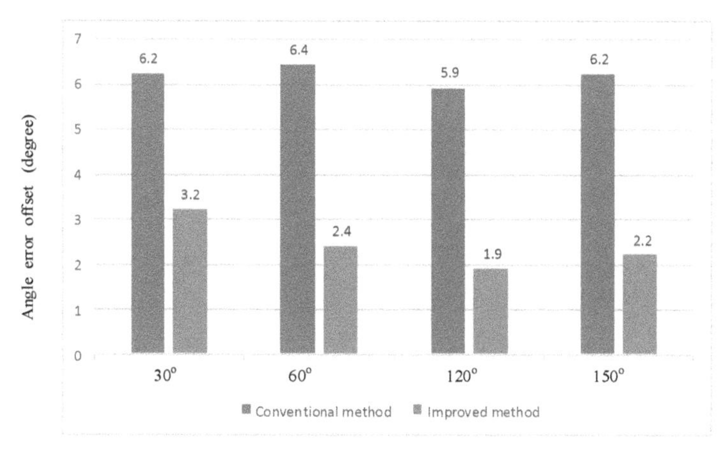

Figure 16. The angle error offset of the two methods.

By summarizing the eight initial position estimation response curves in Figures 14 and 15, the relevant data in Table 3 can be sorted out. In addition, the angle error offset is the size of the offset angle generated on the basis of the actual angle, and the fluctuation range of the angle error is generated on the basis of the angle error offset.

Table 3. Comparison of initial position estimation parameters for two injection methods.

Injection Method	30°		60°		120°		150°	
	Time (s)	error (°)	Time (s)	error (°)	Time (s)	error (°)	Time (s)	error (°)
Conventional method	0.018	±3.3 + 6.2	0.02	±3.4 + 6.4	0.016	±3.2 + 5.9	0.012	±3.6 + 6.2
Improved method	0.022	±3.4 + 3.2	0.032	±3.2 + 2.4	0.023	±2.9 + 1.9	0.017	±3.6 + 2.2

As shown in Figure 16, the estimated angle of the conventional HF square-wave voltage injection method is offset from the actual angle, but this offset is weakened in the improved HF square-wave voltage injection method, in the conventional method, the angle error offset keep within 6.4°, which is caused by the inverter voltage error, in the improved method, the angle error offset keep within 3.2°, it can be seen that the angle error offset is greatly reduced, which is reduced by about 50%, because the improved method reduces the voltage error caused by the nonlinearity of the inverter. It can effectively compensate the voltage error and reduce the influence of the error on the accuracy of rotor position estimation.

4.4. *Estimation of Rotor Position at Low Speed*

4.4.1. Comparison of Rotor Position Estimation Before and After Improvement

In the HF injection sensorless control period, the FOC control period time is 100 microseconds, and the positive voltage injection period and negative voltage injection period are also 100 microseconds. Figure 17 shows the comparison of rotor position estimation before and after improvement. As shown in Figure 17a, in the conventional method, the estimation error results in an offset error of around 5° and an estimated ripple error of ±7°. However, in the improved method of Figure 17b, the offset error is close to 0° and the estimated ripple error is kept within ±5°. The analysis diagram of steady-state estimation error of two methods under different low speed conditions is shown in Figure 18.

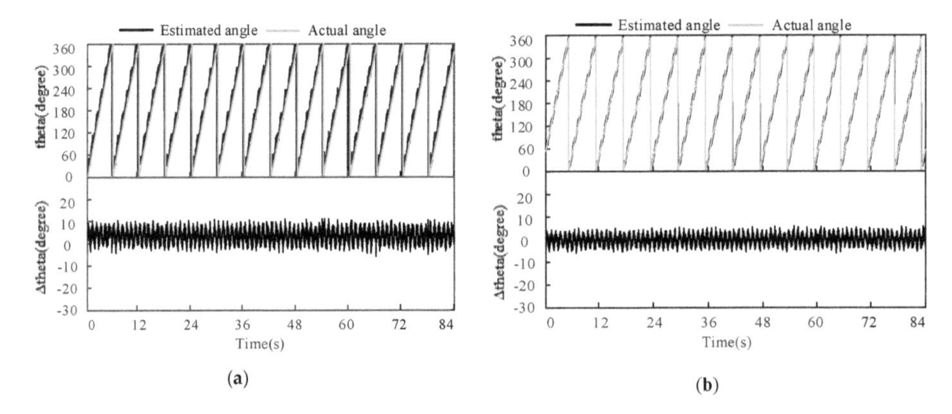

Figure 17. Comparison of rotor position estimation effects between two methods (**a**) Conventional method; (**b**) Improved method.

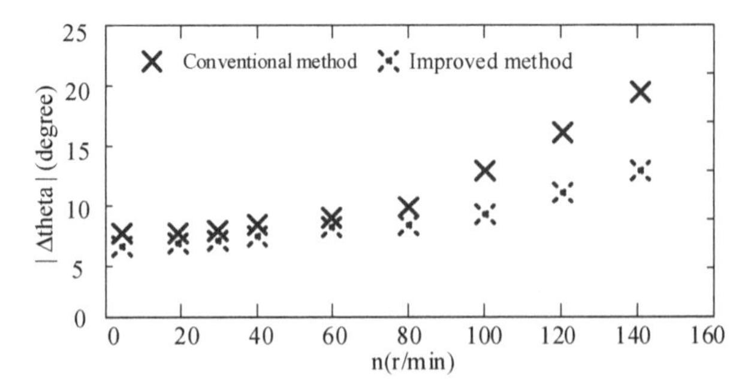

Figure 18. Steady-state estimation error of two injection methods at different speeds.

4.4.2. Rotor Position Observation Experiment Under Forward and Reverse

Figure 19a shows the rotor angle and estimation error waveform of conventional HF square-wave voltage injection method switching back and forth from 5 r/min to −5 r/min. It can be seen that the estimated ripple error can be stabilized within ±10° when the rotation speed is 5 r/min, as shown in Figure 19a. The rotor angle and estimation error waveform of conventional HF square-wave voltage injection method switches back and forth from 20 r/min to −20 r/min, as shown in Figure 19b. It can be seen that the estimated ripple error can be stabilized within ±10° when the rotation speed is 20 r/min, as shown in Figure 19b.

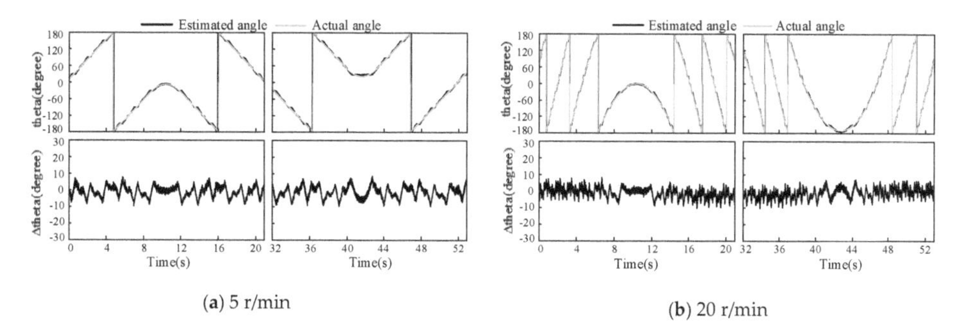

Figure 19. The rotor angle and estimation error waveform of conventional HF square-wave voltage injection method at low-speed operation (**a**) at 5 r/min. (**b**) at 20 r/min.

The rotor angle and estimated error waveform switching back and forth from 5 r/min to −5 r/min and the rotor angle and estimated error waveform switching back and forth from 20 r/min to −20 r/min are shown in Figure 20a,b, respectively. As shown in Figure 20a, it can be seen that the estimated ripple error can be stabilized within ±6° when the rotation speed is 5 r/min. It can be seen that the estimated ripple error can be stabilized within ±8° when the rotation speed is 20 r/min, as shown in Figure 20b.

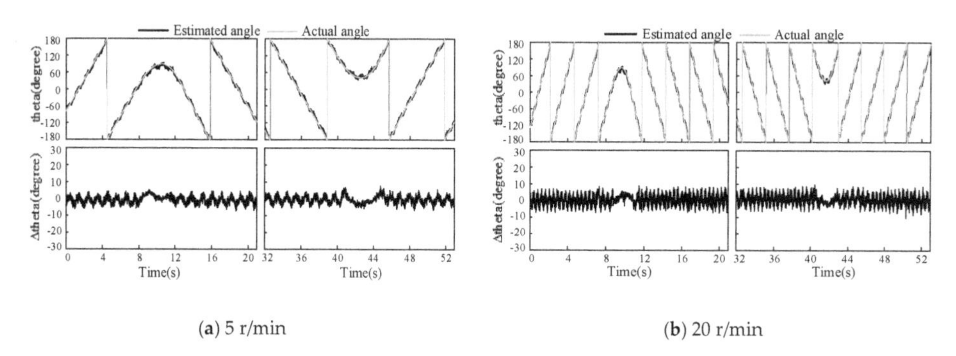

Figure 20. The rotor angle and estimation error waveform of improved HF square-wave voltage injection method at low-speed operation (**a**) at 5 r/min. (**b**) at 20 r/min.

4.5. Judging Compensation of Magnetic Polarity Direction

In this section, the effects of the two magnetic polarity discrimination methods are compared at 30° and 120°, respectively. The magnetic polarity discrimination experiments of the two methods at 30° are shown in Figure 21a,b, respectively. As shown in Figure 21a, the conventional method uses a 20 V DC bias voltage to identify the magnetic polarity. The proposed method uses a 20 V DC bias voltage to superimpose a 16 V HF square-wave voltage signal, as shown in Figure 21b.

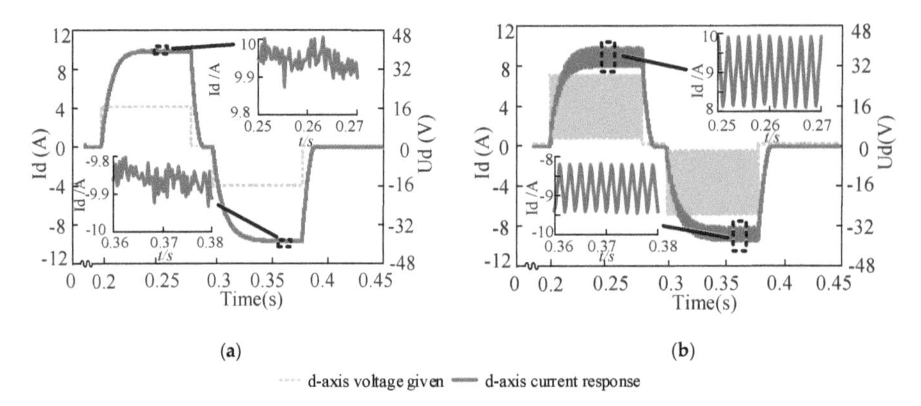

Figure 21. The d-axis current characteristic of the rotor at 30° for magnetic polarity judgment (**a**) conventional method. (**b**) proposed method.

Similarly, the magnetic polarity discrimination experiments of the two methods at 120° are shown in Figure 22a,b, respectively. The experimental data are compared in detail, as shown in Figures 23 and 24. We can divide the process into two stages. When t is between 0 and 0.18 s, the first stage period is the from standstill to startup, that is no procedure and no equipment to be performed at standstill, and the rotor is fixed initially in the startup period. Then, during the second stage, when t is between 0.18 and 0.38 s, the magnetic polarity judgment is performed in order to ensure the accuracy of rotor position estimation. The magnetic polarity identification method proposed in this paper is a combination of HF square-wave voltage injection method and DC bias voltage. In the process of magnetic polarity identification, the HF square-wave voltage injection method has also been updating its angle.

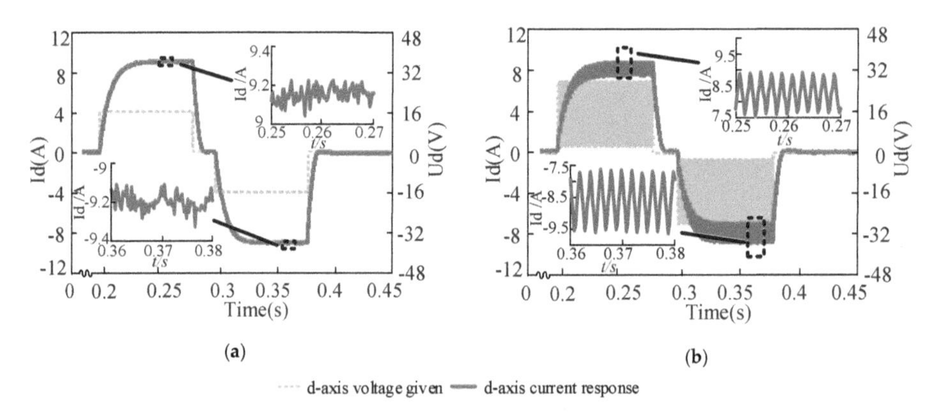

Figure 22. The d-axis current characteristic of the rotor at 120° for magnetic polarity judgment (**a**) conventional method. (**b**) proposed method.

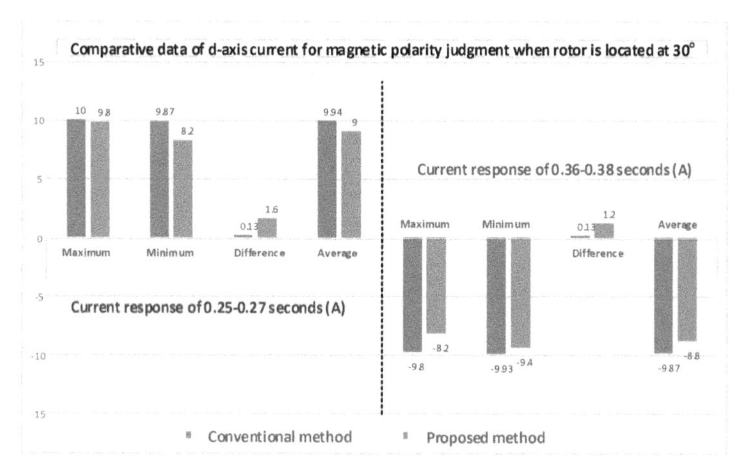

Figure 23. Comparative data of d-axis current characteristics for magnetic polarity judgment when rotor is located at 30°.

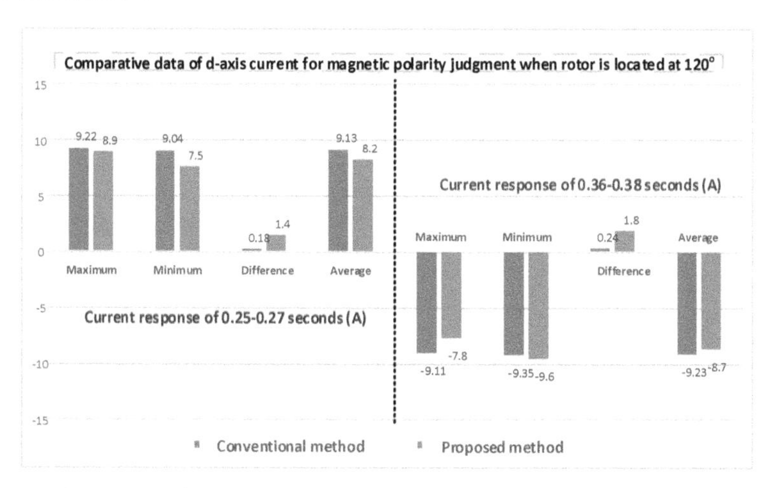

Figure 24. Comparative data of d-axis current characteristics for magnetic polarity judgment when rotor is located at 120°.

Therefore, the method of magnetic polarity identification proposed in this paper is not only suitable for the case where the motor is stationary, but also suitable for the free running condition of the motor. As can be seen from the Figures 21 and 22, the effect of only injecting DC bias voltage on the initial position observation is not very obvious either from the difference of the extreme value of the d-axis current response or from the average value of the steady-state current of the d-axis. If the burr of the current is large, it is easy to affect the judgment result. In 50 repetitive experiments, the probability of method (a) judging errors was about 10%. For method (b), the discrimination is more obvious from the difference of high frequency response of current. In 50 repetitive experiments, the probability of misjudgment was 0, which directly proved that the improved direction judgment of magnetic polarity method proposed in this paper is effective.

5. Conclusions

In this paper, a HF square-wave voltage injection scheme-based position sensorless control of IPMSM in the low-speed range and a new method of magnetic polarity detection in zero-speed range are proposed. The strategy realizes the switch between zero-speed algorithm and low-speed algorithm for sensorless control of IPMSM. Since the proposed method is an improvement on the conventional square-wave injection method, the method is applicable to AC motors with salient pole effect that can be applied in the conventional method. Furthermore, the voltage error caused by the nonlinearity of the inverter are compensated to improve the accuracy of rotor position estimation. In addition, based on the principle analysis of d-axis magnetic circuit characteristics, a method for determining the direction of magnetic polarity of d-axis two-opposite DC voltage offset by uninterruptible square-wave injection is proposed, which is fast in the convergence rate of magnetic polarity detection Compared with the conventional method, the new method is more distinct and the success rate is higher. Finally, comprehensive simulation and experimental results verify that the improved HF square-wave voltage injection method and the detection of magnetic polarity method perform faster and more accurately compared with the conventional method.

Author Contributions: Conceptualization, S.W. and J.Z.; methodology, S.W. and J.Z.; software, K.Y.; validation, S.W., J.Z. and K.Y.; formal analysis, W.S.; data curation, K.Y.; writing—original draft preparation, S.W. and J.Z.; writing—review and editing, S.W. and J.Z. All authors gave advice for the manuscript.

Funding: This work was supported in part by the National Key Research and Development Program of China under Grant 2018YFB0104801.

Conflicts of Interest: The authors declare no conflict of interest.

References

1. Onambele, C.; Elsied, M.; Mpanda Mabwe, A.; El Hajjaji, A. Multi-Phase Modular Drive System: A Case Study in Electrical Aircraft Applications. *Energies* **2018**, *11*, 5. [CrossRef]
2. Zhang, C.; Guo, Q.; Li, L. System Efficiency Improvement for Electric Vehicles Adopting a Permanent Magnet Synchronous Motor Direct Drive System. *Energies* **2017**, *10*, 2030. [CrossRef]
3. Hua, W.; Zhou, K.L. Investigation of a Co-Axial Dual-Mechanical Ports Flux-Switching Permanent Magnet Machine for Hybrid Electric Vehicles. *Energies* **2015**, *8*, 14361–14379. [CrossRef]
4. Wang, S.; Yang, K.; Chen, K. An Improved Position-Sensorless Control Method at Low Speed for PMSM Based on High-Frequency Signal Injection into a Rotating Reference Frame. *IEEE Access* **2019**, *7*, 86510–86521. [CrossRef]
5. Zhang, X.; Hou, B.; Mei, Y. Deadbeat Predictive Current Control of Permanent-Magnet Synchronous Motors with Stator Current and Disturbance Observer. *IEEE Trans. Power Electron.* **2017**, *32*, 3818–3834. [CrossRef]
6. Park, Y.; Sul, S. Sensorless Control Method for PMSM Based on Frequency-Adaptive Disturbance Observer. *IEEE J. Emerg. Sel. Top. Power Electron.* **2014**, *2*, 143–151. [CrossRef]
7. Zhang, G.; Wang, G.; Xu, D.; Zhao, N. ADALINE-Network-Based PLL for Position Sensorless Interior Permanent Magnet Synchronous Motor Drives. *IEEE Trans. Power Electron.* **2016**, *31*, 1450–1460. [CrossRef]
8. Nguyen, H.Q.; Yang, S. Rotor Position Sensorless Control of Wound-Field Flux-Switching Machine Based on High Frequency Square-Wave Voltage Injection. *IEEE Access* **2018**, *6*, 48776–48784. [CrossRef]
9. Wang, T.; Zhang, H.; Gao, Q.; Xu, Z.; Li, J.; Gerada, C. Enhanced Self-Sensing Capability of Permanent-Magnet Synchronous Machines: A Novel Saliency Modulation Rotor End Approach. *IEEE Trans. Ind. Electron.* **2017**, *64*, 3548–3556. [CrossRef]
10. Wang, G.; Yang, L.; Yuan, B.; Wang, B.; Zhang, G.; Xu, D. Pseudo-Random High-Frequency Square-Wave Voltage Injection Based Sensorless Control of IPMSM Drives for Audible Noise Reduction. *IEEE Trans. Ind. Electron.* **2016**, *63*, 7423–7433. [CrossRef]
11. Wang, G.; Yang, L.; Zhang, G.; Zhang, X.; Xu, D. Comparative Investigation of Pseudorandom High-Frequency Signal Injection Schemes for Sensorless IPMSM Drives. *IEEE Trans. Power Electron.* **2017**, *32*, 2123–2132. [CrossRef]
12. Wang, G.; Xiao, D.; Zhao, N.; Zhang, X.; Wang, W.; Xu, D. Low-Frequency Pulse Voltage Injection Scheme-Based Sensorless Control of IPMSM Drives for Audible Noise Reduction. *IEEE Trans. Ind. Electron.* **2017**, *64*, 8415–8426. [CrossRef]

13. Almarhoon, A.H.; Zhu, Z.Q.; Xu, P. Improved Rotor Position Estimation Accuracy by Rotating Carrier Signal Injection Utilizing Zero-Sequence Carrier Voltage for Dual Three-Phase PMSM. *IEEE Trans. Ind. Electron.* **2017**, *64*, 4454–4462. [CrossRef]
14. Seilmeier, M.; Piepenbreier, B. Sensorless Control of PMSM for the Whole Speed Range Using Two-Degree-of-Freedom Current Control and HF Test Current Injection for Low-Speed Range. *IEEE Trans. Power Electron.* **2015**, *30*, 4394–4403. [CrossRef]
15. Zhang, G.; Wang, G.; Yuan, B.; Liu, R.; Xu, D. Active Disturbance Rejection Control Strategy for Signal Injection-Based Sensorless IPMSM Drives. *IEEE Trans. Transp. Electrif.* **2018**, *4*, 330–339. [CrossRef]
16. Luo, X.; Tang, Q.; Shen, A.; Zhang, Q. PMSM Sensorless Control by Injecting HF Pulsating Carrier Signal into Estimated Fixed-Frequency Rotating Reference Frame. *IEEE Trans. Ind. Electron.* **2016**, *63*, 2294–2303. [CrossRef]
17. Yoon, Y.; Sul, S.; Morimoto, S.; Ide, K. High-Bandwidth Sensorless Algorithm for AC Machines Based on Square-Wave-Type Voltage Injection. *IEEE Trans. Ind. Appl.* **2011**, *47*, 1361–1370. [CrossRef]
18. Ni, R.; Xu, D.; Blaabjerg, F.; Lu, K.; Wang, G.; Zhang, G. Square-Wave Voltage Injection Algorithm for PMSM Position Sensorless Control with High Robustness to Voltage Errors. *IEEE Trans. Power Electron.* **2017**, *32*, 5425–5437. [CrossRef]
19. Kim, S.; Ha, J.; Sul, S. PWM Switching Frequency Signal Injection Sensorless Method in IPMSM. *IEEE Trans. Ind. Appl.* **2012**, *48*, 1576–1587. [CrossRef]
20. Park, N.; Kim, S. Simple sensorless algorithm for interior permanent magnet synchronous motors based on high-frequency voltage injection method. *IET Electr. Power Appl.* **2014**, *8*, 68–75. [CrossRef]
21. Chen, L.; Götting, G.; Dietrich, S.; Hahn, I. Self-Sensing Control of Permanent-Magnet Synchronous Machines with Multiple Saliencies Using Pulse-Voltage-Injection. *IEEE Trans. Ind. Appl.* **2016**, *52*, 3480–3491. [CrossRef]
22. Hu, J.; Liu, J.; Xu, L. Eddy Current Effects on Rotor Position Estimation and Magnetic Pole Identification of PMSM at Zero and Low Speeds. *IEEE Trans. Power Electron.* **2008**, *23*, 2565–2575. [CrossRef]
23. Jin, X.; Ni, R.; Chen, W.; Blaabjerg, F.; Xu, D. High-Frequency Voltage-Injection Methods and Observer Design for Initial Position Detection of Permanent Magnet Synchronous Machines. *IEEE Trans. Power Electron.* **2018**, *33*, 7971–7979. [CrossRef]
24. Xie, G.; Lu, K.; Dwivedi, S.K.; Rosholm, J.R.; Blaabjerg, F. Minimum-Voltage Vector Injection Method for Sensorless Control of PMSM for Low-Speed Operations. *IEEE Trans. Power Electron.* **2016**, *31*, 1785–1794. [CrossRef]
25. Liu, Y.; Zhou, B.; Li, S.; Feng, Y. Initial Rotor Position Detection of Surface Mounted Permanent Magnet Synchronous Motor. *Proc. CSEE* **2011**, *31*, 48–54.
26. Yan-Jun, Y.U.; Gao, H.W.; Chai, F.; Cheng, S.K. Rotor magnetic polarity detection method for PMSM. *Elect. Mach. Control.* **2011**, *15*, 86–90.
27. Jeong, Y.S.; Lorenz, R.D.; Jahns, T.M.; Sul, S.K. Initial rotor position estimation of an interior permanent-magnet synchronous machine using carrier-frequency injection methods. *IEEE Trans. Ind. Appl.* **2005**, *41*, 38–45. [CrossRef]
28. Raute, R.; Caruana, C.; Staines, C.S.; Cilia, J.; Sumner, M.; Asher, G. Operation of a sensorless PMSM drive without additional test signal injection. In Proceedings of the 2008 4th IET Conference on Power Electronics, Machines and Drives, York, UK, 2–4 April 2008; pp. 616–620.

Article

Wound Synchronous Machine Sensorless Control Based on Signal Injection into the Rotor Winding

Jongwon Choi and Kwanghee Nam *

Department of Electrical Engineering, Pohang University of Science and Technology, 77 Cheongam-Ro, Nam-Gu, Pohang 37673, Korea; jongwon@postech.ac.kr

* Correspondence: kwnam@postech.ac.kr; Tel.: +82-54-279-5628

Received: 29 October 2018; Accepted: 21 November 2018; Published: 24 November 2018

Abstract: A sensorless position scheme was developed for wound synchronous machines. The demodulation process is fundamentally the same as the conventional signal-injection method. The scheme is different from techniques for permanent-magnet synchronous machines, in that it injects a carrier signal into the field (rotor) winding. The relationship between the high-frequency current responses and the angle estimation error was derived with cross-coupling inductances. Furthermore, we develop a compensation method for the cross-coupling effect, and present several advantages of the proposed method in comparison with signal injection into the stator winding. This method is very robust against magnetic saturation because it does not depend on the saliency of the rotor. Furthermore, the proposed method does not need to check the polarity at a standstill. Experiments were performed to demonstrate the improvement in the compensation of cross-coupling, and the robustness against magnetic saturation with full-load operation.

Keywords: core saturation; cross-coupling inductance; wound synchronous machines (WSM); signal injection; position sensorless; high-frequency model

1. Introduction

Although their speed is not so high, electrical vehicles (EVs) and hybrid EVs (HEVs) are steadily growing their share in the market. Today, permanent-magnet synchronous machines (PMSMs) are widely used in traction applications because of their superior power density and high efficiency. However, the permanent-magnet (PM) materials, typically neodymium (Nd) and dysprosium (Dy), are expensive, and their price fluctuates depending on political situations. Therefore, some research has been directed to developing Nd-free motors. Wound synchronous machines (WSM) is a viable alternative to a PMSM. The main advantage of a WSM is that it has an additional degree of freedom in the field-weakening control because the rotor field can be adjusted [1–3].

There are two types of sensorless angle detection techniques: back-EMF-based and signal-injection methods. The former is based on the relative magnitudes of d and q-axis EMFs, whereas the latter is based on the spatial saliency of rotor. The back-EMF-based methods are reliable and superior in the medium- and high-speed regions [4–10]. However, they exhibit poor performance in the low-speed region owing to lack of the "observability" [11].

On the other hand, signal-injection methods work well in the zero-speed region, even with a full load [12–19]. The injection method does not use the magnetic polarity; it requires the use of a polarity-checking method before starting [14,15]. The signal-injection method is not feasible in some saturation regions where the d and q-axis inductances are close to each other [16,17]. The cross-coupling inductance refers to an incremental inductance developed by the current in the quadrature position. Cross-coupling, being another saturation phenomenon, becomes significant as the load increases. Zhu et al. [18] showed that cross-coupling caused an offset error in the angle estimation and proposed a method eliminating it. However, the cross-coupling inductances change considerably depending on the current magnitudes. Therefore, a lookup table should be used for its compensation method. A group of researchers are working together toward developing specialized motors that are suitable for signal injection [19].

Similar to PMSMs, back-EMF-based sensorless methods were developed for WSMs [20–23]. Boldea et al. [22] proposed an active-flux-based sensorless method, and this paper presents good experimental results in low-speed operation with a heavy load. Amit et al. [23] used a flux observer in the stator flux coordinate. Recently, the sensorless signal injection for WSM was published, whose carrier signal is injected to an estimated d-axis, and the response in estimated q-axis current [24,25]. Griffo et al. [24] applied a signal-injection method to a WSM to start an aircraft engine, and presented full-torque operation from zero to a high speed. Rambetius et al. [25] compared two detection methods when a signal was injected into the stator winding: one from a stator winding and the other from the field (rotor) winding.

Signal injection to the rotor winding of the WSM has been reported in recent years [26–29]. Obviously, detected signals from the (stator) d-q axes differ depending on rotor position. The stator voltage responses [26] and current responses [27–29] are checked to obtain rotor angle information using mutual magnetic coupling between the field coil and the stator coil. Using inverse sine function and q-axis current in the estimated frame, the position estimation algorithm was presented for a sensorless direct torque control WSM and it presented experimental results at zero speed [27]. Rambetius and Piepenbreier [28,29], included cross-saturation effects in the high-frequency model and presented the position estimation method using q-axis current in the estimation synchronous frame and linearization. The model was included in the stator but also in the field dynamics. It is a reasonable approach when a voltage as the carrier signal is injected to rotor winding. However, a 3×3 inductance matrix should be handled, and it is pretty complicated and difficult to analyze.

This paper extends the work in [30]. The signal is injected into the rotor winding, and the resulting high-frequency is detected from the stator currents. The effect of the cross-coupling inductances is modeled. Since the field current is modeled as a current source, 2×2 inductance matrix can be obtained. It is easy to calculate the inverse matrix and analyze the effect of the cross-coupling inductance on rotor angle estimation. The dq-axes stator flux linkage are obtained by finite-element analysis. Using the flux linkage, the dq-axes self-inductances and the mutual inductances between the stator and rotor are obtained and analyzed. Then, an offset angle caused by the cross-coupling inductances is straightforwardly derived. Using both dq-axes currents in the misaligned frame and inverse tangent function, the rotor angle estimation algorithm is developed without linearization. The offset angle caused by cross-coupling effect is directly compensated, and the stability issue of the compensation method is analyzed. Furthermore, it explains why injection into the rotor winding is more robust than the existing methods. Finally, experimental results verify that the rotor position is obtained accurately at standstill and very low speed.

This paper is organized as follows. In Section 2, the WSM model is derived with coupling inductances. In Section 3, the current responses in misaligned coordinate are derived and a sensorless method is proposed. Some advantages of the field signal-injection method are presented in comparison with stator signal injection in Section 4. Section 5 presents a performance comparison between the injection methods to stator and rotor. In Section 6, the performance of the sensorless method is demonstrated by experiment. Finally, in Section 7, some conclusions are drawn.

2. Modeling of a WSM

A schematic diagram of a WSM is depicted in Figure 1. Please note that a high-frequency signal is injected into the field winding via a slip ring.

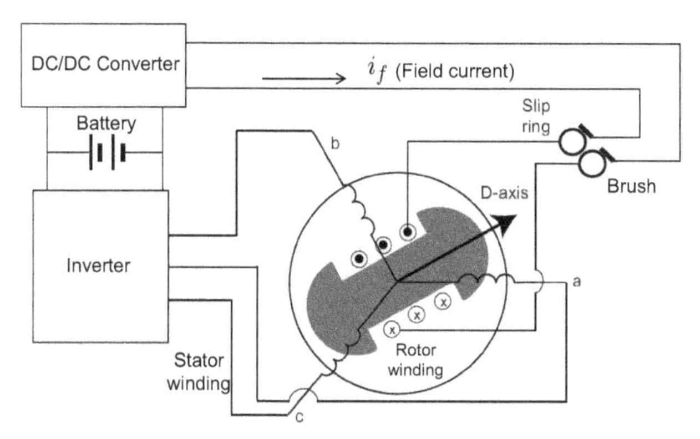

Figure 1. Schematic Diagram of WSM for EV traction.

2.1. WSM Voltage Model

From [31] (in Chapter 11), the voltage equations of WSM are given as

$$v_d^e = r_s i_d^e + \frac{d}{dt}\lambda_d^e - \omega_e \lambda_q^e, \tag{1}$$

$$v_q^e = r_s i_q^e + \frac{d}{dt}\lambda_q^e + \omega_e \lambda_d^e, \tag{2}$$

where v_d^e, v_q^e are the d and q-axis voltages and ω_e is the electrical angular frequency; the superscript e represents the synchronous (rotating) reference frame. Note that $\frac{d}{dt}\lambda_d^e$ and $\frac{d}{dt}\lambda_q^e$ are important terms for sensorless signal-injection methods because the terms are represented as the induced voltage caused by the high-frequency current.

The stator flux linkages of a WSM in the synchronous frame are expressed as $\lambda_d^e(i_d^e, i_q^e, i_f')$, and $\lambda_q^e(i_d^e, i_q^e, i_f')$, where λ_d^e and λ_q^e are non-linear functions [3]; λ_d^e and λ_q^e are the d and q-axes stator flux linkage; i_d^e and i_q^e are the d and q-axes currents; and i_f' is the field current referred to stator. Note that i_f' is represented by $i_f' = \frac{2}{3}\frac{N_s}{N_{fd}} i_f$ [31], where i_f is the field current, N_s and N_{fd} are the number of stator coil and field coil turns. To take account of the cross-coupling effect, the following equations are derived by the chain rule

$$\frac{d\lambda_d^e}{dt} = \underbrace{\frac{\partial \lambda_d^e}{\partial i_d^e}}_{\equiv L_{dd}} \frac{di_d^e}{dt} + \underbrace{\frac{\partial \lambda_d^e}{\partial i_q^e}}_{\equiv L_{dq}} \frac{di_q^e}{dt} + \underbrace{\frac{\partial \lambda_d^e}{\partial i_f'}}_{\equiv L_{df}} \frac{di_f'}{dt} \tag{3}$$

$$\frac{d\lambda_q^e}{dt} = \underbrace{\frac{\partial \lambda_q^e}{\partial i_q^e}}_{\equiv L_{qq}} \frac{di_q^e}{dt} + \underbrace{\frac{\partial \lambda_q^e}{\partial i_d^e}}_{\equiv L_{qd}} \frac{di_d^e}{dt} + \underbrace{\frac{\partial \lambda_q^e}{\partial i_f'}}_{\equiv L_{qf}} \frac{di_f'}{dt}, \tag{4}$$

where L_{dd} and L_{qq} are the self (incremental) inductances, and L_{dq} and L_{qd} are the cross-coupling (incremental) inductances of the stator coil [16]. L_{df} is the mutual (incremental) inductance between the d-axis and the field coils, and L_{qf} is the cross-coupling (incremental) inductance between the stator and field coils. The d-axis inductance can be decomposed as the sum of the mutual and leakage inductances:

$$L_{dd} = L_{df} + L_{ls}, \tag{5}$$

where L_{ls} is the leakage inductance of the stator. Therefore, $L_{dd} \approx L_{df} \gg L_{qf}$.

Substituting (3) and (4) into (1) and (2), the voltage equations are obtained as

$$v_d^e = r_s i_d^e + L_{dd}\frac{di_d^e}{dt} + L_{dq}\frac{di_q^e}{dt} + L_{df}\frac{di_f'}{dt} - \omega_e\lambda_q^e \quad (6)$$

$$v_q^e = r_s i_q^e + L_{qq}\frac{di_q^e}{dt} + L_{qd}\frac{di_d^e}{dt} + L_{qf}\frac{di_f'}{dt} + \omega_e\lambda_d^e. \quad (7)$$

Using (5)–(7), an equivalent circuit can be constructed as shown in Figure 2.

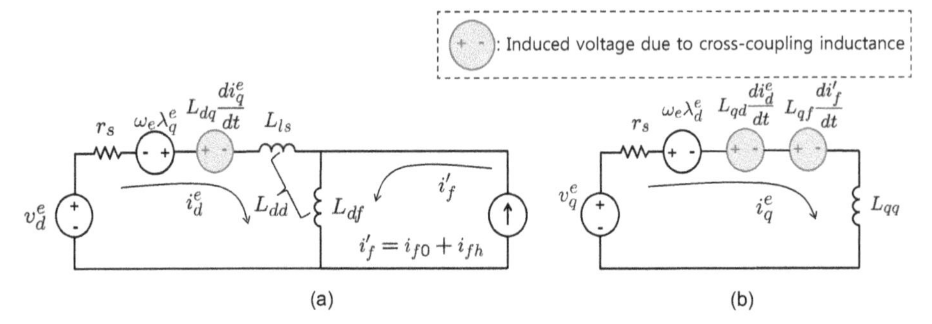

Figure 2. Equivalent circuit of the WSM including cross-coupling inductance: (**a**) d-axis and (**b**) q-axis.

Figure 3a shows a FEM model of WSM used in this experiment. The motor has six salient poles, the continuous rated current is 161 Arms, and the maximum power is 65 kW. The other parameters are listed in Table 1. Figure 3b shows plots of the a-, b-, c-phase flux linkages in the stationary frame as the rotor rotates at 500 r/min, which were obtained through finite-element method (FEM) calculations. The $d-q$ axis flux linkages are obtained using

$$\begin{bmatrix} \lambda_d^e \\ \lambda_q^e \end{bmatrix} = \frac{2}{3}\begin{bmatrix} \cos\theta_e & \sin\theta_e \\ -\sin\theta_e & \cos\theta_e \end{bmatrix}\begin{bmatrix} 1 & -\frac{1}{2} & -\frac{1}{2} \\ 0 & \frac{\sqrt{3}}{2} & -\frac{\sqrt{3}}{2} \end{bmatrix}\begin{bmatrix} \lambda_a \\ \lambda_b \\ \lambda_c \end{bmatrix}, \quad (8)$$

where θ_e is the electrical angle. Figure 4 shows the plots of $\lambda_d^e(i_d^e, i_q^e, i_f)$ and $\lambda_q^e(i_d^e, i_q^e, i_f)$ in the (i_d^e, i_q^e) plane when $i_f = 6$ A. Please note that λ_d^e changes more, i.e., the slope becomes steeper as the d-axis current increases negatively. When i_q^e is under 50 A, i_q^e has little effect on λ_d^e. However, when i_q^e is over 100 A, λ_d^e decreases more as i_q^e increases. It is called "cross-coupling phenomenon". Figure 5 shows a contour of $\lambda_d^e(i_d^e, i_q^e, i_f)$ and $\lambda_q^e(i_d^e, i_q^e, i_f)$ when $i_f = 6$ A or $i_f = 6.5$ A. λ_d^e seems to be proportional to i_f. But, in Figure 5b, λ_q^e with $i_f = 6$ A seems to be the same λ_q^e with $i_f = 6.5$ A when i_q^e is under 50 A. But, it is clear that λ_q^e decreases as i_q^e increases when i_q^e is over 100 A.

Table 1. Parameters of a WSM used in the experiments.

Parameter	Value
Maximum power	65 kW
Maximum torque	123 Nm
Maximum current	161 Arms
Numbers of poles (P)	6 poles
Number of slots	36 slot
Back-EMF coefficient	0.121 Wb
Maximum speed	12,000 r/min
Field current (i_f)	6 A
Number of stator coil turns (N_s)	3 turns
Number of field coil turns (N_{fd})	200 turns

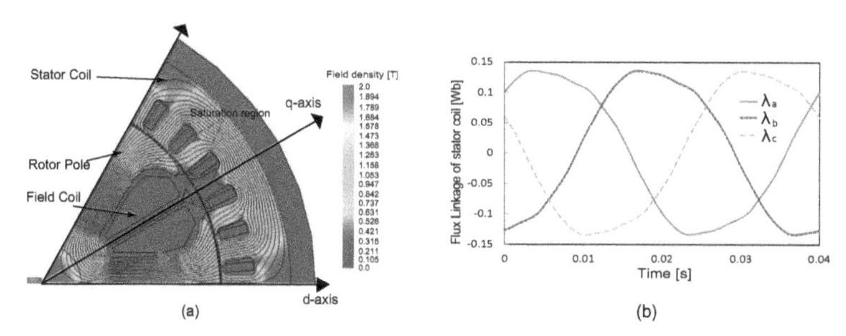

Figure 3. Flux linkage calculation when $i_f = 6$ A: (**a**) Finite-element analysis model and (**b**) flux linkages of stator coils.

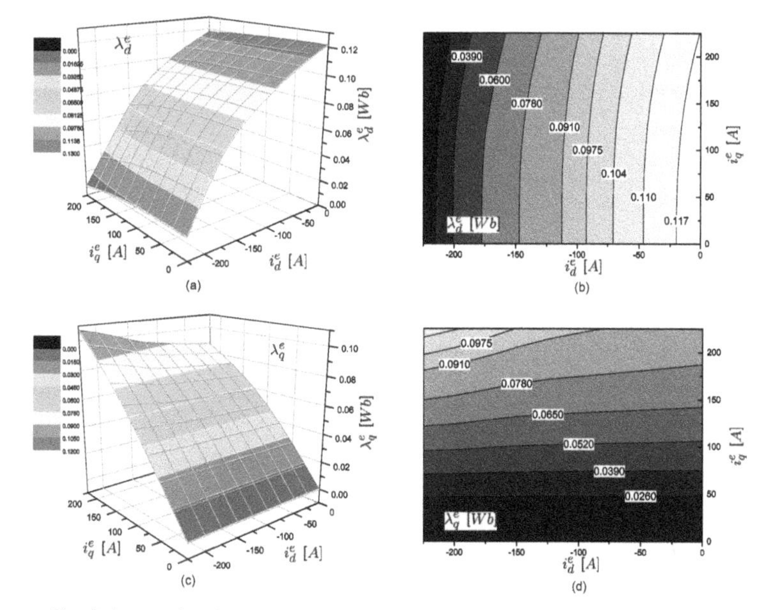

Figure 4. Flux linkages when $i_f = 6$ A (FEM analysis): (**a**) 3-dimensional plot of λ_d^e; (**b**) contour plot of λ_d^e; (**c**) 3-dimensional plot of λ_q^e; and (**d**) contour plot of λ_q^e.

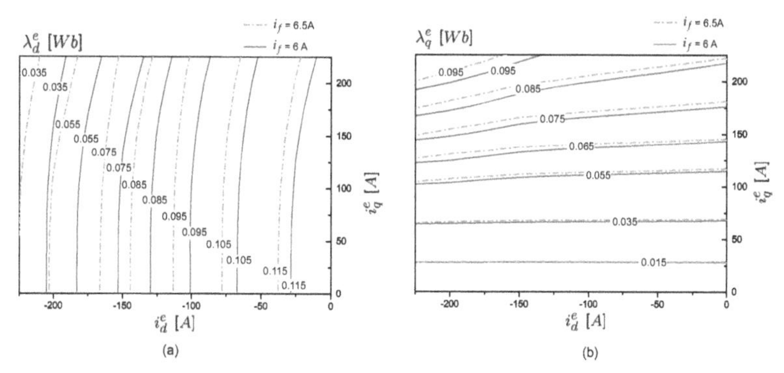

Figure 5. The Influence of the field current on dq-axes stator flux linkage (λ_d^e, λ_q^e): (**a**) Contour plot λ_d^e, and (**b**) contour plot of λ_d^e. (solid and blue line: i_f = 6 A, dash dot and magenta line: i_f = 6.5 A).

2.2. Incremental Inductance Calculations

As shown in [18,32], the incremental inductances are calculated through numerical differentiation:

$$L_{dq} = \frac{\partial \lambda_d^e}{\partial i_d^e} \approx \frac{\lambda_d^e(i_d^e, i_q^e + \Delta i_q^e, i_f) - \lambda_d^e(i_d^e, i_q^e, i_f)}{\Delta i_q^e}$$

$$L_{df} = \frac{\partial \lambda_d^e}{\partial i_f^e} \approx \frac{\lambda_d^e(i_d^e, i_q^e, i_f + \Delta i_f) - \lambda_d^e(i_d^e, i_q^e, i_f)}{\Delta i_f}.$$

The rest of the inductances are obtained similarly. Figure 6 shows L_{dd}, L_{qq}, and L_{dq} under various current conditions. From Figure 6a, L_{dd} is, in general, independent of the q-axis current when i_q is not so high, whereas it depends strongly on the d-axis current, i.e., L_{dd} increases as i_d^e increases negatively. This is because the rotor core becomes free from the saturation caused by the field current. Specifically, the negative d-axis current induces a field in the opposite direction, i.e., it cancels the rotor field.

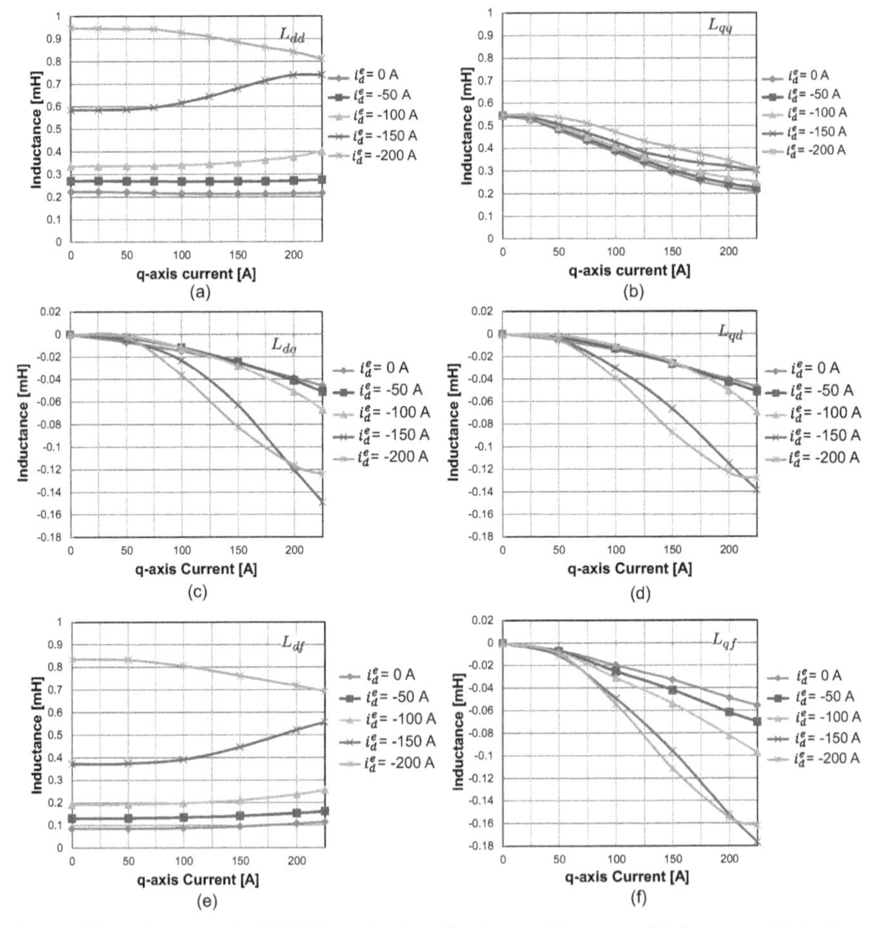

Figure 6. Finite-element method (FEM) data for the self and mutual incremental inductances: (**a**) d-axis self-inductance L_{dd}; (**b**) q-axis self-inductance L_{qq}; (**c**) d-axis and q-axis cross-coupling inductance L_{dq}; (**d**) q-axis and d-axis cross-coupling inductance L_{qd}; (**e**) d-axis and field coil mutual inductance L_{df}; and (**f**) q-axis and field coil cross-coupling inductance L_{qf}.

In addition, also from Figure 6b–d, L_{qq}, L_{dq}, and L_{qd} decrease as i_q^e increases. Figure 6c,f show plots of L_{df} and L_{qf} that show the mutual inductances from the field winding, i.e., the flux variation induced by the field current. Because the high-frequency probing signal is injected into the field coil, L_{df} and L_{qf} play a crucial role in this work. Comparing Figure 6a,c, the shapes of L_{dd} and L_{df} are similar. It is because they are the same except the leakage inductance, as shown in (5). Note from Figure 6c,d,f, the cross-coupling inductances (L_{dq}, L_{qd}, L_{qf}) behave similarly and have negative values. Because of the non-linear magnetic property of core, the core saturation give rise to cross-coupling phenomenon. For instance, when the field current is fixed at 6 A, $\lambda_{d(i_d^e=0, i_q^e=50)} = 0.121345$ Wb and $\lambda_{d(i_d^e=0, i_q^e=200)} = 0.118252$ Wb were obtained in Figure 4b. Despite the same d-axis current, the d-axis flux linkages decrease as i_q^e increases. This is because the q-axis current saturates the core deeper, and it affects d-axis flux linkages to be reduced. Consequently, the cross-coupling inductance between d and q-axis, L_{dq}, can be negative values, the phenomenon is also exhibited in PMSMs [18]. Correspondingly, the q-axis flux linkages decrease as i_f or i_d^e increases in Figures 4d and 5b. Therefore, L_{qd} and L_{qf} have negative values.

When a signal is injected into the field winding, the current responses are monitored in the stator winding via L_{df} and L_{qf}. The use of L_{qf} is different from PMSM signal-injection methods because it provides another signal path to the q-axis besides the one formed by rotor saliency.

3. High-Frequency Model of a WSM

As shown in Figure 1, the WSM has a separate field controller, which feeds i_f to the field winding via a slip ring and brush. A high-frequency carrier is superposed on the field current. Then, the signal is detected on the stator winding, on which the rotor angle is estimated. The field current controller supplies

$$i_f'(t) = I_{f0} + i_{fh}(t) = I_{f0} - I_h \cos\omega_h t, \tag{9}$$

where I_{f0} and I_h are the amplitudes of the dc and ac components, and ω_h is the angular speed of the carrier. Furthermore, the d and q-axis currents can be separated as

$$i_d^e = i_{d0}^e + i_{dh}^e, \tag{10}$$

$$i_q^e = i_{q0}^e + i_{qh}^e, \tag{11}$$

where i_{d0}^e and i_{q0}^e are dc components current, i_{dh}^e and i_{qh}^e are the high-frequency components. Substituting (9)–(11) into (6) and (7), the voltage equations are written as

$$\begin{bmatrix} v_d^e \\ v_q^e \end{bmatrix} = \begin{bmatrix} r_s & 0 \\ 0 & r_s \end{bmatrix} \begin{bmatrix} i_d^e \\ i_q^e \end{bmatrix} + \begin{bmatrix} L_{dd} & L_{dq} \\ L_{qd} & L_{qq} \end{bmatrix} \frac{d}{dt} \begin{bmatrix} i_{d0} + i_{dh}^e \\ i_{q0} + i_{qh}^e \end{bmatrix} + \begin{bmatrix} -\omega_e \lambda_q^e \\ \omega_e \lambda_d^e \end{bmatrix} + \begin{bmatrix} L_{df} \\ L_{qf} \end{bmatrix} \frac{d}{dt}(I_{f0} - I_h \cos\omega_h t). \tag{12}$$

Please note that i_{dh}^e and i_{qh}^e are induced by the high-frequency part of i_f'. From (12), the terms $r_s i_d^e$, $r_s i_q^e$, $-\omega_e\lambda_q^e$, and $\omega_e\lambda_d^e$ are neglected because ω_h is much larger than ω_e, $\omega_h L_{dd} \gg r_s$, and $\omega_h L_{qq} \gg r_s$. Thus, it is following that

$$\begin{bmatrix} v_d^e \\ v_q^e \end{bmatrix} = \begin{bmatrix} L_{dd} & L_{dq} \\ L_{qd} & L_{qq} \end{bmatrix} \frac{d}{dt} \begin{bmatrix} i_{d0}^e + i_{dh}^e \\ i_{q0}^e + i_{qh}^e \end{bmatrix} + \begin{bmatrix} L_{df} \\ L_{qf} \end{bmatrix} \frac{d}{dt}(I_{f0} - I_h \cos\omega_h t). \tag{13}$$

Please note that L_{dd}, L_{qq}, L_{dq}, L_{qd}, L_{df}, and L_{qf} are the incremental inductance in Section 2.2. It means that the inductances are calculated at a specific operating point, $(i_{d0}^e, i_{q0}^e, I_{f0})$. Therefore, all inductances can be assumed the constant value at the operation point. Based on the superposition law at $(i_{d0}^e, i_{q0}^e, I_{f0})$, the high-frequency part is separated as

$$\begin{bmatrix} 0 \\ 0 \end{bmatrix} = \begin{bmatrix} L_{dd} & L_{dq} \\ L_{qd} & L_{qq} \end{bmatrix} \frac{d}{dt} \begin{bmatrix} i_{dh}^e \\ i_{qh}^e \end{bmatrix} + \omega_h \begin{bmatrix} L_{df} \\ L_{qf} \end{bmatrix} I_h \sin \omega_h t. \tag{14}$$

Thus, the solution to (14) is obtained such that

$$\begin{bmatrix} i_{dh}^e(t) \\ i_{qh}^e(t) \end{bmatrix} = I_h \begin{bmatrix} \alpha_A \\ \alpha_B \end{bmatrix} \cos \omega_h t + \begin{bmatrix} i_{dh}^e(0) \\ i_{qh}^e(0) \end{bmatrix}, \tag{15}$$

where

$$\begin{aligned} \alpha_A &= \frac{L_{qq}L_{df} - L_{dq}L_{qf}}{L_{dd}L_{qq} - L_{dq}L_{qd}} \\ \alpha_B &= \frac{-L_{qd}L_{df} + L_{dd}L_{qf}}{L_{dd}L_{qq} - L_{dq}L_{qd}}. \end{aligned}$$

Please note that (15) is the equation in the synchronous (rotor) frame based on the right angle θ_e. As shown in Figure 7, the angle of the misaligned frame is denoted by $\hat{\theta}_e$, and the angle error is defined as

$$\Delta\theta_e = \hat{\theta}_e - \theta_e, \tag{16}$$

where $\Delta\theta_e \in \mathbb{S} \equiv [-\pi, \pi)$.

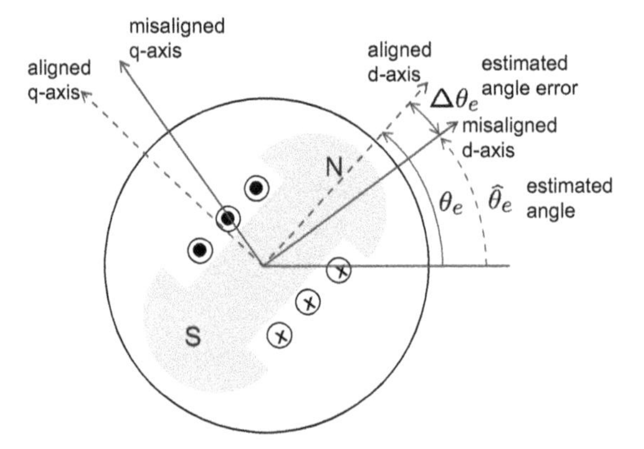

Figure 7. Misaligned *dq*-frame.

It is assumed that current is measured in a misaligned frame, in which the currents are denoted by $i_{dh}^{\hat{e}}$ and $i_{qh}^{\hat{e}}$. The current equation can be transformed to the misaligned frame by the following rotation matrix,

$$\mathbf{R}(\Delta\theta_e) = \begin{bmatrix} \cos(\Delta\theta_e) & \sin(\Delta\theta_e) \\ -\sin(\Delta\theta_e) & \cos(\Delta\theta_e) \end{bmatrix}. \tag{17}$$

By multiplying (17) to (15), we obtain

$$\begin{bmatrix} \hat{i}^e_{dh} \\ \hat{i}^e_{qh} \end{bmatrix} = I_h \begin{bmatrix} \alpha_A \cos\Delta\theta_e + \alpha_B \sin\Delta\theta_e \\ -\alpha_A \sin\Delta\theta_e + \alpha_B \cos\Delta\theta_e \end{bmatrix} \cos\omega_h t$$
$$= I_h\sqrt{\alpha_A^2+\alpha_B^2} \begin{bmatrix} \cos(\Delta\theta_e-\eta) \\ -\sin(\Delta\theta_e-\eta) \end{bmatrix} \cos\omega_h t, \quad (18)$$

where

$$\eta = \tan^{-1}(\frac{\alpha_B}{\alpha_A}) = \tan^{-1}\left(\frac{-L_{qd}L_{df}+L_{dd}L_{qf}}{L_{qq}L_{df}-L_{dq}L_{qf}}\right). \quad (19)$$

Please note that η is an angle offset caused by the cross-coupling inductances. Specifically, $\eta = 0$ when $L_{dq} = L_{qd} = L_{qf} = 0$, which means that η is caused by the cross-coupling.

4. Position Estimation Using Signal Injection into the Rotor Winding

Phase currents are measured and transformed into an estimated frame. Then, a band-pass filter (BPF) is applied to $\hat{i}^e_d$ and $\hat{i}^e_q$ to remove the dc components. To perform synchronous rectification, $\cos\omega_h t$ is multiplied with $\hat{i}^e_{dh}$ and $\hat{i}^e_{qh}$. Then, a low pass filter (LPF) is applied to extract the dc signals X and Y [33]:

$$X \equiv LPF(\hat{i}^e_{dh} \times \cos\omega_h t)$$
$$\approx \frac{I_h}{2}\sqrt{\alpha_A^2+\alpha_B^2}\cos(\Delta\theta_e-\eta) \quad (20)$$
$$Y \equiv LPF(\hat{i}^e_{qh} \times \cos\omega_h t)$$
$$\approx -\frac{I_h}{2}\sqrt{\alpha_A^2+\alpha_B^2}\sin(\Delta\theta_e-\eta). \quad (21)$$

Using the filtered signals, the angle error $\Delta\theta_e$ can be estimated via

$$\Delta\theta_e = -\tan^{-1}\left(\frac{Y}{X}\right) + \eta. \quad (22)$$

Figure 8 shows a block diagram of the signal processing illustrating in the above. A lookup table for η is used to compensate the bias depending on the load condition and current angle. Finally, a phase locked loop (PLL) is employed to obtain an estimate $\hat{\theta}_e$.

The bandwidth of the filter has an impact on the estimation bandwidth. To enhance the performance, the bandwidth of PLL and filter should be increased. However, the high-frequency current (i^e_{dh}, i^e_{qh}) can be contaminated with a noise [34]. In practice, the noise limits the bandwidth of the filter and PLL.

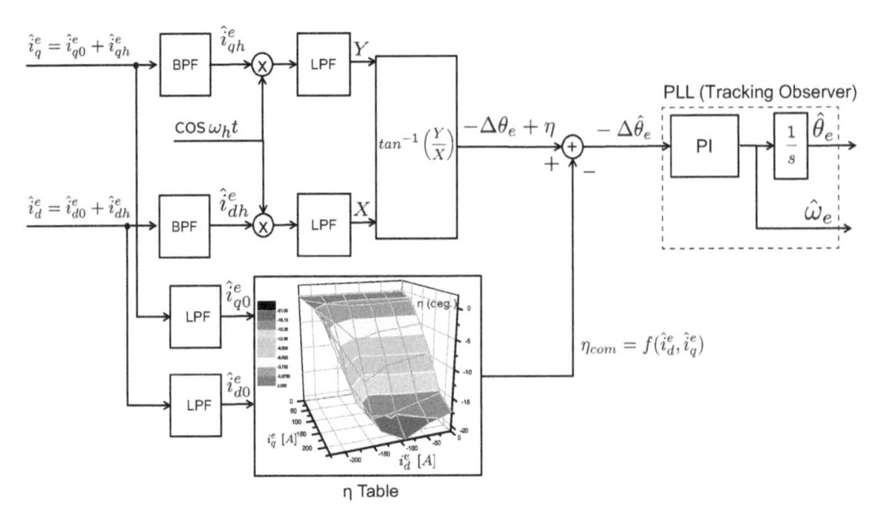

Figure 8. Signal processing block based on the signal-injection method into the field winding.

4.1. The Analysis of Cross-Coupling Offset Angle η

It is clear from (22) that the exact information of η is necessary to obtain an accurate value of $\Delta\theta$. On the other hand, η is a function of L_{dq}, L_{qd}, and L_{qf} in (19), which change nonlinearly with the currents. For practical purposes, it is better to make a lookup table of η over (i_d^e, i_q^e). Figure 9 shows the variations in η when the currents change. In general, the magnitudes of η increases as the core saturation develops. Note also that $|\eta|$ increases with the d-axis current until $i_d^e = -150$ A. However, it has the smallest values when $i_d^e = -200$ A. That situation could be illustrated as follows: The rotor flux generated by the field winding is almost canceled out by the negative d-axis current, $i_d^e = -200$ A. More specifically, note that the ampere-turn of the field winding is $N_{fd} i_f = 200 \times 6$ A−turns, where N_{fd} is the number of turns of the field winding. The ampere-turn of the d-axis stator winding is equal to $\frac{3}{2} \times N_a \times i_a = \frac{3}{2} \times 6 \times \frac{2}{3} \times 200 = 1200$ A-turns, where N_a is the number of turns of a phase winding per pole, and the winding factor is assumed to be unity. That is, they are the same when $i_d^e = -200$ A. Hence, the core is relieved from the saturation induced by the field winding, when $i_d^e = -200$ A. Thus, the non-linear behavior of η is mitigated when $i_d^e = -200$ A.

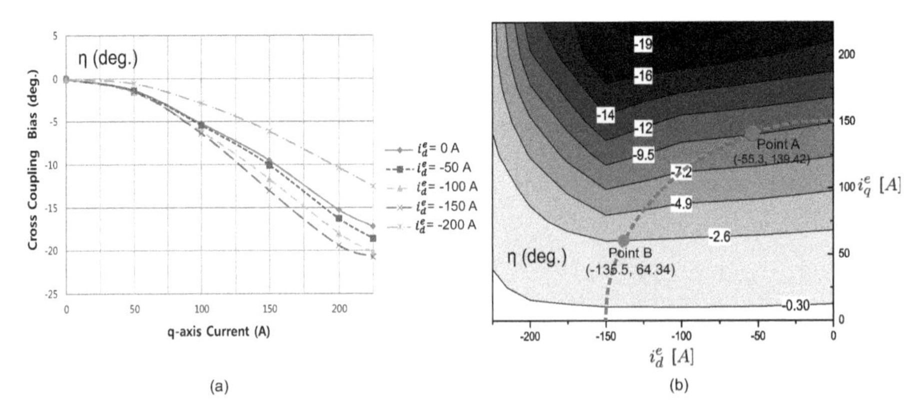

Figure 9. The cross-coupling bias η calculated based on the FEM data: (**a**) Plot of η and (**b**) contour plot of η.

4.2. The Compensation Method for Cross-Coupling Offset Angle η

The cross-coupling bias angle relies on the lookup table, which is the function of i_d^e and i_q^e. In practice, the dq-axes currents in the synchronous reference frame (i_d^e, i_q^e) are unknown in sensorless control. Thus, they are replaced by the dq-axes currents in misaligned frame, ($\hat{i}_d^e$, $\hat{i}_q^e$). However, the inaccurate compensation for η can be made by the difference between (i_d^e, i_q^e) and ($\hat{i}_d^e$, $\hat{i}_q^e$). Therefore, it has an effect on stability of the sensorless observer. To analyze the stability, define the function of η as

$$\eta = f(i_d^e, i_q^e), \tag{23}$$

f is shown in Figure 9. Using (17), the dq-axes currents in misaligned frame are derived as

$$\begin{aligned} \begin{bmatrix} \hat{i}_d^e \\ \hat{i}_q^e \end{bmatrix} &= \begin{bmatrix} \cos\Delta\theta_e & \sin\Delta\theta_e \\ -\sin\Delta\theta_e & \cos\Delta\theta_e \end{bmatrix} \begin{bmatrix} i_d^e \\ i_q^e \end{bmatrix}, \\ &= I_s \begin{bmatrix} -\sin(\beta - \Delta\theta_e) \\ \cos(\beta - \Delta\theta_e) \end{bmatrix}, \end{aligned} \tag{24}$$

where $I_s = \sqrt{i_d^{e\,2} + i_q^{e\,2}}$ and $\beta = \tan^{-1}(-i_d^e/i_q^e)$. Using (24), the compensation offset angle is obtained by $\eta_{com} = f(\hat{i}_d^e, \hat{i}_q^e)$. In Figure 8, $-\Delta\hat{\theta}_e$ is the input of the PLL (tracking filter). Subtracting $\tan^{-1}\frac{Y}{X}$ from the compensation angle η_{com}, $-\Delta\hat{\theta}_e$ can be calculated

$$\begin{aligned} -\Delta\hat{\theta}_e &= -\Delta\theta_e + \eta - \eta_{com} \\ &= -\Delta\theta_e + f(i_d^e, i_q^e) - f(\hat{i}_d^e, \hat{i}_q^e) \\ &= -\Delta\theta_e + f(-I_s\sin\beta, I_s\cos\beta) - f(-I_s\sin(\beta - \Delta\theta_e), I_s\cos(\beta - \Delta\theta_e)) \\ &= -\Delta\theta_e \underbrace{\left[1 + \frac{f(-I_s\sin(\beta - \Delta\theta_e), I_s\cos(\beta - \Delta\theta_e)) - f(-I_s\sin\beta, I_s\cos\beta)}{\Delta\theta_e}\right]}_{\kappa}. \end{aligned} \tag{25}$$

Please note that κ should be the positive value to ensure the stability of the PLL observer, i.e., $\kappa \geq 0$. If $\kappa < 0$, the estimated position error will be amplified. Therefore, our task is to prove that κ is the positive value in the whole operation region. However, it is difficult for analytical demonstration because f is not mathematically represented and is highly non-linear. Figure 9b shows the contour of the cross-bias angle, η. From (24), ($\hat{i}_d^e$, $\hat{i}_q^e$) are rotated by $\Delta\theta_e$ clockwise. It is evident that $f(\hat{i}_d^e, \hat{i}_q^e)$ slightly decreases as $\Delta\theta_e$ increases. It means that κ is maintained over 0, i.e., $\kappa \geq 0$. It is because $\frac{f(\hat{i}_d^e, \hat{i}_q^e) - f(i_d^e, i_q^e)}{\Delta\theta} > -1$. For example, the point B is (i_d^e, i_q^e) and the point A is $(\hat{i}_d^e, \hat{i}_q^e)$ in Figure 9b. β is 64.6° and $\Delta\theta$ is 42.9°. Substituting point A and B into (26), it was obtained as $-\Delta\hat{\theta}_e = -\Delta\theta_e(1 + \frac{-9.5° + 2.6°}{42.9°}) = -0.84\,\Delta\theta_e$. Therefore, the convergence of the estimated angle error can be locally guaranteed due to $\kappa \geq 0$. By contrast, PMSMs have a positive offset angle caused by cross-coupling inductance [15,18] when q-axis current is positive. It shows that the offset angle increases as the q-axis current increases. An offset angle compensation method for PMSM was reported [35], and this paper proposed two different estimation angles: the saliency-based angle and the estimation rotor angle (compensated). Consequently, double-synchronous frames should be used, it causes the increasing of the calculation burden. In comparison, the proposed method has only one estimated synchronous frame.

5. Performance Comparison between Rotor and Stator Injection

Sensorless control performance degrades as the load increases because the inductances vary significantly along with core saturation. Specifically, the saliency ratio decreases, i.e., $L_{dd} \approx L_{qq}$ [17,19]. The accuracy of a sensorless method is determined by signal-to-noise ratio, which the saliency ratio affects. In this section, a common method of injecting a signal into the stator is also considered for the purpose of comparison.

Rambetius et al. studied the WSM model by incorporating the effects of the field winding into the stator. For signal injection into the stator, the following inductances should be used [25]:

$$L_{ddt} = L_{dd} - \frac{3}{2}\frac{L_{df}^2}{(L_{df}+L_{lf})}, \tag{26}$$

$$L_{qqt} = L_{qq} - \frac{3}{2}\frac{L_{qf}^2}{(L_{df}+L_{lf})}, \tag{27}$$

$$L_{dqt} = L_{dq} - \frac{3}{2}\frac{L_{df}L_{qf}}{(L_{df}+L_{lf})}, \tag{28}$$

where $L_{dq} = L_{qd}$ is assumed, L_{lf} is the leakage inductance of field, and L_{ddt}, L_{qqt}, and L_{dqt} are substituted for L_{dd}, L_{qq}, and L_{dq}, respectively. Please note that $L_{ddt} \leq L_{dd}$ and $L_{qqt} \leq L_{qq}$, because the field coil acts as a damper winding and reduces the high-frequency component [25]. The angle error is estimated by

$$\Delta\theta_e \approx \frac{\omega_h}{v_h}\frac{L_{ddt}L_{qqt}}{L_{diff}} LPF(\hat{i}^e_{qh}\sin\omega_h t), \tag{29}$$

where v_h is a high-frequency voltage and $L_{diff} = \frac{L_{qqt}-L_{ddt}}{2}$. It is emphasized that L_{diff} plays a crucial role in the estimation [19]. A smaller error is expected for a larger value of L_{diff}.

Figure 10a shows the loci of constant L_{diff} in the current plane along with the maximum torque per ampere (MTPA) line. As mentioned in the above, L_{diff} decreases as i^e_q increases. In other words, the electromagnetic saliency ratio decreases as the load increases.

On the other hand, the proposed method does not depend on the saliency ratio. When the signal is injected into the field winding, L_{df} plays a similar role as L_{diff}. However, its magnitude is less affected by the current magnitudes. Figure 10b shows the loci of constant L_{df} in the current plane. L_{df} increases as i^e_d increases negatively. Also note that L_{df} increases slightly when i^e_q increases. This can be illustrated by a small increase in L_{df} along with i^e_q as shown in Figure 6c. This supports the robust property of the field coil injection method.

According to the saliency-based method, the angle error is recovered from the term $\sin(2\Delta\theta_e)$. Because "2" is multiplied with $\Delta\theta_e$, polarity check should be carried out. Therefore, before starting the saliency-based sensorless algorithm, a polarity-checking procedure needs to be performed. However, the angle error is estimated from $\sin(\Delta\theta_e - \eta)$ with the field current injection method; therefore, no polarity-checking step is necessary.

Another advantage is that the field current injection method does not undermine the PWM duty of an inverter, which should be used for motor operation. Normally with the stator injection method, approximately 50 Vpeak is used for high-frequency injection for a proper SNR in a 300 V dc-link inverter. However, for the field current injection method, a high-frequency signal is synthesized in a separate dc-dc converter.

For implementation, the field current injection method requires a DSP-based dc-dc converter which can produce a high-frequency signal with a dc bias. Two methods are compared in Table 2.

Table 2. Comparison between the injection methods into stator and field windings.

	Signal Injection into Stator Winding	Signal Injection into Field Winding
Signal amplitude	L_{diff}(Saliency)	L_{df}(Mutual)
Polarity check	Necessary	Not necessary
SNR under core saturation	Small	Large
Signal generation	Inverter	dc-dc converter
Stable region ($\Delta\theta_e$)	<45°	<90°
Implementation	Easy	Medium difficulty

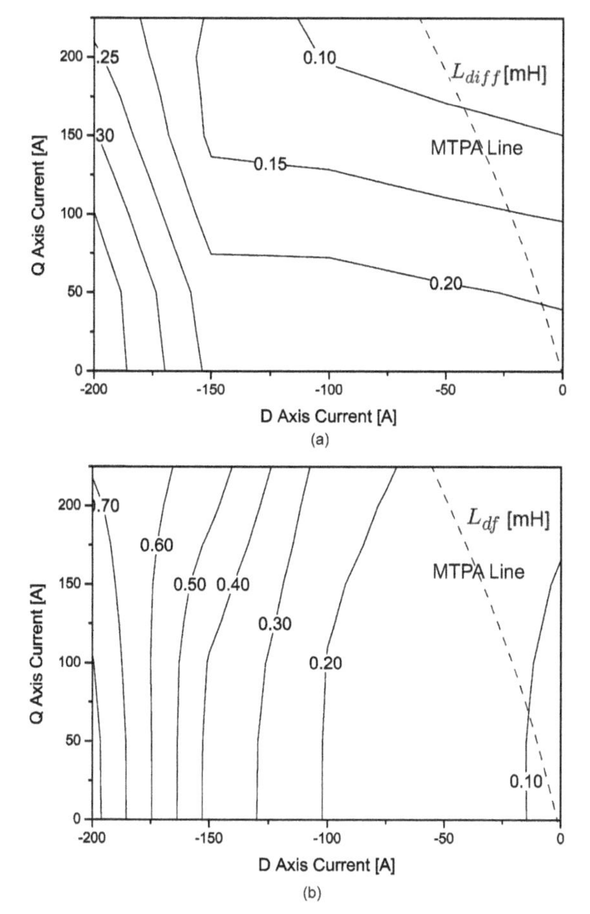

Figure 10. (**a**) L_{diff}, which is important for sensorless control based on signal injection into stator and (**b**) L_{df}, which is important for sensorless control based on signal injection into rotor.

6. Experimental Results

Figure 11 shows the experimental environment consisting of a dynamometer, a test WSM, an inverter, a dc to dc converter to supply the field current, etc. A zero-voltage switching (ZVS) full-bridge topology was used in the dc to dc converter, and operated at 50 kHz. As shown in Figure 12, the inverter dc-link voltage is 360 V and shared with the dc-dc converter. Practically, the field inductance is very large, and the bandwidth of field winding is very low. Consequently, it is difficult for some WSMs to inject high-frequency signal to the field winding. However, the field current supplier (DC-DC converter) is directly connected to the high voltage dc-link of the inverter.Therefore, it is enough to inject the high-frequency signal into the field winding. A carrier signal of 500 Hz, 90 V (peak-to-peak) was superposed on a dc output, which generated a 25 mA (peak-to-peak) current ripple on the dc component, $i_f = 6$ A. In truth, it cannot be guaranteed that I_h is a constant in the whole operation region due to the inductance variations from (9). Fortunately, the impedance between d-axis and rotor winding was not significantly changed in the MTPA operation. Therefore, it may have a small impact on the position estimation. The dynamometer governed the shaft speed, and the WSM was operated in the torque control mode. A Freescale MPC5554 was used as a processor for the inverter control board, and the inverter switching frequency was 8 kHz. The real angle was monitored using a resolver mounted on the WSM shaft.

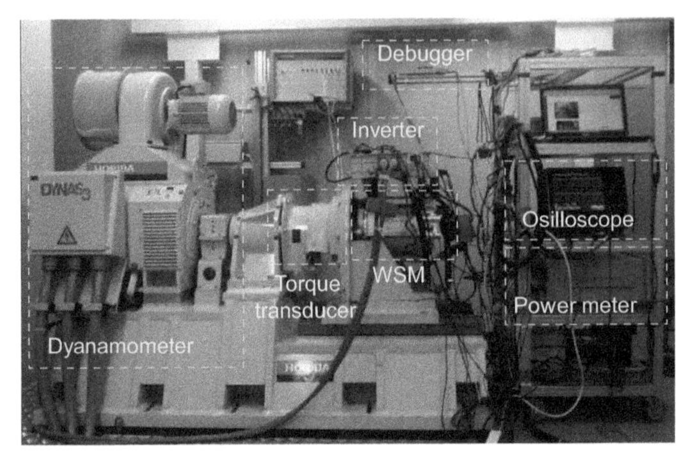

Figure 11. Experiment environment for testing sensorless control: the dynamometer, inverter, debugger, osilloscope, torque transducer, and WSM.

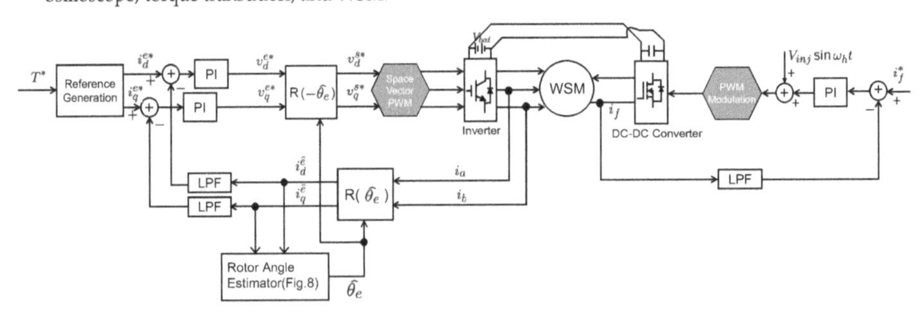

Figure 12. Block diagram of a proposed sensorless control method for a WSM.

The cross-coupling bias angle η is changed depending on the load condition. The bias angle can be directly measured by simple experimental method. The motor was controlled using the real rotor angle from a resolver when the carrier signal is injected into the rotor field. Using the dq-axes currents in the misaligned frame, the estimated angle $\hat{\theta}_e$ can be obtained without the compensation method. Then, the angle offset can be measured by $\eta_{exp} = \hat{\theta}_e - \theta_e$. Figure 13a shows the experimental results of the cross-coupling bias angle. Figure 13b shows η_{exp} and its differences, $\eta_{exp} - \eta$, from the ones computed using the FEM data. Please note that the maximum difference is approximately 6° (ele. degrees).

Figure 14 shows the angle estimate $\hat{\theta}_e$, measured angle θ_e, and their difference $\Delta\theta_e = \hat{\theta}_e - \theta_e$ during the transitions between 0 to 50 r/min and 50 r/min to 0 r/min, when a 28 Nm shaft torque is applied. Please note that the angle error was bounded under ±3.33° (mechanical angle).

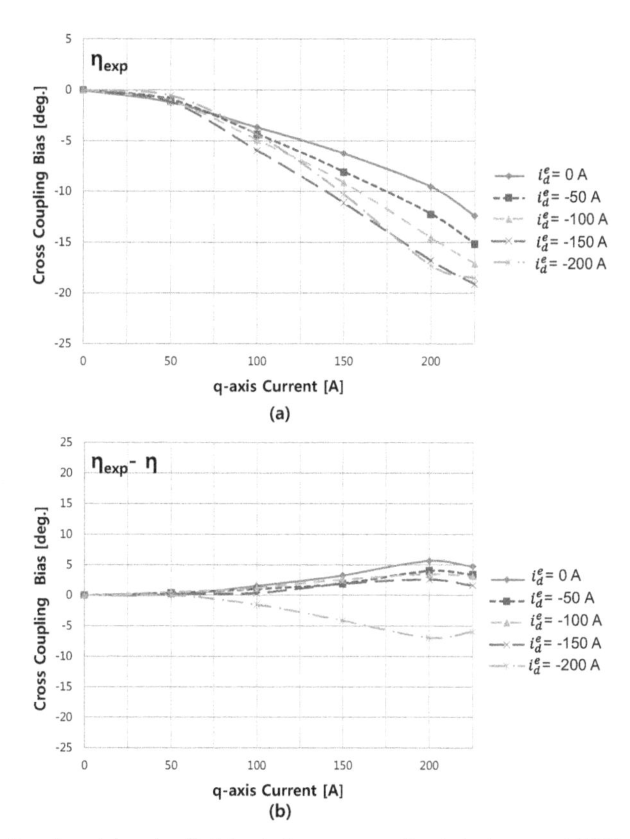

Figure 13. (**a**) Experimental angle offset due to the cross-coupling inductance and (**b**) The differences between the angle offset measured η_{exp} and the angle offset calculated η using FEM data.

Figure 15 shows the current response to a step command $i_q^{e*} = 228$ A when $i_d^e = -10$ A when the motor speed was regulated at a standstill by the dynamometer motor, showing that 123 Nm (1 pu) was produced. However, the torque response appears sluggish owing to a strong filter in CAN communication. Please note that the angle error was regulated below 20°(ele. degrees) under a rated step torque.

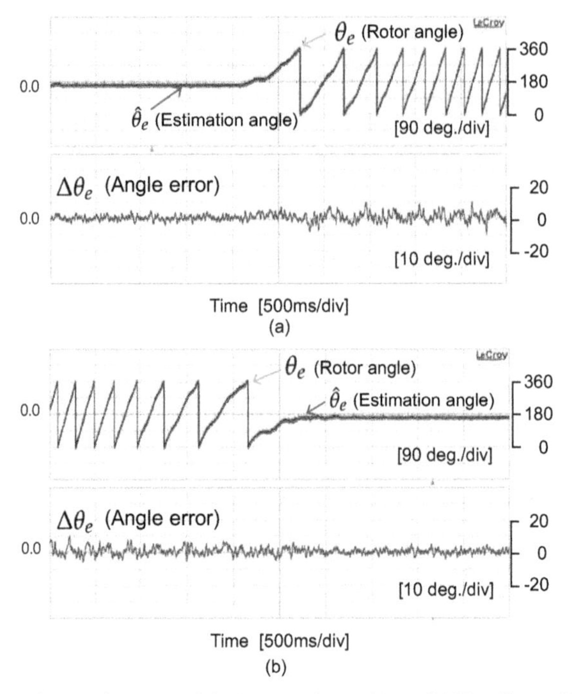

Figure 14. Proposed sensorless control during speed transitions (28 Nm, $i_d^e = -20$ A, $i_q^e = 50$ A): (**a**) 0 → 50 r/min and (**b**) 50 → 0 r/min.

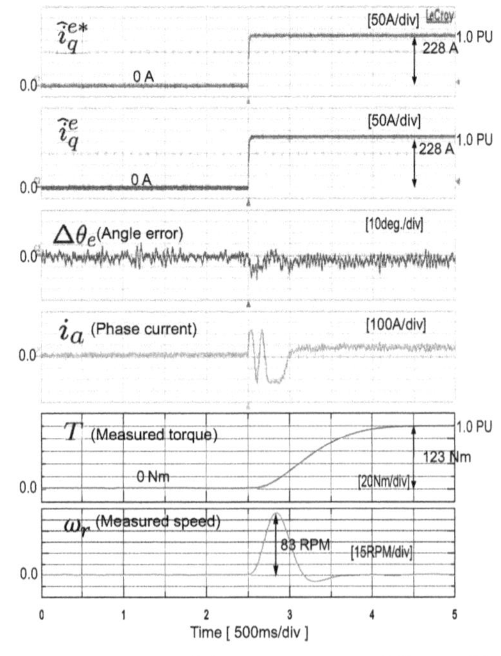

Figure 15. Current responses at a standstill for a q-axis command (0 A → 228 A) when $i_d^e = -10$ A: q-axis current command, q-axis current, angle error, phase current, shaft torque, and shaft speed.

Figure 16 shows the angle error trend at a fixed speed (100 r/min) when the q-axis current increases to a rated point without and with compensation for η, which was caused by cross-coupling. The effectiveness of the cross-coupling compensation was demonstrated, in which the bias error (14° ele. degrees) was monitored without the compensation method whereas no bias error was observed with the compensation method.

Figure 17 shows the experimental result using sensorless control based on signal injection into stator winding. The dyanamometer regulated the WSM speed at 50 r/min. For a light load, the estimated angle error $\Delta\theta_e$ is not over 10° (ele. degrees). However, for a heavy load, the estimated angle error is oscillated between −30° and 5° (ele. degrees) owing to magnetic saturation.

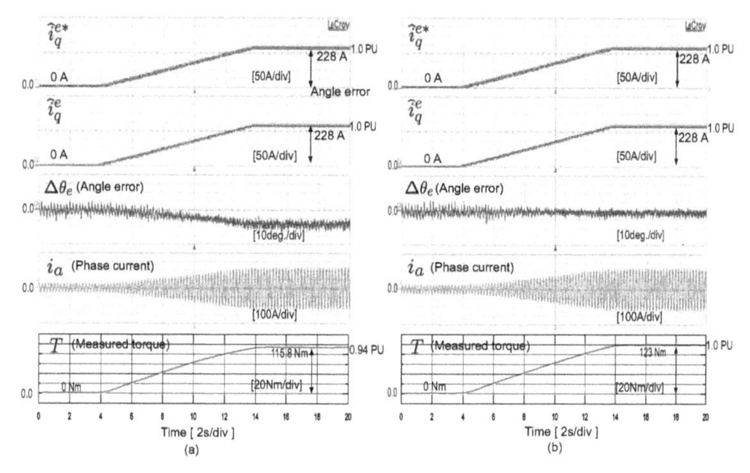

Figure 16. Field current injection method (**a**) without compensation for η caused by cross-coupling and (**b**) with compensation for η: ramp q-axis command (0 A → 228 A), q-axis current response, angle error, phase current, and torque.

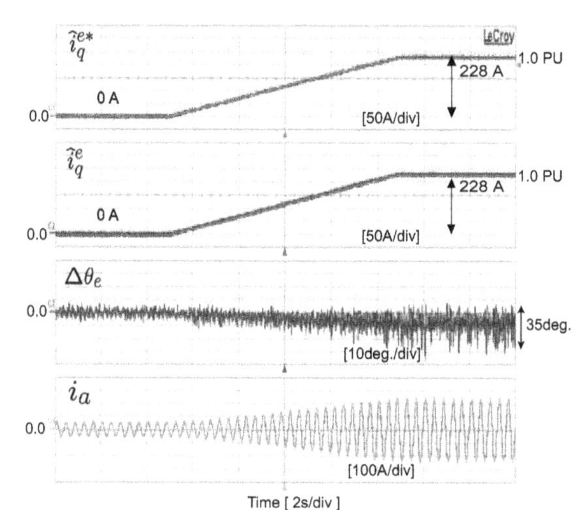

Figure 17. Signal injection into the stator winding without a compensation of cross-coupling effects: response at 50 r/min for a ramp q-axis command (0 A → 228 A) when $i_d^e = -50$ A.

7. Discussion

This study has attempted to investigate sensorless control for a WSM based on signal injection into the field winding. To conclude, we summarize the following contributions of the paper. The sensorless method, which injected a carrier signal into field winding, is not based on the saliency ratio. Therefore, the method does not become unstable caused owing to magnetic saturation phenomenon. In addition, the absolute position angle can be obtained. Both d- and q-axis high-frequency signals were used for angle estimation. Mutual incremental inductances are used to predict and compensate the angle offset caused by the cross-coupling effect. In this work, L_{df} and L_{qf} are significant factors. The algorithm was developed to eliminate the estimation bias caused by the cross-coupling inductance. Experiments were performed for full-torque operation. Furthermore, we obtained an accurate estimated angle in the presence of the cross-coupling effect.

Author Contributions: J.C. and K.N. designed the proposed sensorless algorithm for WSM and J.C. did the experiments and analyzed the data.

Funding: This research received no external funding.

Conflicts of Interest: The authors declare no conflict of interest.

Abbreviations

The following abbreviations are used in this manuscript:

v_d^e, v_q^e	d and q-axes voltage in the synchronous frame
i_d^e, i_q^e	d and q-axes current in the synchronous frame
i_f	Field current
$\lambda_a, \lambda_b, \lambda_c$	Phase a, b, c stator flux linkage
λ_d^e, λ_q^e	d and q-axes stator flux linkage in the synchronous frame
r_s	Stator winding resistance
L_{dd}, L_{qq}	d and q-axes self-inductance
L_{dq}, L_{qd}	d and q-axes cross-coupling inductance
L_{df}	Mutual inductance between d-axis and field coil
L_{qf}	Cross-coupling inductance between q-axis and field coil
L_{ls}	Leakage inductance of stator
L_{lf}	Leakage inductance of field
θ_e	Rotor flux angle
$\hat{\theta}_e$	Estimation rotor flux angle
$\Delta\theta_e$	Estimation rotor flux angle error
ω_e, ω_r	Electrical speed and mechanical speed
ω_h	Angular speed of the carrier signal
η	Angle offset caused by the cross-coupling
η_{com}	Compensation for angle offset
$-\Delta\hat{\theta}_e$	The input of the PLL
T	Shaft torque
WSM	Wound Synchronous Machine
EV	Electrical vehicles
HEV	Hybrid EVs
PMSMs	Permanent-magnet synchronous motors
Nd	Neodymium
Dy	Dysprosium
EMF	Electromotive force
MTPA	Maximum torque per ampere
FEM	Finite-element method
LPF	Low pass filter
BPF	Band-pass filter

References

1. Santiago, J.D.; Bernhoff, H.; Ekergard, B.; Eriksson, S.; Ferhatovic, S.; Waters, R.; Leijon, M. Electrical Motor Drivelines in Commercial All-Electric Vehicles: A Review. *IEEE Trans. Veh. Technol.* **2013**, *61*, 475–484. [CrossRef]
2. Kim, Y.; Nam, K. Copper-Loss-Minimizing Field Current Control Scheme for Wound Synchronous Machines. *IEEE Trans. Power Electron.* **2017**, *32*, 1335–1345. [CrossRef]
3. Friedrich, G.; Girardin, A. Integrated starter generator. *IEEE Trans. Ind. Appl. Mag.* **2009**, *15*, 26–34. [CrossRef]
4. Wang, Y.; Wang, X.; Xie, W.; Dou, M. Full-Speed Range Encoderless Control for Salient-Pole PMSM with a Novel Full-Order SMO. *Energies* **2018**, *11*, 2423. [CrossRef]
5. Morimoto, S.; Kawamoto, K.; Sanada, M.; Takeda, Y. Sensorless control strategy for salient-pole PMSM based on extended EMF in rotating reference frame. *IEEE Trans. Ind. Appl.* **2002**, *38*, 1054–1061. [CrossRef]
6. Chen, G.-R.; Yang, S.-C.; Hsu, Y.-L.; Li, K. Position and Speed Estimation of Permanent Magnet Machine Sensorless Drive at High Speed Using an Improved Phase-Locked Loop. *Energies* **2017**, *10*, 1571. [CrossRef]
7. Wang, M.-S.; Tsai, T.-M. Sliding Mode and Neural Network Control of Sensorless PMSM Controlled System for Power Consumption and Performance Improvement. *Energies* **2017**, *10*, 1780. [CrossRef]
8. Genduso, F.; Miceli, R.; Rando, C.; Galluzzo, G.R. Back EMF Sensorless-Control Algorithm for High-Dynamic Performance PMSM. *IEEE Trans. Ind. Electron.* **2010**, *57*, 2092–2100. [CrossRef]
9. Lee, J.; Hong, J.; Nam, K.; Ortega, R.; Praly, L.; Astolfi, A. Sensorless Control of Surface-Mount Permanent-Magnet Synchronous Motors Based on a Nonlinear Observer. *IEEE Trans. Power Electron.* **2010**, *25*, 290–297.
10. Choi, J.; Nam, K.; Bobtsov, A.A.; Pyrkin, A.; Ortega, R. Robust Adaptive Sensorless Control for Permanent-Magnet Synchronous Motors. *IEEE Trans. Power Electron.* **2017**, *32*, 3989–3997. [CrossRef]
11. Koteich, M.; Maloum, A.; Duc, G.; Sandou, G. Observability analysis of sensorless synchronous machine drives. In Proceedings of the 2015 European Control Conference (ECC), Linz, Austria, 15–17 July 2015; pp. 3560–3565.
12. Tian, L.; Zhao, J.; Sun, J. Sensorless Control of Interior Permanent Magnet Synchronous Motor in Low-Speed Region Using Novel Adaptive Filter. *Energies* **2016**, *9*, 1084. [CrossRef]
13. Gabriel, F.; De Belie, F.; Neyt, X.; Lataire, P. High-Frequency Issues Using Rotating Voltage Injections Intended For Position Self-Sensing. *IEEE Trans. Ind. Electron.* **2013**, *60*, 5447–5457. [CrossRef]
14. Wu, X.; Wang, H.; Huang, S.; Huang, K.; Wang, L. Sensorless Speed Control with Initial Rotor Position Estimation for Surface Mounted Permanent Magnet Synchronous Motor Drive in Electric Vehicles. *Energies* **2015**, *8*, 11030–11046. [CrossRef]
15. Gong, L.M.; Zhu, Z.Q. Robust Initial Rotor Position Estimation of Permanent-Magnet Brushless AC Machines With Carrier-Signal-Injection-Based Sensorless Control. *Trans. Ind. Appl.* **2013**, *49*, 2602–2609. [CrossRef]
16. Li, Y.; Zhu, Z.Q.; Howe, D.; Bingham, C.M. Modeling of Cross-Coupling Magnetic Saturation in Signal-Injection-Based Sensorless Control of Permanent-Magnet Brushless AC Motors. *IEEE Trans. Mag.* **2007**, *43*, 2552–2554. [CrossRef]
17. Zhu, Z.Q.; Gong, L.M. Investigation of Effectiveness of Sensorless Operation in Carrier-Signal-Injection-Based Sensorless-Control Methods. *IEEE Trans. Ind. Electron.* **2011**, *58*, 3431–3439. [CrossRef]
18. Li, Y.; Zhu, Z.Q.; Howe, D.; Bingham, C.M.; Stone, D.A. Improved Rotor-Position Estimation by Signal Injection in Brushless AC Motors, Accounting for Cross-Coupling Magnetic Saturation. *IEEE Trans. Ind. Appl.* **2009**, *44*, 1843–1850. [CrossRef]
19. Sergeant, P.; De Belie, J.; Melkebeek, J. Rotor Geometry Design of Interior PMSMs With and Without Flux Barriers for More Accurate Sensorless Control. *IEEE Trans. Ind. Electron.* **2012**, *59*, 2457–2465. [CrossRef]
20. Li, S.; Ge, Q.; Wang, X.; Li, Y. Implementation of Sensorless Control with Improved Flux Integrator for Wound Field Synchronous Motor. In Proceedings of the 2007 2nd IEEE Conference on Industrial Electronics and Applications, Harbin, China, 23–25 May 2007; Volume 59, pp. 1526–1530.
21. Maalouf, A.; Ballois, S.L.; Idekhajine, L.; Monmasson, E.; Midy, J.; Biais, F. Sensorless Control of Brushless Exciter Synchronous Starter Generator Using Extended Kalman Filter. In Proceedings of the 2009 35th Annual Conference of IEEE Industrial Electronics, Porto, Portugal, 3–5 November 2009; Volume 59, pp. 2581–2586.
22. Boldea, I.; Andreescu, G.D.; Rossi, C. Active Flux Based Motion-Sensorless Vector Control of DC-Excited Synchronous Machines. In Proceedings of the 2009 IEEE Energy Conversion Congress and Exposition, San Jose, CA, USA, 20–24 September 2009; pp. 2496–2503.

23. Jain, A.K.; Ranganathan, V.T. Modeling and Field Oriented Control of Salient Pole Wound Field Synchronous Machine in Stator Flux Coordinates. *IEEE Trans. Ind. Electron.* **2011**, *58*, 960–970. [CrossRef]
24. Griffo, A.; Drury, D.; Sawata, T.; Mellor, P.H. Sensorless starting of a wound-field synchronous starter/generator for aerospace applications. *IEEE Trans. Ind. Electron.* **2012**, *59*, 3579–3587. [CrossRef]
25. Rambetius, A.; Ebersberger, S.; Seilmeier, M.; Piepenbreier, B. Carrier Signal Based Sensorless Control of Electrically Excited Synchronous Machines at Standstill and Low Speed Using The Rotor Winding as a receiver. In Proceedings of the 2013 15th European Conference on Power Electronics and Applications (EPE), Lille, France, 2–6 September 2013; pp. 1–10.
26. Deng, X.; Wang, L.; Zhang, J.; Ma, Z. Rotor Position Detection of Synchronous Motor Based on High-frequency Signal Injection into the Rotor. In Proceedings of the 2011 Third International Conference on Measuring Technology and Mechatronics Automation, Shangshai, China, 6–7 January 2011; pp. 195–198.
27. Zhou, Y.; Long, S. Sensorless Direct Torque Control for Electrically Excited Synchronous Motor Based on Injecting High-Frequency Ripple Current Into Rotor Winding. *IEEE Trans. Energy Convers.* **2015**, *30*, 246–253. [CrossRef]
28. Rambetius, A.; Piepenbreier, B. Comparison of carrier signal based approaches for sensorless wound rotor synchronous machines. In Proceedings of the International Symposium on Power Electronics, Electrical Drives, Automation and Motion, Schia, Italy, 18–20 June 2014; pp. 1152–1159.
29. Rambetius, A.; Piepenbreier, B. Sensorless control of wound rotor synchronous machines using the switching of the rotor chopper as a carrier signal. In Proceedings of the International Symposium on Sensorless Control for Electrical Drives and Predictive Control of Electrical Drives and Power Electronics, Munich, Germany, 17–19 October 2013; pp. 1–8.
30. Choi, J.; Jeong, I.; Jung, S.; Nam, K. Sensorless Control for Electrically Energized Synchronous Motor Based on Signal Injection to Field Winding. In Proceedings of the IECON2013—39th Annual Conference of the IEEE Industrial Electronics Society, Vienna, Austria, 10–13 November 2013; pp. 3120–3129.
31. Selmon, G.R. *Electric Machines and Drives*; Addison Welsley: Boston, MA, USA, 1992; ISBN 0-201-57885-9.
32. Stumberger, B.; Stumberger, G.; Dolinar, D.; Hamler, A.; Trlep, M. Evaluation of Saturation and Cross-Magnetization Effects in Interior Permanent-Magnet Synchronous Motor. *IEEE Trans. Ind. Appl.* **2009**, *48*, 1576–1587.
33. Nam, K.H. *AC Motor Control and Electric Vehicle Application*, 1st ed.; CRC Press: Boca Raton, FL, USA, 2010; ISBN 978-1-49-81963-0.
34. Garcia, P.; Briz, F.; Degner, M.W.; Diaz-Reigosa, D. Accuracy, Bandwidth, and Stability Limits of Carrier-Signal-Injection-Based Sensorless Control Methods. *IEEE Trans. Ind. Appl.* **2009**, *43*, 990–1000. [CrossRef]
35. De Kock, H.W.; Kamper, M.J.; Kennel, R.M. Anisotropy Comparison of Reluctance and PM Synchronous Machines for Position Sensorless Control Using HF Carrier Injection. *IEEE Trans. Power Electron.* **2009**, *24*, 1905–1913. [CrossRef]

 *energies*

Article

Adaptive Robust Control System for Axial Flux Permanent Magnet Synchronous Motor of Electric Medium Bus Based on Torque Optimal Distribution Method

Shuang Wang, Jianfei Zhao *, Tingzhang Liu and Minqi Hua

School of Mechatronic Engineering and Automation, Shanghai University, Baoshan District, Shanghai 200444, China; wang-shuang@shu.edu.cn (S.W.); liutzhcom@oa.shu.edu.cn (T.L.); huaminqi@shu.edu.cn (M.H.)

* Correspondence: jfzhao@shu.edu.cn; Tel.: +86-021-6613-0935

Received: 19 November 2019; Accepted: 7 December 2019; Published: 9 December 2019

Abstract: In this paper, an adaptive robust drive control system for an axial flux permanent magnet synchronous motor of an electric medium-sized bus based on the optimal torque distribution method is studied. The drive control system is mainly divided into two parts. First, a torque distribution method is proposed. The optimal torque distribution method based on particle swarm optimization algorithm is used to increase the high efficiency interval of the system and apply it to the energy feedback braking. Secondly, in order to reduce the nonlinear disturbance of the system and improve the accuracy of the unified control, this paper models and studies the vector system based on adaptive robust control. Finally, the whole drive control system is modeled, simulated and experimented. The simulation and experimental results show that the torque distribution method proposed in this paper can effectively increase the high-efficiency running time of the electric medium bus, and improve the shortcomings of insufficient mileage of the electric medium-sized bus. The use of a current controller based on adaptive robust control improves the control accuracy of the drive system and can effectively suppress the disturbances generated by it.

Keywords: adaptive robust control; AFPMSM; energy feedback; particle swarm optimization; torque optimal distribution method

1. Introduction

The passenger capacity of medium bus is generally 9–20, which is suitable for small and medium-size cities. With the increasing awareness of greenhouse gas emissions, the emergence of electric buses can meet the call for energy conservation and emission reduction [1]. Compared with the internal combustion engine powered medium bus, the electric medium bus have many advantages. For example, they have low vibration noise, simple structure, high power transmission efficiency, easy vehicle layout, and excellent power performance [2,3]. The selection of pure electric medium bus drive motors must meet the vehicle's dynamic requirements, such as maximum speed, acceleration performance, and maximum grade [4]. Among a variety of vehicle drive motors, permanent magnet synchronous motors (PMSMs) have many applications [5,6]. Among them, the axial flux permanent magnet synchronous motors (AFPMSMs) have the advantages of low speed, large torque and high energy density, and are more suitable for use in an electric medium bus with a larger passenger capacity. Double stator-single rotor AFPMSM (two-disc AFPMSM) has better heat dissipation and larger rated torque. Therefore, the two-disc AFPMSM is selected as the driving motor for electric medium bus in this paper.

The two-disc AFPMSM can be equivalent to two PMSMs connected coaxially, and for electric medium-sized buses, there is only one given torque, so this paper involves the problem of multi-motor torque distribution. At present, the commonly used torque distribution method is mainly applied to distributed drive vehicles. Compared with the conventional central direct drive electric vehicles, the drive motors of the various wheels of the distributed drive electric vehicle can be independently controlled, and the torque of each wheel can be distributed in any proportion within its capability range. The energy can be controlled by properly distributing the wheel torque so that the motor works as much as possible in the high efficiency range. The author of [7] proposed a multi-objective optimization method that considers system efficiency and safety for torque distribution. The authors of [8] mainly use torque distribution to enable micro electric vehicles to improve powertrain efficiency. The authors of [9] used the optimal vehicle state estimation method for directional tire torque distribution. In this paper, a torque optimal allocation method is proposed for the purpose of efficiency optimization. The particle swarm optimization algorithm is used to optimize the torque distribution mathematical model to obtain the optimal torque distribution solution. In addition, regenerative braking is one of the most effective ways to extend the durability of electric vehicles [10–12]. In order to further increase the cruising range, this paper applies the previously described optimal torque distribution method to the braking situation.

In addition, due to the large number of passengers in medium bus, some researchers have studied the safety and stability of driving. Authors of [13–16] studied the safety-structure from the structure of the medium bus, among which authors of [13,14] focused on studying the strength of conventional bus structures under operating conditions, authors of [15,16] studied the crashworthiness under rollover accident.

In addition, the research on control systems is mainly divided into two categories, motor design and optimization and motor control. The research on AFPMSM mainly focuses on the optimization of the motor model. The authors of [17] used the combined solution of Maxwell's equations and magnetic equivalent circuits to model the AFPMSM analytically. The authors of [18] used an auxiliary multi-objective optimization algorithm to optimize the design of AFPMSM with dual rotor and single stator. This paper studies the anti-interference and current tracking capabilities of the driving system of medium-sized buses from the perspective of drive control. During the operation of the electric medium bus, due to the complicated operating environment, it will encounter various nonlinear disturbances. Adaptive robust control is used to overcome the effects of nonlinear disturbances, thereby improving tracking accuracy of the current loop. In the study of adaptive robust control, the authors of [19] used adaptive synthesis robust control strategies based on μ synthesis to resist the interference of high-frequency dynamic problems generated by the motor structure mode on linear motor control. The authors of [20] used neural networks to learn adaptive robust controllers to resist interference from unknown factors. The author of [21] used an adaptive robust controller based on extended disturbance observer to improve the control accuracy of linear motors. In this paper, adaptive robust control is applied to the drive control system of AFPMSM for anti-disturbance control.

This paper mainly studies the drive control system of electric medium bus. From the above, the drive control system is mainly divided into two parts. The first part mainly studies the torque distribution method, with the system's highest working efficiency as the distribution target. The second part mainly studies the motor control part. In order to reduce the waveform ripple and improve the system control accuracy, this paper models and studies the adaptive robust control vector system. Finally, the above methods are simulated and the motor experiments and loading experiments are performed, and the results are summarized.

2. Two-Disc AFPMSM Mathematical Model

Compared with the traditional AFPMSM, the AFPMSMs with multi-disc structure improve the overall efficiency of the motor by adjusting the number of stators and rotors running [22]. Therefore, this paper takes the AFPMSM of double-stator-single-rotor structure as the research object, and establishes its mathematical model as the theoretical basis for deducing its control strategy. The structure of the dual-stator-single-rotor AFPMSM (also called double-disc AFPMSM) is shown in Figure 1.

Figure 1. Internal view of double-disc axial flux permanent magnet synchronous motor (AFPMSM).

In order to distinguish the two stators of the AFPMSM, they are respectively defined as the stator 1 and the stator 2. The simplified PMSMs are the motor 1 and the motor 2. For the motor 1 and the motor 2, a d-q axis rotating coordinate system is established, which is d_1-q_1 and d_2-q_2, and the coordinate system rotation speeds are ω_1 and ω_2, and the rotation directions are the same. Since the double disc AFPMSM shares one rotor and is coaxially connected, two mathematical models of the d-q axis rotating coordinate system can be established at the same time. The mathematical model of the AFPMSM is as follows.

Voltage equation:

$$\begin{cases} u_{d1} = R_{s1} i_{d1} + L_{d1} \frac{di_{d1}}{dt} - \omega_1 L_{q1} i_{q1} \\ u_{q1} = R_{s1} i_{q1} + L_{q1} \frac{di_{q1}}{dt} + \omega_1 L_{d1} i_{d1} + \omega_1 \psi_f \\ u_{d2} = R_{s2} i_{d2} + L_{d2} \frac{di_{d2}}{dt} - \omega_2 L_{q2} i_{q2} \\ u_{q2} = R_{s2} i_{q2} + L_{q2} \frac{di_{q2}}{dt} + \omega_2 L_{d2} i_{d2} + \omega_2 \psi_f \end{cases}, \tag{1}$$

Magnetic chain equation:

$$\begin{cases} \psi_{d1} = L_{d1} i_{d1} + \psi_f \\ \psi_{q1} = L_{q1} i_{q1} \\ \psi_{d2} = L_{d2} i_{d2} + \psi_f \\ \psi_{q2} = L_{q2} i_{q2} \end{cases}, \tag{2}$$

Torque equation:

$$\begin{cases} T_{e1} = \frac{3}{2} p [\psi_f i_{q1} + (L_{d1} - L_{q1}) i_{d1} i_{q1}] \\ T_{e2} = \frac{3}{2} p [\psi_f i_{q2} + (L_{d2} - L_{q2}) i_{d2} i_{q2}] \end{cases}, \tag{3}$$

Equation of motion:

$$\begin{cases} T_{e1} - T_{L1} - B\omega_1 = \frac{J}{p} \frac{d\omega_1}{dt} \\ T_{e2} - T_{L2} - B\omega_2 = \frac{J}{p} \frac{d\omega_2}{dt} \end{cases}, \tag{4}$$

In Equations (1)–(4), R_{s1} and R_{s2} are two stator resistances respectively, and u_{d1}, u_{q1}, u_{d2}, and u_{q2} are d-q axis components of the winding voltage vectors of the stator 1 and the stator 2, i_{d1}, i_{q1}, and i_{d2}. i_{q2} is the d-q axis component of the winding current vectors of the stator 1 and stator 2, L_{d1}, L_{q1}, L_{d2}, L_{q2} are the d-q axis components of the winding inductances of the stator 1 and stator 2, ψ_{d1}, ψ_{q1}, ψ_{d2}, ψ_{q2} are the d-q axis components of the winding flux of stator 1 and stator 2, T_{e1} and T_{e2} are the electromagnetic torques of the stator 1 and the stator 2, T_{L1} and T_{L2} are the load torques of the stator 1 and the stator 2, J is the moment of inertia, B is the viscosity coefficient, and T_L is the load torque.

Since the two stators share one rotor, it can be approximated that the two motor modules are coaxially connected, so $\omega_{r1} = \omega_{r2} = \omega_r$ can be obtained. According to the conclusion of coaxial connection, two equivalent motors can now be analyzed under the same d-q reference coordinate system. The two stators are structurally identical and symmetrical, therefore the stator resistances R_{S1} and R_{S2} are equal. According to the uniform air gap of the motor, it can be obtained that the direct-axis inductance and the cross-axis inductance of the two motors are equal. This article uses a hidden-pole motor, so it is also concluded that the inductance of the AC and DC axes is equal. At this time, the electromagnetic torque equation can be rewritten as:

$$T_e = T_{e1} + T_{e2} = \frac{3}{2} p \psi_f \left(i_{q1} + i_{q2} \right), \tag{5}$$

The equation of motion is:

$$T_e - T_L - B\omega_r = \frac{J}{p}\frac{d\omega_r}{dt}, \tag{6}$$

3. Torque Optimal Distribution Method

3.1. Dual Stator AFPMSM Drive System Topolgy

The traditional electric medium bus has only one motor drive system, and the vehicle manager only corresponds to one motor controller [23], and the two are connected by controller area network (CAN) communication. Unlike conventional two-motor electric vehicle drive systems, the dual-station AFPMSM needs to control two sets of stator windings. Although they are driven by separate inverters, the same motor controller can be used, so that the vehicle manager and the motor controller can be directly connected via controller area network (CAN) communication. In order to study the torque distribution method more conveniently, a torque distributor is added between the vehicle manager and the motor controller, and the torque is distributed to the motor through the torque distributor. The topology of the double-disc AFPMSM drive system is shown in Figure 2.

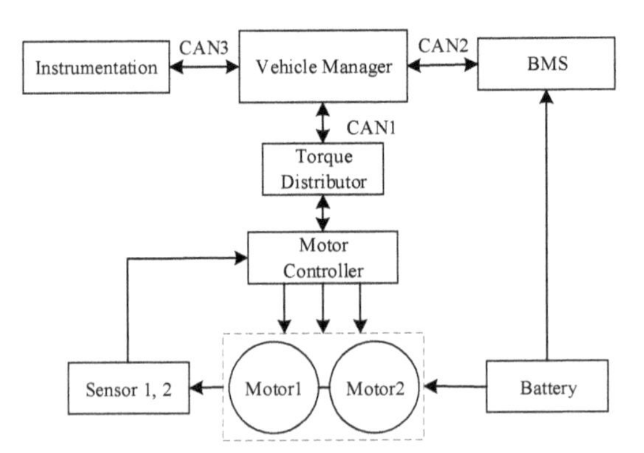

Figure 2. Topology of double-disc AFPMSM drive system for electric medium bus.

The general electric vehicle drive system only has a given torque T_m. The driving motor in this paper can be regarded as two motors after equivalent. Therefore, T_m needs to be allocated. The common method is to evenly distribute torque. In order to save battery power, an optimal torque distribution strategy based on particle swarm optimization is proposed, so that the system can operate in a high efficiency range. The system control block diagram is shown in Figure 3. According to the optimal torque distribution method, the distributed torques of the two motors are obtained, and then the motor control is performed.

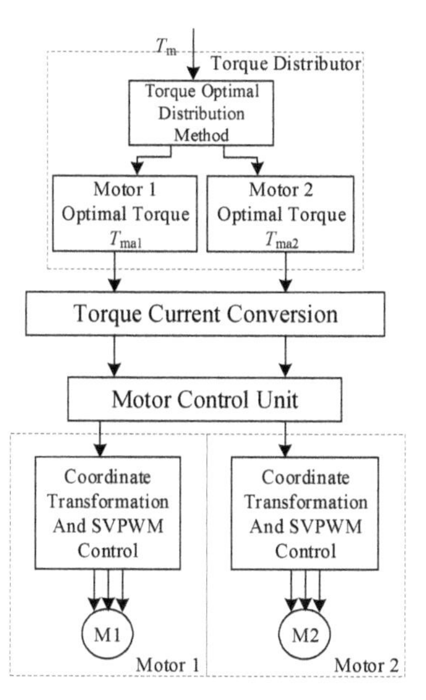

Figure 3. Control block diagram of double-disc AFPMSM drive system for electric medium bus.

3.2. Optimal Torque Distribution Control Method Based on Particle Swarm Optimization in Driving State

The torque optimal distribution strategy based on particle swarm optimization (PSO) is modeled as follows. Taking the double-disc AFPMSM for electric medium bus studied in this paper as an example, the total output torque is T, and the range is [0, 500 Nm]. The output torques of the two motor modules are T_1 and T_2, respectively, and the range is [0, 250 Nm], then:

$$T = T_1 + T_2, \tag{7}$$

Assuming that the mechanical angular velocity of the motor is ω, the output power of the two sets of motor modules is $T_1\omega$ and $T_2\omega$, the input power is P_1 and P_2, and the operating efficiency is η_1 and η_2. Let $T_1 = a_1T$, $T_2 = a_2T$ where $a_1 + a_2 = 1$, $a_1, a_2 \in [0, 1]$. Then the input power of the two motor is:

$$P_1 = \frac{T_1\omega}{\eta_1} = \frac{a_1}{\eta_1}T_1\omega, \tag{8}$$

$$P_2 = \frac{T_2\omega}{\eta_2} = \frac{a_2}{\eta_2}T_2\omega, \tag{9}$$

The total output power of the Motor Module is:

$$P_o = T_1\omega + T_2\omega = T\omega, \tag{10}$$

The total input power is:

$$P_i = P_1 + P_2 = (\frac{a_1}{\eta_1} + \frac{a_2}{\eta_2})T\omega, \tag{11}$$

The total efficiency of the motor module is:

$$\eta = \frac{P_o}{P_i} = \frac{T\omega}{(\frac{a_1}{\eta_1} + \frac{a_2}{\eta_2})T\omega} = \frac{1}{\frac{a_1}{\eta_1} + \frac{a_2}{\eta_2}}, \tag{12}$$

Assume:

$$f(a_1) = \frac{a_1}{\eta_1}, f(a_2) = \frac{a_2}{\eta_2}, \tag{13}$$

It can be known from Equation (10) that the output power is constant during the operation of the electric medium bus and it is only necessary to reduce the total input power to improve the system efficiency. If the maximum value of η is required, that is, the minimum value of $f(a_1) + f(a_2)$ is obtained.

Let

$$A = f(a_1) + f(a_2), \tag{14}$$

Then the problem translates into how the two sets of motor modules are assigned torque ratios for a given torque so that the value of A is minimized.

For Equation (14), when one of the torque distribution ratios a_1 or a_2 is determined, the value of A can be determined. However, the speed and torque at a certain moment are not involved in the Equation (14). In the optimization using the particle swarm optimization algorithm, the optimal distribution must be obtained based on the total given torque and speed. Therefore, the three-dimensional model of $f(a_1)$, the total torque command T, and the current rotational speed n can be obtained by data fitting. Since the two sets of motor modules are identical, the efficiency values are the same under different speeds and torques, so only the total torque command T is required to be the x-axis, and the rotational speed n is the y-axis. The value of $f(a_1)$ is calculated as the z-axis for all torque distribution ratios and corresponding efficiencies at different speeds and torques. The three-dimensional model is shown in Figure 4.When the torque distribution system inputs the torque and the rotational speed at any time, any value of $f(a_1)$ will have a certain value of $f(a_2)$ corresponding to it on the z-axis, so that the value of A under all torque distribution ratios can be calculated. The optimization of the PSO algorithm is to find the smallest one of all fitness functions in the three-dimensional stereogram model, and output the corresponding ratio of a_1 and a_2 to achieve the optimal torque distribution.

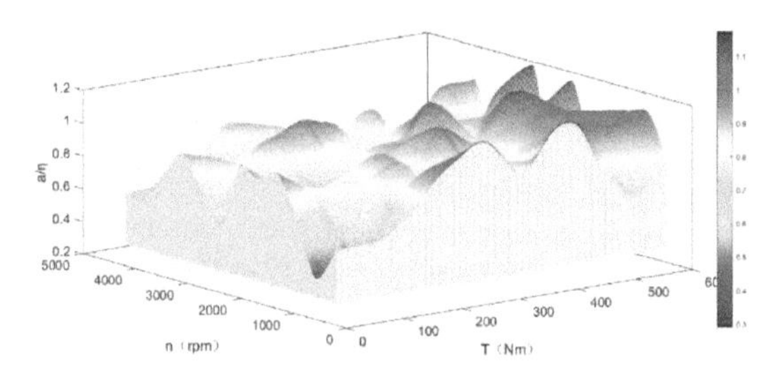

Figure 4. Fitting 3D model under driving state.

Through the analysis above, $A = f(a_1) + f(a_2)$ can be used as fitness function, so the fitness function is designed as follows:

$$\min A = f(a_1) + f(a_2), \tag{15}$$

In a search space of a D-dimensional parameter, the population size of the particles is *Size*. Each particle represents a candidate solution to the solution space, where the position of the *i-th* $(1 \le i \le Size)$ particle in the entire solution space is represented as X_i and the velocity is represented as V_i. The optimal solution generated by the *i-th* particle from the initial to the current iteration number search is the individual extremum p_i, and the current optimal solution of the entire population is *BestS*. *Size* particles are randomly generated, and the position matrix and velocity matrix of the initial population are randomly generated. The learning factors are set as c_1 and c_2, the maximum evolution algebra is G, and g is the current evolutionary algebra. The equation for the velocity and position of a particle in the solution space is as follows:

$$V_i^{kg+1} = w(t) \times V_i^g + c_1 r_1 (p_i^g - X_i^g) + c_2 r_2 (BestS_i^g - X_i^g), \tag{16}$$

$$X_i^{g+1} = X_i^g + V_i^{g+1}, \tag{17}$$

Among them, g = 1,2, ... ,G, I = 1,2, ... ,Size, r_1 and r_2 are random numbers from 0 to 1; c_1 is a local learning factor, and c_2 is a global learning factor, generally c_2 is larger and $w(t)$ is the inertia weight. The particle swarm optimization algorithm has the advantages of strong local search ability, fast calculation speed, and few parameters. However, during the running process, the particle swarm has strong convergence in the local, and it is easy to ignore all and fall into the local optimal solution [24]. In view of the shortcomings of the particle swarm algorithm, the inertia weight $w(t)$ is added to the velocity term, which represents the ability of the particle to update the velocity, which has a great influence on the convergence and accuracy of the whole algorithm. A larger $w(t)$ can improve the global search ability of the algorithm, while a smaller $w(t)$ can improve the local search ability of the algorithm, so the value of $w(t)$ should be decremented during the iterative process, which allows the particle to strike a balance between its search ability and convergence speed. The value of $w(t)$ is determined according to Equation (18).

$$w(t) = w(t)_{\max} - \frac{w_{\max} - w_{\min}}{k_{\max}} \times k, \tag{18}$$

In the equation, w_{max} represents the initial weight, w_{min} represents the final weight, k represents the current iteration number of particles, and k_{max} represents the maximum iteration number of particles. The particle swarm optimization is easy to converge too early and fall into local optimum, which makes it impossible to obtain global optimum solution. Combining with the requirements of speed control for electric vehicles, this paper improves the particle swarm optimization algorithm. The inertia weight is determined by the exponential decrement method, as defined by Equation (19).

$$w = w_{\max} (\frac{w_{\min}}{w_{\max}})^{1/(1+10k/k_{\max})}, \tag{19}$$

In the initial stage, w is larger, and has a strong ability to search in a wide range. In the later stage, w is smaller and has a strong ability to search in a small range, thereby improving the performance of the particle swarm algorithm as a whole. At the same time, in order to avoid premature convergence of the algorithm, the learning factor is dynamically adjusted, as shown in Equation (20):

$$\begin{cases} c_1 = 2 - \sin \frac{k\pi}{k_{\max}} \\ c_2 = 1 + \sin \frac{k\pi}{k_{\max}} \end{cases}, \tag{20}$$

In the early stage of population search, c_1 is larger and c_2 is smaller, which facilitates the particle to learn its own optimal solution and improves the global search ability. In the later stage of population search, c_2 is larger and c_1 is smaller, which facilitates the population to move closer to the global optimal solution and enhances the local optimization performance.

3.3. Energy Feedback Brake Control Based on Optimal Torque Distribution Method

For electric medium bus, in order to further improve the cruising range, the energy feedback brake will be added to the vehicle. It not only saves energy, but also solves the problem that the electric medium bus has a short driving range of one charge, and can also improve the braking performance of the car and reduce the friction loss of the brake pad when the car brakes.

In this paper, the energy feedback brake control is carried out under the condition that the battery is safely charged, the rotation speed is not too low, and the power generation power is in the safe interval. Since the motor provides braking torque in the energy feedback state, the torque optimal control problem of the two sets of motor modules is involved. It is known in the foregoing studies that reasonable torque distribution can improve system efficiency. This conclusion is also applicable in the case of energy feedback. Therefore, in order to further improve the endurance of electric mid-size passenger cars, the optimal torque distribution strategy based on particle swarm optimization is also applied to improve the power generation efficiency of the motor. The basic block diagram of the system incorporating energy feedback is shown in Figure 5. The three-dimensional perspective of the braking mode obtained by the particle swarm optimization algorithm is shown in Figure 6.

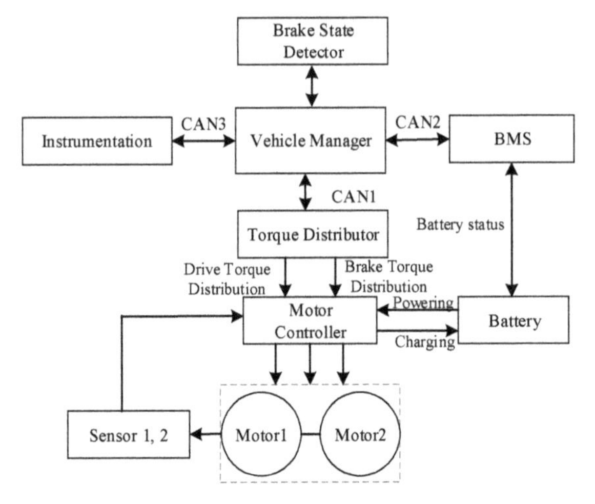

Figure 5. Basic block diagram of the energy feedback system.

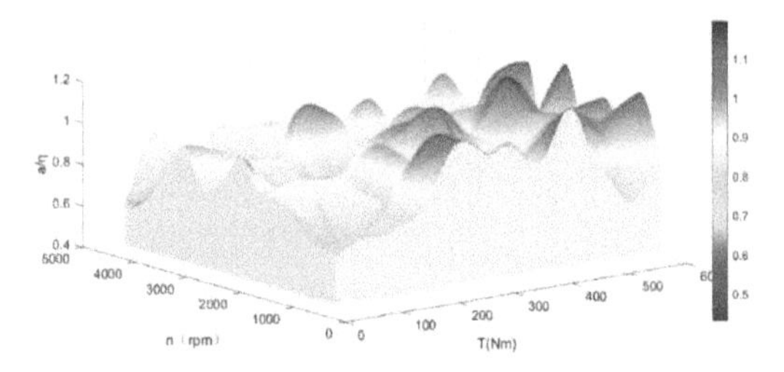

Figure 6. Fitting 3D model in energy feedback state.

4. Electric Medium Passenger Bus Vector Control System Based on Adaptive Robust Current Control

Compared with the application of motors in other aspects, the motor drive system of electric medium-sized buses has higher requirements for the current following ability and control accuracy of the motor. During the operation of electric medium-sized buses, due to the complicated operating environment, it will encounter various non-linear disturbances, resulting in current and torque ripples in the system. Therefore, this paper chooses to study from the perspective of control, and uses adaptive robust control to overcome the effects of nonlinear disturbances, thereby improving the tracking accuracy of the current loop.

According to Equation (1), a voltage model containing non-ideal back-emf can be obtained as shown in Equation (21).

$$L\frac{d}{d_t}i_q = u_q - Ri_q - e_q + \Delta_q, \tag{21}$$

where e_q is the non-ideal back-EMF, and Δq is the sum of the deviation voltages caused by all nonlinear disturbances.

Assume that the non-ideal back EMF model is:

$$e_q = \overline{S}_e K_e, \tag{22}$$

where in $\overline{S}_e$ represents a wave function containing a fundamental wave and a 6th harmonic, K_e represents a back EMF coefficient matrix. Their expressions are shown in Equations (23) and (24), respectively.

$$\overline{S}_e = \frac{3}{2}\omega_e[1 \cos(6\sigma_e)], \tag{23}$$

$$K_e = \left[K_{q1}\ K_{q6}\right]^T, \tag{24}$$

Bringing Equation (22) into Equation (21), the following equation of state can be obtained:

$$L\frac{d}{d_t}i_q = u_q - Ri_q - \overline{S}_e K_e + \Delta_q, \tag{25}$$

To establish a standard adaptive robust control model, let $\dot{x} = \frac{d}{d_t}i_q$, let $\gamma^{\mathrm{T}} = S_e$ and $\sigma = K_e$. Define the amount of virtual control $\overline{u}$ as:

$$\overline{u} = u_q - Ri_q, \tag{26}$$

where u_q is the actual output of the controller and Ri_q is calculated from the known amount and the feedback amount.

Substituting the above assumption and Equation (26) into Equation (25), Equation (26) is obtained as shown below.

$$L\dot{x} = \overline{u} + \gamma^T\sigma + \Delta_q, \tag{27}$$

The adaptive robust current controller designed in this paper makes the output of the double-disc AFPMSM model overcome the effects of non-ideal back EMF and other nonlinear disturbances. The tracking error with the expected value x_d is as small as possible.

It can be seen from the equation of state, Equation (27) that the previously designed double-disc AFPMSM model clearly includes parameter uncertainties and nonlinear disturbances. Adaptive robust control can compensate the uncertainty in the system through the design of adaptive law, and synthesize the robust control law to overcome the influence of nonlinear disturbance, so it is suitable for the design of PMSM current controller. According to Equations (27) and (28) can be obtained.

$$\overline{u} = L\dot{x} - \gamma^T\sigma - \Delta_q, \tag{28}$$

The control law form of the adaptive robust controller is as shown in Equation (29).

$$\overline{u} = u_a + u_r, \tag{29}$$

where u_a is the compensation term for adaptive control, which can be expressed as:

$$u_a = L\dot{x}_d - \gamma^T\hat{\sigma}, \tag{30}$$

where $\dot{x}_d$ is the differential of the expected value of the q-axis current and $\hat{\sigma}$ is the estimated value of the unknown parameter σ.

u_{r} is a robust control term and can be expressed as:

$$u_r = u_{r1} + u_{r2}, \tag{31}$$

In Equation (31), u_{r1} is a linear proportional feedback term and $u_{r1} = -k_{r1}z$, k_{r1} is a proportional coefficient, and $z = x - x_d$ represents a tracking error. According to the control law represented by Equations (28) and (29), the dynamic equation of the system tracking error is:

$$L\dot{z} - u_{r1} = u_{r2} - \left[\gamma^T\widetilde{\sigma} - \Delta_q\right], \tag{32}$$

where $\widetilde{\sigma}$ is the parameter estimation bias and $\widetilde{\sigma} = \hat{\sigma} - \sigma$, u_{r2} is the robust control term. According to the robust control principle, the design requirements are:

$$\begin{cases} zu_{r2} \leq 0 \\ z\left\{u_{r2} - \left[\gamma^T\widetilde{\sigma} - \Delta_q\right]\right\} \leq \varepsilon \end{cases}, \tag{33}$$

In the above equation, ε represents any positive integer, the first condition can be guaranteed to be naturally dissipated, and the second condition indicates that u_{r2} can suppress nonlinear disturbance, modeling error and estimation error of adaptive parameters. There are many ways to select u_{r2} that meet the conditions [19]. The most common one is:

$$\begin{cases} u_{r2} = -\frac{1}{4\varepsilon}h^2 z \\ h = \left|\gamma^T\right|\left|\sigma_{\max} - \sigma_{\min}\right| + \Delta_{q\max} \end{cases}, \tag{34}$$

It can be known from the control law in Equation (29) and the tracking error dynamic in Equation (32) that the tracking performance of the adaptive robust controller depends on the design of the robust control term u_r. Since the adaptive law design is synthesized by tracking error, the parameter projection method is used to modify the adaptive law [20]. Therefore, the adaptive law of adaptive robust control is expressed as:

$$\dot{\hat{\sigma}} = \mathrm{Pr}oj(\Gamma\gamma z), \tag{35}$$

where Γ is a diagonal adaptive law matrix, and $\mathrm{Pr}oj(\lambda)$ is a projection operator, which can be expressed as:

$$\mathrm{Pr}oj(\lambda) = \begin{cases} 0, \text{if} \begin{cases} \hat{\sigma} = \sigma_{\min} \text{ and } \lambda < 0 \\ \hat{\sigma} = \sigma_{\max} \text{and } \lambda > 0 \end{cases} \\ \lambda, \text{others} \end{cases}, \tag{36}$$

When using the control law of Equation (29) and the adaptive law of Equation (35), the adaptive robust current control block diagram is shown in Figure 7.

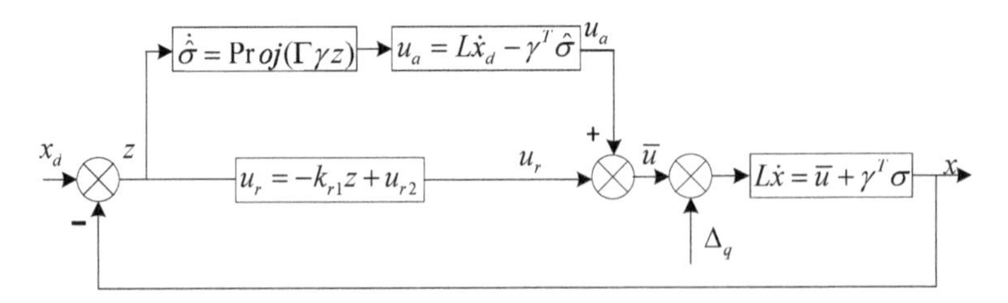

Figure 7. Adaptive robust current control block diagram.

5. Simulation and Experimental Results

5.1. Drive System Simulation

A two-disc AFPMSM simulation system is set up, and the current controller adopts adaptive robust control method and PI control, respectively. Finally, compare the current tracking performance of the two controllers. The current controller adopts the control strategy of $i_d = 0$, the d-axis current adopts PI controller, the q-axis current adopts adaptive robust controller, the given current i_q* is 150 A, and the i_q current response waveform is shown in Figure 8. The three-phase current waveform is shown in Figure 9.

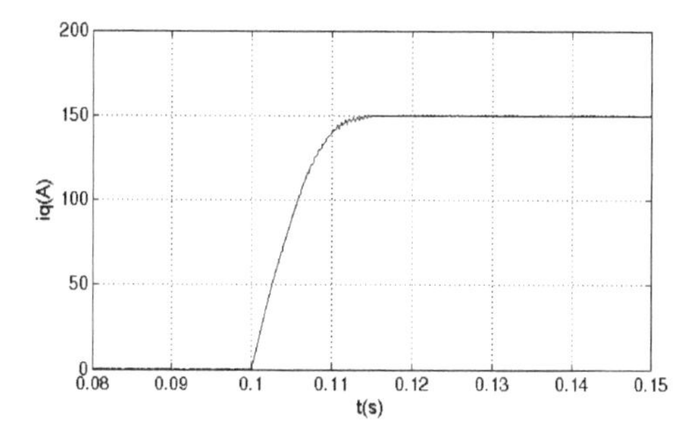

Figure 8. I_q current response waveform based on adaptive robust current controller.

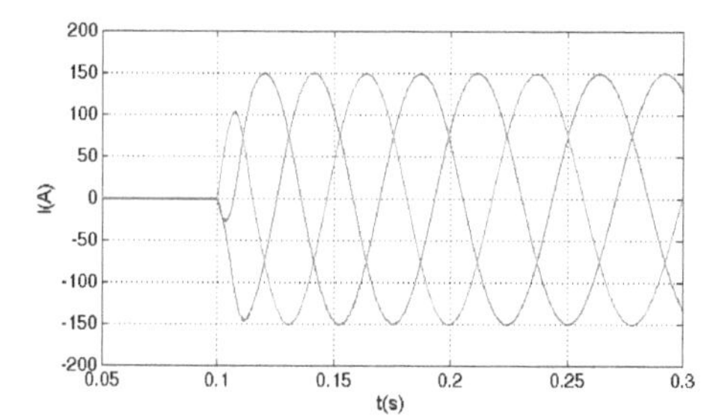

Figure 9. Three-phase current waveform based on adaptive robust current controller.

When the given current i_q* is also 150 A, and the d-axis and q-axis currents all use the PI controller, the i_q current response waveform is shown in Figure 10, and the three-phase current waveform is shown in Figure 11.

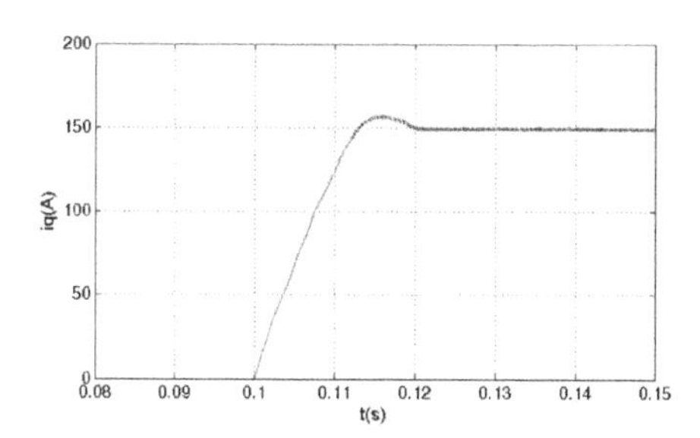

Figure 10. I_q current response waveform based on PI current controller.

It can be seen from the simulation waveform that the control system based on adaptive robust current controller has a response time of about 14 ms from 0 A to 150 A, and the current has almost no overshoot, and the current fluctuation is small at steady state. The control system based on the PI current controller has a response time of approximately 21 ms from 0 A to 150 A and a current overshoot of approximately 6%. The simulation results show that the adaptive robust current controller designed in this paper has better current control performance.

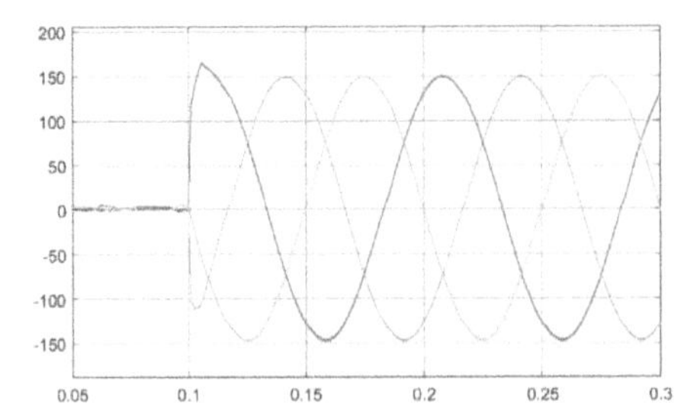

Figure 11. Three-phase current waveform based on PI current controller.

5.2. Experimental Platform Verification Experiment

Figure 12 shows the AFPMSM tow experimental platform. Firstly, the drive test experiment is carried out, and the AFPMSM is used as the drive motor to load. The AC asynchronous motor works in the fixed speed mode, the rotation speed is 1000 rpm. After the rotation speed is stable, the given torque is 150 Nm. The A and B phase current waveforms are shown in Figure 13. The oscilloscope waveform amplitude is about 1.5 V. Then the torque tracking experiment was carried out, and the torque was abruptly changed from 0 Nm to 120 Nm. The torque waveform is shown in Figure 14.

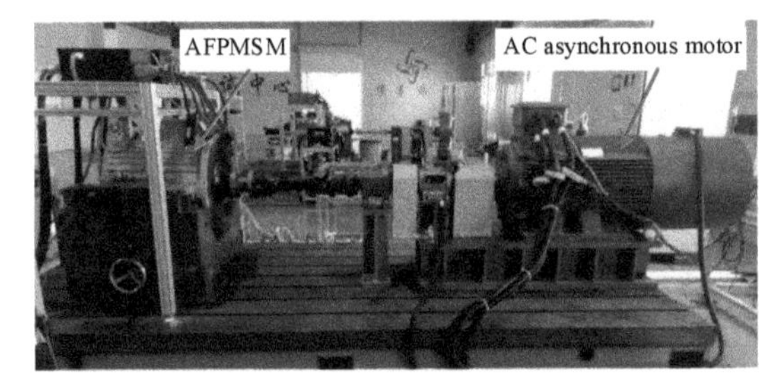

Figure 12. AFPMSM tow experimental platform.

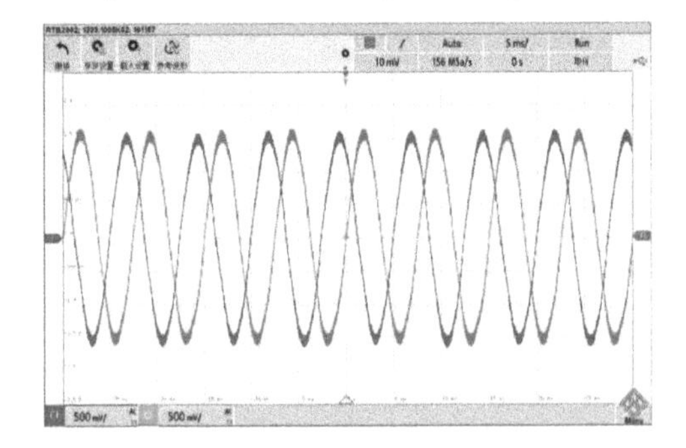

Figure 13. A and B phase current waveforms when the given torque is 150 Nm.

The motor drive system efficiency experiments were carried out under the torque average distribution and the torque optimal distribution control strategy based on PSO. The maximum speed of the motor is 4500 rpm. During the experiment, the motor speed is from 500 rpm to 4500 rpm, and one speed value is taken every 100 rpm for a total of 41 speed values. At every speed value, the output torque is increased from 0 Nm to 600 Nm, and a torque value is selected every 30 Nm. Twenty torque values correspond to 820 efficiency test values. The output torque T, the rotational speed n, the DC-side input voltage U, and the current I of each efficiency point are acquired by a sensor and a storage recording instrument. After calculating the efficiency of the motor drive system, the speed, output torque and efficiency values are finally imported into MATLAB to generate a motor efficiency map. The efficiency map generated by the torque average distribution and the torque optimal distribution control strategy based on the particle swarm optimization algorithm is shown in Figures 15 and 16.

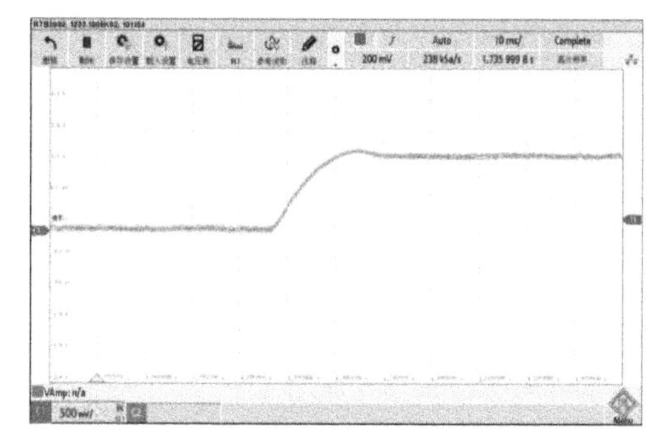

Figure 14. Drive system torque tracking waveform.

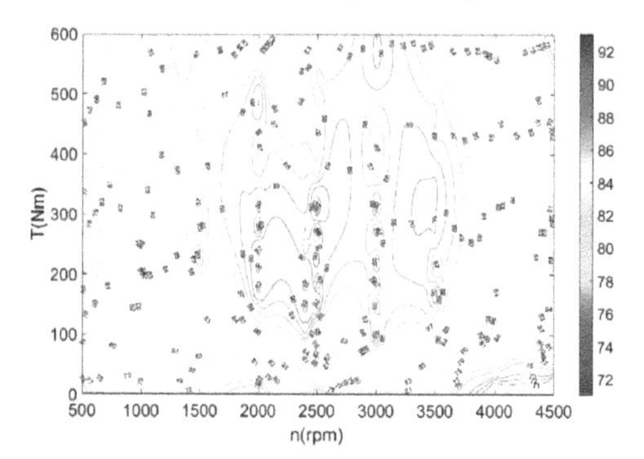

Figure 15. System efficiency map based on traditional torque average distribution strategy.

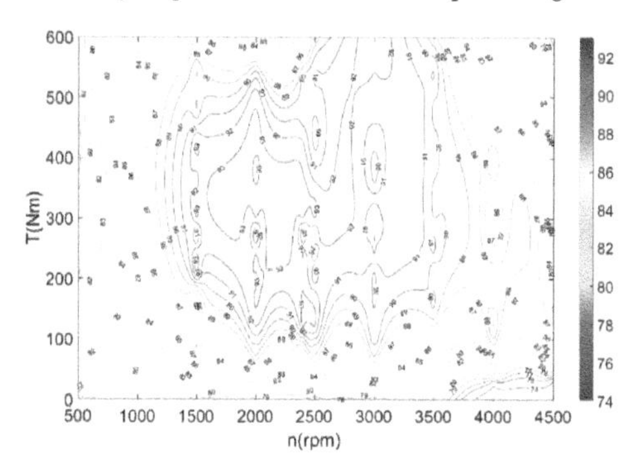

Figure 16. System efficiency map of torque optimal distribution control strategy based on particle swarm optimization.

After comparison, the optimal torque distribution control strategy based on particle swarm optimization algorithm can significantly increase the high efficiency range of AFPMSM. The system efficiency increases by about 15% in the interval of 85% or more, and the system efficiency increases by 20% in the interval of 90% or more. The correctness of the proposed optimal torque distribution control strategy based on particle swarm optimization is verified.

As in the driving state, the efficiency map generated by the torque average distribution and the torque optimal distribution control strategy based on the particle swarm optimization algorithm in the braking situation is as shown in Figures 17 and 18.

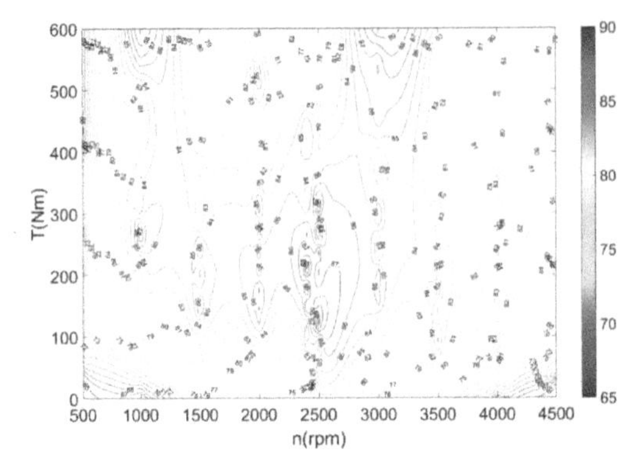

Figure 17. System efficiency map based on traditional torque average allocation strategy in energy feedback state.

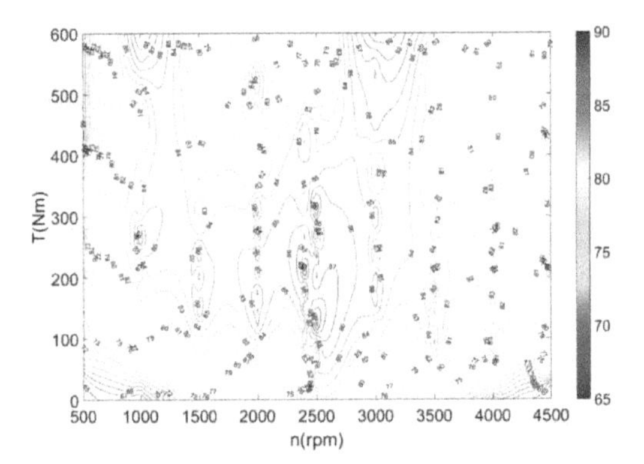

Figure 18. System efficiency map of torque optimal distribution control strategy based on particle swarm optimization in energy feedback state.

After comparison, the optimal torque distribution control strategy based on particle swarm optimization algorithm can significantly increase the high efficiency range of AFPMSM in energy feedback state. The system efficiency is increased by about 25% in the interval above 85%, and the system efficiency increases by 10% in the interval of 90% or more, which verifies the correctness of the optimal torque distribution control strategy in the energy feedback state.

5.3. Vehicle Experiment

As shown in Figure 19, the experimental vehicle is used to simulate the electric medium bus by means of load. The electric mid-size bus simulated in this experiment has an empty load of about 3000 kg. The battery specifications are shown in Table 1. The internal structure of the experimental vehicle is shown in Figure 20. The endurance capability based on the torque average distribution and the optimal torque distribution control strategy based on the particle swarm optimization algorithm is tested on the urban road. The load conditions are no-load and 700 kg (about 10 people). With 300 km as the target cruising range, the data in the table indicates the percentage of completion. The experimental results are shown in Table 2.

Figure 19. Experimental vehicle.

Table 1. The battery specifications.

The Battery Specifications	Specific Parameters
Battery Type	Ternary Polymer Lithium Battery
Battery Capacity	80 kWh
Battery Rated Voltage	384 V
State of Charge	10–100%

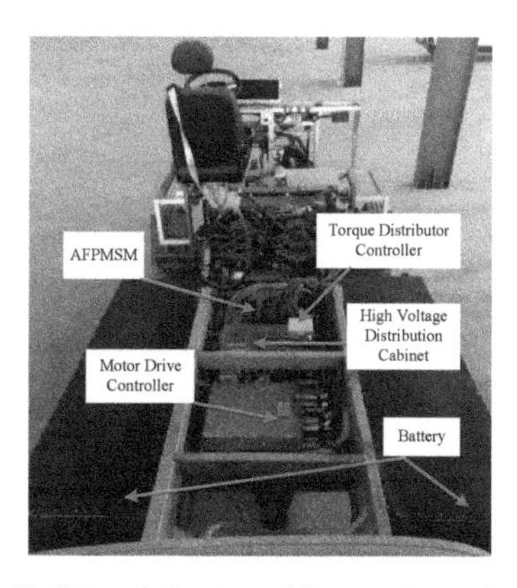

Figure 20. Internal structure of the experimental vehicle.

Table 2. Electric medium-sized passenger car endurance experiment.

Experimental Condition	Torque-Based Average Distribution Method	Optimal Torque Distribution Method
No load	89.67%	97.67%
700 kg load	77.33%	86.00%

Table 2 showed that the optimal torque distribution control strategy based on PSO algorithm can improve the cruising range of 8% under no-load conditions and increase the cruising range by 8.67% under 700 kg load. Therefore, the optimal torque distribution method designed in this paper can effectively improve the system efficiency and improve the endurance.

6. Conclusions

This paper mainly studies the drive control system of electric medium bus. The selected motor is a dual-stator single-rotor AFPMSM, which can be equivalent to two PMSM connected coaxially, therefore the drive control system is divided into two parts: torque distribution and motor control. The first part is the study of the torque distribution method that maximizes the efficiency of the dual-motor system. Simulation and experimental results show that the optimal torque distribution method proposed in this paper can effectively improve system efficiency and endurance. The second part is the motor control part. Simulation and experimental results show that the system based on adaptive robust current controller proposed in this paper has better anti-interference ability and stability than the system based on PI current controller. At the same time, this article also lays the foundation for the subsequent research of multiple modular dual-stator single-rotor combined motor systems, and provides some useful methods and ideas for the drive control system of AFPMSM.

Author Contributions: Conceptualization, S.W., J.Z. and T.L.; methodology, S.W. and J.Z.; software, S.W. and M.H.; validation, J.Z. and T.L.; formal analysis, J.Z. and S.W.; investigation, S.W. and M.H.; resources, J.Z. and T.L.; data curation, S.W. and M.H.; writing—Original draft preparation, J.Z., S.W. and M.H.; writing—Review and editing, J.Z., S.W. and T.L.; visualization, M.H. supervision, J.Z. and S.W.; project administration, J.Z.; funding acquisition, S.W.

Funding: This research was supported by the Shanghai Natural Science Foundation under Grant 19ZR1418600.

Conflicts of Interest: The authors declare no conflict of interest.

References

1. Wang, X.; González, J.A. Assessing feasibility of electric buses in small and medium-sized communities. *Int. J. Sustain. Transp.* **2013**, *7*, 431–448. [CrossRef]
2. Vepsäläinen, K.J.; Tammi, K. Stochastic driving cycle synthesis for analyzing the energy consumption of a battery electric bus. *IEEE Access* **2018**, *6*, 55586–55598.
3. Li, L.; Yang, C.; Zhang, Y.L.; Song, J. Correctional DP-Based Energy Management Strategy of Plug-In Hybrid Electric Bus for City-Bus Route. *IEEE Trans. Veh. Technol.* **2015**, *64*, 2792–2803. [CrossRef]
4. Wang, W.; Zhang, Z.; Shi, J.; Lin, C.; Gao, Y. Optimization of a dual-motor coupled powertrain energy management strategy for a battery electric bus based on dynamic programming method. *IEEE Access* **2018**, *6*, 32899–32909. [CrossRef]
5. Kommuri, S.K.; Defoort, M.; Karimi, H.R.; Veluvolu, K.C. A Robust Observer-Based Sensor Fault-Tolerant Control for PMSM in Electric Vehicles. *IEEE Trans. Ind. Electron.* **2016**, *63*, 7671–7681. [CrossRef]
6. Sant, A.V.; Khadkikar, V.; Xiao, W.; Zeineldin, H.H. Four-Axis Vector-Controlled Dual-Rotor PMSM for Plug-in Electric Vehicles. *IEEE Trans. Ind. Electron.* **2015**, *62*, 3202–3212. [CrossRef]
7. Huang, J.; Liu, Y.; Liu, M.; Cao, M.; Yan, Q. Multi-Objective Optimization Control of Distributed Electric Drive Vehicles Based on Optimal Torque Distribution. *IEEE Access* **2019**, *7*, 16377–16394. [CrossRef]
8. Yuan, X.; Wang, J.; Colombage, K. Torque distribution strategy for a front and rear wheel driven electric vehicle. In Proceedings of the 6th IET International Conference on Power Electronics, Machines and Drives, Bristol, UK, 27–29 March 2012; pp. 1–6.
9. Chen, L.; Chen, T.; Xu, X.; Jiang, H.; Sun, X. Multi-objective coordination control strategy of distributed drive electric vehicle by orientated tire force distribution method. *IEEE Access* **2018**, *6*, 69559–69574. [CrossRef]
10. Xu, G.; Xu, K.; Zheng, C.; Zhang, X.; Zahid, T. Fully Electrified Regenerative Braking Control for Deep Energy Recovery and Safety Maintaining of Electric Vehicles. *IEEE Trans. Veh. Technol.* **2016**, *65*, 1186–1198. [CrossRef]
11. Li, W.; Du, H.; Li, W. Driver intention based coordinate control of regenerative and plugging braking for electric vehicles with in-wheel PMSMs. *IET Intell. Transp. Syst.* **2018**, *12*, 1300–1311. [CrossRef]
12. Heydari, S.; Fajri, P.; Rasheduzzaman, M.; Sabzehgar, R. Maximizing Regenerative Braking Energy Recovery of Electric Vehicles through Dynamic Low-Speed Cutoff Point Detection. *IEEE Trans. Transp. Electrif.* **2019**, *5*, 262–270. [CrossRef]
13. Jain, R.; Tandon, P.; Vasantha, K. Optimization methodology for beam gauges of the bus body for weight reduction. *Appl. Comput. Mech.* **2014**, *8*, 47–62.
14. Croccolo, D.; Agostinis, M.D.; Vincenzi, N. Structural Analysis of an Articulated Urban Bus Chassis via Finite Element Method: A Methodology Applied to a Case Study. *J. Mech. Eng.* **2011**, *57*, 799–809. [CrossRef]
15. Kongwat, S.; Jongpradist, P.; Kamnerdtong, T. Optimization of Bus Body based on Structural Stiffness and Rollover Constraints. In Proceedings of the Asian Congress of Structural and Multidisciplinary Optimization, Nagasaki, Japan, 22–26 May 2016; pp. 22–26.
16. Kunakron-ong, P.; Ruangjirakit, K.; Jongpradist, P. Design and analysis of electric bus structure in compliance with ECE safety regulations. In Proceedings of the 2017 2nd IEEE International Conference on Intelligent Transportation Engineering (ICITE), Singapore, 1–3 September 2017; pp. 25–29.

17. Hemeida, A.; Sergeant, P. Analytical modeling of surface PMSM using a combined solution of Maxwell–s equations and magnetic equivalent circuit. *IEEE Trans. Magn.* **2014**, *50*, 1–13. [CrossRef]
18. Lim, D.K.; Cho, Y.S.; Ro, J.S.; Jung, S.Y.; Jung, H.K. Optimal design of an axial flux permanent magnet synchronous motor for the electric bicycle. *IEEE Trans. Magn.* **2015**, *52*, 1–4. [CrossRef]
19. Chen, N.Z.; Yao, B.; Wang, Q. μ-Synthesis-Based Adaptive Robust Control of Linear Motor Driven Stages with High-Frequency Dynamics: A Case Study. *IEEE/ASME Trans. Mechatron.* **2015**, *20*, 1482–1490. [CrossRef]
20. Wang, Z.; Hu, C.; Zhu, Y.; He, S.; Yang, K.; Zhang, M. Neural Network Learning Adaptive Robust Control of an Industrial Linear Motor-Driven Stage with Disturbance Rejection Ability. *IEEE Trans. Ind. Inform.* **2017**, *13*, 2172–2183. [CrossRef]
21. Yang, Y.; Wang, Y.; Jia, P. Adaptive robust control with extended disturbance observer for motion control of DC motors. *Electron. Lett.* **2015**, *51*, 1761–1763. [CrossRef]
22. Zhao, J.; Hua, M.; Liu, T. Collaborative Optimization and Fault Tolerant Control Method for Multi-disc Permanent Magnet Synchronous Motors for Electric Vehicles. *Proc. CSEE* **2019**, *39*, 386–394.
23. Gan, C.; Jin, N.; Sun, Q.; Kong, W.; Hu, Y.; Tolbert, L.M. Multiport bidirectional SRM drives for solar-assisted hybrid electric bus powertrain with flexible driving and self-charging functions. *IEEE Trans. Power Electron.* **2018**, *33*, 8231–8245. [CrossRef]
24. Bonyadi, M.R.; Michalewicz, Z. Analysis of Stability, Local Convergence, and Transformation Sensitivity of a Variant of the Particle Swarm Optimization Algorithm. *IEEE Trans. Evol. Comput.* **2016**, *20*, 370–385. [CrossRef]

 energies

Article

Online Current Loop Tuning for Permanent Magnet Synchronous Servo Motor Drives with Deadbeat Current Control

Zih-Cing You, Cheng-Hong Huang and Sheng-Ming Yang *

Electrical Engineering, National Taipei University of Technology, Taipei 10608, Taiwan; carefree60024@gmail.com (Z.-C.Y.); yyu124p@gmail.com.tw (C.-H.H.)
* Correspondence: smyang@ntut.edu.tw

Received: 5 August 2019; Accepted: 13 September 2019; Published: 17 September 2019

Abstract: High bandwidths and accurate current controls are essential in high-performance permanent magnet synchronous (PMSM) servo drives. Compared with conventional proportional–integral control, deadbeat current control can considerably enhance the current control loop bandwidth. However, because the deadbeat current control performance is strongly affected by the variations in the electrical parameters, tuning the controller gains to achieve a satisfactory current response is crucial. Because of the prompt current response provided by the deadbeat controller, the gains must be tuned within a few control periods. Therefore, a fast online current loop tuning scheme is proposed in this paper. This scheme can accurately identify the controller gain in one current control period because the scheme is directly derived from the discrete-time motor model. Subsequently, the current loop is tuned by updating the deadbeat controller with the identified gains within eight current control periods or a speed control period. The experimental results prove that in the proposed scheme, the motor current can simultaneously have a critical-damped response equal to its reference in two current control periods. Furthermore, satisfactory current response is persistently guaranteed because of an accurate and short time delay required for the current loop tuning.

Keywords: deadbeat current control; PMSM servo motor drives; auto tuning; parameter identification

1. Introduction

A modern servo motor drive usually includes current, speed, and position control loops. In general, the current loop bandwidth is considerably higher than the bandwidth of the speed and position loops. Therefore, a current loop with a high bandwidth can fundamentally enhance the performance of the servo motor drive.

When the current loop is implemented with a digital signal processor (DSP), because of the limited computation capability, the calculated voltage command requires one control period delay for the pulse width modulation (PWM) module to output voltage to the motor. This time delay causes an underdamped or unstable current response when a proportional–integral (PI) controller is used for motor current regulation [1–4]. The discretized PI controller directly designed in the z-domain has been proposed in [3,4]; however, limited improvement in the current loop bandwidth was achieved and current overshoots were persistent. To eliminate the influences of the time delay, schemes based on the predictive current control [5–11] and deadbeat current control [12–17] have been proposed. The predictive current controller generates the optimal voltage vector by minimizing a specific cost function. This voltage vector allows the motor current to reach its reference value as fast as possible with minimum overshoot. Deadbeat current control is well-known for its zero overshoot, zero steady-state error, and minimum rise time characteristics. Consequently, the motor current can reach its reference value with minimum control periods without overshoot. Compared with predictive current control, deadbeat control is simple to implement and requires less computation. However, its performance is parameter-dependent, as reported in [15–17]. In particular, inductance is sensitive to the current level. Online controller gain tuning is an effective method to mitigate the effects of parameter variations.

Numerous online electrical parameter identification strategies have been proposed. The observer-based methods in [18–21] identify the parameters by converging the error between the sampled and estimated current to zero. In [18], the identified inductance was used for the predictive current controller to improve the robustness of the current loop. Observer-based methods often require long execution times because of the delay of the observer and may encounter stability problems. The authors in [22–24] performed the recursive least-square (RLS) algorithm to identify electrical parameters. The motor model was used to develop the RLS algorithm. Then, the parameters were identified by minimizing the discrepancy between the sampled and calculated current. Although the latency caused by the observer does not exist in RLS-based methods, accurately identifying the parameters in a few control periods is still difficult. In addition, the electrical parameters are generally identified instead of the controller gains in these methods. However, the effect caused by the parameter mismatch can be treated as a disturbance to the current controller. To compensate for this disturbance, the compensation voltage, which was obtained through the disturbance observer in [12,17] and through adaptive control in [25], is added to the current loop. Despite their effectiveness, the schemes in [17,25] involve a complex design procedure to achieve satisfactory performance.

In this study, a deadbeat current controller was designed to enhance the current loop bandwidth for its simple implementation. A novel online current loop tuning strategy is proposed to reduce the effect of parameter variations. The proposed method is simple and effective because the method is directly derived from the discrete-time motor model. In addition, the proposed method directly identifies the gains of the deadbeat controller instead of the electrical parameters. After the controller gains are identified, the gains are averaged to further improve accuracy. Then, the current loop is tuned by updating the deadbeat controller with the average gains.

2. Discrete-Time Motor Model

The stator voltage of a PMSM in the rotor reference frame can be expressed as follows:

$$\begin{bmatrix} v_{qs}^r \\ v_{ds}^r \end{bmatrix} = \begin{bmatrix} r_s + sL_{qs} & \omega_r L_{ds} \\ -\omega_r L_{qs} & r_s + sL_{ds} \end{bmatrix} \begin{bmatrix} i_{qs}^r \\ i_{ds}^r \end{bmatrix} + \begin{bmatrix} \omega_r \lambda_m \\ 0 \end{bmatrix} \tag{1}$$

where v_{qs}^r, v_{ds}^r, i_{qs}^r, and i_{ds}^r are the q- and d-axis voltages and currents, respectively; L_{qs} and L_{ds} are the q- and d-axis inductance, respectively; r_s, ω_r, and λ_m are the phase resistance, rotor electrical speed, and magnet flux, respectively; and s denotes the Laplace operator. When the current loop and PWM function of the PMSM are implemented digitally, a time delay is inevitably introduced. Figure 1 displays the time sequence of the current sampling, voltage command calculation, and voltage command output, where T_s is the sampling period of the motor currents and the control period of the current loop. As depicted in Figure 1, the voltage command is calculated at t_0 and outputs to the PWM module at $t_0 + T_s$. Then, the voltage command is activated by the PWM module and held for one sampling period during $t_0 + T_s$ to $t_0 + 2T_s$. The motor current induced by the corresponding voltage command is then sampled at $t_0 + 2T_s$. Therefore, a time delay of two sampling periods is generated in the current loop. The PWM function and calculation delay can be modeled together by using a zero-order hold involving one sampling period delay. Accordingly, the stator voltage in Equation (1) can be discretized as follows:

$$G_q(z) = \frac{i_{qs}^r(z)}{v_{qs}^r(z)} = Z\left\{e^{-sT_s} \cdot ZOH\left(\frac{1}{L_{qs}s + r_s}\right)\right\} = \frac{B_{mq}z^{-2}}{1 - A_{mq}z^{-1}} \tag{2}$$

$$G_d(z) = \frac{i_{ds}^r(z)}{v_{ds}^r(z)} = Z\left\{e^{-sT_s} \cdot ZOH\left(\frac{1}{L_{ds}s + r_s}\right)\right\} = \frac{B_{md}z^{-2}}{1 - A_{md}z^{-1}} \tag{3}$$

where Z{} is the Z-transform; the model gains A_{mq} and B_{mq} are $e^{-T_s r_s/L_{qs}}$ and $(1 - A_{mq})/r_s$, respectively; and the model gains A_{md} and B_{md} are $e^{-T_s r_s/L_{ds}}$ and $(1 - A_{md})/r_s$, respectively. Because the back-EMF and cross-coupling voltages are approximately constant within one control period, these voltages are assumed to be decoupled from the current controller and are not represented in Equations (2) and (3). The q- and d-axis decoupling voltages, namely v_{qff} and v_{dff}, respectively, are derived from Equation (1) by using the estimated electrical parameters and rotor speed in the following expression:

$$\begin{aligned} v_{qff} &= \omega_r \hat{L}_{ds} i_{ds}^r + \omega_r \lambda_m \\ v_{dff} &= -\omega_r \hat{L}_{qs} i_{qs}^r \end{aligned} \quad (4)$$

where "^" denotes the estimated quantity.

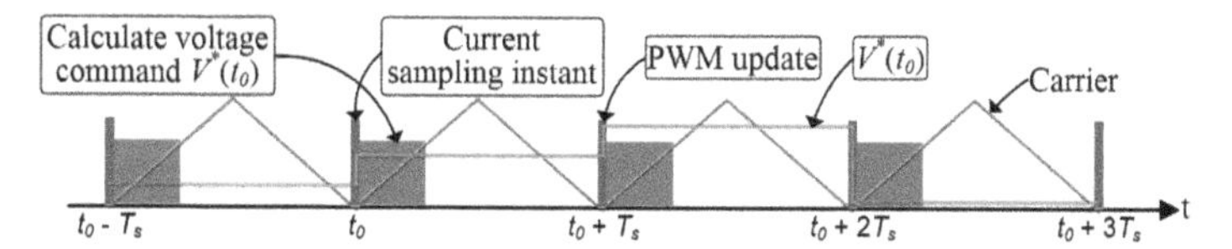

Figure 1. Time sequence for current sampling, voltage command calculation, and output, where PWM is pulse-width modulation.

3. Overall Control System

3.1. Servo Control System

Figure 2 illustrates the overall servo control system, where v_{an}, v_{bn}, and v_{cn} are the phase voltages; θ_m, θ_r, and ω_m denote the mechanical position, electrical angle, and speed, respectively; and "*" denotes the command value. The motor current is regulated using a deadbeat current controller. The classical proportional position with proportional-plus-integral velocity (P-PI) control is implemented to regulate the motor speed and position [26]. The bandwidths of the speed loop and position loop are set as 100 and 10 Hz, respectively. The proposed online current loop tuning algorithm continuously tunes the gains in the current loop to achieve a satisfactory current response. The variables associated with the online tuning are defined in the following text.

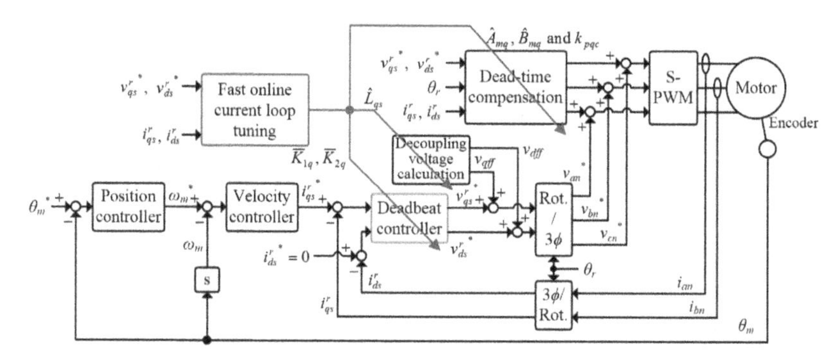

Figure 2. Block diagram of servo control system with the proposed online current loop tuning strategy.

3.2. Dead-Time Compensation

High-performance servo motor drives generally include dead-time compensation. The voltage error caused by the dead time can be measured through the steady-state voltage command and current feedback [27]. The voltage error at different phase currents for the inverter used in this study is depicted in Figure 3. The configuration of the motor drive is listed in Table A1. The voltage error saturates when the magnitude of the phase current is higher than 1 A. After the voltage error is calculated with the phase current feedback, dead-time compensation is performed by adding the voltage error to the current loop, as depicted in Figure 2.

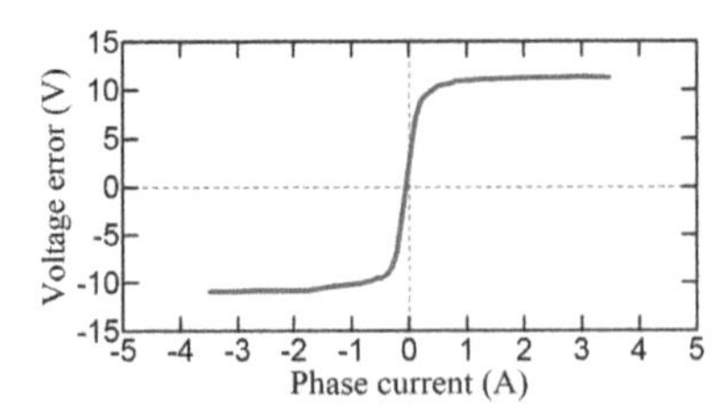

Figure 3. The voltage error caused by the dead-time at different phase current.

The time delay in the current loop can degrade the effectiveness of dead-time compensation. To mitigate the influence of this time delay, the estimated phase current is used to calculate the voltage error. Figure 4 illustrates the proposed current estimator. A PI controller is used to reduce the error between the sampled and estimated current. By ignoring the PI controller, the estimated current can be expressed as follows:

$$\hat{i}^r_{qs}(k) = \hat{B}_{mq} \cdot v^{r}_{qs}{}^*(k) + \hat{A}_{mq} \cdot \hat{i}^r_{qs}(k-1) \tag{5}$$

$$\hat{i}^r_{ds}(k) = \hat{B}_{md} \cdot v^{r}_{ds}{}^*(k) + \hat{A}_{md} \cdot \hat{i}^r_{ds}(k-1) \tag{6}$$

When the parameters are correct, the estimated current approximates the current sampled at the (k + 2)th sampling instant, which is induced by $v^{r*}_{qs}(k)$ and $v^{r*}_{ds}(k)$. An operator z^{-2} is added in the feedback path of the estimator because the estimated current is from two sampling periods before the present sampled current. Then, the estimated phase current can be calculated as follows:

$$\begin{bmatrix} \hat{i}_{an} \\ \hat{i}_{bn} \\ \hat{i}_{cn} \end{bmatrix} = \begin{bmatrix} 1 & 0 \\ -1/2 & -\sqrt{3}/2 \\ -1/2 & \sqrt{3}/2 \end{bmatrix} \begin{bmatrix} \cos\theta_r & \sin\theta_r \\ -\sin\theta_r & \cos\theta_r \end{bmatrix} \begin{bmatrix} \hat{i}^r_{qs} \\ \hat{i}^r_{ds} \end{bmatrix} \tag{7}$$

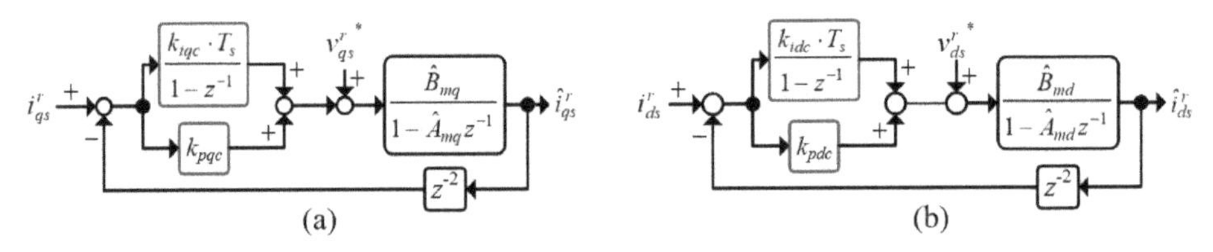

Figure 4. The proposed (**a**) q-axis and (**b**) d-axis current estimator.

The pole-zero cancelation technique is used to design the PI controller of the current estimators. By canceling the plant pole with the controller zero, the proportional gain k_{pqc} and k_{pdc} are calculated as follows:

$$k_{pqc} = k_{iqc} \cdot \hat{A}_{mq} \cdot T_s / \left(1 - \hat{A}_{mq}\right) \tag{8}$$

$$k_{pdc} = k_{idc} \cdot \hat{A}_{md} \cdot T_s / \left(1 - \hat{A}_{md}\right) \tag{9}$$

where k_{iqc} and k_{idc} are the integral gains. Figure 5 shows the damping ratio and bandwidth of the current estimator for various k_{iqc} values. The bandwidth of the current estimator is determined from the integral gain, and the proportional gains are then calculated using Equations (8) and (9).

In this study, the bandwidth of the current estimator was set to 900 Hz because this can simultaneously ensure that the estimated current strictly follows the sampled feedback current and the estimator predicts the sampled current accurately. In addition, the current estimator has a critical-damped response.

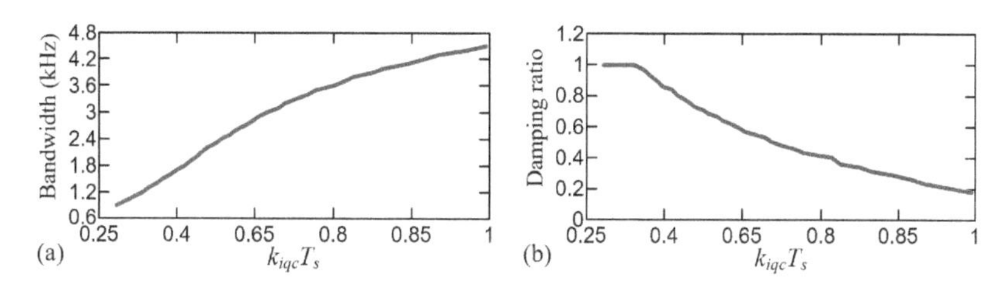

Figure 5. The (**a**) bandwidth and (**b**) damping ratio of the current estimator at different integral gain.

4. Deadbeat Current Controller

When a conventional PI controller is used to regulate the motor current, the time delay in the control loop can cause stability problems because of the degraded phase margin. Consequently, the current loop bandwidth is limited to maintaining an acceptable overshoot on the motor current. To enhance the bandwidth of the current loop to its theoretical maximum, a deadbeat current controller was developed in this study.

Deadbeat Controller Design

Figure 6 depicts the schematics of the q- and d-axis current control loops with the deadbeat controller, where $C_q(z)$ and $C_d(z)$ are the deadbeat controllers for the q- and d-axes, respectively. Deadbeat controller design is conducted entirely in the z-domain. All the closed-loop poles are placed at the origin in the z-domain. The q-axis transfer function is expressed as follows:

$$\frac{i_{qs}^r}{i_{qs}^{r\ *}} = \frac{C_q(z) \cdot G_q(z)}{1 + C_q(z) \cdot G_q(z)} = \frac{h(z)}{z^n} \tag{10}$$

where the numerator $h(z)$ provides an additional degree of freedom for the controller design and n is the number of poles. The q-axis current should strictly follow the command value without a steady-state error. By applying the finite-value theorem to Equation (10), the following result is obtained:

$$\lim_{z=1}\left[(z-1) \cdot \left(\frac{i_{qs}^r}{i_{qs}^{r\ *}} \cdot \frac{z}{z-1}\right)\right] = \lim_{z=1}\frac{h(z)}{z^n} = 1 \tag{11}$$

For convenience, $h(z)$ is set as 1. The difference form of Equation (10) can be derived as follows:

$$i_{qs}^r(k) = i_{qs}^{r\ *}(k-n) \tag{12}$$

The results indicate that the q-axis current lags the command value by n control periods when $h(z) = 1$. Except for the zero steady-state error, the q-axis current can also reach the command value without an overshoot.

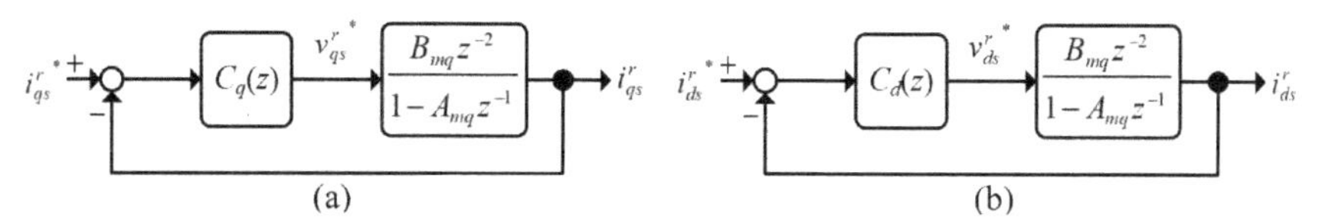

Figure 6. (**a**) q-axis and (**b**) d-axis current loop with the deadbeat controller when the motor is at standstill.

According to Equation (10) and the relation $h(z) = 1$, the deadbeat controller $C_q(z)$ can be derived using the estimated electrical parameters. The deadbeat controller $C_q(z)$ is expressed as follows:

$$C_q(z) = \frac{z^2 - \hat{A}_{mq} \cdot z}{\hat{B}_{mq} \cdot (z^n - 1)} \tag{13}$$

The transfer function of the q-axis voltage command is given as follows:

$$\frac{v_{qs}^{r\ *}}{i_{qs}^{r\ *}} = \frac{z^2 - \hat{A}_{mq} z}{\hat{B}_{mq} z^n} \tag{14}$$

To satisfy the causality, the following inequality must be satisfied:

$$n \geq \deg\left\{z^2 - \hat{A}_{mq} z\right\} = 2 \tag{15}$$

where deg{} denotes the highest order of the polynomial. When n is selected to be 2 and the parameters are perfectly matched, the q-axis current can attain the steady-state and equal the command value in two control periods. In addition, the voltage command can attain the steady state in two control periods after the current command changes. Therefore, $C_q(z)$ is modified as follows:

$$C_q(z) = \frac{z^2 - \hat{A}_{mq} \cdot z}{\hat{B}_{mq} \cdot (z^2 - 1)} \tag{16}$$

The q-axis voltage command at the kth control instant can be derived from (16) as follows:

$$v_{qs}^{r\ *}(k) = v_{qs}^{r\ *}(k-2) + \hat{K}_{1q} \cdot \left(i_{qs}^{r\ *}(k) - i_{qs}^{r}(k)\right) - \hat{K}_{2q} \cdot \left(i_{qs}^{r\ *}(k-1) - i_{qs}^{r}(k-1)\right) \tag{17}$$

where $\hat{K}_{1q} = 1/\hat{B}_{mq}$, and $\hat{K}_{2q} = \hat{A}_{mq}/\hat{B}_{mq}$. Similarly, the deadbeat controller $C_d(z)$ can be derived as follows:

$$C_d(z) = \frac{z^2 - \hat{A}_{md} \cdot z}{\hat{B}_{md} \cdot (z^2 - 1)} \tag{18}$$

The d-axis voltage command at the kth sampling instant is expressed as follows:

$$v_{ds}^{r\ *}(k) = v_{ds}^{r\ *}(k-2) + \hat{K}_{1d} \cdot \left(i_{ds}^{r\ *}(k) - i_{ds}^{r}(k)\right) - \hat{K}_{2d} \cdot \left(i_{ds}^{r\ *}(k-1) - i_{ds}^{r}(k-1)\right) \tag{19}$$

where $\hat{K}_{1d} = 1/\hat{B}_{md}$, and $\hat{K}_{2d} = \hat{A}_{md}/\hat{B}_{md}$. Figure 7 depicts the detailed schematics of $C_q(z)$ and $C_d(z)$ with the decoupling voltage and voltage limitation.

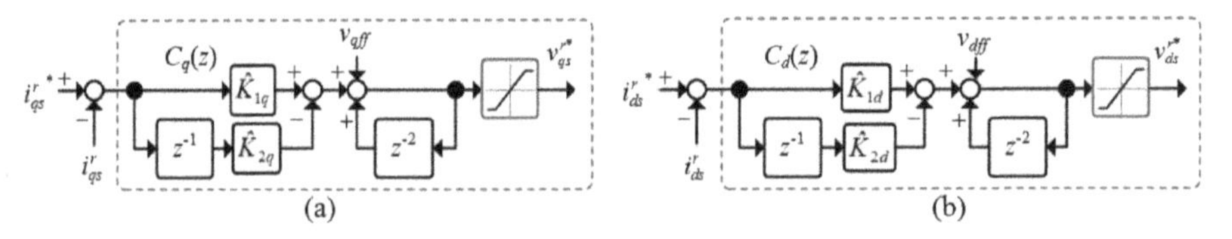

Figure 7. Block diagram of the deadbeat controller with the decoupling voltage and voltage limitation block, (**a**) $C_q(z)$ and (**b**) $C_d(z)$.

5. Simulation Results

A 400-W servo motor was used in the simulation. The motor parameters are listed in Table A2. The drive losses were ignored in the simulation. The voltage command was limited to half of the DC voltage because sinusoidal PWM was implemented, as illustrated in Figure 7. Because the d-axis current is expected to have a similar response as the q-axis current, only the q-axis current simulation results are presented.

5.1. Results with Correct Motor Parameters

Figures 8 and 9 illustrate the q-axis current, current command, and voltage command when the current steps from 0 to 1 A and from 0 to 4 A, respectively, when the motor is at standstill. As depicted in Figure 8, the q-axis current does not exhibit overshoot and is exactly equal to the command value in two control periods after the current command changes. The voltage command is generated immediately after the current command changes. The voltage commands at the kT_s and $(k + 1)T_s$ control periods can be calculated as follows:

$$v_{qs}^{r\ *}(kT_s) \approx L_{qs}\frac{i_{qs}^r((k+2)T_s) - i_{qs}^r((k+1)T_s)}{T_s} = 84.14 \text{ V} \tag{20}$$

$$v_{qs}^{r\ *}((k+1)T_s) \approx r_s \cdot i_{qs}^{r\ *}((k+2)T_s) = 2.1 \text{ V} \tag{21}$$

Note that these command values are less than half of the DC supply.

Conversely, as depicted in Figure 9, the q-axis current increases slowly and requires approximately six control periods to attain the command value because the voltage required for the current to increase from 0 to 4 A in one control period exceeds the command limit. The voltage command saturates several times before the q-axis current reaches its command value, which is in agreement with the control law presented in Equation (17). The actual rise time is dependent on the motor inductance.

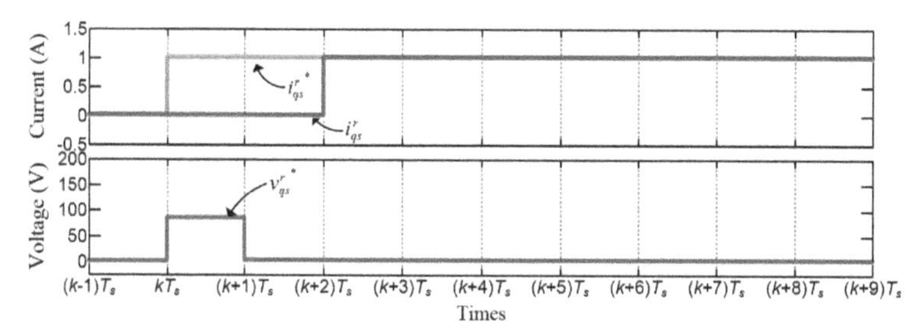

Figure 8. The simulated q-axis current, current command, and voltage command when the current command steps from 0 A to 1 A.

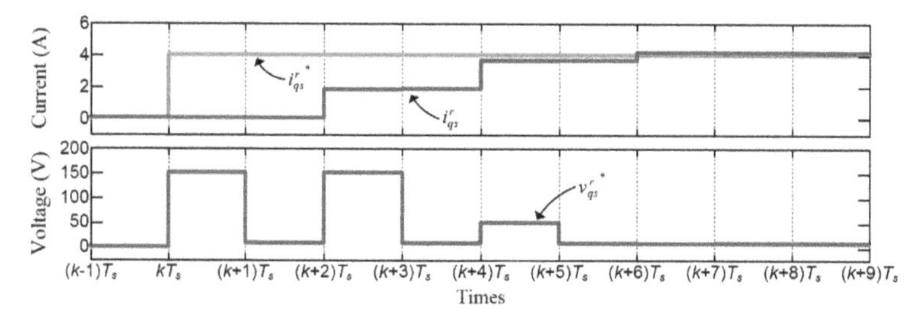

Figure 9. The simulated q-axis current, current command, and voltage command when the current command steps from 0 A to 4 A.

Figure 10 illustrates the frequency response of the deadbeat controller without voltage limitations. The current loop gain is 0 dB at low frequencies and is flat until the Nyquist frequency. This indicates that the current can follow its command without overshoot and steady-state error. However, the phase lag increases with frequency. The phase margin decreases to 0 at 4.575 kHz. Therefore, the maximum theoretical bandwidth of the proposed deadbeat controller is one-fourth of the control frequency.

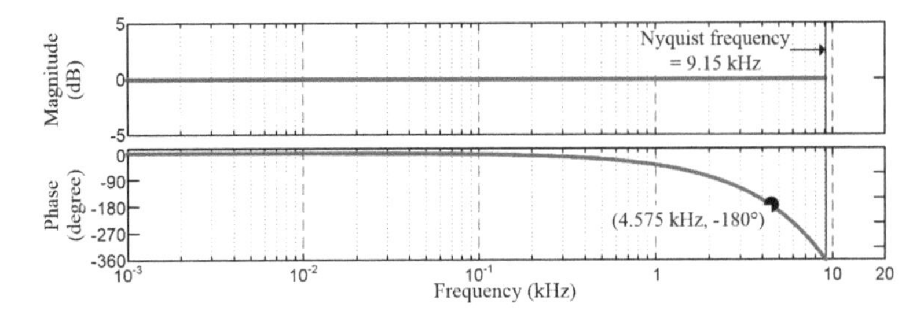

Figure 10. The frequency response of the deadbeat current controller without voltage limitation.

5.2. Results with Parameter Mismatch

The phase resistance and q- and d-axis inductances are required to design a deadbeat controller. The deadbeat controller performance is dependent on the accuracy of the estimated electrical parameters. Figures 11 and 12 display the dominant poles of the q-axis current loop in the z-domain and the corresponding q-axis current response when the current command steps from 0 to 1 A and the phase resistance varies 50% and 150% from its nominal value, respectively. The figures indicate that the poles remain near the origin regardless of the variations in the phase resistance. The q-axis current can still reach its command value in two control periods; however, marginal overshoot is observed. This implies that the influence of the resistance mismatch to the deadbeat controller is trivial. However, the q-axis current depicted in Figure 11 is marginally lower than its command value at the steady state because the resistance is smaller than its nominal value. Conversely, the q-axis current illustrated in Figure 12 is marginally higher than its command value at the steady state because the resistance is larger than its nominal value.

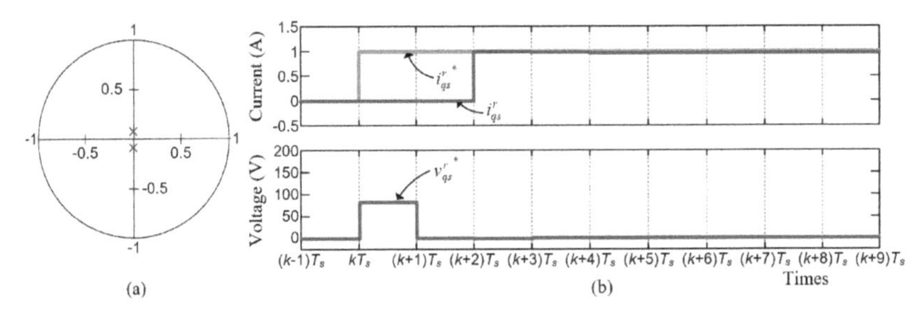

Figure 11. (**a**) The dominant poles and (**b**) simulated q-axis current and voltage command with $\hat{r}_s = 0.5r_s$ when the current command steps from 0 A to 1 A. The motor is at standstill.

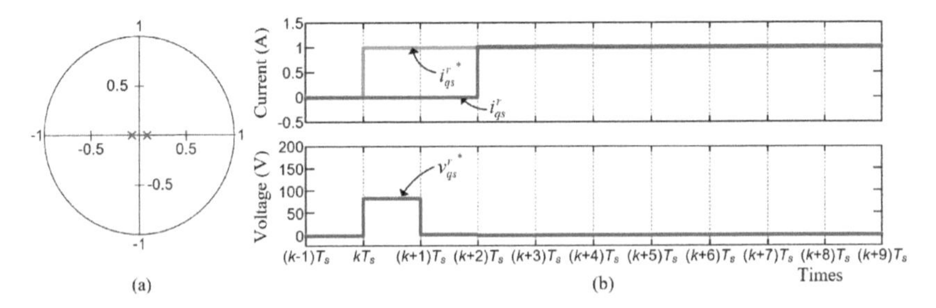

Figure 12. (**a**) The dominant poles and (**b**) simulated q-axis current and voltage command with $\hat{r}_s = 1.5r_s$ when the current command steps from 0 A to 1 A. The motor is at standstill.

Figures 13 and 14 illustrate the dominant poles of the q-axis current loop and the corresponding q-axis current response when the current command steps from 0 to 1 A and the q-axis inductance varies 50% and 120% from its nominal value, respectively. In contrast to the results depicted in Figures 11 and 12, the variations in the inductance considerably deteriorate the system performance. As depicted in Figure 13, the current response becomes overdamped when the inductance is smaller than its nominal value because the poles mitigate toward the unit circle along the real axis. However, in Figure 14, the current response becomes underdamped when the inductance is larger than its nominal value because the poles mitigate toward the unit circle along the imaginary axis. Although no steady-state error is observed, the transient response of the q-axis current is considerably affected. In addition, the current loop can become unstable if the poles mitigate outside the unit circle because of the mismatched inductance.

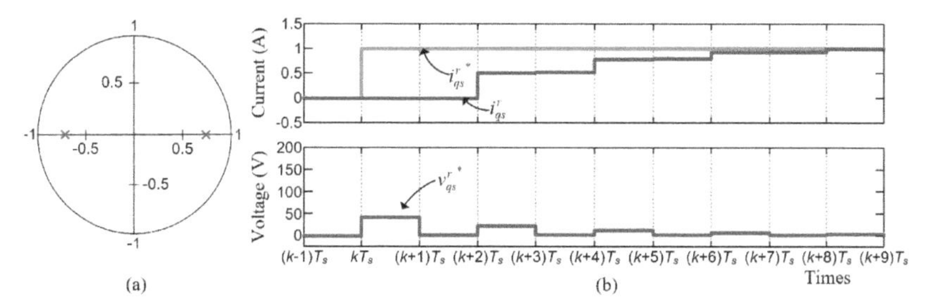

Figure 13. (**a**) The dominant poles and (**b**) simulated q-axis current and voltage command response with $\hat{L}_{qs} = 0.5L_{qs}$ when the current command steps from 0 A to 1 A, the motor is at standstill.

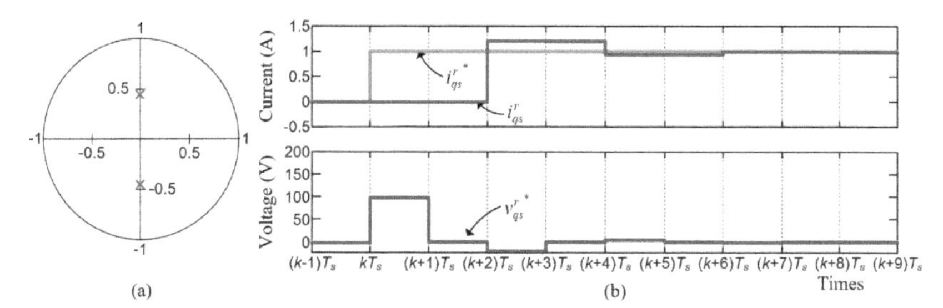

Figure 14. (**a**) The dominant poles and (**b**) simulated q-axis current and voltage command response with $\hat{L}_{qs} = 1.2L_{qs}$ when the current command steps from 0 A to 1 A, the motor is at standstill.

6. Online Current Loop Tuning

The deadbeat controller performance is considerably affected by parameter mismatch because the voltage command is directly related to the voltage drop on the inductance and resistance. In this study, a novel online current loop tuning strategy was developed to preserve the deadbeat controller performance. Only the q-axis current loop is discussed because similar results can be obtained for the d-axis current loop.

6.1. Controller Gain Identification

From Equation (2), the q-axis current sampled at the kth and $(k-1)$th control instants can be expressed as follows:

$$\begin{bmatrix} i_{qs}^r(k) \\ i_{qs}^r(k-1) \end{bmatrix} = \begin{bmatrix} B_{mq} \cdot v_{qs}^r(k-2) + A_{mq} \cdot i_{qs}^r(k-1) \\ B_{mq} \cdot v_{qs}^r(k-3) + A_{mq} \cdot i_{qs}^r(k-2) \end{bmatrix} \tag{22}$$

Equation (22) can be rearranged as follows:

$$\begin{bmatrix} i_{qs}^r(k) & -i_{qs}^r(k-1) \\ i_{qs}^r(k-1) & -i_{qs}^r(k-2) \end{bmatrix} \begin{bmatrix} K_{1q} \\ K_{2q} \end{bmatrix} = \begin{bmatrix} v_{qs}^r(k-2) \\ v_{qs}^r(k-3) \end{bmatrix} \tag{23}$$

where K_{1q} and K_{2q} are defined as $K_{1q} = 1/B_{mq}$ and $K_{2q} = A_{mq}/B_{mq}$, respectively. Because the voltage error caused by the dead-time is satisfactorily compensated, the controller gains K_{1q} and K_{2q} can be reasonably estimated using the command values, which are expressed as follows:

$$\begin{bmatrix} \hat{K}_{1q} \\ \hat{K}_{2q} \end{bmatrix} = \begin{bmatrix} i_{qs}^r(k) & -i_{qs}^r(k-1) \\ i_{qs}^r(k-1) & -i_{qs}^r(k-2) \end{bmatrix}^{-1} \begin{bmatrix} v_{qs}^{r\,*}(k-2) \\ v_{qs}^{r\,*}(k-3) \end{bmatrix} \tag{24}$$

To solve Equation (24), the determinant of the inverse matrix must be a nonzero value. This condition is expressed as follows:

$$\det\left(i_{qs}^r\right) = i_{qs}^r(k-1)^2 - i_{qs}^r(k) \cdot i_{qs}^r(k-2) \neq 0 \tag{25}$$

As presented in Equation (24), the controller gains can be estimated using the sampled currents and voltage commands.

6.2. Estimation Accuracy Improvement

The controller gains cannot be identified in the steady state because $\det\left(i_{qs}^r\right)$ is 0. In addition, although the current ripples caused by the speed and position controller or current sensor noise yield nonzero $\det\left(i_{qs}^r\right)$, these currents cannot be used to identify controller gains because they have a low correlation with the motor parameters and consequently a low signal-to-noise-ratio (SNR). Figure 15a depicts a steady-state q-axis current with a current ripple. Although the current ripple is unpredictable in practice, the ripple is modeled as a square wave with an amplitude of Δi for convenience of analysis. Then, $\det\left(i_{qs}^r\right)$ with the current ripple is calculated as follows:

$$\det\left(i_{qs}^r\right)\Big|_{SS} = \left(i_{qs}^{r\,*} + \Delta i\right)^2 - i_{qs}^{r\,*} \cdot \left(i_{qs}^{r\,*} - \Delta i\right) = 3 i_{qs}^{r\,*} \cdot \Delta i + \Delta i^2 \tag{26}$$

Figure 15b depicts a plot of $\det\left(i_{qs}^r\right)\Big|_{SS}$ versus the q-axis current command when Δi is set as 10% of the command value. It can be seen that $\det\left(i_{qs}^r\right)\Big|_{SS}$ increases with the current level. Therefore, a threshold for $\det\left(i_{qs}^r\right)$ must be set to avoid identification error in the steady state. Accordingly, controller gain identification is performed only when the following condition is satisfied:

$$\left|\det\left(i_{qs}^r\right)\right| > \det_{thres}\left(i_{qs}^r\right) \tag{27}$$

where $\det_{thres}\left(i_{qs}^r\right)$ is the threshold value. In general, the threshold value can be tuned through experiments.

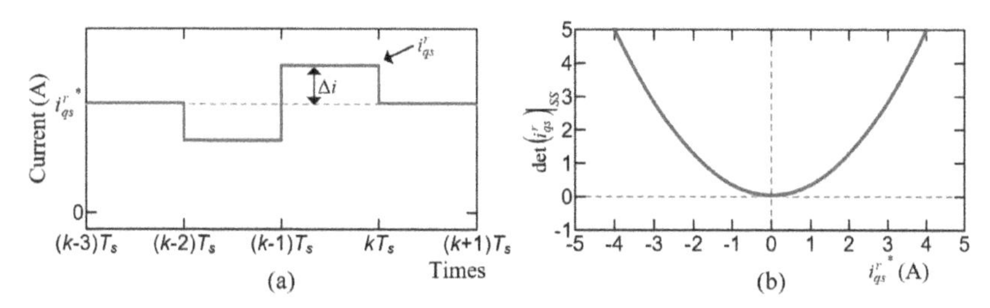

Figure 15. (**a**) Steady-state q-axis current with current ripple, and (**b**) $\det\left(i_{qs}^{r}\right)\Big|_{SS}$ versus current command.

The identification accuracy can be further improved by averaging the controller gains calculated in the last m control periods. In addition, each identified gain is weighted using its $\det\left(i_{qs}^{r}\right)$. The averaged controller gains are calculated as follows:

$$\overline{K}_{1q} = \sum_{y=1}^{m} \hat{K}_{1q,y} \cdot \left|\det_y\left(i_{qs}^{r}\right)\right| / \sum_{y=1}^{m} \left|\det_y\left(i_{qs}^{r}\right)\right| \tag{28}$$

$$\overline{K}_{2q} = \sum_{y=1}^{m} \hat{K}_{2q,y} \cdot \left|\det_y\left(i_{qs}^{r}\right)\right| / \sum_{y=1}^{m} \left|\det_y\left(i_{qs}^{r}\right)\right| \tag{29}$$

Because the average controller gain is dominated by the identified gain with higher $\det\left(i_{qs}^{r}\right)$, the identification accuracy improves.

After the average controller gains are calculated, the model gains can be determined as follows:

$$\hat{A}_{mq} = \overline{K}_{2q} / \overline{K}_{1q} \tag{30}$$

$$\hat{B}_{mq} = 1 / \overline{K}_{2q} \tag{31}$$

Because the effect of resistance variation on the current response is trivial, the estimated q-axis inductance can be approximated as follows:

$$\hat{L}_{qs} \approx -T_s r_s / \ln\left(\hat{A}_{mq}\right) \tag{32}$$

The model gains $\hat{A}_{md}$ and $\hat{B}_{md}$ as well as the estimated d-axis inductance can be obtained similarly. As depicted in Figure 1, the estimated inductances and model gains are used for the decoupling voltage calculation and dead-time compensation, respectively.

6.3. Identification When Voltage Command Is Limited

As illustrated in Figure 7, the stator voltage saturates to a maximum voltage V_{max} as follows:

$$\sqrt{v_{qs}^{r\,2} + v_{ds}^{r\,2}} \leq V_{\max} \tag{33}$$

V_{max} depends on the DC voltage and the dead-time of the inverter. The motor used in this study has almost identical q- and d-axis inductances. Thus, the d-axes current is controlled to 0 to generate the required torque with a minimum stator current. Consequently, the d-axis voltage approximates to the decoupling voltage and the steady-state q-axis voltage can be calculated using Equation (34) when the stator voltage saturates to V_{max}.

$$v_{qs}^{r} = \pm\sqrt{V_{\max}^2 - v_{ds}^{r\,2}} = \pm\sqrt{V_{\max}^2 - v_{dff}^2} \tag{34}$$

Subsequently, the voltage command used to identify the controller gains when the voltage is limited is obtained as follows:

$$v_{qs}^{r\ *} = v_{qs}^{r} - v_{qff} \tag{35}$$

6.4. Gain Update Method

Figure 16 illustrates the timing for identifying and updating the gains in a deadbeat controller and current estimator, where the green bar denotes the execution of the current control and the blue bar denotes the execution of the speed and position control. Because the d-axis current is controlled to 0, only the gains in the q-axis current loop are identified. However, because $L_{qs} \approx L_{ds}$, the gains in $C_d(z)$ and the d-axis current estimator are set equal to the corresponding q-axis values. The controller gains are identified when the current control loop is executed. Because the current control executes eight times faster than the speed and position control, at most eight controller gains are identified before the next speed control is executed. Then, $\overline{K}_{1q}$, $\overline{K}_{2q}$, $\hat{A}_{mq}$, $\hat{B}_{mq}$, $\hat{L}_{qs}$, and k_{pqc} are calculated using Equations (8) and (28)–(32) when the speed control is executed. Subsequently, the gains in the deadbeat controller and the parameters in the current estimator are updated in the next execution of the current control because the motor current generally reaches steady state at this instant. The PI controller in the current estimator is updated two control periods after the model gain is updated because the sampled current is two control periods behind the estimated current.

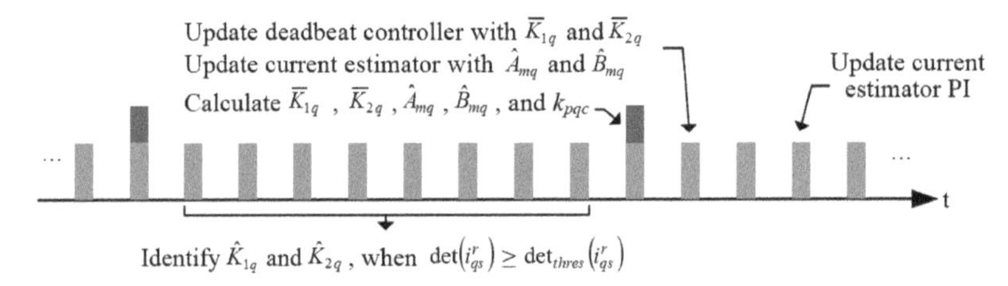

Figure 16. Time sequence of the gain identification, calculation and updating.

7. Experimental Results

A 400-W servo motor was used for experimental verifications. The parameters of the motor are provided in Table A1. Figure 17 illustrates the experimental system. The proposed online current loop tuning scheme is implemented using a Texas Instruments TMS320F28335 DSP. The detailed configuration of the drive is detailed in Table A2. In this study, V_{max} was set to 139 V to account for the losses caused by the dead time. The motor position and speed were measured using an encoder with a resolution of 2500 pulse/rev.

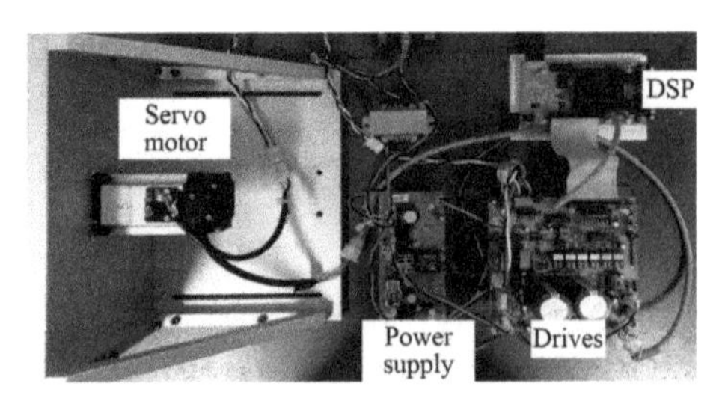

Figure 17. Experimental system.

The experimental results shown in Figures 18–21 were obtained without the online current loop tuning algorithm. Figure 18 illustrates the q-axis current and voltage command response when the current increased from 0 to 1 A and from 0 to 4 A, respectively, when the motor was at standstill. The voltage command was less than the limit for the 1 A step but exceeded the limit for the 4 A step. As depicted in Figure 18a, the q-axis current was exactly two current control periods behind the command value. Furthermore, overshoot and steady state error were not observed for the current. However, the q-axis current presented in Figure 18b required approximately seven control periods to reach the command value because the voltage was limited to 139 V. These results highly concur with the simulation results described in Section 5. Thus, the effectiveness of the deadbeat current controller was verified.

Figure 19a demonstrates the q-axis current, voltage, and speed response when the motor had rotation speeds between −3000 and 3000 rpm. The q-axis current followed the command value closely regardless of the motor speed. Figure 19b,c depicts the amplified views of the situation when the current increased from −4 to 4 A and decreased from 4 to −4 A, respectively. The deadbeat controller produced pulse-wise voltage because the stator voltage was limited. Although the voltage command had an opposite polarity to that of the decoupling voltage, the q-axis current required seven control periods to reach the command value. According to the aforementioned results, the performance of the deadbeat controller was independent of the motor speed. However, a marginal current overshoot is observed in Figures 18b and 19b,c because of the magnetic saturation.

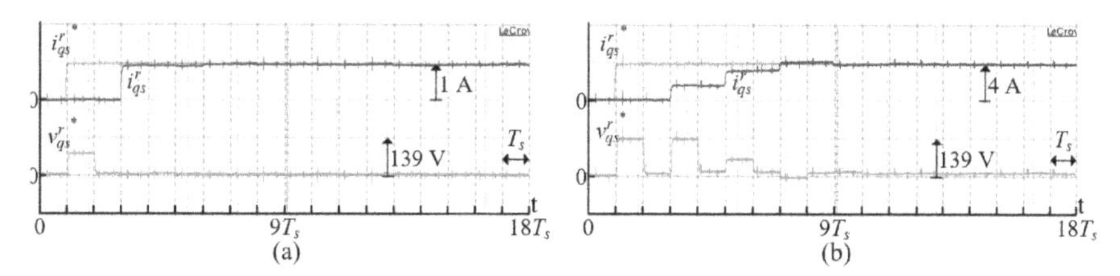

Figure 18. q-axis current and voltage when the motor is at standstill and the current command steps from (**a**) 0 A to 1 A, (**b**) 0 A to 4 A.

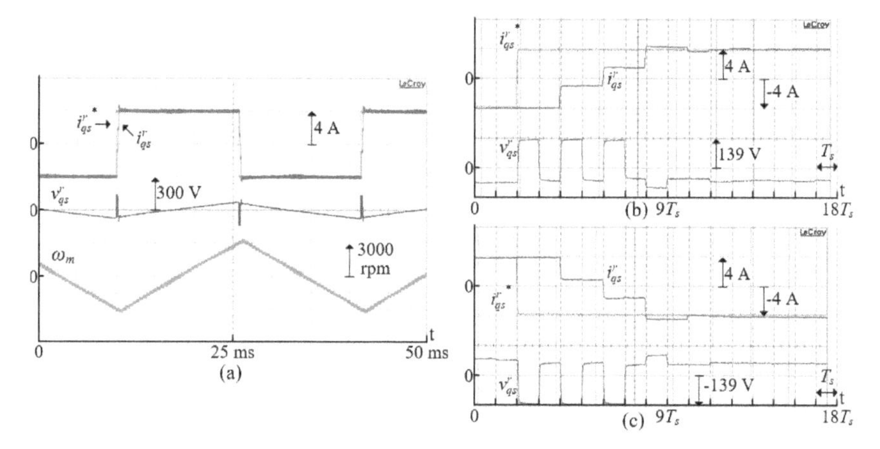

Figure 19. q-axis current, voltage, and speed when the motor cycles between −3000 rpm and 3000 rpm with step current command, (**a**) complete waveform, (**b**) amplified view when current command steps from −4 A to 4 A, (**c**) amplified view when current command steps from 4 A to −4 A.

Figures 20a and 21a illustrate the q-axis current and voltage response when the current increased from 0 to 1 A as the estimated inductance was set as 50% and 120% of its nominal value, respectively. The experiments were performed when the motor was at standstill. As depicted in Figure 20a, when $\hat{L}_{qs} = 0.5L_{qs}$, the q-axis current became overdamped and required seven control periods to reach the command value. However, as depicted in Figure 21a, when $\hat{L}_{qs} = 1.2L_{qs}$, the q-axis current became underdamped and had an observable overshoot. The experimental results are similar to the simulation results presented in Section 5.

Figure 20b,c displays the calculated $\det\left(i_{qs}^{r}\right)$ and controller gains for the transient response depicted in Figure 20a, respectively. Similarly, Figure 21b,c displays the calculated $\det\left(i_{qs}^{r}\right)$ and controller gains for the transient response presented in Figure 21a, respectively. For convenience of observation, the controller gains were normalized by their nominal values. Moreover, only the gains within ±200% of their nominal value are displayed. As depicted in the aforementioned figures, a large current difference resulted in a high $\det\left(i_{qs}^{r}\right)$ magnitude. Consequently, highly accurate gains were obtained because of a superior SNR. In general, the controller gain could be accurately identified for $\left|\det\left(i_{qs}^{r}\right)\right| \geq 0.1$. The maximum error between the identified controller gains and their nominal values were within 16%. Moreover, the proposed identification method could identify the controller gains in one current control period provided $\left|\det\left(i_{qs}^{r}\right)\right|$ was sufficiently large.

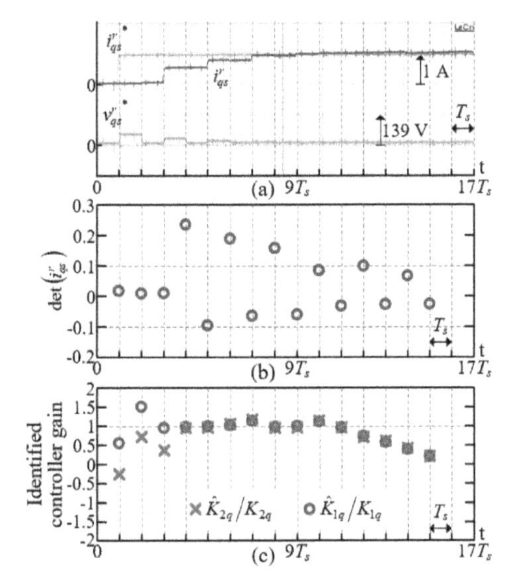

Figure 20. Current command steps from 0 A to 1 A when the motor is at standstill and $\hat{L}_{qs} = 0.5L_{qs}$, (**a**) q-axis current and voltage command, (**b**) $\det\left(i_{qs}^{r}\right)$, (**c**) normalized identified controller gains.

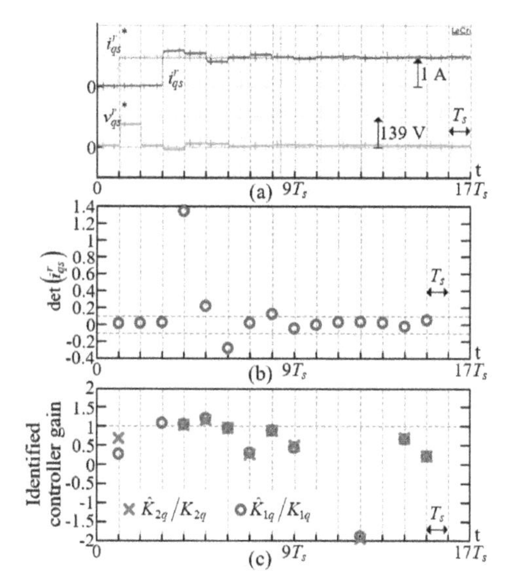

Figure 21. Current command steps from 0 A to 1 A when the motor is at standstill and $\hat{L}_{qs} = 1.2L_{qs}$, (**a**) q-axis current and voltage command, (**b**) $\det(i^r_{qs})$, (**c**) normalized identified controller gains.

Figure 22 displays the speed, position, and current responses when the motor was controlled in the positioning mode. The motor moved forward to 11π and then back to 0. The maximum speed was 3000 rpm, which is the rated speed of the motor. Furthermore, the motor was accelerating and decelerating with its rated current. An observable position error $\theta^*_m - \theta_m$ was obtained only when the motor was accelerating and decelerating. In the following experiments, the waveforms in the acceleration region were amplified to examine the effectiveness of the online current loop tuning scheme. In addition, the lowest bound of $\det_{min}(i^r_{qs})$ for the controller gain calculation was set as 0.2 to ensure sufficient identification accuracy.

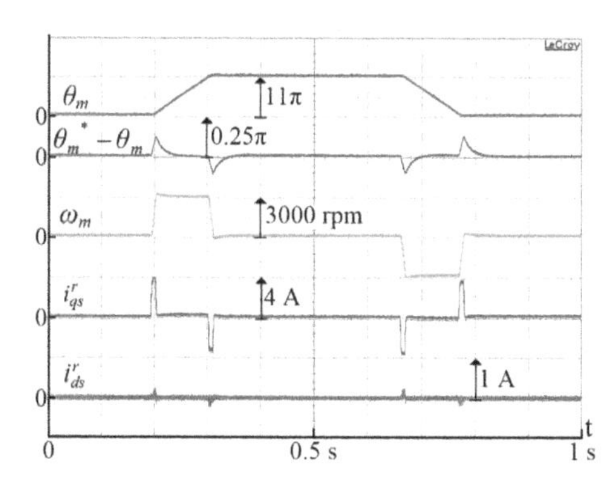

Figure 22. Speed, position, and current waveforms when the motor is controlled in the positioning mode.

Figure 23a, Figure 24a, and Figure 25a depict the current response with $\hat{L}_{qs} = L_{qs}$, $\hat{L}_{qs} = 0.5L_{qs}$, and $\hat{L}_{qs} = 1.2L_{qs}$ respectively, when online current loop tuning was deactivated. Conversely, Figure 23b, Figure 24b, and Figure 25b display the same waveforms but with online current loop tuning activated. The average controller gains were normalized by their nominal value for a clear observation. Figure 23a indicates that even with the correct inductance, overshoot and undershoot were observed for the

q-axis current at high current levels because of the magnetic saturation. By contrast, as indicated in Figure 23b, no apparent overshoot was observed after online current loop tuning was activated.

As depicted in Figure 24a, because the estimated inductance was set to half of the nominal value, the q-axis current response became overdamped. In addition, the d-axis current had a marginal steady-state error. By contrast, as illustrated in Figure 24b, the q-axis current was tuned to reach its reference without overshoot within a speed control period and the d-axis current had no steady state error after online current loop tuning was activated. The q-axis current in Figure 25a exhibits considerable overshoot despite the current level because the q-axis inductance is 20% higher than its nominal value. This caused additional ripples to appear on the d-axis current. However, as depicted in Figure 25b, the overshoot was eliminated within a speed control period after online current loop tuning was activated. It can be observed in Figures 24b and 25b that after the deadbeat controller is tuned by the proposed method, the required sampling period for current to reach its command value is reduced from nine to two sampling periods, and the overshoot on the current is reduced from 0.4 A to 0.09 A. These experimental results verify that the proposed method is effective and can greatly reduce the sensitivity of the deadbeat controller to the variations in inductance.

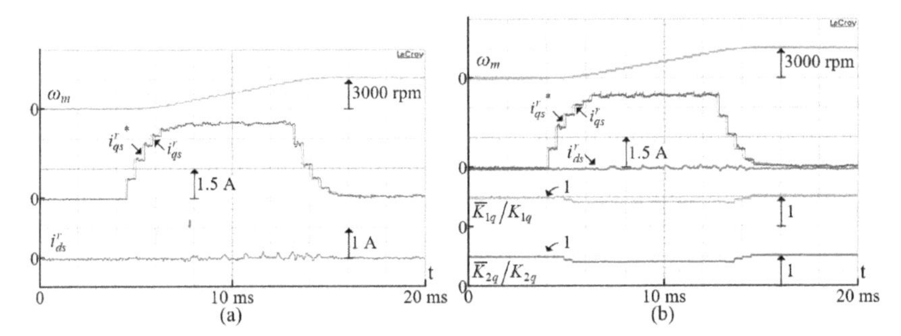

Figure 23. The amplified current response in the acceleration region of Figure 23 with $\hat{L}_{qs} = L_{qs}$ when the online current loop tuning is (**a**) de-activated and (**b**) activated.

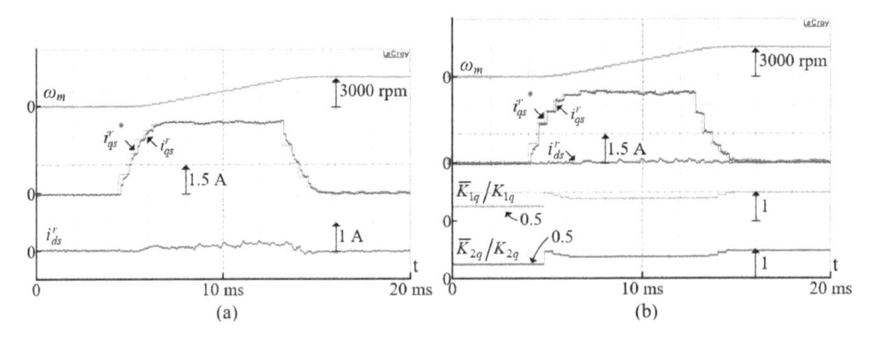

Figure 24. The amplified current response in the acceleration region of Figure 23 with $\hat{L}_{qs} = 0.5L_{qs}$ when the online current loop tuning is (**a**) de-activated and (**b**) activated.

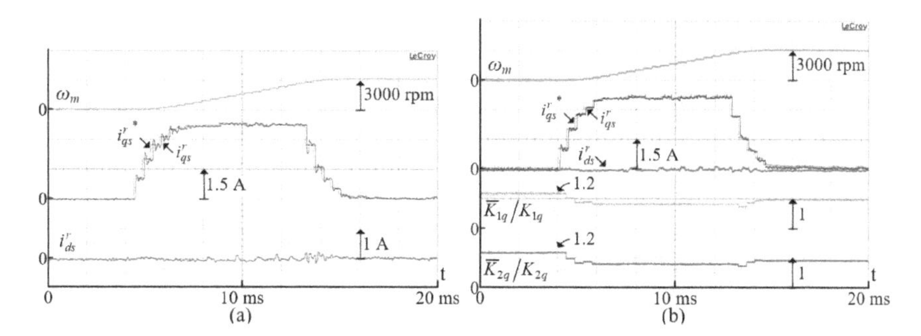

Figure 25. The amplified current response in the acceleration region of Figure 23 with $\hat{L}_{qs} = 1.2L_{qs}$ when the online current loop tuning is (**a**) de-activated and (**b**) activated.

Figure 26 displays the measured and calculated frequency response of the q-axis deadbeat current controller. In the measurements, voltage was within the limit and the motor was at standstill. It can be seen that the current amplitude did not vary with frequency. However, the phase delay gradually increased with frequency. This is because the deadbeat controller was designed to reach its reference in two control periods, and the phase delay for two time periods was small at low frequencies but large at high frequencies.

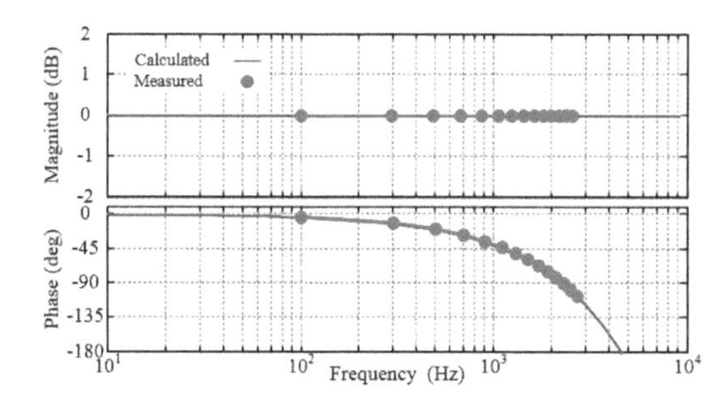

Figure 26. The measured and the calculated frequency response of the deadbeat current controller.

8. Conclusions

In this study, we present an online controller gain tuning scheme for deadbeat current control. The experimental results verify that the motor current can reach its reference value without overshoot in two current control periods with the deadbeat controller and correct parameters. However, the current response can easily become overdamped or underdamped when the controller gains are calculated using incorrectly estimated inductances. The proposed online controller gain tuning scheme is derived on the basis of the discrete-time motor model. The experimental results indicate that the correct controller gains can be identified in one current control period, and the control loop is tuned in a speed control period. Consequently, the deadbeat controller can persistently control the motor current to its reference value in two sampling periods without overshoot irrespective of the inductance variations. Furthermore, the proposed scheme is easy to implement and requires limited computations.

Author Contributions: Conceptualization, Z.-C.Y., C.-H.H., and S.-M.Y.; methodology, Z.-C.Y. and C.-H.H.; software, C.-H.H.; validation, Z.-C.Y., C.-H.H., and S.-M.Y.; formal analysis, Z.-C.Y. and C.-H.H.; investigation, Z.-C.Y. and C.-H.H.; resources, S.-M.Y.; data curation, C.-H.H.; writing—original draft preparation, Z.-C.Y.; writing—review and editing, S.-M.Y.; visualization, Z.-C.Y.; supervision, S.-M.Y.; project administration, S.-M.Y.

Funding: This research received no external funding.

Conflicts of Interest: The authors declare no conflict of interest.

Appendix A

Table A1. Main drive parameters.

	Value	Unit
DC voltage	300	V
Sampling period for current loop (T_s)	55	μs
Sampling period for speed and position loop	440	μs
Dead-time	2	μs

Table A2. Main motor parameters.

	Value	Unit
Rated speed/pole pairs	3000/5	rpm
Rated current	4	A
Magnet flux (λ_m)	0.042	Wb-turns
Stator resistance (r_s)	1.4	Ω
d-axis inductance (L_{ds})	4.46	mH
q-axis inductance (L_{qs})	4.54	mH

References

1. Bocker, J.; Beineke, S.; Bahr, A. On the Control Bandwidth of Servo Drives. In Proceedings of the 13th European Conference Power Electronics Applications, Barcelona, Spain, 8–10 September 2009; pp. 1–10.
2. Huh, K.K.; Lorenz, R.D. Discrete-Time Domain Modeling and Design for AC Machine Current Regulation. In Proceedings of the Conference Record 42th IEEE/IAS Annual Meeting, New Orleans, LA, USA, 23–27 September 2007; pp. 2066–2073.
3. Kim, H.; Degner, M.W.; Guerrero, J.M.; Briz, F.; Lorenz, R.D. Discrete-Time Current Regulator Design for AC Machine Drives. *IEEE Trans. Ind. Appl.* **2011**, *46*, 1425–1435.
4. Yepes, A.G.; Vidal, A.; Malvar, J.; Lopez, O.; Jesus, D.G. Tuning Method Aimed at Optimized Setting Time and Overshoot for Synchronous Proportional-Integral Current Control in Electric Machines. *IEEE Trans. Power Electron.* **2014**, *29*, 3041–3054. [CrossRef]
5. Andersson, A.; Thiringer, T. Assessment of an Improved Finite Control Set Model Predictive Current Controller for Automotive Propulsion Applications. *IEEE Trans. Ind. Electron.* **2019**, *67*, 91–100. [CrossRef]
6. Cortes, P.; Kazmierkowski, M.P.; Kennel, R.M.; Quevedo, D.E.; Rodriguez, J. Predictive Control in Power Electronics and Drives. *IEEE Trans. Ind. Electron.* **2008**, *55*, 4312–4324. [CrossRef]
7. Zhang, Y.; Xu, D.; Liu, J.; Gao, S.; Xu, W. Performance Improvement of Model-Predictive Current Control of Permanent Magnet Synchronous Motor Drives. *IEEE Trans. Ind. Appl.* **2017**, *53*, 3683–3695. [CrossRef]
8. Lin, C.K.; Liu, T.H.; Yu, J.T.; Fu, L.C.; Hsiao, C.F. Model-Free Predictive Current Control for Interior Permanent-Magnet Synchronous Motor Based on Current Difference Detection Technique. *IEEE Trans. Ind. Electron.* **2014**, *61*, 667–681. [CrossRef]
9. Lin, C.K.; Yu, J.T.; Lai, Y.S.; Yu, H.C. Improved Model-Free Predictive Current Control for Synchronous Reluctance Motor Drives. *IEEE Trans. Ind. Electron.* **2016**, *63*, 3942–3953. [CrossRef]
10. Young, H.A.; Perez, M.A.; Rodriguez, J. Analysis of Finite-Control-Set Model Predictive Current Control with Model Parameter Mismatch in a Three-Phase Inverter. *IEEE Trans. Ind. Electron.* **2016**, *63*, 3100–3107. [CrossRef]
11. Ahmed, A.A.; Koh, B.K.; Lee, Y.I. A Comparison of Finite Control Set and Continuous Control Set Model Predictive Control Schemes for Speed Control of Induction Motors. *IEEE Trans. Ind. Informat.* **2018**, *14*, 1334–1346. [CrossRef]
12. Yang, H.; Zhang, Y.; Liang, J.; Xia, B.; Walker, P.D.; Zhang, N. Deadbeat Control Based on a Multipurpose Disturbance Observer for Permanent Magnet Synchronous Motors. *IET Electr. Pwer Appl.* **2018**, *12*, 708–716. [CrossRef]
13. Isermann, R. *Digital Control System*; Springer: Berlin, Germany, 1981.

14. Moon, H.T.; Kim, H.S.; Youn, M.J. A Discrete-Time Predictive Current Control for PMSM. *IEEE Trans. Power Electron.* **2003**, *18*, 464–472. [CrossRef]
15. Walz, S.; Lazar, R.; Buticchi, G.; Liserre, M. Dahlin-Based Fast and Robust Current Control of a PMSM in Case of Low Carrier Ratio. *IEEE Access* **2019**, *7*, 102199–102208. [CrossRef]
16. Yang, S.M.; Lee, C.H. A Deadbeat Current Controller for Field Oriented Induction Motor Drives. *IEEE Trans. Power Electron.* **2002**, *17*, 772–778. [CrossRef]
17. Zhang, X.; Hou, B.; Mei, Y. Deadbeat Predictive Current Control of Permanent-Magnet Synchronous Motors with Stator Current and Disturbance Observer. *IEEE Trans. Power Electron.* **2017**, *32*, 3818–3834. [CrossRef]
18. Zhang, X.; Zhang, L.; Zhang, Y. Model Predictive Current Control for PMSM Drives with Parameter Robustness Improvement. *IEEE Trans. Power Electron.* **2019**, *34*, 1645–1657. [CrossRef]
19. Boileau, T.; Leboeuf, N.; Babak, N.M.; Farid, M.T. Online Identification of PMSM Parameters: Parameter Identifiability and Estimator Comparative Study. *IEEE Trans. Ind. Appl.* **2011**, *47*, 1944–1957. [CrossRef]
20. Liu, K.; Zhu, Z.Q.; Zhang, Q.; Zhang, J. Influence of Nonideal Voltage Measurement on Parameter Estimation in Permanent-Magnet Synchronous Machine. *IEEE Trans. Ind. Electron.* **2012**, *59*, 2438–2447. [CrossRef]
21. Hamida, M.A.; Leon, J.D.; Glumineau, A.; Boisliveau, R. An Adaptive Interconnected Observer for Sensorless Control of PM Synchronous Motors with Online Parameter Identification. *IEEE Trans. Ind. Electron.* **2013**, *60*, 739–748. [CrossRef]
22. Ichikawa, S.; Timita, M.; Doki, S.; Okuma, S. Sensorless Control of Permanent-Magnet Synchronous Motors Using Online Parameter Identification Based on System Identification Theory. *IEEE Trans. Ind. Electron.* **2006**, *53*, 363–372. [CrossRef]
23. Inoue, Y.; Yamada, K.; Morimoto, S.; Sanada, M. Effectiveness of Voltage Error Compensation and Parameter Identification for Model-Based Sensorless Control. *IEEE Trans. Ind. Appl.* **2009**, *45*, 213–221. [CrossRef]
24. Feng, G.; Lai, C.; Mukherjee, K.; Kar, N.C. Current Injection-Based Online Parameter and VSI Nonlinearity Estimation for PMSM Drives Using Current and Voltage DC Components. *IEEE Trans. Transp. Electrific.* **2016**, *2*, 119–128. [CrossRef]
25. Mohamed, Y.A.I.; Saadany, E.F.E. Robust High Bandwidth Discrete-Time Predictive Current Control with Predictive Internal Model–A Unified Approach for Voltage-Source PWM Converters. *IEEE Trans. Power Electron.* **2008**, *23*, 126–136. [CrossRef]
26. Yang, S.M.; Lin, K.W. Automatic Control Loop Tuning for Permanent-Magnet AC Servo Motor Drives. *IEEE Trans. Ind. Electron.* **2016**, *63*, 1499–1506. [CrossRef]
27. Shen, G.; Yao, W.; Chen, B.; Wang, K.; Lee, K.; Lu, Z. Automeasurement of the Inverter Output Voltage Delay Curve to Compensate for Inverter Nonlinearity in Sensorless Motor Drives. *IEEE Trans. Power. Electron.* **2014**, *29*, 5542–5553. [CrossRef]

Article

Commutation Torque Ripple Suppression Strategy of Brushless DC Motor Considering Back Electromotive Force Variation

Xinmin Li [1], Guokai Jiang [2], Wei Chen [3], Tingna Shi [4], Guozheng Zhang [1] and Qiang Geng [1,*]

[1] School of Electrical Engineering and Automation, Tianjin Polytechnic University, Tianjin 300387, China; lixinmin@tju.edu.cn (X.L.); zhanggz@tju.edu.cn (G.Z.)
[2] China Automotive Technology and Research Center Co. Ltd., Tianjin 300300, China; jiangguokai@tju.edu.cn
[3] School of Artificial Intelligence, Tianjin Polytechnic University, Tianjin 300387, China; chen_wei@tju.edu.cn
[4] College of Electrical Engineering, Zhejiang University, Hangzhou 310027, China; tnshi@tju.edu.cn
* Correspondence: gengqiang@tju.edu.cn; Tel.: +86-022-8395-5415

Received: 24 April 2019; Accepted: 17 May 2019; Published: 20 May 2019

Abstract: This paper presents a commutation torque ripple suppression strategy for brushless DC motor (BLDCM) in the high-speed region, which considers the back electromotive force (back-EMF) variation during the commutation process. In the paper, the influence of actual back-EMF variation on the torque and outgoing phase current during the commutation process is analyzed. A modified smooth torque mechanism is then reconstructed considering the back-EMF variation, based on which a novel torque ripple suppression strategy is further designed. Compared with the traditional strategy which controls the chopping duty cycle relatively smoothly in the commutation process, the proposed strategy dynamically regulates the chopping duty cycle, which makes it show a gradual decrease. This strategy can suppress the commutation torque ripple even in a long commutation process, and broaden the speed range of the commutation torque ripple reduction. Under the experimental conditions of this paper, the proposed strategy can effectively reduce the commutation torque ripple in the high-speed region, and avoid the outgoing phase current cannot be reduced to zero. The experimental results verify the correctness of the theoretical analysis and the feasibility of the proposed strategy.

Keywords: brushless DC motor; commutation torque ripple; back electromotive force

---

1. Introduction

The brushless DC motor (BLDCM) has advantages of simple structure, high power density, and high reliable operation [1–5]. However, the windings of the motor have inductances, and the transient process appears during the current exchanged between two phases, and this process will cause commutation torque ripple, which may reach about 50% of the average torque of the BLDCM [6–8]. If the commutation torque ripple is not suppressed by specific reduction strategy, the vibration and noise of motors will increase, and the promotion application of BLDCM will be restricted in a field that has strict requirements for torque ripple and noise [9–11]. When the BLDCM operates in its high-speed range, limited by the output voltage of the inverter, the incoming phase and the outgoing phase current are difficult to change rapidly, so the motor will generate greater commutation torque ripple in this region [12–14].

Pulse width modulation (PWM) techniques can be employed to suppress the commutation torque ripple by maintaining the non-commutation phase current constant [15–18], where only the outgoing phase modulation technique is applicable to the high-speed region in light of the relationship between the back electromotive force (back-EMF) and DC-link voltage. A voltage compensation method is proposed in [15] to control the incoming phase and the outgoing phase current slopes by the same degree. A three-segments modulation strategy is used in [16], where the action time of each segment is acquired based on the minimum commutation time. An integral sliding mode current controller is introduced to enhance the robustness of commutation torque ripple suppression in [17]. A three-phase PWM modulation technique is proposed to suppress commutation torque ripple in [18], where the torque observer and the calculation of commutation process are not necessary.

DC-link voltage boost techniques can also be adopted to reduce the commutation torque ripple by adding a DC–DC converter in front of the voltage source inverter (VSI) [19–23]. In [20], the single ended primary inductor converter (SEPIC) is introduced to adjust the required DC-link voltage during the commutation process. In [22], the Buck converter is added, by regulating the amplitude of the out-voltage and the current during the commutation interval, so the commutation torque can be reduced partly. In [23], a Z-source inverter is used to boost the DC-link voltage by the shoot-through vectors.

Normally, the torque during commutation process is considered to be proportional to the non-commutation phase current when the back-EMF is assumed to be constant. However, a constant non-commutation phase current will still generate a torque ripple under the actual back-EMF variation during the commutation process [24]. Furthermore, the torque ripple is related to the stator current and the commutation duration, which is much heavier in a long commutation process for the high-speed region.

A commutation torque ripple suppression strategy considering the back-EMF variation is designed in this paper for a high-speed region. Section 2 introduces the traditional strategy assuming the back-EMFs constant. Section 3 studies the impact of the back-EMF change on the torque control performance and the normal end of commutation in the traditional strategy, and then Section 4 designs a new strategy in light of the reconstructed smooth torque mechanism. In Section 5, the above theoretical analysis is experimentally verified. Conclusions come in Section 6.

2. Traditional Commutation Torque Ripple Suppression Strategy Ignoring Back-EMF Variation

The equivalent model of the BLDCM drive system is shown in Figure 1, where S_i and D_i, $i \in \{1, 2, \ldots, 6\}$, are metal-oxide-semiconductor field-effect transistor (MOSFET) and its anti-parallel diode respectively; N is the neutral point of three phases windings.

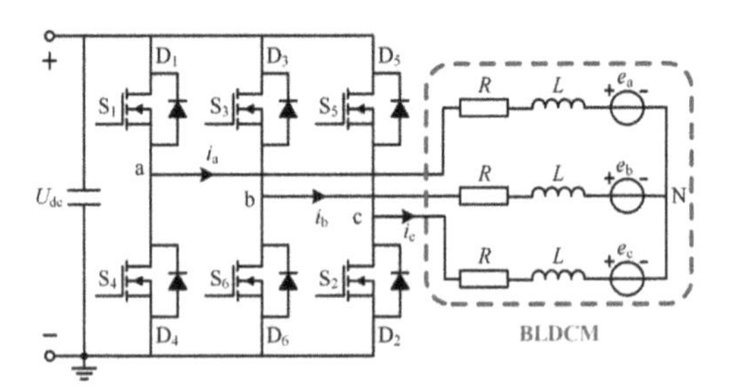

Figure 1. The equivalent model of the BLDCM drive system.

Assuming the three-phase stator windings are symmetrical, simultaneously neglecting the mutual inductance, and as shown in Figure 1 the three-phase windings are star-connected, the terminal voltage model can be expressed as

$$\begin{bmatrix} u_a \\ u_b \\ u_c \end{bmatrix} = \begin{bmatrix} R & 0 & 0 \\ 0 & R & 0 \\ 0 & 0 & R \end{bmatrix} \begin{bmatrix} i_a \\ i_b \\ i_c \end{bmatrix} + \begin{bmatrix} L & 0 & 0 \\ 0 & L & 0 \\ 0 & 0 & L \end{bmatrix} \frac{d}{dt} \begin{bmatrix} i_a \\ i_b \\ i_c \end{bmatrix} + \begin{bmatrix} e_a \\ e_b \\ e_c \end{bmatrix} + u_N \tag{1}$$

where u_a, u_b, and u_c are terminal voltages; i_a, i_b, and i_c are stator currents; e_a, e_b and e_c are phase back-EMFs, and u_N is the neutral point voltage of the motor.

The electromagnetic torque is given by

$$T_e = \frac{e_a i_a + e_b i_b + e_c i_c}{\omega_m} \tag{2}$$

where ω_m is the rotor mechanical angular velocity.

A BLDCM is normally driven by a six-step mode, where normal conduction periods and commutation periods exist. In a normal conduction period, only two windings are energized, and the other one is floating. In a commutation period, three-phase windings are all energized due to the stator inductance and the limited inverter voltage.

Figure 2 shows the modulation method for the high-speed region, where the shaded area and the non-shaded area represent commutation and normal conduction periods, respectively. The outgoing phase modulation starts to be used in the commutation period when Hall sectors change, and the non-commutation modulation is switched to be applied in the normal conduction period as the outgoing phase current is reduced to zero. The duty cycle of the chopping switch in the normal conduction period and the commutation period are denoted as d_{norm} and d_{cmt}, respectively; as shown in Figure 2, two duty cycles are not the same, and the value relation between these two duty cycles is $d_{cmt} < d_{norm}$. Taking the commutation process of $a^+c^- \rightarrow b^+c^-$ as an example, the commutation torque ripple in high-speed region is discussed as follows.

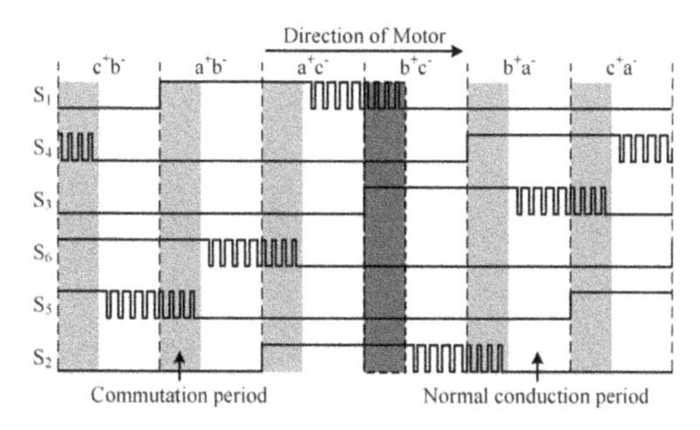

Figure 2. Modulation in high-speed region.

According to Figures 1 and 2, the terminal voltages are satisfied as $u_a = d_{cmt}U_{dc}$, $u_b = U_{dc}$ and $u_c = 0$ in the commutation process of $a^+c^- \rightarrow b^+c^-$, where d_{cmt} is the duty cycle of the commutation process. Substituting it into (1) and considering $i_a + i_b + i_c = 0$ in star-connected motors, the current equation can be obtained as

$$\begin{cases} 3L\frac{di_a}{dt} = (2d_{cmt} - 1)U_{dc} - (2e_a - e_b - e_c) - 3Ri_a \\ 3L\frac{di_b}{dt} = (2 - d_{cmt})U_{dc} - (2e_b - e_a - e_c) - 3Ri_b \\ 3L\frac{di_c}{dt} = -(1 + d_{cmt})U_{dc} - (2e_c - e_a - e_b) - 3Ri_c \end{cases} \tag{3}$$

If the back-EMF variation is neglected, the phase back-EMFs will be

$$\begin{cases} e_a = E \\ e_b = E \\ e_c = -E \end{cases} \tag{4}$$

where E is the magnitude of phase back-EMF.

By substituting (4) into (2), the torque can be simplified as

$$T_\mathrm{e} = \frac{Ei_\mathrm{a} + Ei_\mathrm{b} - Ei_\mathrm{c}}{\omega_\mathrm{m}} = -\frac{2Ei_\mathrm{c}}{\omega_\mathrm{m}} = -\frac{60}{\pi}k_\mathrm{e}i_\mathrm{c} \tag{5}$$

According to (5), the torque is proportional to the non-commutation current during the commutation process on the assumption that back-EMF is unchanged, which means keeping the non-commutation current constant can effectively suppress the commutation torque ripple.

Substitute (4) into (3), the non-commutation current can be expressed as

$$3L\frac{\mathrm{d}i_\mathrm{c}}{\mathrm{d}t} = -(1 + d_\mathrm{cmt})U_\mathrm{dc} + 4E - 3Ri_\mathrm{c} \tag{6}$$

To maintain the current unchanged, namely $\mathrm{d}i_\mathrm{c}/\mathrm{d}t = 0$, the required duty cycle should be

$$d_\mathrm{cmt} = \frac{4E - 3Ri_\mathrm{c}}{U_\mathrm{dc}} - 1 = \frac{4E + 3RI}{U_\mathrm{dc}} - 1 \tag{7}$$

3. Effect of Back-EMF Variation on Commutation Torque Ripple Suppression

The traditional high-speed commutation torque ripple suppression strategies usually neglect the change of back-EMF in the commutation process, and suppresses the commutation torque ripple by controlling the non-commutation phase current constant. The phase currents and back-EMFs in the commutation process of $a^+c^- \rightarrow b^+c^-$ are shown in Figure 3, where I is the current amplitude before the commutation, t_cmt is the commutation duration, namely the time from commutation initial time to the moment that the outgoing phase reduces to zero, t_Hall is the Hall sector's period.

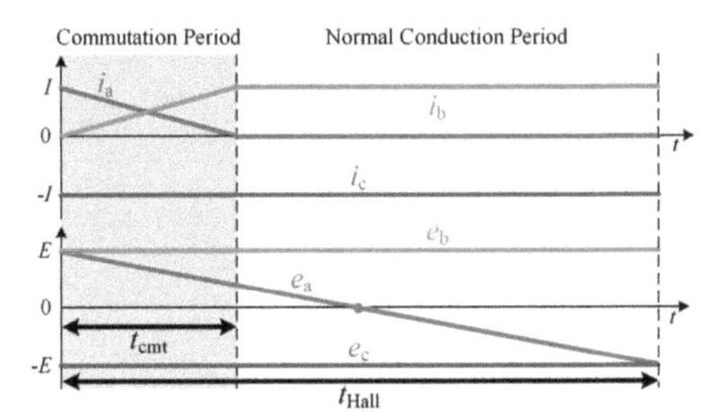

Figure 3. The variation of back-EMF during the commutation process.

It can be seen from Figure 3 that the actual back-EMFs during the commutation process are

$$\begin{cases} e_\mathrm{a} = E - 2Et/t_\mathrm{Hall} \\ e_\mathrm{b} = E \\ e_\mathrm{c} = -E \end{cases}, \quad (0 \le t \le t_\mathrm{cmt}) \tag{8}$$

As the motor speed increases, the Hall sector's period gradually decreases, while the commutation duration is prolonged due to the limited DC-link voltage. Based on (8), when the commutation duration accounts for a large proportion of the Hall sector's period, there is a significant variation in the outgoing phase back-EMF during the commutation process. Therefore, the effect of back-EMF variation on the torque ripple suppression in the traditional strategy is analyzed as follows.

3.1. Effect of Back-EMF Variation on Torque

The actual back-EMF variation affects the suppression effect of commutation torque ripple. By substituting (8) into (2), the torque equation with back-EMF variation can be expressed as

$$T_{\mathrm{e}} = -\frac{2Ei_{\mathrm{c}}}{\omega_{\mathrm{m}}} - \frac{2Ei_{\mathrm{a}}}{\omega_{\mathrm{m}}}\frac{t}{t_{\mathrm{Hall}}} \tag{9}$$

According to (9), it can be seen that the torque during the commutation process is not only related to the non-commutation current, but also to the outgoing phase current. The smooth torque condition can be derived from (9) as

$$\frac{\mathrm{d}T_{\mathrm{e}}}{\mathrm{d}t} = -\frac{2E}{\omega_{\mathrm{m}}}\frac{\mathrm{d}i_{\mathrm{c}}}{\mathrm{d}t} - \frac{2Et}{\omega_{\mathrm{m}}t_{\mathrm{Hall}}}\frac{\mathrm{d}i_{\mathrm{a}}}{\mathrm{d}t} - \frac{2Ei_{\mathrm{a}}}{\omega_{\mathrm{m}}t_{\mathrm{Hall}}} \tag{10}$$

If the traditional duty cycle (7) is employed to suppress the torque ripple, the smooth torque mechanism (10) can be simplified in light of (3) and (8) as

$$\frac{\mathrm{d}T_{\mathrm{e}}}{\mathrm{d}t} = \frac{-2E}{3L\omega_{\mathrm{m}}t_{\mathrm{Hall}}}[3Lt\frac{\mathrm{d}i_{\mathrm{a}}}{\mathrm{d}t} + 3Li_{\mathrm{a}} - 2Et] = \frac{-2E}{3L\omega_{\mathrm{m}}t_{\mathrm{Hall}}}[3L\frac{\mathrm{d}(ti_{\mathrm{a}})}{\mathrm{d}t} - 2Et] \tag{11}$$

Define T_{e0} as the torque at the initial moment of commutation. The torque during the commutation process can be obtained by integrating (11) as

$$T_{\mathrm{e}}(t) = T_{\mathrm{e0}} + \frac{2Et[Et - 3Li_{a}(t)]}{3L\omega_{\mathrm{m}}t_{\mathrm{Hall}}} \tag{12}$$

Based on (12), the torque waveform during the commutation process is shown in Figure 4. It can be seen that the torque change is zero only when $t = 3Li_{\mathrm{a}}/E$, and a large torque ripple occurs at the end of commutation process, namely the moment i_{a} just reduces to zero.

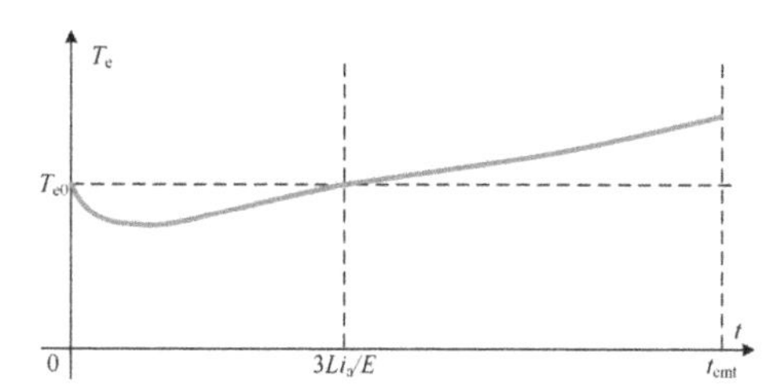

Figure 4. The effect of back-EMF variation on the torque during the commutation process.

3.2. Effect of Back-EMF Variation on Outgoing Phase Current

The actual back-EMF variation also affects the normal turn-off of outgoing phase current. If the traditional duty cycle (7) is used to suppress the torque ripple in high-speed region, the current differential equation can be acquired by substituting (8) into (3) as

$$L\frac{\mathrm{d}i_{\mathrm{a}}}{\mathrm{d}t} = -Ri_{\mathrm{a}} - (U_{\mathrm{dc}} - 2E - 2RI) + \frac{4E}{3t_{\mathrm{Hall}}}t \tag{13}$$

To simplify the calculation, outgoing phase current term Ri_{a} can be neglected as the resistance is small and the outgoing phase is gradually reducing to zero. The outgoing phase current can be obtained by integrating (13) as

$$i_{\mathrm{a}} = I - \frac{1}{L}mt + \frac{2E}{3Lt_{\mathrm{Hall}}}t^2 = \frac{2E}{3Lt_{\mathrm{Hall}}}(t - \frac{3t_{\mathrm{Hall}}m}{4E})^2 + I - \frac{3t_{\mathrm{Hall}}m^2}{8EL} \tag{14}$$

where $m = U_{dc} - 2E - 2RI$.

According to (14), the minimum value of outgoing phase current is $i_{a\_min} = I - 3t_{Hall}m^2/(8EL)$, which happens at $t = 3mt_{Hall}/(4E)$. If the minimum value is less than or equal to zero, the commutation process can be successfully finished as the outgoing phase current reduces to zero, as shown in the gray line in Figure 5. If the minimum value is larger than zero, the commutation process cannot be successfully finished as the outgoing phase current never reduces to zero, as shown in the black line in Figure 5.

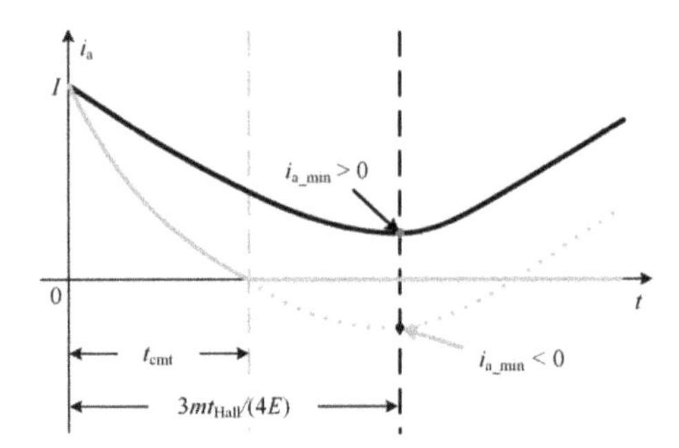

Figure 5. The effect of back-EMF variation on the outgoing phase current.

Therefore, the condition for the normal end of the commutation process is

$$i_{a\_min} = I - \frac{3t_{Hall}m^2}{8EL} \leq 0 \tag{15}$$

Since the variables in (15) are all larger than zero, it can be further simplified as

$$m \geq \sqrt{\frac{8ELI}{3t_{Hall}}} \tag{16}$$

Set p as the pole pairs and n as the motor speed, the Hall sector's period can be expressed as $t_{Hall} = 10/(np)$. Substitute it into (16), the speed range for the normal end of the commutation process in the traditional strategy is

$$n \leq \frac{U_{dc} - 2RI}{2(k_e + \sqrt{\frac{pk_eLI}{15}})} = \frac{f_1}{g_1} \tag{17}$$

where $k_e = E/n$ is phase back-EMF coefficient; f_1 and g_1 are numerator and denominator of the critical speed in the traditional strategy.

4. Proposed Commutation Torque Ripple Suppression Strategy Considering Back-EMF Variation

According to (10), the smooth torque mechanism with back-EMF variation is reconstructed as

$$t_{Hall}\frac{di_c}{dt} + t\frac{di_a}{dt} + i_a = 0 \tag{18}$$

Substitute (3) and (4) into (18), and it can be presented as

$$(2t - t_{Hall})d_{cmt}U_{dc} - (U_{dc} + 4E + 3Ri_a)t - (U_{dc} - 4E + 3Ri_c)t_{Hall} + \frac{4Et^2}{t_{Hall}} + 3Li_a = 0 \tag{19}$$

For commutation torque ripple suppression with back-EMF variation, the duty cycle is

$$d_{\text{cmt}} = \frac{1}{(2t - t_{\text{Hall}})U_{\text{dc}}}[(U_{\text{dc}} + 4E + 3Ri_{\text{a}})t - \frac{4Et^2}{t_{\text{Hall}}} + (U_{\text{dc}} - 4E + 3Ri_{\text{c}})t_{\text{Hall}} - 3Li_{\text{a}}] \tag{20}$$

If this duty cycle is employed, the torque gradient can always be zero over the whole commutation process, even with the changing back-EMF, and the torque ripple will be thus suppressed completely.

Moreover, the outgoing phase current in the proposed strategy is analyzed as follows. By substituting (20) into (3), it can be

$$L\frac{di_{\text{a}}}{dt} = -\frac{t_{\text{Hall}}}{t_{\text{Hall}} - 2t}(a + bi_a) \tag{21}$$

where $a = U_{\text{dc}} - 2E + 2Ri_{\text{c}} \geq 0$, $b = R - 2L/t_{\text{Hall}}$.

If $b > 0$, the current in the interval during commutation process ($t \in [0, 0.5t_{\text{Hall}}]$) satisfies

$$L\frac{di_{\text{a}}}{dt} \leq -\frac{t_{\text{Hall}}a}{t_{\text{Hall}} - 2t} \tag{22}$$

By integrating (22), it can be presented as

$$i_{\text{a}} \leq I + \frac{at_{\text{Hall}}}{2L}\ln(1 - \frac{2t}{t_{\text{Hall}}}) \tag{23}$$

As t gradually increases to $0.5t_{\text{Hall}}$, $\ln(1 - 2t/t_{\text{Hall}})$ will approach to negative infinity. Therefore, i_{a} can be reduced to zero in $t \in [0, 0.5t_{\text{Hall}}]$, and the commutation process can be successfully turned off.

If $b < 0$, the current in the interval during commutation process ($t \in [0, 0.5t_{\text{Hall}}]$) satisfies

$$L\frac{di_{\text{a}}}{dt} \leq -\frac{t_{\text{Hall}}}{t_{\text{Hall}} - 2t}(a + bI) \tag{24}$$

By integrating (24), it can be presented as

$$i_{\text{a}} \leq I + \frac{t_{\text{Hall}}(a + bI)}{2L}\ln(1 - \frac{2t}{t_{\text{Hall}}}) \tag{25}$$

As t gradually increases to $0.5t_{\text{Hall}}$, $\ln(1 - 2t/t_{\text{Hall}})$ will also approach negative infinity. Therefore, if $a + bI \geq 0$, the outgoing phase current can be reduced to zero and the commutation process will end successfully. If $a + bI < 0$, i_{a} will be not decreased but increased as $di_{\text{a}}/dt > 0$ in light of (21), and the commutation cannot be shut off normally.

Based on the above analysis, the condition for $a + bI \geq 0$ is

$$U_{\text{dc}} - 2E + 2Ri_{\text{c}} + RI - \frac{2LI}{t_{\text{Hall}}} \geq 0 \tag{26}$$

By ignoring the non-commutation phase current variation, the condition can be further simplified to be

$$U_{\text{dc}} - 2E - RI - \frac{2LI}{t_{\text{Hall}}} \geq 0 \tag{27}$$

According to (27), the speed range for the normal end of the commutation process in the proposed strategy is

$$n \leq \frac{U_{\text{dc}} - RI}{2k_{\text{e}} + \frac{pLI}{5}} = \frac{f_2}{g_2} \tag{28}$$

where f_2 and g_2 are numerator and denominator of the critical speed in the proposed strategy.

To compare the critical speeds of the traditional strategy and the proposed strategy, it can be obtained by subtracting the denominator of (28) from that of (17) as

$$g_1 - g_2 = \sqrt{\frac{4pLI}{15}}(\sqrt{k_e} - \sqrt{\frac{3pLI}{20}}) = \sqrt{\frac{4pLI}{15n}}(\sqrt{E} - \sqrt{\frac{9}{2\pi}\omega_e LI}) \tag{29}$$

where $w_e = \pi pn/30$ is the electrical angular velocity of the motor.

Since the winding inductance voltage is generally much smaller than the phase back-EMF, $g_1 - g_2 > 0$ is satisfied in light of (29). Moreover, $f_1 - f_2 < 0$ is also met based on (17) and (28). Consequently, $f_1/g_1 - f_2/g_2 < 0$ can be derived, which means the speed range of the proposed strategy is larger than that of the traditional strategy.

Above all, if $b > 0$ is satisfied under the rated speed, the outgoing phase current can be reduced to zero over the full speed range. if $b < 0$ is met under the rated speed, the proposed strategy still broadens the speed range for commutation torque ripple suppression compared with the traditional strategy.

5. Experimental Results

The control block of the proposed strategy is shown in Figure 6. In the normal conduction period, the proportional integral (PI) controller is employed to control the current, where i_p is the amplitude of the conduction phase current. In the commutation period, the duty cycle is obtained from (20) to guarantee $dT_e/dt = 0$ during the commutation process. The period switch is introduced to select the duty cycle and the modulation in normal conduction period or commutation period.

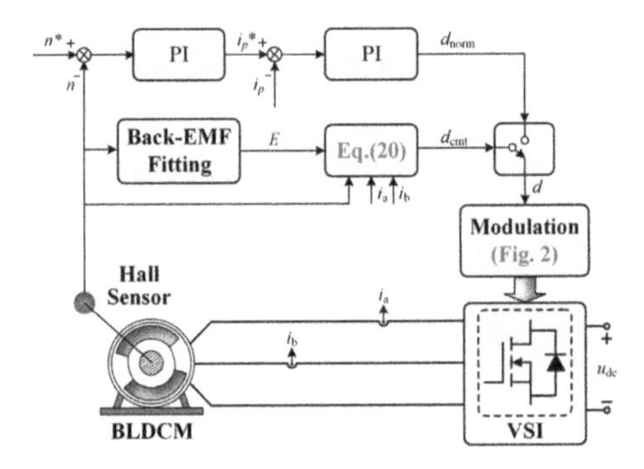

Figure 6. The control block of the proposed strategy.

To verify the correctness of the theoretical analysis and the effectiveness of the proposed strategy, the experimental system is established as shown in Figure 7.

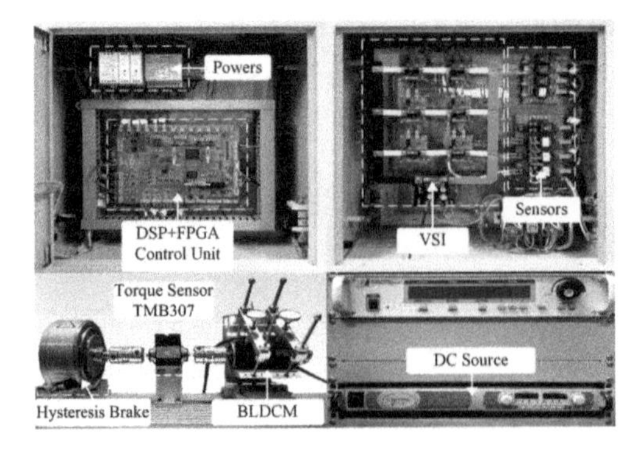

Figure 7. Experimental system.

In the experimental system, the torque sensor TMB307, whose measuring range is 10N·m and resolution is less than 0.01N·m, is used to measure the output mechanical torque of BLDCM. The control unit adopts the hybrid architecture of DSP (TMS320F28335) and FPGA (EP1C6Q240C8). MOSFET is manufactured by IR Corporation as IRFB4310-ZGPBF, whose switching frequency is 20kHz. The phase current is measured using the current sensor CSM025A. The measured torque, speed and duty cycle waveforms are output by the D/A converter. The rotor type of the experimental BLDCM is surface mounted, whose parameters are shown in Table 1.

Table 1. Parameters of BLDCM.

Parameter	Symbol	Value	Unit
Rated voltage	U_N	24	V
Rated current	I_N	14	A
Rated torque	T_N	3.2	N·m
Rated speed	n_N	600	r/min
Back-EMF coefficient	k_e	0.013	V/(r/min)
Phase resistance	R	0.2415	Ω
Phase inductance	L	0.387	mH
Pairs of pole	p	4	/

Based on IEC 60034-20-1, the torque ripple rate K_{rT} is defined as

$$K_{rT} = \frac{T_{high} - T_{low}}{T_{high} + T_{low}} \times 100\% \quad (30)$$

where T_{high} and T_{low} are the maximum and minimum values of the torque, respectively.

In the traditional strategy, the duty cycle is calculated by (7). Except for it, the modulations and the controller in the normal conduction period are all the same as that in the proposed strategy.

Figure 8 shows the waveforms of the motor under the rated load at 500 r/min, where the torque ripple rates of the traditional strategy and the proposed strategy are 7.644% and 4.376% respectively. It can be seen that the torque ripple is reduced to a certain degree in the traditional strategy. However, there is still a large torque fluctuation during the commutation process. Since the back-EMF variation is considered in the proposed strategy, the torque ripple rate is lower than 60% of the one in the traditional strategy.

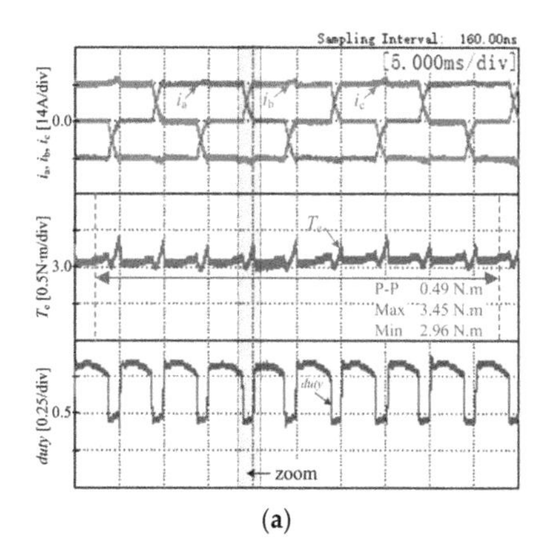

(a)

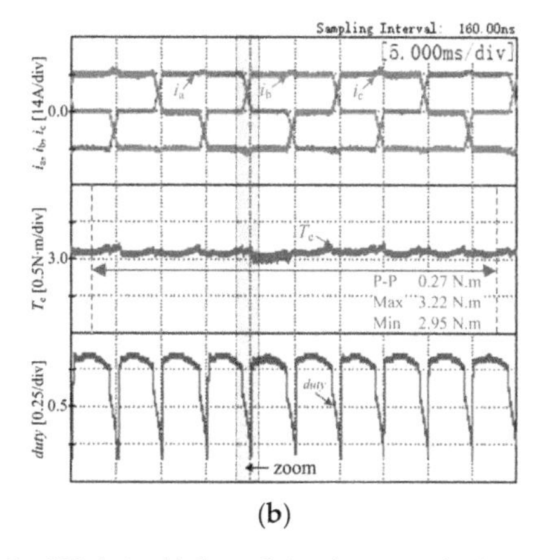

(b)

Figure 8. Experimental waveforms under the rated load at 500 r/min. (**a**) the traditional strategy. (**b**) the proposed strategy.

To account for the commutation process in detail, the enlarged waveforms of the dash area in Figure 8 is presented in Figure 9. In Figure 9a, the outgoing phase current is decreased to zero for about 1.2 ms and the motor torque is reduced first and then increased to a larger value during the commutation process with the constant duty cycle, which verifies the theoretical analysis in Figure 4. In Figure 9b, the outgoing phase current is decreased to zero for about 1 ms and the motor torque is almost constant with the reduced duty cycle.

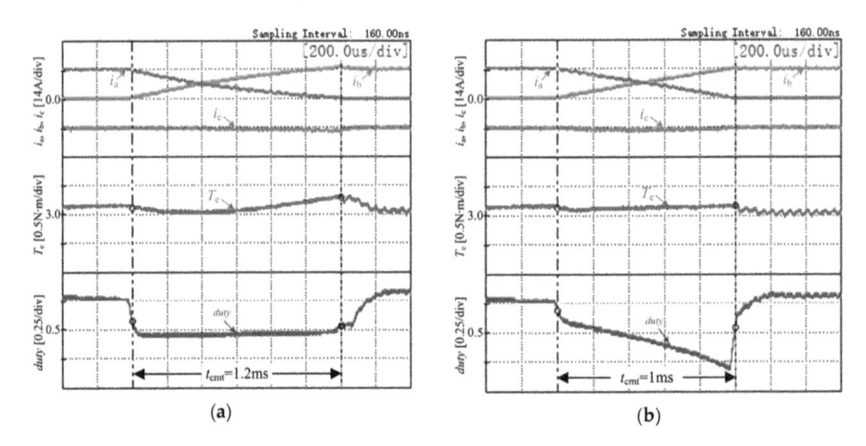

Figure 9. Enlarged waveforms of the commutation process under the rated load at 500 r/min. (**a**) the traditional strategy. (**b**) the proposed strategy.

Figure 10 shows the waveforms of the motor under the rated load at 550 r/min, where the torque ripple rates of the traditional strategy and the proposed strategy are 14.928% and 4.685% respectively. The actual measured critical speed of the traditional strategy is 501 r/min, which is basically same as the theoretical value 497 r/min calculated by (17). When the motor speed is larger than the critical speed, in the traditional strategy the outgoing phase current will not be reduced to zero and the other currents are distorted during the commutation process, which generates a large torque fluctuation. As the experimental motor's parameters meet b > 0 in (21) in light of Table 1, it will never happen over the full speed range in the proposed strategy that the outgoing phase current cannot be reduced to zero.

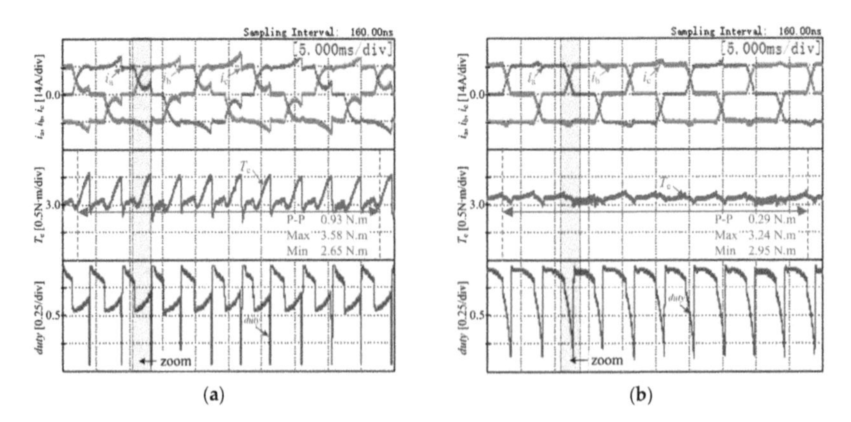

Figure 10. Experimental waveforms under the rated load at 550 r/min. (**a**) the traditional strategy. (**b**) the proposed strategy.

To account for the commutation process in detail, the enlarged waveforms of the dash area in Figure 10 is presented in Figure 11. In Figure 11a, the reason for the current's distortion is that the waveform of outgoing phase current is approximately concave parabola and its minimum value is larger than zero, which causes the failure end of the commutation process. In this experiment, when the commutation time exceeds 2.5 ms, it is considered a failure and the commutation process is forcibly terminated. In addition, the torque is first decreased and then increased to a large value, which aggravates the torque ripple. These experimental results also verify the theoretical analysis in Figure 5. In Figure 11b, it can be seen that the proposed strategy can reduce the outgoing phase current to zero with the back-EMF variation considered, where the normal end of commutation can be acquired.

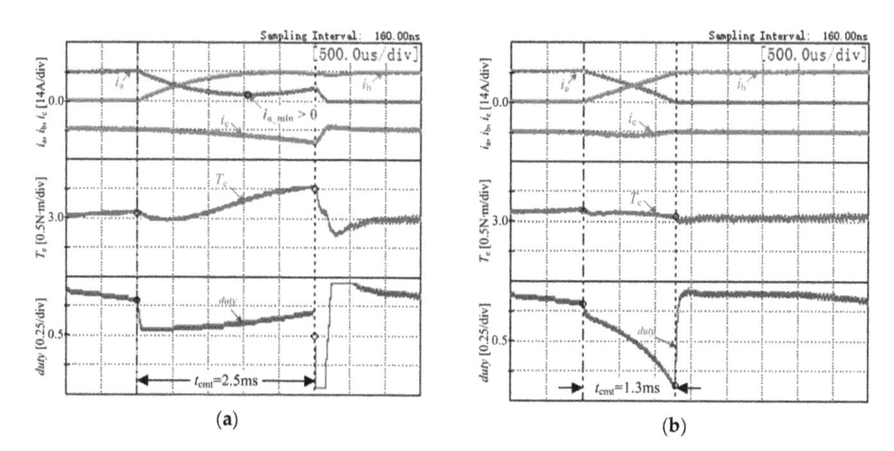

Figure 11. Enlarged waveforms of the commutation process under the rated load at 550 r/min. (**a**) the traditional strategy. (**b**) the proposed strategy.

When the motor is operating under rated conditions (motor speed is 600 r/min, and load torque is 3.2 N·m), owing to the motor rated speed 600r/min is higher than the critical speed 497 r/min, the outgoing phase current will no longer be reduced to zero with the traditional strategy, and the traditional strategy is obviously invalid for rated conditions of the motor. Figure 12 shows the experimental waveforms with the proposed strategy under the rated load at 600 r/min. As shown in Figure 12, the torque ripple rate is 7.792%, which indicates the proposed strategy can suppress the torque ripple effectively, and the phenomenon that the outgoing phase current cannot be reduced to zero does not appear under rated conditions by the proposed strategy.

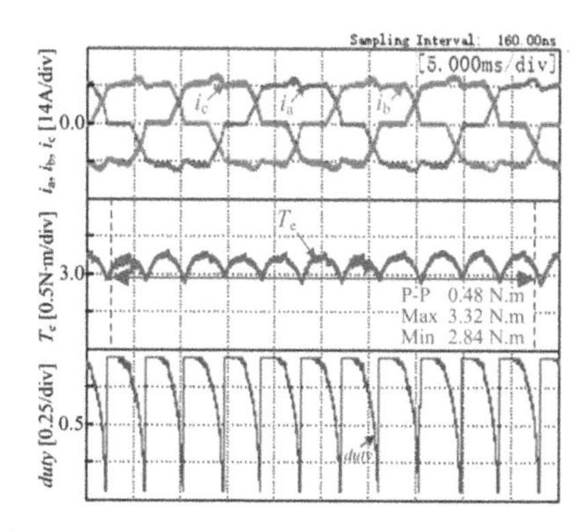

Figure 12. Experimental waveforms with the proposed strategy under the rated load at 600 r/min.

This paper considers the variation of the back-EMF in the commutation period, and the proposed strategy dynamically regulates the chopping duty cycle according to the variation trend of back-EMFs of three phases. Hence, the back-EMF coefficient is one of the important input parameters for the controller. To measure the back-EMF coefficient, driving by another motor at the speed 200 r/min, Figure 13 shows the measured back-EMFs of three phases in the open circuit windings. The back-EMF coefficient k_e = 0.013V/(r/min) listed in Table 1 can be calculated by this experimental result.

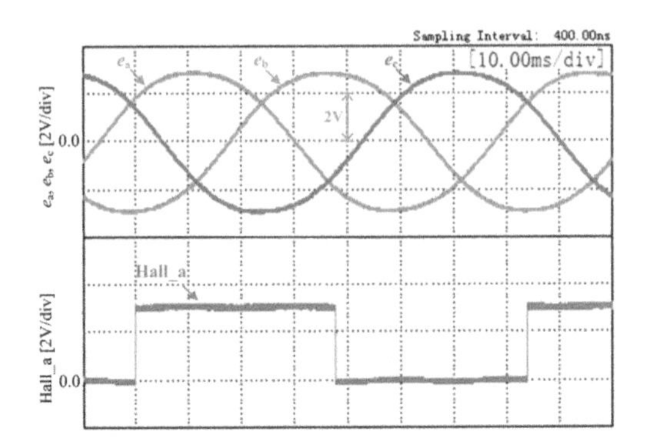

Figure 13. The measured three phases back-EMF and Hall signal of phase-a at 200r/min.

6. Conclusions

This paper aims to reduce the commutation torque ripple of BLDCM in its high-speed region, simultaneously broadening the valid speed range of the torque ripple reduction strategy without increasing the DC voltage. This paper firstly analyzes the problems of the traditional commutation torque ripple suppression for the high-speed region in a long commutation duration. A novel commutation torque ripple reduction strategy is then proposed based on a reconstructed smooth torque mechanism with back-EMF variation considered. Compared with the traditional strategy which controls the chopping duty cycle relatively smoothly in the commutation process, the proposed strategy dynamically regulates the chopping duty cycle, which make it show a gradual decrease. The proposed strategy has the following contributions:

1. The effect of back-EMF variation on torque and outgoing phase current during the commutation process is theoretically analyzed, and the speed range for the normal end of commutation is further deduced.
2. The smooth torque mechanism is reconstructed with the back-EMF variation considered, based on which a new commutation torque ripple suppression strategy is further designed. The proposed strategy can guarantee the torque gradient to be zero ($dT_e/dt = 0$) even under the changing back-EMF during the commutation process, which can more effectively reduce the commutation torque ripple.
3. The proposed strategy broadens the speed range for commutation torque ripple reduction, reducing the risk of phase current distortion and enhancing the reliability of the motor operation.

Author Contributions: Conceptualization, W.C., T.S., X.L and G.J.; methodology, X.L. and G.J.; software, G.J. and X.L.; validation, X.L., G.Z. and G.J.; formal analysis, G.Z. and Q.G.; writing—original draft preparation, G.J. and X.L.; writing—review and editing, G.Z., Q.G. and T.S.; funding acquisition, X.L., Q.G. and W.C.

Funding: This research was jointly supported in part by National Natural Science Foundation of China under Grant 51807141 and 51577126, in part by The Natural Science Foundation of Tianjin (18JCQNJC74200) and in part by The Science & Technology Development Fund of Tianjin Education Commission for Higher Education (2018KJ208).

Conflicts of Interest: The authors declare no conflict of interest.

References

1. Zhu, Z.Q.; Howe, D. Electrical Machines and Drives for Electric, Hybrid, and Fuel Cell Vehicles. *Proc. IEEE* **2007**, *95*, 746–765. [CrossRef]
2. Xia, C.; Jiang, G.; Chen, W.; Shi, T. Switching-Gain Adaptation Current Control for Brushless DC Motors. *IEEE Trans. Ind. Electron.* **2016**, *63*, 2044–2052. [CrossRef]
3. Ustun, O.; Kivanc, O.; Senol, S.; Fincan, B. On Field Weakening Performance of a Brushless Direct Current Motor with Higher Winding Inductance: Why Does Design Matter. *Energies* **2018**, *11*, 3119. [CrossRef]
4. Yoon, K.; Baek, S. Robust Design Optimization with Penalty Function for Electric Oil Pumps with BLDC Motors. *Energies* **2019**, *12*, 153. [CrossRef]
5. Priyadarshi, N.; Padmanaban, S.; Mihet-Popa, L.; Blaabjerg, F.; Azam, F. Maximum Power Point Tracking for Brushless DC Motor-Driven Photovoltaic Pumping Systems Using a Hybrid ANFIS-FLOWER Pollination Optimization Algorithm. *Energies* **2018**, *11*, 1067. [CrossRef]
6. Kamalapathi, K.; Priyadarshi, N.; Padmanaban, S.; Holm-Nielsen, J.; Azam, F.; Umayal, C.; Ramachandaramurthy, V. A Hybrid Moth-Flame Fuzzy Logic Controller Based Integrated Cuk Converter Fed Brushless DC Motor for Power Factor Correction. *Electronics* **2018**, *7*, 288. [CrossRef]
7. Carlson, R.; Mazenc, M.L.; Fagundes, J.C.D.S. Analysis of torque ripple due to phase commutation in brushless DC machines. *IEEE Trans. Ind. Appl.* **1992**, *28*, 632–638. [CrossRef]
8. Liu, Y.; Zhu, Z.Q.; Howe, D. Commutation-Torque-Ripple Minimization in Direct-Torque-Controlled PM Brushless DC Drives. *IEEE Trans. Ind. Appl.* **2007**, *43*, 1012–1021. [CrossRef]
9. Jung, S.; Kim, Y.; Jae, J.; Kim, J. Commutation Control for the Low-Commutation Torque Ripple in the Position Sensorless Drive of the Low-Voltage Brushless DC Motor. *IEEE Trans. Power Electron.* **2014**, *29*, 5983–5994. [CrossRef]
10. Li, H.; Zheng, S.; Ren, H. Self-Correction of Commutation Point for High-Speed Sensorless BLDC Motor with Low Inductance and Nonideal Back EMF. *IEEE Trans. Power Electron.* **2017**, *32*, 642–651. [CrossRef]
11. Cao, Y.; Shi, T.; Li, X.; Chen, W.; Xia, C. A Commutation Torque Ripple Suppression Strategy for Brushless DC Motor Based on Diode-Assisted Buck–Boost Inverter. *IEEE Trans. Power Electron.* **2019**, *34*, 5594–5605. [CrossRef]
12. Tan, B.; Hua, Z.; Zhang, L.; Fang, C. A New Approach of Minimizing Commutation Torque Ripple for BLDCM. *Energies* **2017**, *10*, 1735. [CrossRef]
13. Jiang, G.; Xia, C.; Chen, W.; Shi, T.; Li, X.; Cao, Y. Commutation Torque Ripple Suppression Strategy for Brushless DC Motors with a Novel Noninductive Boost Front End. *IEEE Trans. Power Electron.* **2018**, *33*, 4274–4284. [CrossRef]
14. Shi, T.; Cao, Y.; Jiang, G.; Li, X.; Xia, C. A Torque Control Strategy for Torque Ripple Reduction of Brushless DC Motor with Nonideal Back Electromotive Force. *IEEE Trans. Ind. Electron.* **2017**, *64*, 4423–4433. [CrossRef]
15. Song, J.H.; Choy, I. Commutation Torque Ripple Reduction in Brushless DC Motor Drives Using a Single DC Current Sensor. *IEEE Trans. Power Electron.* **2004**, *19*, 312–319. [CrossRef]
16. Shi, J.; Li, T. New Method to Eliminate Commutation Torque Ripple of Brushless DC Motor with Minimum Commutation Time. *IEEE Trans. Ind. Electron.* **2013**, *60*, 2139–2146. [CrossRef]
17. Xia, C.; Xiao, Y.; Chen, W.; Shi, T. Torque Ripple Reduction in Brushless DC Drives Based on Reference Current Optimization Using Integral Variable Structure Control. *IEEE Trans. Ind. Electron.* **2014**, *61*, 738–752. [CrossRef]
18. Lin, Y.; Lai, Y. Pulsewidth Modulation Technique for BLDCM Drives to Reduce Commutation Torque Ripple Without Calculation of Commutation Time. *IEEE Trans. Ind. Applicat.* **2011**, *47*, 1786–1793. [CrossRef]
19. Nam, K.Y.; Lee, W.T.; Lee, C.M.; Hong, J.P. Reducing torque ripple of brushless DC motor by varying input voltage. *IEEE Trans. Magn.* **2006**, *42*, 1307–1310. [CrossRef]
20. Shi, T.; Guo, Y.; Song, P.; Xia, C. A New Approach of Minimizing Commutation Torque Ripple for Brushless DC Motor Based on DC–DC Converter. *IEEE Trans. Ind. Electron.* **2010**, *57*, 3483–3490. [CrossRef]
21. Viswanathan, V.; Jeevananthan, S. Approach for torque ripple reduction for brushless DC motor based on three-level neutral-point-clamped inverter with DC-DC converter. *IET Power Electron.* **2015**, *8*, 47–55. [CrossRef]

22. Xiaofeng, Z.; Lu, Z. A New BLDC Motor Drives Method Based on BUCK Converter for Torque Ripple Reduction. In Proceedings of the 2006 CES/IEEE 5th International Power Electronics and Motion Control Conference, Shanghai, China, 14–16 August 2006; pp. 1–4.
23. Li, X.; Xia, C.; Cao, Y.; Chen, W.; Shi, T. Commutation Torque Ripple Reduction Strategy of Z-Source Inverter Fed Brushless DC Motor. *IEEE Trans. Power Electron.* **2016**, *31*, 7677–7690. [CrossRef]
24. Kang, B.H.; Kim, C.J.; Mok, H.S.; Choe, G.H. Analysis of torque ripple in BLDC motor with commutation time. In Proceedings of the IEEE International Symposium on Industrial Electronics (ISIE), Pusan, South Korea, 12–16 June 2001. [CrossRef]

Article

Determining the Position of the Brushless DC Motor Rotor

Krzysztof Kolano

Faculty of Electric Drives and Machines, Lublin University of Technology, 20-618 Lublin, Poland; k.kolano@pollub.pl

Received: 11 March 2020; Accepted: 30 March 2020; Published: 1 April 2020

Abstract: In brushless direct current (or BLDC) motors with more than one pole pair, the status of standard shaft position sensors assumes the same distribution several times for its full mechanical rotation. As a result, a simple analysis of the signals reflecting their state does not allow any determination of the mechanical position of the shaft of such a machine. This paper presents a new method for determining the mechanical position of a BLDC motor rotor with a number of pole pairs greater than one. In contrast to the methods used so far, it allows us to determine the mechanical position using only the standard position sensors in which most BLDC motors are equipped. The paper describes a method of determining the mechanical position of the rotor by analyzing the distribution of errors resulting from the accuracy proposed by the BLDC motor's Hall sensor system. Imprecise indications of the rotor position, resulting from the limited accuracy of the production process, offer a possibility of an indirect determination of the rotor's angular position of such a machine.

Keywords: rotor position; BLDC motor; sensor misalignment

---

1. Introduction

Compared to conventional DC brush motors, BLDC brushless motors using shaft position sensors are becoming increasingly popular in many applications, due to their relatively low cost, high performance and high reliability. The operating issues associated with the control of such machines have been extensively studied under the assumption of the correct operation of the shaft position sensor system, namely Hall sensors, and under the assumption of the symmetry of the signals generated by them [1–4].

Some industrial applications require that the position of the rotor is clearly defined and the information about the rotor is used by the control system during the operation cycle of the device. An example of this is the machine spindle, which must be properly positioned in relation to the cutter hopper, in order for the tools to be automatically picked and replaced.

A common solution to this problem of determining the rotor position is to equip the drive system with additional elements, namely, shaft position sensors, whose resolution is matched to the desired accuracy of its positioning. All kinds of encoders of the motor shaft's absolute position are used for this purpose [5]: absolute position encoders, and incremental encoders with an additional index "I" signal generated once per mechanical rotation.

The presence of Hall sensors, mounted in the motor, led the author to use them as an element defining the mechanical angle of the motor shaft. Determining the position of the BLDC motor rotor with an accuracy of 60 mechanical degrees for motors with one pair of poles is possible according to the formula:

$$\theta_e = p \cdot \theta_m \tag{1}$$

(where p equals the number of pole pairs), and it is determined directly by the state of the shaft position sensors, as the change in the electrical angle δ_e is equal to the change in the mechanical angle δ_m of the motor shaft position – Figure 1a.

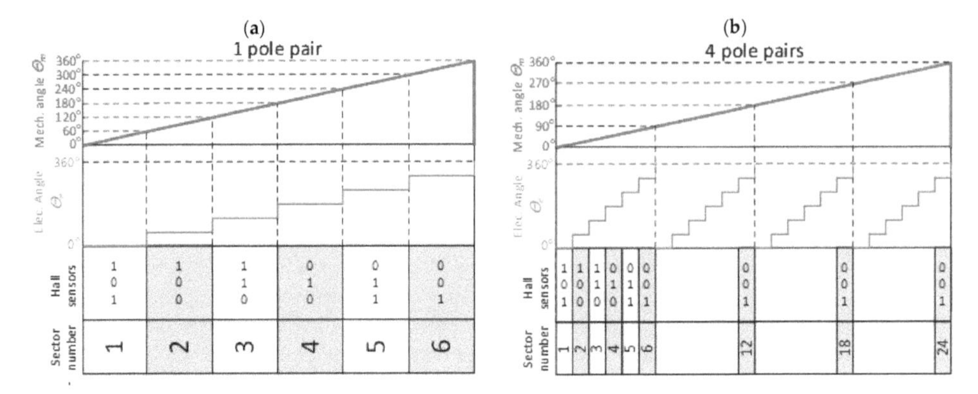

Figure 1. Relationships between the values of mechanical angle θ_m, electrical angle θ_e and the state of Hall sensors for brushless direct current (or BLDC) motors for: (**a**) 1; (**b**) 4 pairs of poles.

However, in machines with the number of $p > 1$ pole pairs ($p = 4$ for most BLDC motors that are present on the market), equipped only with standard Hall sensors, it is not possible to determine the mechanical position of the shaft solely on the basis of the analysis of their state, because the Hall sequence repeats p times per one mechanical revolution, as shown in Figure 1b.

Until now, the solution to this problem has been to install additional elements determining the position of the motor shaft. Unfortunately, this makes it necessary to mechanically modify the typical BLDC motors available on the market.

This article proposes a method that allows us to determine the mechanical position of the rotor with an accuracy equal to 60/p mechanical degrees for BLDC motors with the number of pole pairs greater than one, without the need for additional sensors. This accuracy value is sufficient for rotor positioning in many industrial applications.

2. Errors in the Positioning of Motor Sensors

The disadvantages resulting from the real, different from ideal, arrangement of shaft position sensors are known, and described in detail in the literature [6–13]. This problem is caused by the relatively low accuracy of the motor shaft position determination system during mass production. Researchers' efforts focus on compensating for sensor placement errors by assuming perfect symmetry of the rotating magnetic element (Figure 1 $\delta_{mr} = 0$). This makes it possible to treat the multipolar machine as a machine with one pair of poles [14–16]. Defining the errors in the measurement of the speed and position of the motor shaft, resulting only from the incorrect arrangement of sensors, is a large simplification in the analysis of the operation of the real system for determining the position of the BLDC motor rotor. If the error concerned only the sensor location accuracy (δ_{ms}– Figure 2), the speed measurement errors could be uniquely determined for each combination of Hall sensor states. In fact, these errors result to the same extent from the accuracy of sensor placement (δ_{ms}), as from the accuracy of making the magnetic ring rotating coaxially with the shaft (δ_{mr}); see Figure 2. The angular velocity δ_r of the shaft, measured by the microprocessor system as a value inversely proportional to the time t_s of changing the state of the position sensor values, amounts to:

$$\omega_r = \frac{\pi \pm \delta_{ms} \pm \delta_{mr}}{p \cdot t_s} \tag{2}$$

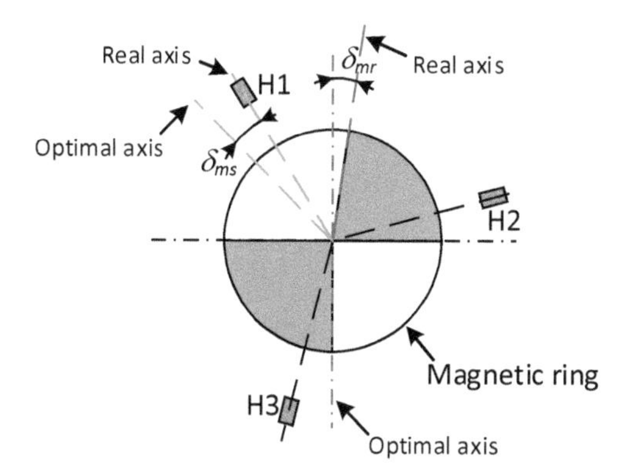

Figure 2. Visualisation of sensor (δ_{ms}) and magnet (δ_{mr}) misalignment in a BLDC shaft sensing system.

If non-zero errors δ_{ms} or δ_{mr} are assumed, the speed values for the individual sectors may vary despite the shaft rotating at a constant speed. This difference can be determined by comparing the measured value with a reference value, measured e.g., by an external measuring system.

In extreme cases, assuming that both the sensor system and the magnetic ring are made with limited accuracy, an error in shaft speed measurement can be calculated for each sector of the BLDC motor (Figure 3), e.g., for a motor with four pole pairs, the number of sectors is 24 per full mechanical rotation of the shaft (Figure 3b).

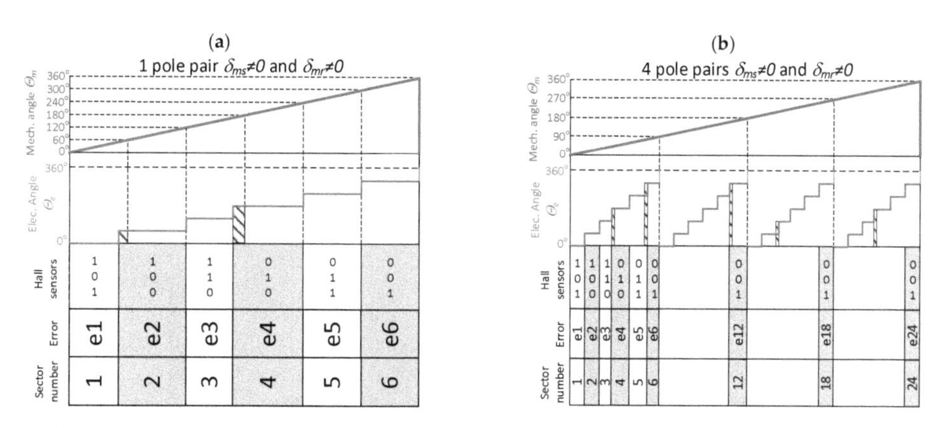

Figure 3. Relationships between the values of mechanical angle θ_m, electrical angle θ_e and the state of Hall sensors for brushless direct current (or BLDC) motors for: (**a**) 1; (**b**) 4 pairs of poles for $\delta_{ms} \neq 0$ and $\delta_{mr} \neq 0$.

For a motor with one pair of poles six unique error values can be assigned to a specific state of Hall sensors (Figure 3a), while for a motor with the number of pole pairs greater than one, these errors cannot be assigned to a specific state of sensors (Figure 3b).

Errors δ_{ms} and δ_{mr} result in switching the keys of the motor stator winding controller in a suboptimal position of the rotor, which in turn leads to the electromagnetic torque pulsation of the motor, increases the amplitude of its current (Figure 4) and the noise level of the motor operation. Boosting the amplitude of the current results in an increase in electrical losses and electromagnetic interference, and forces the constructors of the controller to use in it transistors with a higher rated current, so that they do not suffer thermal and dynamic damage during long-term operation [17].

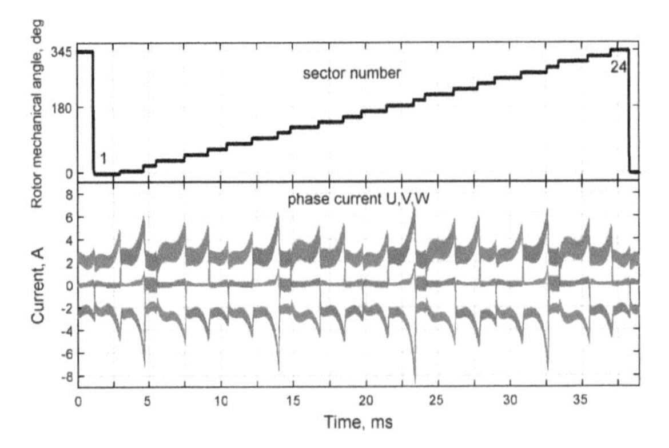

Figure 4. Measurement of phase currents of a BLDC motor (four pole pairs, nominal speed = 3000 rpm/min, nominal power = 380 W) during full mechanical rotation.

3. Laboratory Measurements Taken on Real Motors

In order to confirm the universality of this phenomenon in motors available on the market, it was decided to purchase four types of BLDC motors from four different manufacturers, and to analyze the performance of their shaft positioning systems. For this purpose, an experiment was developed to compare the actual ω_{ref} speed measured by the incremental encoder, with the speed measured using the signal generated by the motor's Hall sensors in an open loop drive system. The measuring functions were executed by the dSpace GmbH (Paderborn, Germany) dSpace MicroLabBox system supported by the Tektronix (Beaverton, Oregon United States) Digital Phosphor Oscilloscope 5054-B (Figure 5). As a measure of the error in determining the speed in individual sectors e_{sec}, a percentage ratio of the ω_{sec} calculated speed to the actual ω_{ref} measured by the reference element, i.e. the incremental encoder, was proposed. Additionally, the values of phase currents of the BLDC motor were measured during the tests.

Figure 5. Research set-up.

One of the aims of the experiment was to determine the influence of sensor distribution error on the pulsation of the actual rotational speed. It was predicted that significant oscillations of the motor phase current amplitude, caused by the premature or delayed activation of the controller transistors, may significantly increase this pulsation. During laboratory tests, it was found that in a wide range of speeds (from 10% to 100% of the n_n rated speed), and for wide ranges of load torque changes (from idle to rated load), the actual speed pulsation is imperceptible (Figure 6), despite significant (up to 80%; Figure 4) phase current fluctuations of the motor.

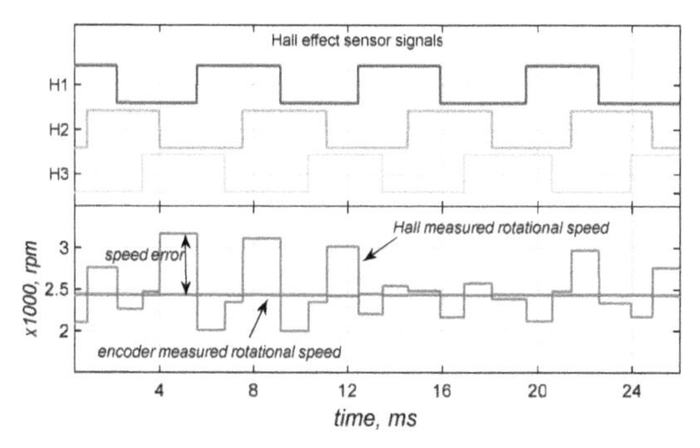

Figure 6. Actual and measured speed of a BLDC motor with factory nonsymmetrical Hall sensor array, for a BLDC motor with four pairs of poles.

The measurements showed that the accuracy of speed determination in the experiment, assumed at 0.5%, was not exceeded. This means that, regardless of errors in the BLDC motor position measurement system, we can assume that its average steady state rotational speed can be considered constant, regardless of the phase current amplitude fluctuations of the motor. This is a very important conclusion, allowing the thesis that, in the steady state the average actual speed for individual motor sectors is equal to the average speed measured during the full mechanical rotation of the shaft. The actual average speed of a full revolution can be measured correctly by any single Hall sensor. This results in the possibility of not using an additional reference speed measurement system (i.e. an additional encoder) in favor of using averaged full revolution speed, which is independent of the sensor placement error [12].

Figure 7 shows the calculated speed error values for the individual e_{sec} sectors for four different BLDC motors with powers ranging from 60 W to 1500 W. All motors had four pairs of poles each, for which 24 sectors of speed measurement can be defined per one mechanical revolution. Each of them corresponds to a mechanical rotation angle equal to $\pi/(3{\cdot}p)$, which gives an angle value for each sector equal to 15 mechanical degrees.

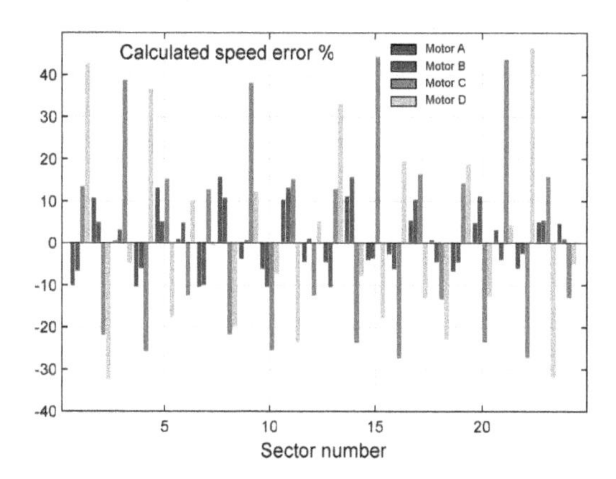

Figure 7. Rotational speed computation error "e_{sec}" in particular sectors of the tested motors.

It can be seen that the values of these errors are significant, and strongly influence the accuracy of motor speed calculation. It is worth emphasizing that all of the tested motors were characterized by a significant error in the arrangement of e_e sensors, which ranged from a few to over 25° of the electric angle, with respect to the optimal location according to the formula $e_e = e_{sec}{\cdot}60°$.

4. Determination of the Absolute Position of the BLDC Motor Rotor

In the course of the work described above, errors in determining the rotational speed for individual sectors of the motors, tested at different rotational speeds, were determined. It was shown that for a wide range of rotational speeds, these errors are almost constant, and what is worth underlining, unique for the specific motor used in the test. This is illustrated in Figure 8 for the motor marked as "D".

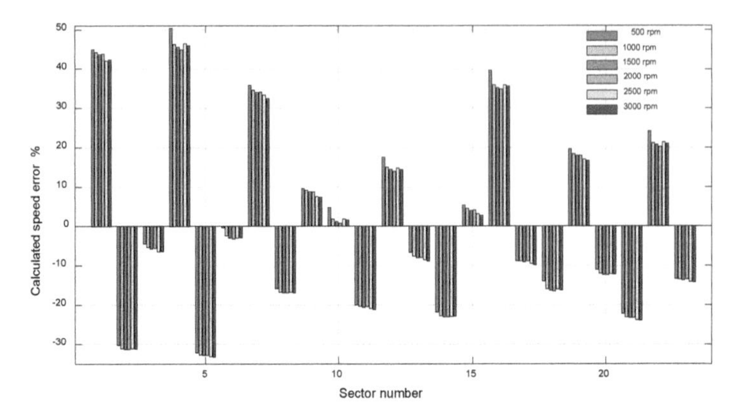

Figure 8. Rotational speed computation errors determined at different rotational speeds of motor D (four pole pairs, nominal speed = 3000 rpm/min, nominal power = 380W).

Figure 9 shows the error distribution of the speed determination for motor "D". Using a standard Hall sensor system causes that during the full rotation of the shaft, the sequence of sensor states is repeated depending on the number of pole pairs of the motor (here, four times).

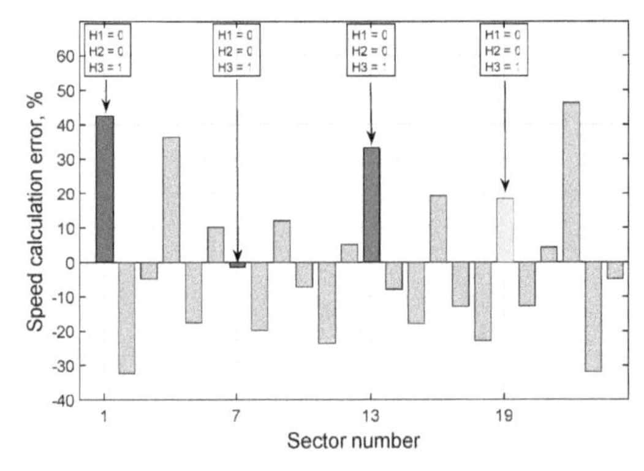

Figure 9. Distribution of errors in determining the e_{sec} speed of the BLDC motor, for motor "D", determined at 2000 rpm.

It can be seen that the value of the speed determination errors for individual sectors varies greatly, creating a sequence unique for a specific motor. This makes it possible to identify the position of the rotor using information about the states of the sensors, with the support of an algorithm that looks for a specific pattern of error distribution. Determination of the rotor position can be done in relation to the reference value of the e_{ref} error distribution, stored in the electrically erasable programmable read-only - EEPROM memory of the motor control device.

The sequence of errors is repeated with the frequency corresponding to the time of full rotation of the shaft by the motor. This sequence can start from different values, which depends only on the initial position of the motor shaft at the moment of starting the drive system.

To simplify the procedure for determining the motor shaft position, only the error distributions corresponding to the same combination of the Hall sensor states can be analyzed, which in the case of the motor under test reduces the number of errors analysed from 24 to only 4.

Figure 10 shows possible error distributions for the selected sensor states (H1 = 0, H2 = 0, H3 = 1). For a motor with four pairs of poles, four different sequences of assigning the error distribution to the absolute position of the motor rotor are possible (depending on the initial position of the rotor after power-up).

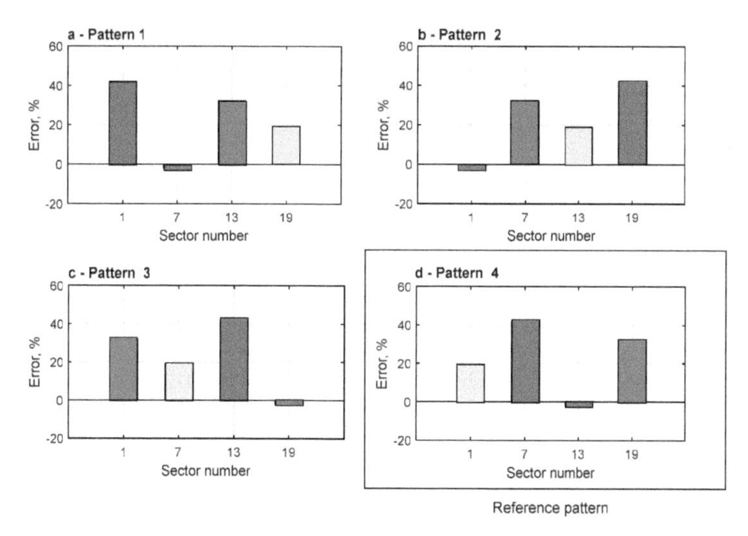

Figure 10. Possible error distributions for determining the rotor speed for the Hall sensor state for different initial rotor positions: (**a–c**); (**d**) for H1 = 0; H2 = 0; H3 = 1; (**d**) reference distribution stored in the memory of the motor controller.

5. Example Implementation

The procedure for determining the absolute position of the rotor is, in the simplest possible implementation, to search for the largest e_{sec} error for the same selected sequence of their state, and to assign its sector number to the sector number of the maximum error of the reference distribution. For example, sectors 1, 7, 13 and 19 (Figure 10) correspond to the sequence (H1 = 0; H2 = 0; H3 = 1). After determining the speed calculation errors, the algorithm compares the results obtained with the reference values stored in the controller's non-volatile EEPROM memory. If the reference error values stored for specific sensor signals reached the maximum value, e.g. for Sector 7, then after searching the newly obtained error distribution, and finding the maximum error value, e.g., for sector 13, one should change the sector number in which the greatest error of speed measurement was found. In this particular case, sector number 13 will be changed by the algorithm determining the position of the shaft to number 7, and the remaining sector numbers will be automatically assigned to the following sectors of Figure 11 by means of software incrementation.

The operation of this algorithm allowed to unambiguously assign the BLDC motor sector number to the angle of the mechanical rotor position by analyzing the reference pattern of error distribution, stored e.g., in the EEPROM memory.

When the differences between the errors of individual sectors are small, i.e., when the shaft position detection system was made precisely at the factory, it is possible to expand the algorithm of searching for similarity between measured values and reference values stored in the EEPROM memory. The minimum error value and even all errors in all sectors can also be analysed.

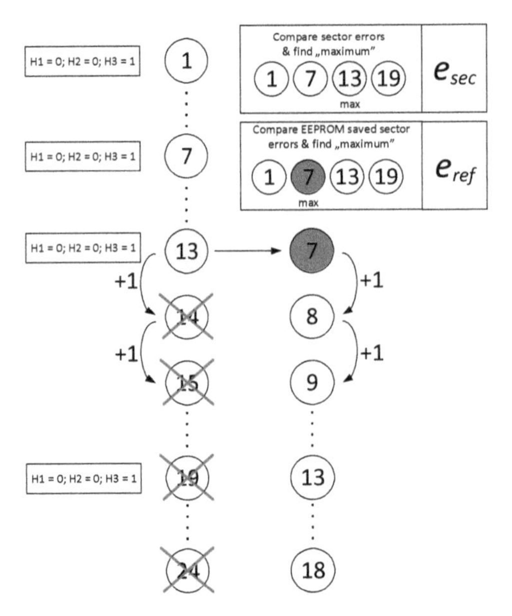

Figure 11. Algorithm for determining the position of the BLDC motor rotor on the basis of the comparison of e_{sec} errors with the reference value of the error distribution e_{ref}, for a given state of the Hall sensors (H1 = 0; H2 = 0; H3 = 1).

6. Conclusions

During the research, four BLDC motors with four pairs of poles each were analysed. Thanks to the proposed method, it was possible to find the mechanical position of a multi-pole BLDC motor, despite the fact that the sequence of the Hall sensor states changes in the same sequence p-times per mechanical revolution. What is very important is that theoretically, if the shaft position monitoring system was made very accurately at the factory, it would not be possible to achieve motor diagnostics results characteristic enough to assign a specific position of the motor rotor to them. Hence, the conclusion that applying the method described in the article can correct a factory's inaccurate fixing of motor components, and result in the possibility of using such machines in drive systems, where the absolute position of the shaft must be determined for the correct operation of the whole system. Another interesting issue is the possibility of identifying a specific electrical machine by analyzing the distribution of errors in speed measurement, using its shaft position sensors. This gives the possibility of the quick, simple and cost-free detection of the replacement of the drive motor, even if it is characterized by the same nominal data. Determination of the angular position of the BLDC motor shaft on the basis of the analysis of the error spectrum allows the application of advanced algorithms for the correction of commutation errors, which result from errors in the position of sensors in multi-pole BLDC motors [3]. This contributes to the reduction of the noise emitted by the drive as well as the use of semiconductor connectors with a lower rated current for the controller.

Funding: This research received no external funding.

Conflicts of Interest: The authors declare no conflict of interest.

References

1. Giulii Capponi, F.; De Donato, G.; Del Ferraro, L.; Honorati, O.; Harke, M.C.; Lorenz, R.D. AC brushless drive with low-rersolution hall-effect sensors for surface-mounted PM machines. *IEEE Trans. Ind. Appl.* **2006**, *42*, 526–535. [CrossRef]
2. Hui, T.S.; Basu, K.P.; Subbiah, V. Permanent magnet brushless motor control techniques. In Proceedings of the National Power Engineering Conference, Bangi, Malaysia, 15–16 December 2003; pp. 133–138.
3. Kolano, K. Improved Sensor Control Method for BLDC Motors. *IEEE Access* **2019**. [CrossRef]
4. Pillay, P.; Krishnan, R. Modeling, simulation, and analysis of permanent-magnet motor drives, Part II: The brushless DC motor drive. *IEEE Trans. Ind. Appl.* **1989**, *25*, 274–279. [CrossRef]
5. Santolaria, J.; Conte, J.; Pueo, M.; Carlos, J. Rotation error modeling and identification for robot kinematic calibration by circle point method. *Metrol. Meas. Syst.* **2014**, *21*, 85–98. [CrossRef]
6. Alaeinovin, P.; Jatskevich, J. Filtering of Hall-Sensor Signals for Improved Operation of Brushless DC Motors. *IEEE Trans. Energy Convers.* **2012**, *27*, 547–549. [CrossRef]
7. Alaeinovin, P.; Chiniforoosh, S.; Jatskevich, J. Evaluating misalignment of hall sensors in brushless DC motors. In Proceedings of the 2008 IEEE Canada Electric Power Conference, Vancouver, BC, Canada, 6–7 October 2008; pp. 1–6.
8. Alaeinovin, P.; Jatskevich, J. Hall-sensor signals filtering for improved operation of brushless DC motors. In Proceedings of the 2011 IEEE International Symposium on Industrial Electronics, Gdansk, Poland, 27–30 June 2011; pp. 613–618.
9. Baszyński, M.; Piróg, S. A Novel Speed Measurement Method for a High-Speed BLDC Motor Based on the Signals From the Rotor Position Sensor. *IEEE Trans. Ind. Inform.* **2014**, *10*, 84–91. [CrossRef]
10. Choi, J.H.; Park, J.S.; Gu, B.G.; Kim, J.H.; Won, C.Y. Position estimation and control of BLDC motor for VVA module with unbalanced hall sensors. In Proceedings of the 2012 IEEE International Conference on Power and Energy (PECon), Kota Kinabalu, Malaysia, 2–5 December 2012; pp. 390–395.
11. Park, J.W.; Kim, J.H.; Kim, J.M. Position Correction Method for Misaligned Hall-Effect Sensor of BLDC Motor using BACK-EMF Estimation. *Trans. Korean Inst. Power Electron.* **2012**, *17*, 246–251. [CrossRef]
12. Kolano, K. Calculation of the brushless dc motor shaft speed with allowances for incorrect alignment of sensors. *Metrol. Meas. Syst.* **2015**, *22*, 393–402. [CrossRef]
13. Samoylenko, N.; Han, Q.; Jatskevich, J. Dynamic performance of brushless DC motors with unbalanced hall sensors. *IEEE Trans. Energy Convers.* **2008**, *23*, 752–763. [CrossRef]
14. Lim, J.S.; Lee, J.K.; Seol, H.S.; Kang, D.W.; Lee, J.; Go, S.C. Position Signal Compensation Control Technique of Hall Sensor Generated by Uneven Magnetic Flux Density. *IEEE Trans. Appl. Supercond.* **2008**, *28*, 1–4. [CrossRef]
15. Nerat, M.; Vrančić, D. A Novel Fast-Filtering Method for Rotational Speed of the BLDC Motor Drive Applied to Valve Actuator. *IEEE/ASME Trans. Mechatron.* **2016**, *21*, 1479–1486. [CrossRef]
16. Park, D.H.; Nguyen, A.T.; Lee, D.C.; Lee, H.G. Compensation of misalignment effect of hall sensors for BLDC motor drives. In Proceedings of the 2017 IEEE 3rd International Future Energy Electronics Conference and ECCE Asia (IFEEC 2017—ECCE Asia), Kaohsiung, Taiwan, 3–7 June 2017; pp. 1659–1664. [CrossRef]
17. Zieliński, D.; Fatyga, K. Comparison of main control strategies for DC/DC stage of bidirectional vehicle charger. In Proceedings of the 2017 International Symposium on Electrical Machines (SME), Naleczow, Poland, 18–21 June 2017. [CrossRef]

Article

The Novel Rotor Flux Estimation Scheme Based on the Induction Motor Mathematical Model Including Rotor Deep-Bar Effect

Grzegorz Utrata [1,*], Jaroslaw Rolek [2] and Andrzej Kaplon [2]

1 Institute of Environmental Engineering, Czestochowa University of Technology, Brzeźnicka 60a, 42-200 Czestochowa, Poland
2 Department of Industrial Electrical Engineering and Automatic Control, Kielce University of Technology, Tysiąclecia Państwa Polskiego 7, 25-314 Kielce, Poland
* Correspondence: gutrata@is.pcz.pl; Tel.: +48-500-145-449

Received: 17 June 2019; Accepted: 10 July 2019; Published: 12 July 2019

Abstract: During torque transients, rotor electromagnetic parameters of an induction motor (IM) vary due to the rotor deep-bar effect. The accurate representation of rotor electromagnetic parameter variability by an adopted IM mathematical model is crucial for a precise estimation of the rotor flux space vector. An imprecise estimation of the rotor flux phase angle leads to incorrect decoupling of electromagnetic torque control and rotor flux amplitude regulation which in turn, causes deterioration in field-oriented control of IM drives. Variability of rotor electromagnetic parameters resulting from the rotor deep-bar effect can be modeled by the IM mathematical model with rotor multi-loop representation. This paper presents a study leading to define the unique rotor flux space vector on the basis of the IM mathematical model with rotor two-terminal network representation. The novel rotor flux estimation scheme was validated with the laboratory test bench employing the IM of type Sg 132S-4 with two variants of rotor construction: a squirrel-cage rotor and a solid rotor manufactured from magnetic material S235JR. The accuracy verification of the rotor flux estimation was performed in a slip frequency range corresponding to the IM load adjustment range up to 1.30 of the stator rated current. This study proved the correct operation of the developed rotor flux estimation scheme and its robustness against electromagnetic parameter variability resulting from the rotor deep-bar effect in the considered slip frequency range.

Keywords: deep-bar effect; mathematical model; estimation; induction motors; motor drives

---

1. Introduction

The development of advanced control methods of induction motors (IMs), such as direct and indirect field-oriented control [1,2] or direct torque control [3], have contributed to the widespread use of this type of motor in modern drive systems intended for various applications in industry. In the rotor-flux-orientation, the stator phase currents through the Park's transformation are represented by the field- and torque-producing components. In cases when the rotor flux amplitude is stabilized by the field-producing component of the stator current space vector, IM electromagnetic torque is linearly proportional to the torque-producing component [1]. The decoupling of IM electromagnetic torque control and rotor flux amplitude regulation is realized based on the phase angle of the rotor flux space vector. Since direct measurement of the rotor flux is practically not achievable, development of indirect methods for rotor flux space vector estimation is reported in the world literature, especially model-based methods.

In IM field-oriented control, slip frequency is controlled within the set range of values, except for very short torque transients. With slip frequency changes, rotor electromagnetic parameters vary due to the rotor deep-bar effect. For maintaining the high dynamic performance of the IM rotor-flux-oriented control during torque transients, the accurate representation of rotor electromagnetic parameter variability by an adopted IM mathematical model is required.

Inaccurate representation of this variability by the adopted IM mathematical model, which serves as basis for the rotor flux estimation scheme, leads to an erroneous estimation of the rotor flux space vector. In consequence, the erroneous estimation of the vector components results in deterioration of decoupling effectiveness of electromagnetic torque control and rotor flux amplitude regulation, thus deteriorating the overall performance of the IM rotor-flux-oriented control [4–6]. For this reason, the compensation of the influence of the rotor deep-bar effect on the rotor flux estimation accuracy is important for the rotor-flux-oriented control of squirrel-cage IMs, especially the ones where the rotor bar is large enough to incorporate high rotor current.

Until now, estimation schemes for the rotor flux space vector have been elaborated predominantly on the basis of the IM classical mathematical model with rotor single-loop representation with constant parameters. In order to compensate for the influence of the rotor electromagnetic parameter variability on the rotor flux estimation accuracy, the estimation schemes extended by algorithms enabling tracking variability of rotor electromagnetic parameters were proposed [4,6–15]. These schemes work very well with reference to IMs with squirrel-cage rotors, in which the electromagnetic parameters do not show significant variability resulting from the rotor deep-bar effect. The response of the proposed algorithms for variability tracking of rotor electromagnetic parameters may not be fast enough to follow rapid parameter variability during torque transients. These algorithms were mainly intended to model rotor resistance changes associated with temperature variation [4,6–15].

The variability of rotor electromagnetic parameters resulting from the rotor deep-bar effect can be modeled by the IM mathematical model with rotor multi-loop representation [16–25]. Nevertheless, an estimation scheme of the rotor flux space vector which algorithm would be formulated on the basis of such IM mathematical models has not been developed so far. What is more, the authors of these works [19–22] stated that defining the unique rotor flux space vector in the IM mathematical model with rotor multi-loop representation is not possible, and thus they proposed IM airgap-flux or pseudorotor-flux oriented control, developed with the use of the mathematical model of this type.

The results of simulation and experimental studies presented previously [19–22] indicate very good dynamic performance of the vector-controlled squirrel-cage and double-cage IMs. This fact encouraged us to carry on work on the application of the IM mathematical model with rotor multi-loop representation in the IM rotor-flux-oriented control, since such a control strategy has a simpler structure and a more effective decoupling of electromagnetic torque control and rotor flux amplitude regulation than airgap-flux-oriented control [22].

This paper presents a study which leads to development of the rotor flux estimation scheme on the basis of the IM mathematical model with rotor two-terminal network representation. The overall goal of this work was focused on the accuracy verification of the rotor flux estimation in a slip frequency range corresponding to the IM load adjustment range up to 1.30 of the stator rated current. Thus, the considered slip frequency range exceeded the typical operating range of slip frequency for IM field-oriented control. This study aimed to prove the proper modeling of the electromagnetic parameter variability resulting from the rotor deep-bar effect by the novel rotor flux estimation scheme. Due to the assumed concept of the conducted work, the experimental investigations were realized in an open-loop drive system (without speed feedback or slip compensator), at a fixed setpoint of stator voltages and step commands of load torque. The evaluation of operation accuracy of the developed rotor flux estimation scheme was realized indirectly with the use of the registered shaft torque.

The results of the conducted study point out an improvement of the estimation accuracy of the rotor flux space vector obtained by the scheme developed on the basis of the IM mathematical model with rotor two-terminal network representation, in comparison to the accuracy which was gained by the estimation schemes formulated with the use of the IM classical mathematical model. In particular, this applies to the tested IM characterized by the intense rotor skin effect. Consequently, the obtained results confirm the correct operation of the novel rotor flux estimation scheme and its robustness for electromagnetic parameter variability resulting from the rotor deep-bar effect.

2. Mathematical Models of an Induction Motor

One of the fundamental problems associated with the use of the IM classical mathematical model with constant parameters in IM control algorithms is the variability of motor electromagnetic parameters which is conditioned by changes of motor winding temperature, ferromagnetic core saturation, as well as the rotor deep-bar effect [26]. Figure 1a presents the T-type equivalent circuit corresponding to the IM classical mathematical model expressed in the Laplace-domain (p-domain), in which the rotor resistance R_2 and leakage inductance $L_{\sigma 2}$ are represented by parameters varying as a function of slip frequency ω_2. The variability of rotor electromagnetic parameters resulting from the rotor deep-bar effect can be modeled in the rotor equivalent circuit by a two-terminal network with constant parameters [16–25]. The electromagnetic parameters of such an IM mathematical model can be determined based on the p-domain motor inductance:

$$L_1(p) = \frac{\underline{\Psi}_{1r}(p)}{\underline{I}_{1r}(p)} = \omega_b \frac{Z_{ab}(p)}{p} \tag{1}$$

where $\underline{\Psi}_{1r}(p)$ and $\underline{I}_{1r}(p)$ are the Laplace transforms of the stator flux and current space vectors, respectively, p is a complex frequency, $Z_{ab}(p)$ denotes the p-domain impedance between the terminals "a" and "b" of the IM equivalent circuit presented in Figure 1a, ω_b is the base frequency (Appendix A), and the subscript "r" denotes physical quantity space vectors expressed in an orthogonal coordinate system rotating at the shaft angular velocity ω_{sh}.

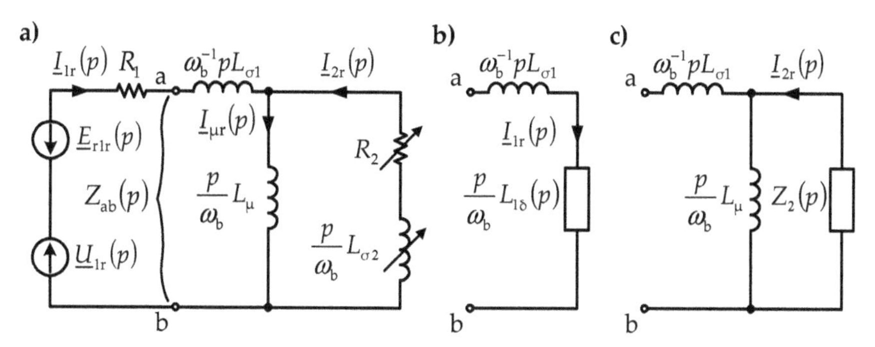

Figure 1. The induction motor (IM) equivalent circuits expressed in the Laplace-domain: (**a**) The IM equivalent circuit with rotor resistance and leakage inductance represented by parameters varying as a function of slip frequency. (**b**) The representation of the p-domain motor inductance by a series connection of the stator leakage inductance and the p-domain inductance associated with the airgap flux. (**c**) The representation of the p-domain inductance associated with the airgap flux by a parallel connection of the magnetizing inductance and the p-domain rotor impedance.

Equation (1), as well as the subsequent equations included in this paper, are expressed in the per-unit (p.u.) system. The base values of the used p.u. system are defined in Appendix A. Moreover, rotor physical quantities and electromagnetic parameters are referred to the stator.

The p-domain motor inductance $L_1(p)$ is a series connection of the stator leakage inductance $L_{\sigma 1}$ and the p-domain inductance associated with the airgap flux $L_{1\delta}(p)$ (Figure 1b):

$$L_1(p) = L_{\sigma 1} + L_{1\delta}(p). \tag{2}$$

The p-domain motor inductance $L_{1\delta}(p)$ can be further represented as a parallel connection of the magnetizing inductance L_μ and the p-domain rotor impedance $Z_2(p)$ (Figure 1c):

$$\frac{\omega_b}{pL_{1\delta}(p)} = \frac{\omega_b}{pL_\mu} + \frac{1}{Z_2(p)}. \tag{3}$$

The p-domain inductance $L_{1\delta}(p)$ can be derived from a solution of Maxwell's differential system of equations which are formulated, for instance, on the basis of an IM multi-layer model [17]. However, the p-domain inductance $L_{1\delta}(p)$ is not directly applicable in an analysis of IM transients due to the lack of possibility for inverse transformation of a Laplace transform including this inductance. The above-mentioned difficulty can be circumvent by the partial fraction decomposition of the inverse p-domain inductance $L_{1\delta}(p)$, which is an irrational function with an infinite number of negative real poles. This, in turn, leads to the rotor mathematical model in the form of a two-terminal network with constant $R_{2(n)}$, $L_{\sigma 2(n)}$ parameters [17]:

$$\frac{\omega_b}{pL_{1\delta}(p)} = \frac{\omega_b}{pL_\mu} + \sum_{n=1}^{\infty} \frac{1}{R_{2(n)} + \frac{1}{\omega_b}pL_{\sigma 2(n)}}. \tag{4}$$

An exact approximation of the reference p-domain inductance $L_{1\delta}(p)$ is obtained with an infinite number of poles of an approximative rational function (an infinite number of parallel connected two-terminals in the rotor mathematical model). For the sake of the desired simplicity of the IM mathematical model, the number of parallel connected two-terminals in the rotor equivalent circuit is limited to N two-terminals, and for achieving the required approximation accuracy of the irrational function $L_{1\delta}(p)$, the $(N + 1)$th residual two-terminal with parameters $R_{2(0)}$, $L_{\sigma 2(0)}$ is included (Figure 2) [17]:

$$\frac{\omega_b}{pL_{1\delta}(p)} = \frac{\omega_b}{pL_\mu} + \sum_{n=1}^{N} \frac{1}{R_{2(n)} + \frac{1}{\omega_b}pL_{\sigma 2(n)}} + \frac{1}{R_{2(0)} + \frac{1}{\omega_b}pL_{\sigma 2(0)}}. \tag{5}$$

The methodology for determination of the residual two-terminal electromagnetic parameters $R_{2(0)}$, $L_{\sigma 2(0)}$ has been described previously [17]. The approximation accuracy of the reference p-domain inductance is evaluated by comparing its frequency characteristic with a characteristic $\underline{L}_{1\delta}(p = j\omega_2)$ resulting from the IM mathematical model with rotor two-terminal network representation.

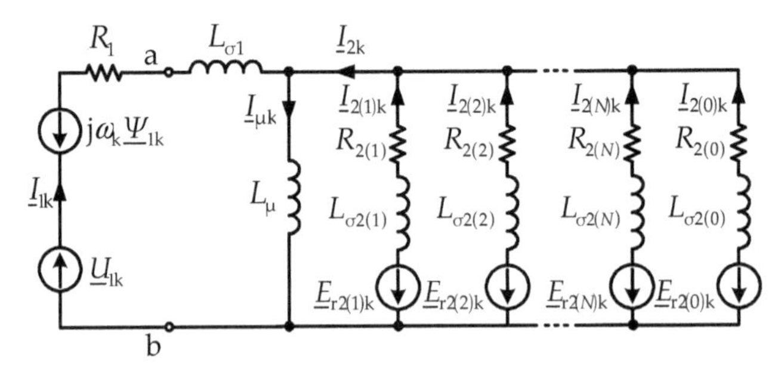

Figure 2. The equivalent circuit of an induction motor with rotor two-terminal network representation.

If the analytical formula describing the p-domain inductance $L_{1\delta}(p)$ is not known, which is the case, for instance, in experimental determination of the reference inductance frequency characteristic (IFCh) $\underline{L}_{1\delta}(\omega_2)$ [25,27], then the "synthetic" electromagnetic parameters of the IM equivalent circuit (Figure 2) can be identified as a result of an approximation of the reference IFCh $\underline{L}_{1\delta}(\omega_2)$ by the characteristic $\underline{L}_{1\delta}(p = j\omega_2)$ derived from the adopted IM mathematical model. In such an approach, the residual two-terminal does not formally occur in Equation (5), and its participation in the approximation accuracy of the p-domain rotor impedance is smaller for larger the numbers N of parallel connected two-terminals in the rotor mathematical model:

$$\frac{1}{Z_2(p)} \cong \sum_{n=1}^{N} \frac{1}{R_{2(n)} + \frac{1}{\omega_b} p L_{\sigma 2(n)}}. \tag{6}$$

In this way, an IM of any rotor construction (e.g., squirrel-cage, double-cage, or solid rotors) can be represented with the use of the mathematical model in which the electromagnetic parameter variability resulting from the rotor deep-bar effect is approximated by the two-terminal network with constant parameters. The process of electromagnetic parameter identification for individual two-terminals in the rotor equivalent circuit, ensuring the required approximation accuracy of the reference IFCh $\underline{L}_{1\delta}(\omega_2)$, can be conveniently performed using selected optimization methods. In such an approach, the number N of the parallel connected two-terminals is determined empirically.

The IM mathematical model with rotor N-loop representation is described by the following system of equations:

$$\underline{U}_{1k} = R_1 \underline{I}_{1k} + T_N \frac{d}{dt} \underline{\Psi}_{1k} + j\omega_k \underline{\Psi}_{1k} \tag{7a}$$

$$\begin{cases} 0 = R_{2(1)} \underline{I}_{2(1)k} + T_N \frac{d}{dt} \underline{\Psi}_{2(1)k} + j(\omega_k - \omega_{sh}) \underline{\Psi}_{2(1)k} \\ \cdots \\ 0 = R_{2(N)} \underline{I}_{2(N)k} + T_N \frac{d}{dt} \underline{\Psi}_{2(N)k} + j(\omega_k - \omega_{sh}) \underline{\Psi}_{2(N)k} \end{cases} \tag{7b}$$

$$\underline{\Psi}_{1k} = L_1 \underline{I}_{1k} + L_\mu \underline{I}_{2k} \tag{7c}$$

$$\begin{cases} \underline{\Psi}_{2(1)k} = L_\mu \left(\underline{I}_{1k} + \underline{I}_{2k} \right) + L_{\sigma 2(1)} \underline{I}_{2(1)k} \\ \cdots \\ \underline{\Psi}_{2(N)k} = L_\mu \left(\underline{I}_{1k} + \underline{I}_{2k} \right) + L_{\sigma 2(N)} \underline{I}_{2(N)k} \end{cases} \tag{7d}$$

$$\underline{I}_{2k} = \sum_{n=1}^{N} \underline{I}_{2(n)k} \tag{7e}$$

$$\frac{d\omega_{sh}}{dt} = \frac{1}{T_M} (T_{em} - T_L) \tag{7f}$$

$$T_{em} = \mathrm{Im}\left(\underline{\Psi}_{1k}^{*} \underline{I}_{1k} \right) \tag{7g}$$

$$\underline{E}_{r1k} = j\omega_k \underline{\Psi}_{1k} \tag{7h}$$

$$\underline{E}_{r2(n)k} = j(\omega_k - \omega_{sh}) \underline{\Psi}_{2(n)k} \tag{7i}$$

where $\underline{U}_{1k}$ and $\underline{E}_{r1k}$ are the stator voltage and electromotive force space vectors, respectively, $\underline{I}_{2(n)k}$, $\underline{\Psi}_{2(n)k}$, and $\underline{E}_{r2(n)k}$ represent the rotor current, flux, and electromotive force, respectively, related to the nth two-terminal in the rotor equivalent circuit, T_{em} and T_L constitute the electromagnetic and load torque, respectively, $T_N = 1/\omega_b$, T_M is the motor mechanical time constant, $j^2 = -1$, * denotes the complex conjugate, and k indicates physical quantity space vectors expressed in an orthogonal coordinate system rotating at an arbitrary angular velocity ω_k.

3. The Novel Estimation Scheme of the Rotor Flux Space Vector

The variability of rotor electromagnetic parameters resulting from the rotor deep-bar effect can be represented in the rotor mathematical model by the two-terminal network with constant $R_{2(n)}$, $L_{\sigma 2(n)}$ parameters. Therefore, the application of such an IM mathematical model in an estimation scheme of the rotor flux space vector is justifiable, especially in the case of an IM characterized by the intense rotor deep-bar effect. However, the rotor flux estimation scheme, which would be based on the IM mathematical model of this type, has not been developed until now. This chapter presents the investigations leading to define the unique rotor flux space vector on the basis of the IM mathematical model with rotor two-terminal network representation.

The current space vector of the nth two-terminal in the rotor equivalent circuit (Figure 2) can be determined with the use of the transformed Equation (7d):

$$\underline{I}^{\bullet}_{2(n)\mathrm{k}} = \frac{1}{L^{\bullet}_{\sigma 2(n)}}\left(\underline{\Psi}^{\bullet}_{2(n)\mathrm{k}} - L_{\mu}\underline{I}_{\mu\mathrm{k}}\right) \tag{8a}$$

$$\underline{I}_{\mu\mathrm{k}} = \underline{I}_{1\mathrm{k}} + \underline{I}_{2\mathrm{k}} \tag{8b}$$

$$\underline{\Psi}_{\mu\mathrm{k}} = L_{\mu}\underline{I}_{\mu\mathrm{k}} \tag{8c}$$

where $\underline{I}_{\mu\mathrm{k}}$ and $\underline{\Psi}_{\mu\mathrm{k}}$ are the magnetizing current and flux space vectors, respectively.

Incorporation of the formulas describing the rotor two-terminal current space vectors (Equation (8a)) to the transformed voltage equation (Equation (7b)) associated with the nth two-terminal in the rotor multi-loop equivalent circuit, the model of the rotor flux space vector related to the nth two-terminal is obtained in the form of:

$$T_{2(n)}\frac{\mathrm{d}}{\mathrm{d}t}\underline{\Psi}_{2(n)\mathrm{k}} = L_{\mu}\underline{I}_{\mu\mathrm{k}} - \underline{\Psi}_{2(n)\mathrm{k}} + \mathrm{j}\omega_{\mathrm{b}}T_{2(n)}(\omega_{\mathrm{k}} - \omega_{\mathrm{m}})\underline{\Psi}_{2(n)\mathrm{k}} \tag{9a}$$

$$T_{2(n)} = \frac{L_{\sigma 2(n)}}{R_{2(n)}}T_{\mathrm{N}} \tag{9b}$$

where $T_{2(n)}$ constitutes the time constant of the nth two-terminal in the rotor equivalent circuit presented in Figure 2.

The magnetizing current space vector $\underline{I}_{\mu\mathrm{k}}$, which is required in Equation (9a), can be determined based on Equation (7c) where the stator flux space vector $\underline{\Psi}_{1\mathrm{k}}$ is obtained with the use of the stator voltage Equation (7a):

$$\underline{I}_{\mu\mathrm{k}} = \frac{1}{L_{\mu}}\left(\underline{\Psi}_{1\mathrm{k}} - L_{\sigma 1}\underline{I}_{1\mathrm{k}}\right). \tag{10}$$

In general, the rotor flux space vector can be expressed as the sum of the magnetizing flux (Equation (8c)) and rotor leakage flux space vectors. Concerning the IM mathematical model with rotor multi-loop representation, the equation takes the following form:

$$\underline{\Psi}_{2\mathrm{k}} = L_{\mu}\underline{I}_{1\mathrm{k}} + \left(L_{\mu} + L_{\sigma 2\mathrm{eq}}\right)\underline{I}_{2\mathrm{k}} \tag{11}$$

where $L_{\sigma 2\mathrm{eq}}$ is the equivalent rotor leakage inductance of the rotor two-terminal network (Figure 2).

The resultant rotor current space vector is the sum of the current space vectors of parallel connected two-terminals in the rotor equivalent circuit (Equation (7e)). Taking into account the formulas describing these current space vectors (Equation (8a)), Equation (11) is as follows:

$$\underline{\Psi}_{2\mathrm{k}} = L_{\mu}\underline{I}_{\mu\mathrm{k}} + L_{\sigma 2\mathrm{eq}}\sum_{n=1}^{N}\frac{\underline{\Psi}_{2(n)\mathrm{k}}}{L_{\sigma 2(n)}} - L_{\mu}\underline{I}_{\mu\mathrm{k}}L_{\sigma 2\mathrm{eq}}\sum_{n=1}^{N}\frac{1}{L_{\sigma 2(n)}}. \tag{12}$$

The magnetizing flux space vector $\underline{\Psi}_{\mu\mathrm{k}}$ has been included in the formulas representing the flux space vectors associated with the individual two-terminals in the rotor multi-loop equivalent circuit (Equation (9a)), thus the magnetizing flux space vector is redundant in Equation (12) for the rotor flux space vector. The first and the third components of the sum in Equation (12), constituting and containing the magnetizing flux space vector, respectively, reduce each other in cases when the inverse equivalent rotor leakage inductance $L_{\sigma 2\mathrm{eq}}$ equals the sum of the inverse leakage inductances of the individual rotor two-terminals:

$$\frac{1}{L_{\sigma 2\mathrm{eq}}} = \sum_{n=1}^{N}\frac{1}{L_{\sigma 2(n)}}. \tag{13}$$

On the basis of the above reasoning, the derived formulas describe the equivalent rotor leakage inductance of the rotor equivalent circuit presented in Figure 2, which result from a parallel connection of leakage inductances of the individual rotor two-terminals:

$$L_{\sigma 2\mathrm{eq}} = \lim_{p \to \infty} \frac{Z_2(p)}{p} \tag{14}$$

and the unique rotor flux space vector of the IM mathematical model with rotor multi-loop representation:

$$\underline{\Psi}_{2\mathrm{k}} = L_{\sigma 2\mathrm{eq}} \sum_{n=1}^{N} \frac{\underline{\Psi}_{2(n)\mathrm{k}}}{L_{\sigma 2(n)}}. \tag{15}$$

According to the above, the voltage–current model of the rotor flux space vector can be formulated. When expressed in the orthogonal coordinate system (α–β) stationary with respect to the stator ($\omega_{\mathrm{k}} = 0$, indicated by s), this model is represented by the following system of equations:

$$\underline{I}^{\mathrm{e}}_{\mu\mathrm{s}} = \frac{1}{L_\mu}\left(\frac{1}{T_\mathrm{N}}\int_0^t \left(\underline{U}_{1\mathrm{s}} - R_1\underline{I}_{1\mathrm{s}}\right)\mathrm{d}t - L_{\sigma 1}\underline{I}_{1\mathrm{s}}\right) \tag{16a}$$

$$\begin{cases} T_{2(1)}\frac{\mathrm{d}}{\mathrm{d}t}\underline{\Psi}^{\mathrm{e}}_{2(1)\mathrm{s}} = L_\mu \underline{I}^{\mathrm{e}}_{\mu\mathrm{s}} - \underline{\Psi}^{\mathrm{e}}_{2(1)\mathrm{s}} + \mathrm{j}\omega_\mathrm{b} T_{2(1)}\omega_\mathrm{m}\underline{\Psi}^{\mathrm{e}}_{2(1)\mathrm{s}} \\ \cdots \\ T_{2(N)}\frac{\mathrm{d}}{\mathrm{d}t}\underline{\Psi}^{\mathrm{e}}_{2(N)\mathrm{s}} = L_\mu \underline{I}^{\mathrm{e}}_{\mu\mathrm{s}} - \underline{\Psi}^{\mathrm{e}}_{2(N)\mathrm{s}} + \mathrm{j}\omega_\mathrm{b} T_{2(N)}\omega_\mathrm{m}\underline{\Psi}^{\mathrm{e}}_{2(N)\mathrm{s}} \end{cases} \tag{16b}$$

$$\underline{\Psi}^{\mathrm{e}}_{2\mathrm{s}} = L_{\sigma 2\mathrm{eq}} \sum_{n=1}^{N} \frac{\underline{\Psi}^{\mathrm{e}}_{2(n)\mathrm{s}}}{L_{\sigma 2(n)}} \tag{16c}$$

where "e" denotes the estimated rotor flux space vector.

Figure 3 presents a schematic diagram of the rotor flux estimation scheme, corresponding to Equations (16a)–(16c). The rotor angular velocity and the stator voltage and current space vector components constitute the input signals of the developed rotor flux estimation scheme. It should also be noted that when the rotor deep-bar effect is represented by the single two-terminal $N = 1$ in the rotor mathematical model, the voltage-current model of the rotor flux space vector remains valid and corresponds, in this case, to the IM classical mathematical model.

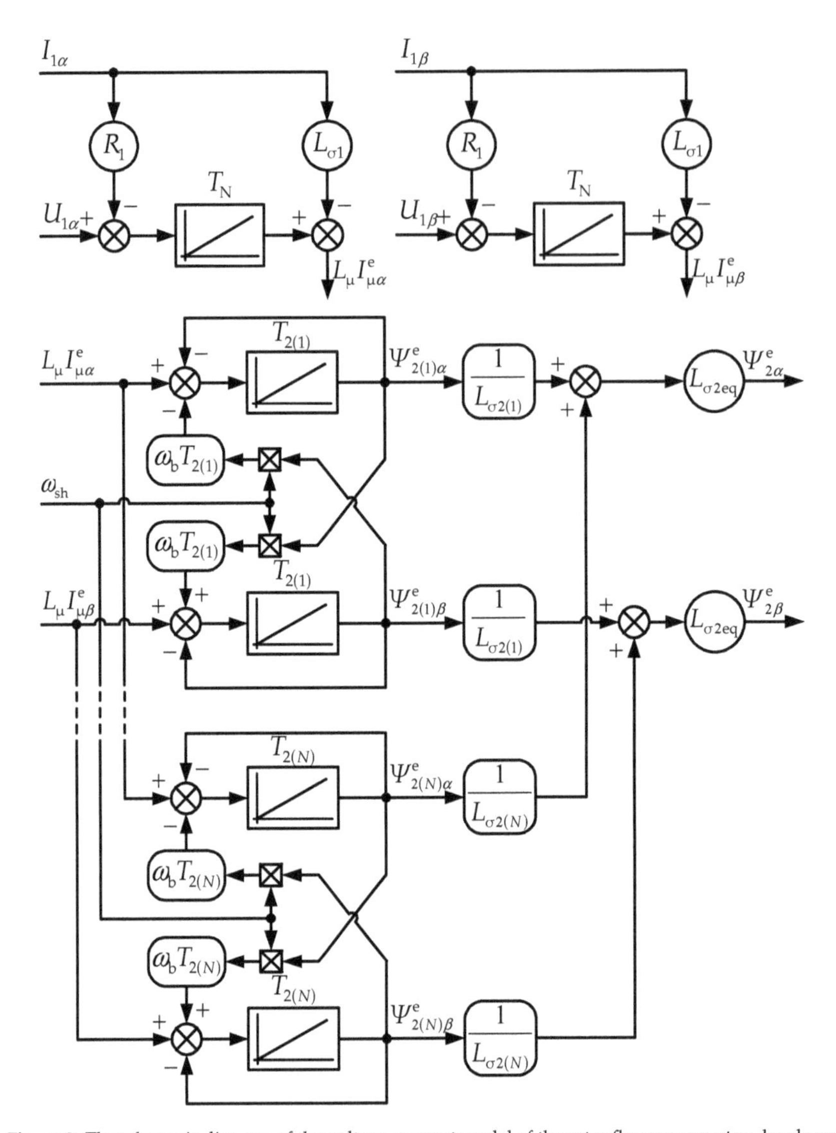

Figure 3. The schematic diagram of the voltage–current model of the rotor flux space vector, developed on the basis of the IM mathematical model with rotor two-terminal network representation, expressed in the orthogonal coordinate system (α–β) stationary with respect to the stator $\omega_k = 0$.

4. Laboratory Tests

4.1. Laboratory Test Bench

The verification of the novel rotor flux estimation scheme was performed with the laboratory test bench, of which a schematic diagram is shown in Figure 4. Figure 5 presents a picture of the laboratory test bench.

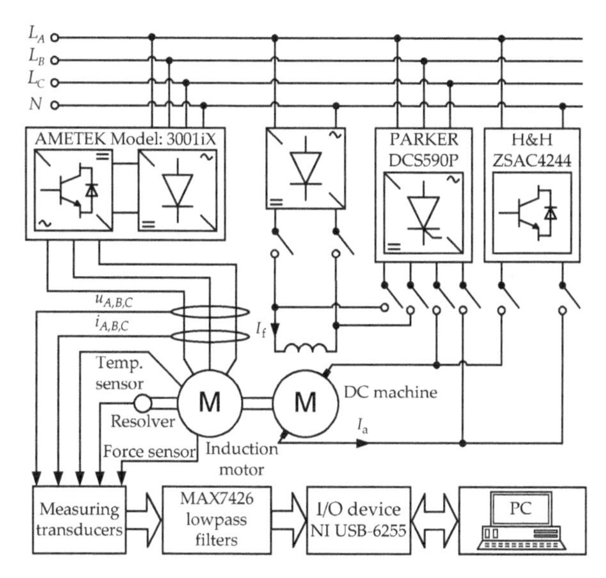

Figure 4. The schematic diagram of the laboratory test bench.

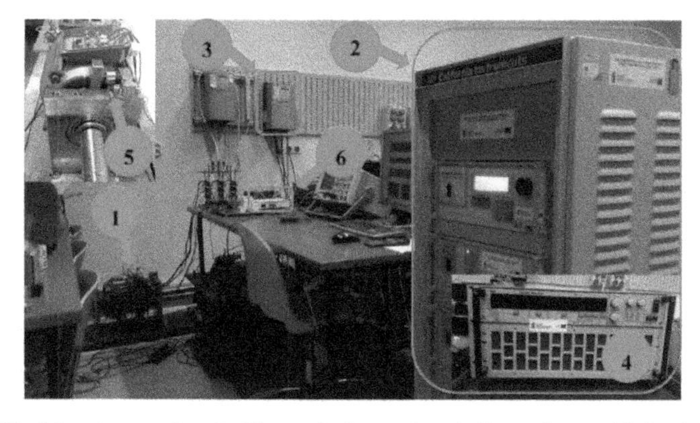

Figure 5. The laboratory test bench. The particular markers indicate: 1. tested induction motor, 2. programmable AC power source, AMETEK model: 3001iX, 3. 4Q thyristor converter, Parker DC590P, 4. electronic AC load ZSAC4244, Höcherl and Hackl GmbH, 5. tensometric force sensor, 6. multifunction I/O device NI USB-6255.

Investigations were conducted on the four-pole IMs of type Sg 132S-4 with a squirrel-cage rotor (CR-IM) and a solid rotor (SR-IM). The cross-section dimensions of the studied rotors are included in Appendix B.

The solid rotor was designed and manufactured only for the purpose of the presented study. This is because of the fact that such a rotor is characterized by the intense skin effect. This feature was conveniently used in the verification of the novel rotor flux estimation scheme. Such an approach allowed us to conduct experimental investigations with the use of low-power IMs, giving the background for the employment of the large squirrel-cage IMs in the next stage of the study. The tested SR-IM is marked by significant slip $s \approx 1$ corresponding to the breakdown torque at the stator supply voltage frequency of $f_1 = 50$ Hz. In order to reduce the breakdown slip of the SR-IM, the new operating points were adopted for both the CR-IM and SR-IM. The rated values of the tested IMs corresponding to the adopted machine operating points were set together in Appendix B.

The tested IMs (marker 1 in Figure 5) were powered by the programmable AC source AMETEK Model: 3001iX (marker 2 in Figure 5). The investigations were carried out at the three setpoints of stator voltage frequency for both considered IMs, maintaining a constant voltage/frequency ratio. Prior to the measurement tests, the CR-IM and SR-IM operated at the given load conditions for a period of time, allowing the stator winding temperature to stabilize at the assumed level. This aimed to reduce the influence of variability of rotor and stator resistances, resulting from winding temperature changes, on the identified electromagnetic parameters of the IM mathematical models and on the accuracy evaluation of the rotor flux space vector estimation.

A separately excited DC machine of type PCMb 54b served as a load for the investigated IMs in the presented study. During the no-load, blocked rotor, and load curve (LC) tests, conducted in order to identify the electromagnetic parameters of the considered IM mathematical models, the DC machine was powered by the 4Q thyristor converter Parker DC590P (marker 3 in Figure 5). Such a solution provided a wide adjustment range of load conditions for the tested IMs, enabling the measurement of demanded physical quantities in the generating, motoring, and ideal no-load modes of machine operation. In turn, the programmable electronic load ZSAC4244 – H&H GmbH (marker 4 in Figure 5) was used to control the armature current of the separately excited DC machine for the verification of the rotor flux estimation scheme. Such an approach enabled shaping of the desired dynamics of IM slip frequency changes in the assumed range, corresponding to the load adjustment range up to 1.30 of the stator rated current for both tested IMs. The slip frequency range considered during the verification of the rotor flux estimation scheme corresponded also to the slip frequency range when the LC tests were conducted. The power rating of the individual devices and DC machine which were used in the experimental investigations are included in Appendix B.

During the laboratory tests, the measurement of stator winding voltages, currents, and temperature, as well as shaft angular velocity, were carried out. Additionally, the shaft torque of the investigated IMs was determined on the basis of the force measurement realized by means of the force sensor (marker 5 in Figure 5). The accuracy class and measuring range of the used force sensor were 0.2 and 5 kN, respectively. The stator currents of the tested IMs were converted into voltage signals by means of non-inductive resistive voltage dividers with an accuracy class of 0.5 and a measuring range of 10 A. Similarly, the stator voltages were scaled by means of voltage dividers with a voltage ratio of 1000:1, composed of non-inductive resistors with an accuracy class of 0.2. The angular velocity measurement was carried with the use of a resolver. Data acquisition (DAQ) was performed by means of the National Instruments USB-6255 high-resolution, multifunction I/O device (marker 6 in Figure 5). The DAQ system was equipped with the MAX7426 5th-order, lowpass, elliptic, switched-capacitor filters. The configuration of the DAQ device and the acquisition of measurement data were carried out in the National Instruments LabView environment.

4.2. The Identification Procedure of Electormagnetic Parametersfor the IM Mathematical Model

The identification process for electromagnetic parameters of the IM mathematical model with rotor two-terminal network representation was conducted in conformity to a procedure described previously [25].The reference IFCh $\underline{L}_1(\omega_2)$ of the considered IMs were determined on the basis of the measurement data derived from the LC test [28] according to the following equations:

$$\frac{\omega_b}{j\omega_1}\left(\underline{Z}_1(\omega_2) - R_1\right) = \underline{L}_1(\omega_2) = L_{1\sigma} + \underline{L}_{1\delta}(\omega_2) \tag{17a}$$

$$\underline{Z}_1(\omega_2) = \frac{\left|\underline{U}_{1\text{ph}}\right|}{\left|\underline{I}_{1\text{ph}}(\omega_2)\right|}\left(\cos(\phi_1(\omega_2)) + j\sqrt{1-[\cos(\phi_1(\omega_2))]^2}\right) \tag{17b}$$

$$\cos(\phi_1(\omega_2)) = \frac{P_1(\omega_2)}{3\left|\underline{U}_{1\text{ph}}\right|\left|\underline{I}_{1\text{ph}}(\omega_2)\right|} \tag{17c}$$

where $\underline{Z}_1(\omega_2)$ is the IM impedance expressed as a function of slip frequency, ω_1 represents the stator voltage angular frequency, $|\underline{U}_{1ph}|$ and $|\underline{I}_{1ph}(\omega_2)|$ constitute the measured root mean squared values of stator phase voltages and currents, respectively, and $P_1(\omega_2)$ and $\cos(\phi_1(\omega_2))$ are the measured stator power and calculated stator power factor, respectively.

Approximation of the reference IFCh $\underline{L}_{1\delta}(\omega_2)$ by means of the frequency-domain inductance $\underline{L}_{1\delta}(p = j\omega_2)$ resulting from the adopted IM mathematical model, for instance, with N parallel connected two-terminals in the rotor equivalent circuit, in the form of the following equations:

$$\frac{1}{\underline{L}_{1\delta}(\omega_2)} \cong \frac{1}{\underline{L}_{1\delta}(p = j\omega_2)} = \frac{1}{L_\mu} + \frac{j\omega_2}{\omega_b} \frac{1}{\underline{Z}_2(p = j\omega_2)} \tag{18a}$$

$$\frac{1}{\underline{Z}_2(p = j\omega_2)} \cong \sum_{n=1}^{N} \frac{1}{R_{2(n)} + \frac{j\omega_2}{\omega_b} L_{\sigma 2(n)}}. \tag{18b}$$

allows determination of the "synthetic" electromagnetic parameters of the IM mathematical model.

The stator phase winding resistance R_1 is identified through the DC line-to-line resistance measurement conducted according to the standards [28,29]. The parameters $L_{1\sigma}$, L_μ, $R_{2(n)}$, and $L_{\sigma 2(n)}$ are subject to the identification process which can be considered as a minimization issue of the evaluation function, defined as the sum of the mean squared errors of the reference characteristic approximation [25]:

$$F\left(\left|\underline{L}_1(\omega_2)\right|, \angle\underline{L}_1(\omega_2)\right) = \sum_{\omega_{2\min}}^{\omega_{2\max}} k_{\mathrm{mod}} \left(\frac{|\underline{L}_1(\omega_2)| - |\underline{L}_1(p=j\omega_2)|}{|\underline{L}_1(\omega_2)|} \right)^2 + \\ + \sum_{\omega_{2\min}}^{\omega_{2\max}} k_{\mathrm{arg}} \left(\angle\underline{L}_1(\omega_2) - \angle\underline{L}_1(p = j\omega_2) \right)^2 \tag{19}$$

where $|\underline{L}_1(\omega_2)|$ and $\angle\underline{L}_1(\omega_2)$ are the modulus and argument of the IM reference IFCh, respectively, $|\underline{L}_1(p = j\omega_2)|$ and $\angle\underline{L}_1(p = j\omega_2)$ constitute the modulus and argument of the frequency-domain IM inductance, respectively, $\omega_{2\min}$ and $\omega_{2\max}$ represent the lower and upper limits of the considered slip frequency range, and k_{mod} and k_{arg} are the weighting factors of the individual components of the evaluation function.

Similarly to the studies presented previously [25], a minimization process of the adopted evaluation function was carried out with the use of the genetic algorithm by means of the Genetic Algorithms for Optimization Toolbox in the Matlab environment. The choice of the genetic algorithm was dictated by the effectiveness of this optimization tool, as indicated in numerous scientific publications concerned the identification of electromagnetic parameters for IM mathematical models [30,31].

The criterion adopted in the identification process of electromagnetic parameters for the CR-IM and SR-IM mathematical models with rotor multi-loop representation, assumed the approximation of the reference IFCh modulus with an error not exceeding 2% in the considered range of slip frequency, while maintaining a possible minimum approximation error of the reference IFCh argument and a minimum number N of two-terminals in a rotor equivalent circuit. The criterion was met with the use of the IM mathematical models with two $N = 2$ and three $N = 3$ parallel connected two-terminals in the cage and solid rotor network representations, respectively. The electromagnetic parameters of the IM mathematical models with multi-loop representation of the cage and solid rotors are denoted as CR-RML and SR-RML, respectively. These parameters are listed in Tables 1 and 2. Tables 1 and 2 also include the electromagnetic parameters of the IM classical mathematical model (T-type equivalent circuit parameters), which were identified based on selected procedures described in the standard 1 [28] (the designations are CR-T Std 1, SR-T Std 1) and standard 2 [29] (the designations are CR-T Std 2, SR-T Std 2) in the vicinity of the adopted operating points of the tested IMs (Appendix B). The stator phase resistance, determined according to the guidelines included in standard 1 [28], after correction to the reference winding temperature of 25 °C, equalled $R_1 = 2.9597\ \Omega$.

Table 1. Electromagnetic parameters (p.u.) of the considered squirrel-cage rotor induction motor (CR-IM) mathematical models. The individual resistances were corrected to the reference winding temperature of 25 °C.

Electromagnetic Parameters	$L_{1\sigma}$	L_{μ}	$R_{2(1)}$	$L_{\sigma 2(1)}$	$R_{2(2)}$	$L_{\sigma 2(2)}$
CR-T Std 1	0.0949	3.0978	0.0309	0.1428	–	–
CR-T Std 2	0.0910	3.1235	0.0335	0.1358	–	–
CR-RML	0.1090	3.0203	0.0395	0.0888	0.1326	1.3294

Table 2. Electromagnetic parameters (p.u.) of the considered solid rotor induction motor (SR-IM) mathematical models. The individual resistances correspond to the average temperature of stator winding of 55 °C registered under the load curve(LC) test.

Electromagnetic Parameters	$L_{1\sigma}$	L_{μ}	$R_{2(1)}$	$L_{\sigma 2(1)}$	$R_{2(2)}$	$L_{\sigma 2(2)}$	$R_{2(3)}$	$L_{\sigma 2(3)}$
SR-T Std 1	0.2597	2.8323	0.1329	0.4550	–	–	–	–
SR-T Std 2	0.1645	3.1736	0.2271	0.2456	–	–	–	–
SR-RML	0.1977	3.4401	0.2852	0.2313	0.4695	2.1032	0.1518	7.4735

As an example, in Figure 6, the SR-IM reference IFCh $\underline{L}_1(\omega_2)$ are set together with the approximative characteristics, which were determined on the basis of the IM mathematical model with single and three parallel connected two-terminals in the solid rotor equivalent circuit. Figure 7 presents the modulus relative errors and the argument absolute errors between the reference and approximative characteristics [25].

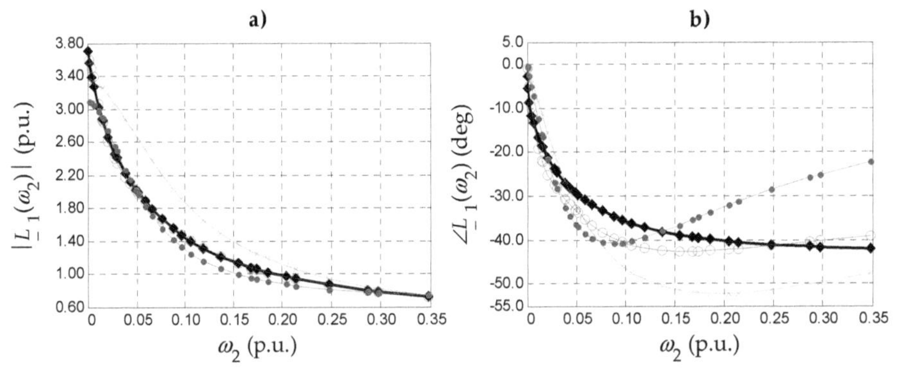

Figure 6. The SR-IM reference inductance frequency characteristic (IFCh) and its approximation by means of the IM mathematical models with single and three two-terminals in the solid rotor equivalent circuit: (**a**) Modulus and (**b**) argument.

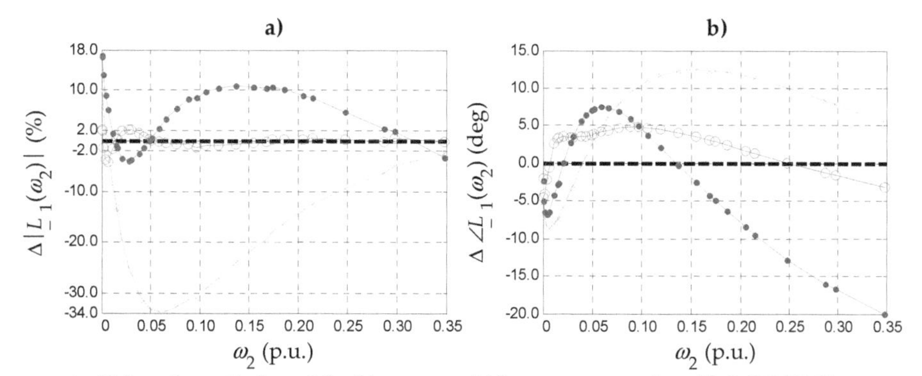

Figure 7. Approximation errors of the SR-IM reference IFCh: (**a**) Modulus relative errors and (**b**) argument absolute errors.

It is worth nothing that the use of the IM mathematical model with three parallel connected two-terminals in the solid rotor equivalent circuit allowed the most accurate approximation of the SR-IM reference IFCh to be achieved, in comparison to the approximation accuracy obtained by the IM classical mathematical model. The required approximation accuracy of the SR-IM reference IFCh modulus was met with argument approximation error not exceeding five degrees in the considered range of slip frequency (Figure 7b). Theoretically, approximation accuracy of the reference IFCh could be further improved by incorporating additional two-terminals in the rotor mathematical model, but the adopted criterion also assumed their minimal number. This resulted from the desirable simplicity of the rotor flux estimation scheme, which is intended for the IM rotor-field-oriented control implementation. Ultimately, the approximation of the SR-IM reference IFCh realized by the IM mathematical model with three parallel connected two-terminals in the solid rotor equivalent circuit was considered to be sufficiently accurate for the purpose of the rotor flux estimation.

4.3. Estimation of the Rotor Flux Space Vector

In the presented study, the estimation accuracy of the rotor flux space vector was evaluated in the slip frequency range corresponding to the IM load adjustment range up to 1.30 of the stator rated current. The range of slip frequency exceeding the typical operating range of slip frequency for the field-oriented controlled IM, was adopted in this study for the robustness verification of the novel rotor flux estimation scheme against electromagnetic parameter variability resulting from the rotor deep-bar effect. This is the reason why the developed rotor flux estimation scheme was not employed in the rotor-flux-oriented control at this stage of the study. Since the direct measurement of rotor flux is not realized in practice, the verification of the rotor flux estimation accuracy was conducted indirectly, based on the reference quantities registered with the laboratory test bench.

During the study, the every endeavor was made to conduct the measurements of the reference quantities in a manner to assure the minimal possible measurement uncertainty. Due to the high resolution of the DAQ device and the precision of the sensors and measuring transducers, the registered shaft torque T_{sh} and the determined power losses ΔP were considered as the reference quantities in the verification of the developed rotor flux estimation scheme. The verification investigations used the fact that the shaft torque can be determined as the quotient of the motor shaft power and angular velocity:

$$T^{\mathrm{e}}_{\mathrm{sh}} = \frac{P^{\mathrm{e}}_{\mathrm{sh}}}{\omega_{\mathrm{sh}}} \tag{20a}$$

$$P_{sh}^{e} = T_{em}^{e}\omega_{sh} - \Delta P \tag{20b}$$

where T_{sh}^{e} and T_{me}^{e} represent the estimated shaft and electromagnetic torque, respectively and P_{sh}^{e} is the estimated shaft power.

The power losses ΔP occurring in Equation (20b) were determined according to Equations (21a)–(21c) based on the measurement data derived from the LC tests. In order to eliminate the necessity to split up individual components of the power losses ΔP at the considered load conditions of the tested IMs, core losses were not erased from the IM input power whilst calculating the airgap power P_{δ} (Equation (21c)). Such an approach is in line with the adopted IM mathematical model described by Equations (7a)–(7i), in which the resistance associated with core losses is not included. In the presented study, the power losses ΔP were considered as any power losses determining the difference between the power transferred to the shaft $(1-s)P_{\delta}$ and the shaft power P_{sh}:

$$\Delta P(s) = (1-s)P_{\delta}(s) - P_{sh}(s) \tag{21a}$$

$$P_{sh}(s) = \frac{\omega_1}{\omega_b}(1-s)T_{sh}(s) \tag{21b}$$

$$P_{\delta}(s) = \frac{3}{2}\mathrm{Re}\left(\underline{U}_1\underline{I}_1^{*}(s)\right) - \frac{3}{2}R_1\left|\underline{I}_1(s)\right|^2 \tag{21c}$$

where $P_{\delta}(s)$constitutes the airgap power and s represent the motor slip.

Figure 8 presents the variability of the power losses, the shaft power, and the power transferred to the shaft expressed as a function of angular velocity of the tested CR-IM (Figure 8a) and SR-IM (Figure 8b), at the stator voltage frequency of $\omega_1 = 1.0$ (p.u.). Due to slight changes of the CR-IM power losses in the considered slip frequency range (Figure 8a), a constant value of these losses $\Delta P = 0.0863$ (p.u.) was assumed in the verification studies of the novel rotor flux estimation scheme. In relation to the SR-IM, on account of significant changes of the power losses in the considered slip frequency range (Figure 8b), the power loss variability was approximated by means of the second order polynomial:

$$\Delta P(\omega_{sh}) = -0.6551\omega_{sh}^2 + 0.8076\omega_{sh} - 0.0232. \tag{22}$$

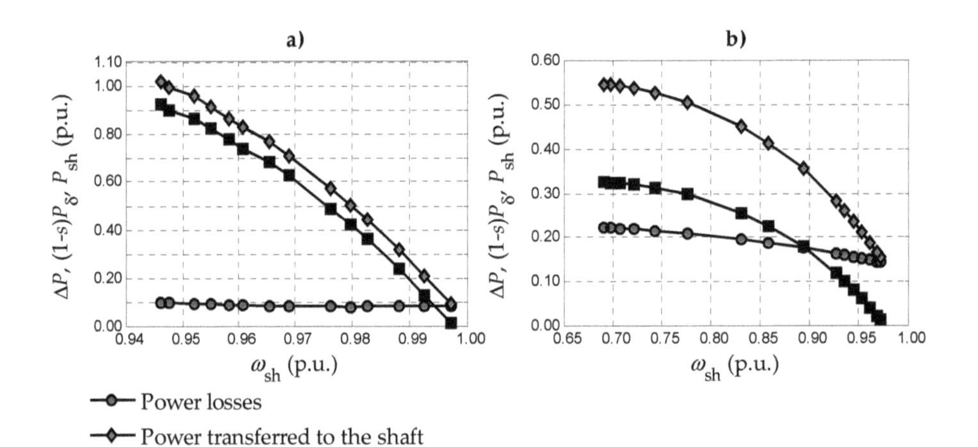

Figure 8. The variability of the power losses, the power transferred to the shaft and the shaft power expressed as a function of angular velocity of the investigated: (**a**) CR-IM and (**b**) SR-IM.

The electromagnetic torque required in Equation (20b) was determined with the use of the estimated rotor flux space vector in conformity with the following equations:

$$T_{\rm em}^{\rm e} = \frac{L_\mu}{L_{\rm 2eq}} {\rm Im}\left(\underline{I}_{\rm 1k} \left(L_{\sigma 2 \rm eq} \sum_{n=1}^{N} \frac{\underline{\Psi}_{2(n)\rm k}}{L_{\sigma 2(n)}} \right)^{*} \right) \tag{23a}$$

$$L_{\rm 2eq} = L_\mu + L_{\sigma 2\rm eq}. \tag{23b}$$

The estimation precision of the registered shaft torque was verified based on the absolute estimation errors:

$$\Delta T_{\rm sh}^{\rm e} = T_{\rm sh} - T_{\rm sh}^{\rm e}. \tag{24}$$

Additionally, Tables 3 and 4 present the maximum and mean absolute errors of the registered shaft torque estimation which were determined in accordance with the following equations:

$$\max\left|\Delta T_{\rm sh}^{\rm e}\right| = \max\left|T_{\rm sh} - T_{\rm sh}^{\rm e}\right| \tag{25a}$$

$${\rm M}\left|\Delta T_{\rm sh}^{\rm e}\right| = \frac{1}{n}\sum_{i=1}^{n}\left|T_{{\rm sh},i} - T_{{\rm sh},i}^{\rm e}\right| \tag{25b}$$

where $T_{{\rm sh},i}$ and $T_{{\rm sh},i}^{\rm e}$ represent the ith samples of the registered and estimated shaft torque, respectively, and n is the number of samples.

For the sake of comparison, the shaft torque estimated through the use of the so called full order open-loop flux observer [32] was also considered in the presented verification. Research results presented previously [32] indicate that this rotor flux estimation scheme is characterized by limited sensitivity to erroneous identification or variability of the rotor electromagnetic time constant in a wide range of slip frequency changes, in comparison to the commonly known current model of the rotor flux space vector. The full order open-loop flux observer was formulated on the basis of the IM classical mathematical model, and is represented by the following system of equations:

$$T_{\rm N}\frac{\rm d}{{\rm d}t}\underline{I}_{\rm 1s}^{\rm e} = \frac{1}{\sigma L_1}\left[\underline{U}_{\rm 1s} - \left(R_1 + \left(\frac{L_\mu}{L_2}\right)^2 R_2\right)\underline{I}_{\rm 1s}^{\rm e} + \frac{L_\mu}{L_2}\left(\frac{R_2}{L_2} - {\rm j}\omega_{\rm m}\right)\underline{\Psi}_{\rm 2s}^{\rm e}\right] \tag{26a}$$

$$T_2\frac{\rm d}{{\rm d}t}\underline{\Psi}_{\rm 2s}^{\rm e} = L_\mu \underline{I}_{\rm 1s}^{\rm e} - \underline{\Psi}_{\rm 2s}^{\rm e} + {\rm j}\omega_{\rm b} T_2 \omega_{\rm m}\underline{\Psi}_{\rm 2s}^{\rm e} \tag{26b}$$

$$T_2 = \frac{L_2}{R_2}T_{\rm N} \tag{26c}$$

where T_2 is the rotor electromagnetic time constant.

The verification of the rotor flux estimation schemes additionally includes the shaft torque, which was estimated with the help of the elaborated voltage–current model (Equations (16a)–(16c)) and formulated on the basis of the IM classical mathematical model.

Figure 9 presents a block diagram of the algorithm used in the accuracy evaluation of the rotor flux space vector estimation. The algorithm was implemented in the Matlab environment.

In the presented study, the investigated estimation schemes of the rotor flux space vector were fed by the registered angular velocity and stator voltages and currents. This case corresponded to the operation of the tested estimation schemes in an IM drive with angular velocity measurements. Due to the similar accuracy of the shaft torque estimation, obtained through the considered estimation schemes of the rotor flux space vector at each setpoint of stator voltage frequency for both tested IMs, this paper presents the study results conducted at the nominal stator voltage frequency ω_1 = 1.0 (p.u.).

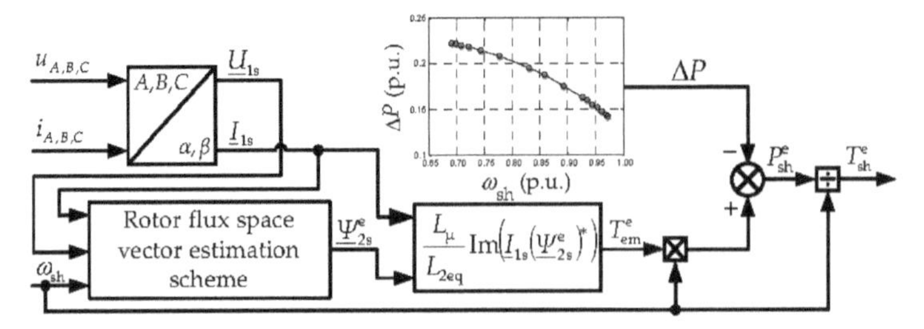

Figure 9. The block diagram of the algorithm used in the accuracy evaluation of the rotor flux space vector estimation.

Figure 10a presents the registered SR-IM shaft torque together with the shaft torque generated with the use of the tested rotor flux estimation schemes which were formulated on the basis of the IM mathematical model with single and three parallel connected two-terminals in the rotor equivalent circuit. The absolute errors of the registered shaft torque estimation are shown in Figure 10b, whereas Table 3 lists the maximal and mean absolute estimation errors. Figure 10 includes the shaft torque obtained with the use of the rotor flux estimation schemes based on the IM classical mathematical model in the structure of which the electromagnetic parameters SR-Std 2 were applied. The use of these parameters allowed for a more accurate estimation of the registered SR-IM shaft torque in relation to the estimation accuracy acquired by means of the employed schemes with the electromagnetic parameters SR-Std 1 (Table 3).

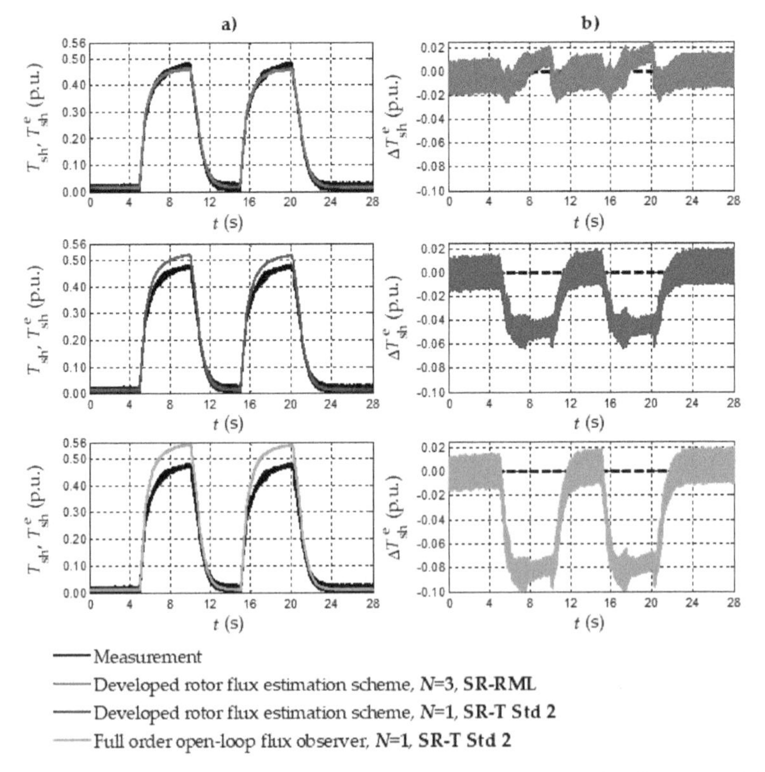

Figure 10. The SR-IM shaft torque estimation: (**a**) The registered and estimated shaft torque and (**b**) the absolute errors of the registered shaft torque estimation.

Table 3. Maximum and mean absolute errors (p.u.) of the registered shaft torque estimation of the tested SR-IM.

Rotor Flux Estimation Schemes	Full Order Open-Loop Flux Observer (26a)–(26c)		Voltage-Current Model (16a)–(16c) $N = 1$		Voltage-Current Model (16a)–(16c) $N = 3$
Electromagnetic parameters	SR-T Std 1	SR-T Std 2	SR-T Std 1	SR-T Std 2	SR-RML
$\max\lvert\Delta T^{e}_{sh}\rvert$	0.3191	0.0986	0.2877	0.0631	0.0262
$M\lvert\Delta T^{e}_{sh}\rvert$	0.0845	0.0346	0.0712	0.0209	0.0075

Figure 11a presents the registered and estimated shaft torque of the tested CR-IM, whereas Figure 11b shows the absolute estimation errors. The maximal and mean absolute estimation errors are listed in Table 4.

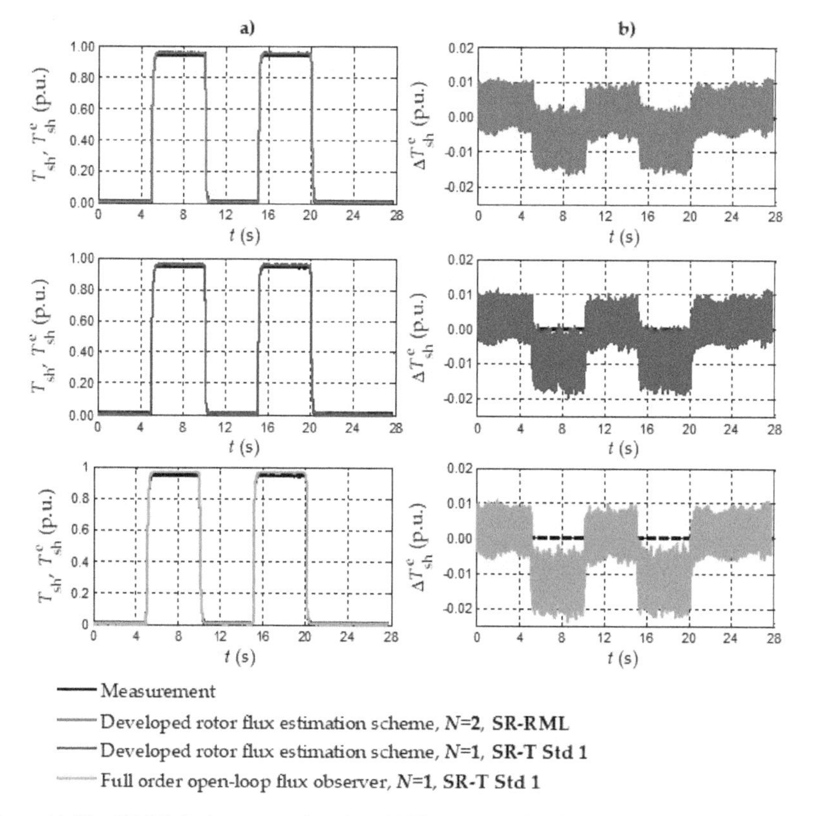

Figure 11. The CR-IM shaft torque estimation: (**a**) The registered and estimated shaft torque and (**b**) the absolute errors of the registered shaft torque estimation.

Table 4. Maximum and mean absolute errors (p.u.) of the registered shaft torque estimation of the tested CR-IM.

Rotor Flux Estimation Schemes	Full Order Open-Loop Flux Observer (26a)–(26c)		Voltage-Current Model (16a)–(16c) $N = 1$		Voltage-Current Model (16a)–(16c) $N = 2$
Electromagnetic parameters	CR-T Std 1	CR-T Std 2	CR-T Std 1	CR-T Std 2	CR-RML
$\max\lvert\Delta T^{e}_{sh}\rvert$	0.0239	0.0345	0.0196	0.0211	0.0164
$M\lvert\Delta T^{e}_{sh}\rvert$	0.0065	0.0105	0.0056	0.0063	0.0047

As regards the CR-IM, a more accurate estimation of the registered shaft torque was achieved using the electromagnetic parameters CR-T Std 1 in the considered estimation schemes based on the IM classical mathematical model, in comparison to when the electromagnetic parameters CR-T Std 2 were applied (Table 4). For this reason, the shaft torque generated through the rotor flux estimation schemes described by Equations (26a)–(26c) and (16a)–(16c) (single two-terminal rotor representation $N = 1$) with the electromagnetic parameters CR-T Std 1 are included in Figure 11.

5. Conclusions

This paper presents a novel estimation scheme of the rotor flux space vector which has been developed on the basis of the IM mathematical model with rotor multi-loop representation. In regards to the tested SR-IM, the use of the rotor flux estimation scheme, in which the rotor skin effect was modeled by three parallel connected two-terminals in the rotor equivalent circuit, enabled a multiple reduction of the registered shaft torque estimation errors in relation to the estimation errors obtained through the considered estimation schemes based on the IM classical mathematical model (Table 3). The absolute estimation errors of the registered SR-IM shaft torque achieved by using the elaborated voltage–current model ($N = 3$) did not exceed the level of ±0.0262 (p.u.) (Figure 10b, Table 3). The results of the presented study indicate considerable improvement in the accuracy of the rotor flux space vector estimation of the tested SR-IM, which was obtained by the estimation scheme elaborated on the IM mathematical model with rotor two-terminal network representation, in comparison with the estimation precision acquired by the schemes formulated on the IM classical mathematical model.

It should also be noted that even for the tested CR-IM, which does not show a substantial deep-bar effect, the superiority of the novel rotor flux estimation scheme (with two parallel connected two-terminals $N= 2$ in the rotor equivalent circuit) over the estimation schemes based on the IM classical mathematical model can be observed (Figure 11, Table 4).

The results of the conducted study indicate that the developed voltage–current model enables accurate estimation of the rotor flux of IMs characterized by intense deep-bar effect, in the operating range of the slip frequency. Considering the above, the novel rotor flux estimation scheme can be applied for the rotor-flux-oriented control of the IMs with any rotor construction, including squirrel-cage, double-cage, and solid rotors. Moreover, the elaborated voltage–current model of the rotor flux space vector can be employed as the adjustable model of the Model Reference Adaptive System based estimator for speed-sensorless IM drive applications.

Future work will include experimental studies of the IM rotor-flux-oriented control with the novel rotor flux estimation scheme.

Author Contributions: Conceptualization, G.U. and J.R.; Formal analysis, G.U.; Investigation, J.R.; Methodology, G.U. and J.R.; Supervision, A.K.; Validation, G.U. and J.R.; Visualization, G.U.; Writing—original draft, G.U., J.R. and A.K.

Funding: This work was supported by the Polish Ministry of Science and Higher Education under research projects: BS/MN-401-312/15 and 03.0.00.00/2.01.01.0001MNSP.E.19.001.

Conflicts of Interest: The authors declare no conflict of interest.

Appendix A

In the presented study, the total apparent electrical power was adopted as base apparent power (input voltampere base).The base values are defined in accordance to the contents of Table A1:

Table A1. The per-unit system base values.

Base Quantity	Symbol	Unit	Formula
Apparent power	S_b	voltampere (V·A)	$3 \cdot U_{1ph} \cdot I_{1ph}$
Frequency	ω_b	radian per second (rad/s)	$2 \cdot \pi \cdot f_{1N}$
Angular velocity	ω_{mb}	radian per second (rad/s)	$2 \cdot \pi \cdot f_{1N} \cdot (p_p)^{-1}$
Magnetic flux	Ψ_b	volt second per radian (V·s/rad)	$U_{1ph} \cdot (2 \cdot \pi \cdot f_{1N})^{-1}$
Impedance	Z_b	ohm (Ω)	$U_{1ph} \cdot (I_{1ph})^{-1}$
Inductance	L_b	henry per radian (H/rad)	$U_{1ph} \cdot (I_{1ph} \cdot 2 \cdot \pi \cdot f_{1N})^{-1}$
Torque	T_b	newton meter per radian (N·m/rad)	$3 \cdot U_{1ph} \cdot I_{1ph} \cdot p_p \cdot (2 \cdot \pi \cdot f_{1N})^{-1}$

Where U_{1ph} and I_{1ph} are the nominal stator phase voltage and current, respectively, f_{1N} stands for the nominal frequency of stator voltages, and p_p is the number of pole pairs.

Appendix B

Figure A1 presents the cross-section dimensions (millimeters) of the tested squirrel-cage (Figure A1a) and solid (Figure A1b) rotors.

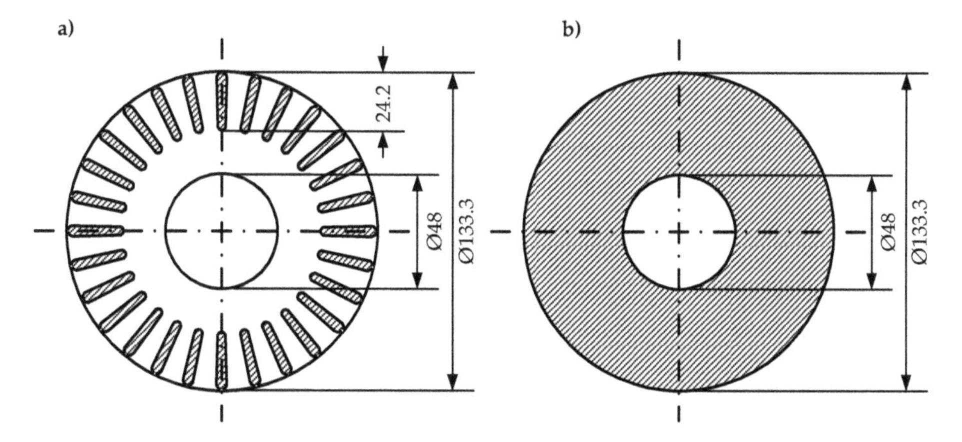

Figure A1. The cross-section dimensions of: (**a**) CR-IM and (**b**) SR-IM.

The rated values of the tested IMs corresponding to the adopted machine operating points are included in Table A2. These rated values were determined for the purpose of the presented study and aimed to reduce the breakdown slip of the SR-IM. The rated values were settled so as to maintain approximately equal stator flux amplitudes of the CR-IM and SR-IM, bearing in mind the limitations resulting from rated values of the programmable AC source (AMETEK Model: 3001iX) powering the investigated IMs.

Table A2. The rated values of the CR-IM and SR-IM corresponding to the adopted operating points.

Rating	Unit	CR-IM	SR-IM
Output power	kilowatt (kW)	2.358	1.992
Stator voltage	volt (V)	400 (wye)	391 (delta)
Stator frequency	hertz (Hz)	50	85
Stator current	ampere (A)	4.536	7.785
Torque	newton meter (N·m)	15.53	9.39
Rotational speed	revolution per minute (r/m)	1450	2030
Power factor	(-)	0.8819	0.6698
Efficiency	(-)	0.8525	0.5641
Stator flux	weber (Wb)	0.973	0.995

Table A3. The power rating of the individual devices and DC machine employed within the laboratory test bench.

Name	Description	Unit	Rated Power
AMETEK Model: 3001iX	programmable AC source	kilowatt (kW)	9.0
DC590P	4Q thyristor converter	kilowatt (kW)	7.5
ZSAC4244 – H&H GmbH	programmable electronic load	kilowatt (kW)	4.2
PCMb 54b	separately excited DC machine	kilowatt (kW)	6.5

References

1. Blaschke, F. The principle of fields-orientation as applied to the transvector closed loop control system for rotating-field machines. *Siemens Rev.* **1972**, *34*, 217–220.
2. Hasse, K. Drehzahlregelverfahren fur schnelle Umkehrantriebe mit stromrichtergespeisten Asynchron-Kurzschlusslaufermotoren. *Regelungstechnik* **1972**, *20*, 60–66. [CrossRef]
3. Takahashi, I.; Noguchi, T. A new quick-response and high efficiency control strategy of an induction motor. *IEEE Trans. Ind. Appl.* **1986**, *5*, 820–827. [CrossRef]
4. Garces, L.J. Parameter adaption for the speed-controlled static ac drive with a squirrel-Cage induction motor. *IEEE Trans. Ind. Appl.* **1980**, *2*, 173–178. [CrossRef]
5. Matsuo, T.; Lipo, T.A. A rotor parameter identification scheme for vector-controlled induction motor drives. *IEEE Trans. Ind. Appl.* **1985**, *3*, 624–632. [CrossRef]
6. Krishnan, R.; Doran, F.C. Study of parameter sensitivity in high-performance inverter-fed induction motor drive systems. *IEEE Trans. Ind. Appl.* **1987**, *4*, 623–635. [CrossRef]
7. Orlowska-Kowalska, T. Application of extended Luenberger observer for flux and rotor time-constant estimation in induction motor drives. *IEE Proc. (Control Theory Appl.)* **1989**, *6*, 324–330. [CrossRef]
8. Garcia Soto, G.; Mendes, E.; Razek, A. Reduced-order observers for rotor flux, rotor resistance and speed estimation for vector controlled induction motor drives using the extended Kalman filter technique. *IEE Proc. (Electr. Power Appl.)* **1999**, *3*, 282–288. [CrossRef]
9. Marčetić, D.P.; Vukosavić, S.N. Speed-sensorless AC drives with the rotor time constant parameter update. *IEEE Trans. Ind. Electron.* **2007**, *5*, 2618–2625. [CrossRef]
10. Proca, A.B.; Keyhani, A. Sliding-mode flux observer with online rotor parameter estimation for induction motors. *IEEE Trans. Ind. Electron.* **2007**, *2*, 716–723. [CrossRef]
11. Salmasi, F.R.; Najafabadi, T.A.; Maralani, P.J. An adaptive flux observer with online estimation of DC-link voltage and rotor resistance for VSI-based induction motors. *IEEE Trans. Power Electron.* **2010**, *5*, 1310–1319. [CrossRef]
12. Wang, S.; Dinavahi, V.; Xiao, J. Multi-rate real-time model-based parameter estimation and state identification for induction motors. *IET Electr. Power Appl.* **2013**, *1*, 77–86. [CrossRef]
13. Zaky, M.S.; Metwaly, M.K. Sensorless torque/speed control of induction motor drives at zero and low frequencies with stator and rotor resistance estimations. *IEEE J. Emerg. Sel. Top. Power Electron.* **2016**, *4*, 1416–1429. [CrossRef]
14. Kim, J.K.; Ko, J.S.; JangHyeon, L.; Lee, Y.K. Rotor flux and rotor resistance estimation using Extended Luenberger-Sliding Mode Observer (ELSMO) for three phase induction motor control. *Can. J. Electr. Comput. Eng.* **2017**, *3*, 181–188. [CrossRef]
15. Yang, S.; Ding, D.; Li, X.; Xie, Z.; Zhang, X.; Chang, L. A novel online parameter estimation method for indirect fieldoriented induction motor drives. *IEEE Trans. Energy Convers.* **2017**, *4*, 1562–1573. [CrossRef]
16. Babb, D.S.; Williams, J.E. Network analysis of A-C machine conductors. *Trans. Am. Inst. Electr. Eng.* **1951**, *2*, 2001–2005. [CrossRef]
17. Paszek, W.; Kaplon, A. Induction machine with anisotropic multilayer rotor modelling the electromagnetic and the electrodynamic states of a symmetrical machine with deep bar cage in solid iron rotor core. *Electromagn. Fields Electr. Eng.* **1988**, 205–210. [CrossRef]
18. Levy, W.; Landy, C.H.; McCulloch, M.D. Improved models for the simulation of deep bar induction motors. *IEEE Trans. Energy Convers.* **1990**, *2*, 393–400. [CrossRef]

19. De Doncker, R.W. Field-oriented controllers with rotor deep bar compensation circuits. *IEEE Trans. Ind. Appl.* **1992**, *5*, 1062–1071. [CrossRef]
20. Healey, R.C.; Williamson, S.; Smith, A.C. Improved cage rotor models for vector controlled induction motors. *IEEE Trans. Ind. Appl.* **1995**, *4*, 812–822. [CrossRef]
21. Smith, A.C.; Healey, R.C.; Williamson, S. A transient induction motor model including saturation and deep bar effect. *IEEE Trans. Energy Convers.* **1996**, *1*, 8–15. [CrossRef]
22. Seok, J.K.; Sul, S.K. Pseudorotor-flux-oriented control of an induction machine for deep-bar-effect compensation. *IEEE Trans. Ind. Appl.* **1998**, *3*, 429–434. [CrossRef]
23. Monjo, L.; Kojooyan-Jafari, H.; Córcoles, F.; Pedra, J. Squirrel-cage induction motor parameter estimation using a variable frequency test. *IEEE Trans. Energy Convers.* **2015**, *2*, 550–557. [CrossRef]
24. Benzaquen, J.; Rengifo, J.; Albánez, E.; Aller, J.M. Parameter estimation for deep-bar induction machines using instantaneous stator measurements from a direct startup. *IEEE Trans. Energy Convers.* **2017**, *2*, 516–524. [CrossRef]
25. Rolek, J.; Utrata, G. An identification procedure of electromagnetic parameters for an induction motor equivalent circuit including rotor deep bar effect. *Arch. Electr. Eng.* **2018**, *2*, 279–291. [CrossRef]
26. McKinnon, D.J.; Seyoum, D.; Grantham, C. Investigation of parameter characteristics for induction machine analysis and control. In Proceedings of the Second International Conference on Power Electronics, Machines and Drives, Edinburgh, UK, 31 March–2 April 2004. [CrossRef]
27. Willis, J.R.; Brock, G.J.; Edmonds, J.S. Derivation of induced motor models from standstill frequency response tests. *IEEE Trans. Energy Convers.* **1989**, *4*, 608–615. [CrossRef]
28. *2013 Rotating Electrical Machines—Part 28: Test Methods for Determining Quantities of Equivalent Circuit Diagrams for Three-Phase Low Voltage Cage Induction Motors*; Standard PN-EN 60034-28; Polski Komitet Normalizacyjny: Warszawa, Poland, 2013.
29. *Standard Test Procedure for Polyphase Induction Motors and Generators*; IEEE Standard 112™-2004; IEEE Power Engineering Society: New York, NY, USA, 4 November 2004. [CrossRef]
30. Alonge, F.; D'ippolito, F.; Ferrante, G.; Raimondi, F.M. Parameter identification of induction motor model using genetic algorithms. *IEE Proc. Control Theory Appl.* **1998**, *6*, 587–593. [CrossRef]
31. Kumar, P.; Dalal, A.; Singh, A.K. Identification of three phase induction machines equivalent circuits parameters using multi-objective genetic algorithms. In Proceedings of the 2014 International Conference on Electrical Machines (ICEM), Berlin, Germany, 2–5 September 2014. [CrossRef]
32. Jansen, P.L.; Lorenz, R.D. A physically insightful approach to the design and accuracy assessment of flux observers for field oriented induction machine drives. *IEEE Trans. Ind. Appl.* **1994**, *1*, 101–110. [CrossRef]

Article

Robust Speed Controller Design Using H_infinity Theory for High-Performance Sensorless Induction Motor Drives

Ahmed A. Zaki Diab [1,*], Abou-Hashema M. El-Sayed [1], Hossam Hefnawy Abbas [1] and Montaser Abd El Sattar [2]

[1] Electrical Engineering Department, Faculty of Engineering, Minia University, Minia 61111, Egypt; dr_mostafa555@yahoo.com (A.-H.M.E.-S.); hosamhe@yahoo.com (H.H.A.)

[2] El-Minia High Institute of Engineering and Technology, Minia 61111, Egypt; mymn2013@yahoo.com

* Correspondence: a.diab@mu.edu.eg; Tel.: +20-102-177-7925

Received: 31 January 2019; Accepted: 9 March 2019; Published: 12 March 2019

Abstract: In this paper, a robust speed control scheme for high dynamic performance sensorless induction motor drives based on the H_infinity (H_∞) theory has been presented and analyzed. The proposed controller is robust against system parameter variations and achieves good dynamic performance. In addition, it rejects disturbances well and can minimize system noise. The H_∞ controller design has a standard form that emphasizes the selection of the weighting functions that achieve the robustness and performance goals of motor drives in a wide range of operating conditions. Moreover, for eliminating the speed encoder—which increases the cost and decreases the overall system reliability—a motor speed estimation using a Model Reference Adaptive System (MRAS) is included. The estimated speed of the motor is used as a control signal in a sensor-free field-oriented control mechanism for induction motor drives. To explore the effectiveness of the suggested robust control scheme, the performance of the control scheme with the proposed controllers at different operating conditions such as a sudden change of the speed command/load torque disturbance is compared with that when using a classical controller. Experimental and simulation results demonstrate that the presented control scheme with the H_∞ controller and MRAS speed estimator has a reasonable estimated motor speed accuracy and a good dynamic performance.

Keywords: Sensorless; induction motors; H_infinity; drives; vector control; experimental implementation

---

1. Introduction

The development of effective induction motor drives for various applications in industry has received intensive effort for many researchers. Many methods have been developed to control induction motor drives such as scalar control, field-oriented control and direct torque control, among which field-oriented control [1–5] is one of the most successful and effective methods. In field orientation, with respect to using the two-axis synchronously rotating frame, the phase current of the stator is represented by two component parts: the field current part and the torque-producing current part. When the component of the field current is adjusted constantly, the electromagnetic torque of the controlled motor is linearly proportional to the torque-producing components, which is comparable to the control of a separately excited DC motor. The torque and flux are considered as input commands for a field-oriented controlled induction motor drive, while the three-phase stator reference currents after a coordinated transformation of the two-axis currents are considered as the output commands. To achieve the decoupling control between the torque and flux currents components, the three-phase currents of the induction motor are controlled so that they follow their reference current commands through the use of current-regulated pulse-width-modulated (CRPWM) inverters [2–8]. Moreover, the controls of the rotor magnetic flux level and the electromagnetic torque are entirely decoupled using an additional outer feedback speed loop. Therefore, the control scheme has two loops; the inner loop of the decoupling the currents components of flux and torque, and the outer control loop which controls the rotor speed and produces the reference electromagnetic torque. Based on that, the control of an induction motor drive can be considered as a multi feedback-loop control problem consisting of current control and speed-control loops. The classical Proportional-Integral (PI) controller is frequently used in a speed-control loop due to its simplicity and stability. The parameters of the PI controller are designed through trial-and-error [3–5]. However, PI controllers often yield poor dynamic responses to changes in the load torque and moment of inertia. To overcome this problem of classical PI controllers and to improve the dynamic performance, various approaches have been proposed in References [6] and [7]. The classical two-degree-of-freedom controller (phase lead compensator and PI controller) [6] was used for indirect vector control of an induction motor drive. However, the parameters of this controller are still obtained through trial-and-error to reach a satisfactory performance level.

In Reference [8], the authors presented a control scheme for the induction motors drives based on fuzzy logic. The proposed control scheme has been applied to improve the overall performance of an induction motor drive system. This controller does not require a system model and it is insensitive to external load torque disturbances and information error. On the other hand, the presented control scheme suffers from drawbacks such as large oscillations in transient operation. Moreover, the control system requires an optical encoder to measure the motor speed [8].

A linear quadratic Gaussian controller was applied in References [9] and [10] to regulate motor speed and improve the motor's dynamic performance. The merits of this controller are as follows: fast response, robustness and the ability to operate with available noise data. However, this controller's drawbacks are that it needs an accurate system model, does not guarantee a stability margin and requires more computation.

Recently, H_infinity (H_∞) control theory has been widely implemented for its robustness against model uncertainty perturbations, external disturbances and noise. Some applications of this technique in different systems, such as permanent magnet DC motors [11], switching converters [12] and synchronous motors [13], have been reported. Moreover, researchers have worked to apply the H_∞ controllers in the induction machines drives. In Reference [14], a control scheme based on the H_∞ is presented for control the speed of the induction motors. However, the control system is validated only through the simulation results. Additionally, the speed sensor which used to measure the rotor speed is reduced the control system reliability and also its cost. In Reference [15], a vector control scheme for the induction machines based on H_∞ has been designed and experimentally validated. However, the authors used a sensor to measure the rotor speed. Additionally, a comparison between the performances of the sensor vector-control scheme of induction motor based on the PI controller and is presented. The results show the priority of the H_∞ control scheme rather than the PI controller. The main drawbacks of the control system of Ref. [15] were that speed sensor data simulation verification results were not included. Another control scheme based on the H_∞ has been presented in Reference [16]. The introduced control scheme in Ref. [16] has many drawbacks such as the need for a speed encoder, and the control law which is based on the linear parameter varying (LPV) should be updated online which increases the cost of implementation. However, validation of the control scheme has been carried out based on only a simulation using the MATLAB/Simulink (2014a, MathWorks, Natick, MA, USA) package. The authors of this paper recommended future work to eliminate the speed sensor and to minimization the implementation time. An interesting research work about the application of induction machines drives has been presented in Reference [17]. The control scheme is applied for Electric Trains application. The control system suffers from reliability reduction because of presence of the speed sensor and also the increasing of the implementation time because of the time which is needed to reach the solution of the Riccati equation. From the previous discussion, further research work is required to enhance the dynamic performance of the induction motor drives with the application of the H_∞ control theory. Moreover, the application of sensorless algorithms with the H_∞ based induction machine drives is an essential research point. Furthermore, the experimental implementation of the induction machines drives based on H_∞ is required for greater validation of the control scheme. Additionally, the major aspect of an H_∞ control is to synthesize a feedback law that forces the closed-loop system to satisfy a prescribed H_∞ norm constraint. This aspect achieves the desired stability and tracking requirements.

In this paper, a robust speed controller design for high-performance sensor-free induction motor drives based on the H_∞ theory is proposed. The proposed speed controller is used to achieve both robust stability and good dynamic performance even under system parameter variations. It can withstand disturbances well and ignores system noise. Moreover, it is simple to implement and has a low computational cost. Additionally, this paper formulates the design problem of an H_∞ controller in a standard form with an emphasis on the selection of the weighting functions that reflect the robustness and performance goals of motor drives. The motor speed is estimated based on a presented model-reference adaptive system (MRAS). The estimated motor speed is used as a control signal in a sensor-free field-oriented control mechanism for induction motor drives. To demonstrate the effectiveness of the proposed controller, the motor speed response following a step-change in speed command and load torque disturbance is compared with that when using a classical controller. The presented experimental and simulation results demonstrate that the proposed control system achieves reasonable estimated motor speed accuracy and good dynamic performance.

2. Mathematical Model of an Induction Motor

The induction motor can be modeled in the following mathematical differential equations represented in the rotating reference frame [8,9]:

$$\begin{bmatrix} v_{ds} \\ v_{qs} \\ 0 \\ 0 \end{bmatrix} = \begin{bmatrix} R_s + pL_s & -\omega_s L_s & pL_m & -\omega_s L_m \\ \omega_s L_s & R_s + pL_s & \omega_s L_m & pL_m \\ pL_m & -s\omega_s L_m & R_r pL_r & -s\omega_s L_r \\ s\omega_s L_m & pL_m & s\omega_s L_r & R_r + pL_r \end{bmatrix} \begin{bmatrix} i_{ds} \\ i_{qs} \\ i_{dr} \\ i_{qr} \end{bmatrix} \quad (1)$$

where, p is the differential operator, d/dt, $L^n{}_m$ is the equivalent magnetizing inductance and s represents the difference between the synchronous speed and the rotor speed and it refers to slip. The self-inductances of the motor can be represented as the following:

$$\begin{array}{l} L^n{}_s = L^n{}_m + L_{ls} \\ L^n{}_r = L^n{}_m + L_{lr} \end{array},$$

where L_{ls} and L_{lr} are the stator and rotor leakage reactances, respectively.

The torque equation, in this case, is expressed as

$$T_e = \frac{3}{2} P \frac{L^n{}_m}{L^n{}_r} (i_{qs}\psi_{dr} - i_{ds}\psi_{qr}) \quad (2)$$

The previous Equation (2) is to calculate the electromagnetic torque of the motor as a function of the stator currents components, rotor flux components, pole pairs P and rotor and magnetizing inductances. Moreover, the rotor flux linkage can be written as the following equations [10]:

$$\psi_{dr} = L^n{}_r i_{dr} + L^n{}_m i_{ds}$$

$$\psi_{qr} = L^{n}{}_{r} i_{qr} + L^{n}{}_{m} i_{qs}.$$

The equation of motion is

$$J_m \frac{d\omega_r}{dt} + f_d \omega_r + T_l = T_e \quad (3)$$

where J_m is the moment of inertia, ω_r is the angular speed the rotor shaft, f_d is to express the damping coefficient, T_e indicates the electromagnetic torque of the induction motor and T_l indicates the load torque.

For achieving the finest decoupling between the ds- and qs-axis currents components, the two components of the rotor flux can be written as the following:

$$\psi_{qr} = 0 \text{ and } \psi_{dr} = \psi_r \quad (4)$$

Based on operational requirements, when the rotor flux is set to a constant, the equation of the electromagnetic torque Equation (2) will be as follows:

$$T_e = K_T i_{qs} \quad (5)$$

where

$$K_T = \frac{3P}{2} \frac{L^{n}{}_{m}}{L^{n}{}_{r}} \psi_r.$$

Equation (3) can be rewritten in the s-domain as follows:

$$\omega_r = G_p(s)(T_e(s) - T_l(s)) \quad (6)$$

where

$$G_p(s) = \frac{1/J_m}{s + f_d/J_m} \quad (7)$$

A block diagram representing an indirect vector-controlled induction motor drive is shown in Figure 1. The diagram consists mainly of three sub-models; a model for an induction motor under load when considering the core-loss, a hysteresis current-controlled pulse-width-modulated (PWM) inverter, and vector-control technique followed by a coordinate transformation and an outer speed-control loop.

In the vector-control scheme of Figure 1, the currents $i^*{}_{ds}$ and $i^*{}_{qs}$ are, respectively, the magnetizing and torque current components commands.

Where $I^*{}_s = \sqrt{i^{*2}{}_{ds} + i^{*2}{}_{qs}}$, $\theta^*{}_t = \tan^{-1}(i^*{}_{qs}/i^*{}_{ds})$, and $\omega^*{}_s = \omega^*{}_{sl} + \omega_r$ (the * refers to the command value). The stator current commands of phase "a", which is the reference current command for the CRPWM inverter, is presented in References [3–6].

$$i^*{}_{as} = I^*{}_s \cos(\omega^*{}_{st} + \theta^*{}_t) \quad (8)$$

The commands for the other two stator phases are defined below. Referring to Figure 1, the slip speed command is calculated by

$$\omega^*{}_{sl} = \frac{1}{T_r} \frac{i^*{}_{qs}}{i^*{}_{ds}} \quad (9)$$

The torque current component command, $i^*{}_{qs}$, is obtained from the error of the speed, which applied to a speed controller provided that $i^*{}_{ds}$ remains constant according to the operational requirements.

According to the above-mentioned analysis, the dynamic performance of the entire drive system, described in Figure 1, can be represented by the control system block diagram in Figure 2. This block diagram calls for accurate K_T parameters and the transfer function blocks of $G_p(s)$. In this paper, the speed controller $K(s)$ is designed using H_∞ theory to eliminate the problems inherent to classical controllers.

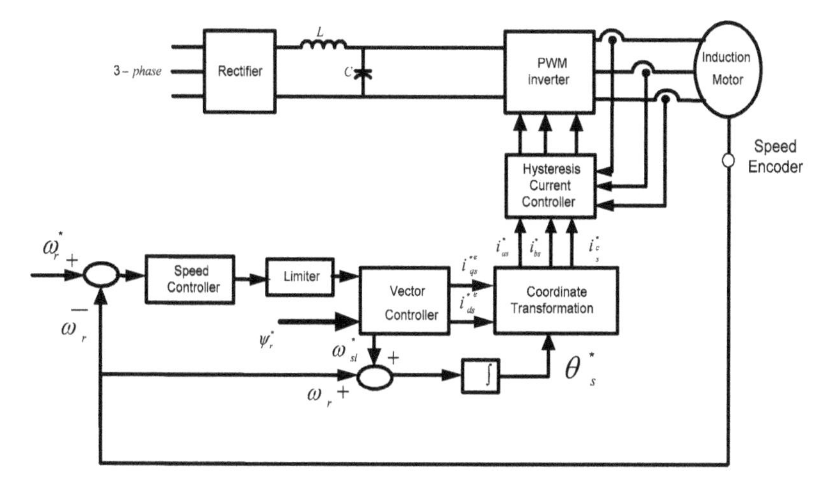

Figure 1. Block diagram of an indirect field-oriented (IFO)-controlled induction motor drive.

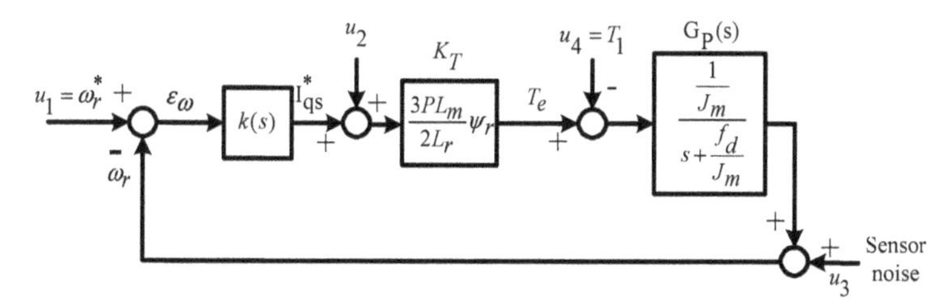

Figure 2. Block diagram of the speed-control system of an IFO-controlled induction motor drive.

3. Design of the Proposed Controller Based on H_∞ Theory

The proposed controller is designed to achieve the following objectives:

(1) Minimum effect of the measurement noise at high frequency
(2) Maximum bounds on closed-loop signals to prevent saturation
(3) Minimum effect of load disturbance rejection, reducing the maximum speed dip
(4) Asymptotic and good tracking for sudden changes in command signals, in addition to a rapid and excellent damping response
(5) Survivability against system parameter variations.

The H_∞ theory offers a reliable procedure for synthesizing a controller that optimally verifies singular value loop-shaping specifications [11–17]. The standard setup of the H_∞ control problem consists of finding a static or dynamic feedback controller such that the H_∞ norm (a standard quantitative measure for the size of the system uncertainty) of the closed-loop transfer function below a given positive number under the constraint that the closed-loop system is internally stable.

H_∞ synthesis is performed in two stages:

i. Formulation: the first stage is to select the optimal weighting functions. The proper selections of the weighting functions give the ability to improve the robustness of the system at different operation condition and varying the model parameters. Moreover, this to reject the disturbance and noises besides the parameter uncertainties.

ii. Solution: The transfer function of the weights has been updated to reach the optimal configuration. In this paper, the MATLAB optimization toolbox in the Simulink is used to determine the best weighting functions.

Figure 3 illustrates the block diagram of the H_∞ design problem, where G(s) is the transfer function of the supplemented plant (nominal plant Gp(s)) plus the weighting functions that represent the design features and objectives. u_1 is the control signal and w is the exogenous input vector, which generally comprises the command signals, perturbation, disturbance, noise and measurement interference; and y_1 is the controller inputs such as commands, measured output to be controlled, and measured disturbance; its components are typically tracking errors and filtered actuator signal; and z is the exogenous outputs; "error" signals to be minimized.

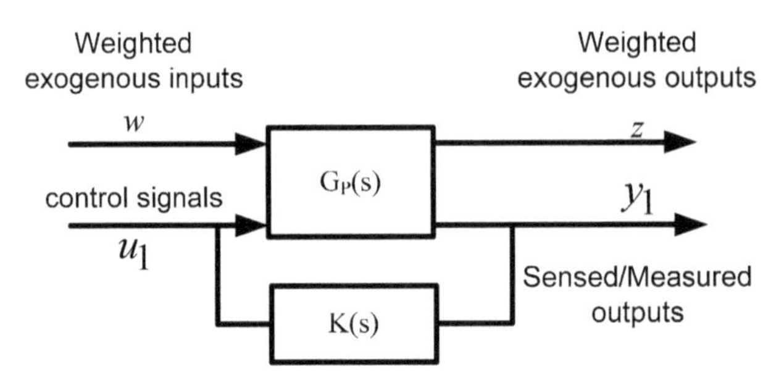

Figure 3. General setup of the H_∞ design problem.

The objective of this problem is to design a controller K(s) for the augmented plant G(s) to have desirable characteristics for the input/output transfer based on the information of $y1$ (inputs to the controller k(s)) to generate the control signal $u1$. Therefore, the design and selection of the K(s) should counteract the influence of w and z. As a conclusion, the H_∞ design problem can be subedited as detecting an equiponderating feedback control law $u1$ (s) = K(s) y_1(s) to neutralizes the effect of w and z and so to minimize the closed loop norm from w to z.

In the proposed control system that includes the H_∞ controller, one feedback loop is designed to adjust the speed of the motor, as given in Figure 4. The nominal system G_P(s) is augmented with the weighting transfer functions $W_1(s)$, $W_2(s)$ and $W_3(s)$, which penalize the error signals, control signals, and output signals, respectively. The selection of the appropriate weighting functions is the quintessence of the H_∞ control. The wrong weighting function may cause the system to suffer from poor dynamic performance and instability characteristics.

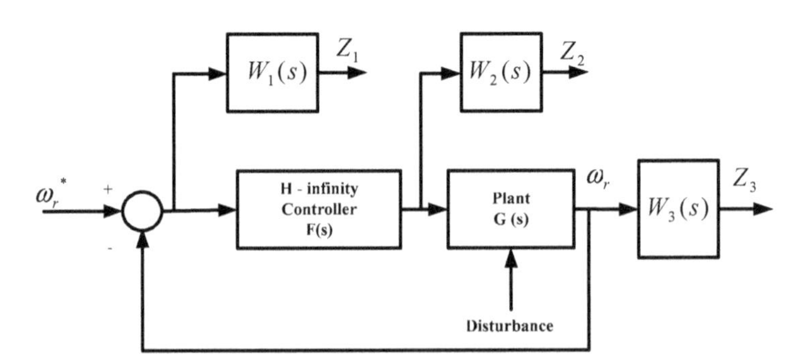

Figure 4. Simplified block diagram of the augmented plant, including the H_∞ controller.

Consider the augmented system shown in Figure 3. The following set of weighting transfer functions are selected to represent the required robustness and operation objectives:

A good choice for $W_1(s)$ is helpful for achieving good input reference tracking and good disturbance rejection. The matrix of the weighted error transfer function Z_1, which is needed for regulation, can be driven as follows:

$$Z_1 = W_1(s)\left\lfloor \omega_{ref} - \omega_r \right\rfloor.$$

A proper selection for the second weight $W_2(s)$ will assist in excluding actuator saturation and provide robustness to plant supplemented disturbances. The matrix of the weighted control function Z_2 can be expressed as:

$$Z_2 = W_2(s) \cdot u(s),$$

where $u(s)$ is the transfer function matrix of the control signal output of the H_∞ controller.

Additionally, a proper selection for the third weight $W_3(s)$ will restrict the bandwidth of the closed loop and achieve robustness to plant output multiplicative perturbations and sensor noise attenuation at high frequencies. The weighted output variable can be provided as:

$$Z_3 = \omega_r\, W_3(s).$$

In summary, the transfer functions of interest that determine the behavior of the voltage and power closed-loop systems are:

(a) Sensitivity function: $S = [I + G(s) \cdot K(s)]^{-1}$,

where G(s) and F(s) are the transfer functions of the nominal plant and the H_∞ controller, respectively, while I is the identity matrix. Therefore, when S is minimized at low frequencies, it will secure perfect tracking and disturbance rejection.

(b) Control function: $C = K(s)\ [I + G(s) \cdot K(s)]^{-1}$.

Minimizing C will preclude saturation of the actuator and acquire robustness to plant additional disturbances.

(c) Complementary function: $T = I - S$.

Minimizing T at high frequencies will ensure robustness to plant output multiplicative perturbations and achieve noise attenuation.

4. Robust Speed Estimation Based on MRAS Techniques for an IFO Control

The using of speed encoder in induction machines drives spoils the ruggedness and simplicity of the induction motor. Moreover, the speed sensor increases the cost of the induction motor drives. To eliminate the speed sensor, the calculation of the speed may be based on the coupled circuit equations of the motor [18–34]. The following explanation and analysis of the stability of the Model Reference Adaptive System (MRAS) speed estimator. In this work, the stability estimator is proven based on Popov's criterion. The measured stator voltages and currents have been used in a stationary reference frame to describe the stator and rotor models of the induction motor. The voltage model (stator model) and the current model (rotor equation) can be written as the following in the stationary reference frame $\alpha - \beta$ [18–29]:

The voltage model (stator equation):

$$p\begin{bmatrix} \psi_{\alpha r} \\ \psi_{\beta r} \end{bmatrix} = \frac{L_r}{L_m}\left(\begin{bmatrix} V_{\alpha s} \\ V_{\beta s} \end{bmatrix} - \begin{bmatrix} (R_s + \sigma L_s p) & 0 \\ 0 & (R_s + \sigma L_s p) \end{bmatrix}\begin{bmatrix} i_{\alpha s} \\ i_{\beta s} \end{bmatrix}\right) \tag{10}$$

The current model (rotor equation):

$$p\begin{bmatrix} \psi_{\alpha r} \\ \psi_{\beta r} \end{bmatrix} = \begin{bmatrix} (-1/T_r) & -\omega_r \\ \omega_r & (-1/T_r) \end{bmatrix} \begin{bmatrix} \psi_{\alpha r} \\ \psi_{\beta r} \end{bmatrix} + \frac{L_m}{T_r} \begin{bmatrix} i_{\alpha s} \\ i_{\beta s} \end{bmatrix} \tag{11}$$

Figure 5 illustrates an alternative way of observing the rotor speed using MRAS. Two independent rotor flux observers are constructed to estimate the components of the rotor flux vector: one based on Equation (10) and the other based on Equation (11). Because Equation (10) does not involve the quantity ω_r, this observer may be regarded as a reference model of the induction motor, while Equation (11), which does involve ω_r, may be regarded as an adjustable model. The states of the two models are compared and the error between them is applied to a suitable adaptation mechanism that produces the observed $\hat{\omega}_r$ for the adjustable model until the estimated motor speed tracks well against the actual speed.

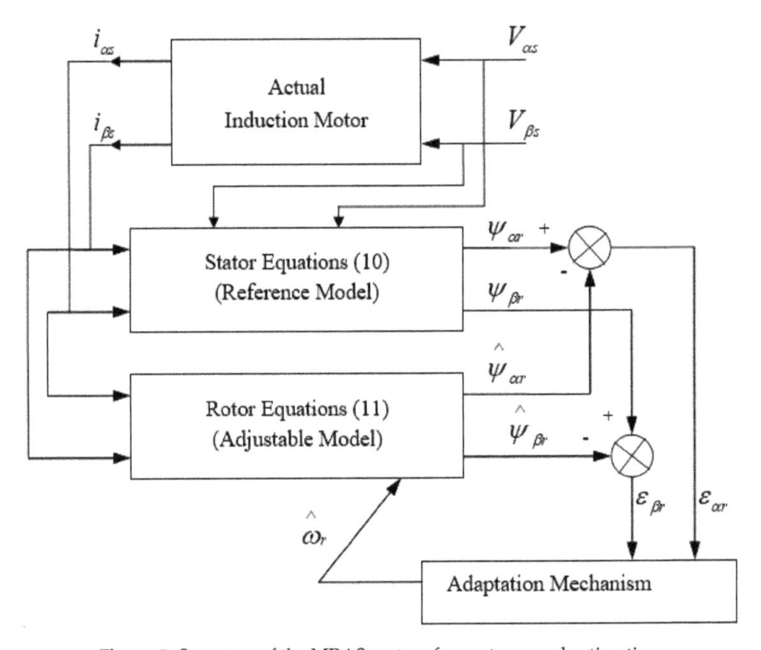

Figure 5. Structure of the MRAS system for motor speed estimation.

When eliciting an adaptation mechanism, it is adequate to initially act as a constant parameter of the reference model. By subtracting Equation (11) for the adjustable model from the corresponding equations belonged to the reference model Equation (10) for the rotor equations, the following equations for the state error can be obtained:

$$p\begin{bmatrix} \varepsilon_{\alpha r} \\ \varepsilon_{\beta r} \end{bmatrix} = \begin{bmatrix} (-1/T_r) & -\omega_r \\ \omega_r & (-1/T_r) \end{bmatrix} \begin{bmatrix} \varepsilon_{\alpha r} \\ \varepsilon_{\beta r} \end{bmatrix} + \begin{bmatrix} -\hat{\psi}_{\beta r} \\ \hat{\psi}_{\alpha r} \end{bmatrix} (\omega_r - \hat{\omega}_r) \tag{12}$$

that is,

$$p\,[\varepsilon] = [A_r]\,[\varepsilon] - [W] \tag{13}$$

Because $\hat{\omega}_r$ is a function of the state error, Equations (12) and (13) represent a non-linear feedback system, as shown in Figure 6.

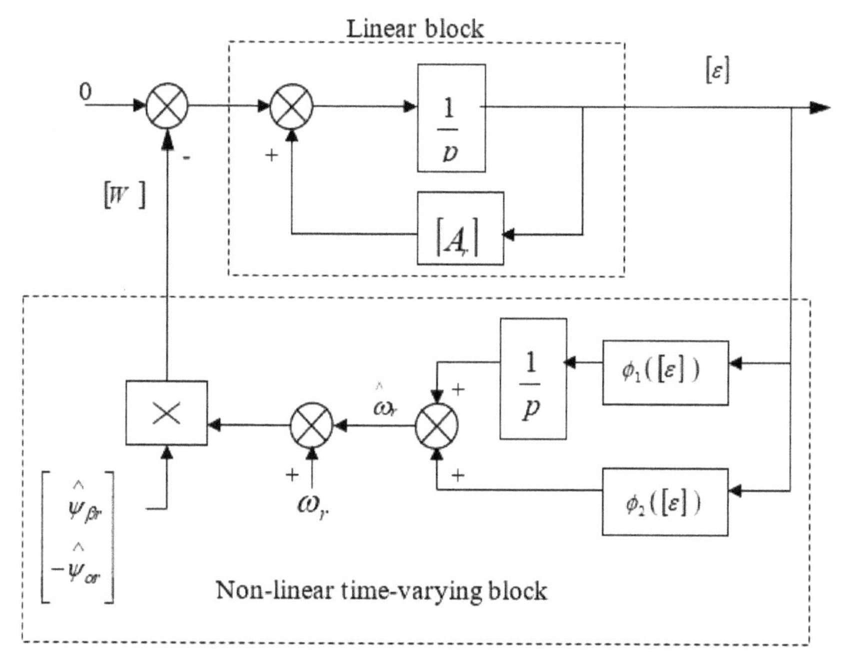

Figure 6. Representation of MRAS as a non-linear feedback system.

According to Landau, hyperstability is confirmed as long as the linear time-invariant forward-path transfer matrix is precisely positive real and that the non-linear feedback (which includes the adaptation mechanism) comply with Popov's hyperstability criterion. Popov's criterion demands a bounded negative limit on the input/output inner product of the non-linear feedback system. Assuring this criterion leads to the following candidate adaptation mechanism [31–34]:

Let

$$\hat{\omega}_r = \phi_2([\varepsilon]) + \int_0^t \phi_1([\varepsilon])d\tau \tag{14}$$

then, Popov's criterion presupposes that

$$\int_0^{t_1} [\varepsilon]^T [W]\, dt \geq -{\gamma_0}^2 \ \text{ for all } t_1 \geq 0 \tag{15}$$

where ${\gamma_0}^2$ is a positive constant. Substituting for $[\varepsilon]$ and $[W]$ in this inequality using the definition of $\hat{\omega}_r$, Popov's criterion for the system under study will be

$$\int_0^{t_1} \left\{ \left[\varepsilon_{\alpha r}\, \hat{\psi}_{\beta r} - \varepsilon_{\beta r}\, \hat{\psi}_{\alpha r}\right] \left[\omega_r - \phi_2([\varepsilon]) - \int_0^t \phi_1([\varepsilon])\, d\tau\right] \right\} dt \geq -{\gamma_0}^2 \tag{16}$$

A proper solution to this inequality can be realized via the following well-known formula:

$$\int_0^{t_1} k(p\, f(t))\, f(t)\, dt \geq -\frac{1}{2} k\, f(0)^2, \ k \succ 0 \tag{17}$$

Using this expression, Popov's inequality is satisfied by the following functions:

$$\phi_1 = K_P(\varepsilon_{\beta r}\hat{\psi}_{\alpha r} - \varepsilon_{\alpha r}\hat{\psi}_{\beta r}) = K_P(\psi_{\beta r}\hat{\psi}_{\alpha r} - \psi_{\alpha r}\hat{\psi}_{\beta r}).$$

$$\phi_2 = K_I(\varepsilon_{\beta r}\hat{\psi}_{\alpha r} - \varepsilon_{\alpha r}\hat{\psi}_{\beta r}) = K_I(\psi_{\beta r}\hat{\psi}_{\alpha r} - \psi_{\alpha r}\hat{\psi}_{\beta r}).$$

Figure 7 illustrates the block diagram of the MRAS. The outputs of the two models the rotor flux components. Moreover, the measured stator voltages and currents are in the stationary reference frame have been applied to be as the inputs of MRAS.

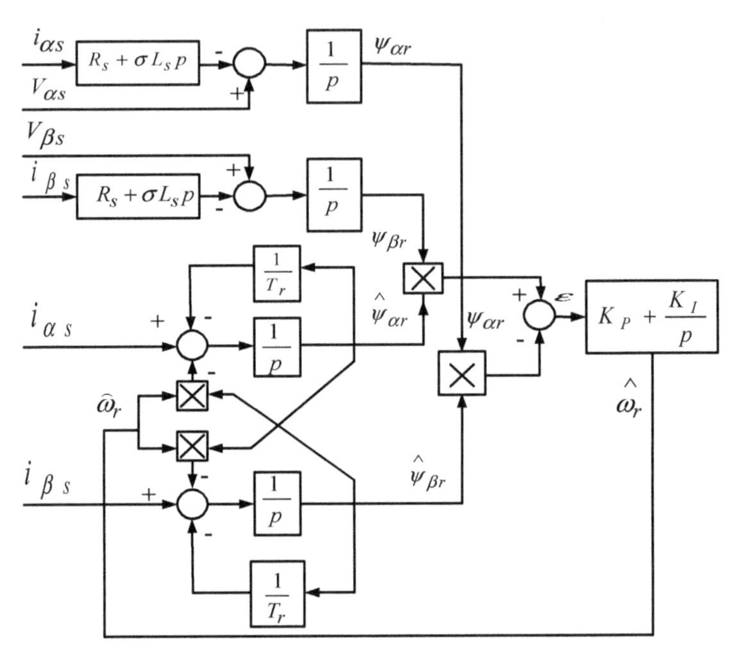

Figure 7. Block diagram of the Model Reference Adaptive System (MRAS) speed estimation system.

5. Proposed Sensor-Free Induction Motor Drive for High-Performance Applications

Figure 8 shows a block diagram of the proposed controller sensor-free induction motor drive system. The control system is composed of a robust controller based on H_∞ theory, hysteresis controllers, a vector rotator, a digital pulse with modulation (PWM) scheme for a transistor bridge voltage source inverter (VSI) and a motor speed estimator based on MRAS. The speed controller makes speed corrections by assessing the error between the command and estimated motor speed. The speed controller is used to generate the command q-component of the stator current i^*_{qs}. The vector rotator and phase transform in Figure 8 are used to transform the stator current components command to the three-phase stator current commands (i^*_{as}, i^*_{bs} and i^*_{cs}) using the field angle position $\hat{\theta}_r$. The field angle is obtained by integrating the summation of the estimated speed and slip speed. The hysteresis current control compares the stator current commands to the actual currents of the machine and switches the inverter transistors in such a way as to obtain the desired command currents. Moreover, the MRAS rotor speed estimator can observe the rotor speed $\hat{\omega}_r$ based its inputs of measured stator voltages and currents. This estimated speed is fed back to the speed controller. Additionally, the estimated speed is added to the slip speed, and the sum is integrated to obtain the field angle θ_s.

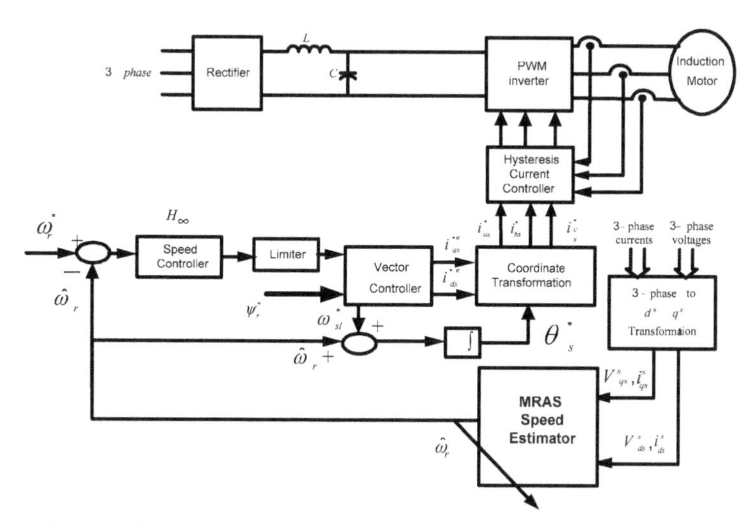

Figure 8. The block diagram of the proposed sensorless induction motor drive based on H_∞ theory.

6. Simulation and Experimental Results

6.1. Simulation Results

The complete block diagram of the field-oriented induction motor drive with the proposed H_∞ and MRAS speed estimator has been presented in Figure 8. The proposed system has been simulated using MATLAB/Simulink under different operating conditions. The parameters and data specifications of the entire system used in the simulation are given in Table 1. The following set of weighting functions are obtained based on the application of the optimization toolbox in MATLAB/Simulink to achieve the proposed robustness and operation objectives:

$$W_1 = \gamma_1 \frac{s + 1e - 4}{s + 0.001}, \; W_2 = \gamma_2 \frac{s + 1e - 4}{s + 10}, \; W_3 = \gamma_3,$$

where $\gamma_1 = 12$, $\gamma_2 = 0.00001$, $\gamma_3 = 0$.

Table 1. Parameters and data specifications of the induction motor.

Rated Power (W)	180	Rated Voltage (V)	220
Rated current (A)	1.3–1.4	Rated frequency (Hz)	60
R_s (Ω)	11.29	R_r (Ω)	6.11
L_s (H)	0.021	L_r (H)	0.021
L_m (H)	0.29	Rated rotor flux, (wb)	0.3
J (kg.m^2)	0.00940	Rated speed (rpm)	1750

The transient behavior of the proposed sensorless control system is evaluated by applying and removing the motor-rated torque (1 N.m), as shown in Figure 9. Figure 9a shows the performance of the proposed control scheme with H_∞ controller. While Figure 9b illustrates the performance of sensorless induction motor (IM) drive based on the conventional PI controller for the comparison purpose. Figure 9 shows that the motor speed can be effectively estimated and accurately tracks the actual speed when using the proposed sensorless scheme. Moreover, the figure shows that the performance of the two controllers of the PI controller and the proposed H_∞ controller have acceptable dynamic performance. Furthermore, the figure also indicates that a fast and precise transient response to motor torque is achieved with the H_∞ controller. Additionally, the stator phase current matches the value of the application and removal of the motor-rated torque.

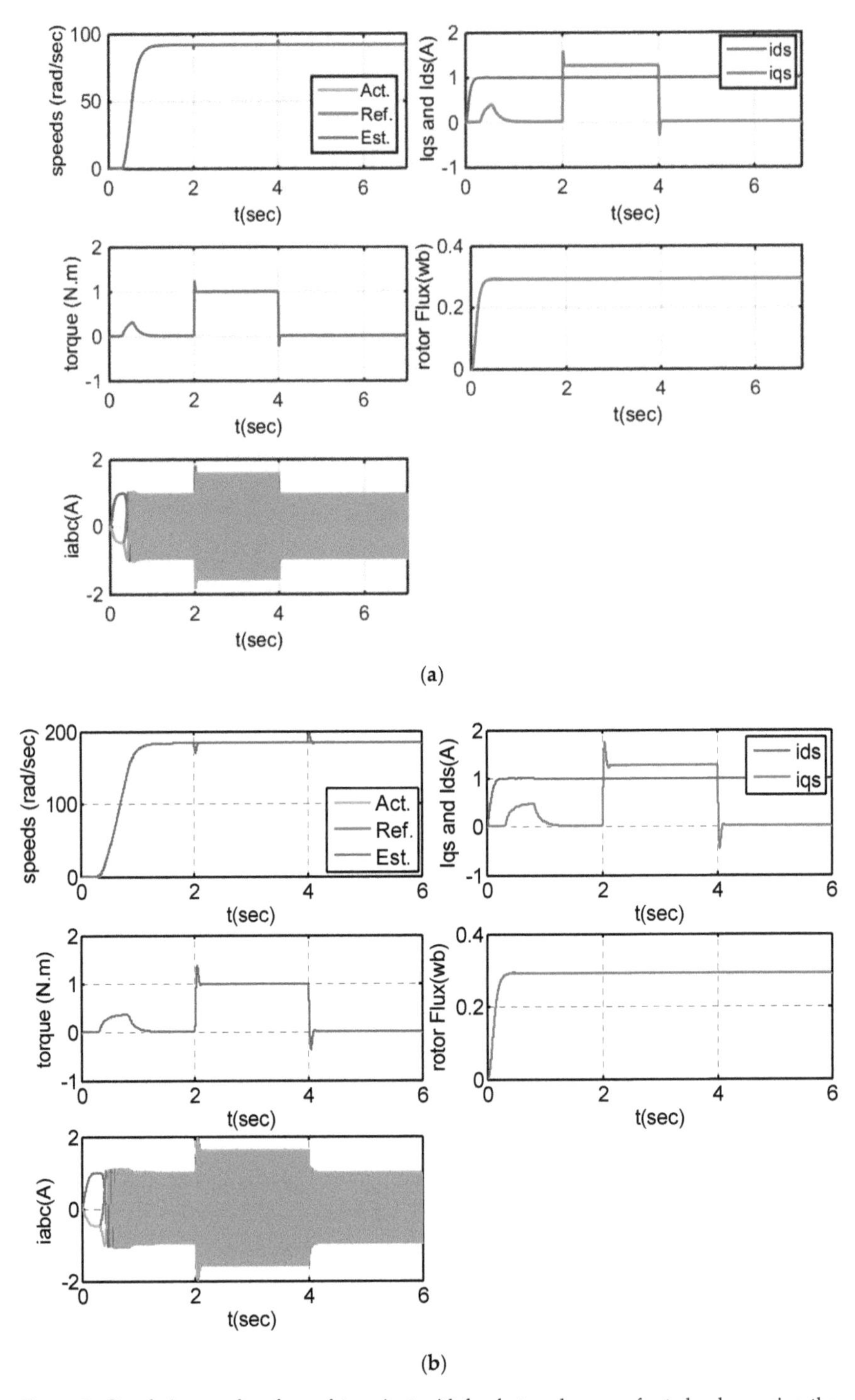

Figure 9. Simulation results of speed transient with load step changes of rated values using the proposed sensorless drive system; (**a**) with H_∞ controller and (**b**) with PI controller.

For more validating of the H_∞ controller, Figure 10 describes the dynamic response of the control scheme based on the proposed H_∞ controller versus the PI controller when the system is subjected to a step change in the load torque. From the figure, the results show an acceptable dynamic performance of the control scheme with the PI controllers and the H_∞ against the step change in the load torque. However, the application of the H_∞ controller causes improvements in the dynamic response of the rotor speed. Clearly, the PI controller requires a long rise time compared to the H_∞ controller and the speed response has a larger overshoot with respect to the H_∞ controller. The reasons for these results of the priority of the H_∞ based induction motor drive may be because the H_∞ controller has been designed taking the weighting function for the disturbance. The discussion of these results can be more discussed as follows: The fixed parameter controllers such as PI controllers are developed at nominal operating points. However, it may not be suitable under various operating conditions. However, the real problem in the robust nonlinear feedback control system is to synthesize a control law which maintains system response and error signals to within prespecified tolerances despite the effects of uncertainty on the system. Uncertainty may take many forms but among the most significant are noise, disturbance signals, and modeling errors. Another source of uncertainty is unmodeled nonlinear distortion. Consequently, researchers have adopted a standard quantitative measure the size of the uncertainty, called the H_∞. Therefore, the dynamic performance of the control scheme with the H_∞ controller is improved rather than the acceptable performance with the PI controller.

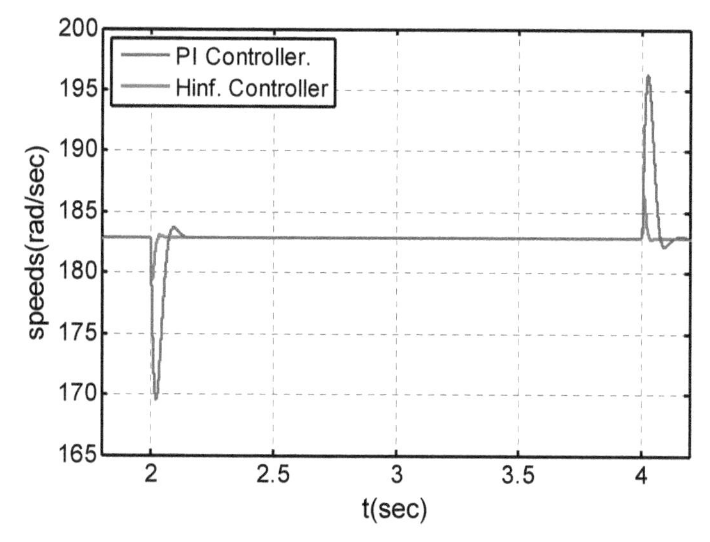

Figure 10. Response of the H_∞ controller versus the PI controller.

Figure 11 shows the actual and estimated speed transient, motor torque and stator phase current during acceleration and deceleration at different speeds. The estimated speed agreed satisfactorily with the actual speed. Additionally, a small deviation occurs from the actual speed before reaching steady-state and subsequently tracking quickly towards the command value. The motor torque response exhibits good dynamic performance. Figure 11 also shows the stator phase current during acceleration. From these simulation results, the proposed sensorless drive system is capable of operating at high speed, as illustrated in Figure 11.

In the last case of the study, the transient performance of the sensorless induction machine drive is examined by reversing the rotor speed. Figure 12 shows that induction motor drive based on the proposed robust controller has high dynamic performance at the reversing the rotor speed from 150 rad/s. Moreover, the speed estimator MRAS can accurately observe the rotor speed.

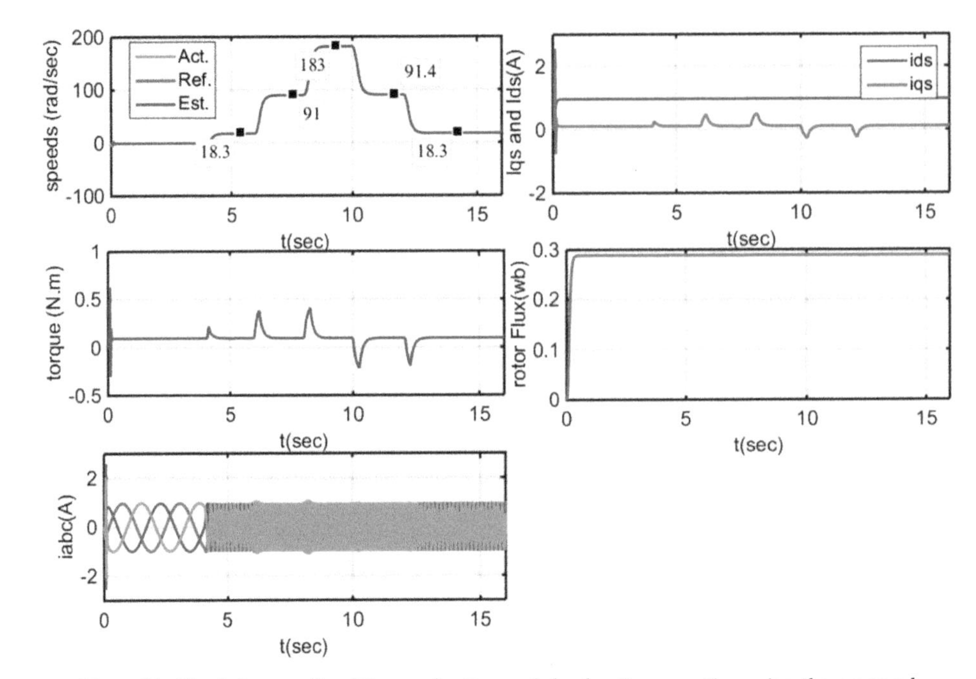

Figure 11. Simulation results of the acceleration and deceleration operation using the proposed sensorless drive system.

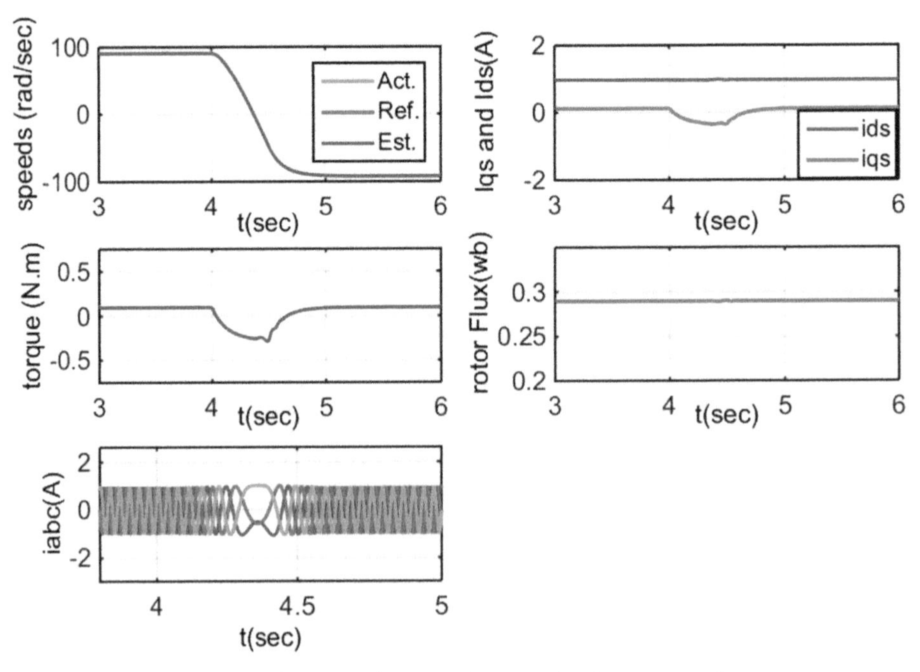

Figure 12. System performance when reversing the motor speed.

6.2. Practical Results

Figure 13 shows the experimental setup for the configured drive system. The drive system includes an induction motor; with the same parameters and data specification of the induction motor which used for simulating the proposed control scheme; linked with a digital control board (TMDSHVMTRPFCKIT from Texas Instruments with a TMS320F28035 control card) [32,35]. The complete control scheme has been programmed in the package of Code Composer Studio CCS from Texas Instruments.

Figure 13. A photo of the experimental setup for induction motor drive.

To validate the effectiveness of the sensorless vector control of the induction motor drive, the experiments were accomplished at different values of the reference speed. Figures 14–16 show samples of the results when the proposed system is tested at reference speeds of 0.2 pu and 0.4 pu; the base speed is assumed to be 3600 rpm (So, when the reference speed is 0.5 pu, it is mean the speed equals 1800 rpm). The results proved that the drive system effectively works at an extensive range of speeds. In addition, the actual and estimated speeds have coincided. Moreover, from Figure 16, it is obviously seen that the current of the motor is sinusoidal.

Another case of study has been tested for more evaluating of the control scheme. In this case of study, the reference speed has been reversed from 0.4 pu to 0.2 pu in the reverse direction in a ramp variation. The results of this case of study are shown in Figures 17 and 18. Figure 17 shows the speed response of the control scheme. Moreover, the phase current is shown in Figure 18. The results show the control scheme has a good dynamic performance.

The last case of study has assumed many ramp changes in the references including reversing the speed. The rotor speed response is shown in Figure 19. The results have been plotted with the aid of CCS package and Digital Signal Processing (DSP). The results show the control scheme has a good dynamic performance.

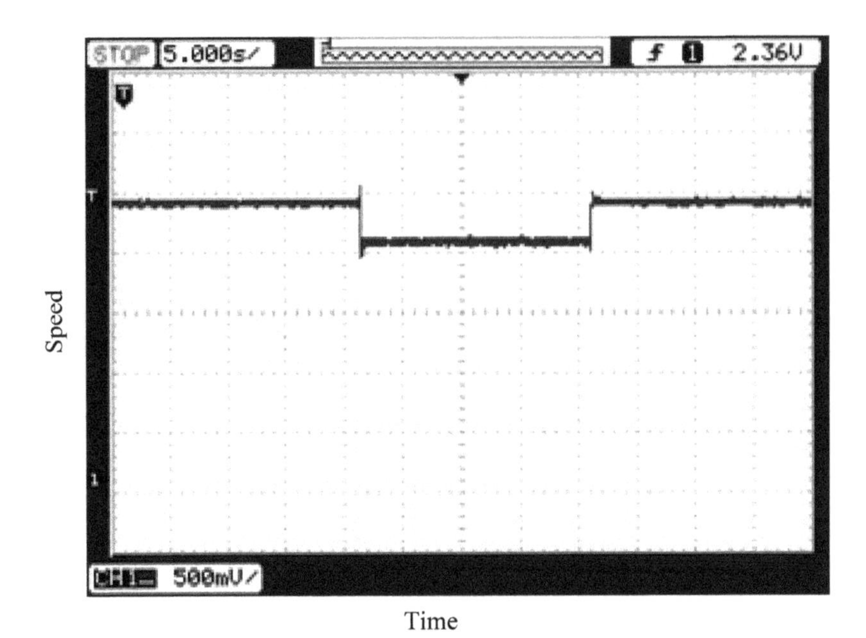

Figure 14. The transient performance of the entire drive system under variable speed from 0.4 pu to 0.2 pu to 0.4 pu.

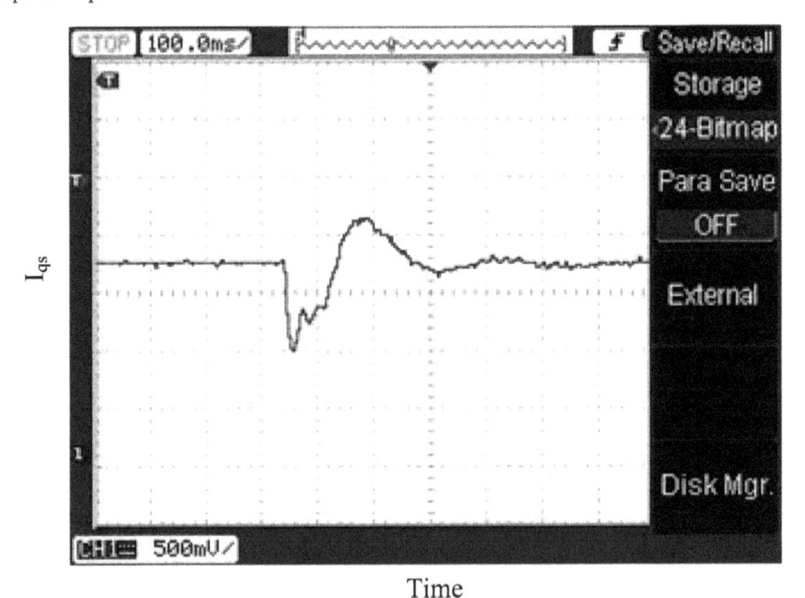

Figure 15. The transient performance of I_{qs} of the entire drive system under variable speed from 0.4 pu to 0.2 pu.

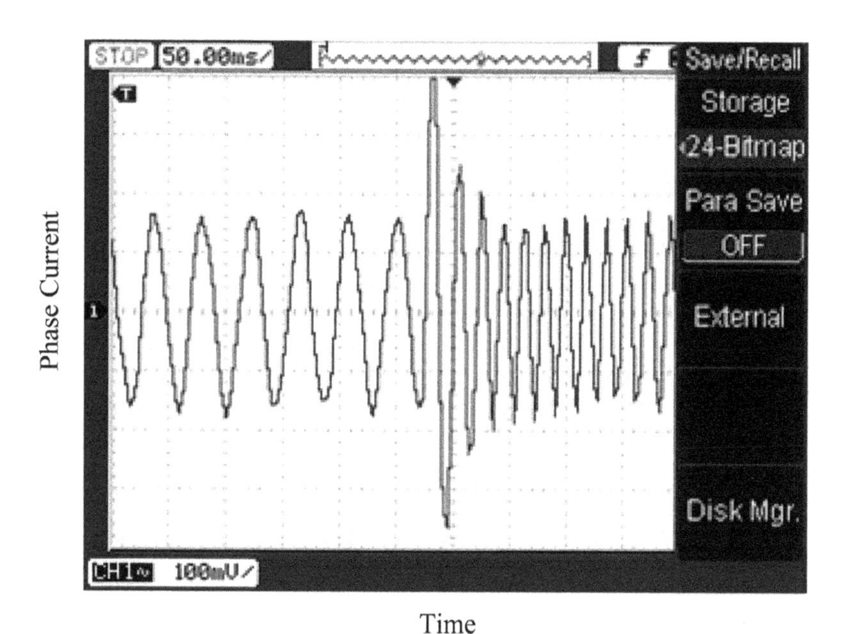

Figure 16. Current of phase (**a**) at reference speeds of 0.2 pu to 0.4 pu.

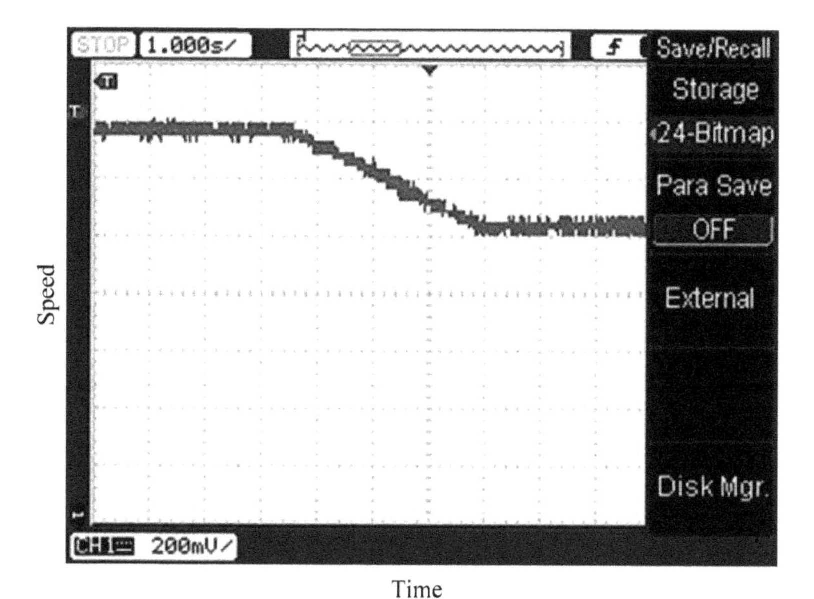

Figure 17. The transient performance of Iq_S of the entire drive system under variable speed from 0.4 pu to 0.2 pu.

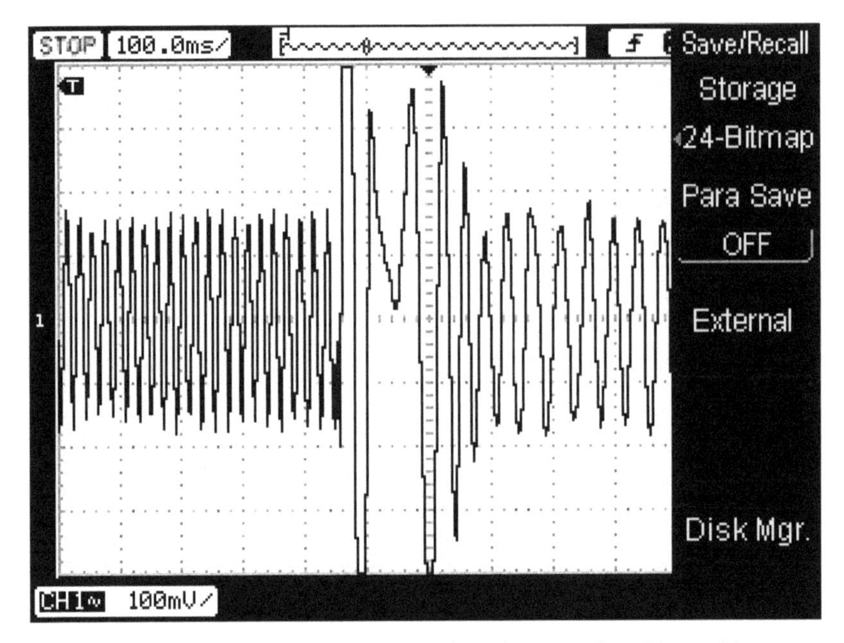

Figure 18. Current of phase (**a**) for the case of speed reversing from 0.4 pu to 0.2 pu.

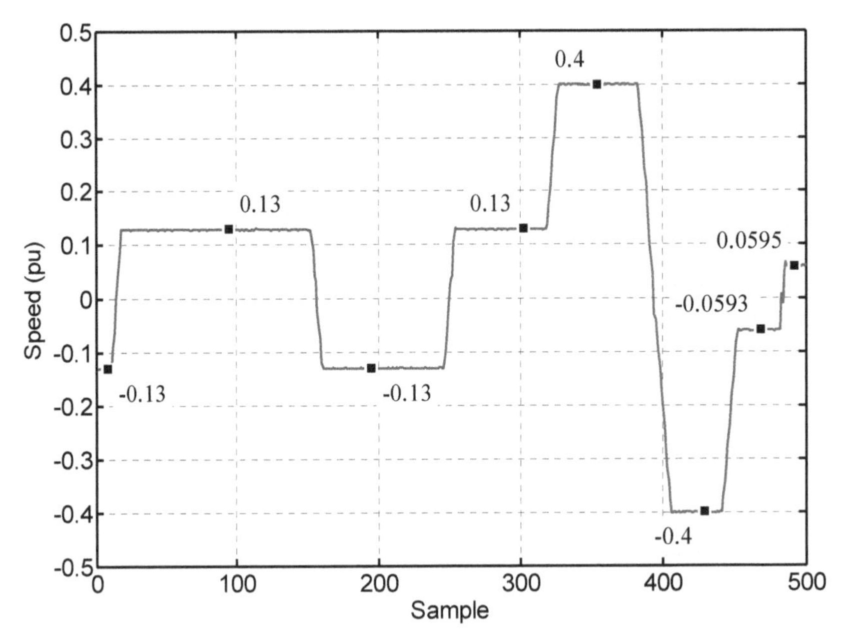

Figure 19. The transient performance of multi-variation in the rotor speed with reversing (The experimental data and measurements have been collected with the aid of the DSP and plotted using Matlab plot tool).

7. Conclusions

In this paper, H_∞ theory has been proposed for designing an optimal robust speed controller for a field-oriented induction motor drive. The design problem of the H_∞ controller was explained and derived in standard form with an assertion on the choice of weighting functions, which fulfills the optimal robustness and performance of the drive system. The proposed control strategy has many advantages: it is robust to plant uncertainties, and has a simple implementation and a fast response. Moreover, a robust motor speed estimator based on the MRAS is presented that estimates motor speed accurately for a sensorless IFO control system. The validation of the induction motor drive was performed using both simulated and experimental implementations. The main conclusions that can be drawn from the results in this study are as follows:

(1) The effectiveness of the considered induction motor drive system with the proposed controller has been demonstrated.

(2) Compared with a PI classical controller, the response of the proposed controller shows a reduced settling time in the case of a sudden change of the speed command in addition to smaller values of the maximum speed dip and overshoot as a result of the application and removal of stepped changes in load torque.

(3) The proposed controller achieved robust performance under stepped speed change commands or changes in load torque even when the parameters of the controlled system were varied.

(4) The forward-reverse operation of the drive is obtained by the robust MRAS speed estimator and guarantees the stability of the proposed sensorless control to the system at a speed of zero. Moreover, the presented speed estimator provides an accurate speed estimation regardless of the load conditions.

(5) Both simulated and real-world experimental results demonstrate that the proposed control drive system is capable of working at a wide range of motor speeds and that it exhibits good performance in both dynamic and steady-state conditions.

Further research work should consider the nonlinearity of the induction machine parameters tacking saturation and/or iron losses into consideration. Additionally, recent optimization techniques may be applied to determine the optimal weight functions for designing the controller. Moreover, the operation range should be expanded to study and analyse the operation of the control scheme in the field weakening region. Moreover, the estimation of the machine parameters may be an interesting research point for future work for improving the overall performance of the control scheme and speed estimator.

Author Contributions: A.A.Z.D. and H.H.A. developed the idea and the simulation models. A.A.Z.D. performed the experiments and analyzed the data. A.A.Z.D., A.-H.M.E.-S. and H.H.A. wrote the paper. A.A.Z.D., A.-H.M.E.-S. and M.A.E.S. contributed by drafting and making critical revisions. All the authors organized and refined the manuscript to its present form.

Funding: This work was supported by Minia University, Egypt.

Conflicts of Interest: The authors declare no conflicts of interest.

References

1. Vas, P. *Vector Control of AC Machines*; Oxford University Press: Oxford, UK, 1990; pp. 122–215.
2. Krishnan, R.; Doran, F.C. Study of parameter Sensitivity in High Performance Inverter Fed Induction Motor Drive Systems. *IEEE Trans. Ind. Appl.* **1987**, *IA-23*, 623–635. [CrossRef]
3. Vas, P. *Parameter Estimation, Condition Monitoring, and Diagnosis of Electrical Machines*; Oxford Science Publications: Oxford, UK, 1993.
4. Toliyat, H.A.; Levi, E.; Raina, M. A Review of RFO Induction Motor Parameter Estimation Techniques. *IEEE Trans. Energy Convers.* **2003**, *18*, 271–283. [CrossRef]

5. Minami, K.; Veles-Reyes, M.; Elten, D.; Verghese, G.C.; Filbert, D. Multi-Stage Speed and Parameter Estimation for Induction Machines. In Proceedings of the Record 22nd Annual IEEE Power Electronics Specialists Conference (IEEE PESC'91), Cambridge, MA, USA, 24–27 June 1991; pp. 596–604.
6. Roncero-Sánchez, P.L.; García-Cerrada, A.; Feliú, V. Rotor-Resistance Estimation for Induction Machines with Indirect-Field Orientation. *Control Eng. Pract.* **2007**, *15*, 1119–1133. [CrossRef]
7. Karayaka, H.B.; Marwali, M.N.; Keyhani, A. Induction Machines Parameter Tracking from Test Data Via PWM Inverters. In Proceedings of the Record of the 1997 IEEE Industry Applications Conference Thirty-Second IAS Annual Meeting, New Orleans, LA, USA, 5–9 October 1997; pp. 227–233.
8. Godoy, M.; Bose, B.K.; Spiegel, R.J. Design and performance evaluation of a fuzzy-logic-based variable-speed wind generation system. *IEEE Trans. Energy Convers.* **1997**, *33*, 956–964.
9. Munteau, I.; Cutululis, N.A.; Bratcu, A.I.; Ceanga, E. Optimization of variable speed wind power systems based on a LQG approach. In Proceedings of the IFAC Workshop on Control Applications of Optimisation—CAO'03 Visegrad, Visegrad, Hungary, 30 June–2 July 2003.
10. Hassan, A.A.; Mohamed, Y.S.; Yousef, A.M.; Kassem, A.M. Robust control of a wind driven induction generator connected to the utility grid. *Bull. Fac. Eng. Assiut Univ.* **2006**, *34*, 107–121.
11. Attaiaence, C.; Perfetto, A.; Tomasso, G. Robust postion control of DC drives by means of H_∞ controllers. *Proc. IEE Electr. Power Appl.* **1999**, *146*, 391–396. [CrossRef]
12. Naim, R.; Weiss, G.; Ben-Yakakov, S. H_∞ control applied to boost power converters. *IEEE Trans. Power Electron.* **1997**, *12*, 677–683. [CrossRef]
13. Lin, F.-J.; Lee, T.; Lin, C. Robust H_∞ controller design with recurrent neural network for linear synchronous motor drive. *IEEE Trans. Ind. Electron.* **2003**, *50*, 456–470.
14. Pohl, L.; Vesely, I. Speed Control of Induction Motor Using H_∞ Linear Parameter Varying Controller. *IFAC-PapersOnLine* **2016**, *49*, 74–79. [CrossRef]
15. Kao, Y.-T.; Liu, C.-H. Analysis and design of microprocessor-based vector-controlled induction motor drives. *IEEE Trans. Ind. Electron.* **1992**, *39*, 46–54. [CrossRef]
16. Prempain, E.; Postlethwaite, I.; Benchaib, A. A linear parameter variant H_∞ control design for an induction motor. *Control Eng. Pract.* **2002**, *10*, 633–644. [CrossRef]
17. Rigatos, G.; Siano, P.; Wira, P.; Profumo, F. Nonlinear H-infinity feedback control for asynchronous motors of electric trains. *Intell. Ind. Syst.* **2015**, *1*, 85–98. [CrossRef]
18. Riberiro, L.A.D.; Lima, A.M.N. Parameter Estimation of Induction Machines Under Sinusoidal PWM Excitation. *IEEE Trans. Energy Convers.* **1999**, *14*, 1218–1223. [CrossRef]
19. Moonl, S.; Keyhani, A. Estimation of Induction Machines Parameters from Standstill Time-Domain Data. *IEEE Trans. Ind. Appl.* **1994**, *30*, 1609–1615. [CrossRef]
20. Lima, A.M.N.; Jacobina, C.B.; Filho, E.B.D. Nonlinear Parameter Estimation of Steady-state Induction Machine Models. *IEEE Trans. Ind. Electron.* **1997**, *44*, 390–397. [CrossRef]
21. Yang, G.; Chin, T.H. Adaptive-Speed Identification Scheme for a Vector-Controlled Speed Sensorless Inverter-Induction Motor Drive. *IEEE Trans. Ind. Appl.* **1993**, *29*, 820–825. [CrossRef]
22. Kubota, H.; Matsuse, K. Speed Sensorless Field-Oriented Control of Induction Motor with Rotor Resistance Adaptation. *IEEE Trans. Ind. Appl.* **1994**, *30*, 1219–1224. [CrossRef]
23. Suwankawin, S.; Sangwongwanich, S. Design Strategy of an Adaptive Full Order Observer for Speed Sensorless Induction Motor Drives Tracking Performance and Stabilization. *IEEE Trans. Ind. Electron.* **2006**, *53*, 96–119. [CrossRef]
24. Al-Tayie, J.; Acarnley, P. Estimation of Speed, Stator Temperature and Rotor Temperature in Cage Induction Motor Drive Using the Extended Kalman Filter Algorithm. *Proc. IEE Electr. Power Appl.* **1997**, *144*, 301–309. [CrossRef]
25. Barut, M.; Bogosyan, O.S.; Gokasan, M. Switching EKF Technique for Rotor and Stator Resistance Estimation in Speed Sensorless Control of Induction Motors. *Energy Convers. Manag.* **2007**, *48*, 3120–3134. [CrossRef]
26. Jingchuan, L.; Longya, X.; Zhang, Z. An Adaptive Sliding-Mode Observer for Induction Motor sensorless Speed Control. *IEEE Trans. Ind. Appl.* **2005**, *41*, 1039–1046.
27. Rashed, M.; Stronach, A.F. A Stable Back-EMF MRAS-Based Sensorless Low Speed Induction Motor Drive Insensitive to Stator Resistance Variation. *IEE Proc. Electr. Power Appl.* **2004**, *151*, 685–693. [CrossRef]

28. Mohamed, Y.S.; El-Sawy, A.M.; Zaki, A.A. Rotor Resistance Identification for Speed Sensorless Vector Controlled Induction Motor Drives Taking Saturation Into Account. *J. Eng. Sci. Assiut Univ.* **2009**, *37*, 393–412.
29. Middeton, R.H.; Goodwin, G.C. *Digital Control and Estimation*, 1st ed.; Prentice-Hall, Inc.: Englewood Cliffs, NJ, USA, 1990; Volume 1.
30. Blasco-Gimenez, R.; Asher, G.; Summer, M.; Bradley, K. Dynamic Performance Limitations for MRAS Based Sensorless Induction Motor Drives. Part 1: Stability Analysis for the Closed Loop Drive. *Proc. IEE-Electr. Power Appl.* **1996**, *143*, 113–122. [CrossRef]
31. Diab, A.A.; Khaled, A.; Elwany, M.A.; Hassaneen, B.M. Parallel estimation of rotor resistance and speed for sensorless vector controlled induction motor drive. In Proceedings of the 2016 17th International Conference of Young Specialists on Micro/Nanotechnologies and Electron Devices (EDM), Erlagol, Russia, 30 June–4 July 2016.
32. Diab, A.A. Real-Time Implementation of Full-Order Observer for Speed Sensorless Vector Control of Induction Motor Drive. *J. Control Autom. Electr. Syst.* **2014**, *25*, 639–648. [CrossRef]
33. Diab, A.A.Z. Implementation of a novel full-order observer for speed sensorless vector control of induction motor drives. *Electr. Eng.* **2016**, *99*, 907–921. [CrossRef]
34. Diab, A.A.Z. Novel robust simultaneous estimation of stator and rotor resistances and rotor speed to improve induction motor efficiency. *Int. J. Power Electron.* **2017**, *8*, 267–287. [CrossRef]
35. Texas Instruments C2000 Systems and Applications Team. High Voltage Motor Control and PFC (R1.1) Kit Hardware Reference Guide, v. 2. 2012. Available online: http://www.ti.com/tool/TMDSHVMTRPFCKIT (accessed on 31 January 2019).

Review

Control Strategies for Induction Motors in Railway Traction Applications

Ahmed Fathy Abouzeid [1], Juan Manuel Guerrero [1], Aitor Endemaño [2], Iker Muniategui [2], David Ortega [2], Igor Larrazabal [2] and Fernando Briz [1,*]

1 Electrical, Electronic, Computers and Systems Engineering, University of Oviedo, 3204 Oviedo, Spain; abouzeidahmed@uniovi.es (A.F.A.); guerrero@uniovi.es (J.M.G.)
2 Ingeteam Power Technology S.A.—Traction, 48170 Zamudio, Spain; aitor.endemano@ingeteam.com (A.E.); iker.muniategui@ingeteam.com (I.M.); David.Ortega@ingeteam.com (D.O.); igor.larrazabal@ingeteam.com (I.L.)
* Correspondence: fbriz@uniovi.es

Received: 30 December 2019; Accepted: 3 February 2020; Published: 6 February 2020

Abstract: This paper analyzes control strategies for induction motors in railway applications. The paper will focus on drives operating with a low switching to fundamental frequency ratio and in the overmodulation region or six-step operation, as these are the most challenging cases. Modulation methods, efficient modes of operation of the drive and the implications for its dynamic performance, and machine design will also be discussed. Extensive simulation results, as well as experimental results, obtained from a railway traction drive, are provided.

Keywords: railway traction drives; induction motor drives; high-speed drives; maximum torque per ampere; overmodulation and six-step operation

1. Introduction

Despite being one of the most energy-efficient means for mass transportation (see Figure 1) [1], there is pressure to develop a more efficient, reliable, cheap, and compact railway traction system, which should be achieved without compromising customer satisfaction.

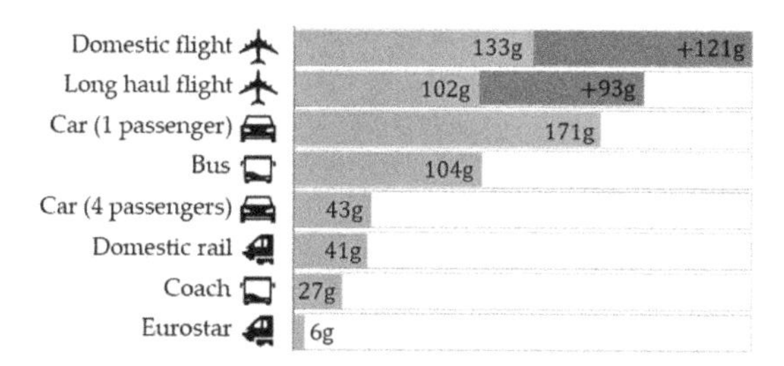

Figure 1. Emissions per passenger per km from different modes of transport. From the UK Department for Business, Energy and Industrial Strategy 2019 Government Greenhouse Gas Conversion Factors [1].

Three-phase induction motors (IMs) were adopted in the 1990s for traction systems in railways replacing DC machines [2] due to their increased robustness and reduced cost and maintenance requirements. In addition, precise control of the IM torque/speed is perfectly possible thanks to the development of new power devices and digital signal processors, combined with the advances in AC-driven control methods. Furthermore, the inherent slip of IM allows multiple motors to be fed from a single inverter, even if they rotate at different speeds due to differences in wheel diameters. As a result, the voltage-source inverter-fed IM drive (VSI-IM) is currently the preferred option in traction systems for railways [3]. While Permanent Magnet Synchronous Machines (PMSM) have also been considered and can be found in several traction systems, cost and reliability concerns intrinsic to this type of machine, mainly due to magnets, have so far prevented their widespread use [4].

Rolling stock can be classified according to the power level of the traction system, ranging from several tens of kW for Light Rail systems, to several MW for High-Speed Trains (HST) and Heavy Rail Locomotives [5]. Traction systems can be concentrated or distributed. In concentrated systems, one or more locomotives pull unmotorized coaches. On the contrary, distributed traction systems use Electric Multiple Units (EMU), i.e., self-propelled carriages. Both options have advantages and disadvantages. EMUs can provide a superior performance in terms of the acceleration and deceleration times, adhesion effort, and transport capacity. However, passenger comfort, maintenance, and pantograph operation can be compromised in this case [6,7]. For the case of HST, European manufacturers have predominantly adopted the concentrated traction option, while the distributed option has been preferred by Japanese manufacturers [8].

The two main elements in a traction system are the electric motor and the inverter. The development of a cost-effective traction system for a given application involves a complex, iterative process to decide the number of traction motors, motor size, inverter rated power, cooling system, etc. Once the physical elements of the traction system have been defined, the control and modulation strategies need to be defined. Additionally, in this case, a complex iterative process can be required as the traction system must comply with a number of requirements. These include those imposed by the desired train performance (e.g., torque-speed characteristic, maximum torque and speed, acceleration/deceleration times, etc.), electric drive performance (e.g., machine and inverter efficiency, temperature limits, maximum torque ripple, etc.), existing standards (e.g., electromagnetic interference, acoustic noise, etc.), and so on. However, these targets will often be in conflict. The reduction of inverter losses requires low switching frequencies, which in turn result in higher losses and large torque pulsations in the motor, and can also compromise the dynamic response or even the stability of the drive. Especially challenging is the operation of the traction drive at high speeds. The large back-electromotive force, in this case, forces the inverter to operate in the overmodulation region, including square-wave modes. The control operates in this case with a reduced (or even no) voltage margin and large distortions in the currents, which can further deteriorate the drive performance.

Figure 2 shows a schematic representation of the main blocks involved in the operation of a traction drive. The drive will normally receive a torque command coming from outer control loops (e.g., the train driver or speed control loop). From the torque command and in the operating condition of the machine, a flux command is derived; different criteria can be followed for this purpose, as shown in Figure 2. Torque and flux are controlled by the inner control loops; a number of solutions are available for this purpose. Inner control loops will provide the voltage command to the Voltage Source Inverter (VSI) feeding the machine, with selection of the modulation method being of the highest importance.

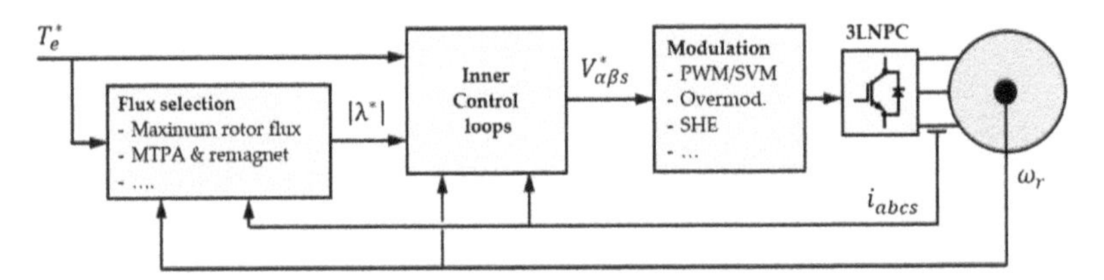

Figure 2. Main blocks of a traction drive.

This paper presents a review of the different aspects involved in the control of IM motor drives for railway applications. Section 2 reviews the IM motor model, including a discussion on the machine characteristics. Section 3 discusses control strategies, with a special focus on their suitability for use at high speed and low switching frequencies, as this is the most frequent and challenging mode of operation for traction drives. Modulation is discussed in Section 4. Section 5 discusses efficient modes of operation and remagnetization strategies. Sections 6 and 7 provide simulation and experimental results, respectively. Section 8 summarizes the conclusions.

2. Induction Motor Model and Machine Characteristics

2.1. Induction Motor Model

Complex vectors allow a compact, insightful dynamic representation of the physical effects occurring in AC machines, i.e., the relationships among electromagnetic variables (voltages, currents, and fluxes) and shaft variables (torque and speed) [9]. Equations (1)–(4) show the electromagnetic complex vector equations describing the squirrel cage induction machine in a synchronous reference frame rotating at the flux angular frequency ω_e, where v_{dqs} denotes the stator voltage; i_{dqs} and i_{dqr} are the stator and rotor currents, respectively; λ_{dqs} and λ_{dqr} represent the stator and rotor fluxes, respectively; R_s and R_r are the stator and rotor resistances, respectively; L_s, L_r, and L_m are the stator, rotor, and mutual inductances, respectively; ω_r is the rotor angular speed in electrical units; and p is the derivative operator.

$$v_{dqs} = R_s i_{dqs} + p\lambda_{dqs} + j\omega_e \lambda_{dqs} \tag{1}$$

$$0 = R_r i_{dqr} + p\lambda_{dqr} + j(\omega_e - \omega_r)\lambda_{dqr} \tag{2}$$

$$\lambda_{dqs} = L_s i_{dqs} + L_m i_{dqr} \tag{3}$$

$$\lambda_{dqr} = L_m i_{dqs} + L_r i_{dqr} \tag{4}$$

The electromagnetic torque T_e can be expressed as the cross product of stator and rotor currents (5). P is the number of pole-pairs, and "*Im*" and "‡" denote the imaginary part and complex conjugate, respectively.

$$T_e = \frac{3}{2} P L_m Im\left\{ i_{dqs} i^{\ddagger}_{dqr} \right\} \tag{5}$$

Equations (1)–(4) can be particularized for the case when the d-axis is aligned with the rotor flux, i.e., $\lambda_{dqr} = \lambda_{dr} = \lambda_r$, which is the base of rotor field-oriented control (RFOC). The stator voltage equation in scalar form is, in this case (6), the rotor flux dynamics being given by (7), where τ_r is the rotor time constant and σ is the leakage factor.

$$\left.\begin{aligned} v_{ds} &= R_s i_{ds} + \sigma L_s p i_{ds} - \omega_e \sigma L_s i_{qs} + \tfrac{L_m}{L_r} p\lambda_r \\ v_{qs} &= R_s i_{qs} + \sigma L_s p i_{qs} + \omega_e \sigma L_s i_{ds} + \omega_e \tfrac{L_m}{L_r} \lambda_r \end{aligned}\right\} \tag{6}$$

$$\tau_r \frac{d\lambda_r}{dt} + \lambda_r = L_m i_{ds}\ ;\ \tau_r = \frac{L_r}{R_r}\ ;\ \sigma = 1 - \frac{L_m^2}{L_s L_r} \tag{7}$$

The torque Equation (5) can be rewritten as (8) in this case. Other forms of the torque equation can be obtained by combining (2)–(4) and (5) and will be the basis of different control strategies, as will be discussed in further sections.

$$T_e = \frac{3}{2} P \frac{L_m}{L_r} \lambda_r i_{qs} \tag{8}$$

2.2. Machine Characteristics

Traction drives commonly receive a torque command from an outer control loop, which is responsible for speed control. The maximum torque that can be produced at a given speed will essentially depend on the current limits of the machine and power converter (due to losses) and on the maximum flux, which is limited by saturation and the available DC link voltage. For most IM designs, the maximum voltage and field weakening occur at the same speed, i.e., field weakening is a direct consequence of reaching the voltage limit. This is shown schematically in Figure 3 (continuous line case). For rotor speeds $\omega_r < \omega_1$, the machine operates with a rated flux and current, with the voltage increasing proportionally to the rotor speed, mainly due to the back-emf. If $\omega_r > \omega_1$, the flux, and consequently torque, must be decreased. The current (Figure 3b) can still be maintained at its rated value until the machine enters field-weakening region II (not shown in the figure) [10]. Therefore, for the machine denoted as conventional in Figure 3, region ① corresponds to a constant torque operation, while regions ②+③ have constant power.

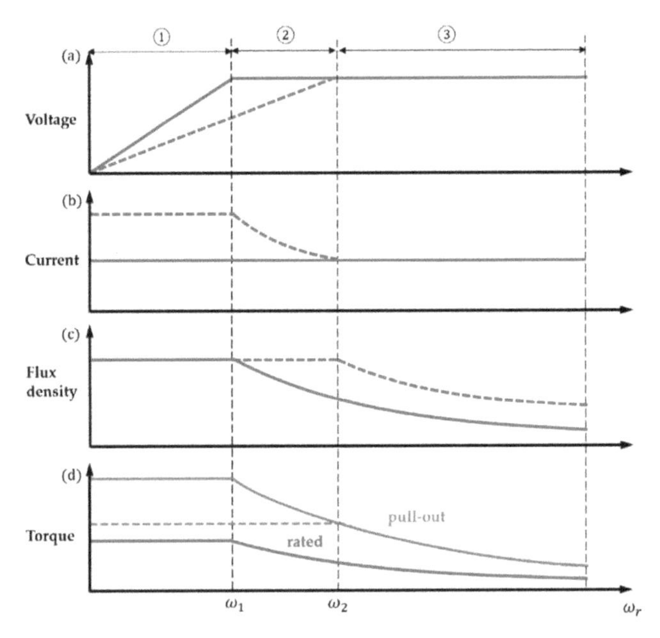

Figure 3. Conventional (–) and extended full flux range (- -) induction motor (IM) design behavior: (**a**) Stator voltage magnitude; (**b**) Stator current magnitude; (**c**) Flux density; (**d**) Electromagnetic torque (rated&pull-out). Both machines are designed to provide the same torque vs. speed characteristic and have the same voltage limit.

IM designs for railway traction are often aimed at reducing the size of the machine, which can be desirable or even imperative due to room constraints. For this purpose, the voltage characteristic of the conventional design in Figure 3 can be modified by rewinding the stator, varying the number of turns, and gauging the wire [10,11]. If the modification is made such that $N_2 < N_1$, with N_1 and N_2 being the number of turns for the conventional and modified designs, respectively, and the active conductor area in each stator slot remaining unchanged, i.e., $N_1 \cdot S_1 = N_2 \cdot S_2$, and S_1 and S_2 being the area of the conductor for the conventional and modified designs, respectively, both machines should be able to produce the same amount of torque, as the total current circulating within the stator slots and the rest of the machine dimensions are the same in both cases [11]. Since the number of turns has been reduced, the voltage vs. speed characteristic is also modified. As seen in Figure 3a (dashed line), for $\omega_r = \omega_1$, the machine is far from its voltage limit. It can also be observed that for $\omega_r < \omega_1$, the current of the modified machine design is N_1/N_2 larger than for the conventional design. This does not imply an increase of joule losses, as the wire in the modified design is thicker, and the current density is the same in both cases. Since, at $\omega_r = \omega_1$, the modified machine operates well below its voltage limit, there is no need to decrease the flux at this point; instead, the nominal air gap flux density can be maintained until $\omega_r = \omega_2$ (region ② in Figure 3), i.e., the full flux range is extended. The fact that the flux weakening region is reduced while the torque characteristic remains unchanged enables a reduction of the stator current for $\omega_r > \omega_1$, as can be readily deduced from (8). Consequently, assuming that the dimensions of the machine do not change, the extended full flux range design in Figure 3 would allow a significant decrease of the current density in regions ② and ③ (i.e., at high train speeds) and consequently of Joule losses, i.e., would be more efficient compared to the conventional design.

However, the design with an extended full flux range offers other possibilities. The torque of an IM can be written as (9), where V_{rotor} is the active volume of the rotor, J is the stator surface current density, B is the air gap flux density, $\varnothing$ is the angle between J and B vectors, and k_1 is a constant which depends on the machine winding design [11].

$$T_e = k_1 \cdot V_{rotor} \cdot J \cdot B \cdot cos(\varnothing) \quad (9)$$

As the extended full flux range design provides higher flux densities at high speeds and the current density J remains constant, it is possible to reduce the volume of the rotor, and consequently the size of the machine, without affecting the torque production capability, i.e., the extended full flux design in Figure 3 will be smaller.

It must be noted, however, that redesigning the machine brings drawbacks that must also be considered. First, the size of the inverter is increased, as the current that the semiconductors must handle is increased by a factor of N_1/N_2, while the voltage and power remain unaffected. However, this penalty is not so relevant nowadays thanks to the latest developments in power devices [10]. Second, the pull-out torque in the low-speed region is significantly decreased, as shown in Figure 3d [11], which must be considered to guarantee that the machine meets the application requirements.

3. Overview of Control Methods for Three-Phase Induction Machines

This section discusses control strategies for IMs in railway applications. The drives must be able to perform properly from zero to relatively high rotational frequencies. On the other hand, the switching frequencies are often limited to several hundred Hz due to the switching losses of high-power semiconductor devices. At low rotational frequencies, the switching to fundamental frequency ratio is still relatively large and the inverter will operate far from its voltage limit. On the contrary, operation at high speeds is characterized by a reduced switching to fundamental frequency ratio and a reduced (or even inexistent) voltage margin in the inverter. Due to this, both control and modulation strategies are often dynamically modified, depending on the IM speed. The following discussion will primarily focus on the most challenging high-speed case.

Control methods for IMs can be classified into scalar and vector types, as shown in Figure 4. Scalar methods are derived from the machine equivalent circuit in a steady-state. Consequently, they can operate properly in applications in which fast changes in the operating conditions of the machine (torque, speed, flux, ...) are not required. On the contrary, vector control methods are based on the dynamic equations of the machine, which, combined with proper control loops, allow the machine's torque capabilities to be fully exploited, without surpassing machine or power converter limits. Both types of methods are briefly discussed in the following. It must be noted, however, that the borderline between scalar- and vector-based methods is sometimes blurred, as there have been several proposals to enhance the dynamic response of scalar methods by adding control loops based on dynamic models.

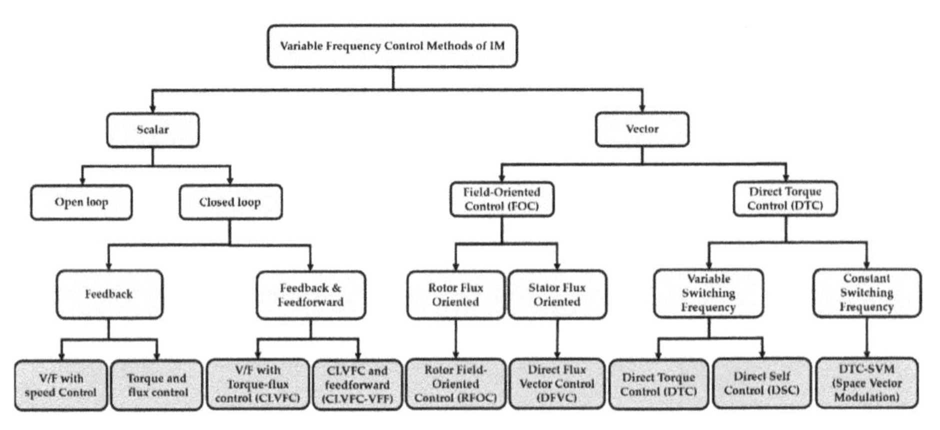

Figure 4. Control methods for IMs.

3.1. Scalar-Based Control

3.1.1. Open-loop V/F

Open-loop V/F varies the stator voltage magnitude proportional to the frequency. This results in an (almost) constant flux. While simple, V/F control has some relevant limitations. The rotor speed is not precisely controlled due to slip. Additionally, an incorrect voltage to frequency ratio, voltage drop in the stator resistance, variations of the DC link voltage feeding the inverter, etc., will result in incorrect flux levels, eventually modifying the operating point of the machine from the desired value.

3.1.2. V/F with Feedback Control

Closed-loop speed control with slip regulation (Figure 5) has been widely used in IM traction drives [12]. Speed error generates the slip command ω_{sl}^* through a Proportional-Integrator (PI) controller, which, when added to the measured speed, provides the angular frequency of the stator voltage ω_e^*.

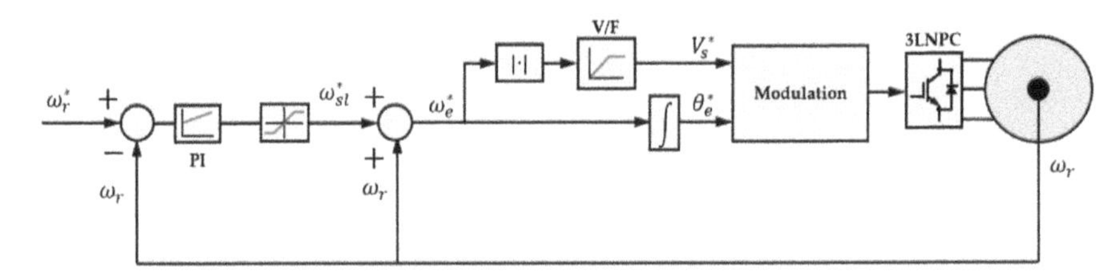

Figure 5. V/F with speed control scheme.

Flux and torque control loops can be used instead of the V/F ratio to obtain the desired stator voltage magnitude and angle (Figure 6). Torque and flux can be estimated from the (commanded) stator voltages and the (measured) stator currents using the voltage model (10); "^" indicates estimated variables/parameters. The pure integrator in (10) is replaced in practice by a first-order system to avoid the drift problems derived from the integrator infinite gain at DC [13]. The torque is obtained using (11).

$$\hat{\lambda}_{\alpha\beta s} = \int \left(V_{\alpha\beta s}^* - \hat{R}_s i_{\alpha\beta s}\right) dt \tag{10}$$

$$\hat{T}_e = \frac{3}{2} P\left(\hat{\lambda}_{\alpha s} i_{\beta s} - \hat{\lambda}_{\beta s} i_{\alpha s}\right) \tag{11}$$

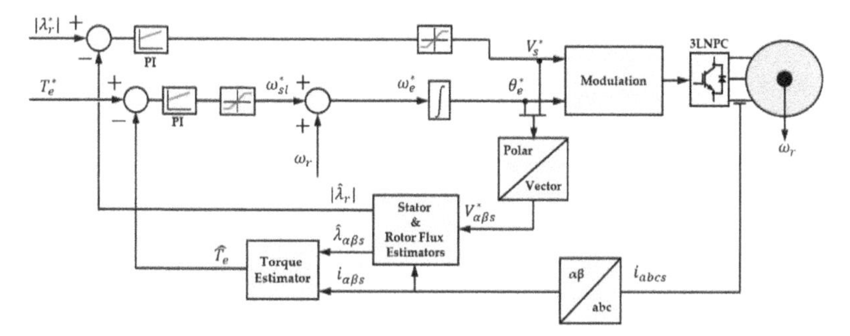

Figure 6. Torque-flux scalar control scheme.

The methods in Figures 5 and 6 are relatively simple to implement, with the second enabling precise control of the machine's operating point in a steady-state. A further advantage of scalar methods is that operation near or at the inverter voltage limit is relatively easy to achieve. However, the fact that coupling between flux and torque is not considered for the control design requires a very slow dynamic response to avoid over currents and torque pulsations.

3.1.3. Torque/Flux Scalar Control with Feedforward

The dynamic response of the closed-loop V/F control scheme in Figure 6 can be enhanced by adding two feedforward terms, as can be seen in Figure 7. The first uses the desired V/F characteristic to provide the base value of the stator voltage magnitude $V^*_{s\_vf}$, with the rotor flux regulator providing the incremental voltage required to track the desired rotor flux with no error. The second provides the base value for the slip $\omega^*_{sl\_ff}$, which is obtained from the desired torque and the estimated rotor flux using (12). The torque regulator corrects the slip so that the desired torque is followed with no error.

$$\omega^*_{sl\_ff} = \frac{2}{3}\frac{1}{P}\frac{\hat{R}_r}{|\hat{\lambda}_r|^2}T^*_e \tag{12}$$

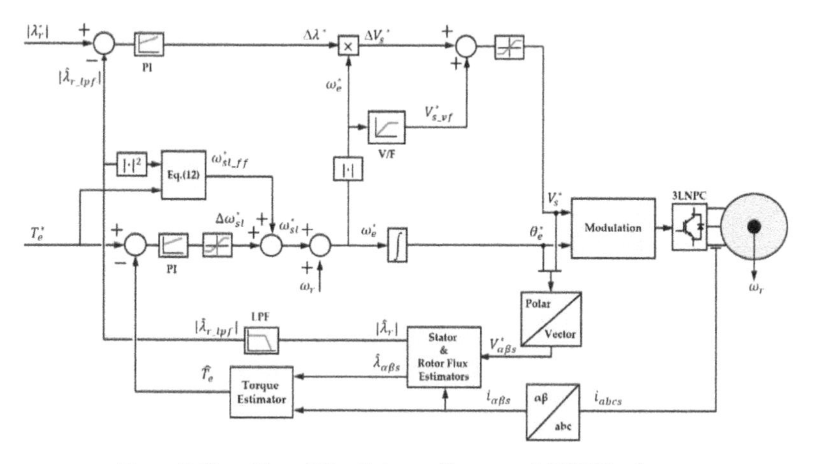

Figure 7. Closed-loop V/F with torque/flux control (CLVFC) scheme.

Due to the fact that the voltage command magnitude and phase angle are independently controlled, flux and torque controllers must be tuned for relatively low bandwidths. A dynamic response eventually relies on the accuracy of the feedforward terms. As for the scheme in Figure 6, the scheme in Figure 7 can easily operate in the field-weakening region, including that of six-step.

An alternative approach for the implementation of the feedforward action is to use the machine *d-q* model in the rotor flux reference frame. The desired d- and q-axis currents are first obtained from the commanded torque and rotor flux using (7) and (8) (see Figure 8). The d- and q-axis stator voltages required to achieve the desired currents are the middle terms in (13), which are obtained from (6).

$$\left.\begin{array}{l} v^*_{ds\_ff} = \hat{R}_s i^*_{ds} + \hat{\sigma}\hat{L}_s p i^*_{ds} - \omega^*_e \hat{\sigma}\hat{L}_s i^*_{qs} + \frac{\hat{L}_m}{\hat{L}_r} p \lambda^*_{dr} \cong -\omega^*_e \hat{\sigma}\hat{L}_s i^*_{qs} \\ v^*_{qs\_ff} = \hat{R}_s i^*_{qs} + \hat{\sigma}\hat{L}_s p i^*_{qs} + \omega^*_e \hat{\sigma}\hat{L}_s i^*_{ds} + \omega^*_e \frac{\hat{L}_m}{\hat{L}_r} \lambda^*_{dr} \cong \omega^*_e \hat{\sigma}\hat{L}_s i^*_{ds} + \omega^*_e \frac{\hat{L}_m}{\hat{L}_r} \lambda^*_{dr} \end{array}\right\} \quad (13)$$

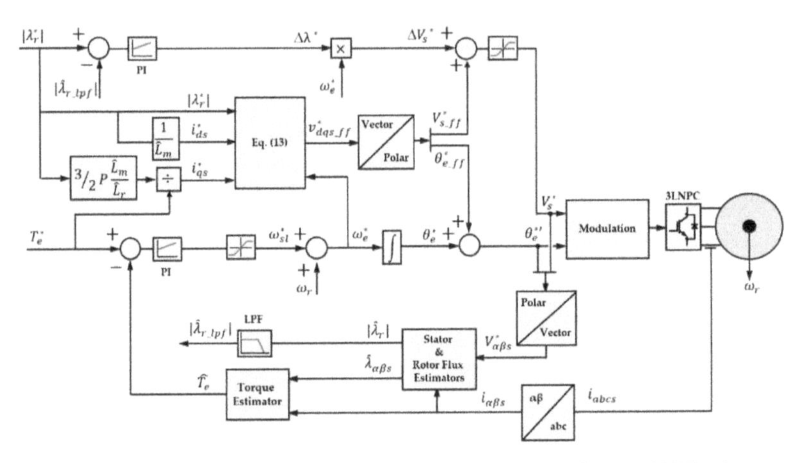

Figure 8. Closed loop V/F with torque/flux control and feedforward (CLVFC&FF) scheme.

Ideally, (13) will produce the voltage needed to obtain the desired torque and rotor flux with no error. However, there are a number of issues to consider. Mismatch between model and actual parameters must be expected and will produce errors in the feedforward voltages. In addition, (13) includes derivatives which are problematic in practice. It is noted, however, that the signals affected by the derivative (13) are (clean) commanded variables, i.e., do not involve (noisy) measured variables. Furthermore, the torque derivative will be limited by the application, meaning that the derivative of q-axis current and flux commands will be limited too. Finally, the terms depending on the stator resistance will have a reduced weight considering that the control is only intended to operate at a high speed. Based on the previous considerations, the feedforward voltage can be safely simplified to form the right hand of (13). The resulting block diagram is shown in Figure 8, with the feedforward term being either the complete or simplified voltage equation in (13).

3.2. Vector-Based Control

Vector control methods are aimed at directly manipulating the IM fields and torque. These methods are based on well-known *d-q* models. Field-Oriented Control (FOC) represents flux and torque as a function of stator currents in a synchronous reference frame, with high-bandwidth current regulators being used to provide the voltage command to the inverter. Alternatively, Direct Torque Control (DTC) methods implement torque and flux controllers which directly provide the IGBT gate signals for the inverter, i.e., without the explicit control of stator currents.

3.2.1. Rotor Field-Oriented Control (RFOC)

RFOC (see Figure 9) is one of the most popular options for the high-performance control of IM drives [14,15], although its discussion is beyond the scope of this paper. RFOC is often used in HST at relatively low speeds, the inverter operates in the linear region and with an adequate switching to fundamental frequency ratio. However, its use at high speeds presents multiple problems, including the lack of a voltage margin in the inverter for proper operation of the current regulator, distortions in the currents due to overmodulation, and delays intrinsic to the reduced switching frequency. The modification of RFOC to enable operation at the voltage limit was discussed in [16].

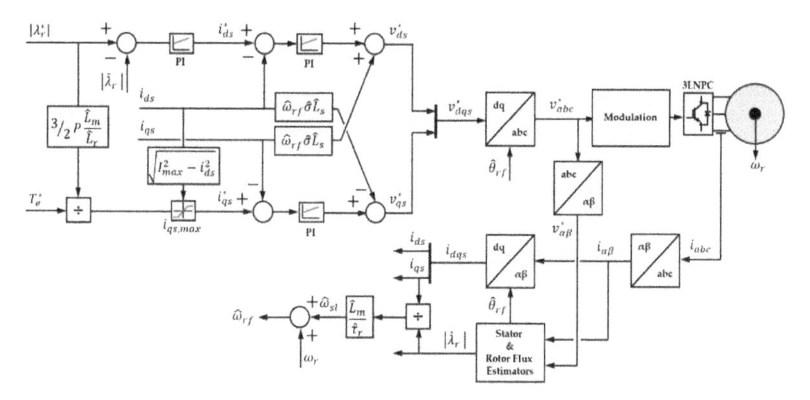

Figure 9. Rotor field-oriented control (RFOC) scheme.

3.2.2. Direct Flux Vector Control (DFVC)

DFVC [17] is a stator-flux-oriented control approach. By writing the voltage Equation (1) in stator flux, reference frame (14) can be obtained. It can be observed from (14) that the stator flux variation can be regulated through the d-axis voltage, and the torque is then controlled through the q-axis current (15), with a current regulator being used for this purpose. The DFVC scheme is shown in Figure 10.

$$\left.\begin{array}{l} v_{ds} = \hat{R}_s i_{ds} + p\hat{\lambda}_{ds} \\ v_{qs} = \hat{R}_s i_{qs} + \hat{\omega}_{sf}\hat{\lambda}_{ds} \end{array}\right\} \quad (14)$$

$$T_e = \frac{3}{2} P \hat{\lambda}_{ds} i_{qs} \quad (15)$$

Stator flux $\alpha\beta$-components are estimated from the voltage-model-based flux estimator. The synchronous frequency can be obtained from the estimated stator flux and back-emf (16) [18], avoiding the use of stator flux angle derivative and time-consuming trigonometric functions.

$$\hat{\omega}_{sf} = p\hat{\theta}_{sf} = \frac{d}{dt}\left[tan^{-1}\left(\frac{\hat{\lambda}_{\beta s}}{\hat{\lambda}_{\alpha s}}\right)\right] = \frac{\hat{\lambda}_{\alpha s}\cdot\hat{e}_{\beta s} - \hat{\lambda}_{\beta s}\cdot\hat{e}_{\alpha s}}{|\hat{\lambda}_s|^2} \quad (16)$$

At low speeds, DFVC can operate either with rated stator flux or a maximum torque per ampere (MTPA) strategy to improve the efficiency. Above the base speed, flux is reduced according to (17), where V_{max} is the maximum output voltage of the inverter, which depends on the available DC-link voltage and the modulation method. Operation in overmodulation is feasible, but a voltage margin must be preserved for proper operation of the q-axis current regulator, meaning that operation with a maximum output voltage (i.e., six-step) is not possible. Furthermore, operation in overmodulation forces a reduction of the current regulator bandwidth to mitigate the effects of the resulting current harmonics in this case. Therefore, current regulator gains may need to be adapted with machine speeds.

$$\lambda_s^* \leq \frac{V_{max} - \hat{R}_s i_{qs}}{|\hat{\omega}_{sf}|} \quad (17)$$

It is finally noted that Figure 10 includes a mechanism to limit the torque angle δ between the stator and rotor fluxes so that it is smaller than the pull-out torque angle of $\delta = 45$ electrical degrees [17].

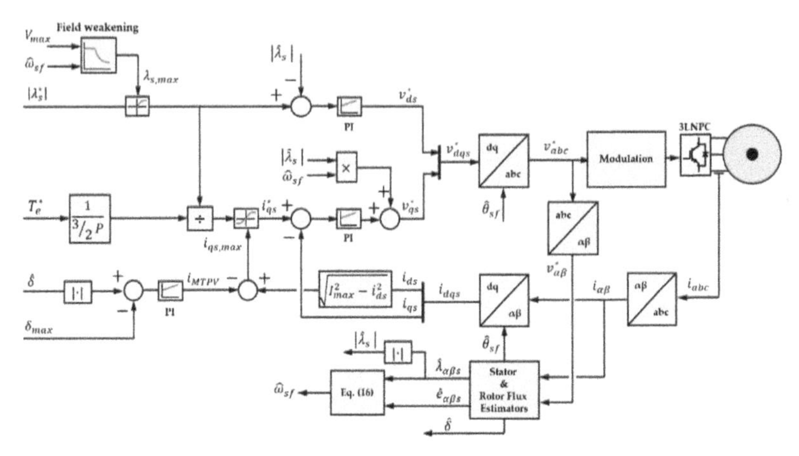

Figure 10. Direct Flux Vector Control (DFVC) scheme.

3.2.3. Direct Torque Control (DTC)

IM torque can be expressed as (18), with δ being the torque angle. DTC methods control torque by controlling the stator flux magnitude and angle with respect to rotor flux. Stator flux is controlled through the stator voltage (19) (stator resistance neglected), with V_s being the inverter output voltage vector.

$$T_e = \frac{L_m}{\sigma L_s L_r} \lambda_s \lambda_r sin(\delta) \quad (18)$$

$$\lambda_s = \int V_s dt \quad (19)$$

Switching-Table-Based (ST-DTC) was introduced by Takahashi and Noguchi [19] in the mid-1980s. Two hysteresis controllers are used to control the stator flux and torque directly. The hysteresis control signals are sent to a look-up table to select the voltage vectors required to achieve high dynamics (see Figure 11). The fact that the switching frequency is not defined and operation in overmodulation and six-step is not straightforward makes this method inadequate for high-power railway drives [20].

Direct-Self Control (DSC) was proposed by Depenbrock [21] for high-power drives. Three hysteresis controllers determine the voltage applied to the machine by comparing a flux magnitude command with the estimated flux for each phase, and a two-level hysteresis torque controller determines the amount of zero voltage (see Figure 12). DSC produces a hexagonal stator flux trajectory, which enables a smooth transition into overmodulation and eventually six-step. However, hexagonal flux trajectories make DSC problematic below ≈30% of the base speed, and remedial actions can be found in [22,23].

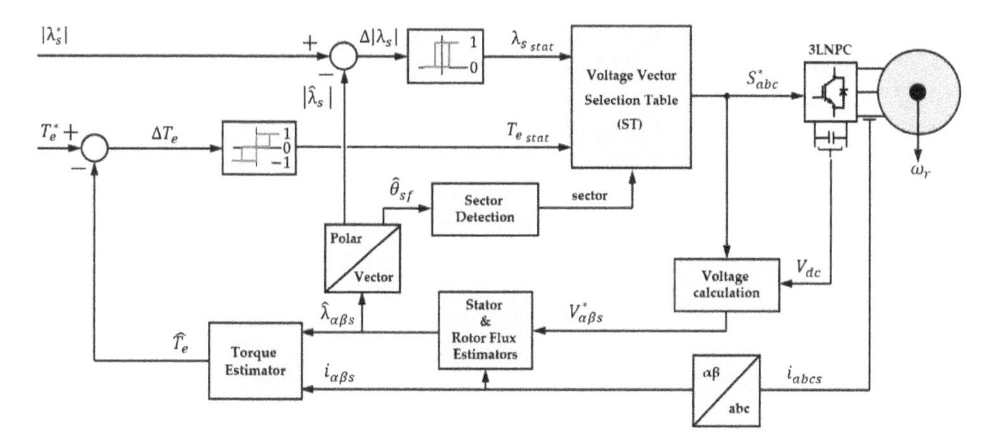

Figure 11. Switching-Table-Based Direct Torque Control (ST-DTC) scheme.

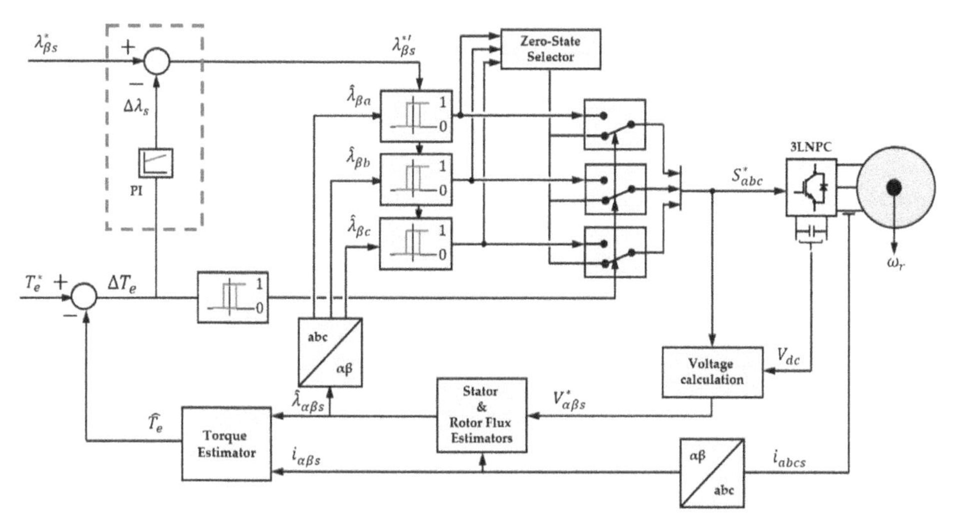

Figure 12. Direct-Self Control (DSC) scheme.

Several modifications have been proposed to overcome the limitations of DTC methods [24]. DTC with a constant switching frequency calculates the required stator voltage vector over a sampling period to achieve the desired torque and stator flux. The voltage vector is synthesized using Space-Vector Modulation (SVM), and these methods are often referred to as DTC-SVM. In the implementation in Figure 13, a PI controls the torque through the torque angle [25]. The stator flux angle is obtained from the estimated rotor flux angle and the commanded torque angle. The stator voltage vector command $V_{\alpha\beta s}^*$ employed to cancel the stator flux error $\Delta\lambda_{\alpha\beta s}^*$ at the end of the next sampling period Δt is obtained as:

$$V_{\alpha\beta s}^* = \frac{\Delta\lambda_{\alpha\beta s}^*}{\Delta t} + \hat{R}_s i_{\alpha\beta s}. \tag{20}$$

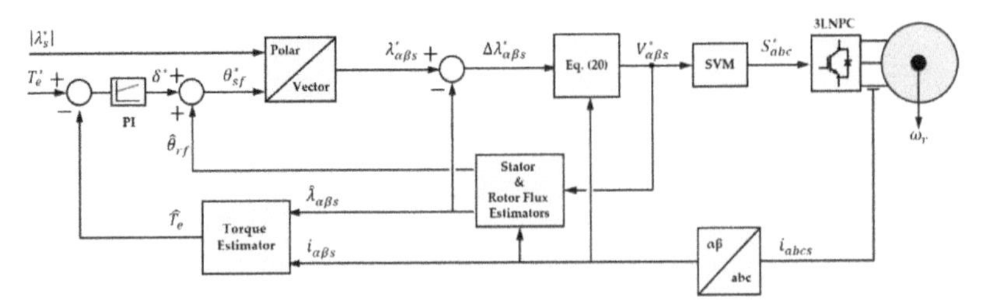

Figure 13. Direct-Torque Control Space-Vector Modulation (DTC-SVM) block diagram.

The scheme in Figure 13 is easy to implement and retains the fast dynamics of DTC if the inverter operates in the linear region. However, voltage distortions intrinsic to overmodulation can result in magnitude and phase deviations of the actual stator flux vector, leading to instability problems. Furthermore, (20) effectively cancels the flux error for relatively small values of Δt, but can result in large steady-state errors in the case of low switching frequencies. DTC-SVM suffers from the same limitation as ST-DTC when operating in overmodulation and six-step, which raises concerns on their use for high-power, high-speed railway traction drives. A predictive term for mitigating the stator flux delay and extending the operation to six-step was proposed in [26]. However, this was at the price of a significant complexity increase.

3.3. Control Strategies Summary

Table 1 summarizes the main conclusions for the control methods discussed in this section, including controlled variables and the easiness of operation at low speeds, overmodulation (high speed), and the transition to six-step. Regarding the dynamic response, it is important to note that the torque ramp is normally limited in railway traction. Consequently, not only the maximum dynamic response (e.g., the minimum time required to respond to a step-like torque command) is relevant, but also the capability of the drive to meet the maximum torque ramp requested by the application, especially when the machine operates at a high speed in the field-weakening region. CLVFC, CLVFVC&FF, DFVC, and DTC-SVM have been selected as a representative subset of the methods in Table 1, and their behavior will be analyzed by means of simulation in Section 6.

Table 1. Summary of the presented control schemes for traction applications.

Properties/ Performance	V/Hz with Feedback			FOC		DTC		
	V/Hz (Figure 5)	**CLVFVC (Figure 7)**	**CLVFVC&FF (Figure 8)**	**RFOC (Figure 9)**	**DFVC (Figure 10)**	**DTC (Figure 11)**	**DSC (Figure 12)**	**DTC-SVM (Figure 13)**
Reference frame	λ_r	λ_r	λ_r	λ_r	λ_s	SRF.	SRF	λ_r
Controlled variables	ω_r †	λ_r; T_e	λ_r; T_e	λ_r&i_{ds}; i_{qs}	λ_s; i_{qs}	λ_s; T_e	λ_s; T_e	λ_s; T_e
Defined switching frequency	Yes	Yes	Yes	Yes	Yes	No	No	Yes
Low speed (linear mod.)	✓	✓	✓	✓	✓	✗	✗	✓
High speed (overmodulation)	✓	✓	✓	—	—	✗	✓	✗
Six-step operation	✓	✓	✓	✗	✗	✗	✓	✗
Dynamic response ††	✗/✗	✗/—	—/✓	✓/—	✓/—	✓/✗	✓/✓	✓/✗

✓: favorable; —: neutral; ✗: unfavorable; "SRF" stands for stationary reference frame. †: Implementation of an outer speed control loop for the rest of the methods is straightforward. ††: (1) maximum torque dynamic response/ (2) capability to provide 3 kNm/s in the overmodulation region.

4. Modulation Techniques

High-power traction drives usually operate with low switching frequencies (<1 kHz) to reduce switching losses. This results in significant current and consequently torque ripples, which can have implications for mechanical transmission stress, train comfort, standards compliance, etc. Trading-off switching losses and torque pulsations is a challenge for the selection of modulation methods. Furthermore, modulation and control strategies often change with the output frequency. Figure 14 shows an example of this [27]. Asynchronous Pulse-Width Modulation (PWM) is used at low speeds, changing to synchronous modulation with Selective Harmonic Elimination (SHE) and finally single pulse modes as the speed increases. The three options are briefly described in the following, and are particularized for a three-level Neutral-Point-Clamped (3L-NPC) scheme [28], as this is the configuration used in this project.

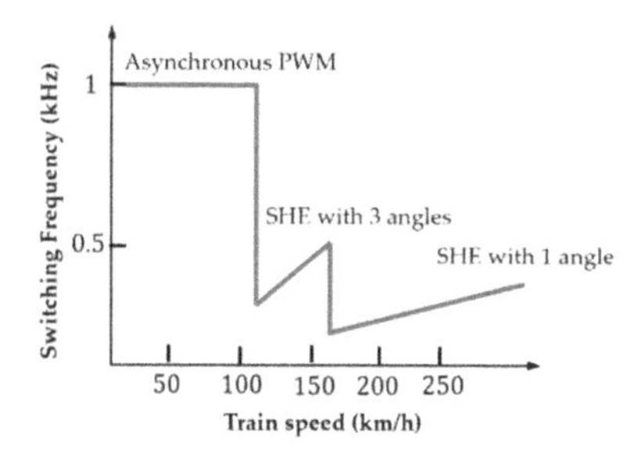

Figure 14. Modulation technique vs. train speed for a High-Speed Train (HST).

4.1. Asynchronous Modulation

Carrier-Based Pulse-Width Modulation (PWM) or Space-Vector Modulation (SVM) can be used at low speeds. The first compares the reference voltages V^*_{abc} with two carriers, as shown in Figure 15a. A level-shifted carrier is normally preferred as it results in a lower voltage harmonic content [28]. A common-mode (homopolar) voltage should be added to fully use the available DC link voltage. Space-Vector Modulation (SVM) for three-level inverters shares the same basic principles as that for two-level inverters, but 24 active voltage and three zero vectors are available. The implementation of SVM is shown in Figure 15b. It typically consists of three steps: (1) sector identification, (2) region identification, and (3) the selection of an appropriate switching sequence. Redundant states are used to balance DC link capacitor voltages. SVM offers the same DC voltage utilization as the PWM with a homopolar voltage, and it has a larger computational burden, but makes better use of the redundant states [29,30].

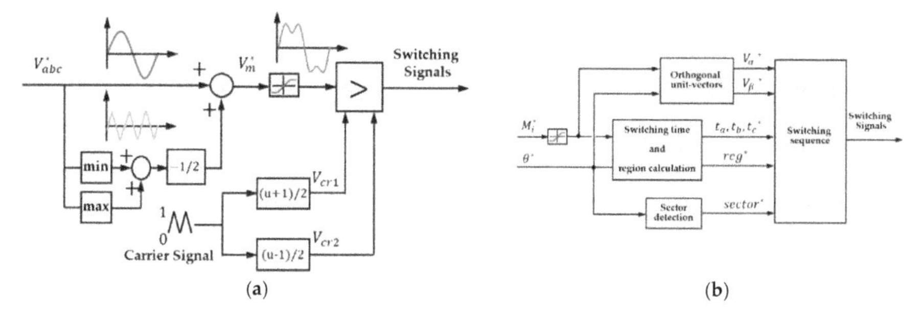

Figure 15. Asynchronous modulation techniques: (**a**) Pulse-Width Modulation (PWM) with triple harmonic injection; (**b**) Space-Vector Modulation (SVM).

4.2. Synchronous Modulation—Selective Harmonic Elimination (SHE)

SHE performs a predefined number of commutations per quarter of the fundamental cycle. Commutations are synchronized with the fundamental wave. Commutation angles are pre-calculated via Fourier analysis [31], with the aim of eliminating specific harmonics of the output voltage. An example of SHE with three switching angles is shown in Figure 16a. With three angles, it is possible to cancel two harmonics of the output voltage (typically the 5th and 7th), in addition to controlling the magnitude of the fundamental voltage. As the speed increases, SHE changes to one pulse mode (Figure 16b) to reduce switching losses. SHE implementation is schematically shown in Figure 16c.

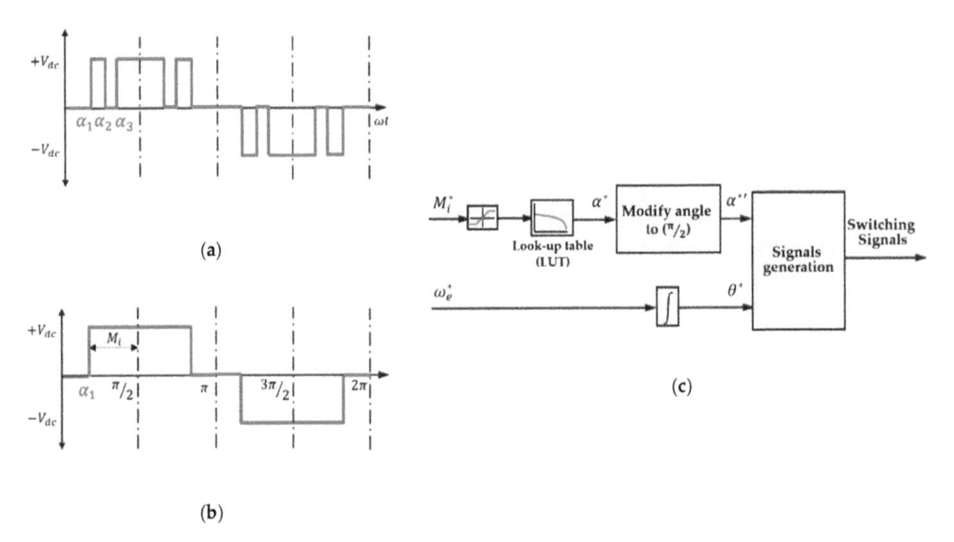

Figure 16. Selective Harmonic Elimination (SHE) for three-level Neutral-Point-Clamped (3L-NPC): (**a**) Phase voltage for the case of three switching angles; (**b**) phase voltage for the case of one switching angle; (**c**) Selective Harmonic Elimination (SHE) block diagram.

5. Operation with Reduced Flux and Remagnetization Strategies

Electric drives in high-speed traction applications can work for certain periods of time with light loads. It is possible in this case to decrease the flux level to reduce the stator current and consequently Joule losses, which is commonly termed MTPA [32]. However, operating with reduced flux levels will penalize the dynamic response of the drive. If a torque increase is demanded, the machine must be remagnetized first. The remagnetization time is determined by the rotor time constant (7) and applied magnetizing current. Due to the relatively large values of the rotor time constant, fast torque changes of torque are not feasible. It must be noted, however, that fast torque changes are not desirable for traction applications, as they might exert stress on the mechanical transmission, produce wheel slip, and raise comfort concerns. The maximum torque-allowed gradient will depend on the application. For the machine considered in this paper, it has a value of 3 kNm/s.

Figure 17 shows two possible remagnetization strategies. RFOC principles are used for the discussion. It is noted, however, that equivalent strategies can be used with other control methods by simply transforming *d*-*q* axis current commands into other commands, e.g., stator flux and slip.

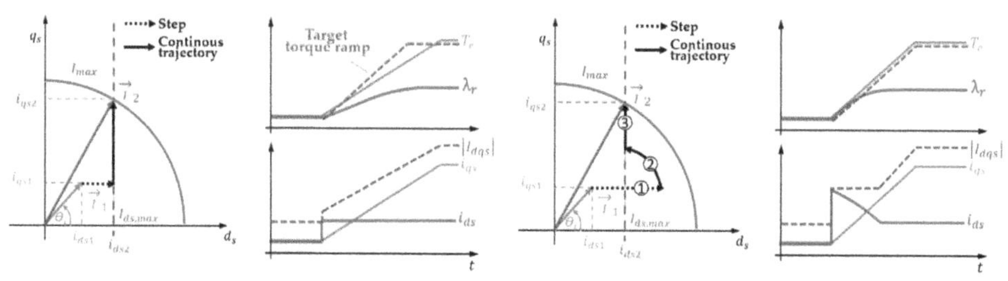

(**a**) Profile 1: step-like d-axis, ramp-like q axis (**b**) Profile 2: 3kNm/s torque ramp, minimum current

Figure 17. Remagnetization (**a**) using the rated d-axis current and (**b**) target torque gradient. Left: Current trajectories in the *d-q* plane; right: torque, rotor flux, and current trajectories vs. time.

The strategy in Figure 14a uses a step-like d-axis current command. While simple, this results in a slow remagnetization, with the 3 kNm/s target not being achieved. The option in Figure 14b is derived from the method described in [33] and is aimed at providing a target torque ramp with the smallest possible current during the remagnetization process. This reduces the stress in the power devices and the risk of surpassing their current limit. This strategy will be used for the simulation results in the next section.

6. Simulation Results

Selected control methods from Section 3 have been evaluated by means of simulation using MATLAB/Simulink. IM parameters for the base speed are given in Table 2. The simulation model implements asynchronous SVM with a switching frequency of 1 kHz at low speeds and SHE at high speeds, as shown in Figure 14.

Table 2. Specifications of the induction motor at base speed ω_{base} (extended full flux range design).

Variable	Value	Unit
DC-link voltage	3600	V
Rated Power	1084	kW
Rated Voltage (L-L, rms)	2727	V
Pole-pairs (P)	2	Pole
Stator resistance (R_s)	55.38	mΩ
Stator inductance (L_s)	26.45	mH
Torque	3241	Nm
Speed	3194	rpm

Since the main focus of this paper is high-speed operation, only results at high speed using SHE are provided in this section. Infinite inertia is assumed. Consequently, the rotor speed remained constant throughout the simulation. This assumption is realistic and has no effect on the conclusions. Profile 2 in Figure 17b was used during remagnetization. The maximum torque ramp was limited to 3 kNm/s, which was imposed by the application. Simulation results are shown in Figure 18.

The most remarkable difference is the slowest transient response of CLVFC due to dynamic limitations intrinsic to scalar control. The dynamic response is seen to improve and be comparable to the other methods when the feedforward defined by (13) is used (CLVFC&FF in Figure 18b).

DFVC and DTC-SVM are seen to provide similar dynamic responses to CLVFC&FF. Regarding DFVC, it must be noted that to achieve proper operation in the overmodulation region, the q-axis current regulator bandwidth was reduced in the range of ten times to avoid a current regulator reaction to low-order current harmonics due to the non-linear operation of the inverter. The need to dynamically adapt the gains of the current regulator in the high-speed region is an obvious concern.

It can be observed that DTC-SVM suffers from a steady-state error in the controlled flux due to the low sampling frequency (Δt) when SHE is used in the inverter. This results in an increase in the load angle. This could lead to overcurrent or instability if the load angle is not monitored.

Figure 19 summarizes the performance in a steady state for the four control methods, i.e., once the machine is providing its maximum torque. The steady-state error in the flux for DTC-SVM is seen to affect the modulation index and slip. This will eventually affect the machine loss distribution, which is a concern as traction motors can be required to operate close to their thermal limit. CLVFC and CLVFC&FF are seen to have a higher torque error compared to DFVC, but with little impact on the modulation index and slip. It is noted that a torque error in the range of 1% is perfectly assumable.

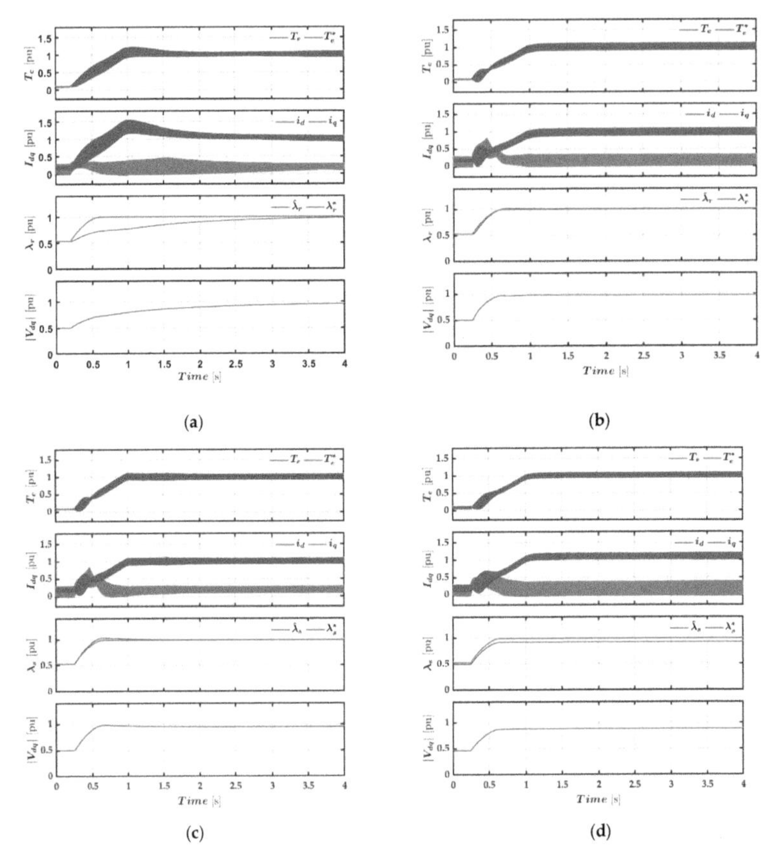

Figure 18. Simulation results of using (**a**) CLVFC, (**b**) CLVFC&FF, (**c**) DFVC, and (**d**) DTC-SVM control methods with SHE. Rotor speed $\omega_r = 1.328\omega_{base}$; torque was increased from 10% (i.e., with the machine operating with reduced flux in MTPA) to 100%. From top to bottom: commanded and actual torque; d- and q-axis currents; commanded and estimated flux (can be stator or rotor flux, depending on the method); and output voltage magnitude. All the variables are shown in p.u.

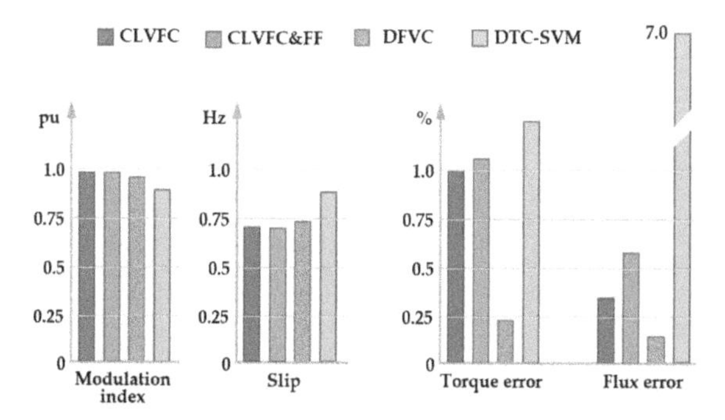

Figure 19. From left to right: modulation index, slip, torque error, and error in the flux being controlled for the four control methods being considered, once the machine has reached a steady state, i.e., is its maximum torque. Torque and slip have been low-pass filtered to eliminate the harmonic content produced by SHE modulation.

It can be concluded that CLVFC&FF is more adequate compared to the simulated schemes at high speeds due to its high dynamics, and the controllers are not affected by low-order harmonics resulting from a square-wave operation, i.e., six-step, as in the case of DFVC, and are simple to implement.

7. Experimental Results

A schematic diagram of the high-power traction system test bench is shown in Figure 20a. It consists of two identical IMs and converters connected back-to-back, which are supplied from a High-Voltage (HV) DC power supply. The power converter module (see Figure 20b) consists of a three-phase, three-level Neutral-Point Clamped (NPC) inverter feeding the IMs. Single-phase inverters feed auxiliary loads, such as cooling systems, control power supply units, etc. A DC-DC chopper is implemented for dissipative braking and DC bus overvoltage protection. A specially designed traction transformer is used to filter off catenary harmonics and allow the interconnection of the different converters. A 100 Hz (2f) filter is included in the DC bus. The overall experimental test rig is shown in Figure 20c.

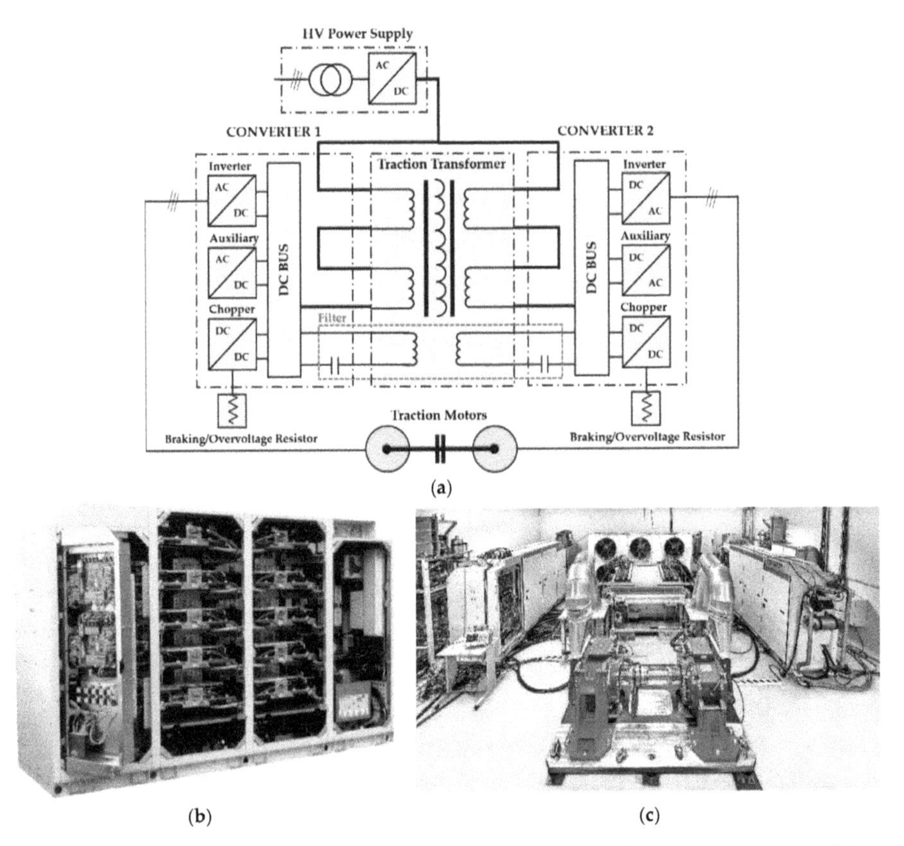

Figure 20. High-power traction test bench: (**a**) Schematic diagram; (**b**) power converter module (INGETRAC); and (**c**) overall view of the laboratory setup.

Preliminary experimental results for a full-scale HS traction drive are presented in the following. The control uses RFOC at low speeds and CLVFC at high speeds. The main system parameters are the same as those used in the simulation shown in Table 2. The torque-flux characteristic of the motor is of the type named as the extended full flux range in Figure 3.

Figure 21a shows the rotor speed, modulation index, commanded and estimated torques, estimated rotor flux, and magnitude of the stator current vector during an acceleration (left) and deceleration (right) process. Figure 21b shows the spectrogram of the stator current vector. For $\omega_r < 0.12$ p.u., RFOC-SVM with a switching frequency of 850 Hz is used; the switching frequency increases to 1 kHz for $0.12 < \omega_r < 0.94$ p.u. For $\omega_r > 0.94$ p.u., CLVFC combined with SHE with one switching angle is used. Changes in the modulation method are readily observable in the spectrogram of Figure 21b, and are aim to trade-off switching losses and torque ripple. The control is seen to precisely follow the commanded torque and rotor flux in the whole speed range. It is noted that the changes in the estimated rotor flux observed in the flux-weakening region respond to changes in the corresponding command (not shown in the figure). Transitions between the different control and modulation strategies can be a challenge due to the high power and low switching frequencies. However, as can be observed from Figure 21, such transitions are satisfactory, i.e., the spikes observed in the currents are perfectly acceptable and do not represent a risk for the power devices. Implementation of the other control methods and remagnetization strategies discussed in Sections 3 and 6 is ongoing.

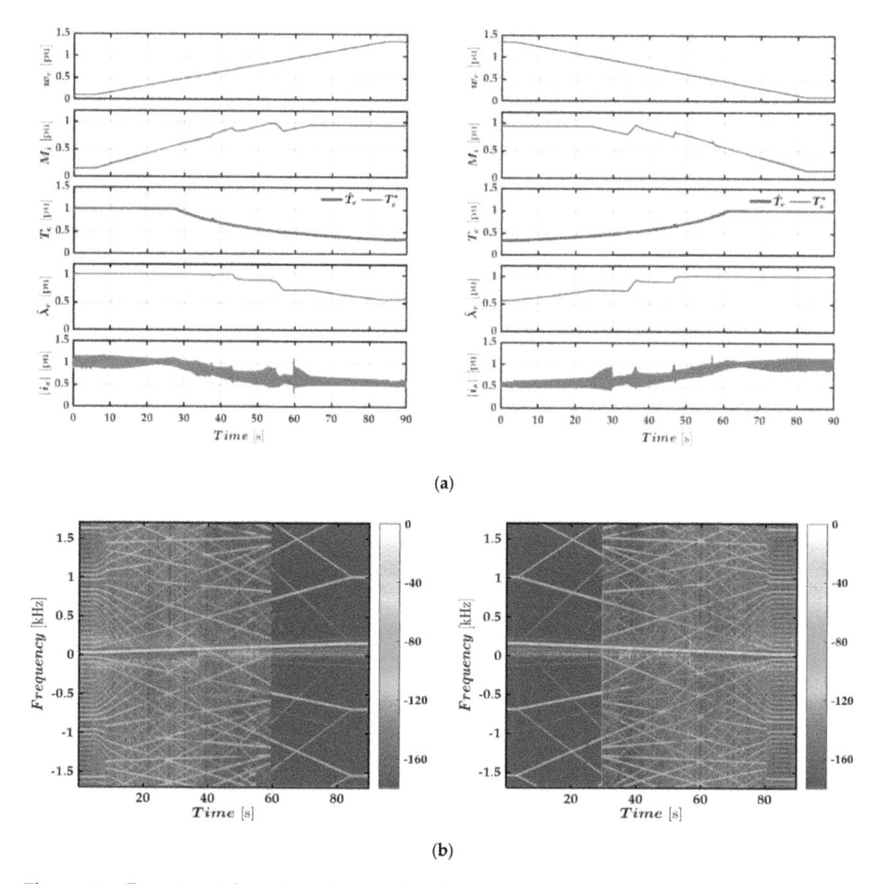

Figure 21. Experimental results. Acceleration (left)-deceleration (right) tests between $\omega_r = 0.1$ and $\omega_r = 1.328$ p.u. (**a**) From top to bottom: rotor speed, modulation index, commanded and estimated torques, estimated rotor flux, and magnitude of the stator current vector; (**b**) stator current vector spectrogram.

8. Conclusions

In this paper, a comparative analysis of scalar and vector control strategies for railway traction applications has been presented, with a special focus on their operation at high speeds.

HSTs normally use medium-voltage, high-power IMs. Rotor flux Oriented Control (RFOC) has been widely adopted at low and medium speeds. However, high fundamental frequencies intrinsic to high-speed operations, combined with the need to reduce inverter losses, force the inverter to operate with reduced switching frequencies and a high modulation index or even at the six-step limit. These limitations seriously compromise the performance of RFOC at high speeds. A common practice is to use RFOC at low speeds, rather than switch to strategies able to perform properly under severe voltage constraints at high speeds.

Methods considered for the analysis included different types of Closed-loop Voltage/Frequency (V/F), Field-Oriented Control (FOC), and Direct-Torque Control (DTC) strategies. Four different control strategies have been selected and tested by means of simulation, namely, Closed Loop V/F with flux/torque Control (CLVFC), CLVFC with feedforward (CLVFC&FF), Direct Flux Vector Control (DFVC), and Direct-Torque Control Space-Vector Modulation (DTC-SVM). The modulation methods that have been considered are PWM/SVM, SHE, and six-step. Their advantages include the easiness of operation with a high modulation index, including six-step; switching frequency; and dynamic response to both torque change demands and rotor flux change demands during remagntization.

The CLVFC&FF method described in Section 3.1.3 and the remagnetization strategy discussed in Section 5 are the original contributions of this paper.

It was concluded from the simulation results that CLVFC, CLVFC&FF, and DFVC provide similar performances. However, DFVC requires a modification of the q-axis current controller gains when the drive enters the overmodulation region. Specifically, CLVFC&FF proposed in this paper operates properly with a high modulation index, including six-step, and provides a good dynamic response during remagnetization.

Preliminary experimental results using CLVFC in a full-scale traction drive have been provided, which are in good agreement with the simulation results, and confirm the viability of this strategy. Implementation of the other strategies, including remagnetization, is ongoing.

Author Contributions: Conceptualization, F.B. and D.O.; methodology, F.B., J.M.G. and D.O.; formal analysis, A.F.A., J.M.G. and F.B.; simulation and programing, A.F.A., A.E. and I.M.; validation, A.F.A., A.E. and I.M.; writing—original draft preparation, A.F.A.; writing—review and editing F.B. and J.M.G.; supervision, F.B., D.O. and I.L.; All authors have read and agreed to the published version of the manuscript.

Funding: This research was funded by the European Commission H2020 under grant UE-18-POWER2POWER-826417; The Spanish Ministry of Science, Innovation and Universities under grant MCIU-19-PCI2019-103490; and by the Government of Asturias under grant IDI/2018/000188 and FEDER funds.

Conflicts of Interest: The authors declare no conflict of interest.

References

1. Nikolas Hill, E.K. *2019 Government Greenhouse Gas Conversion Factors for Company Reporting: Methodology Paper for Emission Factors Final*; Government UK: London, UK, 2019.
2. Hill, R.J. Electric railway traction. Part 2: Traction drives with three-phase induction motors. *Power Eng. J.* **1994**, *8*, 143–152. [CrossRef]
3. El-Refaie, A.M. Motors/generators for traction/propulsion applications: A review. *IEEE Veh. Technol. Mag.* **2013**, *8*, 90–99. [CrossRef]
4. Buyukdegirmenci, V.T.; Bazzi, A.M.; Krein, P.T. Evaluation of induction and permanent-magnet synchronous machines using drive-cycle energy and loss minimization in traction applications. *IEEE Trans. Ind. Appl.* **2014**, *50*, 395–403. [CrossRef]
5. Ronanki, D.; Singh, S.A.; Williamson, S.S. Comprehensive Topological Overview of Rolling Stock Architectures and Recent Trends in Electric Railway Traction Systems. *IEEE Trans. Transp. Electrif.* **2017**, *3*, 724–738. [CrossRef]
6. McGean, T.J. Developing IEEE rail transit vehicle standards. In Proceedings of the 1998 ASME/IEEE Joint Railroad Conference, Philadelphia, PA, USA, 16 April 1998; pp. 95–105.
7. Drofenik, U.; Canales, F. European trends and technologies in traction. In Proceedings of the 2014 International Power Electronics Conference IPEC-Hiroshima—ECCE Asia, Hiroshima, Japan, 18–21 May 2014; pp. 1043–1049.
8. Sato, K.; Yoshizawa, M.; Fukushima, T. Traction systems using power electronics for Shinkansen High-speed Electric Multiple Units. In Proceedings of the 2010 International Power Electronics Conference IPEC—ECCE Asia, Sapporo, Japan, 21–24 June 2010; pp. 2859–2866.
9. Krause, P.C.; Thomas, C.H. Simulation of Symmetrical Induction Machinery. *IEEE Trans. Power Appar. Syst.* **1965**, *84*, 1038–1053. [CrossRef]
10. Ikeda, R.; Yusya, S.; Kondo, K. Study on Design Method for Increasing Power Density of Induction Motors for Electric Railway Vehicle Traction. In Proceedings of the 2019 IEEE International Electric Machines & Drives Conference (IEMDC), San Diego, CA, USA, 12–15 May 2019; pp. 1545–1550.
11. Buhrkall, L. Traction System Case Study. In Proceedings of the 2008 IET Professional Development Course on Electric Traction Systems, Manchester, UK, 3–7 November 2008; pp. 45–63.
12. Bose, B.K. *Modern Power Electronics and AC Drives*; Prentice-Hall, Inc.: Upper Saddle River, NJ, USA, 2002.

13. Holtz, J. Sensorless control of induction machines—With or without signal injection? *IEEE Trans. Ind. Electron.* **2006**, *53*, 7–30. [CrossRef]
14. Blaschke, F. The principle of field orientation as applied to the new TRANSVECTOR closed loop control system for rotating field machines. *J. Chem. Inf. Model.* **1972**, *34*, 217–220.
15. Kubota, H.; Matsuse, K. Speed sensorless field-oriented control of induction motor with rotor resistance adaptation. *IEEE Trans. Ind. Appl.* **1994**, *30*, 1219–1224. [CrossRef]
16. Kwon, Y.-C.; Kim, S.; Sul, S.-K. Six-Step Operation of PMSM With Instantaneous Current Control. *IEEE Trans. Ind. Appl.* **2014**, *50*, 2614–2625. [CrossRef]
17. Pellegrino, G.; Bojoi, R.I.; Guglielmi, P. Unified direct-flux vector control for AC motor drives. *IEEE Trans. Ind. Appl.* **2011**, *47*, 2093–2102. [CrossRef]
18. Novotny, D.W. Implementation of Direct Stator Flux Orientation Control on a Versatile dsp Based System. *IEEE Trans. Ind. Appl.* **1991**, *27*, 694–700.
19. Takahashi, I.; Noguchi, T. A New Quick-Response and High-Efficiency Control Strategy of an Induction Motor. *IEEE Trans. Ind. Appl.* **1986**, *IA-22*, 820–827. [CrossRef]
20. Casadei, D.; Serra, G.; Stefani, A.; Tani, A.; Zarri, L. DTC drives for wide speed range applications using a robust flux-weakening algorithm. *IEEE Trans. Ind. Electron.* **2007**, *54*, 2451–2461. [CrossRef]
21. Depenbrock, M. Direct self-control (DSC) of inverter fed induktion machine. In Proceedings of the 1987 IEEE Power Electronics Specialists Conference, Blacksburg, VA, USA, 21–26 June 1987; pp. 632–641.
22. Steimel, A. Direct self-control and synchronous pulse techniques for high-power traction inverters in comparison. *IEEE Trans. Ind. Electron.* **2004**, *51*, 810–820. [CrossRef]
23. Spichartz, M.; Heising, C.; Staudt, V.; Steimel, A. Indirect Stator-Quantities Control as Benchmark for Highly Dynamic Induction Machine Control in the Full Operating Range. In Proceedings of the 14th International Power Electronics and Motion Control Conference EPE-PEMC, Ohrid, Macedonia, 6–8 September 2010; pp. T3–T13.
24. Buja, G.S.; Kazmierkowski, M.P. Direct Torque Control of PWM Inverter-Fed AC Motors—A Survey. *IEEE Trans. Ind. Electron.* **2004**, *51*, 744–757. [CrossRef]
25. Rodríguez, J.; Pontt, J.; Silva, C.; Kouro, S.; Miranda, H. A novel direct torque control scheme for induction machines with space vector modulation. In Proceedings of the 2004 IEEE 35th Annual Power Electronics Specialists Conference (IEEE Cat. No.04CH37551), Aachen, Germany, 20–25 June 2004; Volume 2, pp. 1392–1397.
26. Tripathi, A.; Khambadkone, A.M.; Panda, S.K. Stator flux based space-vector modulation and closed loop control of the stator flux vector in overmodulation into six-step mode. *IEEE Trans. Power Electron.* **2004**, *19*, 775–782. [CrossRef]
27. Yano, M.; Iwahori, M. Transition from Slip-Frequency Control to Vector Control for Induction Motor Drives of Traction Applications in Japan. In Proceedings of the Fifth International Conference on Power Electronics and Drive Systems, Singapore, 17–20 November 2003; pp. 1246–1251.
28. Rodriguez, J.; Bernet, S.; Steimer, P.K.; Lizama, I.E. A survey on neutral-point-clamped inverters. *IEEE Trans. Ind. Electron.* **2010**, *57*, 2219–2230. [CrossRef]
29. Wang, F. Sine-triangle versus space-vector modulation for three-level PWM voltage-source inverters. *IEEE Trans. Ind. Appl.* **2002**, *38*, 500–506. [CrossRef]
30. Attique, Q.M. A Survey on Space-Vector Pulse Width Modulation for Multilevel Inverters. *CPSS Trans. Power Electron. Appl.* **2017**, *2*, 226–236. [CrossRef]
31. Al-Hitmi, M.; Ahmad, S.; Iqbal, A.; Padmanaban, S.; Ashraf, I. Selective harmonic elimination in awide modulation range using modified Newton-raphson and pattern generation methods for a multilevel inverter. *Energies* **2018**, *11*, 458. [CrossRef]

32. Wasynczuk, O.; Sudhoff, S.D.; Corzine, K.A.; Tichenor, J.L.; Krause, P.C.; Hansen, I.G.; Taylor, L.M. A maximum torque per ampere control strategy for induction motor drives. *IEEE Trans. Energy Convers.* **1998**, *13*, 163–169. [CrossRef]
33. Popov, A.; Lapshina, V.; Briz, F.; Gulyaev, I. Estimation of the Required Voltage for Improved MTPA Algorithm. In Proceedings of the 2018 X International Conference Electrical Power Drive Systems (ICEPDS), Novocherkassk, Russia, 3–8 October 2018; pp. 1–4.

Article

Comparison of an Off-Line Optimized Firing Angle Modulation and Torque Sharing Functions for Switched Reluctance Motor Control

Peter Bober and Želmíra Ferková *

Faculty of Electrical Engineering and Informatics, Technical University of Kosice, 04200 Kosice, Slovakia; peter.bober@tuke.sk
* Correspondence: zelmira.ferkova@tuke.sk

Received: 17 April 2020; Accepted: 11 May 2020; Published: 12 May 2020

Abstract: In this paper, a comparison of the simple firing angle modulation method (FAM) and the more advanced torque sharing function (TSF)-based control of switched reluctance motor (SRM) is presented. The off-line procedure to tailor and optimize the parameters of chosen methods for off-the-shelf SRM is explained. Objective functions for optimization are motor efficiency, torque ripple, and integral square error. The off-line optimization uses a finite element method (FEM) model of the SRM. The model was verified by measurement on the SRM. Simulation results showed that FAM has comparable efficiency to TSF, but has a much higher value of torque ripple. The presented off-line procedure can be used for single or multi-objective optimization.

Keywords: switched reluctance motor; torque sharing functions; finite element method; firing angle modulation; torque ripple; efficiency; optimization

---

1. Introduction

The principle of Switched Reluctance Motor (SRM) is known from the 1830s when the first attempts to build a usable machine were made by many constructors and inventors [1,2]. One of them was Robert Davidson who, in 1838, developed a reluctance motor to power an electric locomotive at the Edinburgh-Glasgow railway [3]. The early attempts suffer from inadequate mechanical switches and poor electromagnetic and mechanical design. The SRM waited 150 years for the theoretical foundations to be developed and new power electronics devices to appear. A work of Lawrenson et al. [4] lays general foundations for the practical design of a family of switched reluctance motors in 1980. Another thirty years were needed to achieve progress in the field of power electronics and powerful microcontrollers. Currently, many companies are producing SRM in the world, and the worldwide market is expected to grow by roughly 5.2% over the next five years [5]. The construction of SRM can be tailored to broad application areas. Report [5] showed that 22.95% of the SRM market demand is in the Automobile Industry, 19.43% in the Appliance Industry, and 39.28% in Industrial Machinery in 2016.

Typical SRM construction has salient poles on both the stator and the rotor. The windings are only on the stator, and phase electric current is unidirectional. Reluctance torque produced by the change in inductance as a function of the angle creates a rotary movement. The advantages resulting from this simple construction are low cost, robustness, high efficiency, and wide speed range up to 150,000 rpm [6]. Each energized phase winding generates torque for a certain rotor angle interval. Therefore, SRM control depends heavily on rotor position information to turn-on and turn-off the phase current at the chosen angle. Torque ripples that occur during the transition from one winding to another are the major drawback of SRMs. The main goal of ongoing research is a torque ripple reduction and efficiency improvement by the optimization of the structure and magnetic design of the motor [6–10], and by adopting advanced motor control [9,11–17].

An overview of SRM control strategies is presented in [9]. The methods are grouped into (1) current and angle modulation, (2) average torque control (ATC) and direct torque control (DTC), (3) torque sharing function (TSF)-based control, (4) feedback linearization control (FBL), (5) iterative learning control (ILC), and (6) intelligent control (IC).

Current and angle modulation control methods are focused on defining turn-on and turn-off angles to lower torque ripple. In addition, the current controller can maintain constant current or follow a current profile specified in look-up tables. ACT and DTC use estimated average or instant torque in the control loop. TSFs define the torque reference for each phase so that their sum is equal to the output torque reference. FBL control algorithm compensates nonlinear characteristics of the motor by utilizing state feedback in the closed-loop system. ILC learns iteratively from the difference between required and actual phase current and motor torque. IC uses self-learning and the adaptive ability of fuzzy logic, genetic algorithms, and neural networks to optimize current profiles either on-line or off-line.

The majority of described SRM control methods, especially in the automotive industry, are complex and require high computational power. Authors in [14] use 1 GHz CPU and FPGA running at 100 MHz. The EPM570 Intel CPLD—Complex Programmable Logic Devices—with a frequency of 150–300 MHz and dsPIC30f6010A MPU is used in [15]. High-dynamic four-quadrant switched reluctance drive based on DITC published in [16] uses DSP/FPGA rapid-prototyping platform. This computational power is needed for torque and speed estimation, look-up tables, vibration and noise suppression, and precise current control, where high sampling frequency for low inductances is required. In the coming years, the price for computing power will fall. Nevertheless, the growing SRM market has the potential for simple and cheap drives where the hardware cannot support complex control methods. It can be assumed that not all SRMs available on the market will be designed optimally. The question then arises as to what characteristics the drive with such a motor and a simple control method will have.

Two control methods applied to the commercially available switched reluctance motor were chosen to investigate in this research. The first method is a conventional control with Firing Angle Modulation (FAM) and constant current profile similar to [18,19]. This method uses on-line turn-on and turn-off angle calculation and simple hysteresis current controller with a low demand on computing power. No look-up tables nor torque estimation is needed.

The second method is more complex TSF-based control that uses special functions to distribute torque between the outgoing and incoming motor phase. According to the evaluation in [20], the cubic, sinusoidal, and exponential torque sharing functions result in similar minimum torque ripples, which are much smaller than the torque ripple produced by the linear torque sharing function. A minimum effective rate-of-change of flux linkage was used in [11,21] as an objective function that should be minimized by selected TSF to extend the torque-speed range. Research results in [11,13,20–23] show that different objective functions lead to different TSF. Therefore, multi-objective optimization with weighted criterion [11,13,19] or the Pareto-optimal approach [22] are used.

Many published works [6–8,11,13,19–23] use two to four objective functions to compare or optimize control methods or the structural design of the motor. This reduction of problem formulation is necessary to handle the complexity of a full industrial design process and can be considered as one stage in the multi-stage optimization process. Authors in [10] address this complexity and present a two-step procedure for a motor design where different performance measures (such as manufacturing cost and iron weight) are included along with efficiency and torque ripple. Here some general optimization frameworks, e.g., ARTAP published in [24], are helpful since various domain-specific numerical solvers are needed for this type of motor design task.

The main aim of this paper is to present a comparison of two selected methods that control the off-the-shelf switched reluctance motor. The procedure to tailor and optimize the parameters of the chosen methods is explained. Objective functions for optimization are motor efficiency, torque ripple, and integral square error between instant and average torque. The offline optimization uses the finite element method (FEM) model of the SRM. The model is verified by comparing simulation results to the motor measurement.

2. Modeling and Control of SRM

2.1. Switched Reluctance Motor and Power Converter

Control methods are tailored and optimized for the SRM that is commercially available on the market. It is a small 200 W three-phase 12/8 pole switched reluctance motor. Rated voltage is 120 V, and the maximum phase current is 6 A. The geometry of the motor is shown in Figure 1a. Each stator phase winding consists of four coils connected in a series-parallel manner. The width of the stator and rotor pole is almost the same. Therefore, inductance profile has a sharp peak in an aligned position, and transition from a motor to generator mode is fast.

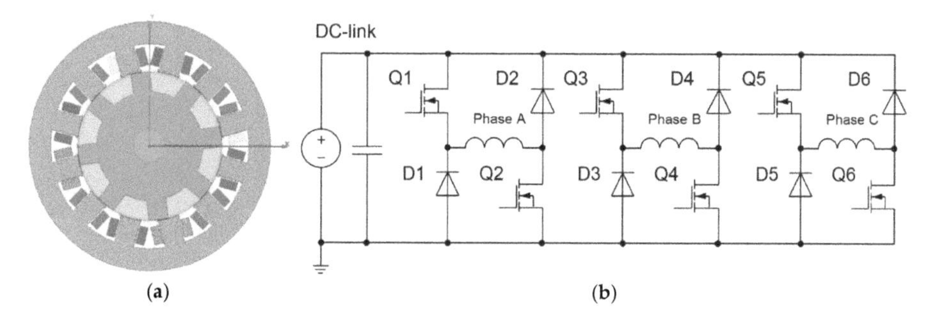

Figure 1. (**a**) Geometry of the switched reluctance motor; (**b**) standard asymmetric H-bridge converter topology.

The experimental setup uses a power converter that was built for measurement purposes. It has a standard asymmetric H-bridge topology to drive three phases of SRM (Figure 1b). A stabilized power supply supplies a DC-link of the power converter. The voltage can vary from 30 V to 120 V.

2.2. FEM Model

The FEM model of SRM was built in ANSYS Maxwell according to the dimensions of the motor and electrical measurement. When creating the model, a compromise solution was chosen between the mesh size and the computational time. The total number of elements is 11,110 in the 2D model. The mesh is denser in the air gap. The additional leakage inductance models the effect of the end winding leakage flux. The magnetic vector potential is zero on the outer circumference of the yoke. The comparison of the simulation results with the electrical measurement led to the selection of the B-H characteristic corresponding to the electrical steel M400.

The SRM model is driven by three-phase asymmetric H-bridge created in ANSYS Simplorer. The control algorithms are also programmed in ANSYS Simplorer. Figure 2 shows magnetic field distribution in the FEM model of the studied motor for aligned rotor position 22.5° (a) and for 7.2° (b). Figure 2c shows simulated motor characteristics for a maximum phase current of 6 A and 120 V DC-link voltage. The FEM model verification is described in Section 3.2.

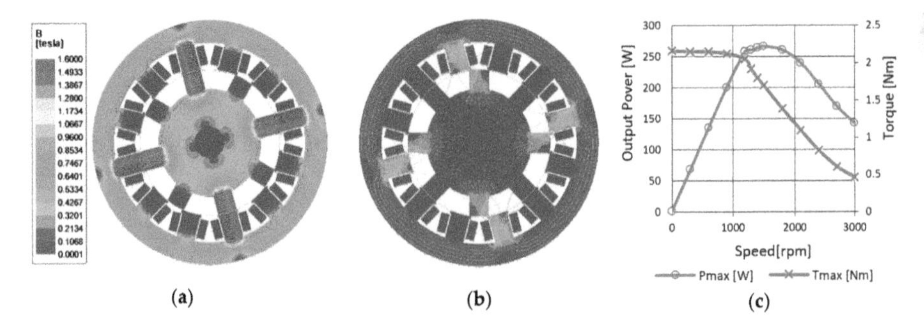

Figure 2. Magnetic field distribution in the finite element method (FEM) model, phase current 6 A: (**a**) aligned position, rotor angle 22.5 degrees; (**b**) rotor angle 7.2 degrees; (**c**) Power–Speed and Torque–Speed characteristics.

2.3. *SRM Control Methods*

The switched reluctance motor is controlled by two methods. The first method is a Firing Angle Modulation, and the second one is a control method with Torque Sharing Functions.

2.3.1. Firing Angle Modulation

The firing angle modulation method is quite straightforward and can be easily implemented. The asymmetric H-bridge converter allows controlling the electric current in each stator phase winding independently by using a simple hysteresis controller. Each motor phase creates torque only in a limited angle interval, and the phases alternate one after the other. Therefore, the current is switched on at the angle θ_{on} and switched off at the angle θ_{off}. The hysteresis current controller controls the current level inside the angle interval θ_{on} and θ_{off}. Figure 3a shows simulated current waveforms of one phase as a function of a rotor angle for different speeds. The shape of the phase current cannot be fully controlled at an available DC-link voltage as it depends on the inductance profile and sampling frequency of the hysteresis controller. The current rising is fast because the phase inductance is low. At the end of the current impulse, the inductance is highest as the stator and rotor poles are aligned. Therefore, the electric current decay is slow, and it can create negative torque, especially at high speeds. Average motor torque corresponds to the current reference I_{ref} and, of course, to the θ_{on} and θ_{off} values. The value of I_{ref} is an output of the speed controller. Figure 3b shows the block diagram of the firing angle modulation method. The procedure to calculate firing angles θ_{on} and θ_{off} is described in Section 2.3.2.

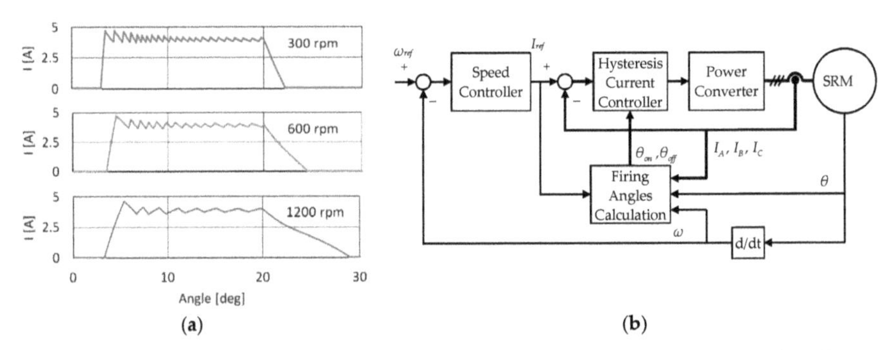

Figure 3. (**a**) Current waveform at speed 300, 600, and 900 rpm, 120 V; (**b**) block diagram of the firing angle modulation method.

2.3.2. Firing Angles Calculation

The phase current generates torque in a rotor angle region, where the phase inductance is changing due to a rotor position change. The electric current rise time is defined by the inductance and the applied voltage. Therefore, the turn-on angle θ_{on} shall precede the increase in inductance to start to generate the desired torque from the very beginning. If the current reaches the specified level too early, it creates a copper loss only. If the current does not reach the specified level at a defined angle, the torque is less than it could be. An early turn-off angle θ_{off} causes torque to drop. However, a late turn-off of the current causes a negative torque to be generated.

A calculation of the turn-on angle θ_{on} is given in [16,17] based on the assumption, that the inductance profile has constant minimum inductance, then the inductance starts to rise at angle θ_m. The value of the current should reach the reference value at this point, and θ_{on} can be calculated as follows:

$$\theta_{on} = \theta_m - 6L_u I_{ref} \frac{n}{V_{DC}}, \tag{1}$$

where θ_{on} is turn-on angle (degrees), θ_m is rotor angle (degrees) where the inductance begins to rise, L_u is the inductance in unaligned position, I_{ref} is a reference current from speed controller, n is rotor speed (rpm), and V_{DC} is the supply voltage. This equation neglects the actual shape of the inductance profile. Therefore, a set of simulation experiments was planned and performed to obtain a more precise formula for turn-on angle calculation. An extrapolation of collected data leads to the formula:

$$\theta_{on} = 7 - \left(0.463 + 8.88e^{-4}n + 1.2e^{-6}n^2\right)\left(\frac{I_{ref0}}{I_{max}}\right)^2, \tag{2}$$

where I_{ref0} is a reference current from speed controller at angle 0°, and I_{max} is a maximum reference current.

The formula for calculation of turn-off angle is:

$$\theta_{off} = 22.5 - \frac{0.2 I_{15} 2\pi\, n}{60 V_{DC}}, \tag{3}$$

where I_{15} is a phase winding current at angle 15°, and V_{DC} is a voltage applied on winding inductance at turn-off time. This formula was derived from the linear approximation of the current fall time. Equations (2) and (3) give angle values for first phase A. Angle values for phases B and C are shifted by 15° and 30°, respectively. Formulas (2) and (3) are valid only for the given SRM and used power converter with a DC-link voltage of 120 V.

2.3.3. Torque Sharing Functions

The motor torque created by consecutive phases is overlapping. Torque sharing function defines a required torque profile for each energized phase that the total motor torque has a small ripple. The shape of TSF can be chosen arbitrarily or is calculated according to selected criteria. Figure 4 shows sinusoidal TSF defined on the interval (0, θ_p) as follows [10]:

$$TSF(\theta) = \begin{cases} 0, & (0 \le \theta < \theta_{on}) \\ f_{up}(\theta), & (\theta_{on} \le \theta < \theta_{on} + \theta_{ov}) \\ T_c & (\theta_{on} + \theta_{ov} \le \theta < \theta_{off}) \\ f_{dn}(\theta), & (\theta_{off} \le \theta < \theta_{off} + \theta_{ov}) \\ 0, & (\theta_{off} + \theta_{ov} \le \theta < \theta_p) \end{cases}, \tag{4}$$

$$f_{up}(\theta) = \frac{T_c}{2}{}_c\left[1 - \cos\left(\frac{\theta - \theta_{on}}{\theta_{ov}}\pi\right)\right], \tag{5}$$

$$f_{dn}(\theta) = T_c - f_{up}\left(\theta + \theta_{ov} - \theta_{off}\right), \tag{6}$$

where θ is the rotor angle, $\theta_p = 2\pi/8$ is the rotor pole pitch, θ_{on} is the turn-on angle, θ_{ov} is the overlap angle, $\theta_{off} = \theta_{on} + 2\pi/24$ is the turn-on angle of the next phase, T_c is the torque command, $f_{up}(\theta)$ is the rising, and $f_{dn}(\theta)$ the declining part of the TSF shape.

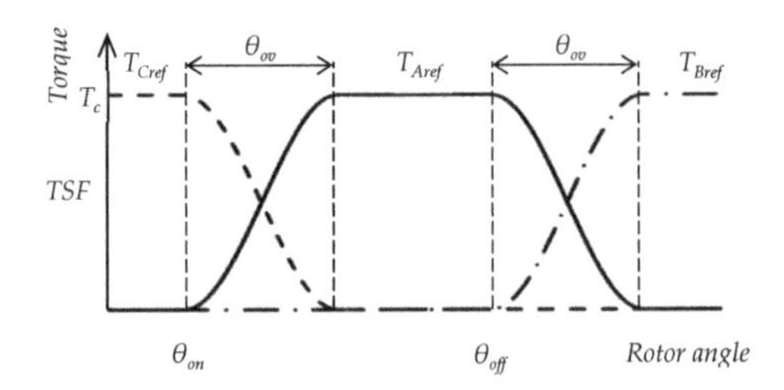

Figure 4. Sinusoidal torque sharing function.

The block diagram of used TSF-based control is shown in Figure 5. The torque command T_c from the *Speed Controller* is translated into three torque profiles T_{Aref}, T_{Bref}, and T_{Cref} for each motor phase in the *TSF* block using Equations (4)–(6). The *Torque to Current* block translates the torque profile to current reference utilizing a look-up table obtained from the FEM model. A *Hysteresis Current Controller* is used to control phase currents. Optimal TSF parameters θ_{on} and θ_{ov} are calculated by functions $f_1(T_c,\omega)$ and $f_2(T_c,\omega)$ from torque command T_c and actual rotor speed ω. The next section explains how to construct functions f_1 and f_2.

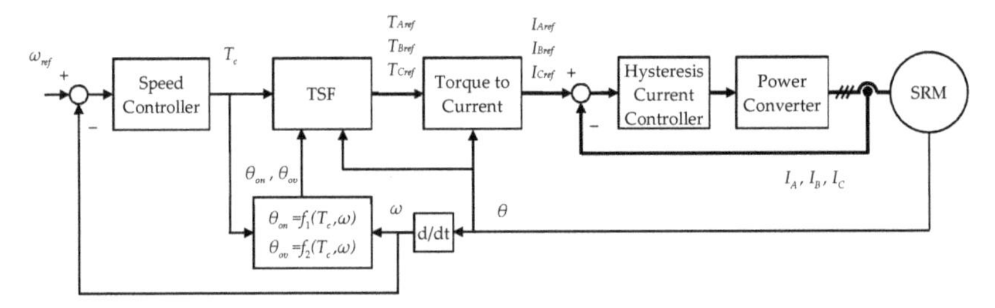

Figure 5. Block diagram of the switched reluctance motor (SRM) control using torque sharing function (TSF) with optimized parameters θ_{on} and θ_{ov}.

2.3.4. Optimization of TSF

The objective function needs to be defined to formulate an optimization problem. Three criteria are selected in this research to find optimal values of parameters θ_{on} and θ_{ov}:

- Efficiency η

$$\eta = \frac{\textit{Mechanical output power}}{\textit{Electrical input power}} \tag{7}$$

- Relative torque ripple T_{rip}

$$T_{rip} = \frac{T_{max} - T_{min}}{T_{avg}} \tag{8}$$

where T_{max}, T_{min}, and T_{avg} are the maximum torque, the minimum torque, and average torque, respectively.

- Integral square error criterion *ISE*

$$ISE = \sqrt{\frac{1}{t_2 - t_1}\int_{t_1}^{t_2}\left(T - T_{avg}\right)^2 dt} \tag{9}$$

where T, and T_{avg} are the instantaneous torque and average torque, respectively, measured in a time-interval (t_1, t_2). This objective function is a measure of overall torque oscillation.

Simulation experiments calculate the numerical value of an objective function. One simulation run is needed for each individual combination of parameters θ_{on} and θ_{ov}. Calculations for all combinations in all operating points are time-consuming. Therefore, a workaround is used: the shape of the objective function is replaced by the interpolation function calculated from a limited set of combinations of θ_{on} and θ_{ov} values. Then, the extreme of the interpolated function determines the optimal parameters for one operating point defined by torque and speed. The procedure described above is repeated to obtain a sufficient number of function points of $f_1(T_c,\omega)$ and $f_2(T_c,\omega)$. Functions obtained in this way are approximations of the optimal solution. Results from optimization are presented in Section 3.3.1.

The described off-line optimization procedure is an alternative to other optimization methods, such as the genetic algorithm. Its advantage is that, unlike genetic algorithms, it does not need new FEM computation when the objective function changes. Instead, the new values of the objective functions are calculated from the interpolated functions for which the computation time is significantly shorter. Once the interpolated functions have been enumerated, no further FEM computation is needed to apply different types of multi-objective optimization.

It must be said that the ability of the current controller to follow the reference current derived from the TSF is limited, especially at high speed. Therefore, the phase torque may not track the TSF. However, described off-line optimization involves the influence of the current controller to the actual phase torque. There is a possibility to reduce the torque ripple further using a more advanced current controller. Authors in [17] propose a predictive pulse width modulation (PWM) current control method that serves this purpose and a modified proportional–derivative (PD) controller that lowers the transient response overshoot is described in [25].

3. Results

3.1. Experimental Setup

The experimental setup consists of 200 W three-phase 12/8 pole switched reluctance motor connected to an induction machine that serves as load. Torque sensor *KISTLER 4520A20* with *CoMo torque evaluation instrument 4700BP0UA* was used to measure the torque, speed, and mechanical power on the shaft. *Infratek 106A Power Analyzer* measured input power. *STM32F303RET6* microcontroller, with the control loop cycle time 50 µs, controlled the asymmetric H-bridge power converter.

3.2. FEM Model Verification

The FEM model of SRM was verified by comparing measured and calculated motor characteristics. Figure 6a shows measured and calculated inductance profiles for currents 1.5 A, 6 A, and 9 A [26]. The more significant deviation occurs at the current of 9 A, but this is not a problem as the operating currents are up to 6 A. Figure 6b shows measured and calculated torque profiles for phase currents 2 A, 4 A, and 6 A. Phase currents from the measurement and FEM model are shown in Figure 7a. The behavior of the FEM hysteresis current controller slightly differs from the real one because it depends on the controller sampling, hysteresis-band, and the current sensor response time. Figure 7b shows the measured and calculated efficiency of the drive. Mechanical and power converter losses are included. Despite the above differences, the authors believe that the match between model and measurement is sufficient to allow meaningful conclusions to be drawn from the simulation results.

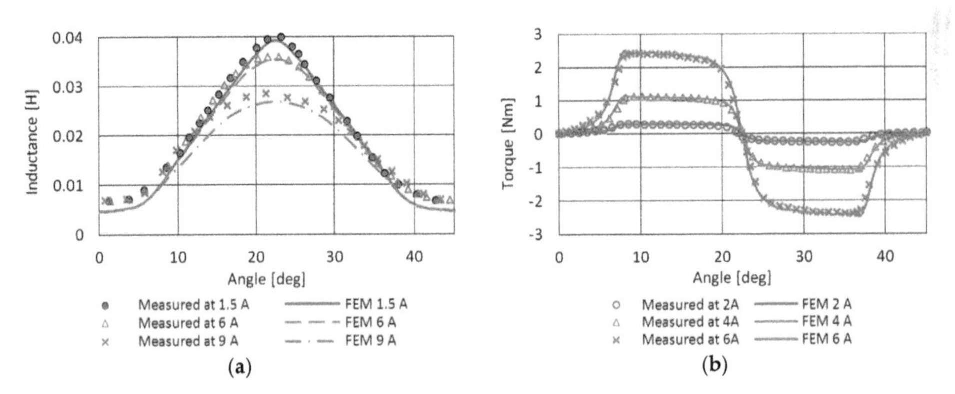

Figure 6. (**a**) Comparison of measured and calculated inductance profiles; (**b**) measured and calculated torque profile.

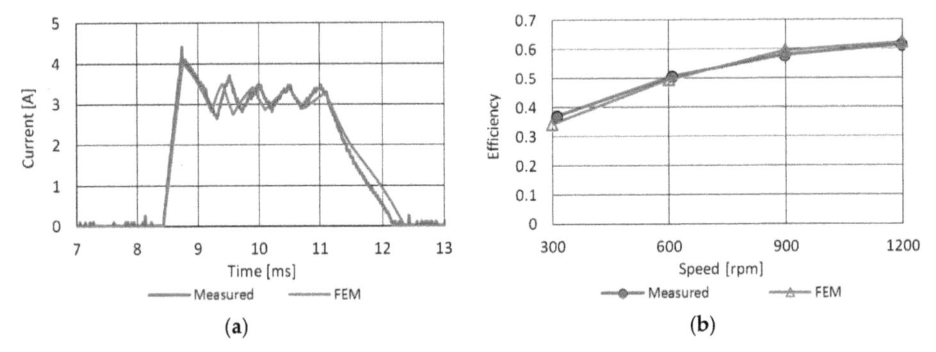

Figure 7. (**a**) Comparison of measured and calculated phase current, $\theta_{on} = 0°$, $\theta_{off} = 19.25°$, speed is 1200 rpm, DC-link voltage is 100 V, load torque is 1 Nm; (**b**) measured and calculated motor efficiency for constant angles $\theta_{on} = 0°$, $\theta_{off} = 17.6°$, mechanical and power converter losses are included.

3.3. *Simulation Results*

3.3.1. Calculation of Optimal TSF Parameters

Results of preliminary simulation experiments were used to determine the range and number of values for speed, torque, θ_{on}, and θ_{ov} parameters. Then, simulation runs for each parameter combination were calculated, and the values of selected objective functions were recorded. The Matlab piece-wise interpolating polynomials function uses recorded values to calculate interpolation function for each objective. The filled contour graph in Figure 8 visualizes the motor efficiency, relative torque ripple, and integral square error criterion for two operational points: torque 1 Nm, speed 600 rpm (Figure 8a), and 1200 rpm (Figure 8b). The data tips show values of θ_{on} and θ_{ov} parameters for the specified operating point that are optimal according to the corresponding objective function. Figure 9 shows optimal values of θ_{on} and θ_{ov} as a function of speed separately in two graphs for speeds 300, 600, 900, and 1200 rpm. Simulated motor torque for TSF-based control with optimal parameter settings according to three objective functions is shown in Figure 10a for speed 600 rpm, and in Figure 10b for speed 1200 rpm.

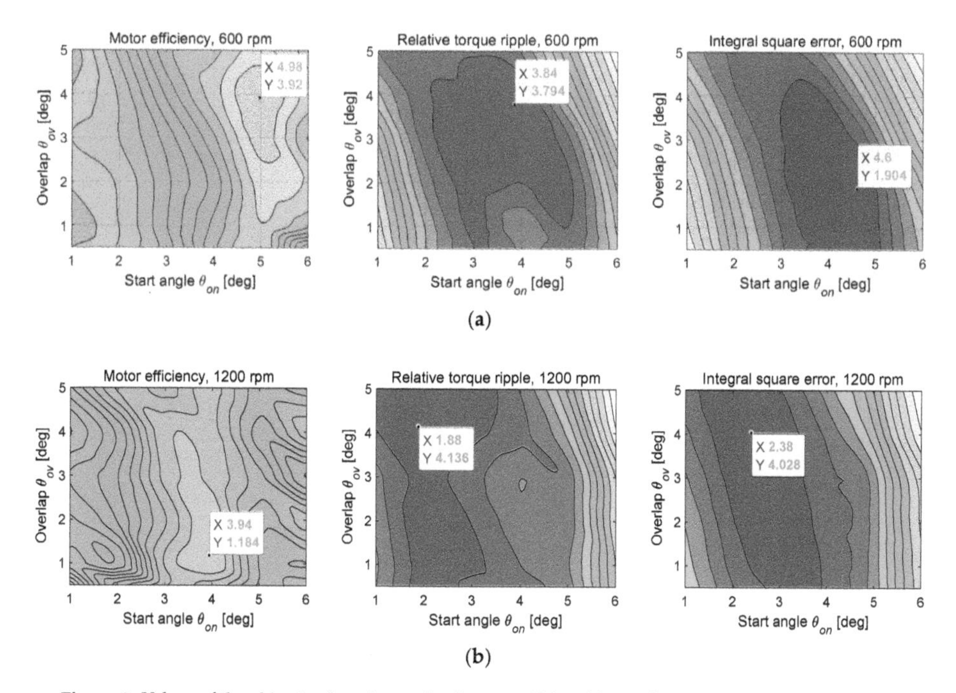

Figure 8. Values of the objective functions at load torque 1 Nm: (**a**) speed 600 rpm; (**b**) speed 1200 rpm.

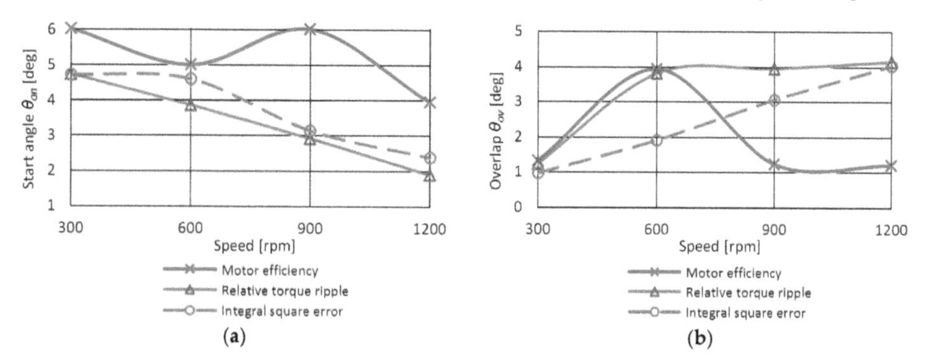

Figure 9. Optimal parameter values at load torque 1 Nm: (**a**) start angle θ_{on}; (**b**) overlap θ_{ov}.

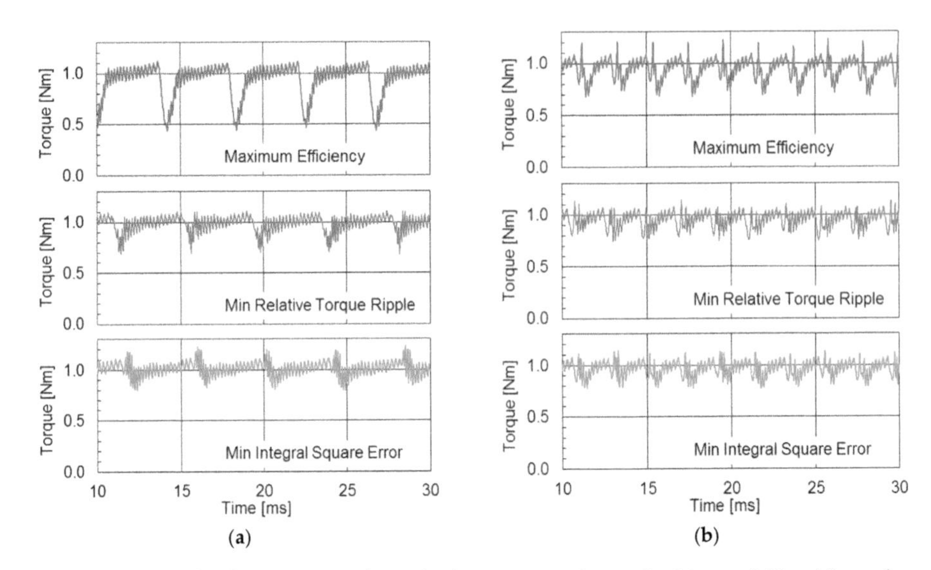

Figure 10. Simulated motor torque for optimal parameter values at load torque 1 Nm: (**a**) speed 600 rpm; (**b**) speed 1200 rpm.

3.3.2. Comparison of FAM and TSF-based control

The comparison of the firing angle modulation method and TSF-based control method according to maximal efficiency, minimal relative torque ripple, and minimum integral square error criteria is presented in Figure 11. Each graph shows the value of the corresponding objective function for four cases. The first three cases are TSF motor control methods optimized according to the defined objective functions. The fourth case is the FAM method, where θ_{on} and θ_{off} are calculated according to Equations (2) and (3), respectively.

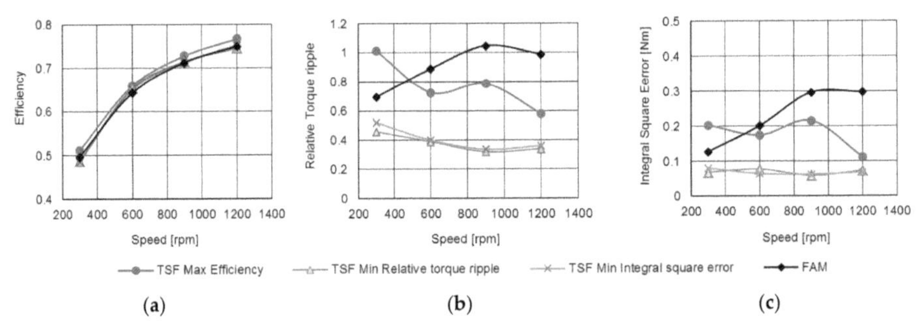

Figure 11. Comparison of the firing angle modulation (FAM) method and TSF at load torque 1 Nm: (**a**) efficiency; (**b**) relative torque ripple; (**c**) integral square error.

Figure 11a shows that the control with TSF optimized for efficiency has the highest efficiency value, but the differences between the other control methods are small, and the efficiency of the FAM is only 1.7% lower (Figure 11a). As shown in Figure 11b, the control with TSF optimized for relative torque ripple, and TSF optimized for integral square error has a small value of relative torque ripple, unlike the FAM and TSF, optimized for efficiency, that has a large value of relative torque ripple. Similar results are displayed in Figure 11c for the integral square error criterion.

Inconsistency in the calculated data appeared in Figure 11c. The value of integral square error criterion (ISE) should be the smallest one for TSF optimized for ISE. However, TSF optimized for relative torque ripple has the lowest value of ISE for 300 and 900 rpm. This unexpected finding shows that the accuracy of a numerical solution is limited.

4. Discussion and Conclusions

This paper presents tailoring and a comparison of two methods to control a commercially available small switched reluctance motor, whose construction is not optimized for the minimization of torque ripple. Furthermore, a new approach to off-line evaluation of objective functions for motor control optimization that reduce computational time is presented.

The first control method is a firing angle modulation, which can be easily implemented on a low-cost microcontroller if the price of the drive is critical for a certain market segment. The second one is a control method with Torque Sharing Functions, which is more complex and uses torque-angle-current look-up tables calculated in advance by FEM.

Objective functions of the control methods optimization are maximal efficiency, minimal relative torque ripple, and minimum integral square error criterion.

The research shows that FAM has the same overall efficiency as TFS optimized for the torque ripple and TSF optimized for integral square error, but FAM has a high torque ripple. The results of simulation experiments show that the presented off-line optimization procedure can find optimal parameters matching the selected objective. However, it is not possible to achieve the lowest torque ripples and the highest efficiency at the same time. This is in line with the results of other authors [11,19,20,22]. Multi-objective optimization should be used to solve the problem. Herein, the results from the presented off-line optimization procedure can serve this purpose. Once the set of simulation runs is executed and the values of each objective function are calculated, another multi-objective optimization, e.g., Pareto-optimal approach, can be applied.

To conclude, the efficiency of FAM control is comparable with a more advanced TSF method, and it makes sense to use it in cost-effective drives where the high torque ripple does not affect the device, e.g., pumps and fans. More advanced SRM control must be used if low torque ripple is required. The presented off-line optimization procedure can be used for single or multi-objective optimization.

The main contribution of this paper is the quantitative comparison of two control methods for SRM that can help in decision making in the process of designing a commercial drive for existing off-the-shelf switched reluctance motors. The second contribution is the explanation of how to tailor and optimize the parameters of control methods using FEM. Presented off-line calculations with the objective function interpolation reduce the computational time. Once the objective functions have been enumerated, no further FEM computation is needed to apply different types of multi-objective optimization.

In future research, the attention will be given to on-line optimization where the objective function value is obtained from the direct measurement of efficiency, vibration, or acoustic noise.

Author Contributions: Conceptualization, P.B. and Ž.F.; methodology, P.B. and Ž.F.; software ANSYS Maxwell and Simplorer, Ž.F.; software Matlab, P.B.; validation, P.B. and Ž.F.; formal analysis, Ž.F.; investigation, P.B. and Ž.F.; writing P.B.; funding acquisition, Ž.F. All authors have read and agreed to the published version of the manuscript.

Funding: This work was supported by the Slovak Research and Development Agency under the contract No. APVV-16-0206 and APVV-15-0750.

Conflicts of Interest: The authors declare no conflict of interest.

Abbreviations and Symbols

ATC	Average Torque Control
CPLD	Complex Programmable Logic Devices
DITC	Direct Instantaneous Torque Control
DSP	Digital Signal Processor
DTC	Direct Torque Control

ISE	Integral Square Error criterion
FAM	Firing Angle Modulation method
FBL	Feedback Linearization control
FEM	Finite Element Method
FPGA	Field-Programmable Gate Array
IC	Intelligent Control
ILC	Iterative Learning Control
MPU	Micro Processor Unit
rpm	revolution per minute [r/min]
SRM	Switched Reluctance Motor
TSF	Torque Sharing Function
$f_1(T_c,\omega)$	The function for optimal value of θ_{on} parameter of TSF
$f_2(T_c,\omega)$	The function for optimal value of θ_{ov} parameter of TSF
$f_{dn}(\theta)$	The declining part of the TSF shape
$f_{up}(\theta)$	The rising part of the TSF shape
I_{15}	The phase current at angle 15°
I_A, I_B, I_C	The instantaneous current in phase A, B, and C
$I_{Aref}, I_{Bref}, I_{Cref}$	The reference current for phase A, B, and C
I_{max}	The maximum current reference
I_{ref}	The reference current
I_{ref0}	The reference current at angle 0°
L_u	The inductance in unaligned position
n	The rotor speed in [rpm]
$T_{Aref}, T_{Bref}, T_{Cref}$	The torque profile for motor phase A, B, and C
T_{avg}	The average torque
T_c	The torque command
T_{max}	The maximum torque on the interval
T_{min}	The minimum torque on the interval
T_{rip}	The relative torque ripple
V_{DC}	The supply voltage
η	The efficiency
θ	The rotor angle
θ_m	The rotor angle where the inductance begins to rise
θ_{off}	The turn-off angle
θ_{on}	The turn-on angle
θ_{ov}	The overlap angle
θ_p	The rotor pole pitch
ω	The rotor angular speed
ω_{ref}	The rotor angular speed reference

References

1. Santo, A.E.; Calado, M.R.; Cabrita, C.M. Sliding Mode Position Controller for a Linear Switched Reluctance Actuator. In *Sliding Mode Control*; IntechOpen: London, UK, 2011; pp. 181–202. [CrossRef]
2. Ahn, J.-W.; Lukman, G.F.L. Switched Reluctance Motor: Research Trends and Overview. *China Electrotech. Soc. Trans. Electr. Mach. Syst.* **2019**, *2*, 339–347. [CrossRef]
3. Jarvis, R. Davidson's locomotive: How did he do it? *Eng. Sci. Educ. J.* **1996**, *5*, 281–288. [CrossRef]
4. Lawrenson, P.; Stephenson, J.; Blenkinsop, P.; Corda, J.; Fulton, N. Variable-speed switched reluctance motors. *IEE Proc. B Electr. Power Appl.* **1980**, *127*, 253. [CrossRef]
5. More, A. Switched Reluctance Motors Market 2019 Industry Size by Global Major Companies Profile, Competitive Landscape and Key Regions 2025. 2019. Available online: https://www.theexpresswire.com/pressrelease/Switched-Reluctance-Motors-Market-2019-Industry-Size-by-Global-Major-Companies-Profile-Competitive-Landscape-and-Key-Regions-2025-360-Research-Report_10273615 (accessed on 21 March 2020).

6. Kozuka, S.; Tanabe, N.; Asama, J.; Chiba, A. Basic characteristics of 150,000r/min switched reluctance motor drive. In Proceedings of the IEEE PES General Meeting, Pittsburgh, PA, USA, 20–24 July 2008; pp. 1–4. [CrossRef]
7. Miller, T. Optimal design of switched reluctance motors. *IEEE Trans. Ind. Electron.* **2002**, *49*, 15–27. [CrossRef]
8. Belhadi, M.; Krebs, G.; Marchand, C.; Hannoun, H.; Mininger, X. Geometrical optimization of SRM on operating mode for automotive application. *Electr. Eng.* **2017**, *100*, 303–310. [CrossRef]
9. Gan, C.; Wu, J.; Sun, Q.; Kong, W.; Li, H.; Hu, Y. A Review on Machine Topologies and Control Techniques for Low-Noise Switched Reluctance Motors in Electric Vehicle Applications. *IEEE Access* **2018**, *6*, 31430–31443. [CrossRef]
10. Lin, C.-H.; Hwang, C.-C. High Performances Design of a Six-Phase Synchronous Reluctance Motor Using Multi-Objective Optimization with Altered Bee Colony Optimization and Taguchi Method. *Energies* **2018**, *11*, 2716. [CrossRef]
11. Xue, X.; Cheng, K.-W.E.; Ho, S.L. Optimization and Evaluation of Torque-Sharing Functions for Torque Ripple Minimization in Switched Reluctance Motor Drives. *IEEE Trans. Power Electron.* **2009**, *24*, 2076–2090. [CrossRef]
12. Evangeline, J.S.; Kumar, S.S.; Jayakumar, J. Torque modeling of Switched Reluctance Motor using LSSVM-DE. *Neurocomputing* **2016**, *211*, 117–128. [CrossRef]
13. Ye, W.; Ma, Q.; Zhang, P. Improvement of the Torque-Speed Performance and Drive Efficiency in an SRM Using an Optimal Torque Sharing Function. *Appl. Sci.* **2018**, *8*, 720. [CrossRef]
14. Zhang, M.; Bahri, I.; Mininger, X.; Vlad, C.; Xie, H.; Berthelot, E. A New Control Method for Vibration and Noise Suppression in Switched Reluctance Machines. *Energies* **2019**, *12*, 1554. [CrossRef]
15. Cheng, H.; Chen, H.; Yang, Z. Average torque control of switched reluctance machine drives for electric vehicles. *IET Electr. Power Appl.* **2015**, *9*, 459–468. [CrossRef]
16. Fuengwarodsakul, N.H.; Menne, M.; Inderka, R.; De Doncker, R. High-Dynamic Four-Quadrant Switched Reluctance Drive Based on DITC. *IEEE Trans. Ind. Appl.* **2005**, *41*, 1232–1242. [CrossRef]
17. Cai, H.; Wang, H.; Li, M.; Shen, S.; Feng, Y.; Zheng, J. Torque Ripple Reduction for Switched Reluctance Motor with Optimized PWM Control Strategy. *Energies* **2018**, *11*, 3215. [CrossRef]
18. Bose, B.K.; Miller, T.J.E.; Szczesny, P.M.; Bicknell, W.H. Microcomputer Control of Switched Reluctance Motor. *IEEE Trans. Ind. Appl.* **1986**, *22*, 708–715. [CrossRef]
19. Hamouda, M.; Szamel, L. *Reduced Torque Ripple based on a Simplified Structure Average Torque Control of Switched Reluctance Motor for Electric Vehicles*; Institute of Electrical and Electronics Engineers (IEEE): Piscataway, NJ, USA, 2018.
20. Xue, X.D.; Cheng, K.W.E.; Cheung, N.C. Evaluation of torque sharing functions for torque ripple minimization of switched reluctance motor drives in electric vehicles. In Proceedings of the 2008 Australasian Universities Power Engineering Conference, Sydney, NSW, Australia, 14–17 December 2008; pp. 1–6.
21. Ye, J.; Bilgin, B.; Emadi, A. An Extended-Speed Low-Ripple Torque Control of Switched Reluctance Motor Drives. *IEEE Trans. Power Electron.* **2014**, *30*, 1457–1470. [CrossRef]
22. Li, H.; Bilgin, B.; Emadi, A. An Improved Torque Sharing Function for Torque Ripple Reduction in Switched Reluctance Machines. *IEEE Trans. Power Electron.* **2018**, *34*, 1635–1644. [CrossRef]
23. Ro, H.-S.; Lee, K.-G.; Lee, J.-S.; Jeong, H.-G.; Lee, K.-B. Torque Ripple Minimization Scheme Using Torque Sharing Function Based Fuzzy Logic Control for a Switched Reluctance Motor. *J. Electr. Eng. Technol.* **2015**, *10*, 118–127. [CrossRef]
24. Panek, D.; Orosz, T.; Karban, P. Artap: Robust Design Optimization Framework for Engineering Applications. In Proceedings of the 2019 Third International Conference on Intelligent Computing in Data Sciences (ICDS), Marrakech, Morocco, 28–30 October 2019; pp. 1–6.
25. Sovicka, P.; Rafajdus, P.; Vavrus, V. Switched reluctance motor drive with low-speed performance improvement. *Electr. Eng.* **2019**, *102*, 27–41. [CrossRef]
26. Ferkova, Z.; Suchy, L.; Cernohorsky, J. Measurement of switched reluctance motor parameters. In Proceedings of the 2017 19th International Conference on Electrical Drives and Power Electronics (EDPE), Dubrovnik, Croatia, 4–6 October 2017; pp. 287–290. [CrossRef]

Article

A Novel Magnet-Axis-Shifted Hybrid Permanent Magnet Machine for Electric Vehicle Applications

Ya Li, Hui Yang *, Heyun Lin, Shuhua Fang and Weijia Wang

School of Electrical Engineering, Southeast University, Nanjing 210096, China; seueelab_ly@163.com (Y.L.); hyling@seu.edu.cn (H.L.); shfang@seu.edu.cn (S.F.); seueelab_wwj@163.com (W.W.)
* Correspondence: huiyang@seu.edu.cn; Tel.: +86-25-837-4169

Received: 15 January 2019; Accepted: 13 February 2019; Published: 16 February 2019

Abstract: This paper proposes a novel magnet-axis-shifted hybrid permanent magnet (MAS-HPM) machine, which features an asymmetrical magnet arrangement, i.e., low-cost ferrite and high-performance NdFeB magnets, are placed in the two sides of a "▽"-shaped rotor pole. The proposed magnet-axis-shift (MAS) effect can effectively reduce the difference between the optimum current angles for maximizing permanent magnet (PM) and reluctance torques, and hence the torque capability of the machine can be further improved. The topology and operating principle of the proposed MAS-HPM machine are introduced and are compared with the BMW i3 interior permanent magnet (IPM) machine as a benchmark. The electromagnetic characteristics of the two machines are investigated and compared by finite element analysis (FEA), which confirms the effectiveness of the proposed MAS design concept for torque improvement.

Keywords: hybrid permanent magnet; interior permanent magnet (IPM) machine; magnet-axis-shifted; reluctance torque

1. Introduction

Due to their high torque/power density, high efficiency and excellent flux weakening capability, interior permanent magnet (IPM) machines are considered as competitive candidates for electric vehicles (EVs) [1]. In order to improve the reluctance torque and reduce the magnet usage, multi-layer IPM machines are widely employed in EV applications, such as the BMW i3 traction machine [2]. However, for the conventional IPM machines, the optimum current angle for maximizing reluctance and permanent magnet (PM) torques basically differs by a 45° electrical angle, which results in a relatively low utilization ratio of the two torque components. Consequently, in order to deal with this issue, hybrid rotor [3–7], dual rotor [8] and asymmetrical permanent magnet (PM)-assisted synchronous reluctance machines [9] have been recently developed. The constant power-maintaining capabilities of the hybrid rotor configurations are investigated by adopting the parameter equivalent circuit method, which shows that the hybrid rotor topologies have more degrees of freedom for a given constant power operating range [10]. Moreover, the theoretical analysis demonstrates that the PM usages of synchronous machines can be reduced by about 10% with the reluctance axis shifted by a displacement angle of about 60° [11]. The hybrid synchronous machines with a displaced reluctance axis are comparatively studied with conventional pure PM and electrically excited synchronous machines [12], which demonstrates that the hybrid topologies exhibit higher torque and high-efficiency operating range. In addition, the effects of shifting the PM axis with respect to the reluctance axis in PM machines are investigated [13], showing that the asymmetric salient PM machine exhibits higher torque and constant power speed range [14]. Nevertheless, the hybrid and dual rotor machines suffer from complicated structures, while the latter asymmetrical one is characterized by shifts of both magnet and reluctance axes that require relatively sophisticated computational design efforts.

Recently, in order to reduce the use of the rare-earth NdFeB magnets, the hybrid PM concept has been proposed and developed in rotor PM [15,16] and stator PM [17–22] configurations. Compared with the structure of conventional spoke-type magnets, the proposed hybrid PM topology exhibits better field weakening capability and lower total cost [15]. Besides, compared with a double-layer PM structure, the U-shaped configuration has good irreversible demagnetization withstanding capability [16]. Due to the variable magnetization state of the low-coercive-force AlNiCo magnets, the flexible air-gap flux adjustment and wide operating range with high efficiency can be readily achieved in stator hybrid PM machines [16–22]. A novel magnet-axis-shifted hybrid PM (MAS-HPM) machine combined with the asymmetric and hybrid PM concepts is proposed in this paper.

The purpose of this paper is to propose an MAS-HPM machine for torque performance improvement. The proposed configuration features an asymmetrical PM arrangement, i.e., low-cost ferrite and high-performance NdFeB magnets, which significantly reduces the difference of the optimum current angle for maximizing PM and reluctance torques. Hence, the torque capability can be further improved. In order to validate the merits of the magnet-axis-shift (MAS) effect, the IPM machine of an BMW i3 vehicle is used as a benchmark. The basic electromagnetic characteristics of the two machines are comparatively investigated, which confirms the validity of the proposed MAS design concept.

2. Machine Topologies and Magnet-Axis-Shift Principle

2.1. Machine Topologies

The topologies of the benchmark 2016 BMW i3 IPM machine and the proposed MAS-HPM machine are shown in Figure 1a,b, respectively. The main design parameters are tabulated in Table 1. It should be noted that the proposed machine shares the same inverter power ratings, stator structure, active stack length and air-gap length with the BMW i3 IPM machine. Meanwhile, in order to make a fair comparison, the rare-earth PM usages are identical in the two structures. The main difference between the two machines lies in the fact that two kinds of PM, i.e., low-cost ferrite and high-performance NdFeB magnets, are simultaneously employed in the developed MAS-HPM machine to achieve the MAS effect. The total costs of the magnets are given in Table 1. Due to the additional ferrite magnets, the proposed machine has a slightly higher total cost of magnets than the BMW i3 IPM counterpart. However, compared with the BMW i3 IPM machine, the ratio of the peak torque to the total cost of magnets in the MAS-HPM configuration is increased by about 7.81%, which indicates that the torque capability can be improved by 7.81% at the same cost of magnets.

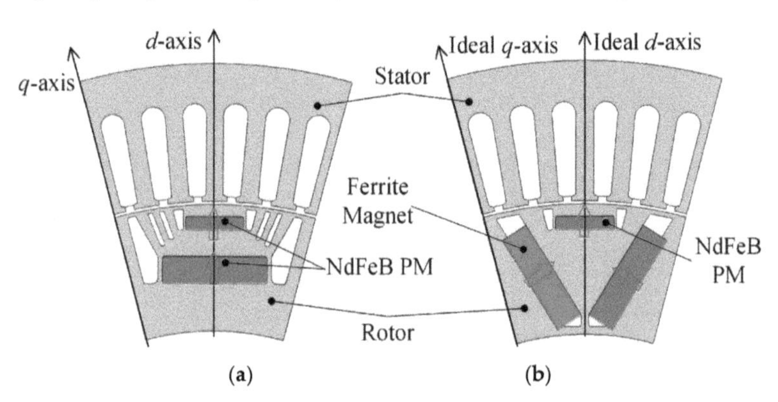

Figure 1. Machine topologies. (**a**) BMW i3 interior permanent magnet (IPM) machine. (**b**) The proposed magnet-axis-shifted hybrid permanent magnet (MAS-HPM) machine.

Table 1. Main design parameters of the machines.

<table>
<tr><th>Items</th><th>BMW i3 IPM</th><th>MAS-HPM</th></tr>
<tr><td>Stator slot number</td><td colspan="2">72</td></tr>
<tr><td>Rotor pole pair number</td><td colspan="2">6</td></tr>
<tr><td>Stator outer radius (mm)</td><td colspan="2">121</td></tr>
<tr><td>Air-gap length (mm)</td><td colspan="2">0.7</td></tr>
<tr><td>Rotor outer radius (mm)</td><td colspan="2">89.3</td></tr>
<tr><td>Active stack length (mm)</td><td colspan="2">132</td></tr>
<tr><td>Peak current (A)</td><td colspan="2">530</td></tr>
<tr><td>Steel grade</td><td colspan="2">TKM330-35</td></tr>
<tr><td>NdFeB grade</td><td colspan="2">N35EH</td></tr>
<tr><td>Ferrite magnet grade</td><td>-</td><td>AC-12</td></tr>
<tr><td>NdFeB PM volume (mm^3)</td><td colspan="2">24,816</td></tr>
<tr><td>Ferrite magnet volume (mm^3)</td><td>-</td><td>19,565</td></tr>
<tr><td>Total cost of magnets (¥)</td><td>51.3</td><td>53.1</td></tr>
<tr><td>Peak torque (Nm)</td><td>269.34</td><td>300.51</td></tr>
<tr><td>Peak torque/total cost of magnets (Nm/¥)</td><td>5.25</td><td>5.66</td></tr>
<tr><td>Working temperature (°C)</td><td colspan="2">100</td></tr>
</table>

The d and q-axes' equivalent electrical circuits are illustrated in Figure 2. In the synchronous reference frame, the voltage equations for the PM synchronous machine are expressed as:

$$\begin{cases} u_d = Ri_d - \omega\psi_q \\ u_q = Ri_q + \omega\psi_d \end{cases}, \tag{1}$$

where R is the stator resistance, ω is the electric frequency, ψ_d and ψ_q are the d and q-axes' flux linkages, respectively. i_d and i_q are the d and q-axes' current, respectively.

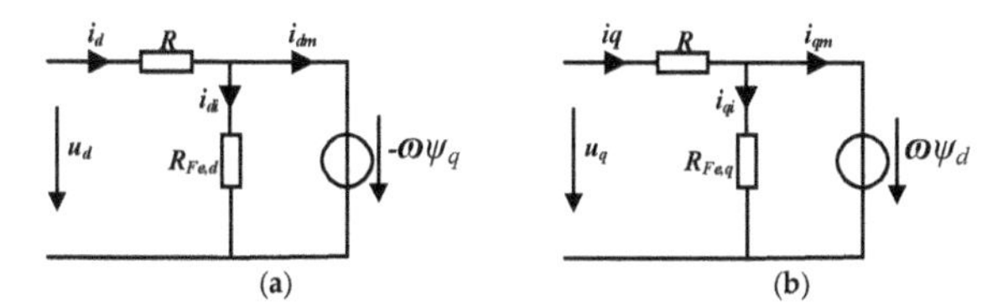

Figure 2. d and q-axes' equivalent electrical circuits. (**a**) d-axis. (**b**) q-axis.

By the application of Kirchhoff's voltage and current laws to both d and q-axes, the four equations can be obtained as:

$$\begin{cases} R_{Fe,d}i_{di} + \omega\psi_q = 0 \\ -R_{Fe,q}i_{qi} + \omega\psi_d = 0 \\ i_d - i_{di} - i_{dm} = 0 \\ i_q - i_{qi} - i_{qm} = 0 \end{cases}, \tag{2}$$

where $R_{Fe,d}$ and $R_{Fe,q}$ are the iron losses resistances in d and q-axes, respectively. i_{di} and i_{qi} are the iron losses currents in d and q-axes, respectively. i_{dm} and i_{qm} are the d and q-axes' magnetization currents, respectively.

2.2. MAS Principle

The total torque T_{total} of an IPM machine, including the PM torque T_{PM} and the reluctance torque T_r, can be expressed as [23]:

$$T_{total} = \frac{3}{2}p_r\psi_f i_s \cos\beta + \frac{3}{4}p_r(L_d - L_q)i_s^2 \sin 2\beta = T_{PM} + T_r, \tag{3}$$

$$T_{PM} = \frac{3}{2}p_r\psi_f i_s \cos\beta, \tag{4}$$

$$T_r = \frac{3}{4} p_r (L_d - L_q) i_s^2 \sin 2\beta, \tag{5}$$

where p_r, ψ_f, i_s, L_d and L_q are the rotor pole pair number, the PM flux linkage, the phase current and the d- and q-axes' inductances, respectively. β is the current angle, which is defined as the angle between the phase current and open-circuit back electro-motive force (EMF) [24].

From Equations (3)–(5), it can be found that the optimum current angle for T_r is theoretically twice that for T_{PM}. If the difference between the optimum current angles for maximizing the two kinds of torques can be reduced, the torque capability of the machine will be improved. To achieve this goal, this paper proposes an asymmetrical PM arrangement by employing the HPM configuration, i.e., low-cost ferrite and high-performance NdFeB magnets, which is termed as the MAS effect. In this case, the magnet axis is shifted while the reluctance axis is unchanged due to the symmetrical rotor configuration. Thus, the difference of the current angles γ_s when both T_{PM} and T_r reach the maximum can be reduced, which can be defined as:

$$\gamma_s = \beta_{PM} - \beta_R, \tag{6}$$

where β_R and β_{PM} are the optimum current angles for the reluctance and PM torques, respectively.

The flux density distributions of the two machines are calculated by finite element analysis (FEA) and illustrated in Figure 3. It can be seen that the d-axis shifted by an angle α_s in the proposed machine under the open-circuit condition, as shown in Figure 3a, which confirms the MAS effect. The reluctance d and q-axes are not changed in the two machines, as shown in Figure 3b, which is mainly attributed to the design of the symmetrical flux barriers in the two rotor configurations. The flux density distributions of the two machines at the peak current load condition are given in Figure 3c. Due to dual excited armature windings and PMs, the two machines under the load condition have higher flux densities than at other operating conditions.

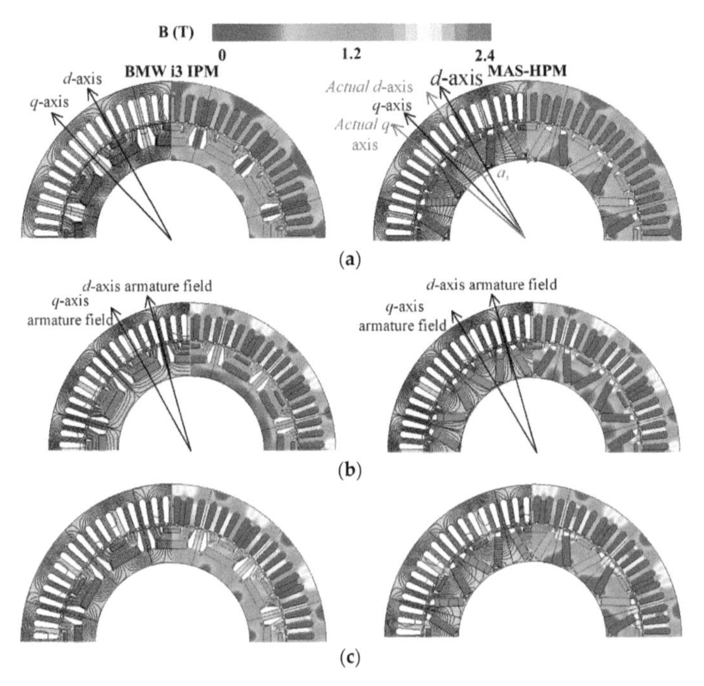

Figure 3. Flux density distributions of the two machines. (**a**) Open-circuit. (**b**) Reluctance axis with only armature windings excited. (**c**) Peak current load condition.

To clearly understand the MAS effect, the open-circuit air-gap flux density waveforms are given in Figure 4. Compared with the *d*-axis in the BMW i3 IPM machine, the displacement of the actual *d*-axis occurred in the proposed topology, which means that the magnet and reluctance axes grow closer by using the HPM configuration. Consequently, the resultant current angles for optimizing the reluctance and PM torques are closer, which enables the torque improvement. Moreover, the fundamental amplitude of the air-gap flux density in the MAS-HPM machine are found to be 53.70% higher than that of the BMW i3 IPM machine, as reflected in Figure 4b. Due to the asymmetrical PM configuration, larger high-order harmonics of the air-gap flux density are observed in the MAS-HPM machine.

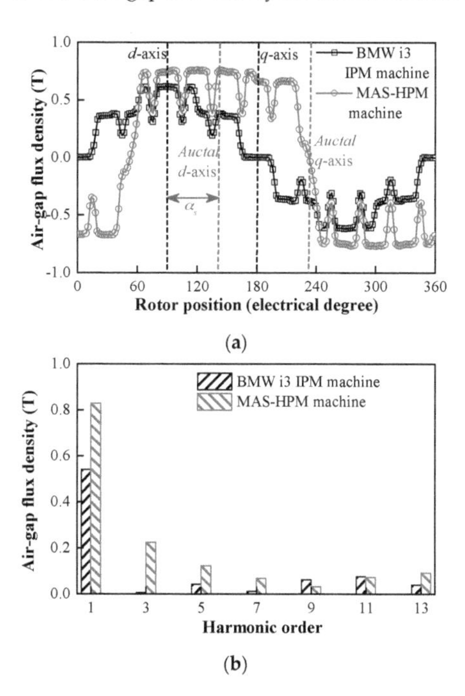

Figure 4. Open-circuit air-gap flux density. (**a**) Waveforms. (**b**) Spectra.

3. Electromagnetic Performance Comparison

In order to validate the MAS effect, the basic electromagnetic characteristics of the proposed MAS-HPM machine are comparatively studied with those of the BMW i3 IPM machine in this section. In order to reduce the computational time, 1:12 scale models are adopted for the two machines. The simulation time is 2.5 h.

3.1. Open-Circuit Performance

The back EMF waveforms of the two investigated machines are shown in Figure 5. Compared with the BMW i3 IPM machine, the proposed configuration exhibits a 53.54% higher back-EMF fundamental amplitude, which indicates that the magnet torque can be effectively improved by using the HPM configuration. In addition, the cogging torque waveforms of the two machines are shown in Figure 6, which experience the same periods due to the same numbers of stator slots and rotor poles. Because the air-gap flux density contains larger high-order harmonics, as shown in Figure 3b, the MAS-HPM structure has a higher cogging torque amplitude. The ratios of the cogging torque amplitudes to the corresponding peak torque values in BMW i3 IPM and MAS-HPM machines are 0.73% and 2.04%, respectively, which are lower than the acceptable value of 2.5%.

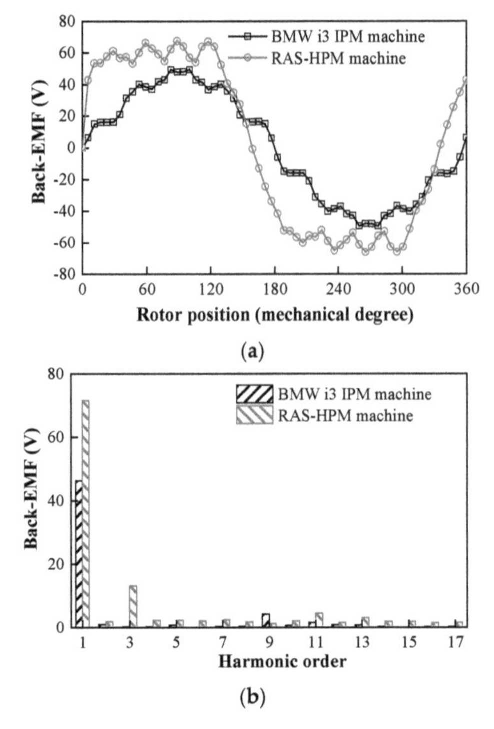

Figure 5. Back electro-motive force (EMF) waveforms at 3000 rpm. (**a**) Waveforms. (**b**) Spectra.

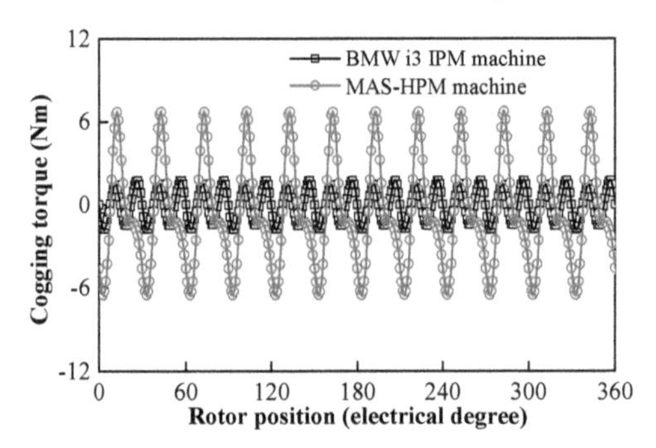

Figure 6. Cogging torque waveforms.

3.2. *Torque Characteristics*

The torque versus current angle characteristics of the two machines are illustrated in Figure 7. The PM and reluctance torques are separated by using the frozen permeability method [25]. It can be seen that the γ_s of the proposed MAS-HPM machine is smaller than that of the BMW i3 machine. As a result, a higher torque capability can be obtained in the HPM case, as evidenced in Figure 8. Moreover, due to the MAS effect, the ripple patterns of the PM and reluctance torques of the proposed machine are different, which results in a torque ripple offset effect. Hence the HPM configuration exhibits 55.99% lower torque ripple than the BMW i3 IPM machine, as shown in Figure 8b. The average torques versus phase current curves of the two machines are shown in Figure 9. It can be observed that the MAS-HPM machine has a higher torque capability regardless of the applied loads. As a whole, the feasibility of the proposed MAS-HPM design for torque performance improvement is clearly confirmed.

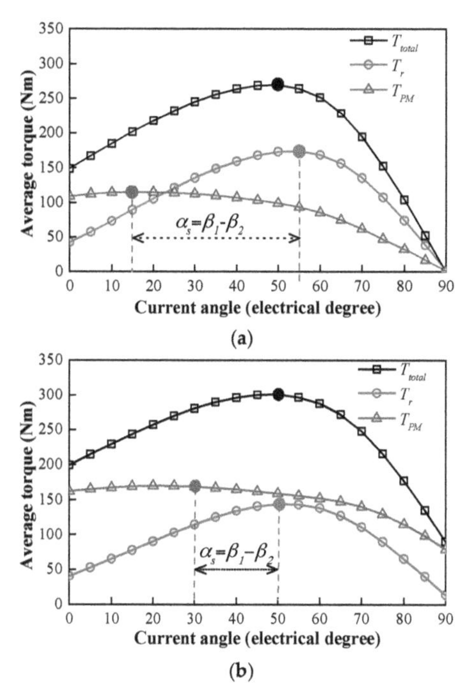

Figure 7. Torque versus current angle characteristics. (**a**) BMW i3 IPM machine. (**b**) MAS-HPM machine.

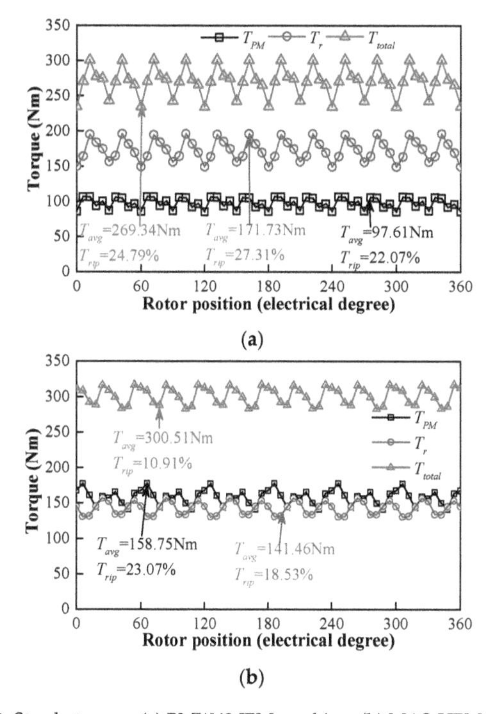

Figure 8. Steady torque. (**a**) BMW i3 IPM machine. (**b**) MAS-HPM machine.

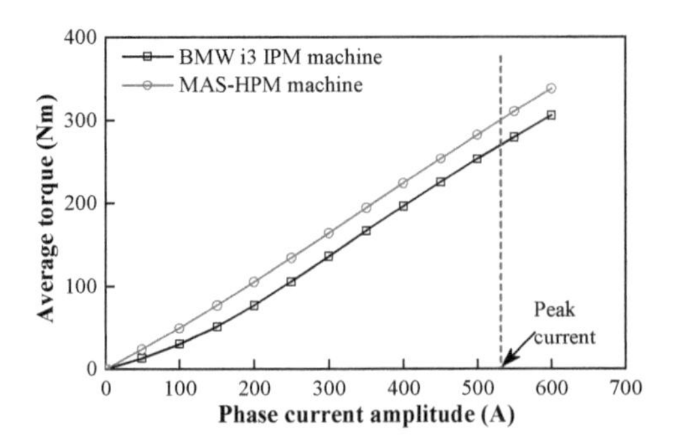

Figure 9. Average torque versus phase current curves.

3.3. Torque/Power versus Speed Curves

The torque and power versus speed curves of the two machines are illustrated in Figure 10. It can be seen that the MAS-HPM machine exhibits higher torque and power than the BMW i3 IPM machine over the whole operating range, consequently achieving a better high-speed constant power-maintaining capability.

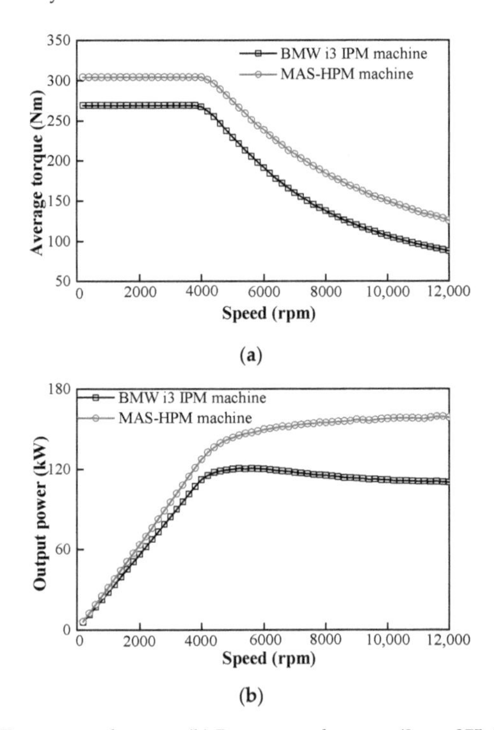

Figure 10. (**a**) Torque–speed curves. (**b**) Power–speed curves. (I_{rms} = 375 A, U_{dc} = 360 V).

3.4. Irreversible Demagnetization

The flux density distributions of the magnets are illustrated in Figure 11. When the working temperature is set as 100 °C, the knee points of ferrite and NdFeB magnets are −0.15 and −0.6 T, respectively. It can be observed that the irreversible demagnetization of ferrite and NdFeB magnets does not occur. In order to quantitatively illustrate the flux density variations of magnets, five typical points are selected in three magnets, as shown in Figure 11. The corresponding flux density variations of the typical five points on magnets are given in Figure 12. It can be seen that the working points of ferrite and NdFeB magnets are greater than the respective knee points, which indicates that good demagnetization withstanding capability can be achieved.

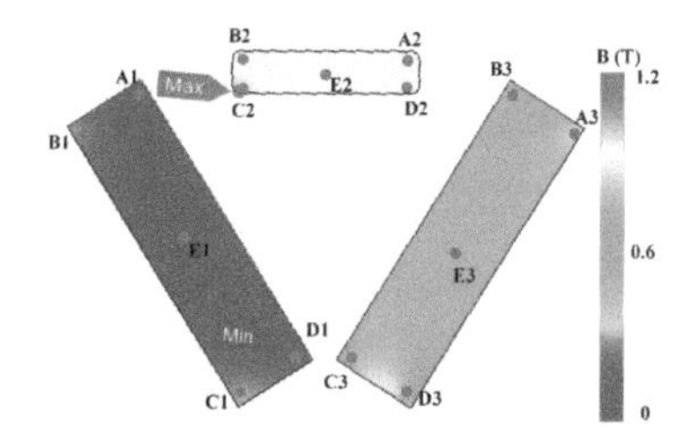

Figure 11. Flux density distributions of ferrite and NdFeB magnets.

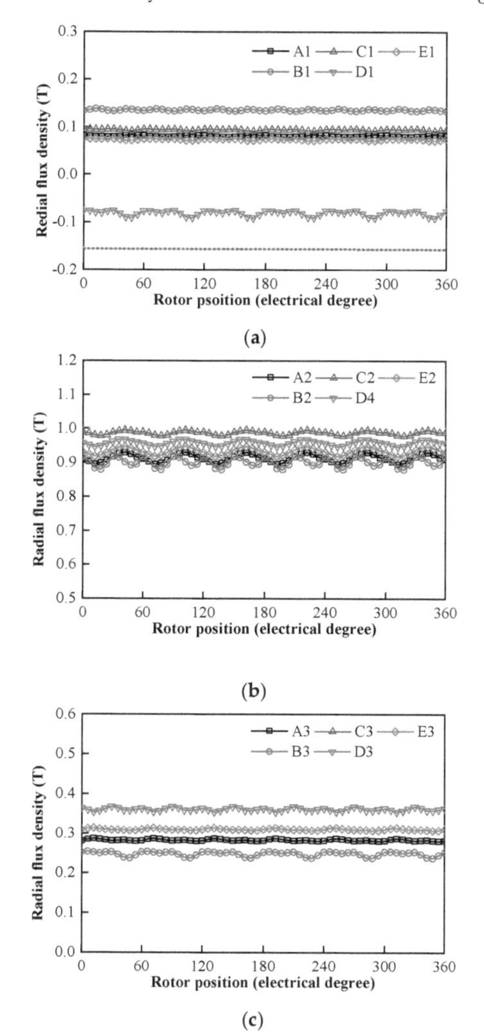

Figure 12. Variations of the working points of the five points on the ferrite and NdFeB magnets. (**a**) Ferrite magnet. (**b**) First-layer NdFeB magnet. (**c**) Second-layer NdFeB magnet.

3.5. Rotor Mechanical Analyses

The rotor mechanical strengths of the two machines are investigated at the maximum speed of 12,000 rpm in this section. The von Mises stress maps are shown in Figure 13. It can be observed that the peak stress of the MAS-HPM machine (268.8 MPa) is slightly lower than that of the BMW i3 IPM machine (282.4 MPa), which are both lower than the threshold yield value (396 MPa). Due to the differences in mesh subdivision, it can be observed that the mismatch between the maximal values occurs at the two sides of the symmetrical configurations. However, the difference in stress values between the points of the symmetrical structure is very small and thus negligible, as shown in Figure 13. As a result, it was confirmed that the proposed rotor configuration can withstand a larger centrifugal force at the maximum speed of 12,000 rpm.

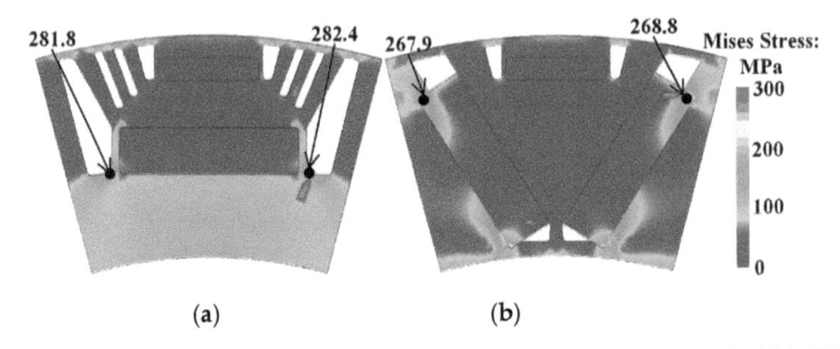

Figure 13. Rotor von Mises stress distributions at the maximum speed (12,000 rpm). (**a**) BMW i3 IPM machine. (**b**) MAS-HPM machine.

3.6. Loss and Efficiency

The two machines have the same stators and windings, which indicates that the same copper losses can be achieved. The iron loss and efficiency are calculated and illustrated in Figures 14 and 15, respectively. The iron loss p_i, eddy-current loss p_e and hysteresis loss p_h in the laminated core are calculated as follows [26]:

$$p_i = p_h + p_e, \tag{7}$$

$$p_e = \sum_n \left\{ \int_{iron} K_e D (nf)^2 (B_{r,n}^2 + B_{\theta,n}^2) dv \right\}, \tag{8}$$

$$p_h = \sum_n \left\{ \int_{iron} K_h D (nf) (B_{r,n}^2 + B_{\theta,n}^2) dv \right\}, \tag{9}$$

where K_e and K_h are the experimental constants obtained by the Epstein frame test of the core material, D is the density of the steel sheets, f is the fundamental frequency, B_r, n and $B_{\theta,n}$ are the radial and tangential components of the flux density at each finite element.

The copper loss p_{cu} and efficiency η can be calculated by:

$$p_{cu} = 3R_a I^2 \tag{10}$$

$$\eta = \frac{\omega T}{\omega T + p_i + p_{cu}} \times 100\%, \tag{11}$$

where R_a, I, and ω are the armature winding resistance, the phase current and the mechanical angular velocity, respectively.

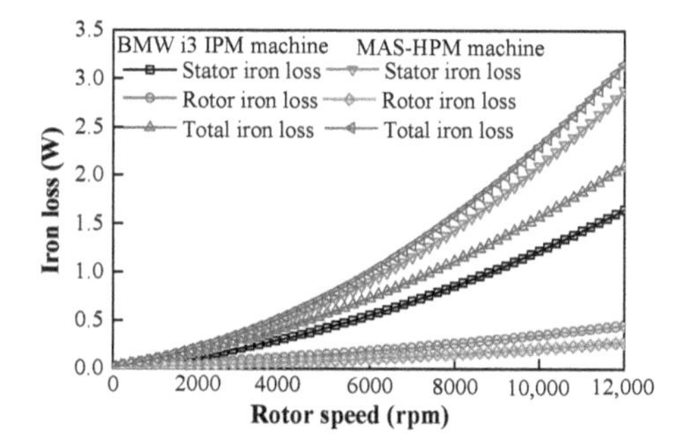

Figure 14. Iron losses versus speed curves.

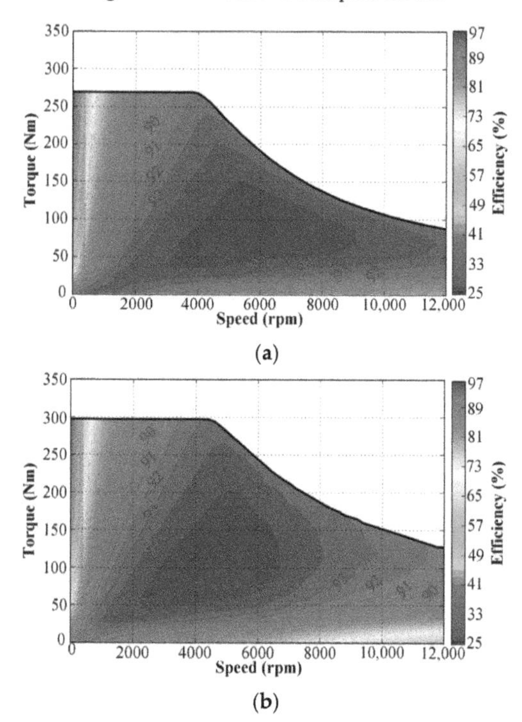

Figure 15. Efficiency maps. (**a**) BMW i3 IPM machine. (**b**) MAS-HPM machine.

The iron losses of the two machines under different speeds are given in Figure 14. It can be observed that the stator iron losses dominate the total iron losses in both machines at the rated load. The iron losses of the two machines are very close when the speed is lower than 4000 rpm. However, due to higher harmonics, the HPM structure produces a larger iron loss than the BMW i3 IPM machine when the speed exceeds 4000 rpm. Furthermore, the efficiency maps of the two cases are illustrated in Figure 15. The maximum efficiency of the proposed MAS-HPM machine (95.79%) is slightly higher than that of the BMW i3 IPM (95.57%). Due to the higher iron loss in high speed range, the proposed structure shows a relatively lower efficiency when the speed exceeds 10,000 rpm. Nevertheless, the MAS-HPM machine still exhibits a similar operating range when the efficiency is higher than 93%.

4. Conclusions

A novel MAS-HPM machine is proposed to achieve a higher torque capability and a wider high-efficiency operation range for EV applications in this paper. The basic electromagnetic characteristics of the proposed MAS-HPM machine and the benchmark BMW i3 IPM machine are comprehensively investigated and compared by FEA. Due to the MAS effect, the difference between the optimal current angles of maximizing the magnet and reluctance torques is reduced. In addition, it was found that the back-EMF and total torque of the proposed MAS-HPM machine can be effectively improved, compared with the conventional BMW i3 IPM machine. Moreover, the proposed machine shows lower peak mechanical stress, better field-weakening capability, higher peak efficiency and comparable high-efficiency operating range, which confirms the effectiveness of the proposed MAS design concept for performance improvement. However, due to higher harmonics, the proposed MAS-HPM configuration has higher cogging torque and iron losses in a high speed operating range.

Author Contributions: Conceptualization, Y.L. and H.Y.; methodology, Y.L. and H.Y.; software, Y.L. and W.W.; validation, Y.L., H.Y. and H.L., formal analysis, Y.L. and H.Y.; investigation, Y.L. and H.Y.; resources, Y.L. and H.Y.; data curation, Y.L. and H.Y.; writing—original draft preparation, Y.L. and H.Y.; writing—review and editing, Y.L., H.Y., H.L. and S.F.; visualization, Y.L. and H.Y.; supervision, H.Y., H.L., and S.F.; project administration, H.Y.; funding acquisition, H.Y. and H.L.

Funding: This work was jointly supported in part by National Natural Science Foundations of China under Grant 51377036, in part by Natural Science Foundation of Jiangsu Province for Youth (BK20170674), in part by the Fundamental Research Funds for the Central Universities (2242017K41003) and in part by the Postgraduate Research and Practice Innovation Program of Jiangsu Province (KYCX18_0096).

Conflicts of Interest: The authors declare no conflict of interest.

References

1. Zhu, Z.Q.; Howe, D. Electrical Machines and Drives for Electric, Hybrid, and Fuel Cell Vehicles. *Proc. IEEE* **2007**, *95*, 745–765. [CrossRef]
2. Burress, T.; Rogers, S.A.; Ozpineci, B. *FY 2016 Aannual Progress Report for Electric Drive Technologies Program*; Oak Ridge National Laboratory Department of Energy: Oak Ridge, TN, USA, 2016; pp. 196–207. Available online: https://www.energy.gov/sites/prod/files/2017/08/f36/FY16%20EDT%20Annual%20Report_FINAL (accessed on 5 February 2019).
3. Chalmers, B.J.; Musaba, L.; Gosden, D.F. Variable-Frequency Synchronous Motor Drives for Electric Vehicles. *IEEE Trans. Ind. Appl.* **1996**, *32*, 896–903. [CrossRef]
4. Chalmers, B.J.; Akmese, R.; Musaba, L. Design and Field-weakening Pperformance of Permanent Magnet/Reluctance Motor with Two-part Rotor. *IEE Proc.-Electr. Power Appl.* **1998**, *145*, 133–139. [CrossRef]
5. Chen, X.; Gu, C.; He, X.; Shao, H. Experimental Research on the ALA+SPM Hybrid Rotor Machine. In Proceedings of the International Conference on Electrical Machines and Systems, Beijing, China, 20–23 August 2011; pp. 1–4.
6. Yang, H.; Li, Y.; Lin, H.; Zhu, Z.Q.; Lyu, S.; Wang, H.; Fang, S.; Huang, Y. Novel Reluctance Axis Shifted Machines with Hybrid Rotors. In Proceedings of the Energy Conversion Congress and Exposition (ECCE), Cincinnati, OH, USA, 1–5 October 2017; pp. 2362–2367.
7. Liu, G.; Xu, G.; Zhao, W.; Du, X.; Chen, Q. Improvement of Torque Capability of Permanent-Magnet Motor by Using Hybrid Rotor Configuration. *IEEE Trans. Energy Convers.* **2017**, *32*, 953–962. [CrossRef]
8. Li, Y.; Bobba, D.; Sarlioglu, B. Design and Optimization of a Novel Dual-Rotor Hybrid PM Machine for Traction Application. *IEEE Trans. Ind. Electron.* **2018**, *65*, 1762–1771. [CrossRef]
9. Zhao, W.; Chen, D.; Lipo, T.A.; Kwon, B. Performance Improvement of Ferrite-Assisted Synchronous Reluctance Machines Using Asymmetrical Rotor Configurations. *IEEE Trans. Magn.* **2015**, *51*, 8108504. [CrossRef]
10. Randi, S.A.; Astier, S. Parameters of Salient Pole Synchronous Motor Drives with Two-Part Rotor to Achieve a Given Constant Power Speed Range. In Proceedings of the IEEE 32nd Annual Power Electronics Specialists Conference, Vancouver, BC, Canada, 17–21 June 2001; pp. 1673–1678.

11. Winzer, P.; Doppelbauer, M. Theoretical Analysis of Synchronous Machines with Displaced Reluctance Axis. In Proceedings of the International Conference on Electrical Machines (ICEM), Berlin, Germany, 2–5 September 2014; pp. 641–647.
12. Winzer, P.; Doppelbauer, M. Comparison of Synchronous Machine Designs with Displaced Reluctance Axis Considering Losses and Iron Saturation. In Proceedings of the IEEE International Electric Machines & Drives Conference (IEMDC), Coeur d'Alene, ID, USA, 10–13 May 2015; pp. 1801–1807.
13. Alsawalhi, J.Y.; Sudhoff, S.D. Effects of Positioning of Permanent Magnet Axis Relative to Reluctance Axis in Permanent Magnet Synchronous Machines. In Proceedings of the IEEE Power and Energy Conference at Illinois (PECI), Champaign, IL, USA, 20–21 February 2015; pp. 1–8.
14. Alsawalhi, J.Y.; Sudhoff, S.D. Design Optimization of Asymmetric Salient Permanent Magnet Synchronous Machines. *IEEE Trans. Energy Convers.* **2016**, *31*, 1315–1324. [CrossRef]
15. Zhu, X.; Zhang, X.W.C.; Wang, L.; Wu, W. Design and analysis of a spoke-type hybrid permanent magnet motor for electric vehicles. *IEEE Trans. Magn.* **2017**, *53*, 8208604. [CrossRef]
16. Jeong, C.L.; Hur, J. Design technique for PMSM with hybrid type permanent magnet. In Proceedings of the IEEE International Electric Machines and Drives Conference (IEMDC), Miami, FL, USA, 21–24 May 2017; pp. 1–6.
17. Li, G.J.; Zhu, Z.Q. Hybrid excited switched flux permanent magnet machines with hybrid magnets. In Proceedings of the 8th International Conference on Power Electronics, Machines and Drives (PEMD 2016), Glasgow, UK, 19–21 April 2016; pp. 1–6.
18. Yang, H.; Lin, H.; Zhu, Z.Q.; Wang, D.; Fang, S.; Huang, Y. A variable-flux hybrid-PM switched-flux memory machine for EV/HEV applications. *IEEE Trans. Ind. Appl.* **2016**, *52*, 2203–2214. [CrossRef]
19. Yang, H.; Zhu, Z.Q.; Lin, H.; Zhan, H.L.; Hua, H.; Zhuang, E.; Fang, S.; Huang, Y. Hybrid-excited switched-flux hybrid magnet memory machines. *IEEE Trans. Magn.* **2016**, *52*, 8202215. [CrossRef]
20. Yang, H.; Zhu, Z.Q.; Lin, H.; Wu, D.; Hua, H.; Fang, S.; Huang, Y. Novel high-performance switched flux hybrid magnet memory machines with reduced rare-earth magnets. *IEEE Trans. Ind. Appl.* **2016**, *52*, 3901–3915. [CrossRef]
21. Yang, H.; Zhu, Z.Q.; Lin, H.; Fang, S.; Huang, Y. Synthesis of hybrid magnet memory machines having separate stators for traction applications. *IEEE Trans. Veh. Technol.* **2018**, *67*, 183–195. [CrossRef]
22. Yang, H.; Lin, H.; Zhu, Z.Q.; Fang, S.; Huang, Y. A dual-consequent-pole Vernier memory machine. *Energies* **2016**, *9*, 134. [CrossRef]
23. Jahns, T.M.; Kliman, G.B.; Neumann, T.W. Interior permanent-magnet synchronous motors for adjustable-speed drives. *IEEE Trans. Ind. Appl.* **1986**, *IA-22*, 738–747. [CrossRef]
24. Gieras, J.F. *Permanent Magnet Motor Technology: Design and Application*, 3rd ed.; CRC Press: Boca Raton, FL, USA, 2009.
25. Chu, W.Q.; Zhu, Z.Q. Average torque separation in permanent magnet synchronous machines using frozen permeability. *IEEE Trans. Magn.* **2013**, *49*, 1202–1210. [CrossRef]
26. Yamazaki, K.; Abe, A. Loss investigated of interior permanent-magnet motors considering carrier harmonics and magnet eddy currents. *IEEE Trans. Ind. Appl.* **2009**, *45*, 659–665. [CrossRef]

 *energies* 

Article

Design and Analysis of Outer Rotor Permanent-Magnet Vernier Machines with Overhang Structure for In-Wheel Direct-Drive Application

Dong Yu, Xiaoyan Huang *, Lijian Wu and Youtong Fang

College of Electrical Engineering, Zhejiang University, Hangzhou 310027, China; 11410025@zju.edu.cn (D.Y.); lijianwu@zju.edu.cn (L.W.); youtong@zju.edu.cn (Y.F.)
* Correspondence: xiaoyanhuang@zju.edu.cn; Tel.: +86-8795-3134

Received: 26 February 2019; Accepted: 28 March 2019; Published: 1 April 2019

Abstract: This paper presents a novel outer rotor permanent-magnet vernier machine (PMVM) for in-wheel direct-drive application. The overhang structures of the rotor and flux modulation pole (FMP) are introduced. The soft magnetic composite (SMC) was adopted in the FMP overhang to allow more axial flux. The 3-D finite element analysis (FEA) was carried out to prove that the proposed machine can effectively utilize the end winding space to enhance the air-gap flux density. Hence the PMVM can offer 27.3% and 14.5% higher torque density than the conventional machine with no overhang structure and the machine with only rotor overhang structure, respectively. Nevertheless, the efficiency of the proposed machine is slightly lower than the conventional ones due to the extra losses from the overhang structures.

Keywords: permanent-magnet vernier machine; in-wheel direct-drive; outer rotor; overhang; soft magnetic composite

---

1. Introduction

The vernier machine was first proposed in the form of the vernier reluctance machine without any attention at that time [1]. Then, with the rapid development of the permanent magnet (PM) materials, the permanent-magnet vernier machine (PMVM) has been highlighted in recent years in various applications requiring high torque and power density, such as the in-wheel direct-drive system for the electric vehicle (EV) [2–6]. The PMVM can effectively utilize the magnetic flux harmonics to achieve higher output torque at a low speed due to the "magnetic gearing effect". Nevertheless, a relatively poor power factor (PF) is inevitable because of its inherent nature. Thus, an inverter with higher power rating is required, which will increase the size and cost of the whole system. The 3-D finite element analysis (FEA) and experiment results indicated that by replacing the surface-mounted PM with the spoke-type PM, the output torque capability and PF of the PMVMs can be significantly improved [7–10]. In addition, the PMVM with concentrated winding and the multitooth flux modulation poles (FMPs) can effectively shorten the length of the end winding, reduce the copper loss, and consequently improve the efficiency of the machine [11–14].

The PMVM with only the rotor overhang structure received more attention recently for applications with limited space in axial direction. By utilizing the end winding space, the rotor overhang structure can enhance the air-gap flux density and therefore the torque density. Both the analytical method and the finite element method (FEM) were carried out to analyze the effect of this structure [15–18]. Nevertheless, the conventional overhang structures only increase the axial stack length of the rotor. The effect could be compromised when the FMP remains the same as in the conventional design.

In this paper, a novel outer rotor PMVM with the overhang structure of the rotor and FMP will be proposed for the in-wheel direct-drive application. In Section 2, the operation principle and topology of the proposed machine are introduced. In Section 3, 3-D FEM simulations are carried out to verify the design of the proposed machine structure.

2. Operation Principle and Machine Configurations

The modulation principle is based on the "magnetic gearing effect", which means that a low-speed rotation of the PMs will cause dramatic variations in flux. Furthermore, the flux will interact with the high-speed rotating field generated by armature windings to produce the torque [19,20]. The FMPs are the key to cause these two rotating fields' modulation in a PMVM. To take advantage of the magnetic gearing effect, the number of FMPs (Z_s), the armature winding fundamental pole pair (p_s), and the PM pole pair (p_r) should meet the following equation:

$$p_r = Z_s \pm p_s \tag{1}$$

Figure 1 shows the configurations of the three PMVMs, M I, M II, and the proposed M III. All of them have the outer rotor structure and surface-mounted PMs. Each machine has the same Z_s, p_r, and p_s combination of 36/28/8 and their stator teeth are split into two FMPs, which makes the stator feasible to adopt concentrated winding to shorten the length of end winding. The double-layer concentrated windings with Y connection were applied in this design as shown in Figure 1g. In order to transmit torque at different speeds, the fundamental space harmonics velocity in the stator should be G_r times higher than that in the rotor, which is

$$G_r = \frac{\omega_s}{\omega_r} = \frac{p_s - Z_s}{p_s} \tag{2}$$

where ω_s and ω_r are the stator and rotor magnetic field rotating speeds, respectively.

The M II PMVM with the 10 mm bilateral rotor overhang structure is shown in Figure 1b. The length of overhang is determined according to the length of end winding. Apart from the difference in rotor, the stator of M II is also optimized for high torque density; the main differences are listed in Table 1.

In the proposed M III PMVM, both rotor overhang and FMP overhang are implemented to maximize the output torque. The FMPs have the same length of the rotor with overhang structure to provide the shortcut for the flux from the rotor overhang to the stator. Moreover, the FMPs of the proposed machine are partitioned from the stator to reduce the losses and cost of the SMC as shown in Figure 1c–f.

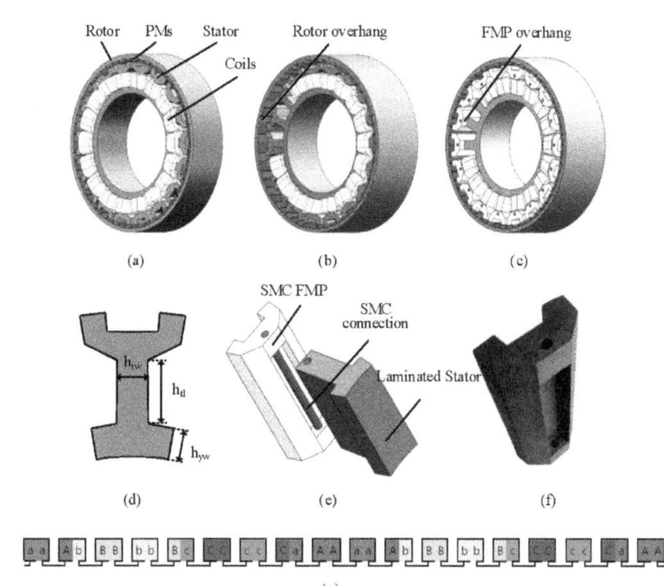

Figure 1. The configurations of the three machines. (**a**) M I: conventional PMVM. (**b**) M II: PMVM with rotor overhang. (**c**) M III: proposed PMVM with rotor and FMP overhang. (**d**) Cross-section of conventional stator tooth. (**e**) Side view of the proposed stator tooth. (**f**) FMP prototype. (**g**) Winding configuration.

Table 1. Main specifications of the three machines.

Parameters	M I	M II	M III
Number of stator slots		18	
Number of FMPs		36	
Number of rotor pole pairs		28	
Number of winding pole pairs		8	
Gear ratio		−28:8	
Rotor outer diameter (mm)		250	
Stator outer diameter (mm)		228	
Stator inner diameter (mm)	140	134	128
Air-gap length (mm)		1	
PM width (mm)		12	
PM thick (mm)		4	
Rotor axial length (mm)	60	80	80
Stator axial length (mm)	60	60	60
FMP axial length (mm)	60	60	80
Stator tooth width h_{tw} (mm)	10.0	11.5	13.0
Stator tooth length h_{tl} (mm)	20.3	22.5	24.4
Stator yoke width h_{yw} (mm)	8.0	9.0	10.0
Stator and rotor material		50ww350	
FMP material	50ww350	50ww350	Somaloy700
PM material	NdFeB38EH	NdFeB38EH	NdFeB38EH
Rated speed (rpm)		600	
Rated current (A)		20	
Current density (A/mm^2)		5	
Slot packing factor		0.6	
Turns per slot		40	
Winding resistance (Ω)		0.088	

The back electromotive force (EMF) and rated load torque increase versus overhang length of M II and M III are shown in Figure 2. It should be noted that when the length of the overhang is under 2 mm, the M II produces higher back EMF and torque than the M III. There are mainly two reasons: one is that when the overhang is short, the air-gap reluctance and the flux leakage of the rotor overhang is relatively small; and the other one is that the permeability of SMC is slightly smaller than the silicon steel as shown in Figure 3. As the length of the overhang increases, the flux linkage must pass through a longer air gap which mitigates the increased air-gap flux density. When the rotor overhang is long enough, the advantage of the proposed structure becomes obvious. The additional FMP overhang can effectively utilize the rotor overhang, which produces higher back EMF and torque in comparison with M II. Since the length of end winding of the machine is around 10.8 mm, a 10 mm overhang length is selected for M II and M III. In addition, the dimension of the stator is optimized for M II and M III to avoid saturation in the stator teeth while maintaining the same current density. Main parameters of the three PMVMs are summarized in Table 1.

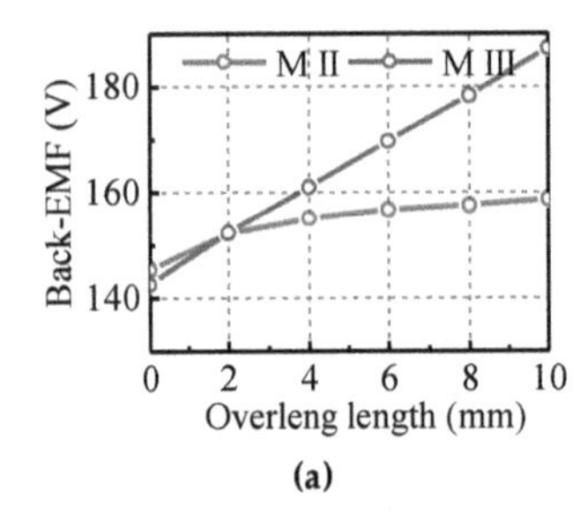

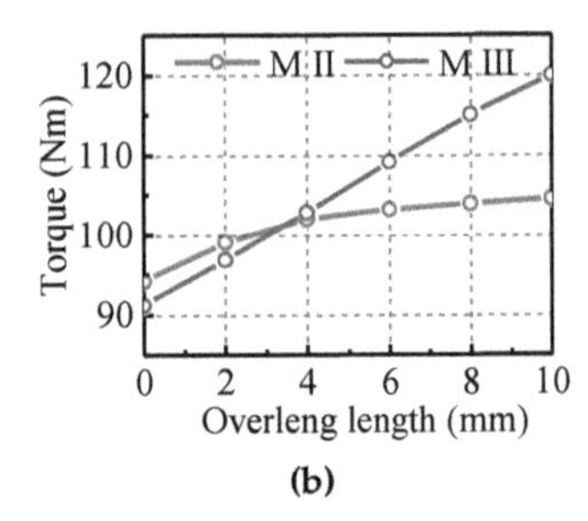

Figure 2. Back EMF (**a**) and torque (**b**) versus overhang length of M II and M III.

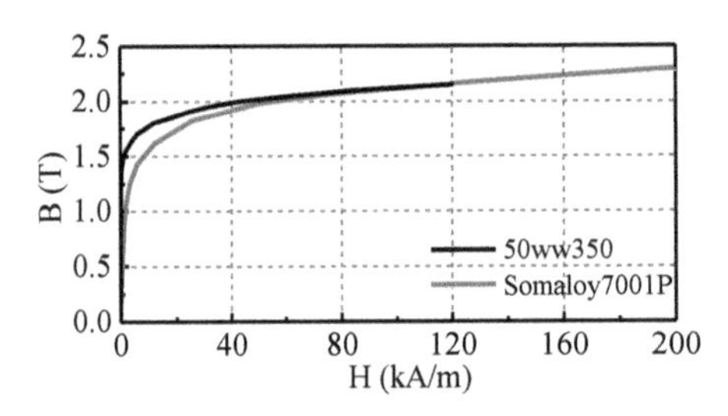

Figure 3. B–H curve of silicon steel and SMC.

3. FEA and Comparisons

As the overhang structure causes asymmetry in the axial direction, the 3-D FEM is essential for further analysis.

3.1. No-Load Characteristics

The no-load back EMF waveforms and their harmonic distributions of the three machines at the rotor speed of 600 rpm are demonstrated and compared in Figure 4. It can be seen that the back EMF amplitude of the proposed M III is the highest, 28.8%and 18.0% higher than the M I and the M II, respectively.

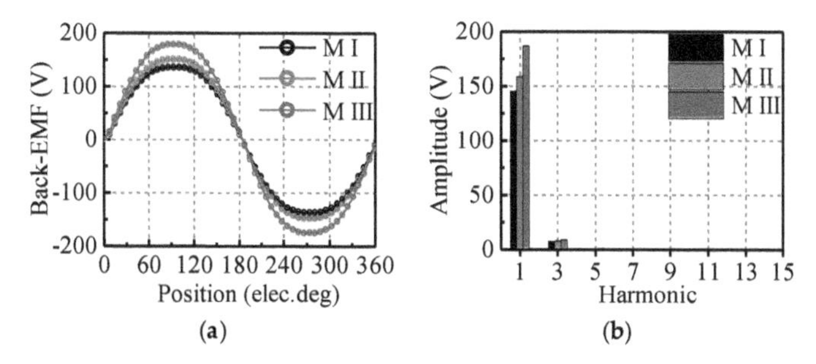

Figure 4. No-load back EMF of the three machines at 600 rpm. (**a**) Waveform. (**b**) Harmonic distributions.

The radial component of the air-gap flux density at the axial position of 0 mm and the corresponding space harmonic spectrum are shown in Figure 5a and c. In addition, the air-gap flux densities of M I and M II at the axial 0 mm position (the middle of the machines) are the same and the 28th harmonic orders are slightly higher than the M III's. This is mainly because of the low permeability of SMC applied to the FMPs, and this also could lead to the difference of back EMF between M II and M III when the overhang length is 0 mm as shown in Figure 2. The amplitude of the 28th harmonic order of air-gap flux density versus axial position are compared in Figure 5b. It should be noted that the air-gap flux density of M II drops apparently at the axial position near the rotor overhang, which is between 30 mm to 40 mm and −30 mm to −40 mm. However, the proposed M III can keep the high air-gap flux density until it reaches the edge of the overhang. As shown in Figure 6a, the PM flux linkage of the rotor overhang passes through the air gap and goes into the stator and contributes to the main flux linkage. However, this effect could be mitigated when the length of the overhang is increasing, and this problem is overcome by adding the FMP overhang. The flux linkage of the rotor overhang can successfully pass into the FMP overhang and then into the stator, in which way the reluctance is much smaller than the M II's, and less flux leakage occurs as shown in Figure 6b. Meanwhile, this also benefits from the SMC, which allows more axial flux. Therefore, the air-gap flux density of M III is improved, which contributes to the increase of back EMF and output torque.

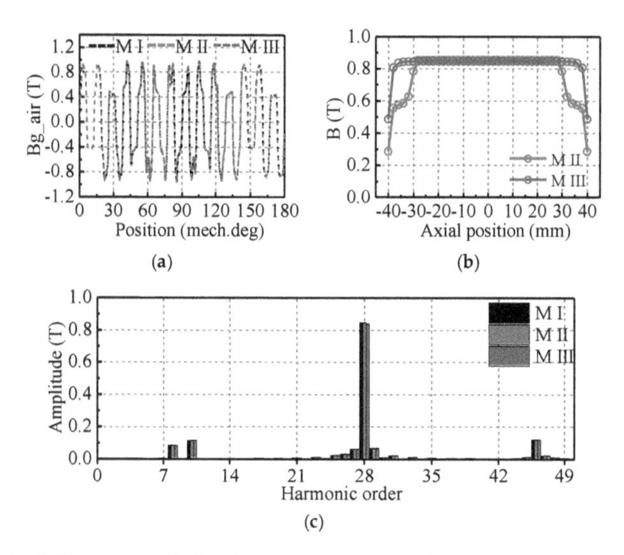

Figure 5. No-load air-gap magnetic flux density. (**a**) 0 mm waveform. (**b**) Air-gap flux density versus axial position. (**c**) 0 mm harmonic distributions.

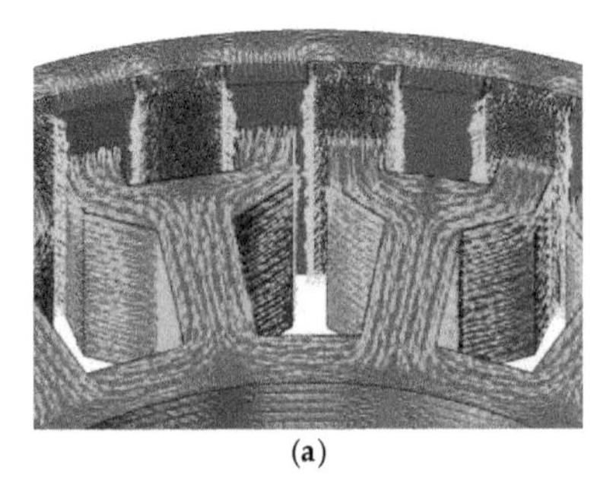

(a)

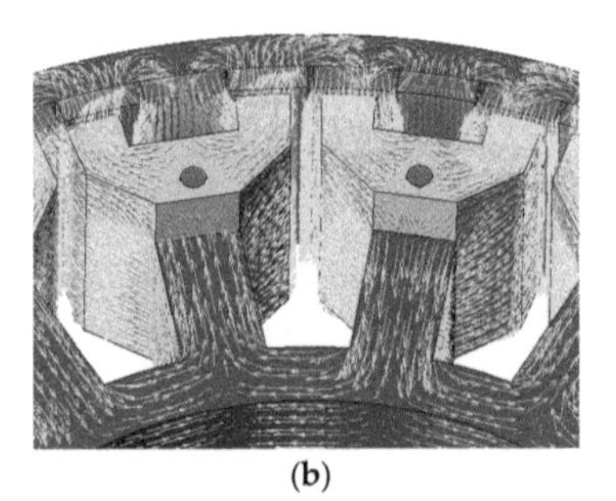

(b)

Figure 6. 3D magnetic flux density vector diagram at no-load condition. (**a**) M II. (**b**) M III.

3.2. On-Load Characteristics

The flux density distributions of the three PMVMs at the rated load (20 Arms current with $i_d = 0$ control) are compared in Figure 7. It can be seen that through the stator size adjustment, the maximum flux density in stator teeth of all the three machines are about 1.7 T, which is a typical value for the silicon steel 50ww350.

The characteristics of the three PMVMs at rated load condition are summarized in Table 2. The output waveforms of the three PMVMs at the rated load condition are shown in Figure 8. The proposed M III can offer higher output torque and power, which is 27.3% higher than M I and 14.5% higher than M II. The torque ripple of all the three machines is relatively low without any additional torque ripple reduction technique adopted, which is 1.52 Nm (1.6%), 1.70 Nm (1.8%), and 1.47 Nm (1.5%), respectively. The total weight (including stator, rotor, PMs, winding, and shell) and volume of the three machines are also calculated, and M III has the highest torque density in terms of Nm/kg and Nm/L.

Table 2. Main rated load performance of the three machines.

Items	M I	M II	M III
Average torque (Nm)	94.0	104.5	119.7
Torque ripple (Nm)	1.52 (1.6%)	1.70 (1.8%)	1.47 (1.5%)
Power (kW)	5.90	6.56	7.52
Loss (W)	412.4	442.2	572.5
Efficiency	93.5%	93.7%	92.9%
Weight (kg)	12.88	14.86	15.76
Torque density (Nm/kg)	7.30	7.03	7.60
Volume (L)	5.54	5.54	5.54
Torque density (Nm/L)	16.97	18.86	21.61

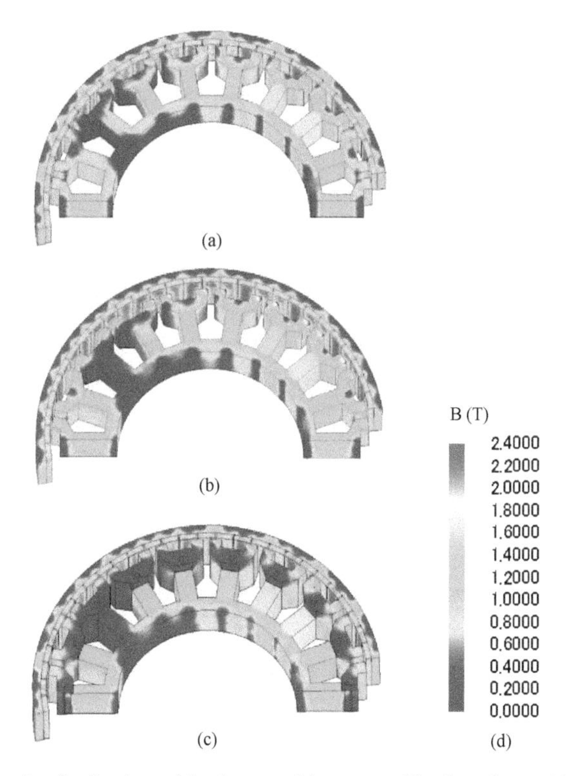

Figure 7. Flux density distributions of the three machines at rated load condition. (**a**) M I. (**b**) M II. (**c**) M III. (**d**) Scale bar.

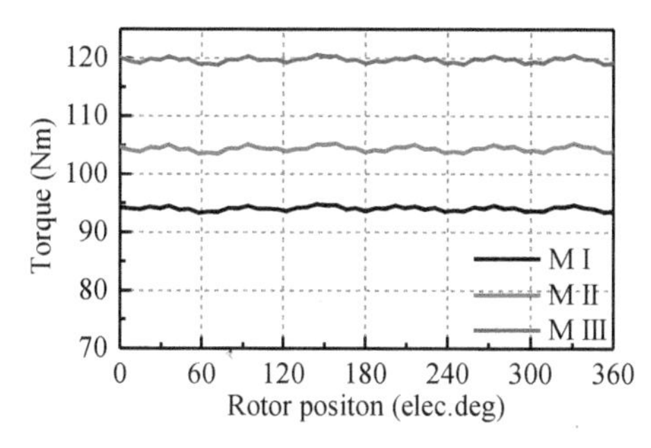

Figure 8. Output torque waveforms of the machines at rated load condition.

The main losses of the three PMVMs at rated load condition are listed in Figure 9. The copper loss of the three PMVMs is almost the same. The PM eddy current loss of these PMVMs cannot be neglected due to the high frequency. The losses in the rotor cores are very small while the stator core losses dominate. The efficiency of the three PMVMs can be approximated as

$$\eta = \frac{P_o}{P_o + P_{\text{loss}}} \tag{3}$$

where P_o is the output power and P_{loss} are the losses listed in Figure 9.

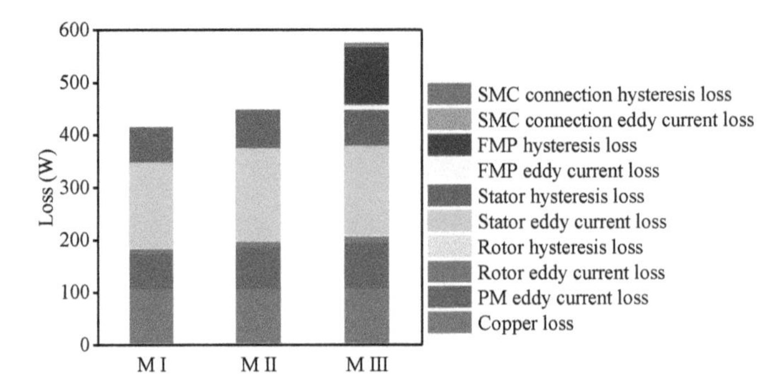

Figure 9. On-load loss comparison of the three PMVMs.

However, the hysteresis loss of the FMPs of M III is relatively high due to SMC inherent material characteristics, which lead to the increase of total loss. Besides the higher cost and relatively higher hysteresis loss of SMC material, only FMPs are made of SMC and the other parts of the stator cores are made of laminations in this paper.

4. Conclusions

In this paper, a novel outer rotor PMVM for in-wheel direct-drive application has been proposed, in which the overhang structure of the rotor and FMP are introduced. The FMP overhang structure is made of SMC to allow more axial flux. Detailed comparisons indicate that the proposed machine can effectively utilize the end winding space to enhance the air-gap flux density and offer higher torque density than the conventional machines with no overhang structure and only rotor overhang structure. Nevertheless, the efficiency of the proposed machine is slightly sacrificed due to the use of SMC.

Author Contributions: Conceptualization, D.Y. and X.H.; methodology, D.Y. and L.W.; software, D.Y. and X.H.; formal analysis, D.Y. and L.W.; data curation, D.Y. and Y.F.; writing, D.Y. and X.H.; supervision, L.W. and Y.F.; funding acquisition, X.H. and Y.F.

Funding: This work was supported by the National Natural Science Foundation of China under Grant 51877196, 51637009, Ningbo Innovation 2020 program, and the Cao Guangbiao High Tech Development Fund of Zhejiang University.

Conflicts of Interest: The authors declare that there is no conflict of interest.

References

1. Lee, C.H. Vernier motor and its design. *IEEE Trans. Power Appl. Syst.* **1963**, *82*, 343–349. [CrossRef]
2. Toba, A.; Lipo, T.A. Generic torque-maximizing design methodology of surface permanent-magnet vernier machine. *IEEE Trans. Ind. Appl.* **2000**, *36*, 1539–1546.
3. Gu, C.; Zhao, W.; Zhang, B. Simplified Minimum Copper Loss Remedial Control of a Five-Phase Fault-Tolerant Permanent-Magnet Vernier Machine under Short-Circuit Fault. *Energies* **2016**, *9*, 860. [CrossRef]
4. Kim, B.; Lipo, T.A. Operation and Design Principles of a PM Vernier Motor. *IEEE Trans. Ind. Appl.* **2014**, *6*, 3656–3663. [CrossRef]
5. Kim, B.; Lipo, T.A. Analysis of a PM vernier motor with spoke structure. *IEEE Trans. Ind. Appl.* **2016**, *52*, 217–225. [CrossRef]
6. Jang, D.; Chang, J. A Novel Design Method for the Geometric Shapes of Flux Modulation Poles in the Surface-Mounted Permanent Magnet Vernier Machines. *Energies* **2017**, *10*, 1551. [CrossRef]
7. Zhang, Y.; Lin, H.; Fang, S.; Huang, Y.; Yang, H.; Wang, D. Air-gap flux density characteristics comparison and analysis of permanent magnet vernier machines with different rotor topologies. *IEEE Trans. Appl. Supercond.* **2016**, *26*, 1–5. [CrossRef]

8. Li, D.; Qu, R.; Lipo, T.A. High-power-factor vernier permanent-magnet machines. *IEEE Trans. Ind. Appl.* **2014**, *50*, 3664–3674. [CrossRef]
9. Li, X.; Chau, K.T.; Cheng, M. Comparative analysis and experimental verification of an effective permanent-magnet vernier machine. *IEEE Trans. Magn.* **2015**, *51*, 1–9.
10. Liu, G.; Chen, M.; Zhao, W.; Chen, Q.; Zhao, W. Design and analysis of five-phase fault-tolerant interior permanent-magnet Vernier machine. *IEEE Trans. Appl. Supercond.* **2016**, *26*, 1–5. [CrossRef]
11. Yang, H.; Lin, H.; Zhu, Z.-Q.; Fang, S.; Huang, Y. A Dual-Consequent-Pole Vernier Memory Machine. *Energies* **2016**, *9*, 134. [CrossRef]
12. Oner, Y.; Zhu, Z.Q.; Wu, L.J.; Ge, X.; Zhan, H.; Chen, J.T. Analytical on-load subdomain field model of permanent-magnet vernier machines. *IEEE Trans. Ind. Electron.* **2016**, *63*, 4105–4117. [CrossRef]
13. Liu, X.; Zhong, X.; Du, Y.; Chen, X.; Wang, D.; Ching, T.W. A New Magnetic Field Modulation Type of Brushless Double-Fed Machine. *IEEE Trans. Appl. Supercond.* **2018**, *28*, 1–5. [CrossRef]
14. Xu, L.; Liu, G.; Zhao, W.; Yang, X.; Cheng, R. Hybrid stator design of fault-tolerant permanent-magnet vernier machines for direct-drive applications. *IEEE Trans. Ind. Electron.* **2017**, *64*, 179–190. [CrossRef]
15. Seo, J.M.; Jung, I.S.; Jung, H.K.; Ro, J.S. Analysis of overhang effect for a surface-mounted permanent magnet machine using a lumped magnetic circuit model. *IEEE Trans. Magn.* **2014**, *50*, 1–7.
16. Yeo, H.K.; Park, H.J.; Seo, J.M.; Jung, S.Y.; Ro, J.S.; Jung, H.K. Electromagnetic and thermal analysis of a surface-mounted permanent-magnet motor with overhang structure. *IEEE Trans. Magn.* **2017**, *53*, 1–4. [CrossRef]
17. Min, S.G.; Bobba, D.; Sarlioglu, B. Analysis of overhang effects using conductor separation method in coreless-type PM linear machines. *IEEE Trans. Magn.* **2018**, *54*, 1–4. [CrossRef]
18. Ishikawa, T.; Sato, Y.; Kurita, N. Performance of novel permanent magnet synchronous machines made of soft magnetic composite core. *IEEE Trans. Magn.* **2014**, *50*, 1–4. [CrossRef]
19. Wn, F.; El-Refaie, A.M. Permanent magnet vernier machine: A review. *IET Electr. Power Appl.* **2019**, *13*, 127–137.
20. Okada, K.; Niguchi, N.; Hirata, K. Analysis of a Vernier Motor with Concentrated Windings. *IEEE Trans. Magn.* **2013**, *49*, 2241–2244. [CrossRef]

Article

Analytical Modeling and Comparison of Two Consequent-Pole Magnetic-Geared Machines for Hybrid Electric Vehicles

Hang Zhao [1,2], Chunhua Liu [1,2,*], Zaixin Song [1,2] and Jincheng Yu [1,2]

1 School of Energy and Environment, City University of Hong Kong, 83 Tat Chee Avenue, Kowloon Tong, Hong Kong, China; zhao.hang@my.cityu.edu.hk (H.Z.); zaixin.song@my.cityu.edu.hk (Z.S.); jincheng.yu@my.cityu.edu.hk (J.Y.)

2 Shenzhen Research Institute, City University of Hong Kong, Nanshan District, Shenzhen 518057, China

* Correspondence: chunliu@cityu.edu.hk; Tel.: +852-34422885

Received: 11 April 2019; Accepted: 13 May 2019; Published: 17 May 2019

Abstract: The exact mathematical modeling of electric machines has always been an effective tool for scholars to understand the working principles and structure requirements of novel machine topologies. This paper provides an analytical modeling method—the harmonic modeling method (HMM)—for two types of consequent-pole magnetic-geared machines, namely the single consequent-pole magnetic-geared machine (SCP-MGM) and the dual consequent-pole magnetic-geared machine (DCP-MGM). By dividing the whole machine domain into different ring-like subdomains and solving the Maxwell equations, the magnetic field distribution and electromagnetic parameters of the two machines can be obtained, respectively. The two machines were applied in the propulsion systems of hybrid electric vehicles (HEVs). The electromagnetic performances of two machines under different operating conditions were also compared. It turns out that the DCP-MGM can reach a larger electromagnetic torque compared to that of the SCP-MGM under the same conditions. Finally, the predicted results were verified by the finite element analysis (FEA). A good agreement can be observed between HMM and FEA. Furthermore, HMM can also be applied to the mathematical modeling of other consequent-pole electric machines in further study.

Keywords: harmonic modeling method; magnetic-geared machine; hybrid electric vehicle; magnetic field; electromagnetic performance; analytical modeling

1. Introduction

The last decade has witnessed rapid developments of magnetic gears (MGs) and electric machines that utilize the magnetic-gearing effect, which are also called magnetic-geared machines (MGMs) [1–3]. Ever since their invention in 2001 [4], MGs have become a research hotspot due to their high efficiency and self-protection characteristics [5–7].

The concept of MGMs is derived from MGs. By substituting stator windings with AC current for one rotating permanent magnet (PM) component, MGMs change one mechanical port of MGs into an electrical port. Thus, the two rotating components of the MGM and its stator windings can be regarded as a combination of a magnetic gear and an electric machine [8–10]. Indeed, with the introduction of another rotating component and the ability to alternate the speed ratio and torque ratio between two rotating components, MGMs have broadened the application scenarios of electric machines [11–13]. A good example is that MGMs can serve as the power split component (PSC) in hybrid electric vehicles (HEVs) to realize energy exchange among the internal combustion engine (ICE), wheels, and battery [14–16]. The ICE and electric machine can provide traction for the wheels independently. The electric machine can work as a generator and a motor. When the electric machine serves as a generator, it can absorb power from the ICE or wheels (depending on working modes) to get the battery charged. When the electric machine serves as a motor, the power flows from the battery to the electric machine to drive the wheels. Hence, the ICE can always work at its highest efficiency to save fuel by alternating the working modes of the electric machine. This application scenario has drawn more and more attention as environmental problems become severe [17]. HEVs do offer a chance to alleviate the exhaust gas emission problem caused by fuel vehicles [18]. Moreover, compared to its counterpart, namely the mechanical gearbox with an electric machine, MGMs not only save space, but also improve efficiency and reduce noise and vibration by eliminating the physical contact of two gear sets [19,20].

Just like permanent magnet synchronous machines (PMSMs), MGMs utilize permanent magnets as the magnetic sources instead of using the electrical excitation method. Thus, the carbon brush structure can be eliminated and the durability of electric machines can be enhanced. However, the rare earth elements make the price of PMs extremely expensive [21]. To solve this problem, a consequent-pole structure can be adopted. The consequent-pole structure can not only reduce the flux linkage, but also improve the torque density [22–24]. Two different topologies of consequent-pole MGMs, i.e., single consequent-pole magnetic-geared machines (SCP-MGM) and dual consequent-pole magnetic-geared machines (DCP-MGM) have been proposed [25], but their mathematical modeling has not been well studied.

Although the MGMs offer many new possibilities for electric machines, their magnetic field distribution is much more complex compared to traditional electric machines with one rotor. Many scholars have focused on the magnetic field distribution calculation of MGMs [26–28]. Yet, to the best of author's knowledge, no literature has studied the magnetic field distribution of consequent-pole MGMs. The introduction of soft magnetic material (SMM) to replace the PM part will make the magnetic field distribution of consequent-pole MGMs even more complicated. Research [29] has solved the magnetic field distribution of a PMSM with PMs inserted into the SMM part, but did not consider of the saturation of SMM. Additionally, the subdomain division method [30] is not suitable for MGMs, since too many subdomains increase the calculation time rapidly. Research [31] has proposed a new harmonic modeling method (HMM) to calculate the magnetic distribution of electric machines. By introducing complex Fourier series and a convolution matrix of permeability, HMM can reduce the number of subdomains to within ten. This is because the total number of these ring-like subdomains will not increase with the increase of modulator pieces and slots.

In this paper, two consequent-pole MGMs were studied using HMM. The paper is organized as follows. Section 2 discusses the configurations and operating principles of consequent-pole MGMs. Mathematical models of SCP-MGM and DCP-MGM considering iron saturation are then proposed and elaborated in Section 3. Finally, the effectiveness of proposed HMM is validated by using finite element analysis (FEA) in Section 4.

2. Configurations and Operating Principles of SCP-MGM and DCP-MGM

When the MGM (either SCP-MGM or DCP-MGM) is applied in HEV, its inner rotor can be connected to the ICE, while the outer rotor can be connected to a permanent magnet synchronous machine (PMSM), which will be further connected to the differential to drive the wheels; the battery provides energy to the windings of both the MGM and PMSM via an inverter. The whole system configuration can be seen in Figure 1. The MGM together with the PMSM can be regarded as the E-CVT in a Toyota Prius. They can cooperate with each other according to different working conditions of HEVs [32]. Briefly speaking, either the torque from the ICE or the torque on the outer rotor driven by AC current can be the prime power to drive the HEV, and they can also work together to enhance the output power. Additionally, the battery can be charged under a regenerative braking state. The concept that a PMSM is added after the CP-MGM is derived from that in E-CVT [33]. The PMSM in Figure 1 is used to regulate the performances of the outer rotor. For instance, it can be used to drag the outer rotor of CP-MGM to a synchronous state (the rated rotating speed) at startup state. In addition, it can deliver extra output torque to the outer rotor shaft if the output torque of the CP-MGM cannot meet the requirement. Since this paper mainly focuses on the operating modes of the CP-MGM, it is reasonable to assume that there is no power flow between the PMSM and the outer rotor shaft at the four steady states mentioned in this paper. In fact, power exchange between the PMSM and the wheels would not affect the conclusion obtained in this paper.

The working principle of MGMs is similar to that of magnetic gears. By adopting a modulator layer, the magnetic field distribution can be changed. Assuming that the pole pair number of the original magnetic field is P_i, and the modulator number is Q, then a novel magnetic field will have a component that has $(Q - P_i)$ pole pairs. Thus, the fundamental structural requirement of an MGM is [4]:

$$P_i + P_s = Q \tag{1}$$

where P_s is the pole pair number of stator windings.

Under steady working conditions, the rotating speed of two rotating rotors and the current frequency f within stator windings should then satisfy:

$$P_i\omega_i - Q\omega_o = P_s w_s = 60f \tag{2}$$

where ω_i, ω_o, and ω_s are the rotating speed of the inner rotor, outer rotor, and the equivalent rotating speed of stator windings.

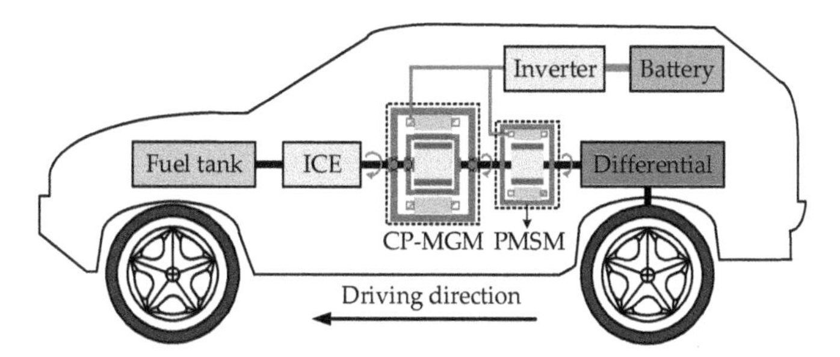

Figure 1. Propulsion system configuration of the consequent-pole magnetic-geared machine (MGM) applied in hybrid electric vehicles (HEVs).

Since the ICE reaches its highest efficiency at the range of ~2000 r/min–3000 r/min, the rotating speed of outer rotor and the current frequency of stator winding must cooperate with the rotating speed of the inner rotor to ensure the highest efficiency of the ICE. However, if the stator windings need to provide energy for the HEV, the rotating speed of the inner rotor must be smaller than that of the outer rotor. Thus, a gearbox must come into service under hybrid mode to reduce the rotating speed of the inner rotor. Therefore, the operation modes of the proposed HEV propulsion system can be divided into four kinds, and their typical operating parameters are shown in Table 1. The rotating speed of the outer rotor is calculated according to the different driving speeds of the HEV, and the current frequency of stator winding is obtained via Equation (2).

Table 1. Operating parameters of MGM under different modes of HEV.

Operation Modes	Rotating Speed of Inner Rotor ω_o	Rotating Speed of Outer Rotor ω_i	Current Frequency f
Pure electric mode	0 r/min	500 r/min	108.3 Hz
Pure mechanical mode	1200 r/min	1015 r/min	0 Hz (DC)
Hybrid mode	1200 r/min	2000 r/min	213.3 Hz
Regenerative braking mode	0 r/min	1000 r/min	216.6 Hz

The topologies of SCP-MGM and DCP-MGM are shown in Figure 2. By substituting SMM for PMs with the same polarity, a consequent-pole structure is obtained. The name "consequent-pole" is due to SMM, and PM appears alternately on the circumferential direction. Although SMM cannot generate a magnetic field itself, it can be easily magnetized to conduct flux lines. Hence, SMM in a consequent-pole structure can be regarded as a magnetic source to some degree. The greatest advantage of using the consequent-pole structure is saving PM material, which is the most expensive material in an electric machine. Both SCP-MGM and DCP-MGM utilize a consequent-pole structure to save PM material. The SMM part in the outer rotor of a DCP-MGM not only works as a consequent-pole structure for the PMs inserted in the outer rotor, it also modulates the magnetic field of the inner rotor. Thus, the P_s-th harmonic component within the DCP-MGM is larger than that of the SCP-MGM. Additionally, the saturation of the DCP-MGM is more severe than that of the SCP-MGM.

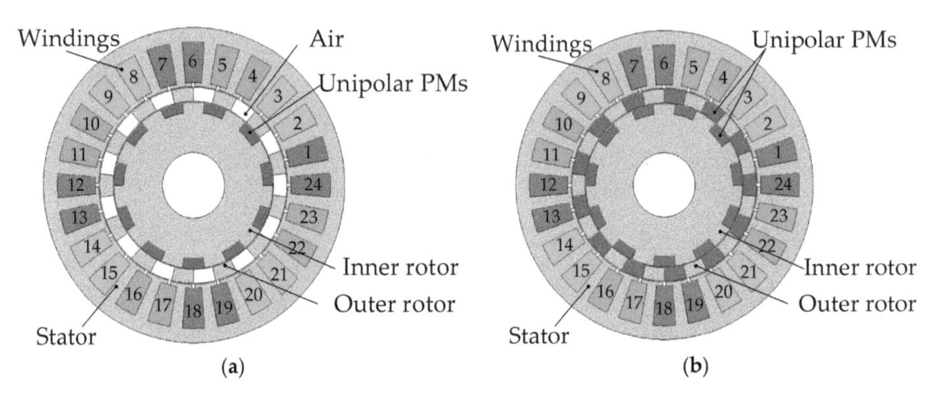

Figure 2. Proposed machine topologies: (**a**) single consequent-pole magnetic-geared machines (SCP-MGMs); (**b**) dual consequent-pole magnetic-geared machines (DCP-MGMs).

3. Mathematical Modeling of SCP-MGMs and DCP-MGMs

3.1. Assumption and Parameter Definition

The machine structure chosen to be studied in this paper was a 24 slot 11 pole-pair SCP-MGM and DCP-MGM, as shown in Figure 2. A few assumptions must be made to simplify the mathematical modeling:

- The geometrical shape of the machine has a radial side and a tangential side;
- The magnetic field distribution is constrained in the 2D plane: the axial component is ignored;
- The machine has infinite axial length, so the end effect is ignored;
- The radial component of the permeability of SMM within a certain region is regarded as a constant;
- Eddy–current effects within SMM and PMs are ignored.

Since there exists a z-direction current within the windings of the studied machines and the machine topology has a circular shape, a vector magnetic potential (VMP) $\boldsymbol{A}_z$ in a polar coordinate is adopted to calculate the magnetic flux density distribution within the machines. The machine structures are then divided into several ring-like regions based on the different material interfaces, as can be seen in Figure 3, where α represents the angle of the inner PM arc, β represents the angle of the slot opening in the modulator, δ is the slot opening angle, and γ is the stator slot angle. The whole machine is divided into ten subdomains: the innermost one (region I) represents the shaft part; region II is the rotor yoke; region III is the inner consequent-pole PM; region IV is the inner air gap; region V is the modulator pieces (it should be noted that for SCP-MGMs, the gap between each two modulator pieces is air, while for DCP-MGM, bipolar PMs are inserted in that gap). Region VI is the outer air gap;

region VII is the stator teeth; region VIII is the stator slots together with windings; region IX is the stator yoke; region X is the outside of the studied machines.

The angular position of the *j*-th PM part of the inner rotor θ_{PM}, the position of the *k*-th modulator piece θ_{Mod}, the position of the *t*-th stator tooth part θ_{tooth}, and the position of the *s*-th stator slot part θ_{slot} can be defined, respectively, as:

$$\theta_{PM} = \varphi_{in} + \frac{j \cdot 2\pi}{P_i} \tag{3}$$

$$\theta_{Mod} = \frac{k \cdot 2\pi}{Q} + \theta_0 - \frac{\beta}{2} \tag{4}$$

$$\theta_{tooth} = t \cdot 2\pi / P \tag{5}$$

$$\theta_{slot} = s \cdot 2\pi / P \tag{6}$$

where φ_{in}, and θ_0 the initial angular positions of the inner rotor and outer rotor, respectively. Due to the symmetrical structure of the inner rotor and outer rotor, φ_{in}, θ_0 has a range of $[0, 2\pi/P_i]$, $[0, 2\pi/Q]$, respectively. Specifically, φ_{in} is defined as zero when the lower edge of the PM in the inner rotor coincides with the positive direction of the angular axis; θ_0 is defined as zero when the center of the slot of the outer rotor (air in SCP-MGM and PM in DCP-MGM) coincides with the positive direction of the angular axis, as shown in Figure 2.

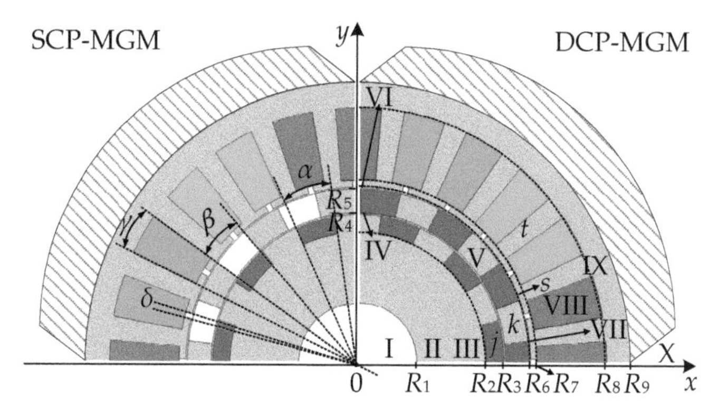

Figure 3. Subdomain divisions of SCP-MGM and DCP-MGM.

The VMP $\vec{A}$, the magnetic flux density $\vec{B}$, the magnetic field strength $\vec{H}$, and the current density distribution $\vec{J}$ in stator windings can be written in vector form as:

$$\vec{A} = A_z(r,\theta) \cdot \vec{u}_z \tag{7}$$

$$\vec{B} = B_r(r,\theta) \cdot \vec{u}_r + B_\theta(r,\theta) \cdot \vec{u}_\theta \tag{8}$$

$$\vec{H} = H_r(r,\theta) \cdot \vec{u}_r + H_\theta(r,\theta) \cdot \vec{u}_\theta \tag{9}$$

$$\vec{J} = J_z(r,\theta) \cdot \vec{u}_z \tag{10}$$

To simplify the solving process, all the parameters related to magnetic field are expressed in terms of complex Fourier series. Thus, the vector amplitude above can be further expressed as:

$$A_z(r,\theta) = \sum_{n=-\infty}^{n=\infty} \hat{A}_{z,n}(r) \cdot e^{-in\theta} \tag{11}$$

$$B_r(r,\theta) = \sum_{n=-\infty}^{n=\infty} \hat{B}_{r,n}(r) \cdot e^{-in\theta} \text{ and } B_\theta(r,\theta) = \sum_{n=-\infty}^{n=\infty} \hat{B}_{\theta,n}(r) \cdot e^{-in\theta} \tag{12}$$

$$H_r(r,\theta) = \sum_{n=-\infty}^{n=\infty} \hat{H}_{r,n}(r) \cdot e^{-in\theta} \text{ and } H_\theta(r,\theta) = \sum_{n=-\infty}^{n=\infty} \hat{H}_{\theta,n}(r) \cdot e^{-in\theta} \tag{13}$$

$$J_z(r,\theta) = \sum_{n=-\infty}^{n=\infty} \hat{J}_{z,n}(r) \cdot e^{-in\theta} \tag{14}$$

where n represents the n-th order coefficient of the corresponding Fourier series. It should be noticed that, in numerical calculation, a reasonable harmonic order N is used to truncate the infinite Fourier series. If N is too small, the Fourier series will have a large error, if N is too large, the calculation time will be rather long.

3.2. Partical Differential Equation Solution

The magnetic field within the machine follows quasistatic Maxwell equations:

$$\nabla \times \vec{H} = \vec{J} \tag{15}$$

$$\nabla \cdot \vec{B} = 0 \tag{16}$$

The relationship between $\vec{B}$ and $\vec{A}$ can be further expressed as:

$$\vec{B} = \nabla \times \vec{A} \tag{17}$$

The radial component and tangential component matrix of magnetic flux density $\vec{B}$ are then obtained in matrix form as [31]:

$$\mathbf{B}_r = \frac{1}{r}\frac{\partial \mathbf{A}_z}{\partial \theta} = -i\frac{1}{r}\mathbf{K}\mathbf{A}_z \tag{18}$$

$$\mathbf{B}_\theta = -\frac{\partial \mathbf{A}_z}{\partial r} \tag{19}$$

where $\mathbf{K}$ represents the harmonic order coefficient diagonal matrix that is related to N, given by:

$$\mathbf{K} = \begin{bmatrix} -N & \cdots & 0 \\ \vdots & \ddots & \vdots \\ 0 & \cdots & N \end{bmatrix} \tag{20}$$

Similar to Equations (11)–(14), the relative permeability of each region can also be expressed in a complex Fourier series form:

$$\mu(\theta) = \sum_{n=-\infty}^{n=\infty} \hat{\mu}_n \cdot e^{-in\theta} \tag{21}$$

Next, based on the relation between $\vec{B}$ and $\vec{H}$, as expressed below:

$$\vec{B} = \mu\vec{H} + \mu_0\vec{M} \tag{22}$$

where $\vec{M}$ is the magnetization vector. The first item on the right is a product of two Fourier series, which can be rewritten in matrix form by using the Cauchy product theorem:

$$\mathbf{B}_r = \mu_{r,\text{cov}}\mathbf{H}_r + \mu_0\mathbf{M}_r \tag{23}$$

$$\mathbf{B}_\theta = \boldsymbol{\mu}_{\theta,\text{cov}}\mathbf{H}_\theta + \mu_0\mathbf{M}_\theta \tag{24}$$

where $\boldsymbol{\mu}_{r,\text{cov}}$ and $\boldsymbol{\mu}_{\theta,\text{cov}}$ are convolution matrices of the radial and tangential components of permeability, respectively. $\mathbf{M}_r$ and $\mathbf{M}_\theta$ are the radial and tangential components of magnetization intensity, respectively. $\mathbf{M}_r$ and $\mathbf{M}_\theta$ can all written in complex Fourier series. The convolution matrix $\boldsymbol{\mu}_{r,\text{cov}}$ can be defined as:

$$\boldsymbol{\mu}_{r,\text{cov}} = \begin{bmatrix} \hat{\mu}_0 & \cdots & \hat{\mu}_{-2N} \\ \vdots & \ddots & \vdots \\ \hat{\mu}_{2N} & \cdots & \hat{\mu}_0 \end{bmatrix} \tag{25}$$

where $\hat{\mu}_n$ is the n-th order coefficient of the Fourier series of corresponding μ.

In Equation (24), $\mathbf{H}_\theta$ is continuous at the interface between two regions, but B_θ is discontinuous at the interface. Hence, the matrix $\boldsymbol{\mu}_{\theta,\text{cov}}$ cannot be settled using Equation (25). Instead, a fast Fourier factorization is applied to calculate $\mu_{\theta,\text{cov}}$, for the sake of keeping the rate of convergence the same for the left and right side of Equation (24) [34]. $\boldsymbol{\mu}_{\theta,\text{cov}}$ can be given by [31]:

$$\boldsymbol{\mu}_{\theta,\text{cov}} = \begin{bmatrix} \hat{\mu}_0^{rec} & \cdots & \hat{\mu}_{-2N}^{rec} \\ \vdots & \ddots & \vdots \\ \hat{\mu}_{2N}^{rec} & \cdots & \hat{\mu}_0^{rec} \end{bmatrix}^{-1} \tag{26}$$

From Equation (26), it can be seen that $\boldsymbol{\mu}_{\theta,\text{cov}}$ is acquired by replacing $\hat{\mu}_n$ with the corresponding n-th order Fourier coefficient of $1/\mu_\theta$ for each element, and there is a matrix inversion outside.

The region V in SCP-MGM is used to illustrated the convolution matrix with respect to relative permeability, as shown in Figure 4.

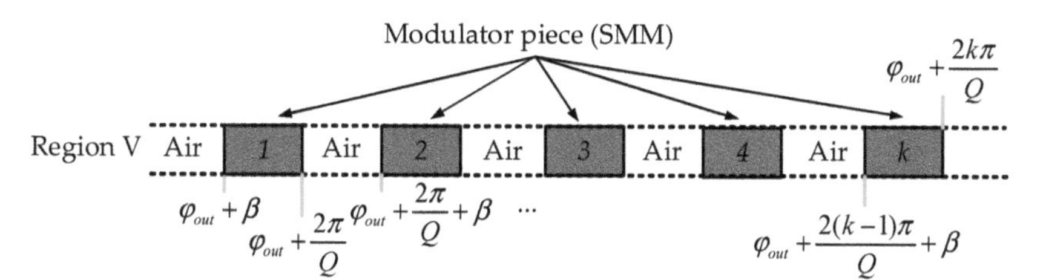

Figure 4. The calculation instance of the convolution matrix with respect to relative permeability.

The relative permeability distribution on the circumferential direction can be expressed as:

$$\mu(\theta) = \begin{cases} \mu_0 & \theta \in [\varphi_{out} + \frac{2(k-1)\pi}{Q}, \varphi_{out} + \frac{2(k-1)\pi}{Q} + \beta) \\ \mu_{iron,k} & \theta \in [\varphi_{out} + \frac{2(k-1)\pi}{Q} + \beta, \varphi_{out} + \frac{2k\pi}{Q}) \end{cases} \tag{27}$$

where $\varphi_{out} = \theta_0 - \frac{\beta}{2}$.

When a Fourier expansion on $[0, 2\pi]$ is applied to Equation (27), the expressions of $\hat{\mu}_k$ and $\hat{\mu}_k^{rec}$ are given by:

$$\hat{\mu}_n = \begin{cases} \sum\limits_{k=1}^{Q} \frac{\mu_0}{2\pi in} e^{in(\frac{2k\pi}{Q}+\theta_0)} (e^{\frac{in\beta}{2}} - e^{-\frac{in\beta}{2}}) + \sum\limits_{k=1}^{Q} \frac{\mu_{iron,k}}{2\pi in} e^{in(\frac{2k\pi}{Q}+\theta_0)} (e^{in(\frac{2\pi}{Q}-\frac{\beta}{2})} - e^{\frac{in\beta}{2}}) n \neq 0 \\ \frac{Q\beta\mu_0 + (\frac{2\pi}{Q}-\beta)\sum\limits_{k=1}^{Q}\mu_{iron,k}}{2\pi} \quad n = 0 \end{cases} \tag{28}$$

$$\hat{\mu}_n^{rec} = \begin{cases} \sum_{k=1}^{Q} \frac{1}{2\pi i n \mu_0} e^{in(\frac{2k\pi}{Q}+\theta_0)} (e^{\frac{in\beta}{2}} - e^{-\frac{in\beta}{2}}) + \sum_{k=1}^{Q} \frac{1}{2\pi i n \mu_{iron,k}} e^{in(\frac{2k\pi}{Q}+\theta_0)} (e^{in(\frac{2\pi}{Q}-\frac{\beta}{2})} - e^{\frac{in\beta}{2}}) n \neq 0 \\ \frac{\frac{Q\beta}{\mu_0} + (\frac{2\pi}{Q} - \beta) \sum_{k=1}^{Q} \frac{1}{\mu_{iron,k}}}{2\pi} n = 0 \end{cases} \quad (29)$$

The convolution matrices $\mu_{r,cov}$ and $\mu_{\theta,cov}$ can then be obtained by substituting Equations (28) and (29) into (25) and (26).

Combining Equations (18)–(26) together, the VMP satisfies:

$$\frac{\partial^2 \mathbf{A}_z^k}{\partial r^2} + \frac{1}{r}\frac{\partial \mathbf{A}_z^k}{\partial r} - \frac{\mathbf{V}\mathbf{A}_z^k}{r^2} = -\mu_{\theta,cov}\mathbf{J}_z - \frac{\mu_0}{r}(\mathbf{M}_\theta + i\mathbf{U}\mathbf{M}_r) \quad (30)$$

where $\mathbf{V} = \mu_{\theta,cov}\mathbf{K}\mu_{r,covc}^{-1}\mathbf{K}$ and $\mathbf{U} = \mu_{\theta,cov}\mathbf{K}\mu_{r,covc}^{-1}$. The derivation process of Equation (30) is given in Appendix A.

Equation (30) is a Cauchy–Euler differential equation system [35]. The general solution of a single differential equation in Equation (30) is given by:

$$y = c_1 r^{v^{\frac{1}{2}}} + c_2 r^{-v^{\frac{1}{2}}} \quad (31)$$

where c_1, c_2 are unknown coefficients. Similarly, the general solution of the differential equation system in Equation (30) can be written in a matrix form, where the new element (i, j) in matrix $r^{\mathbf{V}}$ is defined as:

$$r^{\mathbf{V}}(i,j) = r^{\mathbf{V}(i,j)} \quad (32)$$

Hence, the complementary solution of Equation (30) $\mathbf{A}_z^k|_{com}$ can be written as:

$$\mathbf{A}_z^k\big|_{com} = r^{\mathbf{V}^{\frac{1}{2}}}\mathbf{C}_1 + r^{-\mathbf{V}^{\frac{1}{2}}}\mathbf{C}_2 \quad (33)$$

where $\mathbf{C}_1$ and $\mathbf{C}_2$ are unknown coefficient vectors. Matrix $r^{\mathbf{V}^{\frac{1}{2}}}$ can be factorized as:

$$r^{\mathbf{V}^{\frac{1}{2}}} = \mathbf{P}r^{\boldsymbol{\lambda}}\mathbf{P}^{-1} \quad (34)$$

where $\boldsymbol{\lambda}$ is the eigenvalue matrix of $\mathbf{V}^{1/2}$, and matrix $\mathbf{P}$ is the eigenvector matrix of $\mathbf{V}^{1/2}$. Therefore, Equation (33) can be simplified as:

$$\mathbf{A}_z^k\big|_{com} = \mathbf{P}r^{\boldsymbol{\lambda}}(\mathbf{P}^{-1}\cdot\mathbf{C}_1) + \mathbf{P}r^{-\boldsymbol{\lambda}}(\mathbf{P}^{-1}\cdot\mathbf{C}_2) = \mathbf{P}r^{\boldsymbol{\lambda}}\mathbf{D} + \mathbf{P}r^{-\boldsymbol{\lambda}}\mathbf{E} \quad (35)$$

As for the particular solution of Equation (30), $\mathbf{A}_z^k|_{par}$ is given by:

$$\mathbf{A}_z^k\big|_{par} = r^2\mathbf{F} + r\mathbf{G} \quad (36)$$

where $\mathbf{F} = (\mathbf{V} - 4\mathbf{I})^{-1}\mu_{\theta,cov}\mathbf{J}_z$, $\mathbf{G} = \mu_0(\mathbf{V} - \mathbf{I})^{-1}(\mathbf{M}_\theta + i\mathbf{U}\mathbf{M}_r)$.

The general solution of VMP in Equation (30) is the sum of the complementary solution of Equation (35), and particular solution Equation (36):

$$\mathbf{A}_z^k = \mathbf{A}_z^k\big|_{com} + \mathbf{A}_z^k\big|_{par} \quad (37)$$

The expression of VMP in each subdomain is given in Appendix A.

According to the geometrical characteristic of SCP-MGMs and DCP-MGMs, the magnetization intensity only exists in region III and region V, and the current density distribution only exists in region VIII. Their distribution waveforms and Fourier series coefficients can be seen in Table 1. Additionally, for regions I, II, IV, VI, IX, and X, there only exists one material type, so the coefficients within the convolution matrix are a constant. However, the permeability distributions are different in region III, V, VII, and VIII, as shown in Table 2; their coefficients can be obtained by substituting a, b, and c in Table 3 into the following equations:

$$\hat{\mu}_n = \frac{1}{2\pi}\left(\sum_{k=1}^{c}\int_{a}^{a+b}\mu_0 e^{in\theta}d\theta + \sum_{k=1}^{c}\int_{a+b}^{a+2\pi/c}\mu_{iron}(k)e^{in\theta}d\theta\right) \tag{38}$$

$$\hat{\mu}_n^{rec} = \frac{1}{2\pi}\left(\sum_{k=1}^{c}\int_{a}^{a+b}\frac{e^{in\theta}}{\mu_0}d\theta + \sum_{k=1}^{c}\int_{a+b}^{a+2\pi/c}\frac{e^{in\theta}}{\mu_{iron,k}}d\theta\right) \tag{39}$$

Table 2. Mathematical modeling of magnetic sources within SCP-MGMs and DCP-MGMs.

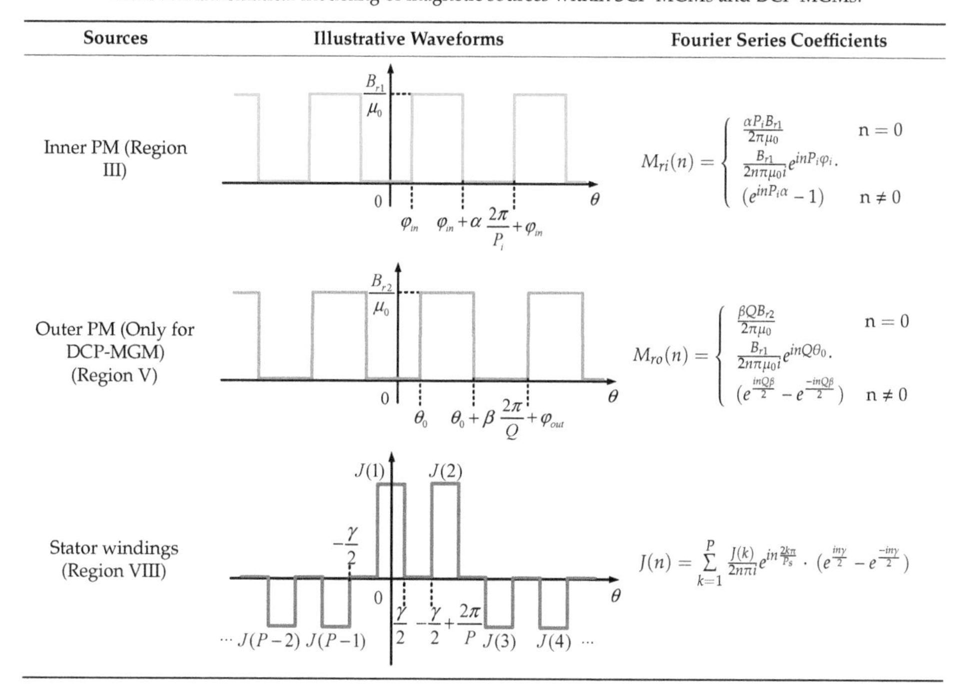

Sources	Illustrative Waveforms	Fourier Series Coefficients
Inner PM (Region III)	$\frac{B_{r1}}{\mu_0}$; 0; φ_{in}; $\varphi_{in}+\alpha$; $\frac{2\pi}{P_i}+\varphi_{in}$; θ	$M_{ri}(n) = \begin{cases} \frac{\alpha P_i B_{r1}}{2\pi\mu_0} & n=0 \\ \frac{B_{r1}}{2n\pi\mu_0 i}e^{inP_i\varphi_i}\cdot(e^{inP_i\alpha}-1) & n\neq 0 \end{cases}$
Outer PM (Only for DCP-MGM) (Region V)	$\frac{B_{r2}}{\mu_0}$; 0; θ_0; $\theta_0+\beta$; $\frac{2\pi}{Q}+\varphi_{out}$; θ	$M_{ro}(n) = \begin{cases} \frac{\beta Q B_{r2}}{2\pi\mu_0} & n=0 \\ \frac{B_{r1}}{2n\pi\mu_0 i}e^{inQ\theta_0}\cdot(e^{\frac{inQ\beta}{2}}-e^{\frac{-inQ\beta}{2}}) & n\neq 0 \end{cases}$
Stator windings (Region VIII)	$J(1)$; $J(2)$; $-\frac{\gamma}{2}$; 0; $\frac{\gamma}{2}$; $-\frac{\gamma}{2}+\frac{2\pi}{P}$; θ; … $J(P-2)$ $J(P-1)$; $J(3)$ $J(4)$ …	$J(n) = \sum_{k=1}^{P}\frac{J(k)}{2n\pi i}e^{in\frac{2k\pi}{P_s}}\cdot(e^{\frac{in\gamma}{2}}-e^{\frac{-in\gamma}{2}})$

Table 3. Mathematical modeling of the permeability distribution of different regions within SCP-MGMs and DCP-MGMs.

Regions	Illustrative Waveforms	Coefficients
III	μ_{iron}; μ_0; 0; a; $a+b$; $a+\frac{2\pi}{c}$; θ	$a=\varphi_{in}, b=\alpha, c=P_i$
V		$a=\varphi_{in}, b=\beta, c=Q$
VII		$a=\delta/2, b=2\pi/P_s-\delta, c=P_s$
VIII		$a=\gamma/2, b=2\pi/P_s-\gamma, c=P_s$

3.3. Bondary Condition Application

At the interfaces between two different subdomains, the radial component of $\vec{B}$ and the tangential component of $\vec{H}$ should be continuous across the boundary. Due to their ring-like shapes, each subdomain only interfaces with two subdomains at most, and region I and X have only one interface. Hence, all the boundary conditions can be written as:

$$\mathbf{B}_r^k(R_k) = \mathbf{B}_r^{k+1}(R_k) \tag{40}$$

$$\mathbf{H}_\theta^k(R_k) = \mathbf{H}_\theta^{k+1}(R_k) \tag{41}$$

where k represents the k-th subdomain of the proposed machine, and $2 \le k \le 9$.

Suppose the subdomain number of the proposed machine is L, and the harmonic order to be calculated is N. By applying the above boundary conditions to each interface of the machines, a system of 2*(L − 1) linear equations with 2*(L − 1) unknowns can be obtained, and each unknown is an (N × 1) vector. The system of linear equations can be further written in matrix form, as below:

$$\mathbf{SX} = \mathbf{T} \tag{42}$$

where the expressions of **S**, **X**, and **T** in this paper are given in Appendix A.

As long as the coefficient matrix **S** is invertible, the unknown vector **X** can be acquired. The numerical solution of Equation (42) can be obtained in the MATLAB software.

3.4. Saturation Consideration of Soft-Magnetic Material

For the SMM in the consequent-pole part, modulator, and stator teeth, the flux line is concentrated, thus, the saturation of the SMM must be considered. The nonlinear B-μ curve of 50JN1300 is given in Figure 4. In HMM, the relative permeability μ is obtained by an iterative method, as shown by the dot lines in Figure 5, where the number "1, 2, 3, 4" means the iteration number. The detailed iterative process is shown in Figure 6. First, an initial value, namely μ_0 = 1500, is assigned to a given region with SMM for the first iteration. In the k-th iteration, the average flux density $\vec{B}$ of a specific region can be obtained by substituting $\overline{\mu}_{i,k}$ and solving the matrix Equation (42). A new average relative permeability $\overline{\mu}_{i,cal}$ in region i can then be acquired. Where i belongs to {III, V, VII, VIII}. The average relative permeability $\mu_{i,k+1}$ for the (k + 1)-th iteration in region i is given as:

$$\mu_{i,k+1} = \frac{\overline{\mu}_{i,k} + \overline{\mu}_{i,kcal}}{2} \tag{43}$$

The iteration will stop only if the iteration time n_i exceeds the maximum number of iterations N_i (N_i is set to be 50 here), or the maximum error Δ in all these regions is below the error requirement ξ, Δ is defined as:

$$\Delta = \max\left\{\frac{\left|\overline{\mu}_{i,k} - \overline{\mu}_{i,cal}\right|}{\overline{\mu}_{i,k}}\right\},\ i \in \{III, V, VII, VIII\} \tag{44}$$

where ξ is set to be 0.05 in this paper. The saturations of SMM in SCP-MGMs and DCP-MGMs are calculated respectively using this method. The relative permeabilities of each region under on-load conditions are shown in Table 4. It can be observed that the saturation of regions V and VII is more severe in DCP-MGMs, since there are PMs inserted in the modulator, thus, there are more flux lines in the modulator and stator teeth.

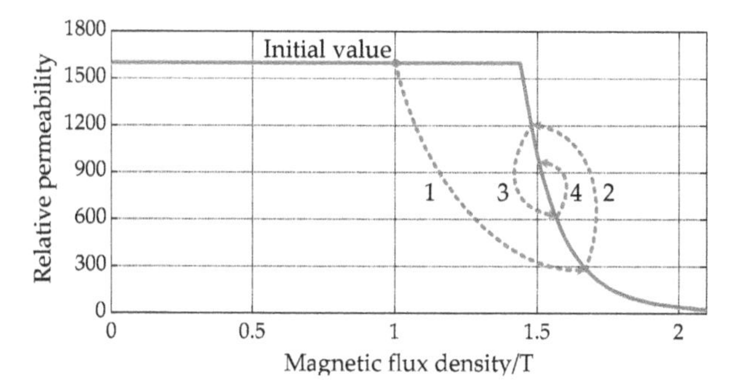

Figure 5. Iterative process illustration.

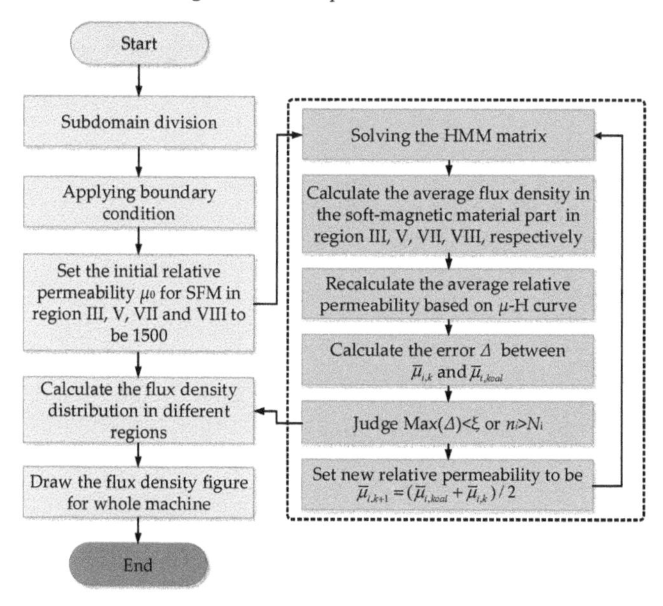

Figure 6. Fast Fourier transform (FFT) of no-load outer-air-gap radial magnetic flux density for SCP-MGMs and DCP-MGMs.

Table 4. Relative permeabilities of different regions within SCP-MGMs and DCP-MGMs.

Regions	SCP-MGM	DCP-MGM
III	27.46	27.92
V	1309.15	1257.49
VII	888.05	838.35
VIII	1013.2	917.24

3.5. Electromagnetic Parameters Calculation

Once the VMP is solved in Equation (42), the related electromagnetic parameters of the two machines can be calculated. The electromagnetic torque of the two rotors of SCP-MGMs and DCP-MGMs can be calculated by using the Maxwell stress tensor.

The flux linkage of each coil side can be given by [36]:

$$\varphi_k = \frac{N_{turn}L}{S_{coil}} \int_{-\frac{\gamma}{2}+\frac{2k\pi}{P}}^{\frac{\gamma}{2}+\frac{2k\pi}{P}} \int_{R_7}^{R_8} A_z^{VIII}(r,\theta) r dr d\theta \tag{45}$$

where N_{turn} is the number of coil turns in each slot, and S_{coil} is the cross section area of a single slot. When all the coils in each phase are in series, the three-phase flux linkage can be written as:

$$\boldsymbol{\psi} = \begin{bmatrix} \psi_A \\ \psi_B \\ \psi_C \end{bmatrix} = \mathbf{C}_{turn} \begin{bmatrix} \varphi_1 & \varphi_2 & \cdots & \varphi_{24} \end{bmatrix} \tag{46}$$

where $\mathbf{C}_{turn}$ is coil-connecting matrix of the proposed machine, the coil number is given in Figure 2, and $\mathbf{C}_{turn}$ can be expressed as:

$$\mathbf{C}_{turn} = \begin{bmatrix} 1 & 0 & 0 & 0 & 0 & -1 & -1 & 0 & 0 & 0 & 0 & 1 & 1 & 0 & 0 & 0 & 0 & -1 & -1 & 0 & 0 & 0 & 0 & 1 \\ 0 & 0 & 0 & 1 & 1 & 0 & 0 & 0 & 0 & -1 & -1 & 0 & 0 & 0 & 0 & 1 & 1 & 0 & 0 & 0 & 0 & -1 & -1 & 0 \\ 0 & -1 & -1 & 0 & 0 & 0 & 0 & 1 & 1 & 0 & 0 & 0 & 0 & -1 & -1 & 0 & 0 & 0 & 0 & 1 & 1 & 0 & 0 & 0 \end{bmatrix} \tag{47}$$

The back electromotive force (EMF) is computed by the derivative of $\boldsymbol{\psi}$ with respect to time:

$$\mathbf{E}_{ABC} = \frac{d\boldsymbol{\psi}}{dt} = \frac{d\boldsymbol{\psi}}{d\theta} \cdot \omega \tag{48}$$

where ω is the rotating speed of the magnetic field in the outer air gap.

The electromagnetic torque is calculated by using a Maxwell stress tensor. Thus, the electromagnetic torque of the inner rotor T_{in} equals the calculus of the Maxwell stress tensor of the inner air gap along the circumferential direction, and the electromagnetic torque of the outer rotor T_{out} equals the algebraic sum of the calculus of the Maxwell stress tensor of both the inner air gap and the outer air gap along the circumferential direction. They can be expressed as:

$$T_{in} = \frac{LR_i^2}{\mu_0} \int_0^{2\pi} B_r^{IV}(R_i,\theta) B_\theta^{IV}(R_i,\theta) d\theta \tag{49}$$

$$T_{out} = \frac{LR_o^2}{\mu_0} \int_0^{2\pi} B_r^{VI}(R_o,\theta) B_\theta^{VI}(R_o,\theta) d\theta - \frac{LR_i^2}{\mu_0} \int_0^{2\pi} B_r^{IV}(R_i,\theta) B_\theta^{IV}(R_i,\theta) d\theta \tag{50}$$

where R_i and R_o are the middle radius of the inner air gap and outer air gap, respectively.

4. Validation and Comparison

4.1. Simulation Environment and Machine Parameters

To make quantitative comparison between the SCP-MGMs and DCP-MGMs, all the geometrical parameters of these two machines should be set as the same, and other parameters, such as the slot filling factor and root mean square value of the winding current should be also set as the same, as shown in Table 5. The analytical prediction of the HMM was carried out using MATLAB, and the FEA model was constructed and run in JMAG software. The FEA model had 35,597 elements and 25,368 nodes; the element size near the air gap was set as 1 mm, to maintain calculation accuracy. It took 4.6 s for the computer to obtain the magnetic field distribution for one step. The computer system configuration was as follows: Processor: Intel Core i7-4790 CPU @ 3.60 GHz; Installed Memory (RAM):

28.0 GB-System type: 64-bit Windows Operating System. Additionally, the mean error percentage ε_1 and maximum difference ε_2 of the two methods was defined as:

$$\varepsilon_1 = \frac{1}{N_c}\sum_{n=1}^{N_c}\left|\frac{V_{FEA,n} - V_{HMM,n}}{V_{FEA,n}}\right| \times 100\% \tag{51}$$

$$\varepsilon_2 = \max\left\{\left|V_{FEA,n} - V_{HMM,n}\right|\right\},\ n \in \{1, 2, \ldots, N_c\} \tag{52}$$

where $V_{HMM,n}$ is the n-th value, calculated using the HMM, and $V_{FEA,n}$ is n-th the value obtained by FEA. N_c is the total number of calculated points.

4.2. Comparison between HMM and FEA

Based on the calculation and simulation, a good agreement between the HMM and FEA can be observed in Figures 7–12 for SCP-MGMs and DCP-MGMs. It was also found that the difference between HMM and FEA at a no-load condition was less compared to that at a load condition. This is due to the truncation of the infinite Fourier series; the high-order components take up a greater proportion at a no-load condition, leading to a larger error for HMM. Additionally, the numerical values calculated via MATLAB were constrained by the computational accuracy. Values exceeding the computational accuracy were ignored, which led to errors of the magnetic field calculation. Generally, if the harmonic order and computational accuracy is improved, the error will decrease. However, the computational time increases rapidly with the increase of harmonic order and computational accuracy. Thus, there is a tradeoff between calculation accuracy and calculation time. In this paper, the computational accuracy of MATLAB was set as 32 bits.

The mean error percentage ε_1 and maximum difference ε_2 for the SCP-MGM and DCP-MGM are listed in Table 6.

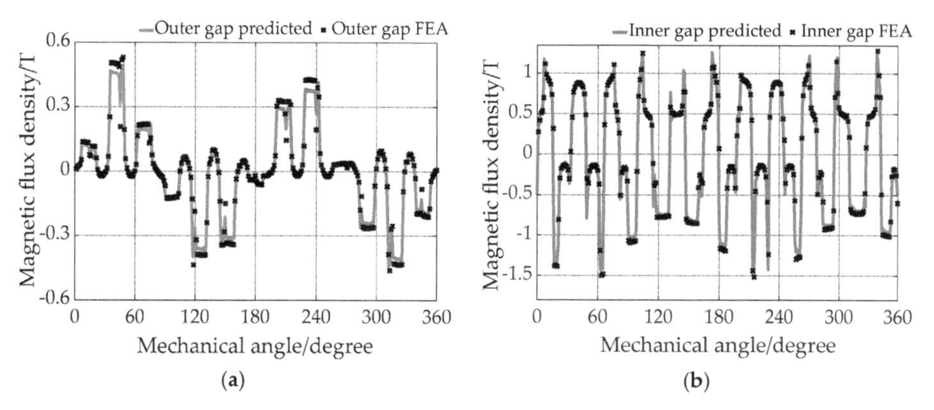

Figure 7. On-load inner and outer air-gap radial magnetic flux density distribution of the SCP-MGM: (**a**) outer air gap; (**b**) inner air gap.

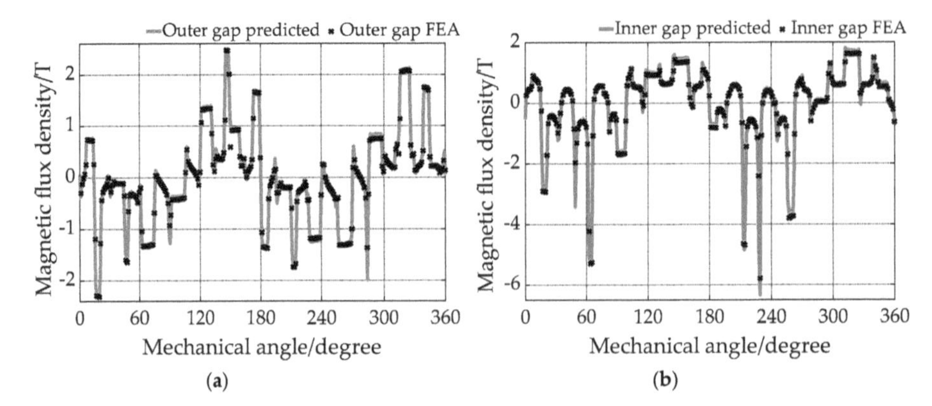

Figure 8. On-load inner and outer air-gap radial magnetic flux density distribution of the SCP-MGM: (**a**) outer air gap; (**b**) inner air gap.

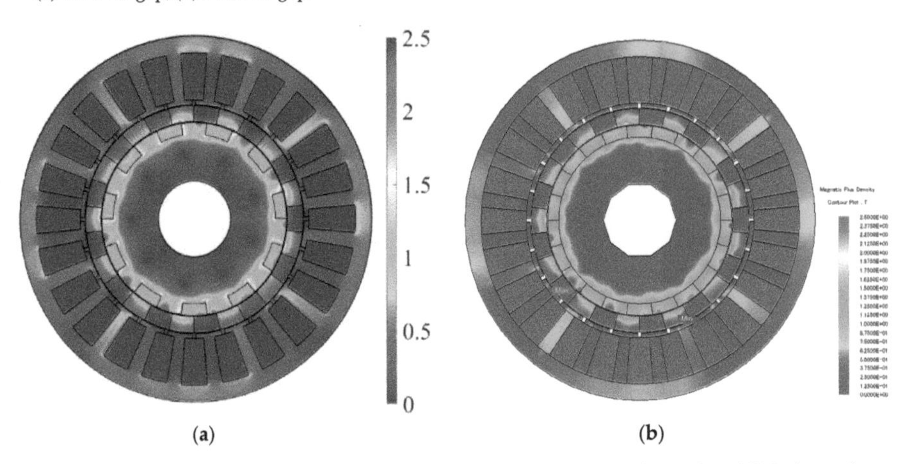

Figure 9. Comparison of on-load magnetic flux density distribution of the SCP-MGM, drawn by harmonic modeling method (HMM) and finite element analysis (FEA): (**a**) HMM; (**b**) FEA.

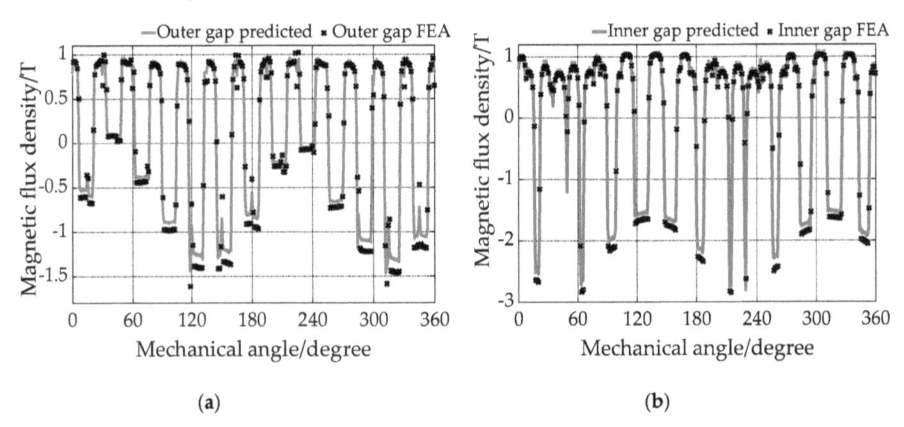

Figure 10. No-load inner and outer air-gap radial magnetic flux density distribution of the DCP-MGM: (**a**) outer air gap; (**b**) inner air gap.

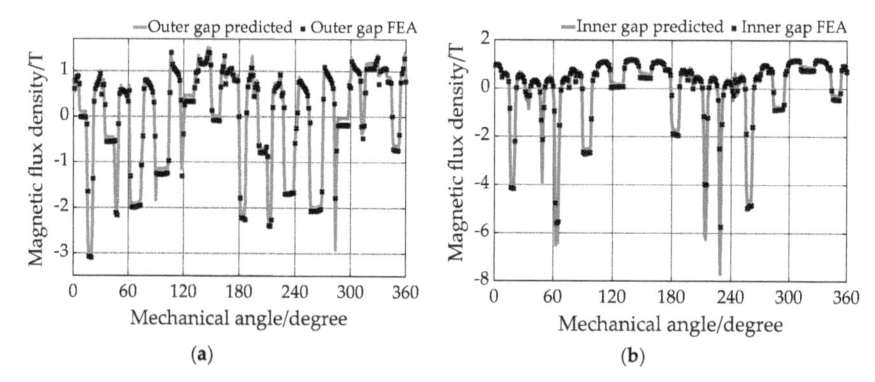

Figure 11. On-load inner and outer air-gap radial magnetic flux density distribution of the DCP-MGM: (**a**) outer air gap; (**b**) inner air gap.

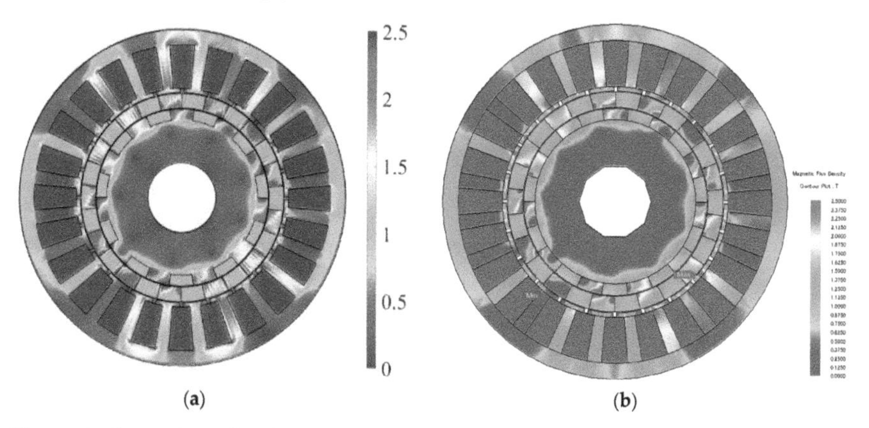

Figure 12. Comparison of on-load magnetic flux density distribution of the DCP-MGM drawn by HMM and FEA: (**a**) HMM; (**b**) FEA.

Table 5. Geometrical parameters of the SCP-MGM and DCP-MGM.

Parameters	Symbols	Values	Units
Number of inner PM pole pairs	P_i	11	-
Number of modulator pieces	Q	13	-
Number of stator slots	P	24	-
Number of stator winding pole pairs	P_s	2	-
Axial length	L	110	mm
Outer radius of shaft	R_1	25	mm
Inner radius of inner PM	R_2	55.5	mm
Outer radius of inner rotor	R_3	63.2	mm
Inner radius of outer rotor	R_4	63.8	mm
Outer radius of outer rotor	R_5	74.4	mm
Inner radius of stator	R_6	75	mm
Radius of stator slot bottom	R_7	78	mm
Outer radius of stator slot	R_8	109	mm
Outer radius of stator	R_9	120	mm
Angle of inner PM arc	α	0.286	rad
Angle of Modulator piece	β	0.242	rad
Angle of slot opening	δ	0.032	rad
Angle of stator slot	γ	0.168	rad
Slot filling factor	Fa	60%	-
Current density in stator windings	I_D	5	A/mm^2

Table 6. Mean error percentage and maximum difference of the magnetic flux density distributions between the FEA and HMM of the SCP-MGM and DCP-MGM.

State	SCP-MGM				DCP-MGM			
	Inner Rotor		Outer Rotor		Inner Rotor		Outer Rotor	
	ε_1	ε_2	ε_1	ε_2	ε_1	ε_2	ε_1	ε_2
No-load	9.3%	0.55 T	15.9%	0.11 T	11.4%	1.06 T	14.4%	0.70 T
On-load	14.2%	1.62 T	14.3%	0.83 T	20.8%	3.11 T	16.7%	1.36 T

A fast Fourier transform (FFT) was executed on the no-load radial component of the magnetic flux density of the outer air gap for both the SCP-MGM and DCP-MGM, as shown in Figure 13. It can be seen that the second harmonic component was much higher for the DCP-MGM, since the inserted PM on the outer rotor produced a second magnetic field after the modulation of the consequent-pole iron part of the inner rotor. Hence, the consequent-pole structure of the inner rotor and outer rotor of the DCP-MGM could modulate the magnetic field generated by its counterpart. The electromagnetic torque of the DCP-MGM was expected to be larger than that of the SCP-MGM under the same working conditions. Additionally, the difference of each frequency component between the HMM and FEA was very small. Specifically, the error percentages of the second harmonic component for the SCP-MGM and the DCP-MGM using HMM and FEA were 4.14% and 2.15%, respectively.

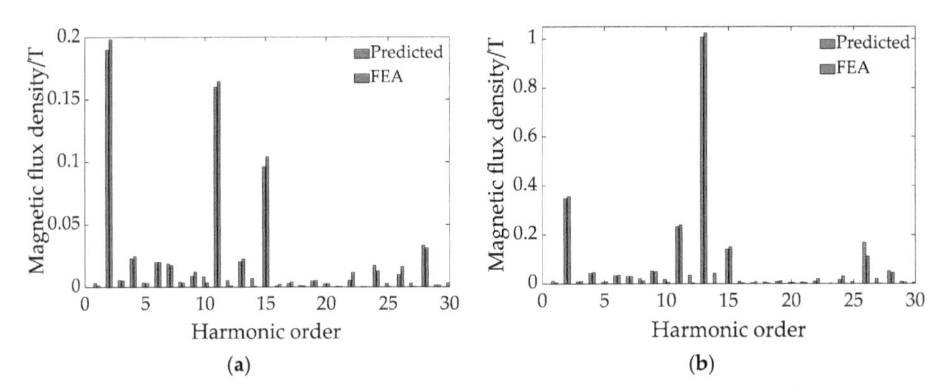

Figure 13. FFT of no-load outer-air-gap radial magnetic flux density for the SCP-MGM and the DCP-MGM: (**a**) SCP-MGM; (**b**) DCP-MGM.

4.3. Electromagnetic Performance Analysis under Different Working Conditions

The electromagnetic performances of the two MGMs are predicted by HMM and simulated in FEA software under different operating conditions, where the rotating speed of two rotors and the current frequency of stator windings are set to cooperate with the practical driving conditions of the HEC, as shown in Table 1. Since the ICE is connected to the inner rotor to provide the power; the outer rotor is connected to the differential, which is further connected to the wheels; the battery is connected to the stator windings, and there can be two-way power flow between the battery and the stator windings. Thus, the power transmission relation among inner rotor, outer rotor and stator windings is equal to the power transmission relation among ICE, wheels and battery. Assume the anti-clock direction is positive, the torque is positive if it's on the anti-clock direction, otherwise it's negative. The power of a component is defined as a positive one when this component is inputting energy; when a component is outputting energy, its power is defined as a negative one.

4.3.1. Back EMF under No-Load Condition

The amplitude of back EMF was determined by the rotating speed of both the inner rotor and the outer rotor. Figure 14 shows the no-load back EMF of the SCP-MGM and the DCP-MGM under the same operating conditions, namely ω_i = 1200 r/min, ω_o = 1500 r/min. It can be seen that there was an error between the predicted back EMF and FEA result; because the back EMF was calculated as the derivative of flux linkage with respect to time, a small error of flux linkage will be amplified on the back EMF. Additionally, the back EMF of the DCP-MGM was larger than that of the SCP-MGM, due to the flux enhancing effect of the PMs inserted into the modulator. The maximum errors for the SCP-MGM and DCP-MGM were 165 V and 520 V, respectively. The average errors of the SCP-MGM and DCP-MGM were 3.7% and 6.9%, respectively.

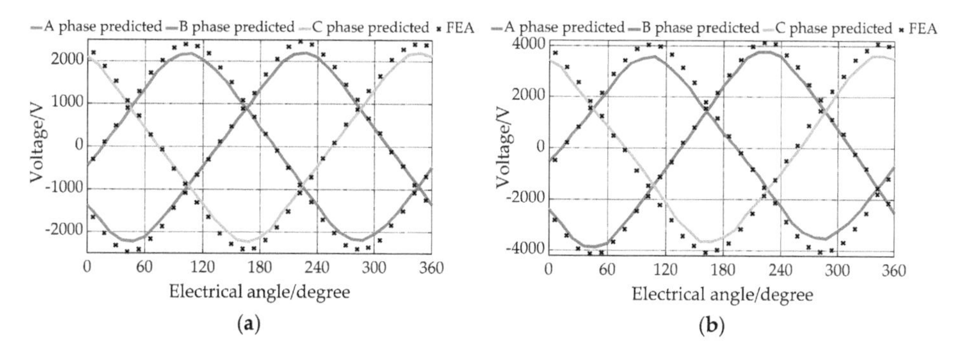

Figure 14. Analytically predicted and FEA simulated back electromotive force (EMF) of the SCP-MGM and DCP-MGM: (**a**) SCP-MGM; (**b**) DCP-MGM.

4.3.2. Pure Electric Mode (Mode 1)

Under this mode, the HEV is at a low driving speed, the inner rotor is locked, and the ICE does not come into service; stator windings only work to provide the power that the HEV needs. Thus, ω_i = 0 r/min, ω_o = 500 r/min. The torque waveforms of the power distribution of the two machines are shown in Figures 15 and 16.

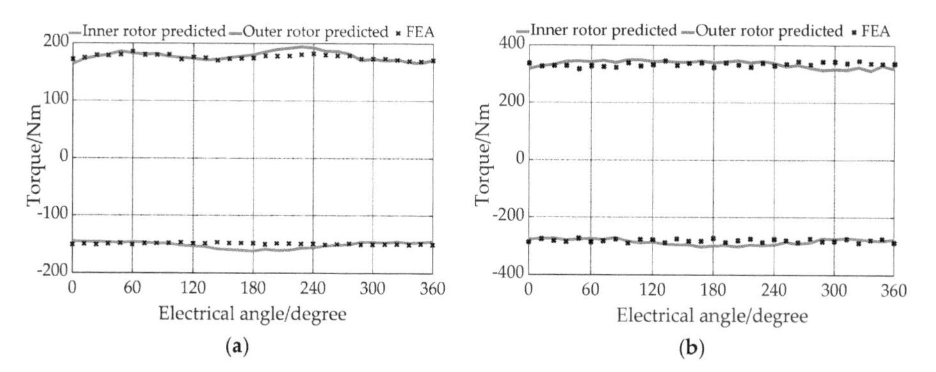

Figure 15. Analytically predicted and FEA simulated torque waveforms of the SCP-MGM and DCP-MGM under pure electric mode: (**a**) SCP-MGM; (**b**) DCP-MGM.

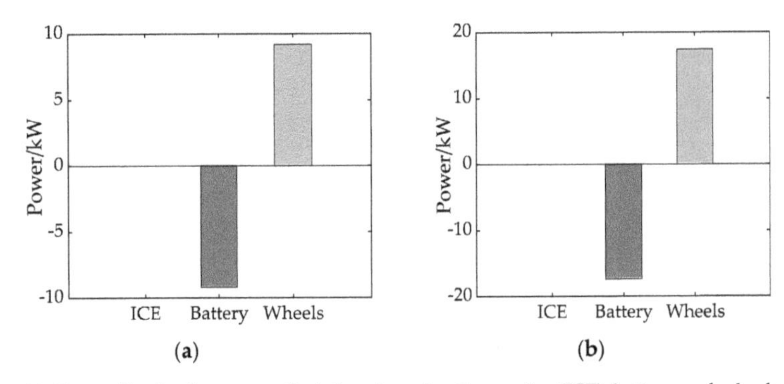

Figure 16. Power distribution among the internal combustion engine (ICE), battery, and wheels under pure electric mode: (**a**) SCP-MGM; (**b**) DCP-MGM.

4.3.3. Pure Mechanical Mode (Mode 2)

Under this mode, the HEV is running at a medium speed, so the battery does not output power anymore and the ICE comes into use. However, to maintain the magnetic field, the stator windings are electrified with DC current [14]. The rotating speeds of two rotors are: ω_i = 1200 r/min, ω_o = 1015 r/min. The torque waveforms of the power distribution of the two machines are shown in Figures 17 and 18.

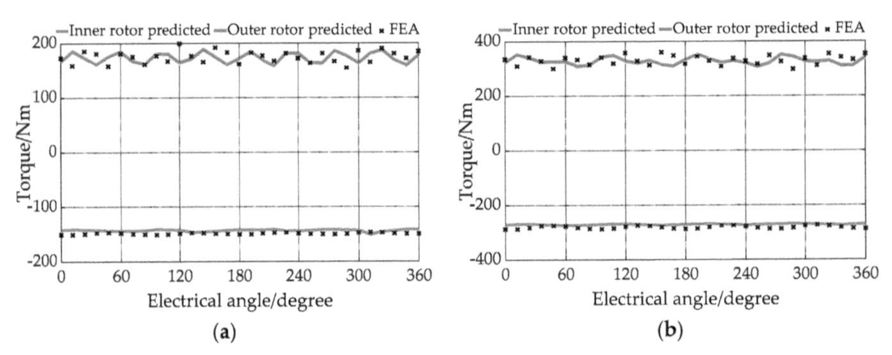

Figure 17. Analytically predicted and FEA simulated torque waveforms of SCP-MGM and DCP-MGM under pure mechanical mode: (**a**) SCP-MGM; (**b**) DCP-MGM.

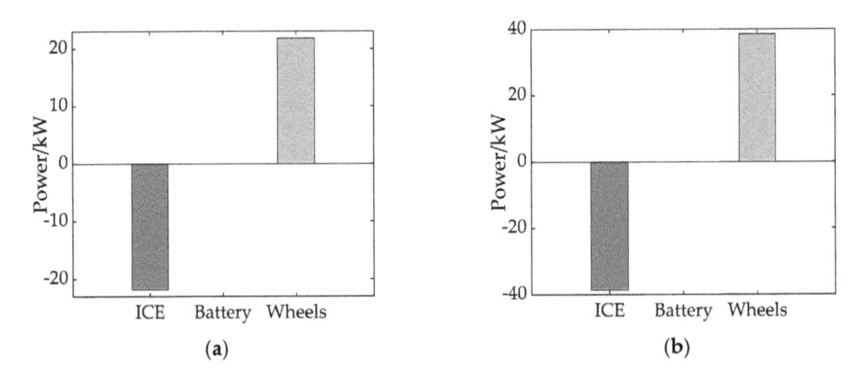

Figure 18. Power distribution among ICE, battery and wheels under pure mechanical mode: (**a**) SCP-MGM; (**b**) DCP-MGM.

4.3.4. Hybrid Mode (Mode 3)

When the HEV needs to further accelerate, the ICE alone is not enough to provide the power that the HEV needs. The SCP-MGM and DCP-MGM can then switch into hybrid mode. In this mode, both ICE and battery provide the energy for the wheels. The rotating speeds of two rotors are: ω_i = 1200 r/min, ω_o = 2000 r/min. The torque waveforms of the power distribution of the two machines are shown in Figures 19 and 20.

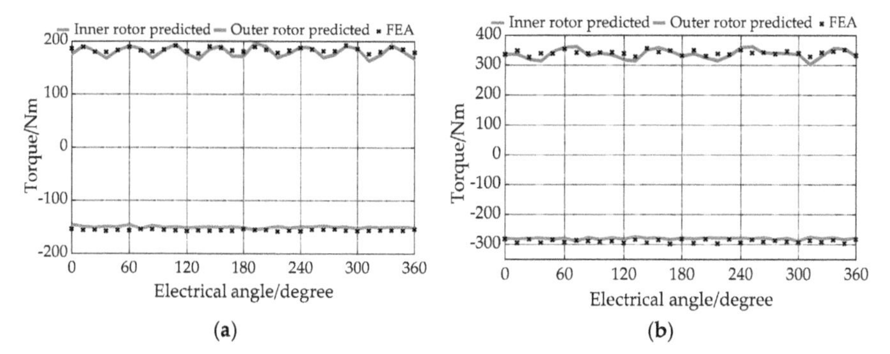

Figure 19. Analytically predicted and FEA simulated electromagnetic torque of SCP-MGM and DCP-MGM under hybrid mode: (**a**) SCP-MGM; (**b**) DCP-MGM.

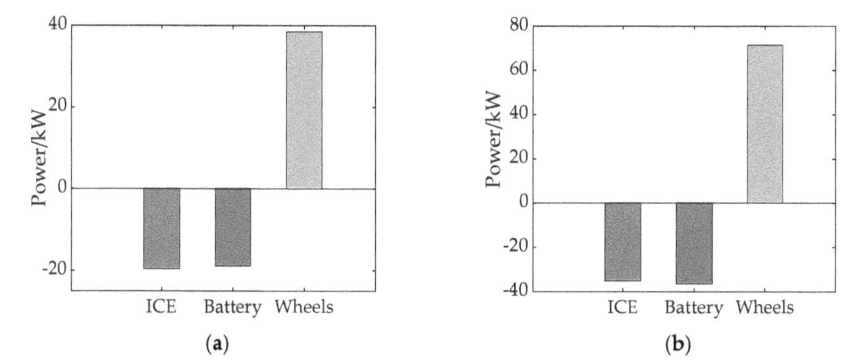

Figure 20. Power distribution among ICE, battery, and wheels under hybrid mode: (**a**) SCP-MGM; (**b**) DCP-MGM.

4.3.5. Regenerative Braking Mode (Mode 4)

When the HEV deaccelerates, the ICE stops working. The magnetic field generated via the stator windings provides a resistance for the wheels, and thus the power flows from the wheels to the battery. At this time, the power flows from the stator windings to the battery. Thus, the battery can be charged under this mode. The rotating speeds of two rotors are: ω_i = 0 r/min, ω_o = 1000 r/min. The torque waveforms of the power distribution of the two machines are shown in Figures 21 and 22.

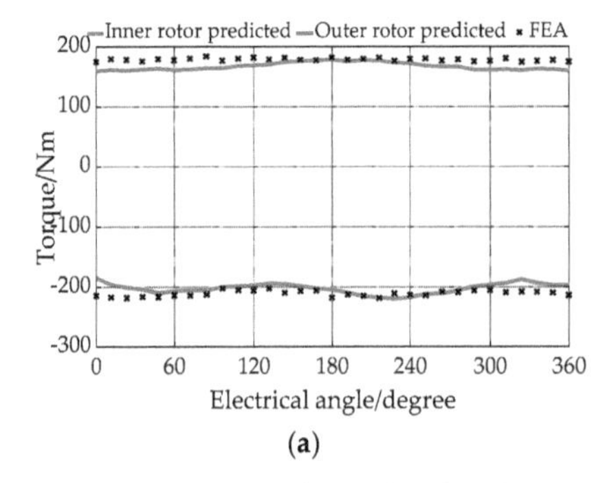

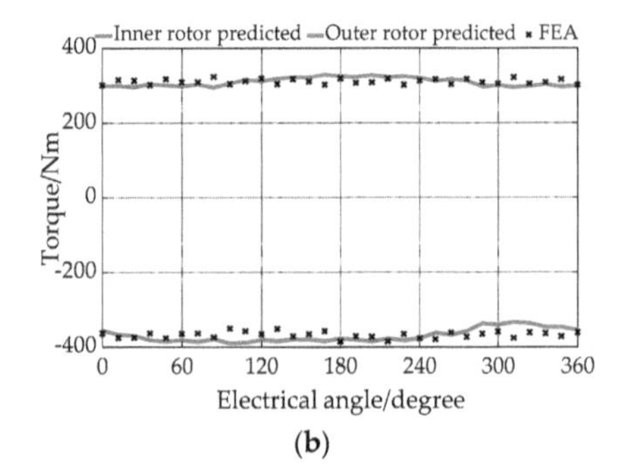

Figure 21. Analytically predicted and FEA simulated back EMF of SCP-MGM and DCP-MGM: (**a**) SCP-MGM; (**b**) DCP-MGM.

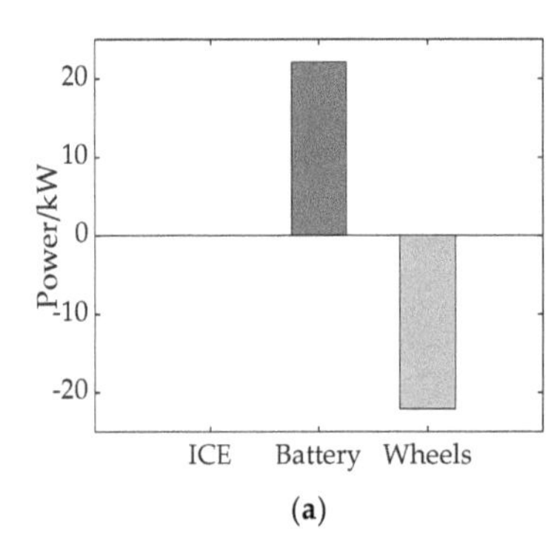

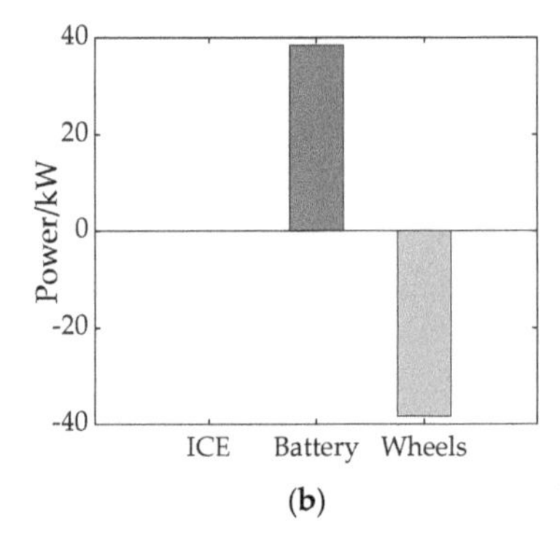

Figure 22. Power distribution among ICE, battery, and wheels under hybrid mode: (**a**) SCP-MGM; (**b**) DCP-MGM.

4.3.6. Quantitative comparison between HMM and FEA

Table 7 shows the mean error percentage and maximum difference of the torque waveforms between FEA and HMM of the SCP-MGM and DCP-MGM under different operating modes. It can be observed that the mean error percentage was below 8%, which is acceptable for the electromagnetic torque prediction of SCP-MGMs and DCP-MGMs. The error of torque prediction was mainly caused by the error of the air-gap magnetic flux density prediction on both the radial and tangential direction, since the torque was calculated using the calculus of the product of the radial component and tangential component of air-gap magnetic flux density.

Table 7. Mean error percentage and maximum difference of the torque waveforms between FEA and HMM of the SCP-MGM and DCP-MGM under different operating modes.

Mode	SCP-MGM				DCP-MGM			
	Inner Rotor		Outer Rotor		Inner Rotor		Outer Rotor	
	ε_1	ε_2	ε_1	ε_2	ε_1	ε_2	ε_1	ε_2
1	3.59%	12.1 Nm	2.64%	14.8 Nm	3.55%	22.5 Nm	4.46%	30.3 Nm
2	3.00%	7.85 Nm	7.39%	37.2 Nm	2.57%	14.9 Nm	5.88%	48.6 Nm
3	2.75%	8.9 Nm	3.61%	15.2 Nm	2.27%	15.3 Nm	3.27%	30.3 Nm
4	5.73%	18.1 Nm	5.06%	30.0 Nm	3.50%	28.4 Nm	4.56%	43.0 Nm

From the above comparisons between the SCP-MGM and DCP-MGM under various working conditions, it can be seen that the torque ratio between outer rotor and inner rotor was about 1.18, which is equal to the pole-pair ratio of outer rotor to inner rotor. The ripple rate of the outer rotor was larger than that of the inner rotor, which can be alleviated by adopting a skewed stator when these machines are applied in practical applications. Additionally, the electromagnetic torque of the DCP-MGM was about 1.88 times of that of SCP-MGM. However, the torque per unit weight of the PM of the DCP-MGM was only about 0.72 of that of SCP-MGM. Considering the fact that the consequent–pole structure already saves half the PM material that would otherwise have been used, the total usage of PM material for DCP-MGM is reasonable. Thus, the DCP-MGM is more suitable to be used in the propulsion systems of HEVs.

4.3.7. Discussion of HMM

The greatest advantage of HMM is that the modeling process is simpler compared to the conventional subdomain method [30]. Since it unitizes Fourier series of relative permeability, the boundary conditions become simpler and the number of unknowns becomes less. For instance, the consequent-pole structure leads to a very complex general solution if the subdomain method is used [29]. Additionally, magnetic saturation can be taken into account to an extent in HMM, and a magnetic flux density distribution figure can be obtained. However, the HMM also has some drawbacks. First, HMM can only provide a simplified magnetic saturation model, since the magnetic saturation varies from point to point in some SMM parts. The relative permeability in the radial direction is also regarded as a constant, but, in reality, it could change. To the authors' knowledge, no existing analytical method can really reflect the magnetic saturation at every point within an electric machine. Secondly, the use of convolution sometimes leads to an extremely large or small value, which further leads to a large error in the final result due to the accuracy limit of the numerical calculation.

5. Conclusions

In this paper, an analytical modeling method for consequent-pole MGMs was proposed and elaborated. Two machine topologies, namely the SCP-MGM and DCP-MGM, were analyzed and compared quantitatively using HMM and FEA. A good agreement was achieved for these two methods. Furthermore, these two machines were embedded into the propulsion system of HEVs under different operating conditions. By inserting extra PMs in the modulator, the electromagnetic torque of the DCP-MGM increased greatly compared to its counterpart. Therefore, the DCP-MGM has the potential to be applied in the propulsion systems of HEVs. Additionally, the proposed HMM can also be applied in the mathematical modeling of other consequent-pole electric machines.

Author Contributions: The work presented in this paper is the output of the research projects undertaken by C.L. In specific, H.Z. and C.L. developed the topic. H.Z. carried out the calculation and simulation, analyzed the results, and wrote the paper. Z.S. gave some suggestions on the calculation process. J.Y. helped to carried out a part of the simulation.

Funding: This research was supported by a grant (Project No. 51677159) from the Natural Science Foundation of China (NSFC), China. Also, the work is supported by a grant (Project No. JCYJ20180307123918658) from the Science Technology and Innovation Committee of Shenzhen Municipality, Shenzhen, China. Moreover, it was also supported by a grant (Project No. CityU 21201216) from Research Grants Council of HKSAR, China.

Conflicts of Interest: The authors declare no conflict of interest.

Appendix A

The expressions of $\mathbf{H}_r$ and $\mathbf{H}_\theta$ are given by:

$$\mathbf{H}_r = -i\frac{1}{r}\mu_{r,\text{cov}}^{-1}\mathbf{K}\mathbf{A}_z - \mu_0\mu_{r,\text{cov}}^{-1}\mathbf{M}_r \text{ and } \mathbf{H}_\theta = -\mu_{\theta,\text{cov}}^{-1}\frac{\partial \mathbf{A}_z}{\partial r} - \mu_0\mu_{\theta,\text{cov}}^{-1}\mathbf{M}_\theta \tag{A1}$$

In polar coordinate, (15) can be simplified as:

$$\nabla \times \mathbf{H} = \frac{1}{r}\frac{\partial}{\partial r}(r\mathbf{H}_\theta) + i\frac{\mathbf{K}\mathbf{H}_r}{r} = \mathbf{J}_z \tag{A2}$$

By substituting (A1) into (A2), one can obtain that:

$$\frac{\partial^2 \mathbf{A}_z^k}{\partial r^2} + \frac{1}{r}\frac{\partial \mathbf{A}_z^k}{\partial r} - \frac{b_{\theta,\mathrm{cov}}\mathbf{K}\boldsymbol{\mu}_{r,\mathrm{covc}}^{-1}\mathbf{K}\mathbf{A}_z^k}{r^2} = -\boldsymbol{\mu}_{\theta,cov}\mathbf{J}_z - \frac{\mu_0}{r}(\mathbf{M}_\theta + i\boldsymbol{\mu}_{\theta,\mathrm{cov}}\mathbf{K}\boldsymbol{\mu}_{r,\mathrm{covc}}^{-1}\mathbf{M}_r) \tag{A3}$$

The vector magnetic potential in region I~X are given in Table A1.

Table A1. The Vector Magnetic Potential Expressions.

Region	VMP Expression	Region	VMP Expression
I	$\mathbf{A}_z^I = \left(\frac{r}{R_1}\right)^{\lvert\mathbf{K}\rvert}\mathbf{D}^I$	II	$\mathbf{A}_z^{II} = \left(\frac{r}{R_2}\right)^{\lvert\mathbf{K}\rvert}\mathbf{D}^{II} + \left(\frac{R_1}{r}\right)^{\lvert\mathbf{K}\rvert}\mathbf{E}^{II}$
III	$\mathbf{A}_z^{III} = \mathbf{P}^{III}\left(\frac{r}{R_3}\right)^{\boldsymbol{\lambda}^{III}}\mathbf{D}^{III} + \mathbf{P}^{III}\left(\frac{R_2}{r}\right)^{\boldsymbol{\lambda}^{III}}\mathbf{E}^{III} + r\mathbf{G}^{III}$	IV	$\mathbf{A}_z^{IV} = \left(\frac{r}{R_4}\right)^{\lvert\mathbf{K}\rvert}\mathbf{D}^{IV} + \left(\frac{R_3}{r}\right)^{\lvert\mathbf{K}\rvert}\mathbf{E}^{IV}$
V	$\mathbf{A}_z^{V} = \mathbf{P}^{V}\left(\frac{r}{R_5}\right)^{\boldsymbol{\lambda}^{V}}\mathbf{D}^{V} + \mathbf{P}^{V}\left(\frac{R_4}{r}\right)^{\boldsymbol{\lambda}^{V}}\mathbf{E}^{V} + r\mathbf{G}^{V}$	VI	$\mathbf{A}_z^{VI} = \left(\frac{r}{R_6}\right)^{\lvert\mathbf{K}\rvert}\mathbf{D}^{VI} + \left(\frac{R_5}{r}\right)^{\lvert\mathbf{K}\rvert}\mathbf{E}^{VI}$
VII	$\mathbf{A}_z^{VII} = \mathbf{P}^{VII}\left(\frac{r}{R_7}\right)^{\boldsymbol{\lambda}^{VII}}\mathbf{D}^{VII} + \mathbf{P}^{VII}\left(\frac{R_6}{r}\right)^{\boldsymbol{\lambda}^{VII}}\mathbf{E}^{VII}$	VIII	$\mathbf{A}_z^{VIII} = \mathbf{P}^{VIII}\left(\frac{r}{R_8}\right)^{\boldsymbol{\lambda}^{VIII}}\mathbf{D}^{VIII} + \mathbf{P}^{VIII}\left(\frac{R_7}{r}\right)^{\boldsymbol{\lambda}^{VIII}}\mathbf{E}^{VIII} + r^2\mathbf{F}$
IX	$\mathbf{A}_z^{IX} = \left(\frac{r}{R_9}\right)^{\lvert\mathbf{K}\rvert}\mathbf{D}^{IX} + \left(\frac{R_8}{r}\right)^{\lvert\mathbf{K}\rvert}\mathbf{E}^{IX}$	X	$\mathbf{A}_z^{X} = \left(\frac{r}{R_9}\right)^{\lvert\mathbf{K}\rvert}\mathbf{E}^{X}$

Where $\mathbf{D}^{\mathrm{I}}, \mathbf{D}^{\mathrm{II}}, \ldots, \mathbf{D}^{\mathrm{IX}}$ and $\mathbf{E}^{\mathrm{I}}, \mathbf{E}^{\mathrm{II}}, \ldots, \mathbf{E}^{\mathrm{X}}$ are (N*1) vector.

The expressions of **X**, **S**, **T** are:

$$\mathbf{X} = \begin{bmatrix} (\mathbf{D}^I)^T & (\mathbf{D}^{II})^T & (\mathbf{E}^{II})^T & (\mathbf{D}^{III})^T & (\mathbf{E}^{III})^T & \cdots & (\mathbf{D}^{IX})^T & (\mathbf{E}^{IX})^T & (\mathbf{E}^{X})^T \end{bmatrix}^T \tag{A4}$$

$$\mathbf{S} = \begin{bmatrix}
\mathbf{K}_{1,1} & \mathbf{K}_{1,2} & \mathbf{K}_{1,3} & 0 & 0 & 0 & 0 & 0 & 0 & 0 & 0 & 0 & 0 & 0 & 0 & 0 & 0 & 0 \\
\mathbf{K}_{2,1} & \mathbf{K}_{2,2} & \mathbf{K}_{2,3} & 0 & 0 & 0 & 0 & 0 & 0 & 0 & 0 & 0 & 0 & 0 & 0 & 0 & 0 & 0 \\
0 & \mathbf{K}_{3,2} & \mathbf{K}_{3,3} & \mathbf{K}_{3,4} & \mathbf{K}_{3,5} & 0 & 0 & 0 & 0 & 0 & 0 & 0 & 0 & 0 & 0 & 0 & 0 & 0 \\
0 & \mathbf{K}_{4,2} & \mathbf{K}_{4,3} & \mathbf{K}_{4,4} & \mathbf{K}_{4,5} & 0 & 0 & 0 & 0 & 0 & 0 & 0 & 0 & 0 & 0 & 0 & 0 & 0 \\
0 & 0 & 0 & \mathbf{K}_{5,4} & \mathbf{K}_{5,5} & \mathbf{K}_{5,6} & \mathbf{K}_{5,7} & 0 & 0 & 0 & 0 & 0 & 0 & 0 & 0 & 0 & 0 & 0 \\
0 & 0 & 0 & \mathbf{K}_{6,4} & \mathbf{K}_{6,5} & \mathbf{K}_{6,6} & \mathbf{K}_{6,7} & 0 & 0 & 0 & 0 & 0 & 0 & 0 & 0 & 0 & 0 & 0 \\
0 & 0 & 0 & 0 & 0 & \mathbf{K}_{7,6} & \mathbf{K}_{7,7} & \mathbf{K}_{7,8} & \mathbf{K}_{7,9} & 0 & 0 & 0 & 0 & 0 & 0 & 0 & 0 & 0 \\
0 & 0 & 0 & 0 & 0 & \mathbf{K}_{8,6} & \mathbf{K}_{8,7} & \mathbf{K}_{8,8} & \mathbf{K}_{8,9} & 0 & 0 & 0 & 0 & 0 & 0 & 0 & 0 & 0 \\
0 & 0 & 0 & 0 & 0 & 0 & 0 & \mathbf{K}_{9,8} & \mathbf{K}_{9,9} & \mathbf{K}_{9,10} & \mathbf{K}_{9,11} & 0 & 0 & 0 & 0 & 0 & 0 & 0 \\
0 & 0 & 0 & 0 & 0 & 0 & 0 & \mathbf{K}_{10,8} & \mathbf{K}_{10,9} & \mathbf{K}_{10,10} & \mathbf{K}_{10,11} & 0 & 0 & 0 & 0 & 0 & 0 & 0 \\
0 & 0 & 0 & 0 & 0 & 0 & 0 & 0 & 0 & \mathbf{K}_{11,10} & \mathbf{K}_{11,11} & \mathbf{K}_{11,12} & \mathbf{K}_{11,13} & 0 & 0 & 0 & 0 & 0 \\
0 & 0 & 0 & 0 & 0 & 0 & 0 & 0 & 0 & \mathbf{K}_{12,10} & \mathbf{K}_{12,11} & \mathbf{K}_{12,12} & \mathbf{K}_{12,13} & 0 & 0 & 0 & 0 & 0 \\
0 & 0 & 0 & 0 & 0 & 0 & 0 & 0 & 0 & 0 & 0 & \mathbf{K}_{13,12} & \mathbf{K}_{13,13} & \mathbf{K}_{13,14} & \mathbf{K}_{13,15} & 0 & 0 & 0 \\
0 & 0 & 0 & 0 & 0 & 0 & 0 & 0 & 0 & 0 & 0 & \mathbf{K}_{14,12} & \mathbf{K}_{14,13} & \mathbf{K}_{14,14} & \mathbf{K}_{14,15} & 0 & 0 & 0 \\
0 & 0 & 0 & 0 & 0 & 0 & 0 & 0 & 0 & 0 & 0 & 0 & 0 & \mathbf{K}_{15,14} & \mathbf{K}_{15,15} & \mathbf{K}_{15,16} & \mathbf{K}_{15,17} & 0 \\
0 & 0 & 0 & 0 & 0 & 0 & 0 & 0 & 0 & 0 & 0 & 0 & 0 & \mathbf{K}_{16,14} & \mathbf{K}_{16,15} & \mathbf{K}_{16,16} & \mathbf{K}_{16,17} & 0 \\
0 & 0 & 0 & 0 & 0 & 0 & 0 & 0 & 0 & 0 & 0 & 0 & 0 & 0 & 0 & \mathbf{K}_{17,16} & \mathbf{K}_{17,17} & \mathbf{K}_{17,18} \\
0 & 0 & 0 & 0 & 0 & 0 & 0 & 0 & 0 & 0 & 0 & 0 & 0 & 0 & 0 & \mathbf{K}_{18,16} & \mathbf{K}_{18,17} & \mathbf{K}_{18,18}
\end{bmatrix} \tag{A5}$$

$$\begin{aligned}\mathbf{T} = \big[\, & 0 \quad 0 \quad (R_2\mathbf{G}_1)^T \quad (R_2\boldsymbol{\mu}_{II}\mathbf{G}_1)^T \quad (-R_3\mathbf{G}_1)^T \quad (-R_3\boldsymbol{\mu}_{IV}\mathbf{G}_1)^T \quad (R_4\mathbf{G}_2)^T \quad (R_4\boldsymbol{\mu}_{IV}\mathbf{G}_2)^T \quad (-R_5\mathbf{G}_2)^T \\ & (-\boldsymbol{\mu}_{VI}R_5\mathbf{G}_2)^T \quad 0 \quad 0 \quad (R_7^2\mathbf{F})^T \quad (2R_7^2\mu_{open,\theta}\mathbf{F})^T \quad (-R_8^2\mathbf{F})^T \quad (-2R_8^2\mu_{IX}\mathbf{F})^T \quad 0 \quad 0 \,\big]^T\end{aligned} \tag{A6}$$

where:

$$\begin{bmatrix} \mathbf{K}_{1,1} & \mathbf{K}_{1,2} & \mathbf{K}_{1,3} \end{bmatrix} = \begin{bmatrix} \mathbf{I} & -\left(\frac{R_1}{R_2}\right)^{\lvert\mathbf{K}\rvert} & -\mathbf{I} \end{bmatrix} \tag{A7}$$

$$\begin{bmatrix} \mathbf{K}_{2,1} & \mathbf{K}_{2,2} & \mathbf{K}_{2,3} \end{bmatrix} = \begin{bmatrix} \boldsymbol{\mu}_{II} & -\boldsymbol{\mu}_{I}\left(\frac{R_1}{R_2}\right)^{\lvert\mathbf{K}\rvert} & \boldsymbol{\mu}_{I} \end{bmatrix} \tag{A8}$$

$$\begin{bmatrix} \mathbf{K}_{3,2} & \mathbf{K}_{3,3} & \mathbf{K}_{3,4} & \mathbf{K}_{3,5} \end{bmatrix} = \begin{bmatrix} \mathbf{I} & \left(\frac{R_1}{R_2}\right)^{|\mathbf{K}|} & -\mathbf{P}^{III}\left(\frac{R_1}{R_2}\right)^{|\mathbf{K}|} & -\mathbf{P}^{III} \end{bmatrix} \tag{A9}$$

$$\begin{bmatrix} \mathbf{K}_{4,2} & \mathbf{K}_{4,3} & \mathbf{K}_{4,4} & \mathbf{K}_{4,5} \end{bmatrix} = \begin{bmatrix} \mu_{PM,\theta} & -\mu_{PM,\theta}|\mathbf{K}|\left(\frac{R_1}{R_2}\right)^{|\mathbf{K}|} & -\mu_{II}\mathbf{P}^{III}\boldsymbol{\lambda}^{III}\left(\frac{R_2}{R_3}\right)^{\boldsymbol{\lambda}^{III}} & \mu_{II}\mathbf{P}^{III}\boldsymbol{\lambda}^{III} \end{bmatrix} \tag{A10}$$

$$\begin{bmatrix} \mathbf{K}_{5,4} & \mathbf{K}_{5,5} & \mathbf{K}_{5,6} & \mathbf{K}_{5,7} \end{bmatrix} = \begin{bmatrix} \mathbf{P}^{III} & \mathbf{P}^{III}\left(\frac{R_2}{R_3}\right)^{\boldsymbol{\lambda}^{III}} & -\left(\frac{R_3}{R_4}\right)^{|\mathbf{K}|} & -\mathbf{I} \end{bmatrix} \tag{A11}$$

$$\begin{bmatrix} \mathbf{K}_{6,4} & \mathbf{K}_{6,5} & \mathbf{K}_{6,6} & \mathbf{K}_{6,7} \end{bmatrix} = \begin{bmatrix} \mu_{IV}\mathbf{P}^{III}\boldsymbol{\lambda}^{III} & -\mu_{IV}\mathbf{P}^{III}\boldsymbol{\lambda}^{III}\left(\frac{R_2}{R_3}\right)^{\boldsymbol{\lambda}^{III}} & \mu_{PM,\theta}|\mathbf{K}|\left(\frac{R_3}{R_4}\right)^{|\mathbf{K}|} & \mu_{PM,\theta}|\mathbf{K}| \end{bmatrix} \tag{A12}$$

$$\begin{bmatrix} \mathbf{K}_{7,6} & \mathbf{K}_{7,7} & \mathbf{K}_{7,8} & \mathbf{K}_{7,9} \end{bmatrix} = \begin{bmatrix} \mathbf{I} & \left(\frac{R_3}{R_4}\right)^{|\mathbf{K}|} & -\mathbf{P}^{V}\left(\frac{R_4}{R_5}\right)^{\boldsymbol{\lambda}^{V}} & -\mathbf{P}^{V} \end{bmatrix} \tag{A13}$$

$$\begin{bmatrix} \mathbf{K}_{8,6} & \mathbf{K}_{8,7} & \mathbf{K}_{8,8} & \mathbf{K}_{8,9} \end{bmatrix} = \begin{bmatrix} \mu_{Mod,\theta}|\mathbf{K}| & -\mu_{Mod,\theta}|\mathbf{K}|\left(\frac{R_3}{R_4}\right)^{|\mathbf{K}|} & -\mu_{IV}\mathbf{P}^{V}\boldsymbol{\lambda}^{V}\left(\frac{R_4}{R_5}\right)^{\boldsymbol{\lambda}^{V}} & \mu_{IV}\mathbf{P}^{V}\boldsymbol{\lambda}^{V} \end{bmatrix} \tag{A14}$$

$$\begin{bmatrix} \mathbf{K}_{9,8} & \mathbf{K}_{9,9} & \mathbf{K}_{9,10} & \mathbf{K}_{9,11} \end{bmatrix} = \begin{bmatrix} \mathbf{P}^{V} & \mathbf{P}^{V}\left(\frac{R_4}{R_5}\right)^{\boldsymbol{\lambda}^{V}} & -\left(\frac{R_5}{R_6}\right)^{|\mathbf{K}|} & -\mathbf{I} \end{bmatrix} \tag{A15}$$

$$\begin{bmatrix} \mathbf{K}_{10,8} & \mathbf{K}_{10,9} & \mathbf{K}_{10,10} & \mathbf{K}_{10,11} \end{bmatrix} = \begin{bmatrix} \mu_{VI}\mathbf{P}^{V}\boldsymbol{\lambda}^{V} & -\mu_{VI}\mathbf{P}^{V}\boldsymbol{\lambda}^{V}\left(\frac{R_4}{R_5}\right)^{\boldsymbol{\lambda}^{V}} & -\mu_{Mod,\theta}|\mathbf{K}|\left(\frac{R_5}{R_6}\right)^{|\mathbf{K}|} & \mu_{Mod,\theta}|\mathbf{K}| \end{bmatrix} \tag{A16}$$

$$\begin{bmatrix} \mathbf{K}_{11,10} & \mathbf{K}_{11,11} & \mathbf{K}_{11,12} & \mathbf{K}_{11,13} \end{bmatrix} = \begin{bmatrix} \mathbf{I} & \left(\frac{R_5}{R_6}\right)^{|\mathbf{K}|} & -\mathbf{P}^{VII}\left(\frac{R_6}{R_7}\right)^{\boldsymbol{\lambda}^{VII}} & -\mathbf{P}^{VII} \end{bmatrix} \tag{A17}$$

$$\begin{bmatrix} \mathbf{K}_{12,10} & \mathbf{K}_{12,11} & \mathbf{K}_{12,12} & \mathbf{K}_{12,13} \end{bmatrix} = \begin{bmatrix} \mu_{Open,\theta}|\mathbf{K}| & \mu_{Open,\theta}|\mathbf{K}|\left(\frac{R_5}{R_6}\right)^{|\mathbf{K}|} & -\mu_{VI}\mathbf{P}^{VII}\boldsymbol{\lambda}^{VII}\left(\frac{R_6}{R_7}\right)^{\boldsymbol{\lambda}^{VII}} & \mu_{VI}\mathbf{P}^{VII}\boldsymbol{\lambda}^{VII} \end{bmatrix} \tag{A18}$$

$$\begin{bmatrix} \mathbf{K}_{13,12} & \mathbf{K}_{13,13} & \mathbf{K}_{13,14} & \mathbf{K}_{13,15} \end{bmatrix} = \begin{bmatrix} \mathbf{P}^{VII}\boldsymbol{\lambda}^{VII} & \mathbf{P}^{VII}\left(\frac{R_6}{R_7}\right)^{\boldsymbol{\lambda}^{VII}} & -\mathbf{P}^{VIII}\left(\frac{R_7}{R_8}\right)^{\boldsymbol{\lambda}^{VIII}} & -\mathbf{P}^{VIII} \end{bmatrix} \tag{A19}$$

$$\begin{bmatrix} \mathbf{K}_{14,12} & \mathbf{K}_{14,13} & \mathbf{K}_{14,14} & \mathbf{K}_{14,15} \end{bmatrix} = \begin{bmatrix} \mu_{Slot,\theta}\mathbf{P}^{VII}\boldsymbol{\lambda}^{VII} & -\mu_{Slot,\theta}\mathbf{P}^{VII}\boldsymbol{\lambda}^{VII}\left(\frac{R_6}{R_7}\right)^{\boldsymbol{\lambda}^{VII}} & -\mu_{Open,\theta}\mathbf{P}^{VIII}\boldsymbol{\lambda}^{VIII}\left(\frac{R_7}{R_8}\right)^{\boldsymbol{\lambda}^{VIII}} & \mu_{Open,\theta}\mathbf{P}^{VIII}\boldsymbol{\lambda}^{VIII} \end{bmatrix} \tag{A20}$$

$$\begin{bmatrix} \mathbf{K}_{15,14} & \mathbf{K}_{15,15} & \mathbf{K}_{15,16} & \mathbf{K}_{15,17} \end{bmatrix} = \begin{bmatrix} \mathbf{P}^{VIII} & \mathbf{P}^{VIII}\left(\frac{R_7}{R_8}\right)^{\boldsymbol{\lambda}^{VIII}} & -\left(\frac{R_8}{R_9}\right)^{|\mathbf{K}|} & -\mathbf{I} \end{bmatrix} \tag{A21}$$

$$\begin{bmatrix} \mathbf{K}_{16,14} & \mathbf{K}_{16,15} & \mathbf{K}_{16,16} & \mathbf{K}_{16,17} \end{bmatrix} = \begin{bmatrix} \mu_{IX}\mathbf{P}^{VIII}\boldsymbol{\lambda}^{VIII} & -\mu_{IX}\mathbf{P}^{VIII}\boldsymbol{\lambda}^{VIII}\left(\frac{R_7}{R_8}\right)^{\boldsymbol{\lambda}^{VIII}} & \mu_{Slot,\theta}|\mathbf{K}|\left(\frac{R_8}{R_9}\right)^{|\mathbf{K}|} & \mu_{Slot,\theta}|\mathbf{K}| \end{bmatrix} \tag{A22}$$

$$\begin{bmatrix} \mathbf{K}_{17,16} & \mathbf{K}_{17,17} & \mathbf{K}_{17,18} \end{bmatrix} = \begin{bmatrix} \mathbf{I} & -\left(\frac{R_8}{R_9}\right)^{|\mathbf{K}|} & -\mathbf{I} \end{bmatrix} \tag{A23}$$

$$\begin{bmatrix} \mathbf{K}_{18,16} & \mathbf{K}_{18,17} & \mathbf{K}_{18,18} \end{bmatrix} = \begin{bmatrix} \mu_{X} & -\mu_{X}\left(\frac{R_8}{R_9}\right)^{|\mathbf{K}|} & \mu_{IX} \end{bmatrix} \tag{A24}$$

References

1. Liu, C. Emerging Electric Machines and Drives—An Overview. *IEEE Trans. Energy Convers.* **2018**, *33*, 2270–2280. [CrossRef]
2. Zhu, Z.Q.; Liu, Y. Analysis of Air-Gap Field Modulation and Magnetic Gearing Effect in Fractional-Slot Concentrated-Winding Permanent-Magnet Synchronous Machines. *IEEE Trans. Ind. Electron.* **2018**, *65*, 3688–3698. [CrossRef]
3. Atallah, K.; Rens, J.; Mezani, S.; Howe, D. A Novel "Pseudo" Direct-Drive Brushless Permanent Magnet Machine. *IEEE Trans. Magn.* **2008**, *44*, 4349–4352. [CrossRef]
4. Atallah, K.; Howe, D. A novel high-performance magnetic gear. *IEEE Trans. Magn.* **2001**, *37*, 2844–2846. [CrossRef]
5. Acharya, V.M.; Bird, J.Z.; Calvin, M. A Flux Focusing Axial Magnetic Gear. *IEEE Trans. Magn.* **2013**, *49*, 4092–4095. [CrossRef]
6. Linni, J.; Chau, K.T.; Gong, Y.; Jiang, J.Z.; Chuang, Y.; Wenlong, L. Comparison of Coaxial Magnetic Gears With Different Topologies. *IEEE Trans. Magn.* **2009**, *45*, 4526–4529. [CrossRef]
7. Holehouse, R.C.; Atallah, K.; Wang, J.B. Design and Realization of a Linear Magnetic Gear. *IEEE Trans. Magn.* **2011**, *47*, 4171–4174. [CrossRef]
8. Liu, C.T.; Chung, H.Y.; Hwang, C.C. Design Assessments of a Magnetic-Geared Double-Rotor Permanent Magnet Generator. *IEEE Trans. Magn.* **2014**, *50*, 1–4. [CrossRef]

9. Wang, L.L.; Shen, J.X.; Luk, P.C.K.; Fei, W.Z.; Wang, C.F.; Hao, H. Development of a Magnetic-Geared Permanent-Magnet Brushless Motor. *IEEE Trans. Magn.* **2009**, *45*, 4578–4581. [CrossRef]
10. Zhu, X.; Chen, L.; Quan, L.; Sun, Y.; Hua, W.; Wang, Z. A New Magnetic-Planetary-Geared Permanent Magnet Brushless Machine for Hybrid Electric Vehicle. *IEEE Trans. Magn.* **2012**, *48*, 4642–4645. [CrossRef]
11. Liu, C.; Chau, K.T. Electromagnetic Design of a New Electrically Controlled Magnetic Variable-Speed Gearing Machine. *Energies* **2014**, *7*, 1539–1554. [CrossRef]
12. Ho, S.L.; Wang, Q.; Niu, S.; Fu, W.N. A Novel Magnetic-Geared Tubular Linear Machine With Halbach Permanent-Magnet Arrays for Tidal Energy Conversion. *IEEE Trans. Magn.* **2015**, *51*, 1–4.
13. Liu, C.; Yu, J.C.; Lee, C.H.T. A New Electric Magnetic-Geared Machine for Electric Unmanned Aerial Vehicles. *IEEE Trans. Magn.* **2017**, *53*, 1–6. [CrossRef]
14. Zhao, H.; Liu, C.; Song, Z.; Liu, S. A Consequent-Pole PM Magnetic-Geared Double-Rotor Machine With Flux-Weakening Ability for Hybrid Electric Vehicle Application. *IEEE Trans. Magn.* **2019**, 1–7. [CrossRef]
15. Sun, L.; Cheng, M.; Zhang, J.W.; Song, L.H. Analysis and Control of Complementary Magnetic-Geared Dual-Rotor Motor. *IEEE Trans. Ind. Electron.* **2016**, *63*, 6715–6725. [CrossRef]
16. Bai, J.; Liu, J.; Zheng, P.; Tong, C. Design and Analysis of a Magnetic-Field Modulated Brushless Double-Rotor Machine—Part I: Pole Pair Combination of Stator, PM Rotor and Magnetic Blocks. *IEEE Trans. Ind. Electron.* **2019**, *66*, 2540–2549. [CrossRef]
17. Chan, C.C. The state of the art of electric, hybrid, and fuel cell vehicles. *Proc. IEEE* **2007**, *95*, 704–718. [CrossRef]
18. Miller, J.M. Hybrid electric vehicle propulsion system architectures of the e-CVT type. *IEEE Trans. Power Electron.* **2006**, *21*, 756–767. [CrossRef]
19. Liu, C.; Chau, K.T.; Zhang, Z. Novel Design of Double-Stator Single-Rotor Magnetic-Geared Machines. *IEEE Trans. Magn.* **2012**, *48*, 4180–4183. [CrossRef]
20. Chen, C.L.; Tsai, M.C. Kinematic and Dynamic Analysis of Magnetic Gear With Dual-Mechanical Port Using Block Diagrams. *IEEE Trans. Magn.* **2018**, *54*, 1–5. [CrossRef]
21. Chen, L.; Hopkinson, D.; Wang, J.; Cockburn, A.; Sparkes, M.; O'Neill, W. Reduced Dysprosium Permanent Magnets and Their Applications in Electric Vehicle Traction Motors. *IEEE Trans. Magn.* **2015**, *51*, 1–4.
22. Baloch, N.; Kwon, B.I.; Gao, Y.T. Low-Cost High-Torque-Density Dual-Stator Consequent-Pole Permanent Magnet Vernier Machine. *IEEE Trans. Magn.* **2018**, *54*, 1–5. [CrossRef]
23. Gao, Y.; Qu, R.; Li, D.; Li, J.; Zhou, G. Consequent-Pole Flux-Reversal Permanent-Magnet Machine for Electric Vehicle Propulsion. *IEEE Trans. Appl. Supercond.* **2016**, *26*, 1–5. [CrossRef]
24. Wang, H.T.; Fang, S.H.; Yang, H.; Lin, H.Y.; Wang, D.; Li, Y.B.; Jiu, C.X. A Novel Consequent-Pole Hybrid Excited Vernier Machine. *IEEE Trans. Magn.* **2017**, *53*, 1–4. [CrossRef]
25. Wang, Q.S.; Niu, S.X.; Yang, S.Y. Design Optimization and Comparative Study of Novel Magnetic-Geared Permanent Magnet Machines. *IEEE Trans. Magn.* **2017**, *53*, 1–4. [CrossRef]
26. Zhang, X.X.; Liu, X.; Chen, Z. Investigation of Unbalanced Magnetic Force in Magnetic Geared Machine Using Analytical Methods. *IEEE Trans. Magn.* **2016**, *52*, 1–4. [CrossRef]
27. Shin, K.H.; Cho, H.W.; Kim, K.H.; Hong, K.; Choi, J.Y. Analytical Investigation of the On-Load Electromagnetic Performance of Magnetic-Geared Permanent-Magnet Machines. *IEEE Trans. Magn.* **2018**, *54*, 1–5. [CrossRef]
28. Djelloul-Khedda, Z.; Boughrara, K.; Dubas, F.; Kechroud, A.; Tikellaline, A. Analytical Prediction of Iron-Core Losses in Flux-Modulated Permanent-Magnet Synchronous Machines. *IEEE Trans. Magn.* **2019**, *55*, 1–12. [CrossRef]
29. Lubin, T.; Mezani, S.; Rezzoug, A. Two-Dimensional Analytical Calculation of Magnetic Field and Electromagnetic Torque for Surface-Inset Permanent-Magnet Motors. *IEEE Trans. Magn.* **2012**, *48*, 2080–2091. [CrossRef]
30. Lubin, T.; Mezani, S.; Rezzoug, A. Analytical Computation of the Magnetic Field Distribution in a Magnetic Gear. *IEEE Trans. Magn.* **2010**, *46*, 2611–2621. [CrossRef]
31. Sprangers, R.L.J.; Paulides, J.J.H.; Gysen, B.L.J.; Lomonova, E.A. Magnetic Saturation in Semi-Analytical Harmonic Modeling for Electric Machine Analysis. *IEEE Trans. Magn.* **2016**, *52*, 1–10. [CrossRef]
32. Bai, J.G.; Zheng, P.; Tong, C.D.; Song, Z.Y.; Zhao, Q.B. Characteristic Analysis and Verification of the Magnetic-Field-Modulated Brushless Double-Rotor Machine. *IEEE Trans. Ind. Electron.* **2015**, *62*, 4023–4033. [CrossRef]

33. Chung, C.-T.; Wu, C.-H.; Hung, Y.-H. Effects of Electric Circulation on the Energy Efficiency of the Power Split e-CVT Hybrid Systems. *Energies* **2018**, *11*, 2342. [CrossRef]
34. Li, L. Use of Fourier series in the analysis of discontinuous periodic structures. *JOSA A* **1996**, *13*, 1870–1876. [CrossRef]
35. Articolo, G.A. *Partial Differential Equations and Boundary Value Problems with Maple*, 2nd ed.; Elsevier: Burlington, MA, USA, 2009; pp. 29–30.
36. Djelloul-Khedda, Z.; Boughrara, K.; Dubas, F.; Ibtiouen, R. Nonlinear Analytical Prediction of Magnetic Field and Electromagnetic Performances in Switched Reluctance Machines. *IEEE Trans. Magn.* **2017**, *53*, 1–11. [CrossRef]

Article

Design and Optimization of a Magnetically Levitated Inductive Reaction Sphere for Spacecraft Attitude Control

Liming Yuan [1,2,†], Jie Zhang [1,2,†], Si-Lu Chen [1,*], Yusheng Liang [1], Jinhua Chen [1], Chi Zhang [1] and Guilin Yang [1]

[1] Zhejiang Key Laboratory of Robotics and Intelligent Manufacturing Equipment Technology, Ningbo Institute of Materials Technology and Engineering, Chinese Academy of Sciences, Ningbo 315201, China; yuanliming@nimte.ac.cn (L.Y.); zhangjie@nimte.ac.cn (J.Z.); liangyusheng@nimte.ac.cn (Y.L.); chenjinhua@nimte.ac.cn (J.C.); zhangchi@nimte.ac.cn (C.Z.); glyang@nimte.ac.cn (G.Y.)

[2] University of Chinese Academy of Sciences, Beijing 100049, China

* Correspondence: chensilu@nimte.ac.cn; Tel.: +86-574-8668-5861

† These authors contributed equally to this work.

Received: 15 March 2019; Accepted: 18 April 2019; Published: 24 April 2019

Abstract: The inductive reaction sphere (RS) brings the benefit of simple, economical, and miniaturized design, and it is capable of multi-DOF torque generation. Thus, it is a suitable choice for the angular momentum exchange actuator in attitude control of micro-spacecrafts. To synthesize symmetric distribution of eddy currents and improve the speed and stability of rotation, a novel 4-pole winding design is proposed. However, the developed simplified analytical model shows that reduced pole number degrades the torque generation. To enhance the output torque of 4-pole RS, its curved cores and electromagnets are redesigned to enable the side teeth to be functional. As the analytical torque model for the RS with the slotted cores is not available, a constrained optimization problem is formulated, and the optimized parameters are calculated based on the prediction model from supported vector machine and finite element analysis. The lab prototypes are developed to validate the proposed design and test the speed performance. The experimental results show that the 4-pole RS prototype obtains a stable rotation over 700 rpm about X, Y and Z axis respectively with the angular momentum of 0.08 kg·m^2/s, being superior to the 6-pole counterpart.

Keywords: reaction sphere; spherical motor; structural design; torque density optimization; support vector machines; finite element method

---

1. Introduction

Besides the orbit control, the attitude control is essential to prevent rollover of the spacecraft and to ensure its antennae consistently directs to a fixed point on the Earth surface [1]. Since spacecrafts are free-floating in space, one way to adjust their attitudes is to transfer the momentum from a rotating actuator to the spacecraft body back and forth. When the actuator changes its rotational speed, the spacecraft will counter-rotate to remain the conservation of angular momentum.

The concept of using a spherical rotor as a momentum exchanger for spacecrafts attitude control starts in the 1960s [2–4]. This is where the name "reaction sphere (RS)" comes from. Compared with other momentum exchange actuators such as the flywheel [5] and the control moment gyroscope [6], the RS can generate 3-axis output torque within one single actuator, thus giving a compact design. In addition, the torques generated in the RS around different axes are naturally decoupled, which simplify the controller design. However, due to the difficulties in designing a working prototype, the RS had not gained much attention for decades. With the rapid development of micro-spacecrafts from the early of this century, the demand for minimization design of the attitude control system has helped the reaction spheres get noticed again [7,8].

A RS can be driven by several single axis motors or a single spherical motor. The former scheme drives the reaction sphere with multiple wheels rotating at desired directions [9–12]. These wheels are independently driven by their corresponding single-axis motors. The later scheme considers the reaction sphere as the rotor of a spherical motor, where the stator of the motor is designed in a spherically symmetric way. No mechanical output shaft exists in these spherical motors. The driving principle can be inductive [13–19], permanent magnet (PM) [7,8,20,21], hysteresis [22,23], variable reluctance [24] and ultrasonic [25,26].

Among all types of reaction spheres, the inductive RSs have the advantages of simple design and high reliability, making them be more competitive in commercial applications. Inductive RSs follow the same driving principle as the traditional single-DOF induction motors. The stator of an induction reaction sphere is excited with alternating currents. Then the generated rotating magnetic fields cut the rotor conductor to produce the induced currents. The interaction between the induced currents on the rotor and the magnetic fields in the air gap produces Lorentz force, which drives the rotor to rotate in the direction of the rotating magnetic fields. Compared with conventional induction motors, inductive RSs feature the capacity of generating multi-DOF rotating magnetic fields.

One class of inductive reaction spheres obtains this capacity via distributed electromagnets [16–18]. A 3D reaction sphere design is proposed in [17], where 3 pairs of electromagnets are put orthogonally in space. A one-DOF inductive RS prototype is developed, with 4 electromagnets for rotation and 1 electromagnet for levitation. With the similar design, a well-developed one-DOF inductive RS prototype is presented in [18]. It reaches the maximum speed of 13,500 rpm and the maximum torque of 0.7 Nm in the performance test experiments. However, this prototype has the disadvantage of large stator. The stator occupies much more space than the rotor. An inductive RS design with 20 distributed electromagnets is proposed in [16]. Subsequently, a one-DOF prototype is developed, but no performance data are reported. The control strategy can be very complicated if the suspension and rotation are driven by the same electromagnets.

The other class of inductive RSs gains this capacity via curved inductors. This type of inductive RSs tends to have a more compact design over the ones with distributed electromagnets, for the coils of curved inductors are extended along the rotor surface but coils of electromagnets are extended radially. The curved inductors in inductive RSs can be arranged to be vortex-like [13,19] and orthogonal [13–15]. The vortex-like arrangement can get the spherical rotor half exposed. This feature makes it be suitable for applications like mobile platforms and robots and the case with no requirement of full angle rotation. The design and control aspects of a 3D spherical induction motor for mobile robots is presented in [19], featuring the vortex-like arranged four curved inductors and the half exposed spherical rotor. The angular velocity and orientation control studies enable a large prototype to rotate along arbitrary axes under the speed of 300 rpm, which validate the feasibility of the vortex-like arrangement. In orthogonal arrangement, the curved inductors go all around the rotor and form the closed structure. It is well fit for the devices working through momentum exchange. Patent [14] proposes a magnetic bearing inductive RS design with three pairs of curved inductors orthogonally arranged. In this design, the AC windings for rotation and the DC windings for suspension are placed on the same teeth of curved cores. No relevant feasibility research is found of this design.

In this paper, a compact design of inductive RS is proposed, featuring 12 curved inductors in orthogonal arrangement for rotating and 6 electromagnets in pairs for magnetic bearing. The overall design enables the RS to generate three-DOF rotation under magnetic levitation. To synthesize symmetric distribution of eddy currents and improve the stability during high-speed rotation, the 4-pole winding is proposed instead of the intuitive 6-pole design. However, the analytical torque model with simplified slotless assumption indicates that reduced pole number degrades the torque generation. To enhance the capability of the torque synthesis in the 4-pole winding design, the slotted iron-core is imposed in design of curve stators. In addition, the electromagnets in the RS are redesigned to be the cross-shape, so that two more slots near the notches of the iron-core curve stator become functional. Such design improves the uniformity of the eddy current and magnetic field induced along the stator circle. As the analytical torque model for the RS with slotted curve stator is not available, a constrained optimization problem is formulated for torque maximization in the RS. From here, the data-based regression algorithms are applied to find the optimal design parameters. Speed test experiments on the developed prototype are conducted to show the superior performance of the optimized 4-pole inductive RS design over the conventional 6-pole counterpart.

2. Structure Design

2.1. Rotor

The attitude adjustment of the spacecraft is realized through the exchange of angular momentum between the rotating sphere and spacecraft body. The rotor of the spherical motor is the carrier of angular momentum. To store more angular momentum, large moment of inertia (MOI) is preferred with a fixed amount of mass. In rotor structural design, MOI of a hollow and uniform sphere is calculated by

$$J = \frac{2}{5}M\frac{R^5 - R_0^5}{R^3 - R_0^3}, \quad 0 \leq R_0 \leq R \tag{1}$$

where J refers to the MOI, M is the mass of the rotor, R and R_0 are the outer and inner radius of the sphere.

Consider the case given the fixed mass and the fixed outer radius. As plotted in Figure 1, it can be seen that the thinner spherical shell has higher MOI. For example, the MOI of a hollow sphere with $0.1R$ thickness is 1.365 times higher than the one with $0.5R$ thickness. In the extreme case, the inertia of a solid sphere is 60% of the spherical shell with its thickness arbitrarily approaching to zero. The brief mathematical proof sees Appendix A. When R_0 increases, the density is expected to be raised given a certain mass. Therefore, materials with high density get priority in machining the rotor. Furthermore, consider the case given the same mass and the same material, it can be concluded that the outer mass gives more MOI than the inner mass through calculation. For example, the inertial of a hollow sphere with radius of R and thickness of $0.1R$ is 3.6 times higher than the solid sphere with radius of $0.647R$. Therefore, hollow sphere is preferred when designing the reaction sphere.

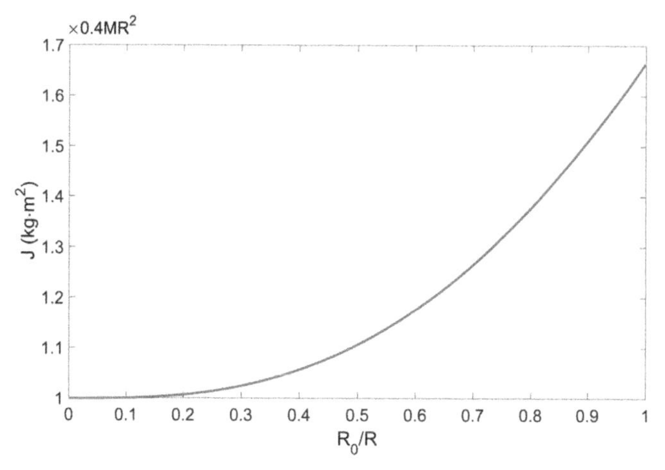

Figure 1. MOI of a hollow sphere with a fixed amount of mass.

Eventually, the rotor is designed as a hollow sphere, consisting of three layers of metals, as shown in Figure 2. The inner layer is made of steel, which is used to form the magnetic circuits with the stator cores through the air gap. The surface of the steel is electroplated with a copper film, forming the middle layer, which is used to raise the conductivity and enlarge the electromagnet torque induced by the eddy currents on the surface of the rotor. Additionally, to prevent the copper from oxidation and leading to conductivity reduction, a thin nickel layer is deposited on the copper by electroplating, forming the outer layer.

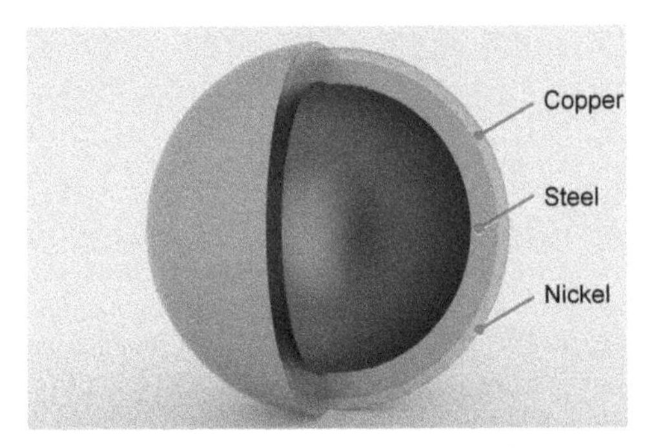

Figure 2. Multilayer spherical shell.

2.2. *Stator*

The stator of the reaction sphere is fixed with the spacecraft body, which consists of the curved stators and the electromagnets. They are responsible for rotation motion and magnetic suspension respectively. The curved cores and the electromagnet cores are used to conduct the magnetic flux generated by the stator coils. To meet the requirement of generating tri-axial torques, the stator must be designed as a spherically symmetric structure. As shown in Figure 3, the stator of the reaction is formed by three circles that are orthogonal to each other. Each circle is composed of 4 curved cores and enables the rotor to rotate about one certain axis. The output torque about any axis can be realized by combining the torques generated by the three circles. On the other hand, in order to eliminate the mechanical friction and extend its working life, six electromagnets are set in the position of the six intersection of three circles to get the spherical rotor magnetically levitated in the center of the stator.

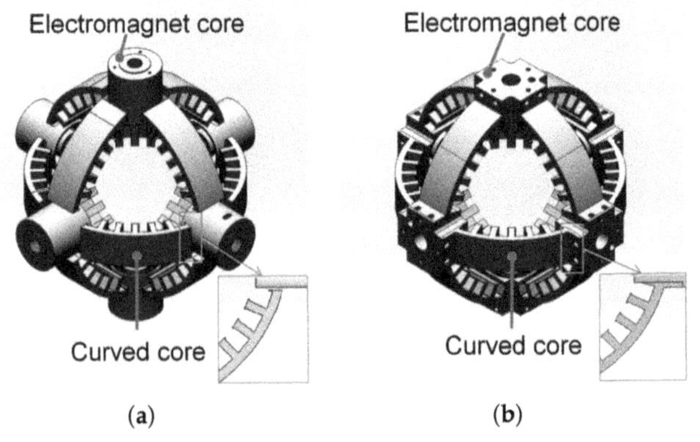

Figure 3. Structural design of the stator cores: (**a**) original design and (**b**) improved design.

In the original design with round-shape electromagnet cores, as shown in Figure 3a, only five complete teeth on the curve iron-core are functional, while the two teeth near the side of the curve stator are not. The notches are formed by connecting three pieces of iron-core curve stator to form one complete stator circle. It causes non-uniformity eddy current and magnetic field induced along the stator circle. To reduce such effect and get the stator magnetic fields utilized in a more sufficient way, the structure of the electromagnet cores are revised to be the cross-shape. In this way, the curved iron-core and electromagnet core can be assembled together and so that two extra teeth on the side are functional, as shown in Figure 3b.

2.3. Winding

For the convenience of speed adjustment and the feature of self-starting, the stator coils are rolled as three-phase windings. In order to generate the equivalent tri-axial torques, the stator should be symmetric in both mechanical and electrical designs. Since the uniaxial plane is divided into four equal parts, as shown in Figure 4, 12 coils are missing in each circle due to the occupation of the electromagnets. All in all, 1/3 of the driving coils are replaced by electromagnet coils.

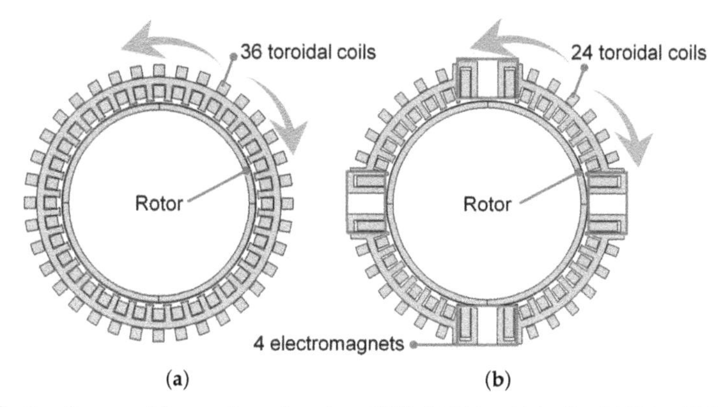

Figure 4. Section diagram of the reaction sphere (one-DOF): (**a**) 36 toroidal coils. (**b**) 24 toroidal coils and 4 electromagnet coils.

As can be seen in Figure 5a,c, the three-phase windings come to two types, 6-pole and 4-pole. 6-pole windings can be regarded as removing 12 coils directly from the complete set of three-phase windings, which does not make any adjustment of the rest windings. As for 4-pole windings, the phase of next coil starts right from the end of the last one so that the windings on the four curved cores can achieve consistency.

With finite element method (FEM) modeling in COMSOL Multiphysics 5.3, the fields and currents distribution on the surface of the rotor can be visually analyzed. As shown in Figure 5b,d, the gray lines stand for the magnetic line of force, the red arrows refer to the induced currents and the surface colors refer to the radius magnetic flux density. In this simulation, copper thickness, steel thickness, ampere turns, tooth thickness and tooth width are set as 0.3 mm, 3 mm, 2.4 kA, 3 mm and 25 mm respectively. In this case, the driving torques of the 6-pole arrangement and 4-pole arrangement are 33.8 mNm and 27.0 mNm, where the former is 1.25 times larger than the latter. However, it can be seen that, in 6-pole scheme, two poles are much larger than the rest four and the field distortions in the regions that face the electromagnets are serious. Also, it induces an asymmetric distribution of eddy currents by the adjacent curved stators. These features increase the difficulty of multi-DOF motion control of this 6-pole reaction sphere. Furthermore, from the radius magnetic flux density in Figure 5b,d, it can be seen that, in 6-pole arrangement it is the electromagnet cores that mainly contribute to the driving torque, while in 4-pole arrangement it is the curved cores that mainly contribute to the driving torque. The 4-pole arrangement makes the curved cores better exploited. Additionally, the case without electromagnets cores is simulated, where the torque of 4-pole scheme declines from 27 mNm to 18.4 mNm, and the torque of 6-pole scheme declines from 33.8 mNm to 9.89 mNm. This phenomenon reflects the generated torque is stronger dependent on the electromagnet cores in 6-pole scheme. The flux generated by this 6-pole arrangement mainly passes through the inner core of the electromagnets, which can interfere the magnetic suspension and increase the risk of magnetic saturation.

In summary, while reducing the output torque (20% reduction rate in case of the Figure 5), 4-pole arrangement brings the benefits in three aspects compared with 6-pole arrangement. Firstly, it generates a more uniform distribution of magnetic field and eddy currents on the surface of the rotor. Secondly, it is beneficial for larger angular momentum storage due to its higher synchronous speed with the same power frequency. Thirdly, it reduces the risk of magnetic saturation when the curved inductors and electromagnets are working at the same time. Thus, the 4-pole scheme is adopted for the reaction sphere windings.

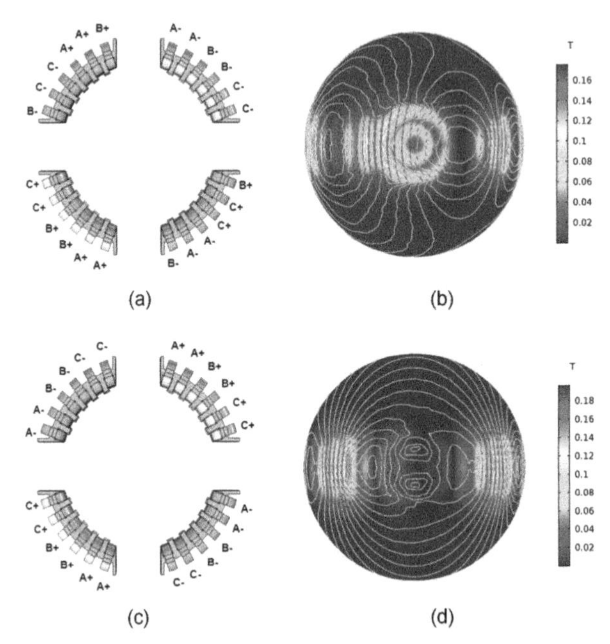

Figure 5. Winding mode and simulation of magnetic field and induced currents on the surface of the rotor: (**a**) 6-pole scheme. (**b**) Simulated distribution of 6-pole scheme. (**c**) 4-pole scheme. (**d**) Simulated distribution of 4-pole scheme.

3. Analytical Model

This section analyzes the magnetic fields on the surface of the rotor, which leads to the solution to electromagnetic driving torque. This also allows us to analyze the generated torque with respect to some of the key design parameters, such as number of poles. Several simplifications are made here. (1) The magnetic permeability of the rotor core is infinite. Thus, no magnetic saturation is encountered. (2) The length of air-gap is evenly distributed. (3) Displacement currents are neglected. Thus, the current law are simplified as Ampere's law.

Generally speaking, the magnetic flux density can be calculated by computing the curl of magnetic vector potential. In spherical coordinates (r, θ, φ), the formula can be expressed as

$$\vec{B} = \frac{1}{r^2 \sin\theta} \begin{vmatrix} \vec{e_r} & r\vec{e_\theta} & r\sin\theta\vec{e_\varphi} \\ \frac{\partial}{\partial r} & \frac{\partial}{\partial \theta} & \frac{\partial}{\partial \varphi} \\ A_r & rA_\theta & r\sin\theta A_\varphi \end{vmatrix},$$

where $\overrightarrow{e_r}$, $\overrightarrow{e_\theta}$ and $\overrightarrow{e_\varphi}$ refer to the unit vectors, A_r, A_θ and A_φ refer to components of the vector potential in spherical coordinate system.

The general torque $\overrightarrow{\tau}$ of the reaction sphere is the vector summation of 3 torques $\overrightarrow{\tau_x}$, $\overrightarrow{\tau_y}$, and $\overrightarrow{\tau_z}$ generated by 3 identical circles of the stator.

$$\overrightarrow{\tau} = \overrightarrow{\tau_x} + \overrightarrow{\tau_y} + \overrightarrow{\tau_z}.$$

The following analysis will take torque about z-axis $\overrightarrow{\tau_z}$ for example. It is the longitudinal magnetomotive force (mmf) that induces the z-axis rotating magnetic field, in which case $A_\varphi = A_r = 0$. Therefore, the magnetic flux density can be simplified as

$$\overrightarrow{B} = -\frac{\overrightarrow{e_r}}{r\sin\theta}\frac{\partial A_\theta}{\partial\varphi} + \frac{\overrightarrow{e_\varphi}}{r}\frac{\partial r A_\theta}{\partial r} \equiv B_r\overrightarrow{e_r} + B_\varphi\overrightarrow{e_\varphi}. \tag{2}$$

In spherical coordinate system, by Ampere's law $\overrightarrow{J} = \Delta \times \overrightarrow{H}$,

$$\overrightarrow{J} = \frac{1}{r^2\sin\theta}\begin{vmatrix} \overrightarrow{e_r} & r\overrightarrow{e_\theta} & r\sin\theta\overrightarrow{e_\varphi} \\ \frac{\partial}{\partial r} & \frac{\partial}{\partial\theta} & \frac{\partial}{\partial\varphi} \\ H_r & rH_\theta & r\sin\theta H_\varphi \end{vmatrix}.$$

In this case, only those currents flowing in θ-direction induce torque about z-axis,

$$J_\theta = -\frac{1}{r\sin\theta}\left(\frac{\partial r\sin\theta H_\varphi}{\partial r} - \frac{\partial H_r}{\partial\varphi}\right) = -\frac{1}{\mu_0 r\sin\theta}\left(\frac{\partial r\sin\theta B_\varphi}{\partial r} - \frac{\partial B_r}{\partial\varphi}\right). \tag{3}$$

Combine (3) and (2),

$$J_\theta = -\frac{1}{\mu_0 r}\frac{\partial^2 r A_\theta}{\partial r^2} - \frac{1}{\mu_0 r^2\sin^2\theta}\frac{\partial^2 A_\theta}{\partial\varphi^2}. \tag{4}$$

Express the magnetic potential A_θ written in complex form $e^{j\omega t}$ [27], where ω is the angular frequency of input currents, j is the imaginary unit. Then,

$$J_\theta = \sigma E = -\sigma\frac{\partial A}{\partial t} = -j\sigma\omega A_\theta. \tag{5}$$

Combine (4) and (5), the differential equation for magnetic field distribution on the surface of the rotor is obtained as

$$\sigma j\omega A_\theta = \frac{1}{\mu_0 r}\frac{\partial^2 r A_\theta}{\partial r^2} + \frac{1}{\mu_0 r^2\sin^2\theta}\frac{\partial^2 A_\theta}{\partial\varphi^2}. \tag{6}$$

To make (6) solvable, apply the the method of separation of variables as

$$A_\theta(r,\theta,\varphi) = X(r)Y(\theta,\varphi). \tag{7}$$

Substitute (7) into (6),

$$\sigma j\omega = \frac{1}{\mu_0 r X}\frac{\partial^2 r X}{\partial r^2} + \frac{1}{\mu_0 r^2\sin^2\theta Y}\frac{\partial^2 Y}{\partial\varphi^2}. \tag{8}$$

With high-order time and space harmonics neglected, it is assumed that $e^{j\omega t}$ only has the first mmf space harmonic. In this way,

$$\frac{1}{Y\sin^2\theta}\frac{\partial^2\theta Y}{\partial\varphi^2} = -p^2, \tag{9}$$

where p is the number of pair pole. The analytical solution to (9) is given as

$$Y = C_1 e^{jp\varphi\sin\theta} + C_2 e^{-jp\varphi\sin\theta}. \tag{10}$$

From (8) and (9), it yields

$$r^2\frac{\partial^2 X}{\partial r^2} + 2r\frac{\partial X}{\partial r} - X(r^2\mu\sigma j\omega + p^2) = 0. \tag{11}$$

Set $X(r,\theta) = \frac{1}{\sqrt{r}}X_1(r,\theta)$. Then, (11) are transformed to modified Bessel's Equation (12) [28],

$$r_1{}^2\frac{\partial^2 X_1}{\partial r_1{}^2} + r_1\frac{\partial X_1}{\partial r_1} - X_1(r_1{}^2 + p^2 + \frac{1}{4}) = 0, \tag{12}$$

where $r_1 = r\sqrt{\mu\sigma j\omega}$. The analytical solution to (12) is as follows:

$$X(r) = \frac{1}{\sqrt{r}}X_1(r,\theta) = \frac{1}{\sqrt{r}}C_3 I_{\sqrt{p^2+\frac{1}{4}}}(r\sqrt{\mu\sigma j\omega}) + \frac{1}{\sqrt{r}}C_4 K_{\sqrt{p^2+\frac{1}{4}}}(r\sqrt{\mu\sigma j\omega}), \tag{13}$$

where $I_p(r_1)$ and $K_p(r_1)$ are the two linearly independent solutions to the modified Bessel's Equation [28]:

$$\begin{aligned} I_\alpha(x) &= \sum_{m=0}^{\infty} \frac{1}{m!\Gamma(m+\alpha+1)}(\frac{x}{2})^{\alpha+2m}, \\ K_\alpha(x) &= \frac{\pi}{2}\frac{I_{-\alpha}(x) - I_\alpha(x)}{\sin\alpha\pi}. \end{aligned} \tag{14}$$

Subsequently, the magnetic flux density can be calculated through (15), with boundary conditions of $H_\varphi|_{r=R_s} = 0$ and $H_{\varphi\text{copper}}|_{r=R_c} = H_{\varphi\text{air}}|_{r=R_c}$, $B_{r\text{copper}}|_{r=R_c} = B_{r\text{air}}|_{r=R_c}$ for continuity, where R_c is the radius of rotor outer surface and R_s is the radius of rotor inner surface. The nickel layer is neglected because it is much thinner than copper layer and steel layer.

$$\begin{aligned} B_r &= -\frac{X(r)}{r\sin\theta}\frac{\partial Y(\theta,\varphi)}{\partial\varphi}, \\ B_\varphi &= \frac{X(r)Y(\theta,\varphi)}{r} + Y(\theta,\varphi)\frac{\partial X(r)}{\partial r}. \end{aligned} \tag{15}$$

Based on the flux density distribution, the driving torque can be evaluated by Lorentz forces induced by eddy currents on the copper layer of the rotor. The electromagnetic torque arising from the rotor steel is negligible [27].

$$T_{eL} = -\int_0^{2\pi}\int_{R_s}^{R_c}\int_0^{\pi} J_\theta B_r r^3 \sin^2\theta \, d\theta dr d\varphi. \tag{16}$$

In case of proposed design, 16 of the 36 teeth in circle are occupied by 4 electromagnets (see Figure 4a). After structural optimization (see Figure 3b), each side tooth of these curved cores can be approximated as a half tooth concerning its combination with the electromagnet core. This helps form 4 equivalent teeth. So the output torque generated by the four curved inductors is 2/3 of the complete electromagnetic torque. Then, substitute (5), (7) and (10) into (16),

$$T_{eLRS} = \frac{2}{3}T_{eL} \tag{17}$$

$$= \frac{2}{3}\int_0^{2\pi}\int_{R_s}^{R_c}\int_0^{\pi} J_\theta \frac{\partial A_\theta}{\partial \varphi} r^2 \sin\theta \, d\theta dr d\varphi \tag{18}$$

$$= \frac{2}{3}\sigma p\omega \int_{R_s}^{R_c} X^2(r) r^2 dr \int_0^{2\pi}\int_0^{\pi} Y^2(\theta,\varphi)\sin^2\theta d\theta d\varphi \tag{19}$$

$$= \frac{\pi^2}{3}\sigma p\omega \int_{R_s}^{R_c} X^2(r) r^2 dr. \tag{20}$$

From (20), it can be seen that higher electrical conductivity and pole numbers can enhance the output torque. So it is reasonable to adopt high conductivity materials like copper as the outer layer of the rotor. Also, the torque is expected to be reduced when choosing the 4-pole winding instead of 6-pole winding. To raise the output torque, the design parameters like tooth thickness and tooth width are required to be optimized in developing the prototypes. However, the analytical torque model (20) does not include these parameters. To authors' best knowledge, all the proposed analytical fields and torque models of inductive reaction spheres are developed based on the assumptions that the stator is slotless [27,29–31]. Otherwise, the differential equation will be too complex to find an analytical solution. Due to the nonlinear dependency of design parameters and the 3-D fields distribution, it is too complicated to derive a purely analytical torque model if the stator contains slots. Hence, data-based modeling technique is used to predict the output torque with design parameters considered to be optimized in the following section.

4. Torque Density Optimization

4.1. Data-Based Torque Model

In this optimization, by dimension constraints, the air gap length is fixed as 1 mm, the outside diameter of the rotor is fixed as 99 mm and the outside diameter of the stator is fixed as 123 mm. The goal is to raise the output torque under the constraints of mass and MOI by searching the optimal tooth thickness (S), tooth width (B), ampere turns (NI), copper thickness (T_{co}) and steel thickness (T_{st}) as shown in Figure 6. Due to the time consuming of FEM calculation, the comprehensive design is not practical. Thus, Taguchi design of five factors five levels is adopted to select the representative samples data necessary for building learning models as shown in Table 1 [32]. Consider the overall size and the process feature, the ranges of T_{co}, T_{st}, NI, S and B are set as 0.1–0.5 mm, 1–5 mm, 1.5–2.7 kA, 2–4 mm and 20–30 mm respectively.

Eight groups of the Taguchi design form the 200 trials with no repeated experiments. Each group consists of 25 samples, which selects the representative samples from the whole parameter space by following the two rules. (1) The value in column has the same number of occurrences. (2) Any pair of values between two arbitrary columns have the same number of occurrences. Subsequently, of all the eight groups, seven groups are used to train the learning model and the rest one is used to test the prediction performance of the model. The output torque is obtained by FEM calculation. To reveal the uneven distribution of the air gap length and raise the accuracy of the calculation, 3D FEM is adopted instead of 2D one. The whole mass is estimated by the CAD model volume and its density. The materials used in FEM simulation are listed in Table 2. The rated current is set to 1.4 A. The power source is 50 Hz.

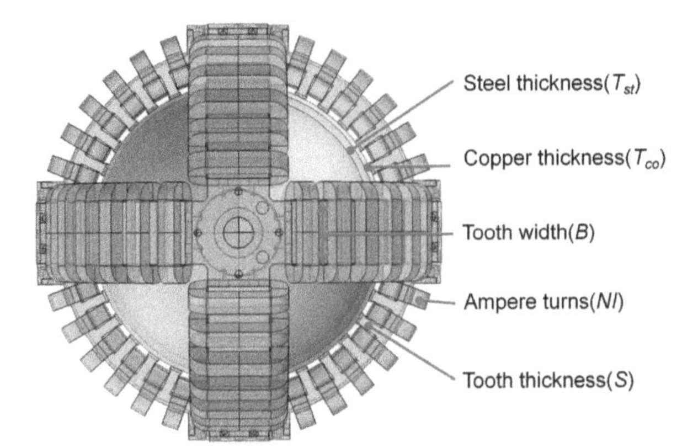

Figure 6. Design parameters to be optimized.

Table 1. The parameters of sample space.

Element	Level 1	Level 2	Level 3	Level 4	Level 5
T_{co} (mm)	0.1	0.2	0.3	0.4	0.5
T_{st} (mm)	1	2	3	4	5
NI (kA)	1.5	1.8	2.1	2.4	2.7
S (mm)	2	2.5	3	3.5	4
B (mm)	20	22.5	25	27.5	30

Table 2. Material components of FEM and lab prototype.

Component	Material
Rotor outer layer	Copper
Rotor inner layer	Low Carbon Steel 1010
Curved core	Silicon Steel NGO M-22
Electromagnet core	Iron
Winding	Copper

The various parameters of the inductive RS have high nonlinear relationships due to the 3-D field distribution and electromagnetic conversion. Fitting neural network (FNN) has the strength of realizing the complex non-linear mapping between the input and the output. So it is suitable for solving the problem of inductive RS torque prediction with complicated internal mechanism. Due to the large calculation and time consumption of 3D finite element simulation, obtaining a large number of training samples are not practical. Support vector machines (SVM) fit well for the small or medium size of training samples. It determines the final results with only a few support vectors. The major redundant samples are eliminated after training. Random forest (RF) can handle data with many features and have high training speed with independent decision trees. This fits the efficient multi-variable optimization of the inductive RS. Also, it can reduce the risk of over-fitting by increasing tree numbers. It is hard to give strict conclusion of application domains for the three methods. In practice, we choose the appropriate method according to the specific situation.

To enhance the prediction performance, the tree numbers of RF, the hidden layer size of the FNN and parameters of the SVM are optimized. As shown in Figure 7, increasing the number of trees generally improves the accuracy of RF and prediction tends to be stable. It reaches the highest average accuracy of 0.0022 Nm with 44 trees. As for fitting neural networks, it takes a row vector of N hidden layer sizes, and a BP training function, and returns a feed-forward neural network with N + 1 layers. Theoretically, multi-layer fitting neural networks can fit any finite input-output relationship well given enough hidden neurons. However, the risk of over fitting increases with the increase of hidden neurons, where it may fit well for the training data but badly for the test data. Thus, the hidden layers should be chosen properly. As shown in the second figure of Figure 7, it reaches the highest average accuracy of 0.0011 Nm with 10 hidden layers. As for SVM, the radial basis function is set as the kernel function considering the nonlinearity of the torque model. The SVM parameter cost (c), radial basis function parameter (g) and epsilon in loss function (p) are chosen to be optimized. This optimization is carried out by genetic algorithms. To avoid over fitting, c, g and p are constrained within 100, 100 and 1 respectively. It shows that the SVM prediction model reaches the highest average accuracy of 0.00093 Nm when c p and g are 70.716, 0.019 and 0.066 respectively.

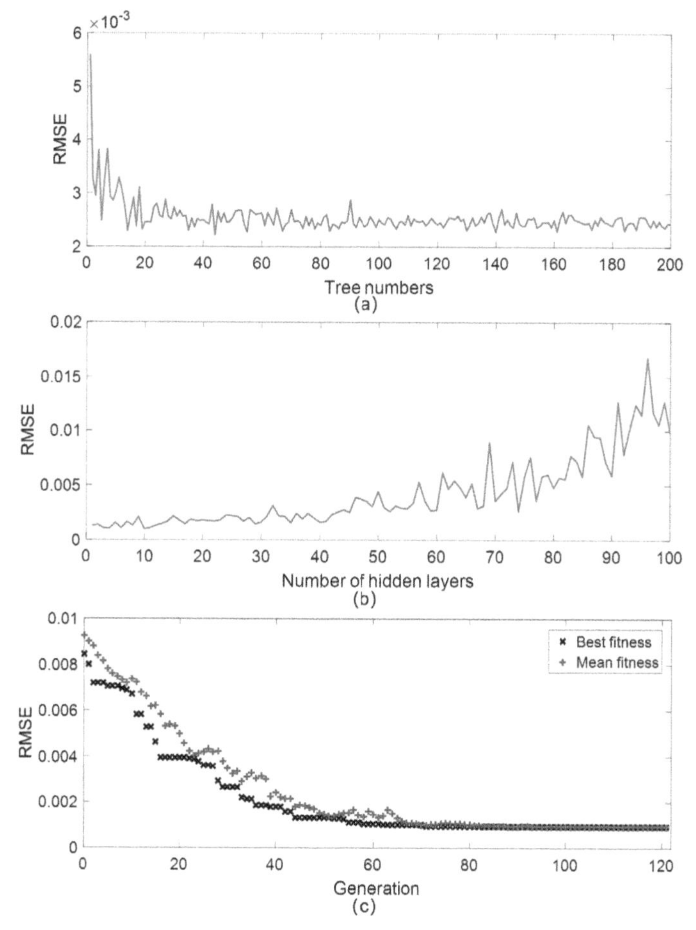

Figure 7. Data-based models parameters optimization: (**a**) RF regression with increasing trees, best root-mean-square error (RMSE) of 0.0022 Nm with 44 trees. (**b**) NN regression with increasing hidden layers, best RMSE of 0.0011 Nm with 10 hidden layers. (**c**) SVM regression with genetic algorithm optimizing parameters, best RMSE of 0.0009 Nm with c (70.716), p (0.019) and g (0.066).

All in all, the regression results of the test set show that the SVM model achieves the least RMSE over the other two in the case of small-sample database. The optimal predictions by RF, FNN and SVM are shown in Figure 8a, with these test samples arranged by Taguchi design of five factors five levels. Figure 8b shows the prediction errors of RF, FNN and SVM for each sample. It can be seen that the

SVM torque model has a stable prediction, with all errors restricted in 2 mNm. Thus, the SVM model is chosen to be the objective function of the optimization in next subsection.

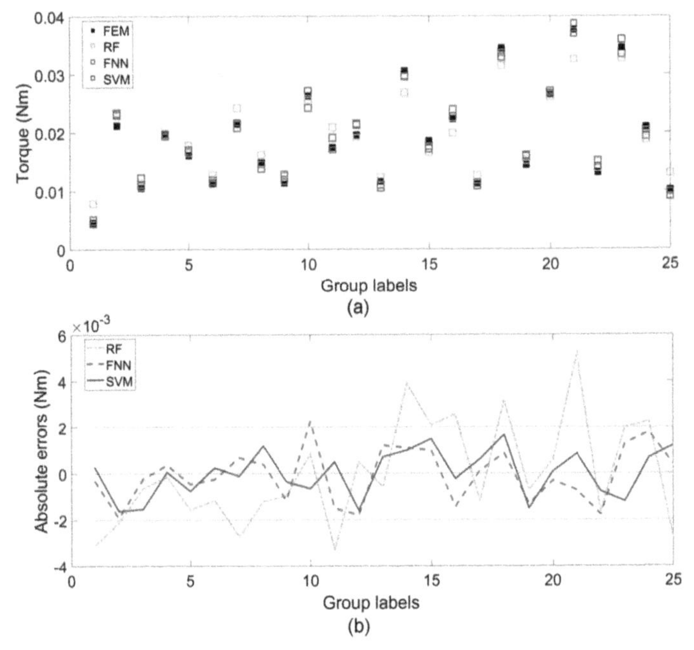

Figure 8. RF, FNN and SVM models prediction performance with test data: (**a**) torque predictions and (**b**) absolute errors.

4.2. Design Parameters Optimization

The reaction sphere design parameters optimization is a multi-parameter optimization problem with constraints. The objective function is the SVM torque prediction model established above. The constraints include the box constraints, linear inequality constraints and nonlinear inequality constraints. The box constraints (22)–(26) are set to be consistent with the range of the sample spaces shown in Table 1. The linear inequality constraints are considered between coil turns, tooth thickness and tooth width: The coil turns and tooth thickness are limited by the tooth pitch (27); The coil turns and tooth width are limited by the fixed stator arc length (28). The parameters of the linear inequality constraints are validated by the lab possessed parts shown in Figure 9. The nonlinear inequality constraints consider the application in micro spacecrafts: the MOI of the rotor is larger than 0.001 kg·m^2 (29) and the whole mass of the actuator is no more than 3 kg (30).

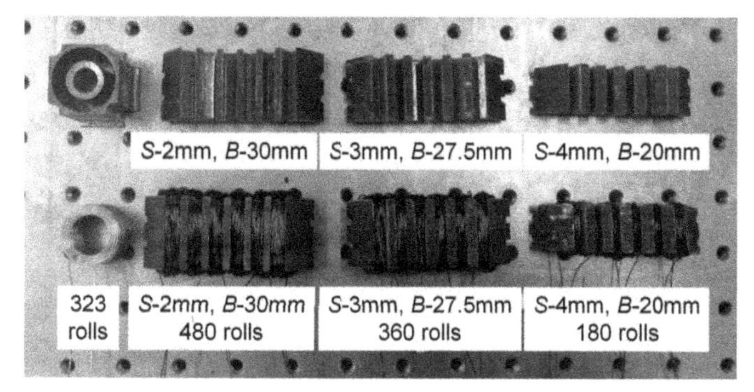

Figure 9. Parts of the reaction sphere prototype for constraints and masses validation.

The mass function $M(T_{co}, T_{st}, NI, S, B)$ evaluates mass of the whole actuator, consisting of rotor, stator core, toroidal winding and electromagnets. The mass of the rotor and stator core is estimated by their measured volume of the digital model and their corresponding densities. The mass of the toroidal winding is calculated by the product of turn number and mass of each turn. The mass of the electromagnets is a constant, which is weighed with prototypes. Several sample components are selected to validate its accuracy as shown in Table 3. Due to the insulating coating, the mass test results of the curved cores are slightly larger than the estimated ones.

Table 3. Mass estimation and experimental validation.

Samples	Estimated Mass (kg)	Tested Mass (kg)
Rotor(T_{co}0.3 mm, T_{st}3 mm)	0.7396	0.7400
Curved core(S2 mm,B30 mm)	0.0795	0.0805
Curved core(S3 mm,B27.5 mm)	0.0813	0.0820
Curved core(S4 mm,B20 mm)	0.0652	0.0660
Curved core(S2 mm,B30 mm), 480 rolls	0.1482	0.1370
Curved core(S3 mm,B27.5 mm), 360 rolls	0.1296	0.1225
Curved core(S4 mm,B20 mm), 180 rolls	0.0759	0.0750
Electromagnets core	\	0.0625
Electromagnets coil	\	0.027

Based on the analysis above, the mathematical description of the optimization problem can be expressed as:

$$\max\ T = \text{svmpredict}(T_{co}, T_{st}, NI, S, B) \tag{21}$$
$$s.t.\quad 0.1\ \text{mm} \leq T_{co} \leq 0.5\ \text{mm}, \tag{22}$$
$$1\ \text{mm} \leq T_{st} \leq 5\ \text{mm}, \tag{23}$$
$$1.5\ \text{kA} \leq NI \leq 2.7\ \text{kA}, \tag{24}$$
$$2\ \text{mm} \leq S \leq 4\ \text{mm}, \tag{25}$$
$$20\ \text{mm} \leq B \leq 30\ \text{mm}, \tag{26}$$
$$NI/33.6 \leq 140 - 20S, \tag{27}$$
$$NI/33.6 \leq 340 - 10S - 8B, \tag{28}$$
$$I_{co} + I_{st} \geq 0.001\ \text{kg} \cdot \text{m}^2, \tag{29}$$
$$M(T_{co}, T_{st}, NI, S, B) \leq 3\text{kg}. \tag{30}$$

This constrained optimization is carried out by interior point methods on MATLAB2016a optimization toolbox, with parameter setting as follows: function tolerance and constraints tolerance:

10^{-6}, max iterations: 1000. Inner max iterations: 100, relative tolerance: 10^{-2}. The optimization results are listed in Table 4. The iteration process is shown in Figure 10. The final T_{co}, T_{st}, NI, S, B for developed 4-pole prototype are rounded as 0.5 mm, 2.4 mm, 2.688 kA, 3 mm and 27.5 mm respectively. The driving torque is 0.0371 Nm (FEM) with the whole mass of 3 kg and MOI of 0.001 kg·m^2.

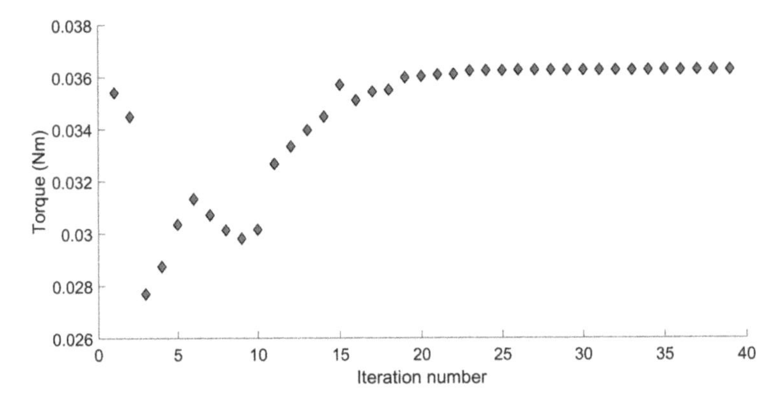

Figure 10. Design parameters optimization by interior point method for better output torque.

Table 4. Constrained optimization results by interior point methods.

T_{co}	T_{st}	NI	S	B	Torque	Density	Moment of Inertia
0.5 mm	2.395 mm	2.7 kA	2.982 mm	27.671 mm	0.0363 Nm	0.0121 Nm/kg	0.001 kg· m^2

5. Feasibility Test with Lab Prototypes

To validate the feasibility of the proposed design, the experimental setup is developed to test the speed performance of the lab prototypes, which is composed of converters, DC power supply, reaction sphere actuators, controllers and amplifiers as shown in Figure 11. The three converters are responsible for driving the reaction sphere actuators in the three orthogonal axis respectively. The DC power supply provides direct currents for amplifiers and controllers for magnetic levitation control.

In this magnetic levitation system, to get the feedback position signal in the closed-loop control, an eddy current displacement sensor made of tiny coils is placed at the center of each electromagnet. These sensors provide the signal of air gap distance to the controller. From here, a negative feedback control loop is set up by two serial-link PD controller to regulate the levitation distance. Specifically, the controller uses STM32F051K6T6 (Suzhou Joint-Tech Mechanical and Electrical Technology Co., LTD, Suzhou, China), which collects the signal of eddy current displacement sensors and generates PWM signal to drive the power amplifier. Each driver amplifies the control signal and drives the two electromagnetic coils on the same axis. What should be mentioned is that the electromagnets about Z-axis need to overcome the gravity of the rotor. They require higher currents compared with the counterparts about the other two axes. A 4-pole reaction sphere prototype and a 6-pole reaction sphere prototype are developed to test the speed performance with the same controllers and drivers. A binary painting strategy for speed measurement is presented as shown in Figure 12. The four silvery white circles on the surface of the rotor are located on the four vertexes of a tetrahedron. In case of measuring the X, Y and Z rotational speed, the laser focus will come across the high reflective areas, which is the nickel protection layer shown in Figure 2, 3 times when the rotor turn around once. Therefore, the output speed can be estimated as the one-third of counts per minute. The frequency of AC is set to 60 Hz in X, Y and Z axis respectively of the 4-pole reaction sphere prototype and in X and Y axis respectively of the 6-pole reaction sphere prototype. The Z axis of the 6-pole reaction sphere is set to 54 Hz, for higher frequency cannot generate a stable rotation with its significantly increasing vibration.

This results from the parts manufacturing and assembling error, as well as the dynamical stability of the sphere about Z-axis rotation under gravity force for 6-pole RS, probably due to its unequal pole. The speed test results are listed in Table 5. It is shown that the 4-pole reaction sphere prototype has the maximum of 775 rpm, 791 rpm and 795 rpm in X, Y and Z axis respectively, which obtains a better speed performance over the 6-pole one as predicted. As for speed uncertainties, the 4-pole prototype shows a relatively smaller fluctuations, which may benefit from the uniform fields distributions.

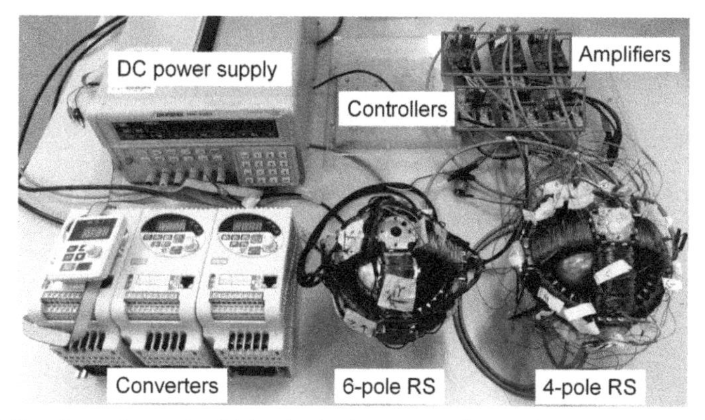

Figure 11. Experimental setup for driving the 4-pole reaction sphere and 6-pole reaction sphere under magnetic levitation.

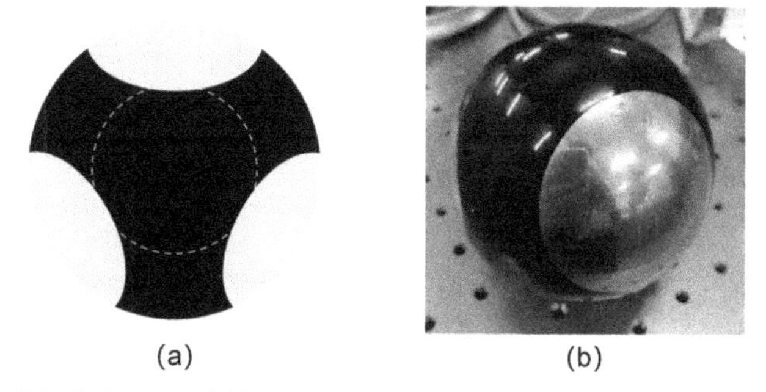

Figure 12. Spherical rotor with black painting for speed counts: (**a**) diagram and (**b**) prototype.

Table 5. Maximum speed of the four-pole reaction sphere prototype and the six-pole reaction sphere prototype.

Axis	Four-Pole Reaction Sphere			Six-Pole Reaction Sphere		
	X (60 Hz)	Y (60 Hz)	Z (60 Hz)	X (60 Hz)	Y (60 Hz)	Z (54 Hz)
Tachometer counts1	2278	2290	2349	1667	1447	963
Tachometer counts2	2351	2413	2365	1712	1322	851
Tachometer counts3	2344	2412	2439	1597	1362	810
Average counts	2324	2372	2384	1659	1377	875
Average speed (rpm)	775	791	795	553	459	292
Speed uncertainty (rpm)	24.3	41.0	30.0	38.3	42.7	51.0
Angular momentum (kg·m^2/s)	0.081	0.083	0.083	0.058	0.048	0.031

The speed control experiment is not so successful. The tested output speed is much less than the synchronization speed (1800 rpm @ 60 Hz). When the frequency is over 60 Hz, the increasing vibration leads to the mechanical collision between the rotor and the stator. One reason lies in the quality of the prototypes. The magnetic levitation demands for the high precision of parts processing and assembling, which is not well guaranteed in the developed prototypes. This results in the non-unique performance in different axes, which is particularly evident between Z axis and the other two axes in the 6-pole prototype.

A more essential reason lies in a drawback of the proposed design: When producing the driving torque, the rotation windings on the curved inductor will generate an extra thrust in the meantime. This extra thrust will push the electromagnets to generate the flux and attract the rotor to be back in the center of the stator. Then, additional damping torques will be induced by the interaction between the rotating rotor and the electromagnets flux, which leads to speed reduction.

6. Conclusions

This paper presents the new design, modeling and parameter optimization methods of a magnetically levitated inductive reaction sphere. It provides a simple, low-cost and miniaturized solution to spacecraft attitude control via momentum exchange. The compact design of electromagnets and curved inductors has the dimensional ratio of 1:1.4 between the rotor and the stator assembly. The simplified, analytical driving torque model is derived based on Ampere's law and separation of variables, showing that the conductivity of the rotor outer layer and pole number of the stator fields contribute to the driving torque. To raise the agility of attitude adjustment of satellites, it is crucial to improve the torque density of this 4-pole inductive RS besides introducing slots in the iron-core curve stator. Here, its curved inductors and electromagnets are redesigned as the cross-shape, allowing two more slots near the notches of the curve stator to be functional. This also improves the uniformity of the induced eddy current and magnetic field along the stator circle. As the full analytical torque model is not available for inductive RS with slotted curve stators, a constrained optimization problem is formulated to maximize the output torque. In addition, the design parameters are efficiently solved by SVM with only small volume of training data. The lab prototype is developed for feasibility test. It shows that the optimized RS can rotate over 700 rpm in X, Y and Z axis with the angular momentum of 0.08 $kg{\cdot}m^2/s$.

Small volume and small mass are two essential issues in reaction spheres design. The proposed compact design enables the developed prototype to have the rotor mass of 0.66 kg and the whole assembly of 3 kg under the dimension constraints of 136 mm × 136 mm × 136 mm. The necessary angular momentum of the actuator is expected to be within 0.1–1 $kg{\cdot}m^2/s$ for actual micro-spacecraft (10–100 kg) attitude control [33,34], and the practical design target is set as 0.4 $kg{\cdot}m^2/s$. If we maintain the design parameters in this paper, the rotor has to spin about 3800 rpm to achieve this, demanding a high level of dynamic balance. Due to the current restraint of motor drivers and the error during part fabrication and assembly, it is not practical to realize such speed with our prototypes. A more practical way is to increase both the output speed and the mass of the steel layer of the rotor appropriately. For example, we can achieve the angular momentum of 0.4 $kg{\cdot}m^2/s$ by increasing the speed to 1500 rpm and the steel thickness of the hollow sphere to 9.5 mm respectively. Remarkably, the design and optimization methods proposed in this paper are still applicable in the practical reaction sphere development.

Author Contributions: L.Y. derived the electromagnetic modeling, optimization method and prepared the manuscript; J.Z. and J.C. proposed the original inductive RS design; S.-L.C. advised the theoretical study, designed the experiments and refined the manuscript; Y.L. and L.Y. assembled the prototypes and carried out experimental test; C.Z. and G.Y. supervised the research throughout.

Funding: This research was funded by Zhejiang provincial public welfare research program(Y19E070003, LGG18E070007) and the innovation team of key components and key technology for the new generation robot(2016B10016).

Conflicts of Interest: The authors declare no conflict of interest.

Appendix A. Proof: (A1) is a Monotonically Increasing Function

The MOI of a hollow and uniform sphere is

$$J = J(R_0) = \frac{2}{5}M\frac{R^5 - R_0^5}{R^3 - R_0^3}, \quad 0 \leq R_0 \leq R, \tag{A1}$$

where R_0 is a variable, R and M are constants, $R_0, R, M \in \mathbb{R}^+$. Take logarithms on both sides of (A1),

$$\ln(J) = \ln\left(\frac{2}{5}M\right) + \ln(R^5 - R_0^5) + \ln(R^3 - R_0^3). \tag{A2}$$

Take the derivative of (A2) with respect to R_0, it yields

$$\frac{J'}{J} = \frac{R_0^2(-5R_0^2R^3 + 2R_0^5 + 3R^5)}{(R^5 - R_0^5)(R^3 - R_0^3)}. \tag{A3}$$

Set $H(R_0) = -5R_0^2R^3 + 2R_0^5 + 3R^5$, so $H(R_0)|_{R_0=R} = 0$. In addition, $\frac{dH}{dR_0} = 10R_0(R_0^3 - R^3) < 0$. In this way, for $R_0 < R$, $H(R_0) > 0$. This yields $J' = JH(R_0)/[(R^5 - R_0^5)(R^3 - R_0^3)] > 0$.

This concludes $J(R_0)$ is monotonically increasing when $R_0 \in [0, R)$. Additionally, $J_{\min} = J(0) = \frac{2}{5}MR_0^2$, $J < J(R) = \lim_{R_0 \to R} \frac{2}{5}M\frac{R^5 - R_0^5}{R^3 - R_0^3} = \frac{2}{3}MR_0^2$, $J(0) = 60\%J(R)$.

References

1. Maini, A.K.; Agrawal, V. *Satellite Technology: Principles and Applications*; John Wiley & Sons: New York, NY, USA, 2011.
2. Ormsby, R.D. Capabilities and limitations of reaction spheres for attitude control. *ARS J.* **1961**, *31*, 808–812. [CrossRef]
3. Haeussermann, W. *The Spherical Control Motor for Three Axis Attitude Control of Space Vehicles*; NASA TM X-50071; NASA: Greenbelt, MD, USA, 1959.
4. Haeussermann, W. Space Vehicle Attitude Control Mechanism. U.S. Patent 3,017,777, 23 June 1962.
5. Bitterly, J.G. Flywheel technology: Past, present, and 21st century projections. *IEEE Aerosp. Electron. Syst. Mag.* **1998**, *13*, 13–16. [CrossRef]
6. Kurokawa, H. *A Geometric Study of Single Gimbal Control Moment Gyros*; Report of Mechanical Engineering Laboratory, no. 175; Agency of Industrial Technology and Science: Tokyo, Japan, 1998.
7. Stagmer, E. Reaction Sphere for Stabilization and Control in Three Axes. U.S. Patent 9,475,592, 25 October 2016.
8. Rossini, L. Electromagnetic mOdeling and Control Aspects of a Reaction Sphere for Satellite Attitude Control. Ph.D. Thesis, EPFL, Lausanne, Switzerland, 2014.
9. Craveiro, A.A.; Sequeira, J.S. Reaction sphere actuator. *IFAC-PapersOnLine* **2016**, *49*, 212–217. [CrossRef]
10. Zhang, M.; Zhu, Y.; Chen, A.L.; Yang, K.M.; Cheng, R. A Maglev Reaction Sphere Driven by Magnetic Wheels. CN Patent 105207430 A, 30 December 2015.
11. Keshtkar, S.; Moreno, J.A.; Kojima, H.; Hernández, E. Design concept and development of a new spherical attitude stabilizer for small satellites. *IEEE Access* **2018**, *6*, 57353–57365. [CrossRef]
12. Takehana, R.; Paku, H.; Uchiyama, K. Attitude control of satellite with a spherical rotor using two-degree-of-freedom controller. In Proceedings of the 2016 7th IEEE International Conference on Mechanical and Aerospace Engineering (ICMAE), London, UK, 18–20 July 2016; pp. 352–357.
13. Hollis, R.L., Jr.; Kumagai, M. Spherical Induction Motor. U.S. Patent 9,853,528, 26 December 2017.
14. Fan, D.; Chun, C.S.; He, Y.; Song, J.; Zhang, N. Magnetization Suspension Inductive Reaction Sphere. CN Patent 105,775,169, 20 July 2016.
15. Isely, W.H. Magnetically Supported and Torqued Momentum Reaction Sphere. U.S. Patent 4,611,863, 16 September 1986.

16. Wampler-Doty, M.P.; Doty, J. A reaction sphere for high performance attitude control. In Proceedings of the 8th Annual CubeSat Developers' Workshop, San Luis Obispo, CA, USA, 20–22 April 2011.
17. Iwakura, A.; Tsuda, S.-I.; Tsuda, Y. Feasibility study on three dimensional reaction wheel. *Proc. Sch. Sci. Tokai Univ. Ser.* **2008**, *33*, 51–57.
18. Kim, D.-K.; Yoon, H.; Kang, W.-Y.; Kim, Y.-B.; Choi, H.-T. Development of a spherical reaction wheel actuator using electromagnetic induction. *Aerosp. Sci. Technol.* **2014**, *39*, 86–94. [CrossRef]
19. Kumagai, M.; Hollis, R.L. Development and control of a three dof spherical induction motor. In Proceedings of the 2013 IEEE International Conference on Robotics and Automation, Karlsruhe, Germany, 6–10 May 2013; pp. 1528–1533.
20. Chételat, O. Torquer Apparatus. U.S. Patent 8,164,294, 24 April 2012.
21. Chabot, J.; Schaub, H. Spherical magnetic dipole actuator for spacecraft attitude control. *J. Guid. Control. Dyn.* **2016**, *39*, 911–915. [CrossRef]
22. Zhou, L.; Nejad, M.I.; Trumper, D.L. *Hysteresis Motor Driven One Axis Magnetically Suspended Reaction Sphere*; American Society for Precision Engineering: Raleigh, NC, USA, 2014.
23. Zhou, L.; Nejad, M.I.; Trumper, D.L. One-axis hysteresis motor driven magnetically suspended reaction sphere. *Mechatronics* **2017**, *42*, 69–80. [CrossRef]
24. Liu, J. Structural Design and Three-Dimensional Magnetic Field Analysis of Magnetic Levitation Switched Reluctance Spherical Motor. Master's Thesis, Yangzhou University, Yangzhou, China, 2011.
25. Paku, H.; Uchiyama, K. Satellite attitude control system using a spherical reaction wheel. In *Applied Mechanics and Materials*; Trans Tech Publ: Zurich, Switzerland, 2015; Volume 798, pp. 256–260.
26. Mashimo, T.; Awaga, K.; Toyama, S. Development of a spherical ultrasonic motor with an attitude sensing system using optical fibers. In Proceedings of the 2007 IEEE International Conference on Robotics and Automation, Roma, Italy, 10–14 April 2007; pp. 4466–4471.
27. Zhu, L.; Guo, J.; Gill, E. Analytical field and torque analysis of a reaction sphere. *IEEE Trans. Magn.* **2018**, *54*, 1–11. [CrossRef]
28. Stegun, A. Handbook of Mathematical Functions. US Government Printing Office: October, 1973. Available online: http://people.math.sfu.ca/~cbm/aands/page_374.htm (accessed on 14 January 2019).
29. Spałek, D. Spherical Induction Motor with Anisotropic Rotor-Analytical Solutions for Electromagnetic Field Distribution, Electromagnetic Torques and Power Losses. 2007. Available online: http://www.compumag.org/jsite/images/stories/TEAM/problem34.pdf (accessed on 15 November 2018).
30. Davey, K.; Vachtsevanos, G.; Powers, R. The analysis of fields and torques in spherical induction motors. *IEEE Trans. Magn.* **1987**, *23*, 273–282. [CrossRef]
31. Davey, K.; Vachtsevanos, G.; Powers, R. Analysis of a spherical induction motor. *Electr. Mach. Power Syst.* **1987**, *12*, 206–223. [CrossRef]
32. Mark A. Tschopp, DOE with MATLAB. Available online: https://icme.hpc.msstate.edu/mediawiki/index.php/DOE_with_MATLAB_3 (accessed on 8 October 2018).
33. Scharfe, M.; Roschke, T.; Bindl, E.; Blonski, D. Design and development of a compact magnetic bearing momentum wheel for micro and small satellites. In Proceedings of the 15th Annual/USU Conference on Small Satellites, Logan, UT, USA, 13–16 August 2001.
34. Grillmayer, G.; Falke, A.; Roeser, H.-P. Technology demonstration with the micro-satellite flying la'pt op. In *Small Satellites for Earth Observation: Selected, Proceedings of the 5th International Symposium of the International Academy of Astronautics, Berlin, Germany, 4–8 April 2005*; Walter de Gruyter: Berlin, Germany 2011; p. 419.

Article

Mitigation Method of Slot Harmonic Cogging Torque Considering Unevenly Magnetized Permanent Magnets in PMSM

Chaelim Jeong [1], Dongho Lee [2] and Jin Hur [3,*]

[1] Department of Industrial Engineering, University of Padova, 35131 Padova, Italy; cofla827@gmail.com
[2] Sungshin Precision Global (SPG) Co., Ltd., Incheon 21633, Korea; leedh38126@gmail.com
[3] Department of Electrical Engineering, Incheon National University, Incheon 22012, Korea
* Correspondence: jinhur@inu.ac.kr; Tel.: +82-10-4024-1728

Received: 29 August 2019; Accepted: 11 October 2019; Published: 14 October 2019

Abstract: This paper presents a mitigation method of slot harmonic cogging torque considering unevenly magnetized magnets in a permanent magnet synchronous motor. In previous studies, it has been confirmed that non-uniformly magnetized permanent magnets cause an unexpected increase of cogging torque because of additional slot harmonic components. However, these studies did not offer a countermeasure against it. First, in this study, the relationship between the residual magnetic flux density of the permanent magnet and the cogging torque is derived from the basic form of the Maxwell stress tensor equation. Second, the principle of the slot harmonic cogging torque generation is explained qualitatively, and the mitigation method of the slot harmonic component is proposed. Finally, the proposed method is verified with the finite element analysis and experimental results.

Keywords: cogging torque; permanent magnet machine; torque ripple; uneven magnets

1. Introduction

The cogging torque is one of the most representative components of torque ripple in the permanent magnet synchronous motor (PMSM). Therefore, studies on the reduction method for the cogging torque have been actively carried out to minimize the torque ripple [1–11]. Those studies on cogging torques have been mainly focused on reducing the cogging effect by modulating the combination of the pole and slot number, the pole arc, the shape of the core, the skew angle, the notching, etc. In general, the results of such studies are based on a simple theoretical analysis, so it is assumed that the magnetic components of the motor are ideal. However, since there are many possible manufacturing errors such as eccentricity, machining error, and unevenly magnetized magnets, motors always contain non-ideal components. As a result of those errors, the measured cogging torque of the actual motor may be very different from what is expected in the simulation [12]. This phenomenon can be a critical issue to those applications that need precision control of the motor and are sensitive to noise and vibration.

For this reason, several studies that take manufacturing errors into consideration have emerged [13–19]. In [13–16], analytical solutions of cogging torque are studied by considering the magnet imperfections, rotor eccentricity, geometrical variation, and magnetizing fixture. In addition, [17] mathematically investigated the cogging torque caused by the simultaneous existence of eccentricities and the uneven magnetization. Those studies have focused on analysis methods of cogging torque by considering manufacturing errors, and they reported that those errors generate additional harmonic components. In [18], they show that the unevenly magnetized permanent magnet (PM) can have a negative impact on applying the cogging torque reduction method (teeth curvature modulation method), leading to additional slot harmonic cogging torque. In [19], it is confirmed that the main contributors, which have the greatest effect on the cogging torque distortion, are the inner radius tolerance of the stator and the tolerance of PM remanence (unevenly magnetized magnets), among many other manufacturing errors. In addition to the aforementioned studies, there are a few studies that have analyzed motor performance in consideration of manufacturing errors, but those studies only handle the phenomena analysis caused by them, and there is a lack of research on the mitigation countermeasures.

Unlike most studies that have only analyzed the effects of manufacturing errors on cogging torque, this paper proposes a method to counteract the influence of unevenly magnetized magnets, which are one of the main contributors of cogging torque distortion [19]. Here, unevenly magnetized magnets mean that each magnet has different magnetic strength. This study is carried out in the following order. First, the relationship between the remanence of each PM and the cogging torque is derived from the basic form of the Maxwell stress tensor equation. Second, the principle of slot harmonic cogging torque generation is explained qualitatively. Based on this principle, a new mitigation method of slot harmonic cogging torque is proposed. This method involves a series of processes that select the position of the PMs, taking into account the remanence deviations of each PM. Finally, the proposed method is verified with a finite element analysis (FEA) and experimental results. Here, note that this study assumed that each magnet is pre-magnetized before the assembly. Therefore, the proposed method is more appropriate for small quantity customized production than mass production. Moreover, based on the principle of the slot harmonic component mitigation condition, it is possible to adjust that the manufacturing tolerance of the magnetization yoke, leading to the alleviation of the influence of the uneven magnetization.

2. Analysis of Cogging Torque in PMSM from a Macroscopic Perspective

Before examining the process of the slot harmonic cogging torque generation caused by the unevenness in magnetic strengths of each pole, we first analyzed the generation of cogging torque from a macroscopic perspective.

2.1. The Relation between the Electromagnetic Force and the Remanence of Magnet

The cogging torque refers to a torque caused by an electromagnetic force generated when the PM's magnetic flux passing through the air gap between a rotor and a stator is concentrated in a path that has a relatively small magnetic reluctance. Therefore, in order to analyze the magnitude of the cogging torque, it is necessary to understand the electromagnetic force. In many studies, according to Maxwell stress tensor theory, the electromagnetic force of the tangential component is defined as follows in a single rotor position [13,20–22]:

$$F_t = \frac{R}{\mu_0} \int_0^{2\pi} B_{rgap}(\phi_r) B_{tgap}(\phi_r) d\phi_r \quad (1)$$

where F_t is the tangential force density, μ_0 is the permeability of free space, R is the radius for which the Maxwell stress tensor is calculated, Φ_r is the space angle at single rotor position, and B_{rgap} and B_{tgap} represent radial and tangential components of magnetic flux density, respectively. Here, if the saturation phenomenon of the magnetic material is ignored, B_{rgap} and B_{tgap} are always proportional to the remanence of the PM (B_r), and the following relationship holds:

$$F_t \propto B_r^2 \quad (2)$$

2.2. The Cogging Torque Caused by Single Pole

The cogging torque according to the electromagnetic force described above can be defined as follows:

$$T_{cog}(\theta, l) = R \int_0^{L_{stk}} F_t(\theta, l) dl \quad (3)$$

where θ is rotation angle of the rotor, L_{stk} is stack length, and l is axial length. Here, if the electromagnetic force is uniformly distributed in the stacking direction, the above equation can be simplified as the following equation:

$$T_{cog}(\theta) = RL_{stk}F_t(\theta) \tag{4}$$

To understand the period of cogging torque and the interaction of harmonic components, we first assumed an example model with one magnet (pole), as in Figure 1. Here, the pole-arc of the PM is 24° and the stator has 12 teeth. In this case, since the influences of eccentricity and shape error are deviated from the subject of this study, the cogging torque can be expressed in the following form of the Fourier series [23]:

$$T_{cog}(\theta) = RL_{stk}\sum_{k=1}^{\infty} F_{tk}\sin(kS\theta) = RL_{stk}\sum_{k=1}^{\infty} F_{tk}\sin(k\theta_s) \tag{5}$$

where θ_s is a slot periodic angle that is calculated by multiplying θ (mechanical angle) with the slot number, F_{tk} is Fourier coefficient (amplitude of force) of the 'k'th harmonic component, and S is the number of the slots.

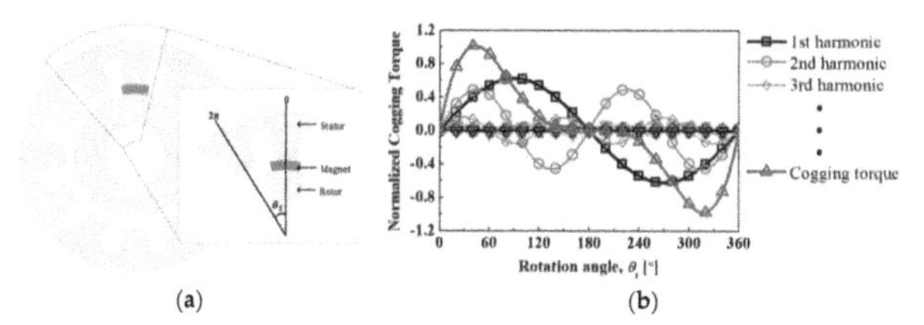

Figure 1. (**a**) The geometry of the example model with one pole and twelve teeth; (**b**) cogging torque harmonic component due to one pole rotation (θ_s).

The cogging torque result of the example model of Figure 1a is shown in Figure 1b with the harmonic component. Here, the magnitude of each harmonic component is normalized to the peak value of the total cogging torque. Since the first harmonic component is the largest, the total cogging torque has the same period as the first harmonic component.

2.3. The Cogging Torque Caused by Multi Poles

Now suppose that we add some magnets to the example model. Figure 2a–c have 8 poles, 10 poles, and 14 poles, respectively. Each added pole has the same remanence, pole-arc, and thickness. In this case, the cogging torque generated in each adjacent pole has a phase difference by a pole pitch. Further, due to the phase difference, the cogging torques generated by each pole interfere with each other. Using the property from Equation (5), the cogging torque caused by multi-poles can be expressed as follows using superposition technique [24,25].

$$\begin{aligned} T_{cog}(\theta_s) &= T_{p1} + T_{p2} + T_{p3} + \cdots + T_{pP} \\ &= RL_{stk}\left\{\sum_{k=1}^{\infty} F_{t1_k}\sin(k\theta_s) + \sum_{k=1}^{\infty} F_{t2_k}\sin\left[k\left(\theta_s + (n_2-1)\tfrac{2\pi S}{P}\right)\right]\right. \\ &\left. + \sum_{k=1}^{\infty} F_{t3_k}\sin\left[k\left(\theta_s + (n_3-1)\tfrac{2\pi S}{P}\right)\right] + \cdots + \sum_{k=1}^{\infty} F_{tP_k}\sin\left[k\left(\theta_s + (n_P-1)\tfrac{2\pi S}{P}\right)\right]\right\} \end{aligned} \tag{6}$$

where T_{pP} is the cogging torque that is generated by each pole, F_{tPk} is the 'k'th harmonic component of the tangential magnetic force density generated by each pole, $n_{2\ldots P}$ are the order of the poles, P is the number of the poles, and S is the number of the slots. Here, the rotation angle of the rotor is expressed in the slot periodic angle.

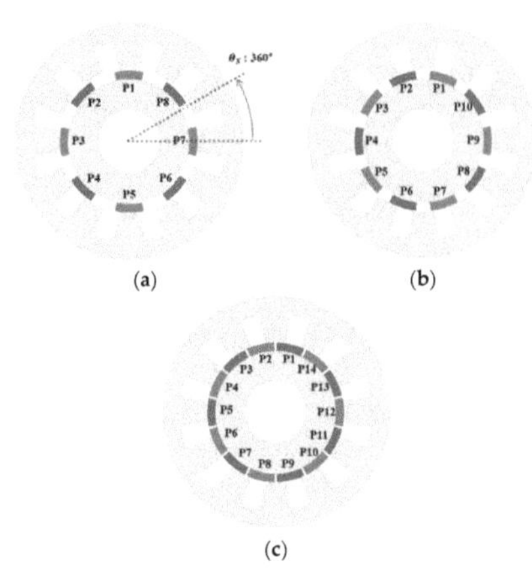

Figure 2. The geometry of the example model with multi poles and twelve teeth for (**a**) 8p/12s, (**b**) 10p/12s, and (**c**) 14p/12s.

Figure 3 shows the cogging torque of each example model case that resulted in the mutual interference of cogging torque for each pole. As can be seen from the figure, depending on the phase difference of each pole, the cogging torque can be increased or decreased by overlapping. Also, the harmonic component of cogging torque is demonstrated in Figure 4. Here, if those harmonic orders are calculated based on the mechanical angle (θ), the most dominant harmonic component has the order of least common multiple (LCM) of the number of slots and the number of poles. This is because the fundamental frequency of the cogging torque is calculated by using the LCM [26]. For example, since the LCM of an 8-pole/12-slot motor is 24, the 24th harmonic becomes as the most dominant harmonic component, as shown in Figure 4. This is a well-known fact of the cogging torque period and has been confirmed once again through this analysis.

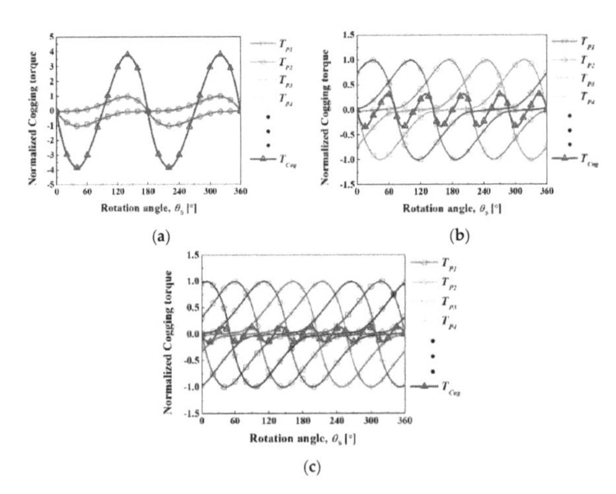

Figure 3. Cogging torque caused by multi poles rotation based on θ_S: (**a**) cogging torque of 8p/12s, (**b**) cogging torque of 10p/12s, and (**c**) cogging torque of 14p/12s.

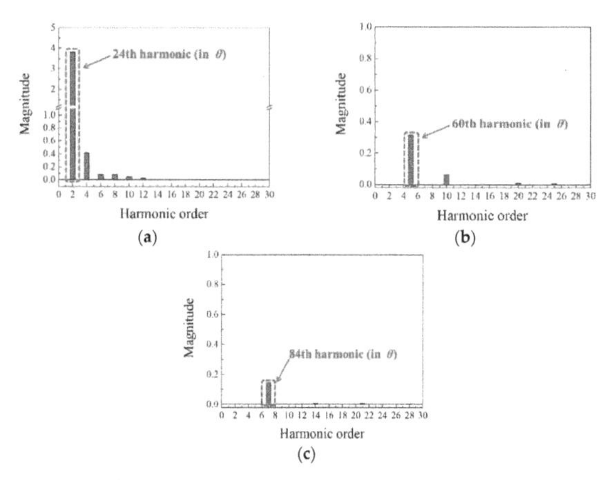

Figure 4. Harmonic spectra of cogging torque caused by multi poles rotation based on θ_S: (**a**) harmonics of 8p/12s, (**b**) harmonics of 10p/12s, and (**c**) harmonics of 14p/12s.

3. Mitigation Method of Slot Harmonic Cogging Torque Component

As the name implies, the slot harmonic component refers to a harmonic component that has the same number of cycles as the number of slots when the rotor rotates 360° (based on θ). Hence, in Equation (6), the first order harmonic component can be defined as the slot harmonic component (because of θ_s = 12θ). Then, the slot harmonic cogging torque can be expressed as follows:

$$\begin{aligned} T_{slot}(\theta_s) = RL_{stk}\Big\{F_{t1_1}\sin(\theta_s) + F_{t2_1}\sin\Big[\theta_s + (n_2-1)\tfrac{2\pi S}{P}\Big] \\ +F_{t3_1}\sin\Big[\theta_s + (n_3-1)\tfrac{2\pi S}{P}\Big] + \cdots + F_{tP_1}\sin\Big[\theta_s + (n_P-1)\tfrac{2\pi S}{P}\Big]\Big\} \end{aligned} \tag{7}$$

As can be seen from the result of Figure 4, when the motor is under the ideal condition and all the magnetic forces of each pole are equal to each other, the slot harmonic component does not exist in all cases. This is because the first harmonic component of each pole is canceled out by the phase difference from the other pole. This is easy to understand the phenomenon with the 8-pole/12-slot and 10-pole/12-slot models.

Substituting the number of poles of 8 and the number of slots of 12 into Equation (7), it is summarized as follows:

$$\begin{aligned} T_{slot}(\theta_s) = RL_{stk}\Big\{F_{t1_1}\sin(\theta_s) + F_{t2_1}\sin\Big[\theta_s + (2-1)\tfrac{2\pi\cdot 12}{8}\Big] \\ +F_{t3_1}\sin\Big[\theta_s + (3-1)\tfrac{2\pi\cdot 12}{8}\Big] + \cdots + F_{t8_1}\sin\Big[\theta_s + (8-1)\tfrac{2\pi\cdot 12}{8}\Big]\Big\} \\ = RL_{stk}\Big[\Big(F_{t1_1} + F_{t3_1} + F_{t5_1} + F_{t7_1}\Big)\sin(\theta_S) + \Big(F_{t2_1} + F_{t4_1} + F_{t6_1} + F_{t8_1}\Big)\sin(\theta_S + \pi)\Big]. \end{aligned} \tag{8}$$

As a result, in the motor of the 8-pole/12-slot, the cogging torque produced by each pole had two phases, and the phase difference was π. Therefore, under ideal conditions, the slot harmonic torque (first harmonic torque of each pole) becomes zero because each pole produces the same amount of magnetic force. Here, through the above equation, the removal condition of the slot harmonic can be more clearly expressed as follows:

$$F_{t1_1} + F_{t3_1} + F_{t5_1} + F_{t7_1} = F_{t2_1} + F_{t4_1} + F_{t6_1} + F_{t8_1} \tag{9}$$

According to this condition, the slot harmonic component is likely to be canceled even if the density of each magnetic force does not exactly coincide. That is, even if the magnetic strength of each pole (the remanence of each magnet) is different, the slot harmonic component may be removed.

In this paper, since the influence of shape error and eccentricity is not considered, the above condition can be rearranged as follows using the relation of (2):

$$B_{r1}^2 + B_{r3}^2 + B_{r5}^2 + B_{r7}^2 = B_{r2}^2 + B_{r4}^2 + B_{r6}^2 + B_{r8}^2 \quad (10)$$

where $B_{r1\ldots rP}$ are the remanence of the magnet in each pole. Under this condition, the slot harmonic component can be mitigated, and all harmonics except the LCM harmonic component will be also mitigated by superposition because the phase difference is equal to the pole pitch.

As in the example of the 8-pole/12-slot motor, by substituting the number of poles of 10 and the number of slots of 12 into Equation (7), it can be summarized as follows:

$$\begin{aligned} T_{slot}(\theta_s) = \; & RL_{stk}[\left(F_{t1_1} + F_{t6_1}\right)\sin(\theta_s) + \left(F_{t2_1} + F_{t7_1}\right)\sin\left(\theta_s + \tfrac{12}{5}\pi\right) \\ & +\left(F_{t3_1} + F_{t8_1}\right)\sin\left(\theta_s + \tfrac{24}{5}\pi\right) + \left(F_{t4_1} + F_{t9_1}\right)\sin\left(\theta_s + \tfrac{36}{5}\pi\right) \\ & +(F_{t5_1} + F_{t10_1})\sin\left(\theta_s + \tfrac{48}{5}\pi\right)]. \end{aligned} \quad (11)$$

Under ideal conditions, the slot harmonic torque (first harmonic torque of each pole) becomes zero because each pole produces the same amount of magnetic force. As a result, the removal condition of the slot harmonic can be expressed as follows:

$$F_{t1_1} + F_{t6_1} = F_{t2_1} + F_{t7_1} = F_{t3_1} + F_{t8_1} = F_{t4_1} + F_{t9_1} = F_{t5_1} + F_{t10_1} \quad (12)$$

$$B_{r1}^2 + B_{r6}^2 = B_{r2}^2 + B_{r7}^2 = B_{r3}^2 + B_{r8}^2 = B_{r4}^2 + B_{r9}^2 = B_{r5}^2 + B_{r10}^2. \quad (13)$$

Through the above examples, it is verified that the slot harmonic cogging torque can be zero, when the motor is under the ideal condition and all the magnetic forces of each pole are equal to each other regardless of the odd or even number of magnet set.

Looking at the process of deriving this condition, consequently, it is important to find poles with the same cogging torque phase so that the sum of the remanence of the magnets is equal to the poles with different phases. By using the equation below, the distance of pole (N), which has the same torque phase with the first (reference) pole, can be calculated in the pole number. This can be simply derived with the number of poles and the number of slots, and by taking into account the relationship between pole pitch and slot pitch.

$$N = \left|\frac{P}{(S-P)}\right| \quad (14)$$

According to Equation (14), N of the 14-pole/12-slot is assigned "7." Therefore, in the case of the 14-pole/12-slot, the torque phases of the first pole and the eighth pole are the same. Therefore, the slot harmonic torque removal condition for each pole/slot combination is derived as follows:

$$\begin{aligned} & B_{r1}^2 + B_{r8}^2 = B_{r2}^2 + B_{r9}^2 = B_{r3}^2 + B_{r10}^2 = B_{r4}^2 + B_{r11}^2 \\ & = B_{r5}^2 + B_{r12}^2 = B_{r6}^2 + B_{r13}^2 = B_{r7}^2 + B_{r14}^2. \end{aligned} \quad (15)$$

In addition to the above condition, in order to make the magnetic flux connected to the winding uniform according to the polarity, the remanence summation of the entire magnets located at the N pole must be equal to that of the S pole. Therefore, the following condition should be also met:

$$B_N(B_{r1} + B_{r3} + B_{r5} + \cdots) = B_S(B_{r2} + B_{r4} + B_{r6} + \cdots) \quad (16)$$

where B_N and B_S are total remanences of the north and south poles.

Considering these conditions when assembling magnets and rotors, the generation of the slot harmonic cogging torque components will be minimized. Hence, the work flow chart of the slot harmonic component mitigation method is shown in Figure 5. Here, the sum of ${B_r}^2$ of the poles with the same cogging torque phase is conveniently referred to as Z. The smaller the difference in Z value for each torque phase is the smaller the slot harmonic size is. In addition, the tolerance is denoted as δ, and it is reasonable to choose this to be larger than the measurement uncertainty of the Gauss value of the magnet in practical.

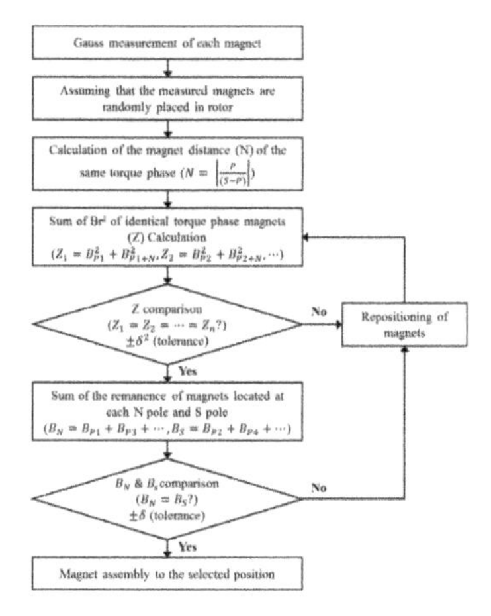

Figure 5. The workflow diagram of a method for mitigating slot harmonic cogging torque before rotor assembly.

4. Verification of the Proposed Method

4.1. Verification Using the Finite Element Analysis (FEA)

In this section, the FEA is conducted to verify the validity of the slot harmonic mitigation method proposed in Figure 5. Two example models are selected for this analysis. The first model is a 6-pole/9-slot interior permanent-magnet motor (IPM), and the other is an 8-pole/12-slot IPM. The geometry with the mesh information and the specification of each model are shown in Figure 6 and Table 1. Here, the saturation point of each core material was adjusted to be lower than the actual property. This is to confirm that the proposed method is still valid under nonlinear material properties.

In order to consider the unevenly magnetized magnet in FEA verification, firstly, the management tolerance on the B_r of the commercial magnets was investigated and is shown in Table 2. Then based on the data, the B_r of each magnet was randomly selected and positioned on the rotor, as shown in Figure 7. The selected B_r results and the magnet position are recorded in Tables 3 and 4. Here, in each table, Case A is each magnet arranged according to its number order, and Case B is where it is arranged according to the proposed method in Figure 5. The change in position of each magnet is shown more clearly in Figure 7. In Tables 3 and 4, it can be seen that the Z comparison and B_N-B_S comparison value are not 'zero,' even in Case B. In fact, since there is very low probability that there can be a magnet arrangement that satisfies this in reality, the tolerances were changed step by step to have the magnet array with the smallest comparison result.

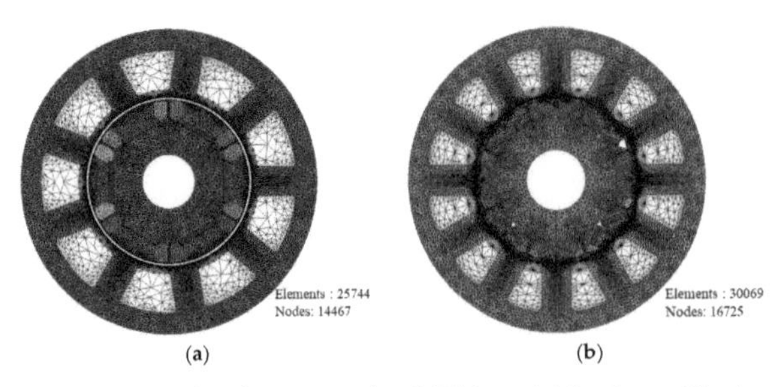

Figure 6. The geometry and mesh information of each FEA model for the verification of the proposed method at (**a**) 6p/9s IPM and (**b**) 8p/12s IPM.

Table 1. Specification of each FEA model for the verification of the proposed method.

Item	6p/9s IPM	8p/12s IPM
Stator outer diameter	100.0 mm	150.0 mm
Rotor outer diameter	54.0 mm	82.0 mm
Stack length	40.0 mm	72 mm
Air gap length	1.0 mm	0.6 mm
Rated power	400 W	5000 W
Rated speed	3500 rpm	2000 rpm
Rated torque	1.1 Nm	23.8 Nm
Rated ph. current	10.3 Arms	120 Arms
Series turn per phase	72	20
Core material	50PN470 (FEM: saturate@1.2T)	50PN470 (FEM: saturate@1.2T)
Magnet material	NMX-36EH	NEOREC 40UH

Table 2. The magnet management tolerance of the manufacturer.

Company	6p/9s IPM	8p/12s IPM
TDK	NEOREC 40UH	1290 ± 30
	NEOREC 40TH	1285 ± 30
	NEOREC 38UX	1250 ± 30
	NEOREC 35NX	1200 ± 30
Hitachi	NMX-43SH	1295 ± 35
	NMX-41SH	1275 ± 35
	NMX-39EH	1235 ± 35
	NMX-36EH	1195 ± 35

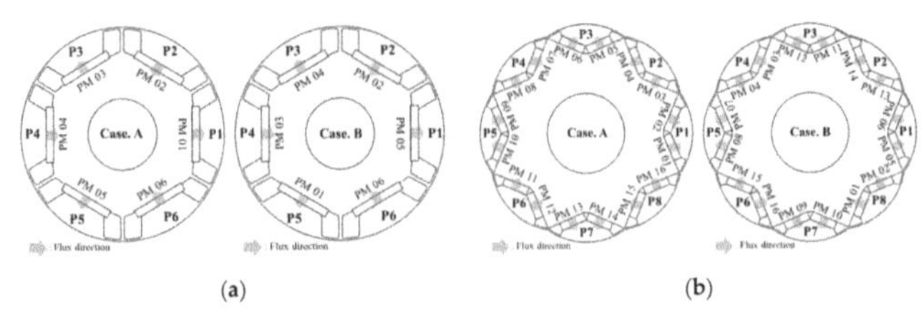

Figure 7. PM position change according to the proposed method for the FEA verification for (**a**) 6p/9s IPM and (**b**) 8p/12s IPM.

Table 3. Random selection of the magnet remanence for the 6p/9s model considering the management tolerance.

Position & Comparison	Case A		Case B	
	PM No.	B_r	PM No.	B_r
P1	PM 01	1.194 T	PM 05	1.226 T
P2	PM 02	1.195 T	PM 02	1.195 T
P3	PM 03	1.198 T	PM 04	1.162 T
P4	PM 04	1.162 T	PM 03	1.198 T
P5	PM 05	1.226 T	PM 01	1.194 T
P6	PM 06	1.172 T	PM 06	1.172 T
Z_1-Z_2	0.212 (T)2		0.042 (T)2	
B_N-B_S	0.089 (T)		0.017 (T)	

Table 4. Random selection of the magnet remanence of 8p/12s model considering the management tolerance.

Position & Comparison	Case A		Case B	
	PM No.	B_r	PM No.	B_r
P1	PM 01	1.265 T	PM 05	1.269 T
	PM 02	1.275 T	PM 06	1.278 T
P2	PM 03	1.290 T	PM 13	1.287 T
	PM 04	1.291 T	PM 14	1.285 T
P3	PM 05	1.269 T	PM 11	1.309 T
	PM 06	1.278 T	PM 12	1.309 T
P4	PM 07	1.294 T	PM 03	1.290 T
	PM 08	1.295 T	PM 04	1.291 T
P5	PM 09	1.278 T	PM 07	1.294 T
	PM 10	1.290 T	PM 08	1.295 T
P6	PM 11	1.309 T	PM 15	1.310 T
	PM 12	1.309 T	PM 16	1.315 T
P7	PM 13	1.287 T	PM 09	1.278 T
	PM 14	1.285 T	PM 10	1.290 T
P8	PM 15	1.310 T	PM 01	1.265 T
	PM 16	1.315 T	PM 02	1.275 T
Z_1-Z_2	−0.483 (T)2		0.009 (T)2	
B_N-B_S	−0.187 (T)		0.004 (T)	

Now, the validity of the proposed method can be verified by examining the variation of the cogging torque harmonic component in each case. The FEA results of the cogging torque harmonic component are shown in Figure 8. Figure 8a shows the result of the 6-pole/9-slot model, and Figure 8b shows the result of the 8-pole/12-slot model. In both models, it can be seen that the slot harmonic component of Case B is much smaller than Case A. Therefore, the proposed method was effective in mitigating the slot harmonic component of cogging torque. Furthermore, it can be seen that the permeability of the core is in a somewhat non-linear region by observing the flux density distribution in Figure 9. Hence, although we ignored the saturation when deriving the method, the result of Figure 8 proves that the proposed method is still effective under the non-linear material characteristics of the ferromagnetic.

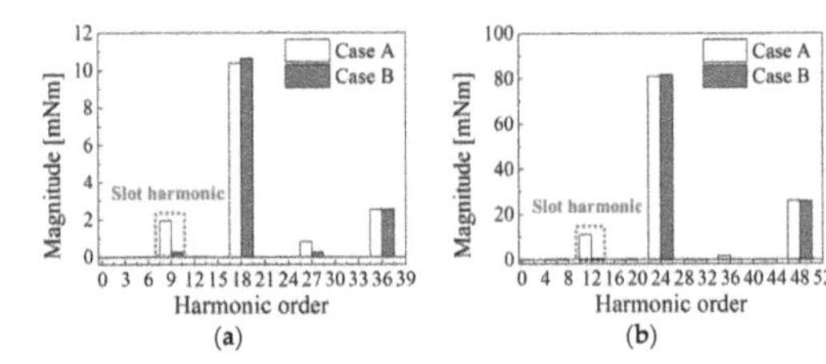

Figure 8. FEA results of cogging torque harmonic component according to each case: (**a**) 6p/9s IPM and (**b**) 8p/12s IPM.

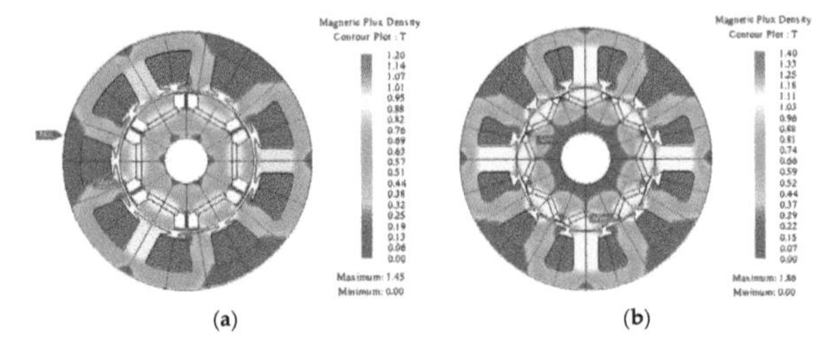

Figure 9. Contour plot of magnetic flux density under no load condition for (**a**) 6p/9s IPM and (**b**) 8p/12s IPM.

4.2. Verification with Experimentation

For the experimental verification, both models in Figure 6 were manufactured, one of each. Figure 10 is a picture of the produced motor. Then, the experiment process was performed as follows.

Figure 10. The manufactured motors for the experiment: (**a**) 6p/9s IPM and (**b**) 8p/12s IPM.

1. The surface Gauss value of each magnet was measured (with ATM 1000, SCMI) in the space, excluding the magnetic substance. Figure 11 shows a picture of the measurement, and the results are written in Tables 5 and 6. The Gauss average value was calculated from the seven measurement points per each magnet, and the measurement uncertainty was calculated by repeating the measurement five times.
2. The position of each magnet was set according to the proposed method. The results are shown in Tables 5 and 6 and in Figure 12 (Case B).
3. These magnets were alternately assembled to the rotor according to the case of each model shown in Figure 12, and the cogging torque according to each case was measured (with ATM-5KA, SUGAWARA) as shown in Figure 13.

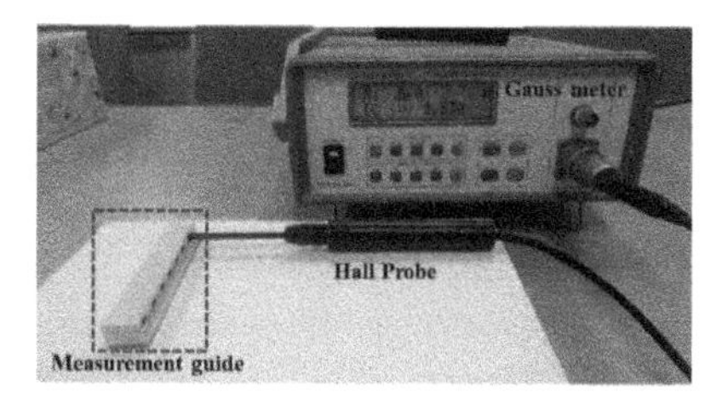

Figure 11. Gauss measurement of the magnet surface.

Table 5. Gauss measurement of each magnet surface of the 6p/9s model and the changes in magnet position according to the proposed method.

Position & Comparison	Case A		Case B	
	PM No.	Gauss Avg.	PM No.	Gauss Avg.
P1	PM 01	194.3 mT	PM 04	198.4 mT
P2	PM 02	198.3 mT	PM 02	198.3 mT
P3	PM 03	196.7 mT	PM 05	196.9 mT
P4	PM 04	198.4 mT	PM 06	199.1 mT
P5	PM 05	196.9 mT	PM 03	196.7 mT
P6	PM 06	199.1 mT	PM 01	194.3 mT
Uncertainty	±0.2%			
Z_1-Z_2	−3113.3 (mT)2		106.9 (mT)2	
B_N-B_S	−7.9 (mT)		0.3 (mT)	

The results of cogging torque measurements are demonstrated in Figure 14. Figure 14a shows the result of the 6-pole/9-slot motor. Case A had a cogging torque of 56.7 $mNm_{pk\text{-}pk}$, and Case B had 55.1 $mNm_{pk\text{-}pk}$. In Figure 14b the 8-pole/12-slot motor showed 227.3 $mNm_{pk\text{-}pk}$ for Case A and 214.8 $mNm_{pk\text{-}pk}$ for Case B. As a result, although the shapes of the stator and rotor of the analyzed motors were already optimized for reducing cogging torque, the cogging torque could be improved more by using the proposed method. The main cause of this cogging difference between Case A and B is due to the slot harmonic component of Case B being smaller than Case A, as can be seen in the FFT result of each cogging torque in Figure 15. Consequently, the validity of the proposed method was confirmed again by the experimental results. Overall, since this method only affects the position of each magnet before assembly, it can be compatible with the conventional cogging torque reduction methods using the teeth curvature and rotor shape modulation.

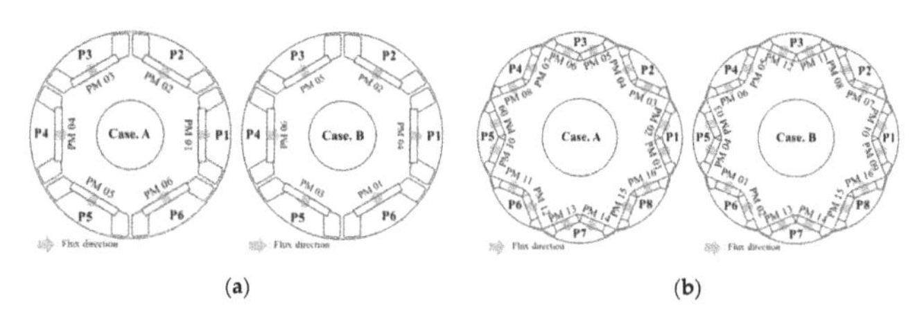

Figure 12. PM position change according to the proposed method for the experimental verification for (**a**) 6p/9s IPM and (**b**) 8p/12s IPM.

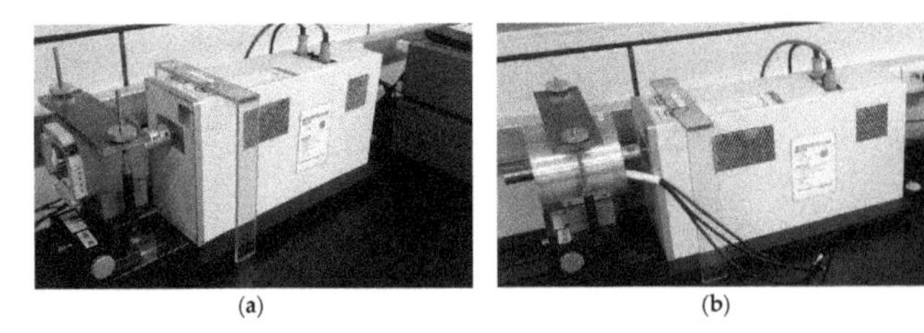

Figure 13. Cogging torque measurement of each motor for (**a**) 6p/9s IPM and (**b**) 8p/12s IPM.

Table 6. Gauss measurement of each magnet surface of the 8p/12s model and the changes in magnet position according to the proposed method.

Position & Comparison	Case A		Case B	
	PM No.	Gauss Avg.	PM No.	Gauss Avg.
P1	PM 01	227.1 mT	PM 09	225.5 mT
	PM 02	228.7 mT	PM 10	226.2 mT
P2	PM 03	233.8 mT	PM 07	231.1 mT
	PM 04	234.0 mT	PM 08	231.5 mT
P3	PM 05	226.3 mT	PM 11	230.6 mT
	PM 06	226.8 mT	PM 12	230.9 mT
P4	PM 07	231.2 mT	PM 05	226.3 mT
	PM 08	231.5 mT	PM 06	226.8 mT
P5	PM 09	225.5 mT	PM 03	233.9 mT
	PM 10	226.2 mT	PM 04	234.0 mT
P6	PM 11	230.6 mT	PM 01	227.1 mT
	PM 12	230.9 mT	PM 02	228.7 mT
P7	PM 13	229.0 mT	PM 13	229.0 mT
	PM 14	229.9 mT	PM 14	229.9 mT
P8	PM 15	234.2 mT	PM 15	234.2 mT
	PM 16	234.3 mT	PM 16	234.2 mT
Uncertainty	±0.1%			
Z_1-Z_2	−10,065.1 (mT)2		133.9 (mT)2	
B_N-B_S	−21.9 (mT)		0.3 (mT)	

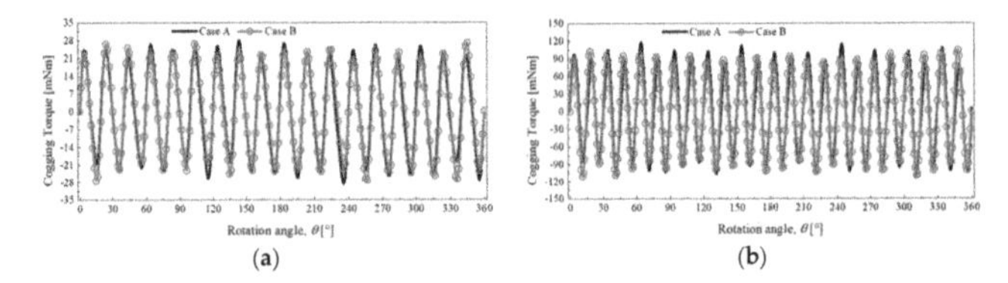

Figure 14. The measured cogging torque of each motor for (**a**) 6p/9s IPM and (**b**) 8p/12s IPM.

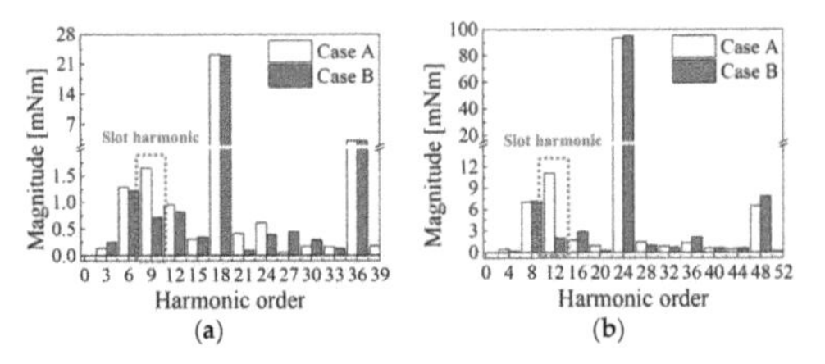

Figure 15. Cogging torque measurement of each motor for (**a**) 6p/9s IPM and (**b**) 8p/12s IPM.

5. Discussion

The reduction effect cannot be clearly seen in the peak-peak comparison of cogging torque in Figure 14. This is because the LCM component is much larger than the slot harmonic component in both cases. In this case, although the slot harmonic component was reduced, as shown in Figure 15, by the proposed method, the effect is not seen much. If the proposed method is applied to a model that is sensitive to the slot harmonic component, the cogging torque can be effectively mitigated, compared with the results of this paper. In other words, the proposed method has a different effect on the mitigation of cogging torque depending on which harmonic component is dominant.

Additionally, since there are some methods to measure the B_r or flux density of PM, the real application for applying the proposed method can be manufactured. Among the measurement methods, the simplest example is using Helmholtz coil. As mentioned in the introduction, the proposed method is more appropriate for small quantity customized production than mass production because the Gauss value of each magnet should be measured before the assembly. In the case of mass production, it is possible that if the manufacturing tolerance of the magnetization yoke is adjusted based on the principle of the slot harmonic component mitigation condition, that the influence of the uneven magnetization can be alleviated.

As described above, there are some limitations to the proposed method. However, it is meaningful that we have dealt with the method to compensate manufacturing tolerance (Uneven PM) that has not been covered in the meantime. Furthermore, this method can prevent an increase in cogging torque caused by unevenly magnetized PMs of motors with a high number of poles. Since small scale customized manufacturing process, which adopts the method of the pre-magnetization of magnets before assembly, cannot adjust and compensate for the unevenness of the PMs, by using the proposed method, it will be possible to ensure the cogging performance of a manufactured motor.

6. Conclusions

In this study, a mitigation method of slot harmonic cogging torque caused by the unevenly magnetized magnet was proposed. This method was drawn through the qualitative analysis of the cogging torque from a macroscopic perspective. As shown in Figure 5, the main process of this method is arranging each magnet according to the non-slot harmonic condition described in Section 3. The validity of the proposed method was verified using FEA and experimentation. Here, the verification was performed by comparing the harmonic components of the cogging torque with and without the proposed method. In this process, it was confirmed that this method is sufficiently effective, even when considering the non-linear material characteristics of the ferromagnetic.

Author Contributions: Conceptualization, C.J. and D.L.; methodology, C.J.; software, D.L.; validation, C.J., and D.L.; formal analysis, C.J.; investigation, C.J. and D.L.; resources, C.J.; data curation, C.J.; writing—original draft preparation, C.J. and D.L.; writing—review and editing, C.J. and J.H.; visualization, C.J.; supervision, J.H.; project administration, J.H.

Funding: This research received no external funding.

Acknowledgments: This work was supported by the Incheon National University under Research Grant 2019-0254 (Corresponding author: Jin Hur).

Conflicts of Interest: The authors declare no conflict of interest.

References

1. Li, G.J.; Ren, B.; Zhu, Z.Q.; Li, Y.X.; Ma, J. Cogging torque mitigation of modular permanent magnet machines. *IEEE Trans. Magn.* **2016**, *52*, 1–10. [CrossRef]
2. Park, Y.U.; Cho, J.H.; Kim, D.K. Cogging torque reduction of single-phase brushless DC motor with a tapered air-gap using optimizing notch size and position. *IEEE Trans. Ind. Appl.* **2015**, *51*, 4455–4463. [CrossRef]

3. Xue, Z.; Li, H.; Zhou, Y.; Ren, N.; Wen, W. Analytical prediction and optimization of cogging torque in surface-mounted permanent magnet machines with modified particle swarm optimization. *IEEE Trans. Ind. Electron.* **2017**, *64*, 9795–9805. [CrossRef]
4. Ren, W.; Xu, Q.; Li, Q.; Zhou, L. Reduction of cogging torque and torque ripple in interior PM Machines with asymmetrical V-type rotor design. *IEEE Trans. Magn.* **2016**, *52*, 1–5. [CrossRef]
5. Dosiek, L.; Pillay, P. Cogging torque reduction in permanent magnet machines. *IEEE Trans. Ind. Appl.* **2007**, *43*, 1565–1571. [CrossRef]
6. Kim, K. A novel method for minimization of cogging torque and torque ripple for interior permanent magnet synchronous motor. *IEEE Trans. Magn.* **2014**, *50*, 793–796. [CrossRef]
7. Wang, D.; Wang, X.; Jung, S. Cogging torque minimization and torque ripple suppression in surface-mounted permanent magnet synchronous machines using different magnet widths. *IEEE Trans. Magn.* **2013**, *49*, 2295–2298. [CrossRef]
8. Wanjiku, J.; Khan, M.A.; Barendse, P.S.; Pilay, P. Influence of slot openings and tooth profile on cogging torque in axial-flux pm machines. *IEEE Trans. Ind. Electron.* **2015**, *62*, 7578–7589. [CrossRef]
9. Kwon, J.; Lee, J.; Zhao, W.; Kwon, B. Flux-switching permanent magnet machine with phase-group concentrated-coil windings and cogging torque reduction technique. *Energies* **2018**, *11*, 2758. [CrossRef]
10. Hwang, M.; Lee, H.; Cha, H. Analysis of torque ripple and cogging torque reduction in electric vehicle traction platform applying rotor notched design. *Energies* **2018**, *11*, 3053. [CrossRef]
11. Dini, P.; Saponara, S. Cogging torque reduction in brushless motors by a nonlinear control technique. *Energies* **2019**, *12*, 2224. [CrossRef]
12. Kim, J.M.; Yoon, M.H.; Hong, J.P.; Kim, S.I. Analysis of cogging torque caused by manufacturing tolerances of surface-mounted permanent magnet synchronous motor for electric power steering. *IEEE Trans. Electr. Power Appl.* **2016**, *10*, 691–696. [CrossRef]
13. Ortega, A.J.P.; Paul, S.; Islam, R.; Xu, L. Analytical model for predicting effects of manufacturing variations on cogging torque in surface-mounted permanent magnet motors. *IEEE Trans. Magn.* **2016**, *52*, 3050–3061. [CrossRef]
14. Qian, H.; Guo, H.; Wu, Z.; Ding, X. Analytical solution for cogging torque in surface-mounted permanent-magnet motors with magnet imperfections and rotor eccentricity. *IEEE Trans. Magn.* **2014**, *50*, 1–15. [CrossRef]
15. Zhou, Y.; Li, H.; Meng, G.; Zhou, S.; Cao, Q. Analytical calculation of magnetic field and cogging torque in surface-mounted permanent-magnet machines accounting for any eccentric rotor shape. *IEEE Trans. Ind. Electron.* **2015**, *62*, 3438–3447. [CrossRef]
16. Lee, C.J.; Jang, G.H. Development of a new magnetizing fixture for the permanent magnet brushless dc motors to reduce the cogging torque. *IEEE Trans. Magn.* **2011**, *47*, 2410–2413. [CrossRef]
17. Song, J.Y.; Kang, K.J.; Kang, C.H.; Jang, G.H. Cogging torque and unbalanced magnetic pull due to simultaneous existence of dynamic and static eccentricities and uneven magnetization in permanent magnet motors. *IEEE Trans. Magn.* **2014**, *50*, 1–9. [CrossRef]
18. Sung, S.J.; Park, S.J.; Jang, G.H. Cogging torque of brushless DC motors due to the interaction between the uneven magnetization of a permanent magnet and teeth curvature. *IEEE Trans. Magn.* **2011**, *47*, 1923–1928. [CrossRef]
19. Ou, J.; Liu, Y.; Qu, R.; Doppelbauer, M. Experimental and theoretical research on cogging torque of PM synchronous motors considering manufacturing tolerances. *IEEE Trans. Ind. Electron.* **2018**, *65*, 3772–3783. [CrossRef]
20. Ionel, D.M.; Popescu, M.; McGilp, M.I.; Miller, T.J.E.; Dellinger, S.J. Assessment of torque components in brushless permanent-magnet machines through numerical analysis of the electromagnetic field. *IEEE Trans. Electr. Power Appl.* **2005**, *41*, 1149–1158. [CrossRef]
21. Zhu, W.; Pekarek, S.; Fahimi, B.; Deken, B.J. Investigation of force generation in a permanent magnet synchronous machine. *IEEE Trans. Energy Convers.* **2007**, *22*, 557–565. [CrossRef]
22. Meessen, K.J.; Paulides, J.J.H.; Lomonova, E.A. Force calculations in 3-D cylindrical structures using fourier analysis and the Maxwell stress tensor. *IEEE Trans. Magn.* **2013**, *49*, 536–545. [CrossRef]
23. Gao, J.; Wang, G.; Liu, X.; Zhang, W.; Huang, S.; Li, H. Cogging torque reduction by elementary-cogging-unit shift for permanent magnet machines. *IEEE Trans. Magn.* **2017**, *53*, 1–5. [CrossRef]

24. Zhu, Z.Q.; Ruangsinchaiwanich, S.; Howe, D. Synthesis of cogging-torque waveform from analysis of a single stator slot. *IEEE Trans. Ind. Appl.* **2006**, *42*, 650–657. [CrossRef]
25. Zhu, Z.Q.; Ruangsinchaiwanich, S.; Chen, Y.; Howe, D. Evaluation of superposition technique for calculating cogging torque in permanent-magnet brushless machines. *IEEE Trans. Magn.* **2006**, *42*, 1597–1603. [CrossRef]
26. Gieras, J.F. *Electrical Machines: Fundamentals of Electromechanical Energy Conversion*; CRC Press: Boca Raton, FL, USA, 2016.

Article

Study on the Electromagnetic Design and Analysis of Axial Flux Permanent Magnet Synchronous Motors for Electric Vehicles

Jianfei Zhao *, Qingjiang Han, Ying Dai and Minqi Hua

School of Mechatronic Engineering and Automation, Shanghai University, Baoshan District, Shanghai 200444, China
* Correspondence: jfzhao@shu.edu.cn; Tel.: +86-021-6613-0935

Received: 30 July 2019; Accepted: 3 September 2019; Published: 6 September 2019

Abstract: In order to provide a complete solution for designing and analyzing the axial flux permanent magnet synchronous motor (AFPMSM) for electric vehicles, this paper covers the electromagnetic design and multi-physics analysis technology of AFPMSM in depth. Firstly, an electromagnetic evaluation method based on an analytical algorithm for efficient evaluation of AFPMSM was studied. The simulation results were compared with the 3D electromagnetic field simulation results to verify the correctness of the analytical algorithm. Secondly, the stator core was used to open the auxiliary slot to optimize the torque ripple of the AFPMSM, which reduced the torque ripple peak-to-peak value by 2%. From the perspective of ensuring the reliability, safety, and driving comfort of the traction motor in-vehicle working conditions, multi-physics analysis software was used to analyze and check the vibration and noise characteristics and temperature rise of several key operating conditions of the automotive AFPMSM. The analysis results showed that the motor designed in this paper can operate reliably.

Keywords: AFPMSM; analytical algorithm; torque ripple; vibration noise; temperature field analysis

---

1. Introduction

Drive motors used in new energy vehicles require frequent start-ups and shutdowns, are subject to large accelerations or decelerations, and require high-speed, low-torque mode operation. Compared with radial flux permanent magnet synchronous motors (RFPMSM) [1–4], axial flux permanent magnet synchronous motors (AFPMSM) have the advantages of compact axial structure and high power density, which are suitable for new energy vehicles. As early as 2005, researchers have studied AFPMSMs. Methods of finite-element analysis and theoretical analysis were combined into a multi-dimensional optimization program to optimize the design of high-power coreless stator AFPM generators in 2005 [5]. In 2006, the author of Reference [6] studied the torque ripple component of permanent magnet motors and proposed a method to minimize the torque ripple of surface permanent magnet motors. In order to ensure the reliability, safety, and ride comfort of the AFPMSM for electric vehicles, it is also possible to find an effective and reliable method from the control strategy. In Reference [7], a sliding mode vector control system based on cooperative control is established to drive AFPMSM. In Reference [8], the current control method is used to drive the AFPMSM, and internal model control was introduced. This paper focuses on the design and optimization of the AFPMSM for electric vehicles.

The design of AFPMSM usually requires 3D finite element method (FEM) analysis. Considering that multiple design variables need to be analyzed simultaneously, the time to design AFPMSM is significantly increased. The analytical algorithms to simplify analysis time are used to solve this problem [9–11]. A surrogate assisted multi-objective optimization algorithm was applied to significantly reduce function calls, effectively reducing motor design time [9]. Under the same magnetic energy of the two models' permanent magnets, an AFPM motor is equivalent to a linear synchronous permanent magnet motor [10]. In order to obtain the key parameters of accurate surface mount permanent magnet synchronous motors, an analysis model based on Maxwell's equation and magnetic equivalent circuit was introduced [11]. The analysis algorithm can effectively reduce the development time when designing the motor. The AFPMSM for electric vehicles has a complicated structure, and the use of FEM will greatly increase the analysis time. For AFPMSM, its structure is complex, and the analytical algorithm can effectively reduce the development time in motor design. Therefore, the electromagnetic design method based on analytical formula studied in this paper has certain advantages.

Similar to RFPMSMs, AFPMSMs also have a torque ripple problem during operation. The existence of torque ripple will cause torsional vibration of the transmission system, which will adversely affect the motor control and torque output quality. Therefore, it is necessary to analyze and optimize its torque characteristics. The authors of Reference [12,13] optimized the torque ripple from the control method. The authors of Reference [14,15] optimized the torque ripple from the motor itself. In this paper, the stator core was used to open the auxiliary slot to optimize the torque ripple.

In recent years, AFPMSM's multi-physics analysis has also made great progress. The authors of Reference [16–18] used the steady-state heat conduction equation method, the T-type thermal network model method, and the AFPMSM full prediction heat transfer coefficient model with magnet geometry parameters to perform thermal analysis on AFPMSM. The authors of Reference [19–22] conducted a detailed study of the AFPMSM's degaussing fault identification. The method was proposed for detecting and locating asymmetric demagnetization defects in AFPMSM, which can be used for real-time condition monitoring of demagnetization defects or as a virtual temperature sensor for magnets. [19]. A time harmonic analysis model was proposed to study the demagnetization and rotor eccentricity of a single stator dual rotor AFPMSM [20]. A forward model was established for demagnetization detection of an AFPMSM. [21]. In addition, the eddy current losses of AFPMSMs have been analyzed in detail in Reference [22–24]. Because electric vehicles require high system efficiency, strong system environment adaptability and low noise, based on these characteristics, the current study analyzed and checked the vibration noise and temperature rise of several key operating conditions of automotive AFPMSMs based on multi-physics analysis software to ensure that the motor can always run smoothly.

In summary, this paper covers the study and design of a dual-stator-single-rotor structure of the new energy vehicle traction AFPMSM (the motor parameters are shown in Table 1; the motor structure diagram is shown in Figure 1). The AFPMSM for electric vehicles was based on a fast electromagnetic design method based on analytical formulas and a refined optimization design simulation technology based on machine-electric-thermal-structure multi-physics analysis, providing a complete solution for designing and optimizing AFPMSM for electric vehicles.

Figure 1. Double stator-single rotor type axial flux permanent magnet synchronous motor (AFPMSM).

Table 1. Motor parameters.

Motor Parameters	Symbol	Value
Pole-pairs	p	6
Slot number	Q	27 × 2
Air gap length	δ	1 mm
Rated torque	T_N	300 Nm
Rated power	P_N	50 kW
Stator outer diameter	D_{so}	250 mm
Phase number	m	3
Rated speed	n	1600 rpm
Permanent magnet Remanence	B_r	1.089 T
Coercivity	HC	2 A/m

2. Magnetic Path Analysis Algorithm for AFPMSM

2.1. Modeling and Analysis of Magnetic Circuit Analysis Algorithm

The AFPMSM has a unique magnetic circuit structure, and its electromagnetic design is quite different from that of the RFPMSM. In this paper, the model of AFPMSM was simplified by the 2D multi-loop equivalent method. Based on this, the magnetic circuit analysis algorithm was used to design the motor and obtain the electromagnetic parameters of the motor. The specific method is as follows. Take the motor stator N different diameters and expand it into a linear motor, and then superimpose the simulation results of N linear motors to get the final result, as shown in Figure 2. The radius D_{iave} taken by the i-th linear motor is as follows:

$$D_{iave} = D_{out} - \frac{2i-1}{N}\frac{D_{out} - D_{in}}{2}, i = 1, 2, \cdots, N \quad (1)$$

where D_{out} and D_{in} are the outer diameter and inner diameter of the stator.

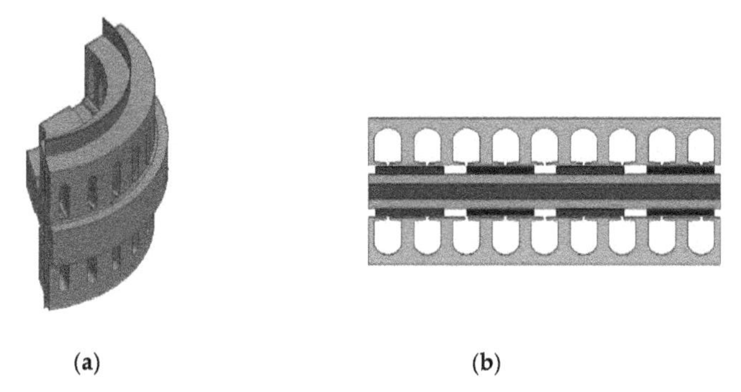

Figure 2. Double stator-single rotor type AFPMSM. (**a**) AFPMSM 3D model. (**b**) Equivalent linear motor model.

When calculating the value of the electromagnetic field under ideal conditions, take $N = 1$. When calculating the loss and other information, take $N > 1$.

Then, use the equivalent magnetic network circuit as shown in Figure 3 to calculate the key electromagnetic parameters of the motor. In Figure 3, R_y is the equivalent reluctance of the stator yoke, R_t is the magnetic reluctance of the tooth, F_l is the magneto motive force generated by the stator winding, R_s is the slot reluctance, R_g is the air gap reluctance, and F_{PM} is the magneto motive force generated by the permanent magnet.

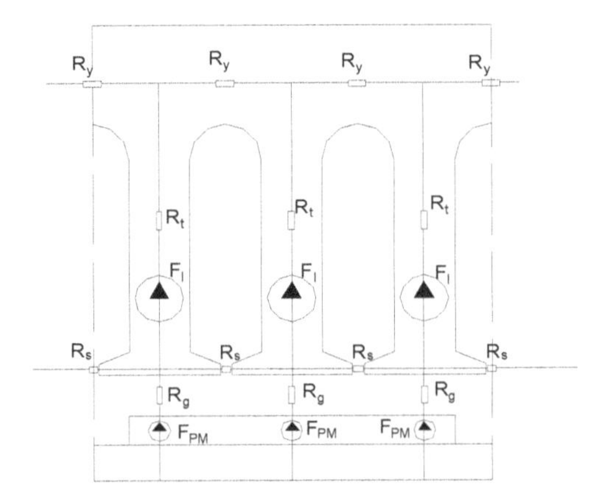

Figure 3. Equivalent magnetic network model of the AFPMSM.

The AFPMSM is equivalent to a linear motor in this paper, and the magnetic circuit analysis method (MCAM) model was used to analyze the air gap magnetic density distribution of AFPMSM [25–28]:

$$B_{mg}(r,x) = \sum_{n=1,3,5,\cdots}^{\infty} \frac{-\frac{8B_r}{n\pi}\sin\left(\frac{\alpha_p(r)n\pi}{2}\right)e^{\frac{-n\pi g'}{\tau_p(r)}}\cos\left(\frac{n\pi x}{\tau_p(r)}\right)}{\left(e^{\frac{-n\pi g'}{\tau_p(r)}}+1\right)+\frac{\mu_{PM}\left(e^{\frac{-n\pi g'}{\tau_p(r)}}+1\right)\left(e^{\frac{2n\pi l_{PM}}{\tau_p(r)}}+1\right)}{\mu_0\left(e^{\frac{2n\pi l_{PM}}{\tau_p(r)}}-1\right)}} \tag{2}$$

where r is the radius and x is the perimeter at different radii and is equal to $2\pi r$. τ_p(r) is the pole pitch of the permanent magnet when the radius is r, and $\tau_p(r) = \pi r/p$. $\alpha_p(r)$ is the polar arc coefficient when the radius is r, and $\alpha_p(r) = w_{PM}(r)/\tau_p(r)$, where $w_{PM}(r)$ is the width of the permanent magnet when the radius is r. l_{PM} is the thickness of the permanent magnet, μ_0 is the vacuum permeability, μ_{PM} is the permanent magnet permeability, and g' is the air gap length.

Since the open slot affects the air gap flux density, the air gap permeability function $\lambda(x)$ is introduced to consider the effect of the open slot on the air gap flux distribution. The air gap magnetic density distribution $B_{mag}(r,x)$ including the slotting effect is

$$B_{mag}(r,x) = \lambda(x)B_{mg}(r,x) \tag{3}$$

when the motor is used as a drive motor; the no-load phase voltage is generated only by the flux generated by the permanent magnet. According to Equations (2) and (3), it can be calculated as:

$$e_{i,PM} = -N_{ph}k_{db}\frac{\Delta\Phi_{i,PM}}{\Delta t} \tag{4}$$

where N_{ph} is the number of turns in series for each phase, K_{db} is the fundamental winding factor, and $\Phi_{i,\,PM}$ is the air gap flux obtained by numerical integration of the air gap flux density distribution,

$$\Phi_{i,PM} = \int_0^{\tau_{p,i}}\int_0^{\frac{l_s}{N}} B_{agap,i}(x)dxdl \tag{5}$$

where $\tau_{p,i}$ is the permanent magnet pole pitch of the i-th equivalent plane, l_s is the length of the stator punch and $l_s = (D_{out}-D_{in})/2$, and $B_{agap,i}(x)$ is the air gap magnetic density function in the i-th equivalent plane.

Therefore, the no-load back electromotive force (EMF) of the AFPMSM is

$$e_{PM}(t) = \sum_{i=1}^{N} e_{i,PM}(t) \tag{6}$$

This paper presents a method to reduce the cogging torque. The cogging torque is calculated from the no-load air gap flux density distribution. Assume that the cogging torque of the axial flux motor unit is

$$T_{cog,i}(\theta) = \frac{\partial W}{\partial \theta} = \frac{\partial\left(\iiint\limits_{V} B_{agap,i}^{2} dV\right)}{2\mu_0 \partial \theta} \tag{7}$$

where V is the air gap volume, θ is the rotor position angle, and W is magnetic energy.

During calculation of the cogging torque, it is assumed that the magnetic flux density distribution of each axial flux motor unit does not change in the radial direction. Analogizing the back EMF calculation, the cogging torque generated by the axial flux motor is

$$T_{cog}(\theta) = \sum_{i=1}^{N} T_{cog,i}(\theta) \tag{8}$$

2.2. Comparative Analysis of MCAM and 3D FEM

As described in Section 2.1, based on Matrix Laboratory programming, the magnetic circuit analysis method compared with the finite element method as follows.

The air gap magnetic density of the motor has a crucial influence on the saturation degree of the motor, the output power requirement, and the loss level. Therefore, it is important to analyze the air gap magnetic density of the AFPMSM. Figure 4 shows that the air-gap magnetic density distribution waveforms calculated by the FEM and the MCAM are basically the same, and the FEM can more accurately consider the cogging and magnetic flux leakage of the motor, so the calculation accuracy is higher than the analytical method. However, the MCAM takes a short time and can be used for screening the initial design of the AFPMSM.

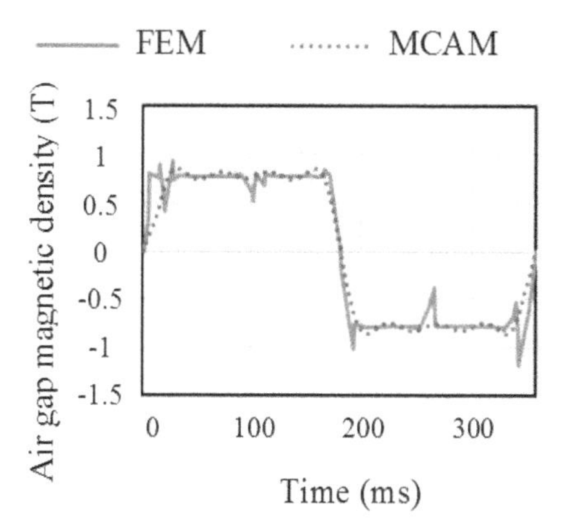

Figure 4. No-load air gap magnetic density of AFPMSM.

Figure 5 is a comparison of the analytical results of the FEM and MCAM of the AFPMSM back EMF. It can be seen from Figure 6a,b that the MCAM has omissions in the analysis of the third harmonic magnetic density, and the analysis results of the fundamental wave and other harmonics are in good agreement with the finite element method.

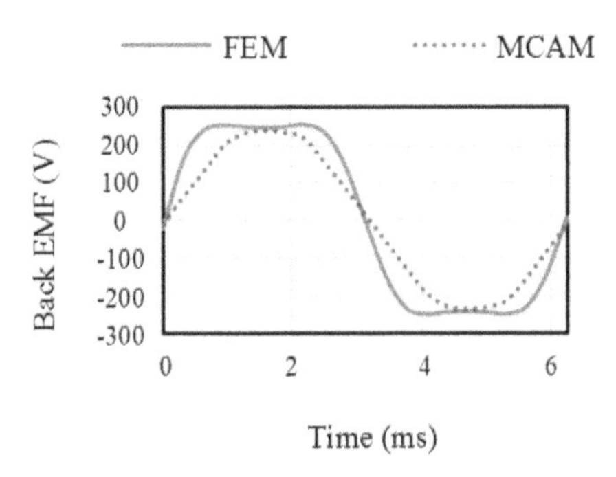

Figure 5. No-load back electromotive force (EMF) of the AFPMSM.

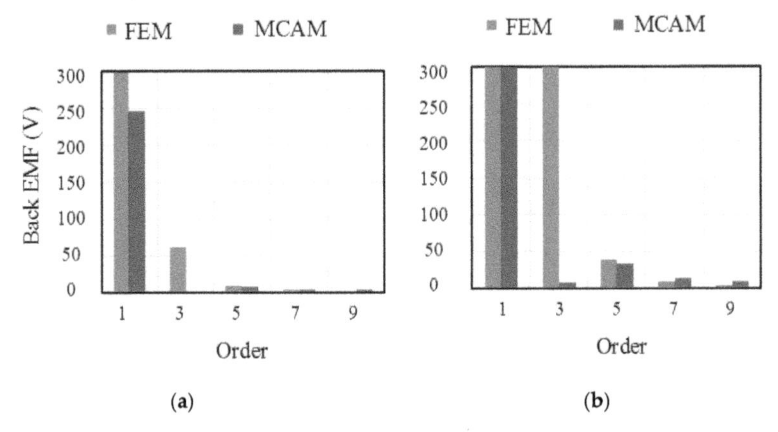

Figure 6. Comparison of back-EMF harmonic analysis results of the AFPMSM. (**a**) Overall picture. (**b**) Partial enlargement picture.

Figure 7 shows that that the MCAM is basically consistent with the analysis of the positioning torque and the finite element analysis results, and the maximum amplitude difference is 0.2 Nm, which satisfies the initial design accuracy requirements of the motor.

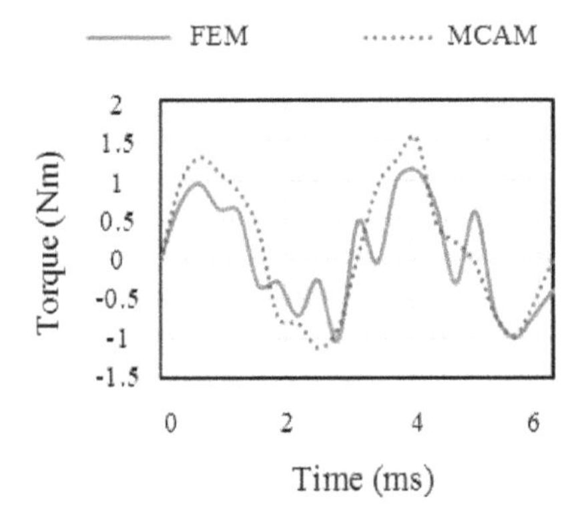

Figure 7. Analysis of positioning torque of AFPMSM.

In summary, the air gap magnetic density, the no-load back EMF, and the positioning torque calculated by the FEM and the MCAM are basically the same. Since the FEM can fully consider the motor cogging effect and the magnetic flux leakage coefficient, the accuracy is higher than the MCAM. However, when the distribution trend is basically the same, the time used by the MCAM is much less than that of the FEM. Therefore, the MCAM studied in this paper can be used to screen the initial design of the AFPMSM.

3. Torque Ripple Optimization of AFPMSM

Electric vehicles generally operate in torque mode. The output of the motor torque is controlled according to the depth at which the driver steps on the throttle. The presence of torque ripple adversely affects the torque output. Therefore, the torque ripple needs to be optimized. The effect of opening the auxiliary slot on the motor stator on the torque ripple is equivalent to increasing the number of armature slots, i.e., changing the pole slot fit. If the position and size of the open auxiliary slot are properly selected, the torque ripple of the motor can be weakened. First, theoretical analysis is used to determine the number of auxiliary slots, slot width, and position distribution on each tooth, and then the slot depth is continuously adjusted in the three-dimensional simulation. The slot depth should suppress the torque ripple as much as possible without affecting the performance of the motor. The number of auxiliary slots is derived below. The motor in this paper has 12 poles and 27 slots. According to the literature [23]:

$$\{i(k+1) | i \in N^{*}\} \tag{9}$$

$$\left\{ i \cdot \frac{2p}{GCD(z, 2p)} \middle| i \in N^{*} \right\} \tag{10}$$

where p is the pole pairs, z is the number of stator slots, and $GCD(z, 2p)$ is the greatest common divisor (GCD) of z and $2p$.

To attenuate torque ripple, the number k of stator toothed auxiliary slots should be such that the intersection of set (9) and set (10) is reduced. The AFPMSM studied in this paper has $p = 6$, $z = 27$, and $k = 2$, so the two sets are $\{3, 6, 9 \ldots\}$ and $\{4, 8, 12 \ldots\}$. Obviously, there is no intersection between the two sets, so the torque ripple is significantly reduced compared to the absence of the auxiliary slot. If $k = 1$, the intersection of the above two sets is $\{4, 8, 12 \ldots\}$. In summary, $k = 2$ and the width of the auxiliary slot is equal to the slot width.

In summary, the motor of the present invention opens two auxiliary slots, the auxiliary slots are evenly distributed on the stator teeth, and the width of the auxiliary slots is the same as the width of the stator slots. The finite element analysis model is shown in Figure 8a, and a partial enlarged view of the auxiliary slot is shown in Figure 8b.

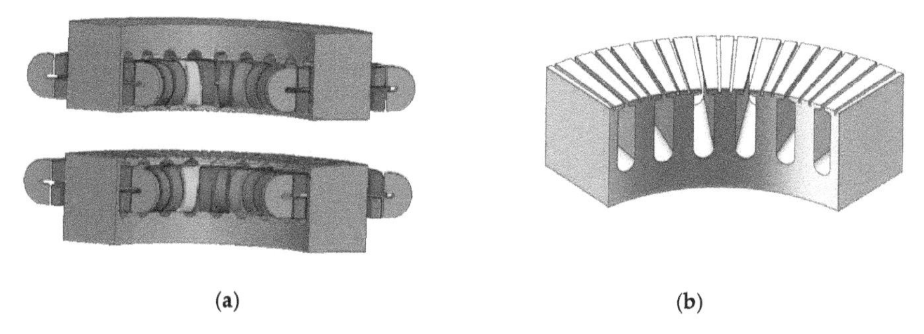

Figure 8. AFPMSM auxiliary slot schematic. (**a**) Finite element analysis model. (**b**) Partial enlarged view of the auxiliary slot.

In order to verify the feasibility of opening the auxiliary slot to suppress the torque ripple, this section analyzes the load back EMF and load torque of constant torque maximum speed operating point and constant power zone maximum speed operating point under rated load and peak load before and after optimization. The maximum speed of the constant torque is 1600 rpm, and the maximum speed of the constant power is 3000 rpm.

3.1. Comparative Analysis of Rated Load Characteristics at 1600 rpm

Under rated load conditions, it can be seen from Figure 9 that the third, fifth, seventh, and ninth harmonics of the load back electromotive force of the auxiliary slot method are weakened when the rotational speed is 1600 rpm; the weakening of the fifth and seventh harmonics is more significant.

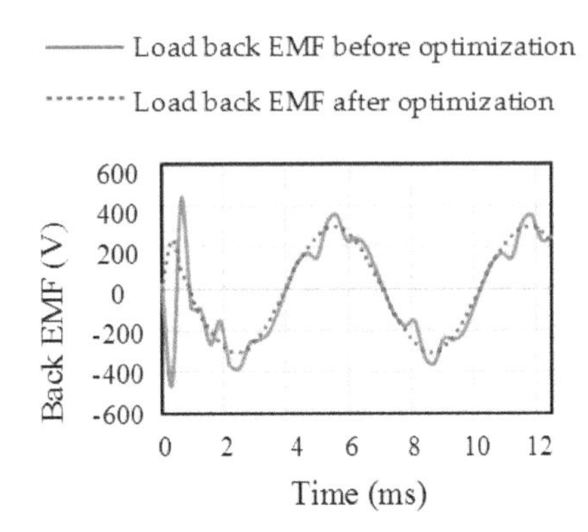

Figure 9. Back EMF of the AFPMSM at the rated load.

Figure 10 shows that when the rotation speed is 1600 rpm, the peak-to-peak load torque of the auxiliary slot method is 10 Nm, and the peak-to-peak value of the load torque without the auxiliary slot method is 16 Nm. The peak-to-peak value of the torque ripple after opening the auxiliary slot is reduced by about 4%, so the optimization method of opening the auxiliary slot at rated load can effectively attenuate the torque ripple under constant torque conditions.

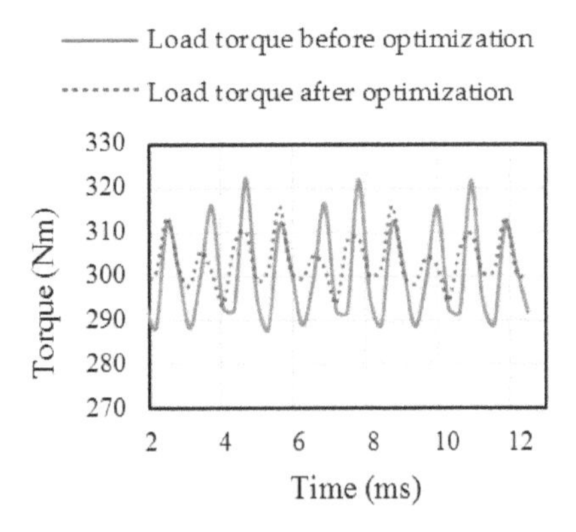

Figure 10. Torque of the AFPMSM at the rated load.

3.2. Comparative Analysis of Peak Load Characteristics at 1600 Rpm

Under the peak load condition, it can be seen from Figure 11 that the seventh and ninth harmonics of the load back electromotive force of the auxiliary slot method are weakened when the rotation speed is 1600 rpm; the third and fifth harmonics are higher than the no auxiliary slot method.

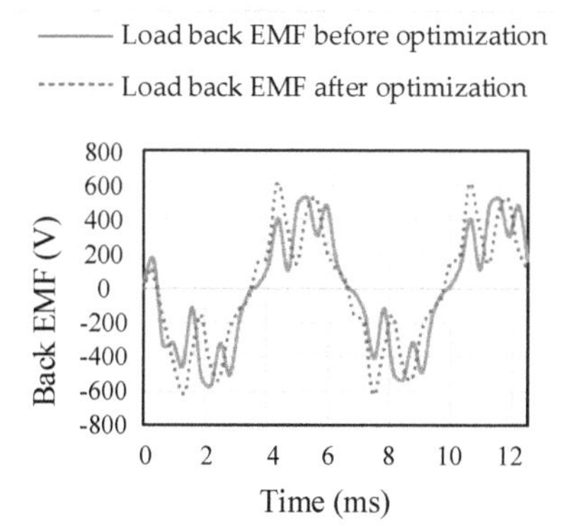

Figure 11. Back EMF of the AFPMSM at the peak load.

Figure 12 shows that the load torque peak-to-peak value of the open auxiliary slot method is 26 Nm when the rotation speed is 1600 rpm, and the load torque peak-to-peak value of the no auxiliary slot method is 32 Nm. The torque ripple of the open auxiliary slot method is reduced by about 2%, so the optimization method of opening the auxiliary slot at the peak load can also effectively attenuate the torque ripple under constant torque conditions.

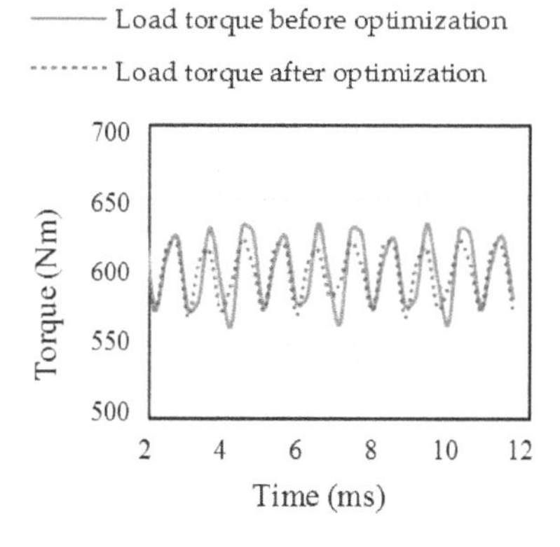

Figure 12. Torque of the AFPMSM at the peak load.

3.3. *Comparative Analysis of Rated Load Power Characteristics at 3000 rpm*

Under the rated load condition, as can be seen from Figure 13, when the rotational speed is 3000 rpm, the fifth, seventh, and ninth harmonics of the load back EMF of the auxiliary slot method are all weakened; the weakening of the fifth, seventh, and ninth harmonics is more significant.

Figure 14 shows that the load torque peak-to-peak value of the auxiliary slot optimization method is 3.5 Nm when the rotation speed is 3000 rpm, and the peak-to-peak value of the load torque before optimization is 10 Nm. The torque ripple of the auxiliary slot method is reduced by about 4%, so the torque ripple optimization method of the auxiliary slot can effectively attenuate the load torque ripple under constant power conditions.

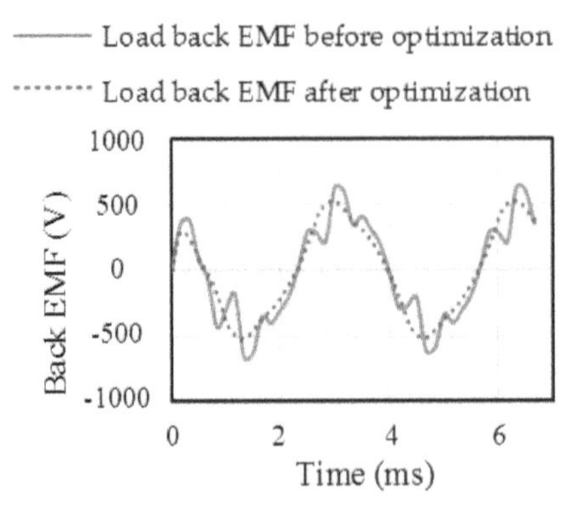

Figure 13. Back EMF of the AFPMSM at the rated load power.

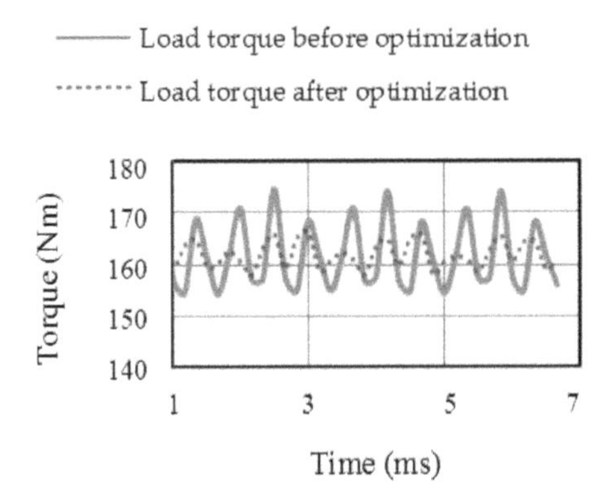

Figure 14. Torque of the AFPMSM at the rated load power.

3.4. Comparative Analysis of Peak Load Power Characteristics at 3000 rpm

Figure 15 shows that the load back EMF waveform is significantly optimized under the peak load power speed of 3000 rpm. Figure 16 shows that the load torque peak-to-peak value of the auxiliary slot optimization method is 10 Nm when the rotation speed is 3000 rpm, and the peak-to-peak value of the load torque before optimization is 16 Nm. The torque ripple of the auxiliary slot method is reduced by about 2%. Therefore, the torque ripple optimization method of the auxiliary slot can effectively attenuate the load torque ripple under constant power conditions and peak load conditions.

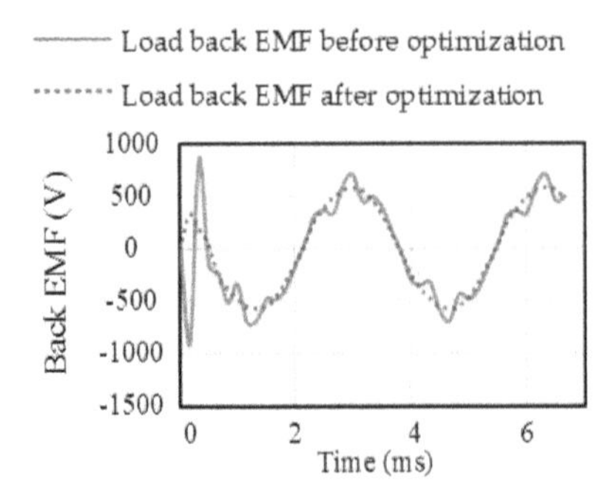

Figure 15. Back EMF of the AFPMSM at the peak load power.

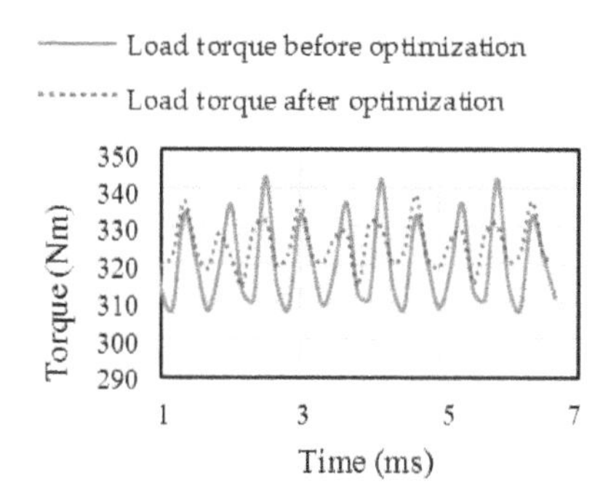

Figure 16. Torque of the AFPMSM at the peak load power.

In summary, at the constant torque maximum speed operating point, the rated load and peak load torque ripple ratio are reduced by 4% and 2%, respectively, before optimization. At the constant power maximum speed operating point, the rated load and peak load torque ripple are reduced by 4% and 2%, respectively, before optimization. Therefore, the AFPMSM opening auxiliary slot can effectively weaken the torque ripple of the motor, which is a reference for scholars studying the axial flux motor optimization design.

4. Multi-Physics Analysis

The traction motor of electric vehicle pursues high density and light weight in its design, and the working environment is complex and changeable. This is easy for causing electromagnetic noise and excessive temperature rise, which affects the reliability, safety and ride comfort of the whole vehicle. Therefore, multi-physics analysis of the motor is required. Controlling the electromagnetic noise within a certain range can improve the ride comfort; in addition, the heating problem of the motor directly affects the safety and reliability of the drive system, so thermal analysis of the motor is necessary. Figure 17 shows the finite element simulation model obtained and optimized above.

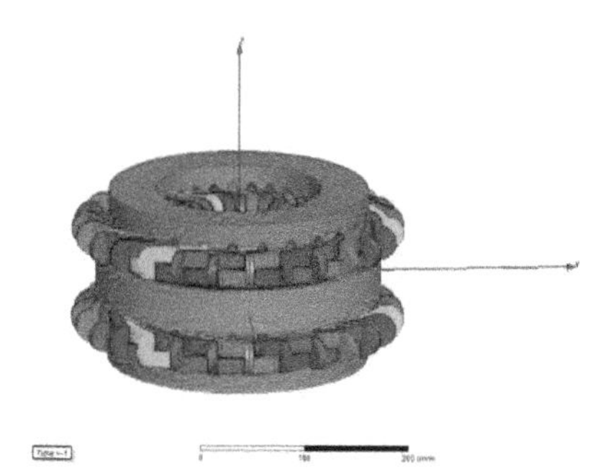

Figure 17. AFPMSM finite element simulation model.

The stator and rotor structure of AFPMSM is different from the traditional RFPMSM. In order to ensure the comprehensiveness of the analysis, the electromagnetic interference generated by the electromagnetic force of the stator and the electromagnetic force of the rotor is analyzed.

4.1. Electromagnetic Noise Generated by the AFPMSM Stator Structure

Firstly, a 3D finite element modal simulation model of the AFPMSM stator structure was established. The stator core structure has an elastic modulus of 205 GPa, a Poisson's ratio of 0.27 and a density of 7305 kg/m^3. Figure 18 shows the stator structure finite element simulation model and mesh segmentation diagram.

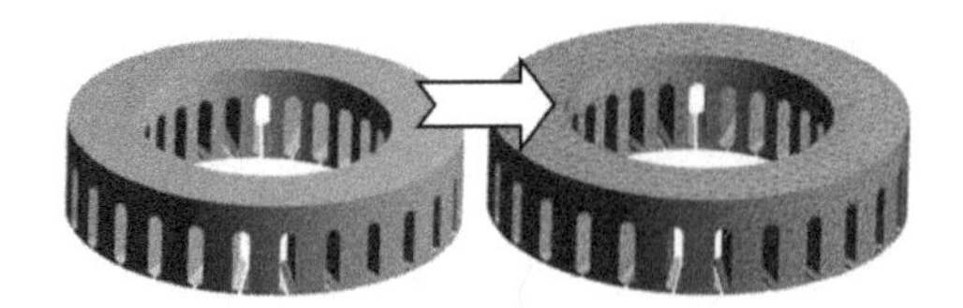

Figure 18. Finite element simulation model of the stator core.

In the finite element modal calculation, the AFPMSM stator model uses the additional mass method to make the winding end equivalent to a ring, thus calculating the natural frequency of the motor stator. The AFPMSM stator structure mode shape obtained by simulation is shown in Figure 19, and the modal frequency is shown in Table 2. It can be seen from Table 2 that using the winding as an additional mass analysis stator mode can increase the stiffness of the stator, increase the frequency range, and reduce the possibility of motor resonance.

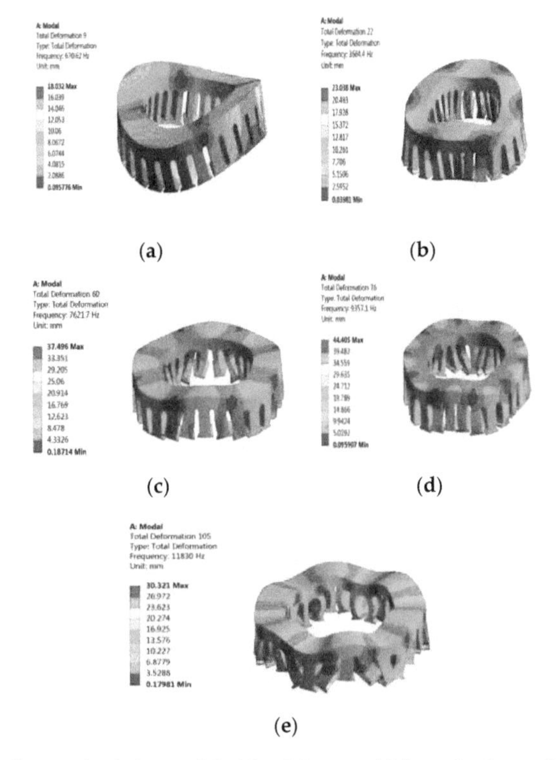

Figure 19. Finite element simulation model of the stator core. (**a**) Second order model. (**b**) Third order model. (**c**) Fourth order model. (**d**) Fifth order model. (**e**) Sixth order model.

Table 2. Modal frequency of stator core (Hz).

Modal Order	Order 2	Order 3	Order 4	Order 5	Order 6
Frequency	671	3664	7621	9357	11,830

The air domain model of motor electromagnetic noise radiation is established by A-weighting. The radiation characteristics of motor noise are analyzed based on the cylindrical air domain model. The results of sound pressure level analysis at each frequency are shown in Figures 20 and 21.

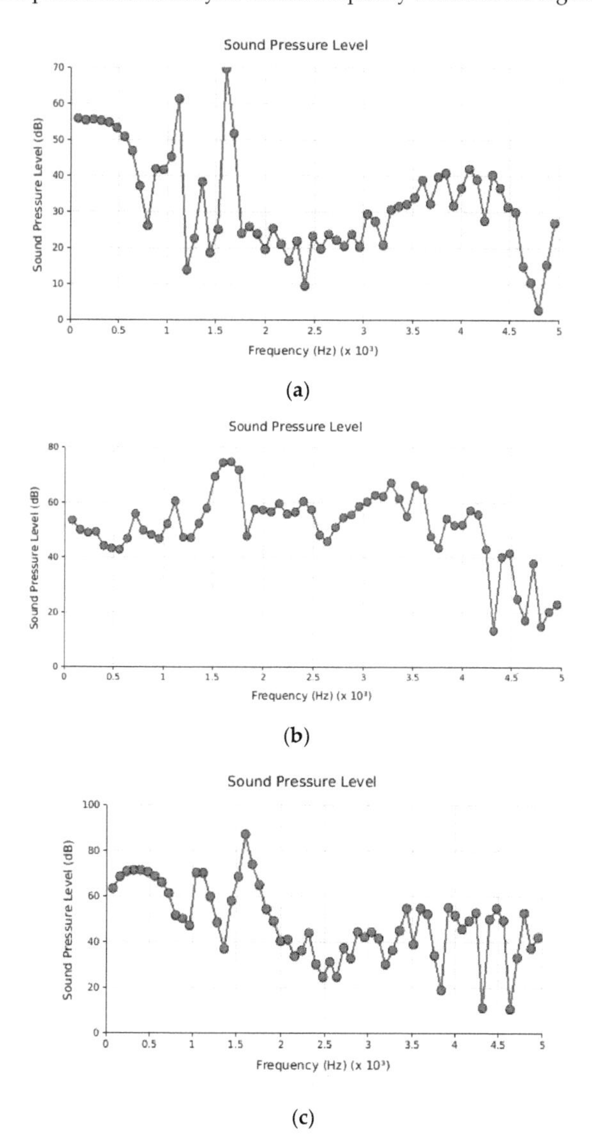

Figure 20. Stator system electromagnetic noise at 1600 rpm. (**a**) Electromagnetic noise at no load. (**b**) Electromagnetic noise at rated load. (**c**) Electromagnetic noise at peak load.

The sound pressure level of the electromagnetic noise at each working point of the motor can be calculated by the sound pressure level summation formula, as shown below:

$$L = 10\lg(\sum_{i=1}^{n} 10^{0.1L_i}) \tag{11}$$

where L_i is the sound pressure level value at the frequency and n is the number of frequency points.

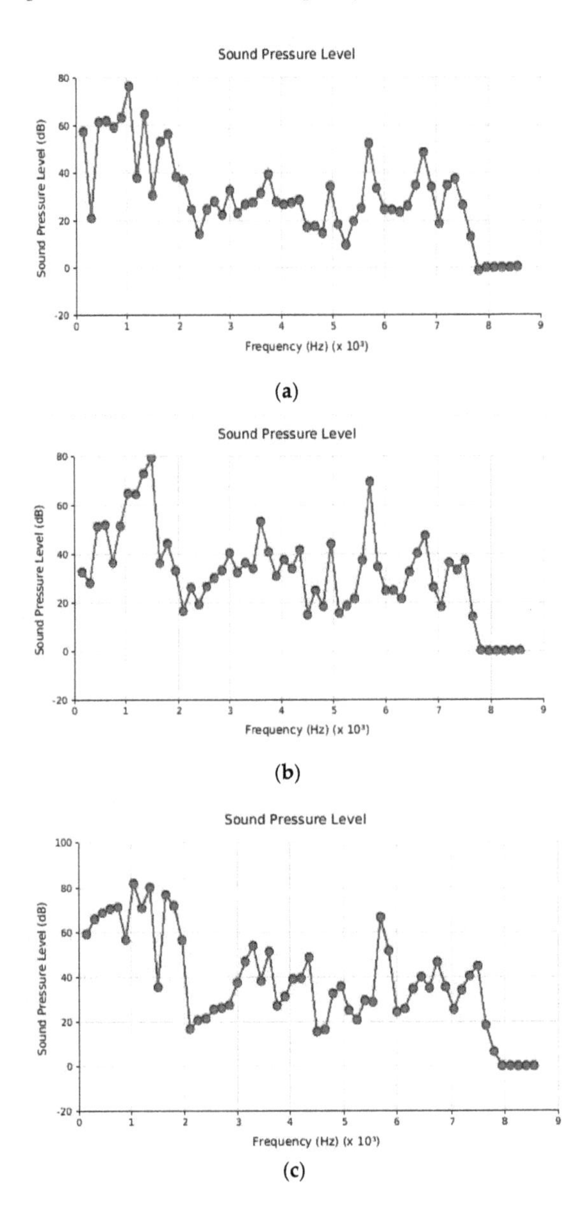

Figure 21. Stator system electromagnetic noise at 3000 rpm. (**a**) Electromagnetic noise at no load. (**b**) Electromagnetic noise at rated load. (**c**) Electromagnetic noise at peak load.

According to the sensitive range of the human ear to frequency, the sound pressure components in the range of 0 to 6000 Hz were selected for summation. The electromagnetic noise of the no-load, rated load, and peak load of the motor 1600 r/min were calculated to be 71 dB, 80 dB, and 88 dB, respectively; the electromagnetic noise of no-load, rated load, and peak load of 3000 r/min are 77 dB, 81 dB, and 86 dB respectively. The peak load noise of the motor 3000 r/min is large, but the torque required by the general vehicle during high-speed cruising is not large, therefore, the electromagnetic noise of the working point has little effect on the performance of the whole vehicle, and certain sound-absorbing or sound-insulating materials can be used for noise reduction processing.

4.2. Electromagnetic Noise Generated by the AFPMSM Rotor Structure

The finite element modal simulation model of the automotive AFPMSM rotor structure was established, and the natural frequency and vibration mode of the rotor structure were obtained, as shown in Figure 22.

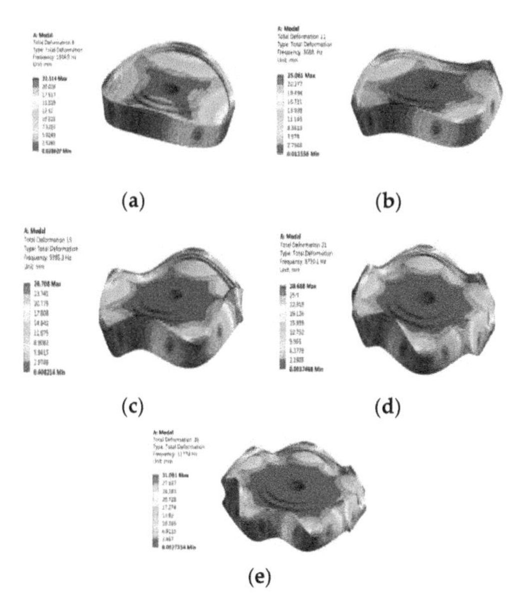

Figure 22. Rotor core axial flux mode shape. (**a**) Second order model. (**b**) Third order model. (**c**) Fourth order model. (**d**) Fifth order model. (**e**) Sixth order model.

The natural frequencies of the second-order mode and the third-order mode of the rotor of the AFPMSM are 1905 Hz and 3688 Hz, respectively, and the generated electromagnetic vibration amplitude is small, so electromagnetic vibration does not cause large vibration noise.

The electromagnetic noise of the rotor structure of the motor was simulated, and the noise sound pressure level results of the respective frequency components were obtained as shown in Figures 23 and 24. Since the peak load electromagnetic noise is greater than the rated load electromagnetic noise, the rotor section only shows the electromagnetic noise under peak load conditions.

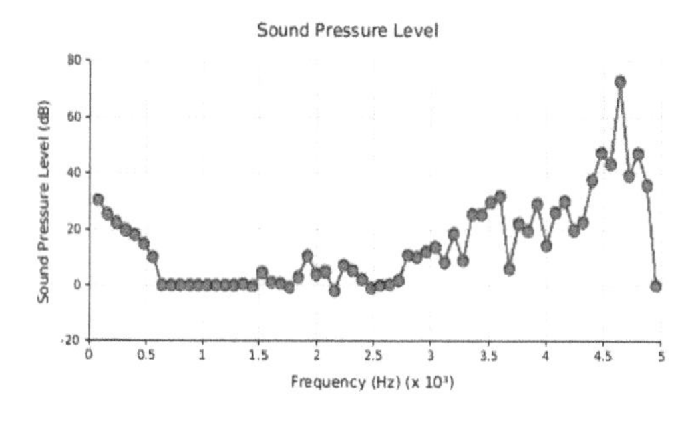

Figure 23. Peak load electromagnetic noise at 1600 rpm.

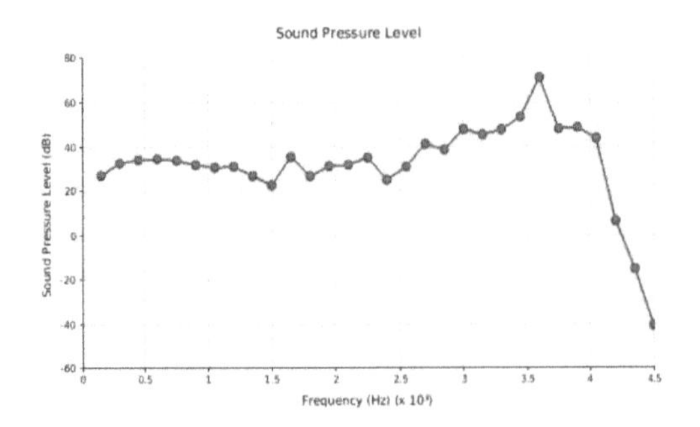

Figure 24. Value-loaded electromagnetic noise at a speed of 3000 rpm.

Calculated by Formula (11), the peak load electromagnetic noise of 1600 r/min is 72 dB; the peak load electromagnetic noise of the motor speed of 3000 r/min is 71 dB. The electromagnetic noise generated by the stator structure is the main source of AFPMSMs.

When the electromagnetic performance of the vehicle motor was evaluated, the magnetic field saturation of the constant torque working point was the largest, and the magnetic field distortion of the working point of the constant power maximum speed was the largest, which is the typical working condition of the traction motor for the vehicle. Therefore, the maximum speed operating point of the constant torque zone and the maximum speed operating point of the constant power zone are often selected as the key working points of the electromagnetic performance. In summary, the rated load and peak load electromagnetic noise of the stator structure at the constant torque operating point are 80 dB and 88 dB, respectively, and the rated load and peak load electromagnetic noise at the constant power maximum speed operating point are 81 dB and 86 dB, respectively. The peak load electromagnetic noise of the rotor structure at the constant torque operating point is 72 dB, and the peak load electromagnetic noise at the constant power maximum speed operating point is 71 dB. According to biology, the sound range that does not affect human health ranges from 0 to 90 dB. The maximum noise of the motor designed in this paper is 88 dB, which is within a reasonable range.

5. Thermal Analysis of AFPMSM

The problem of heating and cooling of the motor for vehicles directly affects the safety and reliability of the drive system. The accurate prediction of the temperature rise of the motor and the design of the efficient cooling system are of great significance to the safety and reliability of the new energy vehicles.

5.1. AFPMSM Temperature Field Equivalent Simplified Calculation Model

To simplify the analysis, an equivalent model of the stator winding was established, introducing an equivalent conductor that replaced the winding copper wire and an equivalent insulating layer of all the insulating materials in the slot, as shown in Figure 25.

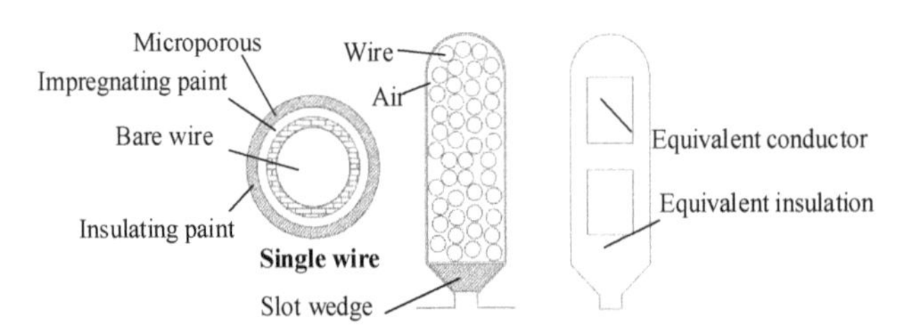

Figure 25. Equivalent winding model.

The equivalent thermal conductivity of the slot insulation is calculated as

$$\lambda_{eq} = \sum_{i=1}^{n} d_i / \left(\sum_{i=1}^{n} \frac{d_i}{\lambda_i} \right) \tag{12}$$

where λ_{eq} is the equivalent thermal conductivity of the insulating material, d_i is the equivalent thickness of each insulating material in the groove, λ_i is the thermal conductivity of the corresponding insulating material.

The AFPMSM in this paper uses a fully enclosed end cap cooling structure. To simplify the analysis, the following assumptions were made:

1. The insulating paint of the winding is evenly distributed, and the winding is completely dip-coated;
2. Since the inter-strand insulation of the armature windings cannot be completely considered and wound, the inner region of the stator slots is equivalently treated. The cross-sectional areas of the upper and lower windings are equal, and the thickness of the insulating layer, the windings and the left and right sides, and the windings of the windings and the upper and lower sides are equal;
3. Ignoring radiation heat dissipation and contact thermal resistance between the rotor and the rotating shaft;
4. The loss of each part of the motor does not change with temperature;
5. The heat generated by the heat source is mainly carried away by the end cover cooling water, and the heat exchange between the outer surface of the casing and the surrounding air is negligible.

The solution domain model of the motor based on the above assumptions is shown in Figure 26.

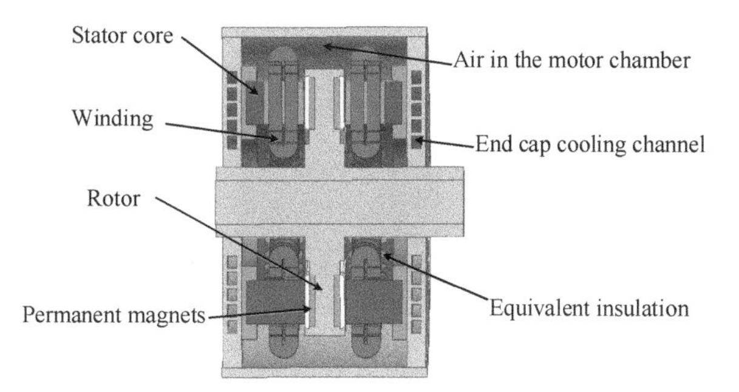

Figure 26. Motor temperature field solution domain model.

The axial flux motor is meshed, and the split model of each part of the motor is shown in Figure 27. The whole motor is divided into 17,346,364 individual cell grids and 3,412,225 nodes.

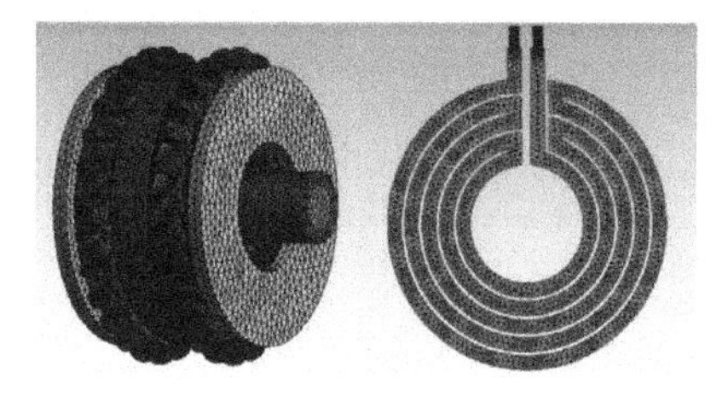

Figure 27. Schematic diagram of the prototype fluid-solid coupling simulation model.

5.2. Temperature Field Analysis under AFPMSM Rated Conditions

Thermal simulation of the motor was conducted based on ANSYS/Workbench. The inlet water flow rate was 0.8 m/s, the ambient temperature was 30 °C, and the inlet coolant temperature was 25 °C. The rated torque operating point of 50 kW and 1600 r/min was selected, and the motor temperature distribution as shown as Figure 28 is when the motor reached the thermal steady state for a long time.

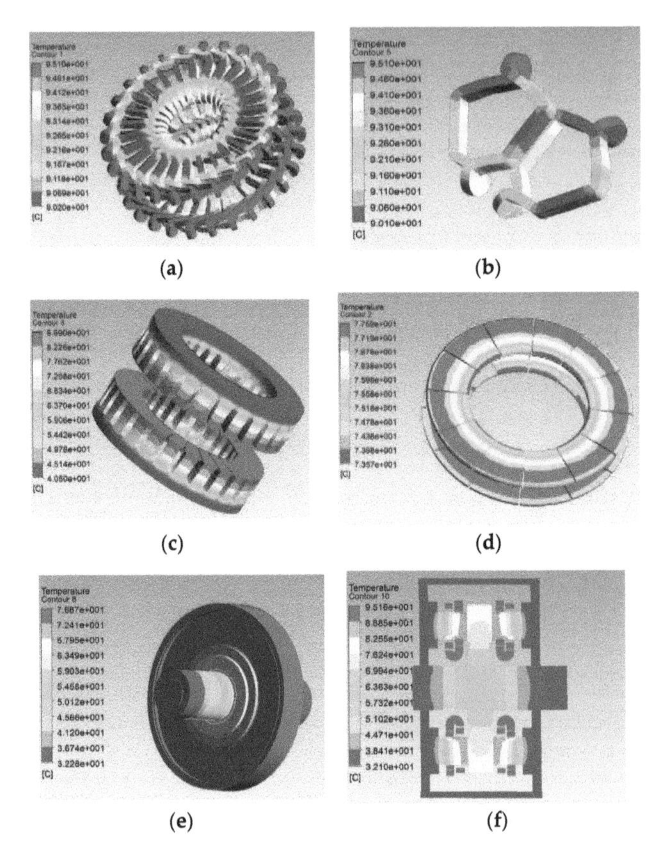

Figure 28. Temperature map of each part of the AFPMSM. (**a**) Winding temperature cloud map. (**b**) Upper and lower winding temperature cloud map. (**c**) Stator core temperature cloud map. (**d**) Permanent magnet temperature map. (**e**) Rotor and shaft temperature cloud map. (**f**)Whole machine section temperature cloud map.

As seen from the thermal simulation results, the temperature rise at the end of the stator winding is high. There is a direct temperature difference between the upper and lower windings in the same slot, and the temperature near the stator yoke winding is lower. The temperature of the stator teeth is significantly higher than the stator yoke. The temperature at the outer diameter of the permanent magnet is significantly higher than the inner diameter. The maximum temperature of the motor appears at the outer end of the winding (near the casing) at 95 °C. The maximum temperature of the stator core is the stator tooth at 87 °C. The maximum temperature of the permanent magnet is at the outer diameter of the permanent magnet at 78 °C. The internal temperature of the whole machine is, from the highest to the lowest, the winding end, the stator tooth, the permanent magnet, the rotor disk, the stator disk, and the casing. The analysis shows that the temperature rise of the motor's long-term rated operation is within a reasonable range, and the motor can operate safely and reliably.

5.3. Temperature Field Analysis of AFPMSM under Peak Operating Conditions

Considering the operating conditions of the actual electric vehicle, the temperature rise of the motor is further checked by operating the peak torque for 60 s after assuming that the rated operation of the motor reaches a thermal steady state. The temperature rise of the motor components over time during the 60 s operation of the motor with peak torque is shown in Figure 29.

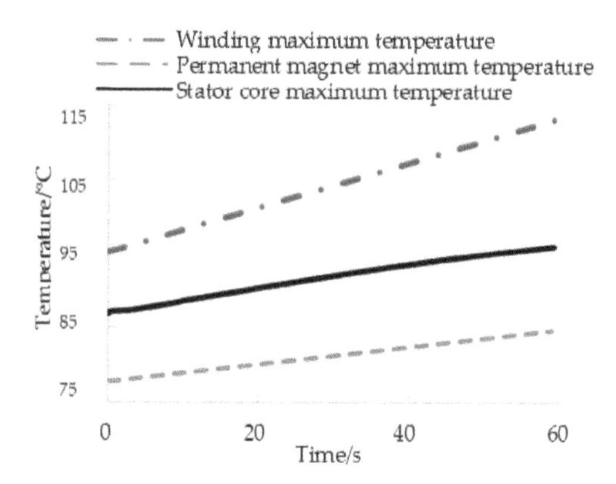

Figure 29. Temperature of main components of the motor changes with time.

Figure 29 shows that the winding temperature rises the fastest when the motor peak torque is running, and the maximum temperature of the winding reaches 113 °C after running for 60 s, which is 18 °C higher than the rated working temperature; the maximum temperature of the stator core is 95.9 °C, which is 9 °C higher than the rated working condition. The maximum temperature of the permanent magnet is 84.68 °C, which is 7 °C higher than the rated working temperature. The winding temperature rises the fastest. Because the copper consumption of the peak torque condition is relatively large, the thermal conductivity of the insulating material is poor, and the generated heat is difficult to dissipate in a short time. In this paper, the motor uses H-class insulation. The results of motor thermal analysis show that the temperature rise of the motor meets the technical requirements.

6. Conclusions

This paper first proposes a 2D multi-loop equivalent method to simplify the AFPMSM model and, based on this, establishes the equivalent magnetic network model and air gap magnetic density distribution function of AFPMSMs. Secondly, by comparing the torque ripple values under multi-operating conditions before and after the open stator auxiliary slot, the feasibility of reducing the AFPMSM torque ripple is verified by the open stator auxiliary slot. Finally, the experimental results of AFPMSM in multi-physics such as vibration, noise, and temperature rise are obtained by finite element software and simulation experiments. The following conclusions can be drawn:

1. The electromagnetic design method based on the analytical method was used to analyze the key electromagnetic parameters such as air gap magnetic density harmonics, positioning torque and no-load back EMF of AFPMSM. The correctness of the analytical calculation was verified by 3D electromagnetic field simulation.
2. The stator core was used to open the auxiliary slot to suppress the torque ripple, and the maximum torque ripple peak-to-peak value was reduced by 2%.
3. Based on the multi-physics simulation software, the vibration and noise characteristics of the motor were analyzed from the electromagnetic force and the modal state. The electromagnetic noise characteristics of the no-load, rated load, and peak load operating points at different speeds were simulated. The noise, vibration and harshness (NVH) of the motor was verified.

4. The thermal simulation based on fluid-structure coupling checked the temperature rise of the AFPMSM after the rated operating point reached the thermal steady state and transient operation for 1 min, which ensured the safety and reliability of the motor running under vehicle working conditions.

In summary, the AFPMSM design and research method proposed in this paper verifies the superior performance of AFPMSM as a motor for vehicles. On the other hand, this paper provides a complete solution for the design and optimization of AFPMSMs for electric vehicles, which can provide relevant researchers with AFPMSM performance reference indicators.

Author Contributions: Conceptualization, J.Z., Y.D., and Q.H.; methodology, J.Z.; software, Q.H. and M.H.; validation, J.Z., Y.D., and Q.H.; formal analysis, J.Z. and Y.D.; investigation, Y.D. and M.H.; resources, J.Z.; data curation, Q.H. and M.H.; writing—original draft preparation, J.Z., Q.H., and Y.D.; writing—review and editing, J.Z. and Q.H.; visualization, Q.H. supervision, J.Z.; project administration, J.Z.; funding acquisition, J.Z.

Funding: This research was supported by 111 project, No. D18003.

Conflicts of Interest: The authors declare no conflict of interest.

References

1. Chen, Y.; Zhang, B. Minimization of the Electromagnetic Torque Ripple Caused by the Coils Inter-Turn Short Circuit Fault in Dual-Redundancy Permanent Magnet Synchronous Motors. *Energies* **2017**, *10*, 1798. [CrossRef]
2. Wang, W.; Wang, W. Compensation for Inverter Nonlinearity in Permanent Magnet Synchronous Motor Drive and Effect on Torsional Vibration of Electric Vehicle Driveline. *Energies* **2018**, *11*, 2542. [CrossRef]
3. Pietrusewicz, K.; Waszczuk, P.; Kubicki, M. MFC/IMC Control Algorithm for Reduction of Load Torque Disturbance in PMSM Servo Drive Systems. *Appl. Sci.* **2019**, *9*, 86. [CrossRef]
4. Jia, Y.F.; Chu, L.; Xu, N.; Li, Y.K.; Zhao, D.; Tang, X. Power Sharing and Voltage Vector Distribution Model of a Dual Inverter Open-End Winding Motor Drive System for Electric Vehicles. *Appl. Sci.* **2018**, *8*, 254. [CrossRef]
5. Wang, R.J. Optimal design of a coreless stator axial flux permanent-magnet generator. *IEEE Trans. Magn.* **2005**, *41*, 55–64. [CrossRef]
6. Aydin, M.; Huang, S.; Lipo, T.A. Torque quality and comparison of internal and external rotor axial flux surface-magnet disc machines. *IEEE Trans. Ind. Electron.* **2006**, *53*, 822–830. [CrossRef]
7. Zhao, J.F.; Hua, M.Q.; Liu, T.Z. Research on a Sliding Mode Vector Control System Based on Collaborative Optimization of an Axial Flux Permanent Magnet Synchronous Motor for an Electric Vehicle. *Energies* **2018**, *11*, 3116. [CrossRef]
8. Darba, A.; Esmalifalak, M.; Barazandeh, E.S. Implementing SVPWM technique to axial flux permanent magnet synchronous motor drive with internal model current controller. In Proceedings of the 2010 4th International Power Engineering and Optimization Conference, Shah Alam, Malaysia, 23–24 June 2010; pp. 126–131.
9. Lim, D.K.; Cho, Y.S.; Ro, J.S.; Jung, S.Y.; Jung, H.K. Optimal Design of an Axial Flux Permanent Magnet Synchronous Motor for the Electric Bicycle. *IEEE Trans. Magn.* **2016**, *52*, 1–4. [CrossRef]
10. Kim, J.S.; Lee, J.H.; Kim, Y.J.; Jung, S.Y. Characteristics analysis method of axial flux permanent magnet motor based on 2-D finite element analysis. *IEEE Trans. Magn.* **2017**, *53*, 1–4. [CrossRef]
11. Hemeida, A.; Sergeant, P. Analytical Modeling of Surface PMSM Using a Combined Solution of Maxwell–s Equations and Magnetic Equivalent Circuit. *IEEE Trans. Magn.* **2014**, *50*, 1–13. [CrossRef]
12. Xia, C.; Zhao, J.; Yan, Y.; Shi, T. A novel direct torque control of matrix converter-fed PMSM drives using duty cycle control for torque ripple reduction. *IEEE Trans. Ind. Electron.* **2013**, *61*, 2700–2713. [CrossRef]
13. Vafaie, M.H.; Dehkordi, B.M.; Moallem, P.; Kiyoumarsi, A. Minimizing torque and flux ripples and improving dynamic response of PMSM using a voltage vector with optimal parameters. *IEEE Trans. Ind. Electron.* **2015**, *63*, 3876–3888. [CrossRef]
14. Wang, D.; Lin, H.; Yang, H.; Wang, K. Cogging torque optimization of flux memory pole-changing permanent magnet machine. *IEEE Trans. Appl. Supercond.* **2016**, *26*, 1–5. [CrossRef]

15. Lai, C.; Feng, G.; Mukherjee, K.; Kar, N.C. Investigations of the influence of PMSM parameter variations in optimal stator current design for torque ripple minimization. *IEEE Trans. Energy Convers.* **2017**, *32*, 1052–1062. [CrossRef]
16. Wahsh, S.; Shazly, J.; Yassin, A. Steady state heat conduction problems of AFPMSM using 3D Finite Element. In Proceedings of the 18th International Middle-East Power Systems Conference, Helwan Univ, ON, Egypt, 27–29 December 2016; pp. 949–953.
17. Chen, Q.; Liang, D.; Gao, L.; Wang, Q.; Liu, Y. Hierarchical thermal network analysis of axial-flux permanent-magnet synchronous machine for electric motorcycle. *IET Electr. Power Appl.* **2018**, *12*, 859–866. [CrossRef]
18. Rasekh, A.; Sergeant, P.; Vierendeels, J. Fully predictive heat transfer coefficient modeling of an axial flux permanent magnet synchronous machine with geometrical parameters of the magnets. *Appl. Thermal Eng.* **2017**, *110*, 1343–1357. [CrossRef]
19. De Bisschop, J.; Sergeant, P.; Dupré, L. Demagnetization fault detection in axial flux PM machines by using sensing coils and an analytical model. *IEEE Trans. Magn.* **2017**, *53*, 1–4. [CrossRef]
20. De Bisschop, J.; Sergeant, P.; Hemeida, A.; Vansompel, H.; Dupre, L. Analytical Model for Combined Study of Magnet Demagnetization and Eccentricity Defects in Axial Flux Permanent Magnet Synchronous Machines. *IEEE Trans. Magn.* **2017**, *53*, 1–12. [CrossRef]
21. De Bisschop, J.; Abdallh, A.; Sergeant, P.; Dupré, L. Identification of Demagnetization Faults in Axial Flux Permanent Magnet Synchronous Machines Using an Inverse Problem Coupled With an Analytical Model. *IEEE Trans. Magn.* **2014**, *50*, 1–4. [CrossRef]
22. Hemeida, A.; Sergeant, P.; Vansompel, H. Comparison of Methods for Permanent Magnet Eddy-Current Loss Computations With and Without Reaction Field Considerations in Axial Flux PMSM. *IEEE Trans. Magn.* **2015**, *51*, 1–11. [CrossRef]
23. Scheerlinck, B.; De Gersem, H.; Sergeant, P. 3D Eddy-Current and Fringing-Flux Distribution in an Axial-Flux Permanent-Magnet Synchronous Machine with Stator in Laminated Iron or SMC. *IEEE Trans. Magn.* **2015**, *51*, 1–4. [CrossRef]
24. Chen, Q.; Liang, D.; Jia, S.; Ze, Q.; Liu, Y. Loss Analysis and Experiment of Fractional-Slot Concentrated-Winding Axial Flux PMSM for EV Applications. In Proceedings of the 10th IEEE Annual Energy Conversion Congress and Exposition, Portland, ON, USA, 23–27 September 2018; pp. 4329–4335.
25. Fei, W.; Luk, P.C.K. Torque ripple reduction of axial flux permanent magnet synchronous machine with segmented and laminated stator. In Proceedings of the 2009 IEEE Energy Conversion Congress and Exposition, San Jose, CA, USA, September 2009; pp. 132–138.
26. Woolmer, T.; McCulloch, M. Analysis of the yokeless and segmented armature machine. In Proceedings of the 2007 IEEE International Electric Machines & Drives Conference, Antalya, Turkey, 3–5 May 2007; pp. 704–708.
27. Aydin, M.; Huang, S.; Lipo, T. Axial flux permanent magnet disc machine: A review. In Proceedings of the Symposium on Power Electronics, Electrical Drives, Automation, and Motion (SPEEDAM), Capri, Italy, 16–18 June 2004; pp. 61–71.
28. Hosseini, S.M.; Agha-Mirsalim, M.; Mirzaei, M. Design, prototyping, and analysis of a low cost axial-flux coreless permanent-magnet generator. *IEEE Trans. Magn.* **2008**, *44*, 75–80. [CrossRef]

MDPI
St. Alban-Anlage 66
4052 Basel
Switzerland
Tel. +41 61 683 77 34
Fax +41 61 302 89 18
www.mdpi.com

Energies Editorial Office
E-mail: energies@mdpi.com
www.mdpi.com/journal/energies

Lightning Source UK Ltd.
Milton Keynes UK
UKHW051538160920
369951UK00007B/100